INTERNATIONAL BACCALAUREATE

Physics
FOURTH EDITION

Gregg Kerr
4th Edition

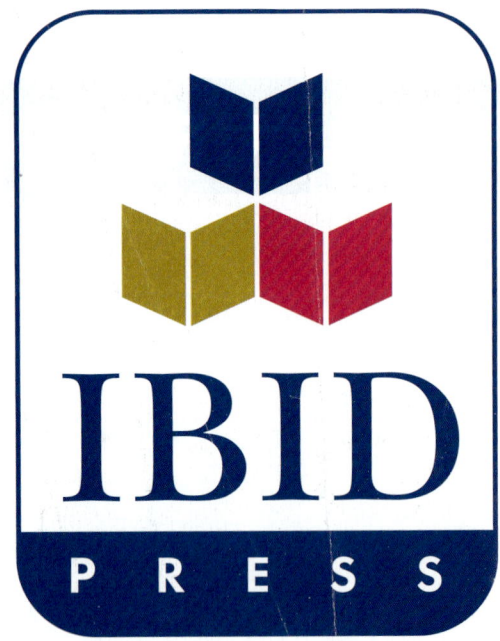

Copyright ©IBID Press, Victoria.

www.ibid.com.au

First published in 2014 by IBID Press, Victoria

Library Catalogue:

Kerr, G.

1. Physics (4th Edition)

2. International Baccalaureate.

Series Title: International Baccalaureate in Detail

ISBN: 978 – 1 – 921917 – 21 – 9

All rights reserved except under the conditions described in the Copyright Act 1968 of Australia and subsequent amendments. No part of this publication may be reproduced, stored in a retrieval system, or transmitted in any form or by any means, without the prior permission of the publishers.

While every care has been taken to trace and acknowledge copyright, the publishers tender their apologies for any accidental infringement where copyright has proved untraceable. They would be pleased to come to a suitable arrangement with the rightful owner in each case.

This material has been developed independently by the publisher and the content is in no way connected with nor endorsed by the International Baccalaureate Organization.

All copyright statements, '© IBO 20014' refer to the Physics guide published by the International Baccalaureate Organization in 2014.

IBID Press express their thanks to the International Baccalaureate Organization for permission to reproduce its intellectual property.

Cover design by Key-Strokes.

Published by IBID Press, 36 Quail Crescent, Melton, 3337, Australia.

Printed by Red Planet.

ELECTRONIC SUPPLEMENT

This text is accompanied by a FREE electronic supplement. This can be obtained from the iTunes Bookstore.

Full details about downloading the supplement can be found on our website:

www.ibid.com.au

The supplement has been designed to be viewed using Apple devices such as the iPad and iPhone.

A pdf version of the supplement will be available as a download from our website. This will be readable on all computer platforms. Most, but not all, features will be operative in the pdf format.

PREFACE TO THE FOURTH EDITION

This fourth edition is based on the revised *International Baccalaureate Physics syllabus* (© IBO 2014) that will be examined for the first time in May 2016. The author, as in the first, second and third editions, has attempted to develop the relevant theory with the objectives of the syllabus assessment statements in mind.

The major changes to the syllabus are the inclusion of several new topics, several new options, the option structure and the inclusion of new assessment statements. In order that physics is not seen as just an academic subject, an effort has been made to link various topics with the role that they play in technology.

Please refer to the *IBO Syllabus Guide* (© IBO 2014) for details about changes to the course.

Although this book might not be regarded as a conventional textbook, many ideas are discussed in detail particularly those ideas that, in the author's experience as a teacher, students find difficult to comprehend on first acquaintance. It is hoped that both students and teachers will find these discussions useful, as they will the worked examples and exercises. However, the exercises and examples supplied do not necessarily reflect the views of the Senior Examining team appointed by the International Baccalaureate. Nor does the book in any way suggest the way and the order in which the syllabus should be taught. In line with I.B policy such decisions are left to the individual teacher, as is the choice of resources that the teacher might wish to use.

The author has obviously tried to be accurate in the presentation of the material and also in the answers to the exercises and examples. However, he strongly welcomes feedback on any errors that may be spotted by student or teacher. Such errors will be corrected in future reprints of this book.

THE AUTHORS

GREGG KERR

Gregg has a Bachelor of Education degree from Charles Sturt University, Australia, and a Master of Science in Education from the State University of New York, USA. He has taught in Germany, Hong Kong, Australia, Brunei, Thailand and is presently Student Welfare Coordinator and Head of Science at the Utahloy International School Guangzhou, China.

Gregg has taught IB Physics since 1988. He has been a member of the Physics Subject Committee, and is presently a Senior Examiner and a Senior Moderator in Physics. He has been an IB Physics workshop leader at conferences in Tokyo, Mumbai, Brisbane, Adelaide, Invercargill, Sydney, Singapore and Chiang Mai. I would like to thank my Bhutanese friend Kinga Tshering for all the suggestions he made to make this edition better. I would especially like to thank Ofelia and Angelica for giving me the support at home while I tore my hair out in writing my contribution to this fourth edition.

PAUL RUTH

IBID Press record with sadness the passing of Paul Ruth. Paul was co-author of previous editions. Whilst he has been unable to contribute directly to this edition, it is appropriate to record his substantial contributions to the series as a whole.

ACKNOWLEDGEMENTS

We wish to acknowledge the advice and assistance of the following people in the development and production of these materials to support the teaching of IB Physics.

AUTHOR

Gregg Kerr

PROOF READER

Richard Marsland

ARTWORK AND GRAPHICS

Charlotte Gelin

LAYOUT AND EDITORIAL

Charlotte Gelin

Contents

Chapter 1   **Measurements and Uncertainties**

1.1	Measurements in Physics	2
1.2	Uncertainties and errors	9
1.3	Vectors and scalars	22

Chapter 2   **Mechanics**

2.1	Motion	33
2.2	Forces	51
2.3	Work, energy and power	65
2.4	Momentum and impulse	74

Chapter 3   **Thermal Physics**

3.1	Thermal Physics	84
3.2	Modelling a gas	100

Chapter 4   **Waves**

4.1	Oscillations	112
4.2	Travelling waves	122
4.3	Wave characteristics	132
4.4	Wave behaviour	132
4.5	Standing Waves	147

Chapter 5   Electricity and Magnetism

5.1	Electric fields	152
5.2	Heating effect of electric currents	165
5.3	Electric cells	178
5.4	Magnetic effects of electric currents	186

Chapter 6   Electricity and Magnetism

6.1	Circular Motion	194
6.2	Newton's law of gravitation	200

Chapter 7   Atomic, Nuclear and Particle Physics

7.1	Discrete energy and radioactivity	209
7.2	Nuclear reactions	221
7.3	The structure of matter	229

Chapter 8   Energy Production

8.1	Energy sources	246
8.2	Thermal energy transfer	273

Chapter 9   Wave Phenomena

9.1	Simple harmonic motion	284
9.2	Single-slit diffraction	289
9.3	Interference	291
9.4	Resolution	299
9.5	Doppler effect	303

Chapter 10 Fields

| 10.1 | Describing fields | 308 |
| 10.2 | Fields at work | 318 |

Chapter 11 Electromagnetic Induction

11.1	Electromagnetic induction	324
11.2	Power generation and transmission	331
11.3	Capacitance	341

Chapter 12 Atomic, Nuclear and Particle Physics

| 12.1 | The interaction of matter with radiation Photons | 348 |
| 12.2 | Nuclear physics | 360 |

Chapter 13 Option A: Relativity

A.1	The beginnings of relativity	366
A.2	Lorentz transformations	369
A.3	Space-time diagrams	375
A.4	Relativistic mechanics	379
A.5	General relativity	382

Chapter 14 Option B: Engineering Physics

B.1	Rigid bodies and rotational dynamics	392
B.2	Thermodynamics	401
B.3	Fluids and fluid dynamics	415
B.4	Forced vibrations and resonance	422

Chapter 15 Option C: Imaging

C.1	Introduction to imaging	428
C.2	Imaging instrumentation	437
C.3	Fibre Optics	446
C.4	Medical imaging	450

Chapter 16 Option D: Astrophysics

D.1	Stellar quantities	466
D.2	Stellar characteristics and stellar evolution	471
D.3	Cosmology	482
D.4	Stellar processes	490
D.5	Further cosmology	494

Glossary   499

Index   520

1. Measurements and Uncertainties

Contents

1.1 – Measurements in physics
1.2 – Uncertainties and errors
1.3 – Vectors and scalars

Essential Ideas

Since 1948, the Système International d'Unités (SI) has been used as the preferred language of science and technology across the globe and reflects current best measurement practice.

Some quantities have direction and magnitude, others have magnitude only, and this understanding is the key to correct manipulation of quantities. This sub-topic will have broad applications across multiple fields within physics and other sciences. © IBO 2014

CHAPTER 1

TOK BACKGROUND

Search the web for a mathematical timeline and it becomes clear that the idea that science is a Western invention is a myth. The foundations of science and technology were beginning to be established centuries before by Arabic, Indian, Mayan and Chinese civilizations, among others. "Teachers are encouraged to emphasize this contribution in their teaching of various topics, perhaps through the use of timeline websites. The scientific method in its widest sense, with its emphasis on peer review, open-mindedness and freedom of thought, transcends politics, religion, gender and nationality." © IBO 2014

So how was distance, mass and time measured over these past centuries, and what has influenced the common language used in science terminology? More to the point, why in this world of internationalism and scientific collaboration have we not been able to implement fully the use of one system of common units throughout the world? One pint or one acre has very little significance to the majority of countries throughout the world.

A derived unit of energy that was in common use in the 1900's was the ergon abbreviated to erg. It was equivalent to 107 joules. I can remember this only because it was the nickname we gave to one of our science teachers, Brother Metcalf. But how was space, mass and time measured in the past and what was the derivation of these units? How did Galileo Galilei (1564 – 1642) measure distance and time to perform his experiments to try and determine the nature of motion? For time measurement, he may have used a candle clock or a water clock or more likely a pendulum clock. And how was distance measured? Common units it his time were the digit (the width of a finger), the inch (the width of a man's thumb), and the cubit (the length from the elbow to the middle finger). However, the degree of accuracy of these units varied considerably from person to person. The accuracy and precision of measuring instruments has greatly improved although the myth that digital devices are better than analogue devices is still out there.

Another common unit to measure area of land is the acre. It was defined as the area of land that could be ploughed by an ox team in a morning. Landowners and property developers would not be too happy in modern times with this measurement given the price of land today.

In the 1800's, the French mathematician Poincaré proposed a thought experiment concerning the absolute nature of matter. Suppose that you went to sleep, and while you were sleeping, everything in the Universe increased a hundred times. By everything he meant electrons, protons, the earth, the stars etc. When you woke up, could you tell if anything has changed? Poincaré argued that there is no experiment that could be performed to prove that you had become bigger. So size is not an absolute quantity because always needs to be measured relative to something else.

"Scientists aim towards designing experiments that can give a "true value" from their measurements, but due to the limited precision in measuring devices, they often quote their results with some form of uncertainty". © IBO 2014

NATURE OF SCIENCE:

Common terminology: Since the 18th century, scientists have sought to establish common systems of measurements to facilitate international collaboration across science disciplines and ensure replication and comparability of experimental findings. (1.6)

Improvement in instrumentation: An improvement in apparatus and instrumentation, such as using the transition of cesium-133 atoms for atomic clocks, has led to more refined definitions of standard units. (1.8)

Certainty: Although scientists are perceived as working towards finding "exact" answers, the unavoidable uncertainty in any measurement always exists. (3.6)

© IBO 2014

1.1 Measurements in Physics

Essential idea: Since 1948, the Système International d'Unités (SI) has been used as the preferred language of science and technology across the globe and reflects current best measurement practice.

Understandings
- Fundamental and derived SI units
- Scientific notation and metric multipliers
- Significant figures
- Orders of magnitude
- Estimation

Fundamental and derived SI units

SI units are those of Le Système International d'Unités adopted in 1960 by the Conférence Générale des Poids et Mesures. They are adopted in all countries for science research and education. They are also used for general measurement in most countries with the USA and the UK being the major exceptions.

Physics is the most fundamental of the sciences in that it involves the process of comparing the physical properties of what is being measured against reference or fundamental quantities, and expressing the answer in numbers and units.

Some quantities cannot be measured in a simpler form, and others are chosen for convenience. They have been selected as the basic quantities and are termed fundamental quantities. Figure 101 lists the fundamental quantities of the SI system together with their respective SI unit and SI symbol.

Quantity	SI unit	Symbol
length	metre	m
mass	kilogram	kg
time	second	s
electric current	ampere	A
thermodynamic temperature	Kelvin	K
amount of substance	mole	mol
luminous intensity	candela	cd*

*not required for this course

Figure 101 Fundamental quantities

Scientists and engineers need to be able to make accurate measurements so that they can exchange information. To be useful, a standard of measurement must be:

1. Invariant in time. For example, a standard of length that keeps changing would be useless.

2. Readily accessible so that it can be easily compared.

3. Reproducible so that people all over the world can check their instruments.

The standard metre, in 1960, was defined as the length equal to 1 650 763.73 wavelengths of a particular orange–red line of krypton–86 undergoing electrical discharge. Since 1983 the metre has been defined in terms of the speed of light. The current definition states that 'the metre is the length of path travelled by light in a vacuum during a time interval of 1/299 792 453 second'.

The standard kilogram is the mass of a particular piece of platinum-iridium alloy that is kept in Sèvres, France. Copies of this prototype are sent periodically to Sèvres for adjustments.

The standard second is the time for 9 192 631 770 vibrations of the cesium-133 atom.

Standards are commonly based upon properties of atoms. It is for this reason that the standard kilogram could be replaced at some future date. When measuring lengths, we choose an instrument that is appropriate to the order of magnitude, the nature of the length, and the sensitivity required. For example, the orders of magnitude (the factor of 10) of the radius of a gold atom, a person's height and the radius of the solar system in metres are 10^{-15}, 10^0 and 10^{12} respectively.

Physical Quantity	Symbol	Name and Symbol SI Unit	Fundamental Units Involved	Derived Units involved
frequency	f or v	hertz (Hz)	s^{-1}	s^{-1}
force	F	newton (N)	$kg\ m\ s^{-2}$	$kg\ m\ s^{-2}$
work	W	joule (J)	$kg\ m^2\ s^{-2}$	Nm
energy	Q, E_p, E_k, E_{elas}	joule (J)	$kg\ m^2\ s^{-2}$	Nm
power	P	watt (W)	$kg\ m^2\ s^{-3}$	$J\ s^{-1}$
pressure	P	pascal (Pa)	$kg\ m^{-1}\ s^{-2}$	$N\ m^{-2}$
charge	Q	coulomb (C)	A s	A s
potential difference	V	volt (V)	$kg\ m^2\ s^{-3}\ A^{-1}$	$J\ C^{-1}$
resistance	R	ohm (Ω)	$kg\ m^2\ s^{-3}\ A^{-2}$	$V\ A^{-1}$
magnetic field intensity	B	tesla (T)	$kg\ s^{-3}\ A^{-1}$	$NA^{-1}\ m^{-1}$
magnetic flux	Φ	weber (Wb)	$kg\ m^2\ s^{-2}\ A^{-2}$	$T\ m^2$
activity	A	becquerel (Bq)	s^{-1}	s^{-1}
absorbed dose	W/m	gray (Gy)	$m^2\ s^{-2}$	$J\ kg^{-1}$

Figure 102 Derived Units

CHAPTER 1

The nature of a person's height is different from that of the radius of a gold atom in that the person's height is macroscopic (visible to the naked eye) and can be measured with, say, a metre stick, whereas the diameter of the atom is microscopic and can be inferred from electron diffraction.

When a quantity involves the measurement of two or more fundamental quantities it is called a derived quantity, and the units of these derived quantities are called derived units. Some examples include acceleration (m s^{-2}), angular acceleration (rad s^{-2}) and momentum (kg m s^{-1} or N s). It should be noted that the litre (L) and the millilitre (mL) are often used for measuring the volume of liquid or the capacity of a container. The litre is a derived unit but not a SI unit. The equivalent SI unit is dm^3.

Some derived units are relatively complex and contain a number of fundamental units. Figure 102 lists the common relevant derived units and associated information.

Sometimes, it is possible to express the units in different derived units. This concept will become clear as the various topics are introduced throughout the course. For example, the unit of momentum can be kg m s^{-1} or N s. The unit of electrical energy could be J or W h or kJ or kWh (kilowatt-hour). In atomic and nuclear physics the unit of energy could be J or eV (electronvolt) where 1 eV = 1.6 × 10^{-19} J.

Note the use of the accepted SI format. For example, the unit for acceleration is written as m s^{-2} and not m/s/s. No mathematical denominators are used but rather inverse numerators are the preferred option.

Scientific notation and metric multipliers

Scientists tend to use scientific notation when stating a measurement rather than writing lots of figures. 1.2 × 10^6 is easier to write and has more significance than 1 200 000. In order to minimise confusion and ambiguity, all quantities are best written as a value between one and ten multiplied by a power of ten.

For example, we have that,

0.06 kg = 6 × 10^{-2} kg

140 kg = 1.4 × 10^2 kg or 1.40 × 10^2 kg depending on the significance of the zero in 140.

132.97 kg = 1.3297 × 10^2 kg

The terms 'standard notation' and 'standard form' are synonymous with scientific notation. The use of prefixes for units is also preferred in the SI system – multiple or submultiple units for large or small quantities respectively. The prefix is combined with the unit name. The main prefixes are related to the SI units by powers of three.

However, some other multiples are used.

1 000 000 000 m = 1 Gm

1 000 000 dm^3 = 1 Mdm^3

0.000 000 001 s = 1 ns

0.000 001 m = 1 µm

The main prefixes and other prefixes are shown in Figure 103 and some of these are also given in the Physics data booklet.

Multiple	Prefix	Symbol	Multiple	Prefix	Symbol
10^{24}	yotta	Y	10^{-1}	deci	d
10^{21}	zetta	Z	10^{-2}	centi	c
10^{18}	exa	E	10^{-3}	milli	m
10^{15}	peta	P	10^{-6}	micro	µ
10^{12}	tera	T	10^{-9}	nano	n
10^{9}	giga	G	10^{-12}	pico	p
10^{6}	mega	M	10^{-15}	femto	f
10^{3}	kilo	k	10^{-18}	atto	a
10^{2}	hecto	h	10^{-21}	zepto	z
10^{1}	deca	da	10^{-24}	yocto	y

Figure 103 Preferred and some common prefixes

Exercise 1.1

1. Which of the following isotopes is associated with the standard measurement of time?

    A. uranium–235
    B. krypton–86
    C. cesium–133
    D. carbon–12

2. Which one of the following lists a fundamental unit followed by a derived unit?

    A. ampere   mole
    B. coulomb   watt
    C. ampere   joule
    D. second   kilogram

3. Which one of the following is a fundamental unit?

   A. Kelvin
   B. Ohm
   C. Volt
   D. Newton

4. Which of the following is measured in fundamental units?

   A. velocity
   B. electric charge
   C. electric current
   D. force

5. The density in g cm^{-3} of a sphere with a radius of 3 cm and a mass of 0.54 kg is:

   A. 2 g cm^{-3}
   B. 2.0 × 10 g cm^{-3}
   C. 0.50 g cm^{-3}
   D. 5.0 g cm^{-3}

6. Convert the following to fundamental S.I. units:

   (a) 5.6 g
   (b) 3.5 μA
   (c) 3.2 dm
   (d) 6.3 nm
   (e) 2.25 tonnes
   (f) 440 Hz

7. Convert the following to S.I. units:

   (a) 2.24 MJ
   (b) 2.50 kPa
   (c) 2.7 km h^{-1}
   (d) 2.5 mm^2
   (e) 2.4 L
   (f) 3.6 cm^3
   (g) 230.1 M dm^3
   (h) 3.62 mm^3

8. Estimate the power of ten for each the following quantities:

   (a) your height in metres
   (b) the mass of a 250 tonne aeroplane in kilograms
   (c) the diameter of a hair in metres
   (d) human life span in seconds.

9. Calculate the distance in metres travelled by a parachute moving at a constant speed of 6 km h^{-1} in 4 min.

10. The force of attraction F in newtons between the Earth with mass M and the Moon with mass m separated by a distance r in metres from their centres of mass is given by the following equation:

    $$F = G M m \, r^{-2}$$

    where G is a constant called the Universal Gravitation constant. Determine the correct SI units of G.

11. Determine the SI units for viscosity η if the equation for the force on a sphere moving through a fluid is:

    $$F = 6\pi\eta r v$$

    where r is the radius of the sphere, v is the speed of the sphere in the fluid.

Significant figures

The concept of **significant figures** may be used to indicate the degree of accuracy or precision in a measurement. Significant figures (sf) are those digits that are known with certainty followed by the first digit that is uncertain.

Suppose you want to find the volume of a lead cube. You could measure the length l of the side of a lead cube with vernier calipers (refer Figure 112). Suppose this length was 1.76 cm. The volume l cm^3 from your calculator reads 5.451776. The measurement 1.76 cm was to three significant figures so the answer can only be to three significant figures. So that the volume = 5.45 cm^3.

The following rules are applied in this book.

1. All non-zero digits are significant. (22.2 has 3 sf).

2. All zeros between two non-zero digits are significant. (1007 has 4 sf).

3. For numbers less than one, zeros directly after the decimal point are not significant. (0.0024 has 2 sf).

4. A zero to the right of a decimal and following a non-zero digit is significant. (0.0500 has 3 sf).

5. All other zeros are not significant. (500 has 1 sf).

Scientific notation allows you to give a zero significance. For example, 10 has 1 sf but 1.00×10^1 has 3 sf.

6. When adding and subtracting a series of measurements, the least accurate place value in the answer can only be stated to the same number of significant figures as the measurement of the series with the least number of decimal places.

For example, if you add 24.2 g and 0.51 g and 7.134 g, your answer is 31.844 g which has increased in significant digits. The least accurate place value in the series of measurements is 24.2 g with only one number to the right of the decimal point. So the answer can only be expressed to 3sf. Therefore, the answer is 31.8 g or 3.18×10^1 g.

CHAPTER 1

7. When multiplying and dividing a series of measurements, the number of significant figures in the answer should be equal to the least number of significant figures in any of the data of the series.

For example, if you multiply 3.22 cm by 12.34 cm by 1.8 cm to find the volume of a piece of wood your initial answer is 71.52264 cm³. However, the measurement with the least number of significant figures is 1.8 cm with 2 sf. Therefore, the correct answer is 72 cm³ or 7.2×10^1 cm³.

8. When rounding off a number, if the digit following the required rounding off digit is 4 or less, you maintain the last reportable digit and if it is six or more you increase the last reportable digit by one. If it is a five followed by more digits except an immediate zero, increase the last reportable digit. If there is only a five with no digits following, increase reportable odd digits by one and maintain reportable even digits.

For example if you are asked to round off the following numbers to two significant numbers

6.42	becomes	6.4
6.46	becomes	6.5
6.451	becomes	6.5
6.498	becomes	6.5
6.55	becomes	6.6
6.45	becomes	6.4

As a general rule, round off in the final step of a series of calculations.

Exercise 1.2

1. A student measures the current in a resistor as 655 mA for a potential difference of 2.0 V. A calculator shows the resistance of the resistor to be 3.053 Ω. Which one of the following gives the resistance to an appropriate number of significant figures?

    A. 3.1 Ω
    B. 3.05 Ω
    C. 3.053 Ω
    D. 3 Ω

2. How many significant figures are indicated by each of the following:

    (a) 1247   (b) 1007
    (c) 0.034   (d) 1.20×10^7
    (e) 62.0   (f) 0.0025
    (g) 0.00250   (h) sin 45.2°
    (i) $\tan^{-1} 0.24$ (j) 3.2×10^{-16}
    (k) 0.0300   (l) 1.0×10^1

3. Express the following in standard notation (scientific notation):

    (a) 1250   (b) 30007
    (c) 25.10   (d) an area of 4 km² in m²
    (e) an object of 12.0 nm2 in m2

4. Calculate the area of a square with a side of 3.2 m.

5. Add the following lengths of 2.35 cm, 7.62 m and 14.2 m.

6. Calculate the volume of a rectangular block 1.52 cm by 103.4 cm by 3.1 cm.

9. A metal block has a mass of 2.0 g and a volume of 0.01 cm³. Calculate the density of the metal in g cm⁻³.

7. Round off the following to three significant figures:

    (a) 7.1249   (b) 2561
    (c) 2001   (d) 21256
    (e) 6.5647

8. Determine the following to the correct number of significant figures:

    (a) $(3.74 - 1.3) \times 2.12 \times 17.65$
    (b) $(2.9 + 3.2 + 7.1) \div 0.134$

12. Add 2.76×10^{-6} cm and 3.4×10^{-5} cm.

Orders of magnitude

The order of magnitude of a number is the power of ten closest to that number. Often, when dealing with very big or very small numbers, scientists are more concerned with the order of magnitude of a measurement rather than the precise value. For example, the number of particles in the Universe and the mass of an electron are of the orders of magnitude of 10^{80} particles and 10^{-30} kg. It is not important to know the exact values for all microscopic and macroscopic quantities because, when you are using the order of magnitude of a quantity, you are giving an indication of size and not necessarily a very accurate value.

The order of magnitude of large or small numbers can be difficult to comprehend at this introductory stage of the course. For example, 10^{23} grains of rice would cover Brazil to a depth of about one kilometre.

The order of magnitude of some relevant lengths in metres (m), masses in kilograms (kg) and times in seconds (s) are given in Figure 104.

MEASUREMENTS AND UNCERTAINTIES

Mass of Universe	10^{50} kg	Height of a person	10^0 m
Mass of Sun	10^{30} kg	1 gram	10^{-3} kg
Extent of the visible Universe	10^{25} m	Wavelength of visible light	10^{-6} m
Mass of the Earth	10^{25} kg	Diameter of an atom	10^{-10} m
Age of the Universe	10^{18} s	Period of visible light	10^{-15} s
One light year	10^{16} m	Shortest lived subatomic particle	10^{-23} s
Human light span	10^9 s	Passage of light across the nucleus	10^{-23} s
One year	10^7 s	Mass of proton	10^{-27} kg
One day	10^5 s	Mass of neutron	10^{-27} kg
Mass of car	10^3 kg	Mass of electron	10^{-30} kg

Figure 104   Range of magnitudes

Examples

1. The number 8 is closer to 10^1 (10) than 10^0 (1). So the order of magnitude is 10^1. Similarly, 10 000 has an order of magnitude of 10^4.

2. However, 4.3×10^3 has an order of magnitude of 10^4. The reason for this is if you use the log button on your calculator, the value of $4.3 \times 10^3 = 10^{3.633}$.

Therefore the order of magnitude is 10^4. So, the normal mathematical rounding up or down above or below 5 does not apply with order of magnitude values. In fact, $100^{0.5} = 3.16$. This becomes our 'rounding' value in determining the order of magnitude of a quantity.

Order of magnitude, for all its uncertainty, is a good indicator of size. Let's look at two ways of calculating the order of magnitude of the number of heartbeats in a human in a lifetime. The average relaxed heart beats at 100 beats per minute. Do you agree? Try the following activity:

Using a timing device such as a wristwatch or a stopwatch, take your pulse for 60 seconds (1 minute). Repeat this 3 times. Find the average pulse rate. Now, using your pulse, multiply your pulse per minute (say 100) × 60 minutes in an hour × 24 hours in a day × 365.25 days in a year × 78 years in a lifetime. Your answer is 4.102×10^9. Take the log of this answer, and you get $10^{9.613}$. The order of magnitude is 10^{10}. Now let us repeat this but this time we will use the order of magnitude at each step:

10^2 beats min^{-1} × 10^2 min h^{-1} × 10^1 h day^{-1} × 10^3 day yr^{-1} × 10^2 yr

The order of magnitude is 10^{10}.

Do the same calculations using your own pulse rate. Note that the two uncertain values here are pulse rate and lifespan. Therefore, you are only giving an estimate or indication. You are not giving an accurate value.

Ratios can also be expressed as differences in order of magnitude. For example, the diameter of the hydrogen atom has an order of magnitude of 10^{-10} m and the diameter of a hydrogen nucleus is 10^{-15} m. Therefore, the ratio of the diameter of a hydrogen atom to the diameter of a hydrogen nucleus is $10^{-10} / 10^{-15} = 10^5$ or five orders of magnitude.

The orders of magnitude of quantities in the macroscopic world are also important when expressing uncertainty in a measurement. This is covered in the next section of this chapter.

Exercise 1.3

1. The order of magnitude of 4 200 000 is:

   A. 10^4 C. 10^6
   B. 10^5 D. 10^7

2. Give the order of magnitude of the following quantities:

   (a)   20 000   (d) 7.4×10^{15}
   (b) 2.6×10^4 (e) 2.8×10^{-24}
   (c) 3.9×10^{17} (f) 4.2×10^{-30}

3. Give the order of magnitude of the following measurements:

   (a)   The mean radius of the Earth, 6 370 000 m.
   (b)   The half-life of a radioactive isotope 0.0015 s.
   (c)   The mass of Jupiter: 1 870 000 000 000 000 000 000 000 000 kg.
   (d)   The average distance of the Moon from the Earth is 380 000 000 m.
   (e)   The wavelength of red light 0.000 000 7 m.

4. The ratio of the diameter of the nucleus to the diameter of the atom is approximately equal to:

   A. 10^{-15} C. 10^{-5}
   B. 10^{-8} D. 10^{-2}

5. What is the order of magnitude of:

   (a) the time in seconds in a year.
   (b) the time for the moon to revolve around the earth in seconds.

6. A sample of a radioactive element contains 6.02×10^{23} atoms. It is found that 3.5×10^{10} atoms decay in one day.

   (a) Estimate the order of magnitude of the number of atoms that disintegrate in one second.
   (b) What is the ratio of the original number of atoms to the number of atoms that remain after one day in orders of magnitude?

Estimation

Many problems in physics and engineering require very precise numerical measurements or calculations. The numbers of significant digits in a measured quantity show the precision of that quantity. When English and French engineers used their excavation machinery to dig the tunnel under the North Sea, they hoped that they would meet at a common point. The laser guidance systems used allowed for a good degree of precision in the digging process. High precision is also required in cancer radiotherapy so that the cancerous cells are killed and the good body cells are not damaged in amounts greater than necessary. Also to our amazement and sadness we have witnessed too often on television the accuracy of laser guided missiles seeking out targets with incredible accuracy.

However, in other applications, estimation may be acceptable in order to grasp the significance of a physical phenomenon. For example, if we wanted to estimate the water needed to flush the toilet in your dwelling in a year, it would be reasonable to remove the lid off the toilet cistern (reservoir for storing water) and seeing whether there are graduations (or indicators) of the water capacity given on the inside on the cistern. When I removed the lid from my cistern, the water was at the 9 L (9 dm^3) mark and when I did a "water saving" flush, the water went to the 6 L mark. A long flush emptied the cistern. Now let's assume there are three people in the house who are using one long flush and five short flushes a day. This makes a total of $(3 \times 9 \text{ dm}^3) + (15 \times 3 \text{ dm}^3) = 72 \text{ dm}^3$ per day or an estimate of 10^2 dm^3 per day. There are 365.25 days in a year or an estimate of 10^3 (using the order of magnitude) days. So the water used by this family would be 2.6×10^4 dm^3 per year or an estimate of 10^4 dm^3. Neither answer is accurate because both answers are only rough estimates.

With practice and experience, we will get a feel for reasonable estimates of everyday quantities. We should be able to estimate approximate values of everyday quantities to the nearest order of magnitude to one or two significant digits. We need to develop a way to estimate an answer to a reasonable value.

Suppose we wanted to estimate the answer to:

$$16 \times 5280 \times 12 \times 12 \times 12 \times 5280$$

This can be estimated as:

$= (2 \times 10^1) \times (5 \times 10^3) \times (1 \times 10^1) \times (1 \times 10^1) \times (1 \times 10^1) \times (5 \times 10^3)$

$= 5 \times 10^{11}$

The calculator answer is 7.7×10^{11}. So our estimate gives a reasonable order of magnitude.

Exercise 1.4

1. A rough estimate of the volume of your body in cm^3 would be closest to:

   A. 2×10^3 C. 5×10^3
   B. 2×10^5 D. 5×10^5

2. Estimate the:

   (a) dimensions of this textbook in cm
   (b) mass of an apple in g
   (c) period of a heartbeat in s
   (d) temperature of a typical room in °C

3. Estimate the answer to:

   (a) $16 \times 5280 \times 5280 \times 5280 \times 12 \times 12 \times 12$
   (b) $3728 \times (470165 \times 10^{-14}) \div 278146 \times (0.000713 \times 10^{-5})$
   (c) $47816 \times (4293 \times 10^{-4}) \div 403000$

4. The universe is considered to have begun with the "Big Bang" event. The galaxies that have moved the farthest are those with the greatest initial speeds. It is believed that these speeds have been constant in time. If a galaxy 3×10^{21} km away is receding from us at 1.5×10^{11} km y^{-1}, calculate the age of the universe in years.

5. Give an estimate of the order of magnitude of the following:

   (a) The length of your arm in mm.
   (b) The quantity of milk you drink in a year in cm^3.
   (c) The mass of your backpack that contains your school materials in g.
   (d) The diameter of a human hair in mm.
   (e) The time you spend at school in a year in minutes.
   (f) The number of people in the country where you live.

1.2 Uncertainties and errors

> **NATURE OF SCIENCE:**
>
> Uncertainties: "All scientific knowledge is uncertain… if you have made up your mind already, you might not solve it. When the scientist tells you he does not know the answer, he is an ignorant man. When he tells you he has a hunch about how it is going to work, he is uncertain about it. When he is pretty sure of how it is going to work, and he tells you, "This is the way it's going to work, I'll bet," he still is in some doubt. And it is of paramount importance, in order to make progress, that we recognize this ignorance and this doubt. Because we have the doubt, we then propose looking in new directions for new ideas." (3.4)
>
> Feynman, Richard P. 1998. *The Meaning of It All: Thoughts of a Citizen-Scientist*. Reading, Massachusetts, USA. Perseus. P 13.
>
> © IBO 2014

Essential idea: Scientists aim towards designing experiments that can give a "true value" from their measurements, but due to the limited precision in measuring devices, they often quote their results with some form of uncertainty.

Understandings:
- Random and systematic errors
- Absolute, fractional and percentage uncertainties
- Error bars
- Uncertainty of gradient and intercepts

Random and systematic errors

Errors can be divided into two main classes, random errors and systematic errors.

Mistakes on the part of the individual such as:

- misreading scales.
- poor arithmetic and computational skills.
- wrongly transferring raw data to the final report.
- using the wrong theory and equations.
- are definite sources of error but they are not considered as experimental errors.

A systematic error causes a random set of measurements to be spread about a wrong value rather than being spread about the accepted value. It is a system or instrument error. Systematic errors can result from:

- badly made instruments.
- low quality instruments.
- an instrument having a zero error, a form of calibration.
- poorly timed actions.
- instrument parallax error.

Many analogue ammeters and voltmeters have a means of adjustment to remove zero offset error. When you click a stopwatch, your reaction time for clicking at the start and at the end of the measurement interval is a systematic error. The timing instrument and you are part of the system.

Systematic errors can, on most occasions, be eliminated or corrected before the investigation is carried out.

Random errors are due to variations in the performance of the instrument and the operator. Even when systematic errors have been allowed for, there exists error. Random errors can be caused by such things as:

- vibrations and air convection currents in mass readings.
- temperature variations.
- misreadings.
- variations in the thickness of a surface being measured (thickness of a wire).
- not collecting enough data.
- using a less sensitive instrument when a more sensitive instrument is available.
- human parallax error (one has to view the scale of the meter in direct line, and not to the sides of the scale in order to minimise parallax error).

As well as obtaining a series of measurements with the correct units for the measurements, an indication of the experimental error or degree of uncertainty in the measurements and the solution is required. The higher the accuracy and precision in carrying out investigations, the lower will the degree of uncertainty be. The meanings of the words accuracy and precision are clearly defined in scientific fields.

Accuracy is an indication of how close a measurement is to the accepted value indicated by the relative or percentage error in the measurement. An accurate experiment has a low systematic error.

Precision is an indication of the agreement among a number of measurements made in the same way indicated by the absolute error. A precise experiment has a low random error.

Suppose a technician was fine-tuning a computer monitor by aiming an electron gun at a pixel in the screen as shown in Figure 105.

Chapter 1

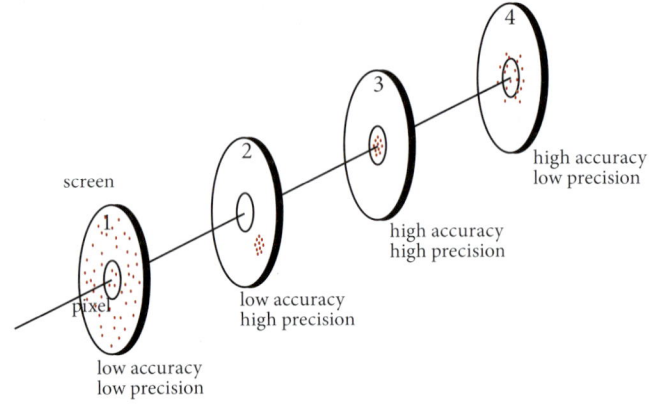

Figure 105   Precision and accuracy

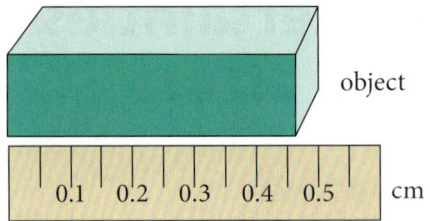

Figure 106   Linear measurement

In case 1 there is low accuracy and precision. The technician needs to adjust the collimator to reduce the scattering of electrons, and to change the magnetic field so the electrons hit the pixel target. In case 2, the electron gun has been adjusted to increase precision but the magnetic field still needs adjustment. In case 3, both adjustments have been made. Can you give an explanation for case four?

Often the random error is not revealed until a large sample of measurements is taken. So taking a required number of readings/samples not only reveals random uncertainty but also helps to reduce it. Consistent experimental procedures can minimise random uncertainty.

Random errors can also be reduced by choosing an instrument that has a high degree of precision. When measuring mass, it would be best to choose a digital balance that can read to 2 decimal places rather than a top pan balance or a digital balance that can read to 1 decimal place. Further reduction of random error can be obtained by reducing variations such as air currents, vibrations, temperature variation, loss of heat to the surroundings.

However, you should be aware that repeating measurements may reduce the random uncertainty but at the same time the systematic error will not be reduced.

Absolute, fractional and percentage uncertainties

The limit of reading of a measurement is equal to the smallest graduation of the scale of an instrument.

The maximum degree of uncertainty of a measurement is equal to half the limit of reading.

When a measuring device is used, more often than not the measurement falls between two graduations on the scale being used. In Figure 106, the length of the block is between 0.4 cm and 0.5 cm.

The limit of reading is 0.05 cm and the uncertainty of the measurement is ± 0.025 cm.

The length is stated as (0.47 ± 0.02) cm. (Uncertainties are given to 1 significant figure).

The smallest uncertainty possible with any measuring device is half the limit of reading. However, most investigations generate an uncertainty greater than this. Figure 107 lists the uncertainty of some common laboratory equipment.

Metre rule	± 0.05 cm
Vernier calipers	± 0.005 cm
Micrometer screw gauge	± 0.005 mm
50 cm^3 measuring cylinder	± 0.2 cm^3
10 cm^3 measuring cylinder	± 0.1 cm^3
Electric balance	± 0.005 g
Watch second hand	± 0.5 s
Digital timer	± 0.0005 s
Spring balance (0–20N)	± 0.1 N
Resistor	± 2%

Figure 107   Equipment uncertainties

Absolute uncertainty is the size of an error and its units. In most cases it is not the same as the maximum degree of uncertainty (as in the previous example) because it can be larger than half the limit of reading. The experimenter can determine the absolute error to be different to half the limit of reading provided some justification can be given. For example, mercury and alcohol thermometers are quite often not as accurate as the maximum absolute uncertainty.

Fractional (relative) uncertainty equals the absolute uncertainty divided by the measurement as follows. It has no units.

$$\text{Fractional uncertainty} = \frac{\text{absolute uncertainty}}{\text{measurement}}$$

Percentage uncertainty is the fractional uncertainty multiplied by 100% to produce a percentage as follows:

MEASUREMENTS AND UNCERTAINTIES

$$= \frac{\text{accepted value} - \text{experimental value}}{\text{accepted value}} \times 100$$

For example, if a measurement is written as (9.8 ± 0.2) m, then there is a

- limit of reading = 0.1 m
- uncertainty = 0.05 m
- absolute uncertainty = 0.2 m
- fractional uncertainty = 0.2/9.8 = 0.02
- and percentage uncertainty = 0.02 × 100% = 2%

Percentage uncertainty should not be confused with percentage discrepancy or percentage difference which is an indication of how much your experimental answer varies from the known accepted value of a quantity. Percentage discrepancy is often used in the conclusion of laboratory reports.

$$= \frac{(\text{accepted value} - \text{experimental value})}{\text{accepted value}} \times 100$$

Note that errors are stated to only one significant figure

1. The arithmetic mean – averaging

When a series of readings are taken for a measurement, then the arithmetic mean of the readings is taken as the most probable answer, and the greatest deviation or residual from the mean is taken as the absolute error.

Study the following data in Table 108 for the thickness of a copper wire as measured with a micrometer screw gauge:

Reading/mm	5.821	5.825	5.820	5.832	5.826	5.826	5.828	5.824
Residual/mm	−0.004	0	−0.005	+0.007	+0.001	+0.001	+0.003	−0.001

Figure 108   Sample measurements

The sum of the readings = 46.602 and so the mean of the readings is 5.825.

Then, the value for the thickness is 5.825 ± 0.007 mm

This method can be used to suggest an appropriate uncertainty range for trigonometric functions. Alternatively, the mean, maximum and minimum values can be calculated to suggest an appropriate uncertainty range. For example, if an angle is measured as 30 ± 2°, then the mean value of sin 30 = 0.5, the maximum value is sin 32 = 0.53 and the minimum value is sin 28 = 0.47. The answer with correct uncertainty range is 0.5 ± 0.03. (Analysis of uncertainties will not be expected for trigonometric or logarithmic functions in examinations).

2. Addition, subtraction and multiplication involving errors

When *adding* measurements, the error in the sum is the sum of the absolute error in each measurement taken. Similarly, when subtracting measurements, add the absolute errors.

- If y = a ± b, then $\Delta y = \Delta a \pm \Delta b$

For example, the sum of (2.6 ± 0.5) cm and (2.8 ± 0.5) cm is (5.4 ± 1) cm.

If you place two metre rulers on top of each other to measure your height, remember that the total error is the sum of the uncertainty of each metre rule. (0.05 cm + 0.05 cm). If there is a zero offset error on an instrument, say a newton balance, you will have to subtract the given reading from the zero error value.
So (25 ± 2.5) N − (2 ± 2.5) equals (23 ± 5)N.

3. Multiplication and division involving errors

When multiplying and dividing measurements, add the fractional or percentage errors of the measurements being multiplied/divided. The absolute error is then the fraction or percentage of the most probable answer.

- If $y = \dfrac{ab}{c}$

   then $\dfrac{\Delta y}{y} = \dfrac{\Delta a}{a} + \dfrac{\Delta b}{b} + \dfrac{\Delta c}{c}$

Example

What is the product of (2.6 ± 0.5) cm and (2.8 ± 0.5) cm?

Solution

First, we determine the product
2.6 cm × 2.8 cm = 7.28 cm²
Fractional error 1 = 0.5/2.6 = 0.192
Fractional error 2 = 0.5/2.8 = 0.179
Sum of the Fractional errors = 0.371 or 37.1%
Absolute error = 0.371 × 7.28 cm² or 37.1% × 7.28 cm² = 2.70 cm²
Errors are expressed to one significant figure = 3 cm²
The product is equal to (7.3 ± 3) cm²

4. Uncertainties and powers

When raising a measurement to the nth power, multiply the percentage uncertainty by *n*, and when extracting the nth root, divide the percentage uncertainty by *n*.

- If $y = a^n$

   then $\dfrac{\Delta y}{y} = \left| n \dfrac{\Delta a}{a} \right|$

11

Chapter 1

For example, if the length x of a cube is (2.5 ± 0.1) cm, then the volume will be given by $x^3 = 15.625$ cm^3. The percentage uncertainty in the volume

$= 3(\frac{0.1}{2.5} \times 100) = 12\%$.

Therefore, 12% of 15.625 = 1.875.
Volume of the cube = (16 ± 2) cm^3.
If $x = (9.0 \pm 0.3)$ m, then $\sqrt{x} = x^{½} = (3.0 \pm 0.15)$ m
$= (3.0 \pm 0.02)$ m.

5. Measuring length with vernier calipers or a micrometer screw gauge

Two length measuring devices with lower uncertainty than the metre rule are vernier calipers and the micrometer screw gauge. The uncertainty of these instruments was given in Figure 109.

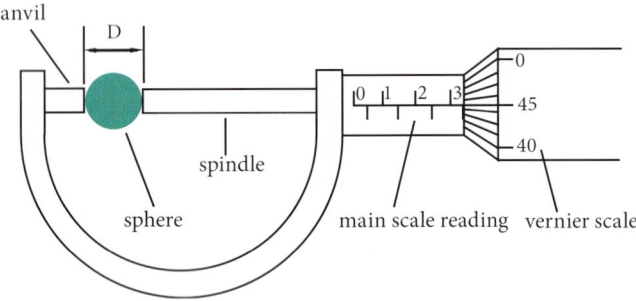

Figure 109 A micrometer screw gauge

In Figure 110, the reading on the micrometer screw gauge is 3.45 mm. You can see that the thimble (on the right of the gauge) is to the right of the 3 mm mark but you cannot see the 3.5 mm mark on the main scale. The vernier thimble scale is close to the 45 mark.

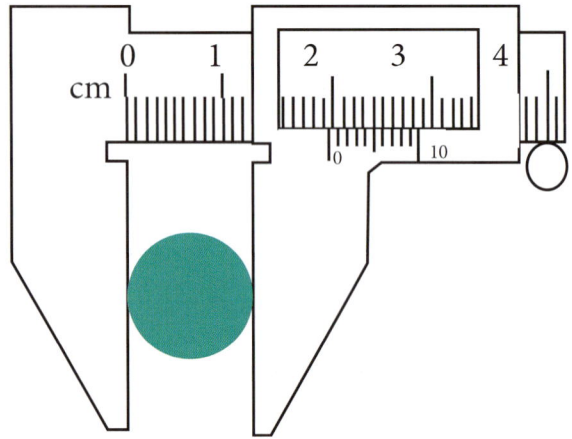

Figure 110 Vernier calipers

In Figure 110, the reading on the vernier calipers is 1.95 cm. The vertical line showing zero on the vernier scale lies between 1.9 cm and 2.0 cm. The vertical graduation on the vernier scale that is aligned to the main scale best is the fifth graduation.

Exercise 1.5

1. Consider the following measured quantities

   (a) 3.00 ± 0.05 m   (b) 12.0 ± 0.3 m

   Which alternative is the best when the accuracy and precision for a and b are compared?

	a	b
A	Low accuracy	Low precision
B	Low accuracy	High precision
C	High accuracy	Low precision
D	High accuracy	High precision

2. A voltmeter has a zero offset error of 1.2 V. This fault will affect:

   A. neither the precision nor the accuracy of the readings.
   B. only the precision of the readings.
   C. only the accuracy of the readings.
   D. both the precision and the accuracy of the readings.

3. A student measures the mass m of a ball. The percentage uncertainty in the measurement of the mass is 5%. The student drops the ball from a height h of 100 m. The percentage uncertainty in the measurement of the height is 7%. The calculated value of the gravitational potential energy of the ball will have an uncertainty of: (use $E_p = mgh$)

   A. 2%   C. 7%
   B. 5%   D. 12%

4. The electrical power dissipation P in a resistor of resistance R when a current I is flowing through it is given by the expression:

   $P = I^2 R$.

   In an investigation, I was determined from measurements of P and R. The uncertainties in P and in R are as shown below.

   $P \pm 4\%$ $R \pm 10\%$

   The uncertainty in I would have been most likely:

   A. 14 %   C. 6 %
   B. 7 %   D. 5 %

5. The mass of the Earth is stated as 5.98 × 10²⁴ kg. The absolute uncertainty is:

   A. 0.005   C. 0.005 × 10²⁴ kg
   B. 0.005 kg   D. 0.005 × 10²⁴

6. If a = (20 ± 0.5) m and b = (5 ± 1) m, then 2a − b should be stated as:

   A. (35 ± 1.5) m   C. (35 ± 0.0) m
   B. (35 ± 2) m   D. (5 ± 2) m

7. How should a calculation result be stated if it is found to be 0.931940 μm with an absolute error of ± 0.0005 μm.

8. This question concerns the micrometer screw gauge in the Figure shown below.

   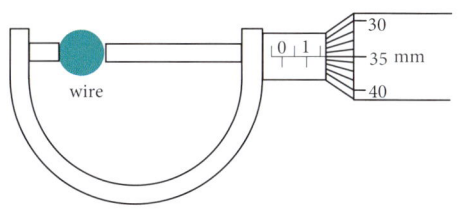

   (a) What is the reading and error on the micrometer?
   (b) The thickness of the wire being measured varies over its length. What sort of error would this be?

9. A student records the following currents in amperes A when the potential difference V across a resistor is 12V:

   0.9 A   0.9 A   0.85 A   0.8 A   1.2 A   0.75 A
   0.8 A   0.7 A   0.8 A   0.95 A

   (a) Would you disregard any of the readings? Justify your answer.
   (b) Calculate the current and its uncertainty.

10. A spring balance reads 0.5 N when it is not being used. If the needle reads 9.5 N when masses are attached to it, then what would be the correct reading to record (with uncertainty)?

11. Five measurements of the length of a piece of string were recorded in metres as:

    1.48   1.46   1.47   1.50   1.45

    Record a feasible length of the string with its uncertainty.

12. A metal cube has a side length of (3.00 ± 0.01) cm. Calculate the volume of the cube.

13. An iron cube has sides (10.3 ± 0.2) cm, and a mass of (1.3 ± 0.2) g. What is the density of the cube in g cm⁻³?

14. The energy E of an α–particle is (4.20 ± 0.03) MeV. How should the value and uncertainty of $E^{-1/2}$ be stated?

15. Suggest an appropriate answer with uncertainty range for sin θ if θ = 6° ± 5°.

Error bars

When an answer is expressed as a value with uncertainty such as (2.3 ± 0.1) m, then the uncertainty range is evident. The value lies between 2.4 (2.3 + 0.1) m and 2.2 (2.3 − 0.1) m. In Physics, we often determine the relationship that exists between variables. To view the relationship, we can perform an investigation and plot a graph of the dependant variable (y–axis) against the independent variable (x–axis). This aspect will be discussed fully in this section.

Consider a spring that has various weights attached to it. As a heavier weight is attached to a spring, the spring extends further from its equilibrium position. Figure 111 shows some possible values for this weight/extension investigation.

Force ± 5 N	100	150	200	250	300
Extension ± 0.2 cm	3.0	4.4	6.2	7.5	9.1

Figure 111   Extension of a spring

When a graph of force versus extension is plotted, the line of best fit does not pass through every point. An error bar can be used to give an indication of the uncertainty range for each point as shown in Figure 111.

In the vertical direction, we draw a line up and down for each point to show the uncertainty range of the force value. Then we place a small horizontal marker line on the extreme uncertainty boundary for the point.

In the horizontal direction, we draw a line left and right for each point to show the uncertainty range of the extension value. Then we place a small vertical marker line on the extreme uncertainty boundary for the point.

CHAPTER 1

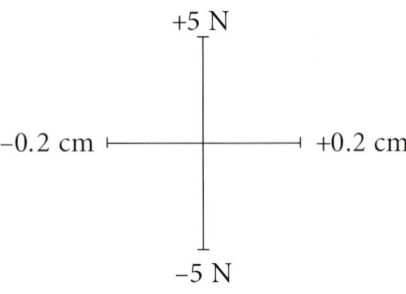

Figure 112   Error bars

When all the points in Figure 113 are plotted on a graph, then the line of best fit with the appropriate error bars is shown in Figure 117. You can see that the line of best fit lies within the error bar uncertainty range. The line of best fit is interpolated between the plotted points. The line of best fit is extrapolated outside the plotted points.

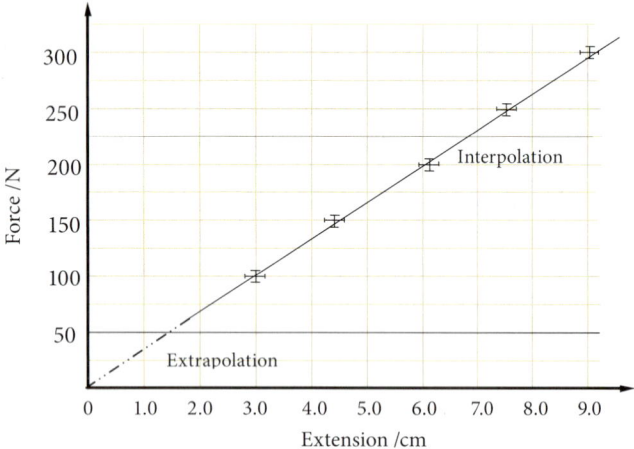

Figure 113   Example of error bars

Error bars will not be expected for trigonometric or logarithmic functions in this course.

Random uncertainty and uncertainties in the slope and intercepts of a straight–line graph

Graphs are very useful for analysing the data that is collected during investigations and is one of the most valuable tools used by physicists and physics students because:

(a) they give a visual display of the relationship that exists between two or more variables.

(b) they show which data points obey the relationship and which do not.

(c) they give an indication of the point at which a particular relationship ceases to be true.

(d) they can be used to determine the constants in an equation relating two variable quantities.

Some of these features are shown in the graphs in Figure 114. Notice how two variables can be drawn on the same axis as in Figure 114 (b).

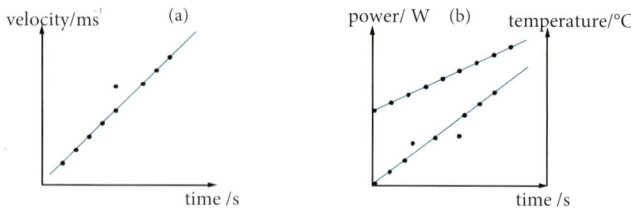

Figure 114   Examples of graphs

1. Choice of axes

A variable is a quantity that varies when another quantity is changed. A variable can be an independent variable, a dependent variable or a controlled variable. During an experiment, an independent variable is altered while the dependent variable is measured.

Controlled variables are the other variables that may be present but are kept constant. For example, when measuring the extension of a spring when different masses are added to it, the weight force is altered and the extension from the spring's original length is measured. The force would be the independent variable plotted on the x-axis and the extension would be the dependant variable plotted on the y-axis. (The extension depends on the mass added). Possible controlled variables would be using the same spring, the same measuring device and the same temperature control.

The values of the independent variable are plotted on the x-axis (the abscissa), and the values of the dependent variable are plotted on the y-axis (the ordinate), as shown in Figure 115.

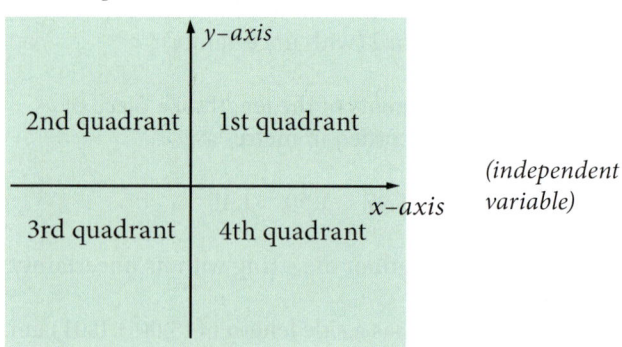

Figure 115   Use of axes

14

It is not always clear which variable is the dependent and which is the independent. When time is involved it is the independent variable. In many electrodynamic and electromagnetic experiments the potential difference (voltage) or the current can be varied to see what happens to the other variable – either could be the independent variable. Most experimental results will be plotted in the first quadrant. However, they can extend over the four quadrants as is the case with aspects of simple harmonic motion and waves, alternating current and the cathode ray oscilloscope to name a few.

When you are asked to plot a graph of displacement against time or to plot a graph of force versus time, the variable first mentioned is plotted on the y-axis. Therefore displacement and force would be plotted on the y-axis in the two given examples.

These days, graphs are quickly generated with graphic calculators and computer software. This is fine for quickly viewing the relationship being investigated. However, the graph is usually small and does not contain all the information that is required, such as error bars. Generally, a graph should be plotted on a piece of 1 mm or 2 mm graph paper and the scale chosen should use the majority of the graph paper. In the beginning of the course, it is good practice to plot some graphs manually. As the course progresses, software packages that allow for good graphing should be explored.

2. Scales

In order to convey the desired information, the size of the graph should be large, and this usually means making the graph fill as much of the graph paper as possible. Choose a convenient scale that is easily subdivided.

3. Labels

Each axis is labelled with the name and/or symbols of the quantity plotted and the relevant unit used. For example, you could write current/A or current (A). The graph can also be given a descriptive title such as 'graph showing the relationship between the pressure of a gas and its volume at constant temperature'.

4. Plotting the points

Points are plotted with a fine pencil cross or as a circled dot. In many cases, error bars are required. Of course, you are strongly recommended to use a graphing software package. These are short lines drawn from the plotted points parallel to the axes indicating the absolute error of the measurement. A typical graph is shown in Figure 116.

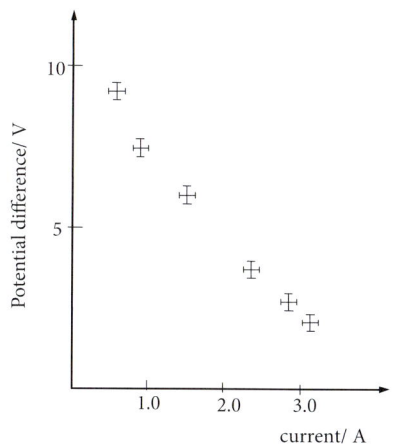

Figure 116   An example of plotting points

5. Lines of best fit

The line or curve of best fit is called the line of best fit. When choosing the line or curve of best fit it is practical to use a transparent ruler. Position the ruler until it lies along the ideal line. Shapes and curves can be purchased to help you draw curves. The line or curve does not have to pass through every point. Do not assume that the line should pass through the origin (0,0) as lines with an x-intercept or y-intercept are common.

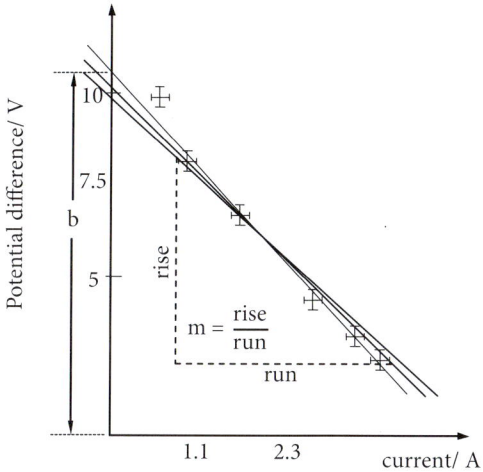

Figure 117   A graph showing the error range

Normally, the line of best fit should lie within the error range of the plotted points as shown in Figure 117. The uncertainty in the slope and intercepts can be obtained by drawing the maximum gradient and minimum gradient lines passing through the error bars. The line of best fit should lie in between these two lines. The uncertainty in the y-intercept can be determined as being the difference in potential difference between the best fit line and the maximum/minimum lines. The uncertainty in the slope can be obtained using the same procedure. However, do not forget that you are dividing. You will therefore have to add the percentage errors to find the final uncertainty.

15

Chapter 1

In the graph, the top plotted point appears to be a data point that could be discarded as a mistake or a random uncertainty.

Area under a straight-line graph

The area under a straight-line graph is a useful tool in Physics. Consider the two graphs of Figure 118.

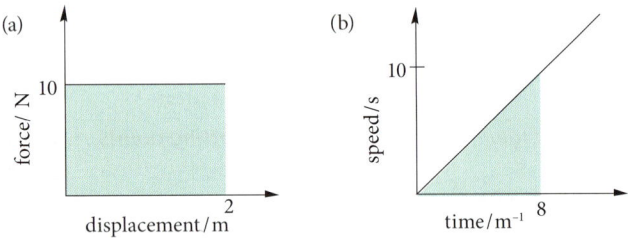

Figure 118 The area under a graph

Two equations that you will become familiar with in Chapter 2 are:

work (J) = force (N) × displacement (m)

distance (m) = speed (m s^{-1}) × time (s)

In these examples, the area under the straight line (Figure 1.22(a)) will give the values for the work done (10 N × 2 m = 20 J).

In Figure 1.22(b), the area enclosed by the triangle will give the distance travelled in the first eight seconds (i.e., ½ × 8 s × 10 m s^{-1} = 40 m).

Graphical analysis and determination of relationships

Straight–line Equation

The 'straight line' graph is easy to recognise and analyse. Often the relationships being investigated will first produce a parabola, a hyperbola or an exponential growth or decay curve. These can be transformed to a straight line relationship (as we will see later).

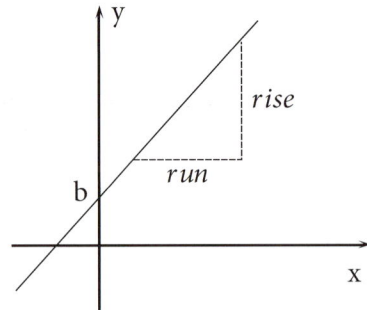

Figure 119 A straight line graph

The general equation for a straight line is:

$y = mx + b$

where y is the dependent variable, x is the independent variable, m is the slope or gradient given by:

$$\frac{\text{vertical rise}}{\text{horizontal run}} = \frac{\Delta y}{\Delta x}$$

and b is the point where the line intersects the y-axis.

In short, an 'uphill' slope is positive and a 'downhill' slope is negative. The value of m has units.

Consider Figure 124 below. The slope of the graph shown can be determined. Past IB markschemes suggest that you should use half the line of best fit when drawing a triangle.

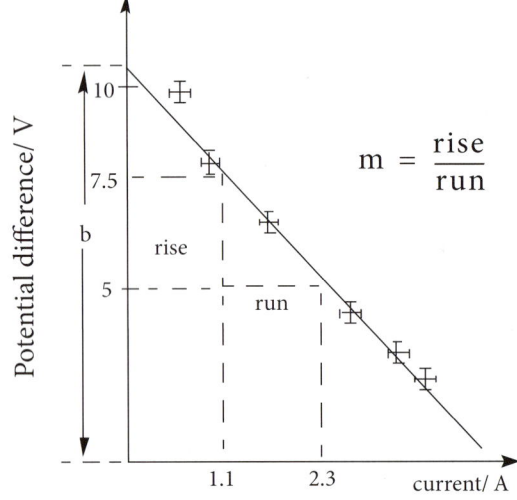

Figure 120 Determining the slope of the graph

$$m = \frac{\text{rise}}{\text{run}} = \frac{\Delta y}{\Delta x} = \frac{5.0 - 7.5}{2.3 - 1.1} = -2.08 \text{ V A}^{-1}$$

The equation for the graph shown is generally given as

$V = \varepsilon - Ir$ or $V = -Ir + \varepsilon$

Because V and I are variables, then m = -r and b = ε.

MEASUREMENTS AND UNCERTAINTIES

If $T = 2\pi\sqrt{\left(\dfrac{l}{g}\right)}$

where T and l are the variables, and 2π and g are constants, then T plotted against l will not give a straight-line relationship. But if a plot of T against $\sqrt{l}$ or T^2 against l is plotted, it will yield a straight line.

These graphs are shown in below.

(i) T vs l (ii) T vs $\sqrt{l}$ (iii) T^2 vs l

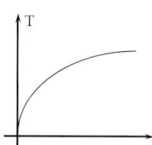 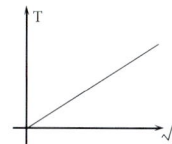 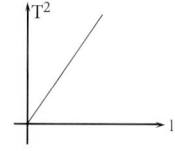

Figure 121   *Some different relationships*

Standard Graphs

1. Linear

The linear graph shows that y is directly proportional to x

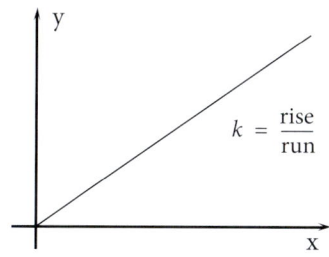

i.e., $y \, \alpha \, x$ or $y = k\,x$ where k is the constant of proportionality.

2. Parabola

The parabola shows that y is directly proportional to x^2. That is, $y \, \alpha \, x^2$ or $y = k\,x^2$ where k is the constant of proportionality.

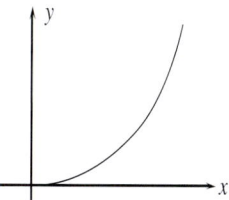 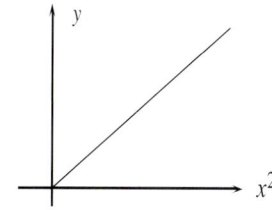

In the equation $s = u\,t + \tfrac{1}{2}\,a\,t^2$, where,

s = displacement in m
u = initial velocity in m s^{-1}
a = acceleration in m s^{-2}
t = time in s

then, $s \, \alpha \, t^2$, $k = \tfrac{1}{2}$ and u = y–intercept

3. Hyperbola

The hyperbola shows that y is inversely proportional to x or y is directly proportional to the reciprocal of x.

i.e., $y \, \alpha \, 1/x$ or $xy = k$

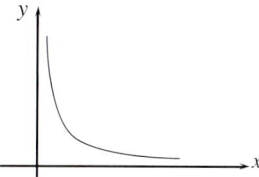 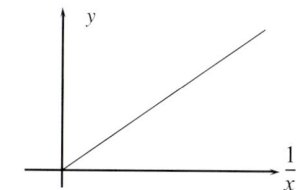

An example of an inverse proportionality is found in relating pressure, P, and volume, V, of a fixed mass of gas at constant temperature

$P \, \alpha \, \dfrac{1}{V} \Rightarrow P = \dfrac{k}{V}$

or $PV = k$ (= constant)

An inverse square law graph is also a hyperbola. The force F between electric charges at different distances d is given by:

$F = \dfrac{k\,q_1 q_2}{r^2}$

A graph of F versus d has a hyperbolic shape, and a graph of F versus $1/r^2$ is a straight line.

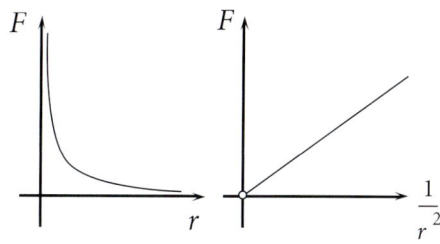

4. Sinusoidal

A sinusoidal graph is a graph that has the shape of a sine curve and its mathematics is unique. It can be expressed using degrees or radians.

The wavelength λ is the length of each complete wave in metres and the amplitude A is the maximum displacement from the x-axis. In the top sinusoidal graph (on page 18) the wavelength is equal to 5 m and the amplitude is equal to 2 m.

The frequency f of the wave is the number of waves occurring in a second measured in hertz (Hz) or s^{-1}. The period T is the time for one complete wave. In the bottom sinusoidal wave, the frequency is 4 Hz, and the period is 0.2 s.

Chapter 1

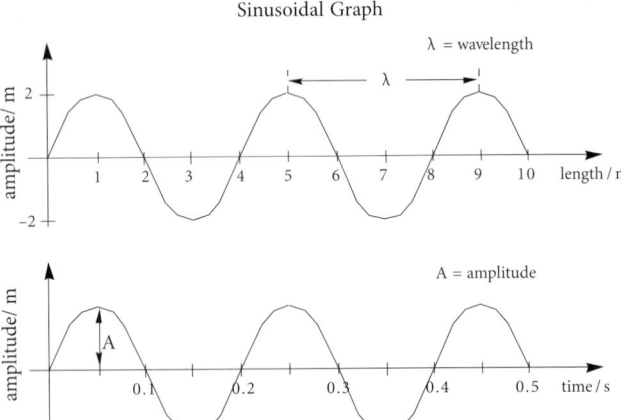

The equations for these graphs will be explored in Chapter 4 when you will study oscillations and simple harmonic motion.

Logarithmic Functions

Exponential and logarithmic graphs

If the rate of change of a quantity over time depends on the remaining amount of matter, the rate of change may well be exponential. Certain elements undergo exponential decay when they decay radioactively. When bacteria reproduce, the change in the number of bacteria over time is given by an exponential growth.

Consider a sample of a material with an original number of atoms N_0 that undergo radioactive decay as shown in Figure 122. It can be shown that the number of atoms N left to decay after a period of time t is given by

$$N = N_0 e^{-kt}$$

From the logarithmic equations given in Appendix 1, it can be shown that

$$\ln N = -kt + \ln N_0$$

Therefore when $\ln N$ is plotted against time the slope of the straight line produced is equal to $-k$.

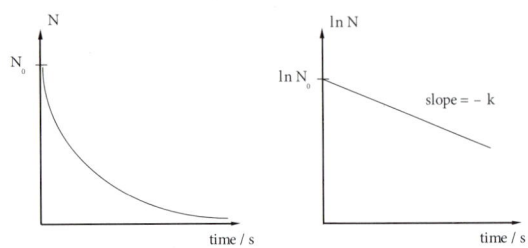

Figure 122   Logarithmic Graphs

Now let us examine a logarithmic function. In thermodynamics, the pressure p versus volume V curve for an adiabatic change at constant temperature is given by the equation

$$pV^\gamma = k \quad \text{(where } \gamma \text{ and k are constants)}$$

If we take the log of both sides then the equation will be

$$\log p + \gamma \log V = \log k$$

Now if the equation is rearranged into the straight line form y = mx + b, we get

$$\log p = -\gamma \log V + \log k$$

If a graph of log p versus log V is plotted, a straight line is obtained with the gradient being equal to γ and the y-intercept being equal to log k.

Exercise 1.6

1. It can be shown that the pressure of a fixed mass of gas at constant temperature is inversely proportional to the volume of the gas. If a graph of pressure versus volume was plotted, the shape of the graph would be:

   A. a straight line.
   B. a parabola.
   C. an exponential graph.
   D. a hyperbola.

2. Newton showed that a force of attraction F of two masses m and M separated by a distance d was given by $F \propto Mm/d^2$. If m and M are constant, a graph of F versus d^{-2} would have which shape?

   A. a parabola
   B. a straight line
   C. a hyperbola
   D. an exponential shape

3. The resistance R_θ of a coil of wire increases as the temperature θ is increased. The resistance R_θ at a temperature can be expressed as $R_\theta = R_0 (1 + \mu\theta)$ where μ is the temperature coefficient of resistance. Given the following data, plot a graph that will allow you to determine R_0 and μ.

RΩ / Ω	23.8	25.3	26.5	28.1	29.7	31.7
θ / °C	15	30	45	60	80	100

18

4. Given that $s = \frac{1}{2}gt^2$ where s is the distance travelled by a falling object in time t, and g is a constant. The following data are provided:

s (m)	5.0	20	45	80
T^2 (s^2)	1.0	4.0	9.0	16.0

Plot a relevant graph in order to determine the value of the constant g.

5. It can be shown that $V = \dfrac{RE}{(R+r)}$ where E and r are constants.

In order to obtain a straight line graph, one would plot a graph of

A. $\dfrac{1}{V}$ against R

B. V against R

C. $\dfrac{1}{V}$ against $\dfrac{1}{R}$

D. V against $\dfrac{1}{R}$

6. The magnetic force F between 2 magnets and their distance of separation d are related by the equation $F = kd^n$ where n and k are constants.

    (a) What graph would you plot to determine the values of the two constants?
    (b) From the graph how could you determine n and k?

7. The intensity I of a laser beam passing through cancer growth cells decreases exponentially with the thickness x of the cancer tissue according to the equation $I = I_0 e^{-\mu x}$, where I_0 is the intensity before absorption and μ is a constant for cancer tissue.

What graph would you draw to determine the values of I_0 and μ?

Uncertainties of gradient and intercepts

Students must be able to determine the uncertainties in the slope and intercepts of a straight line graph. To determine the uncertainty of the gradient, draw the maximum gradient and minimum gradient lines so that they touch the extremities of the first and last error bars as shown in Figure 123.

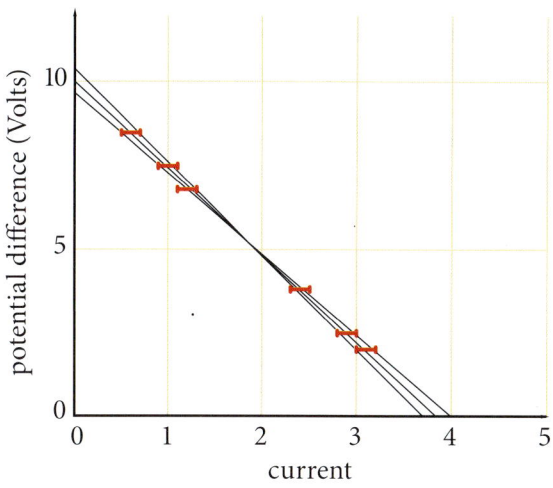

Figure 123 Maximum and minimum lines of best fit

The gradient of the line of best fit is -2.595 V A^{-1}. The maximum gradient is -2.800 V A^{-1} that varies by 0.205 from the best-fit gradient. The minimum gradient is -2.420 V A^{-1} that varies by 0.175 from the best-fit gradient. The greatest residual is 0.205 V A^{-1}. Therefore, a possible answer for the internal resistance could be (2.6 ± 0.2) W.

In this example, the y intercept would be the emf of the electrical source. So the value obtained from the graph could be (10.0 ± 0.4) V.

Example

The schematic diagram in Figure 124 demonstrates an experiment to determine Planck's constant using the photoelectric effect proposed by Albert Einstein in 1905 as outlined in Topic 12 of the syllabus. The wavelength (λ) of light from the light source incident on a metal photo-emissive plate of a photoelectric cell is varied, and the stopping voltage V_s applied across the photoelectric cell is measured.

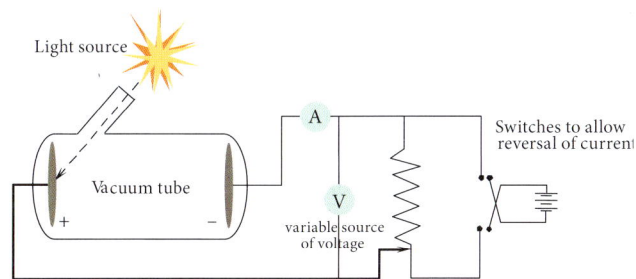

Figure 124 Determining Planck's constant

CHAPTER 1

The following values were obtained for different light radiation colours

Light Radiation Colour	Stopping Voltage Vs ±0.05 V	λ ± 0.3 × 10⁻⁷ m
Red	1.20	6.1
Orange	1.40	5.5
Yellow	1.55	5.2
Green	1.88	4.6
Blue	2.15	4.2
Violet	2.50	3.8

Figure 125 *Data For Planck's constant*

It can be shown that for this experiment

$\frac{hc}{\lambda} = hf = \Phi + eV_s$ where h is Planck's Constant

c is the speed of light constant 3.0×10^8 m s⁻¹

λ is the wavelength in m and f is the frequency in Hz

Φ is the work function.

e is the charge on an electron (1.6×10^{-19} C)

(a) Copy Figure 125, add 2 more columns and complete the frequency and the uncertainties columns for each colour of light radiation in the table.

Because the wavelength is given to two significant figures, the frequency can only be given to two significant figures.

For division, to find the frequency from hc/λ, the relative uncertainty in the frequency has to be calculated for each wavelength. For example, for dark red:

$f = \frac{c}{\lambda} = \frac{3.0^8}{6.1^{-7}} = 1.6^{14}$

the relative uncertainty $= 0.3 \times 10^{-7} \div 6.1 \times 10^{-7} = 0.0492$

the absolute uncertainty $= 0.0492 \times 1.6 \times 10^{14}$

$= \pm 0.07 \times 10^{14}$ Hz

In this case, the absolute uncertainty is not half the limit of reading as the absolute uncertainty of the wavelength was given as $\pm 0.3 \times 10^{-7}$ m. Remember that the minimum possible absolute uncertainty is half the limit of reading which would be $\pm 0.05 \times 10^{-7}$ m.

Light Radiation Colour	Stopping Voltage Vs ±0.05 V	λ ±0.3 × 10⁻⁷m	Frequency × 10¹⁴ Hz	Uncertainty ±10¹⁴ Hz
Red	1.20	6.1	1.6	0.07
Orange	1.40	5.5	1.8	0.09
Yellow	1.55	5.2	1.9	0.1
Green	1.88	4.6	2.2	0.1
Blue	2.15	4.2	2.4	0.2
Violet	2.50	3.8	2.6	0.2

Figure 126 *Data showing uncertainties*

(b) Plot a fully labelled graph with stopping voltage on the vertical axis against the reciprocal of the wavelength on the horizontal axis. Allow for a possible negative y–intercept.

Now you can put in the error bars for each point and label the axis. There will be a negative y–intercept.

Mark in the gradient and the y–intercept.

The required graph is shown in Figure 127. Note the maximum and minimum lines and the line of best fit, the gradient of the straight line of best fit and the value of the negative y-intercept

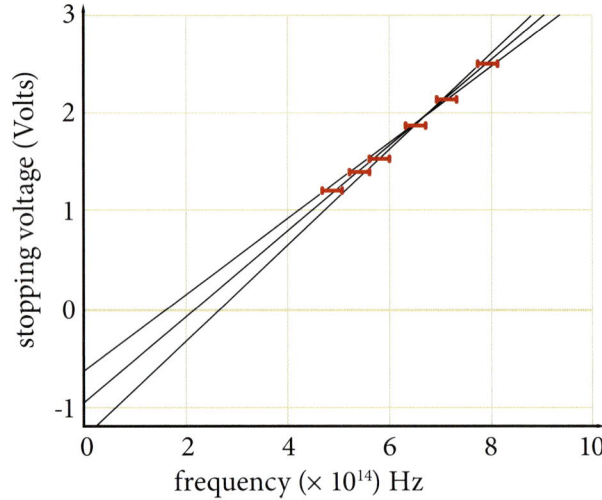

Figure 127 *Graph showing best-fit lines*

(c) Calculate Planck's Constant by graphical means and compare your value with the theoretical value of 6.63×10^{-34} J s.

20

The equation given at the start of this example was:

$$\frac{hc}{\lambda} = hf = \Phi + eV_s$$

If we rearrange this equation in the form y = mx + c, the equation becomes:

$$V_s = \frac{hf}{e} - \frac{\Phi}{e}$$

Therefore, the gradient $= \frac{h}{e} = \frac{2.07\ V}{4.62 \times 10^{14}}\ s^{-1}$

$= 4.5 \times 10^{-15}\ Vs$

$h = \frac{gradient}{e} = 4.5 \times 10^{-15}\ Vs \times 1.6 \times 10^{-19}\ C$

$= 7.2 \times 10^{-34}\ Js$

The accepted value of Planck's constant is $6.63 \times 10^{-34}\ Js$.

The percentage discrepancy $= \frac{7.2 - 6.63}{6.63} \times 100\%$

$= 8.6\ \%$

(d) Determine the minimum frequency of the photoelectric cell by graphical means.

The threshold frequency is the x-intercept

$= (2.2 \pm 0.6) \times 10^{14}\ Hz$

(e) From the graph, calculate the work function of the photoemissive surface in the photoelectric cell in joules and electron-volts.

The y-intercept is equal to $\frac{-\Phi}{e}$

Work function, $\Phi = e \times (y\text{-intercept}) = 1.6 \times 10^{-19}\ C \times -1\ V$
$= 1.6 \times 10^{-19}\ J$

(c) Determine the absolute error of T^2 for each value.

(d) Draw a graph of T^2 against l. Make sure that you choose an appropriate scale to use as much of a piece of graph paper as possible. Label the axes, put a heading on the graph, and use error bars. Draw the curve of best fit.

(e) What is the relationship that exists between T^2 and l?

(f) Are there any outliers?

(g) From the graph determine a value for g.

Length of pendulum ± 0.05 m	Time for 20 oscillations ± 0.2 s	Period T / s²	T²	Absolute error of T²
0.21	18.1			
0.40	25.5			
0.62	31.5			
0.80	36.8			
1.00	40.4			

Exercise 1.7

1. An investigation was undertaken to determine the relationship between the length of a pendulum l and the time taken for the pendulum to oscillate twenty times. The time it takes to complete one swing back and forth is called the period T. It can be shown that

$$T = 2\pi\sqrt{\frac{l}{g}}$$

where g is the acceleration due to gravity.

The data in the table below was obtained.

(a) Copy the table and complete the period column for the measurements. Be sure to give the uncertainty and the units of T.

(b) Calculate the various values for T^2 including its units.

CHAPTER 1

1.3 Vectors and scalars

NATURE OF SCIENCE:

Models: First mentioned explicitly in a scientific paper in 1846, scalars and vectors reflected the work of scientists and mathematicians across the globe for over 300 years on representing measurements in three dimensional space. (1.10)

© IBO 2014

Scalars	Vectors
distance (s)	displacement ($\mathbf{s}$)
speed	velocity ($\mathbf{v}$)
mass (m)	area ($\mathbf{A}$)
time (t)	acceleration ($\mathbf{a}$)
volume (V)	momentum ($\mathbf{p}$)
temperature (T)	force ($\mathbf{F}$)
charge (Q)	torque ($\boldsymbol{\tau}$)
density (ρ)	angular momentum ($\mathbf{L}$)
pressure (P)	flux density ($\boldsymbol{\Phi}$)
energy (E)	electric field intensity ($\mathbf{E}$)
power (P)	magnetic field intensity ($\mathbf{B}$)

Figure 128   Examples of scalar and vector quantities

Essential idea: Some quantities have direction and magnitude, others have magnitude only, and this understanding is the key to correct manipulation of quantities. This sub-topic will have broad applications across multiple fields within physics and other sciences.

Understandings:
- Vector and scalar quantities
- Combination and resolution of vectors

Vector and scalar quantities

Scalars are quantities that can be completely described by a magnitude (size). Scalar quantities can be added algebraically. They are expressed as a positive or negative number and a unit. Some scalar quantities, such as mass, are always positive, whereas others, such as electric charge, can be positive or negative.

Vectors are quantities that need both magnitude and direction to describe them. The magnitude of the vector is always positive. In this textbook, vectors will be represented in heavy print. However, they can also be represented by underlined symbols or symbols with an arrow above or below the symbol. Because vectors have both magnitude and direction, they must be added, subtracted and multiplied in a special way.

When vectors are graphed, the system of coordinates is called a rectangular coordinate system or a Cartesian coordinate system, or simply, a coordinate plane. Vectors in the same plane are said to be co-planar.

Figure 128 lists some examples of scalar and vector quantities.

Addition of vectors

From simple arithmetic it is known that 4 cm + 5 cm = 9 cm
However, in vector context, a different answer is possible when 4 and 5 are added.
For example, 4 cm north (N) + 5 cm south (S) = 1 cm south
Suppose you move the mouse of your computer 4 cm up your screen (N), and then 5 cm down the screen (S), you move the mouse a total distance of 9 cm. This does not give the final position of the arrow moved by the mouse. In fact, the arrow is 1cm due south of its starting point, and this is its displacement from its original position. The first statement adds scalar quantities and the second statement adds two vector quantities to give the resultant vector R.

The addition of vectors which have the same or opposite directions can be done quite easily:

1 N east + 3 N east = 4 N east (newton force)

200 μm north + 500 μm south = 300 μm south (micrometre)

300 m s^{-1} north-east + 400 m s^{-1} south-west = 100 m s^{-1} south west (velocity)

The addition of co-planar vectors that do not have the same or opposite directions can be solved by using scale drawings or by calculation using Pythagoras' theorem and trigonometry.

Vectors can be denoted by boldtype, with an arrow above the letter, or a tilde, i.e., **a**, $\vec{a}$ or $\underline{a}$ respectively. They are represented by a straight line segment with an arrow at

22

the end. They are added by placing the tail of one to the tip of the first (placing the arrow head of one to the tail of the other). The resultant vector is then the third side of the triangle and the arrowhead points in the direction from the 'free' tail to the 'free' tip. This method of adding is called the **triangle of vectors** (see Figure 129).

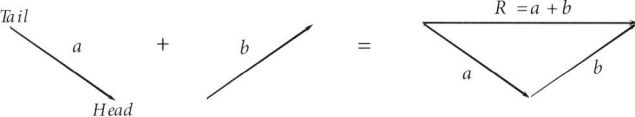

Figure 129   Addition Of vectors

The parallelogram of vectors rule for adding vectors can also be used. That is, place the two vectors tail to tail and then complete a parallelogram, so that the diagonal starting where the two tails meet, becomes the resultant vector. This is shown in Figure 130.

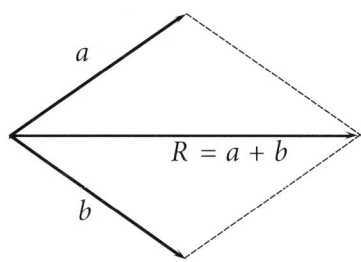

Figure 130   Addition of vectors using parallelogram rule

If more than two co-planar vectors are to be added, place them all head to tail to form a polygon. Consider the three vectors, a, b and c shown in Figure 131. Adding the three vectors produces the result shown in Figure (b).

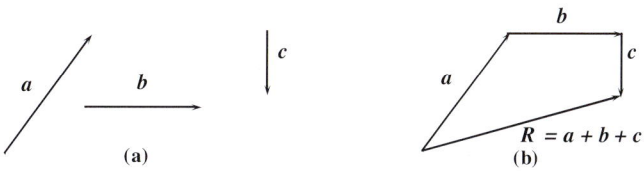

Figure 131   Addition of more than two vectors

Notice then that **a + b + c = a + c + b = b + a + c = . . .** That is, vectors can be added in any order, the resultant vector remaining the same.

Example

On an orienteering expedition, you walk 40 m due south and then 30 m due west. Determine how far and in what direction are you from your starting point.

Solution

Method 1   *By scale drawing*

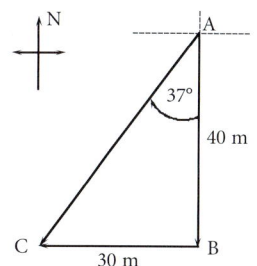

Figure 132   Orienteering

Draw a sketch of the two stages of your journey.

From the sketch make a scale drawing using 1 cm equal to 10 m (1 cm ≡: 10m).

If you then draw the resultant AC, it should be 5 cm in length. Measure ∠CAB with a protractor.
The angle should be about 37°.

Therefore, you are 50 m in a direction south 37° west from your starting point (i.e., S 37° W).

Method 2   *By calculation*

Using Phythagoras' theorem, we have

$AC^2 = 40^2 + 30^2$ ∴ $AC = \sqrt{40^2 + 30^2} = 50$

(taking the positive square root).

From the tan ratio,

$\tan\theta = \dfrac{\text{opposite}}{\text{adjacent}}$ we have $\tan\theta = \dfrac{BC}{AB} = \dfrac{30}{40} = 0.75$

∴ $\tan^{-1}(0.75) = 36.9°$

You are 50 m in a direction south 37° west from your starting point (i.e. S 37° W).

Subtraction of vectors

In Chapter 2, you will describe motion – kinematics. You will learn that change in velocity, Δv, is equal to the final velocity minus the initial velocity, $v - u$. Velocity is a vector quantity so Δv, v and u are vectors. To subtract $v - u$, you reverse the direction of u to obtain $-u$, and then you add vector v and vector $-u$ to obtain the resultant Δv.

That is, $\Delta v = v + (-u)$. Vectors v and u are shown. For $v - u$, we reverse the direction of u and then add head to tail

CHAPTER 1

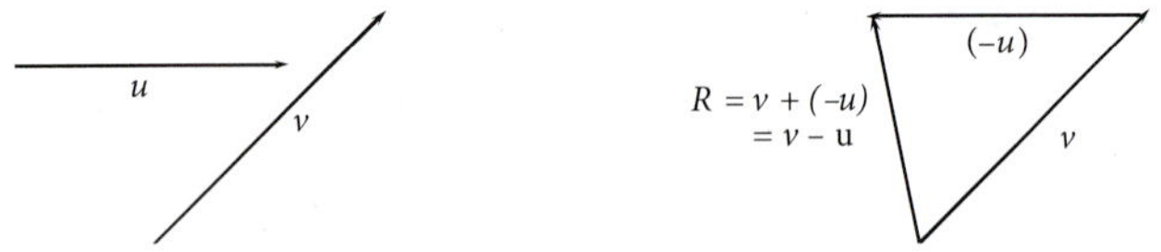

Figure 133 **Subtraction of vectors**

Example

A snooker ball is cued and strikes the cushion of the snooker table with a velocity of 5.0 m s⁻¹ at an angle of 45° to the cushion. It then rebounds off the cushion with a velocity of 5.0 m s⁻¹ at an angle of 45° to the cushion. Determine the change in velocity? (Assume the collision is perfectly elastic with no loss in energy).

Solution

You can solve this problem by scale drawing or calculation. Draw a sketch before solving the problem, then draw the correct vector diagram.

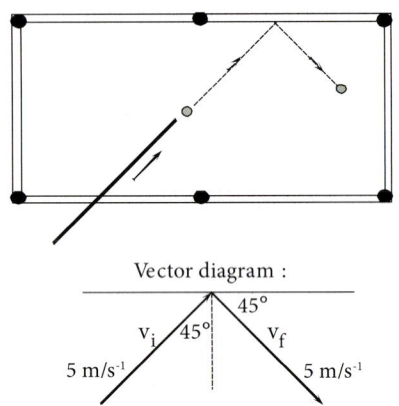

Notice that the lengths of the initial velocity vector v_i, and the final velocity vector, v_f, are equal.

Using the vector diagram above we can now draw a vector diagram to show the change in velocity.

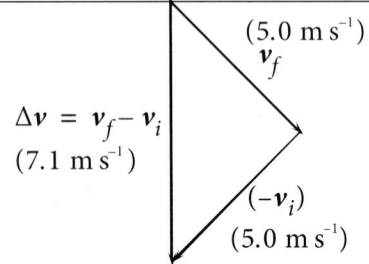

Using the same scale as that used for the 5.0 m s⁻¹ velocity vector, the change in velocity is 7.1 m s⁻¹ at right angles to the cushion.

We could also use Pythagoras' theorem to determine the length (or magnitude) of the change in velocity vector, Δ v:

$$|\Delta v|^2 = |v_f|^2 + |v_i|^2,$$

so that $|\Delta v|^2 = 5^2 + 5^2 = 50$ ∴ $|\Delta v|^2 = \sqrt{50} \approx 7.1$ m s⁻¹

Multiplying vectors and scalars

Scalars are multiplied and divided in the normal algebraic manner, for example:

5 m ÷ 2 s = 2.5 m s⁻¹   2 kW × 3 h = 6 kW h (kilowatt-hours)

A vector multiplied by a scalar gives a vector with the same direction as the vector and magnitude equal to the product of the scalar and the vector.

For example: 3 × 15 N east = 45 N east;
2kg × 15 m s⁻¹ south = 30 kg m s⁻¹ south

Exercise 1.8

Which of the following lines best represents the vector 175 km east (1 cm : 25 km)?

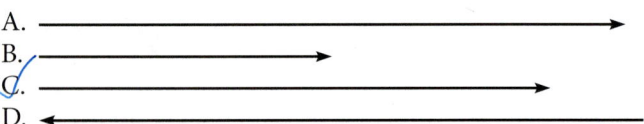

2. Which one of the following is a vector quantity?

   A. Work   C. Acceleration
   B. Speed   D. Pressure

3. Which one of the following is a scalar quantity?

   A. Force   C. Momentum
   B. Velocity   D. Energy

4. The diagram below shows a boat crossing a river with a velocity of 4 m s⁻¹ north. The current flows at 3 m s⁻¹ west.

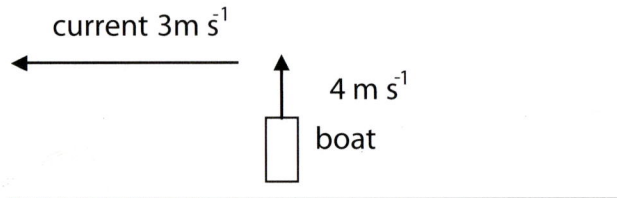

24

The resultant magnitude of the velocity of the boat will be

    A.   3 m s⁻¹   C.   5 m s⁻¹
    B.   4 m s⁻¹   D.   7 m s⁻¹

5. Two vectors with displacements of 10 m north–west and 10 m north–east are added. The direction of the resultant vector is

    A.   south   B.   north-east
    C.   north   D.   north-west

6. Add the following vectors by the graphical method

    (a)   4 m south and 8 m north, *4m north*
    (b)   5 m north and 12 m west, *13 m SW*
    (c)   6.0 N west and 6.0 N north, *12N NE*
    (d)   9.0 m s⁻¹ north + 4.0 m s⁻¹ east + 6.0 m s⁻¹ south. *19 m/s W*

7. Subtract the following vectors by either the graphical method or by calculation

    (a)   2 m east from 5 m east (i.e., 5 m east − 2m east), *3m east*
    (b)   9 m s⁻² north from 4 m s⁻² south, *13 m/s S*
    (c)   4.0 N north from 3.0 N east,
    (d)   3.2 T east from 5.1 T south.

8. Calculate the following products

    (a)   20 m s⁻¹ north by 3 *60 m/s N*
    (b)   12 by 5 N s north 12° east

9. If a cyclist travelling east at 40 m s⁻¹ slows down to 20 m s⁻¹, what is the change in velocity? *20 m/s*

10. Find the resultant of a vector of 5 m north 40° west added to a vector of 8 m east 35° north

Resolution of vectors

The process of finding the components of vectors is called resolving vectors. Just as two vectors can be added to give a resultant vector, a single vector can be split into two components or parts.

The vector 5 m south has a vertical component of 5 m south and a zero horizontal component just as the vector 10 N east has a zero vertical component and a horizontal component of 10 N east.

Suppose you have a vector that is at an angle to the horizontal direction. Then that vector consists of measurable horizontal and vertical components. In Figure 134, the vector F is broken into its components. Note that the addition of the components gives the resultant F.

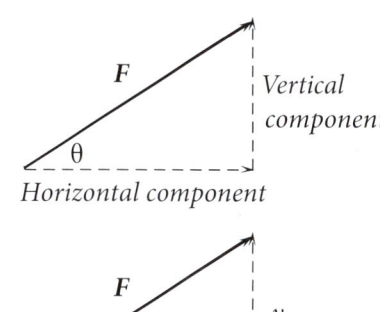
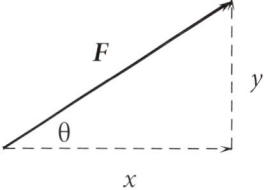

Figure 134   Resolution of vectors

From trigonometry

$$\sin\theta = \frac{\text{opposite}}{\text{hypotenuse}} = \frac{y}{|F|}$$

$$\cos\theta = \frac{\text{adjacent}}{\text{hypotenuse}} = \frac{x}{|F|}$$

This means that the magnitude of the vertical component
$= y = F \sin\theta$

and the magnitude of the horizontal component
$= x = F \cos\theta$

25

CHAPTER 1

Example

A sky rocket is launched from the ground at an angle of 61.0° with an initial velocity of 120 m s⁻¹. Determine the components of this initial velocity?

Solution

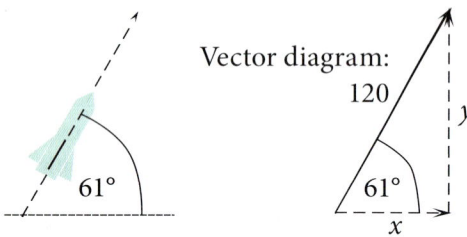

From the vector diagram we have that

$$\sin 61° = \frac{y}{120} \Rightarrow y = 120 \sin 61°$$

$$= 104.954...$$

$$= 1.0 \times 10^2 \text{ m s}^{-1}$$

and $\cos 61° = \frac{x}{120} \Rightarrow x = 120 \cos 61°$

$$= 58.177...$$

$$= 58 \text{ m s}^{-1}$$

That is, the magnitude of the vertical component is 1.1×10^2 m s⁻¹ and the magnitude of the horizontal component is 58 m s⁻¹.

Exercise 1.9

1. The vertical component of a vector of a 4.0 N force acting at 30° to the horizontal is

    A. 4.3 N   C. 4 N
    B. 2 N   D. 8.6 N

2. Calculate the horizontal component of a force of 8.4 N acting at 60.0° to the horizontal.

3. Calculate the vertical and horizontal components of the velocity of a projectile that is launched with an initial velocity of 25.0 m s⁻¹ at an angle of elevation of 65° to the ground.

4. Calculate the easterly component of a force of 15 N south-east.

5. Calculate the vector whose components are 5.0 N vertically and 12 N horizontally.

6. Calculate F in the diagram below if the sum of all the forces is zero.

    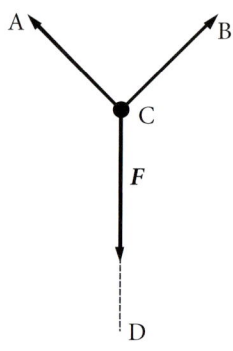

    AC = 2 N   BC = 2 N and
    ∠ACD = 135°   ∠BCD = 135°

7. Calculate the acceleration of a small object down a frictionless plane that is inclined at 30.0° to the horizontal. Take the acceleration due to gravity g equal to 9.81 ms⁻².

8. Calculate the resultant force of all the forces acting on a point object in the diagram below.

    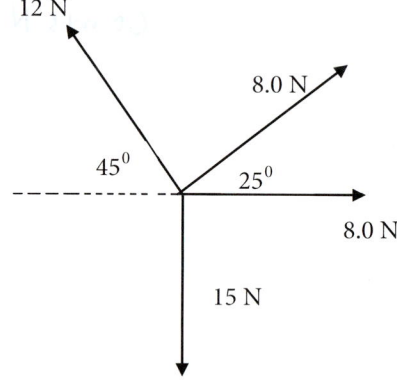

MEASUREMENTS AND UNCERTAINTIES

APPENDIX

Mathematical requirements

During this course you should experience a range of mathematical techniques. You will be required to develop mathematical skills in the areas of arithmetic and computation, algebra, geometry and trigonometry, and graphical and vector analysis for both external and internal assessment.

Mathematical sentences

$=$	is equal to
$/$	divided by or in units of
$<$	is less than
$>$	is greater than
$\propto$	is proportional to
$\approx$	is approximately equal to
Δx	a small difference between two values of x
$\|x\|$	the absolute value of x

Geometry

Indices

1. $a^x \times a^y = a^{x+y}$
2. $a^x \div a^y = \dfrac{a^x}{a^y} = a^{x-y}$
3. $(a^x)^y = a^{x \times y}$
4. $a^x \times b^x = (a \times b)^x$
5. $a^0 = 1, 1^x = 1, 0^x = 0 \ (x \neq 0), \sqrt[x]{a} = a^{1/x}$

Logarithms

1. $\log x + \log y = \log(x \times y), x > 0, y > 0.$
2. $\log x - \log y = \log\left(\dfrac{x}{y}\right), x > 0, y > 0.$
3. $x \log y = \log y^x, y > 0.$
4. $a^x = y \Leftrightarrow x = \log_a y$

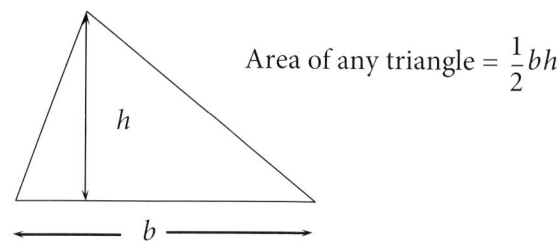

Area of any triangle $= \dfrac{1}{2}bh$

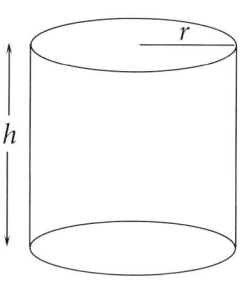

Area of a hollow cylinder $= 2\pi rh$
Surface area of a cylinder $= 2\pi r(h+r)$
Volume of a cylinder $= \pi r^2 h$

Circumference $= 2\pi r$
Area of a circle $= \pi r^2$

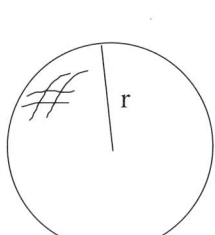

Volume of a sphere $= \dfrac{4}{3}\pi r^3$
Surface area of a sphere $= 4\pi r^2$

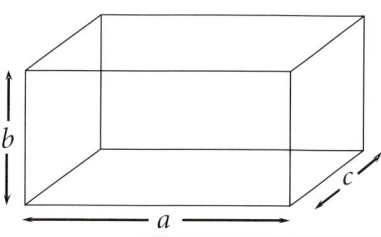

Surface area of a cuboid $= 2(ab + bc + ac)$

CHAPTER 1

Trigonometry

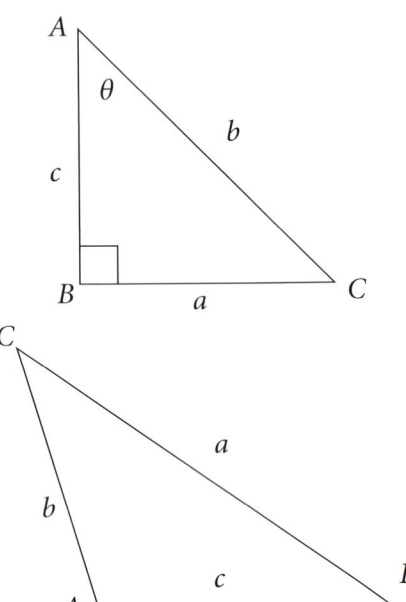

$\sin\theta = \dfrac{\text{opposite}}{\text{hypotenuse}} = \dfrac{a}{b}$

$\cos\theta = \dfrac{\text{adjacent}}{\text{hypotenuse}} = \dfrac{c}{b}$

$\tan\theta = \dfrac{\text{opposite}}{\text{adjacent}} = \dfrac{a}{c}$

$\tan\theta = \dfrac{\sin\theta}{\cos\theta}, \ \cos\theta \neq 0$

For very small angles, $\sin\theta \approx \tan\theta \approx \theta$, $\cos\theta \approx 1$

Sine rule: $\dfrac{a}{\sin A} = \dfrac{b}{\sin B} = \dfrac{c}{\sin C}$

Cosine rule: $a^2 = b^2 + c^2 - 2bc \times \cos A$

Area of triangle: $A = \tfrac{1}{2}ab \times \sin C$

Identities:

$\sin^2 A + \cos^2 A = 1$

$\sin(A - B) + \sin(A + B) = 2\sin A \sin B$

$\sin A + \sin B = 2\sin\left[\dfrac{(A+B)}{2}\right] \times \cos\left[\dfrac{(A-B)}{2}\right]$

Angular measure

Angles are measured in radians. One radian is the angle subtended by an arc with length equal to the radius. If $s = r$, then $\theta = s/r$.

Note then, that 2π rad = 360°, and 1 rad = 57.3°

Exercise 1.10

1. Convert $\dfrac{13}{17}$ to a decimal and to a percentage.

2. Use a calculator to find 3.6^3 and log 120.

3. Make y the subject of the equation if $x = 2y - 6$.

4. Make v the subject of the equation given that:

   $F = \dfrac{mv^2}{r}$

5. Make g the subject of the equation given that:

   $T = 2\pi\sqrt{\left(\dfrac{l}{g}\right)}$

6. Solve for x and y in the following simultaneous equations:

   $2x + 4y = 18$
   $x - y = -1$

7. Calculate the following:

   (a) $16^2 + 16^3$
   (b) $25^{1.5}$
   (c) $(\sqrt{2})^4$
   (d) $(\sqrt{3})^{-2}$

8. Evaluate the following

   (a) $\log_4 64$
   (b) $\log_{10} 0.01$

9. Find the circumference and area of a circle of radius 0.8 cm.

10. Calculate the volume and surface area of a sphere of radius 0.023 m.

11. How many radians are there in:

    A. 270°
    B. 45°

12. If $\sin 2\theta = 1$ then what is θ equal to?

28

Greek Symbols

The Greek alphabet is commonly used in Physics for various quantities and constants. The capital and small letters and their names are given here for your convenience:

Letters		Name
A	α	alpha
B	β	beta
Γ	γ	gamma
Δ	δ	delta
E	ε	epsilon
Z	ζ	zeta
H	η	eta
Θ	θ	theta
I	ι	iota
K	κ	kappa
M	μ	mu
N	ν	nu
Ξ	ξ	xi
O	ο	omicron
Π	π	pi
P	ρ	rho
Σ	σ	sigma
T	τ	tau
Φ	φ	phi
X	χ	chi
Ψ	ψ	psi
Ω	ω	omega

This page is intentionally left blank

2. Mechanics

Contents

2.1 – Motion

2.2 – Forces

2.3 – Work, energy, and power

2.4 – Momentum and impulse

Essential Ideas

Motion may be described and analysed by the use of graphs and equations.

Classical physics requires a force to change a state of motion, as suggested by Newton in his laws of motion.

The fundamental concept of energy lays the basis upon which much of science is built.

Conservation of momentum is an example of a law that is never violated. © IBO 2014

Chapter 2

TOK BACKGROUND

The late Richard Feynman described the process of Physics as akin to observing a vast chess game in which the boundaries of the chessboard cannot be seen. Furthermore, we have no idea why the game is being played or by whom. Nor do we know when the game started, nor will we ever see the end of the game. We don't know the rules of the game and our problem is to figure them out. By careful observation over a period of time we might, for example, discover the rule that governs the move of the bishops and if we are really clever we might even find the rule which governs the movement of the knights. Occasionally something really odd might happen like two white queens appearing on the board at the same time. All our subsequent observations had led us to the conclusion that this could not be the case.

The chessboard in this analogy is the Universe and the chess pieces are the matter in the Universe. The rules that we discover are the laws of Physics and the observation we make of the pieces are the experiments that we carry out to establish the laws of Physics. The rules give the "how" and not the "why". In other words they do not tell us why the pieces move but they help us understand the manner in which they move. And so it is with Physics. We will never know for example, why when we push something it moves. However, we can give a very good description of how it will move under different circumstances. Physics is the science that describes how the Universe "works".

Physics falls into two main categories. There is the Physics before 1926 – Classical Physics – and there is the Physics after 1926- Quantum Physics. Most of the Physics that is studied in an IB course is Classical Physics. However, it is important to realise that ultimately our description of how the Universe works must be understood in terms of Quantum Physics because we know this to be (so far.) the "correct Physics". So you might ask why do we spend so much time in teaching you the "wrong" Physics? Well, it's not quite as bad as it sounds. For example, if we apply the laws of Classical Physics to the behaviour of electrons in solids we get the wrong answer. The laws of Quantum Physics give the right answer. On the other hand if we apply the laws of Classical Physics and the laws of Quantum Physics to the behaviour of billiard balls both give the right answer. However, using Quantum Physics in this situation is rather like taking the proverbial sledgehammer to crack a walnut. In many of the situations that we encounter, Classical Physics will give us the right answer and so for this reason, and the fact that Quantum Physics is not easy to grasp on first acquaintance, we spend a lot of time teaching students Classical Physics.

If we "plot" the speed of things against size, then we can see the sort of areas pertinent to each of the main areas of Physics. Refer to Figure 201.

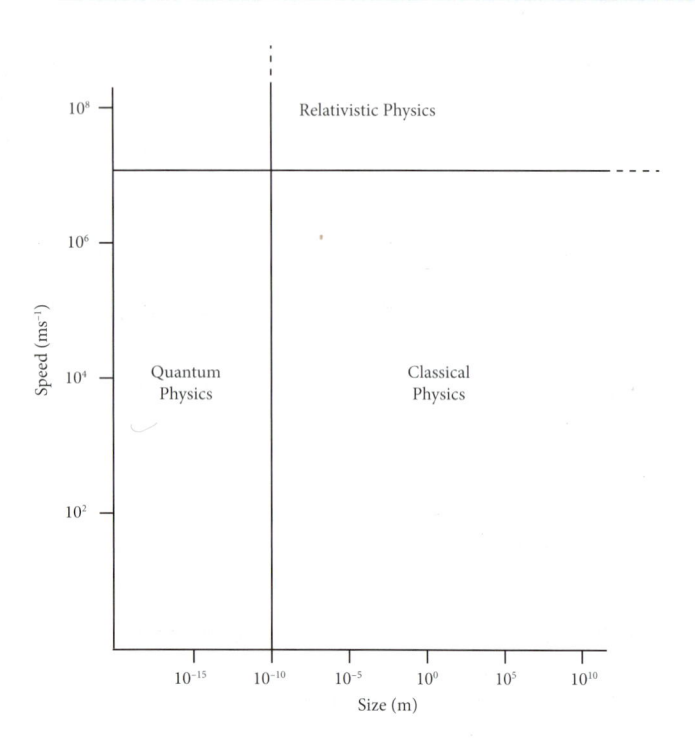

Figure 201   The different areas of Physics

You will note that there is a region that overlaps both Quantum and Classical Physics. This is Relativistic Physics and is the Physics we have to use when we are dealing with speeds close to that of the speed of light. So there is Relativistic Quantum Physics and Relativistic Classical Physics. The latter is discussed in more detail in Option A.

The two great "pillars" upon which Classical Physics rests are Newtonian Mechanics and Electromagnetism. Mass and electric charge are the two basic properties that we associate with all matter and Newtonian mechanics essentially deals with mass and electromagnetism essentially deals with charge. The two corresponding pillars upon which Quantum Mechanics rest are Quantum Mechanics and Quantum Electrodynamics. Bridging both Quantum and Classical Physics is Relativistic Physics and Thermodynamics. This latter subject essentially deals with the relationship between heat and work and also such interesting questions as "how can order arise from disorder?"

The Figure 202 summarises the essential branches of Physics and also gives the appropriate syllabus reference in the IB Physics Guide.

MECHANICS

```
                        Physics
                           |
              Relativity and Thermodynamics
                        A 3, 8, B
                       /         \
              Classical Physics    Quantum Physics
              /         \             /         \
    Newtonian Mechanics  Electromagnetism  Quantum mechanics  QED
         (mass)            (charge)            7, 12, C,       7
       1, 2, 4, 6, 8, 9.1
                              |
       ┌──────────────────────┼──────────────────────┐
   Electrostatics        Magnetostatics        Electromagnetism
  (charge at rest)    (charge in uniform      (charge in accelerated
       5, 10            motion) 5, 10.         motion) 5, 11, C
```

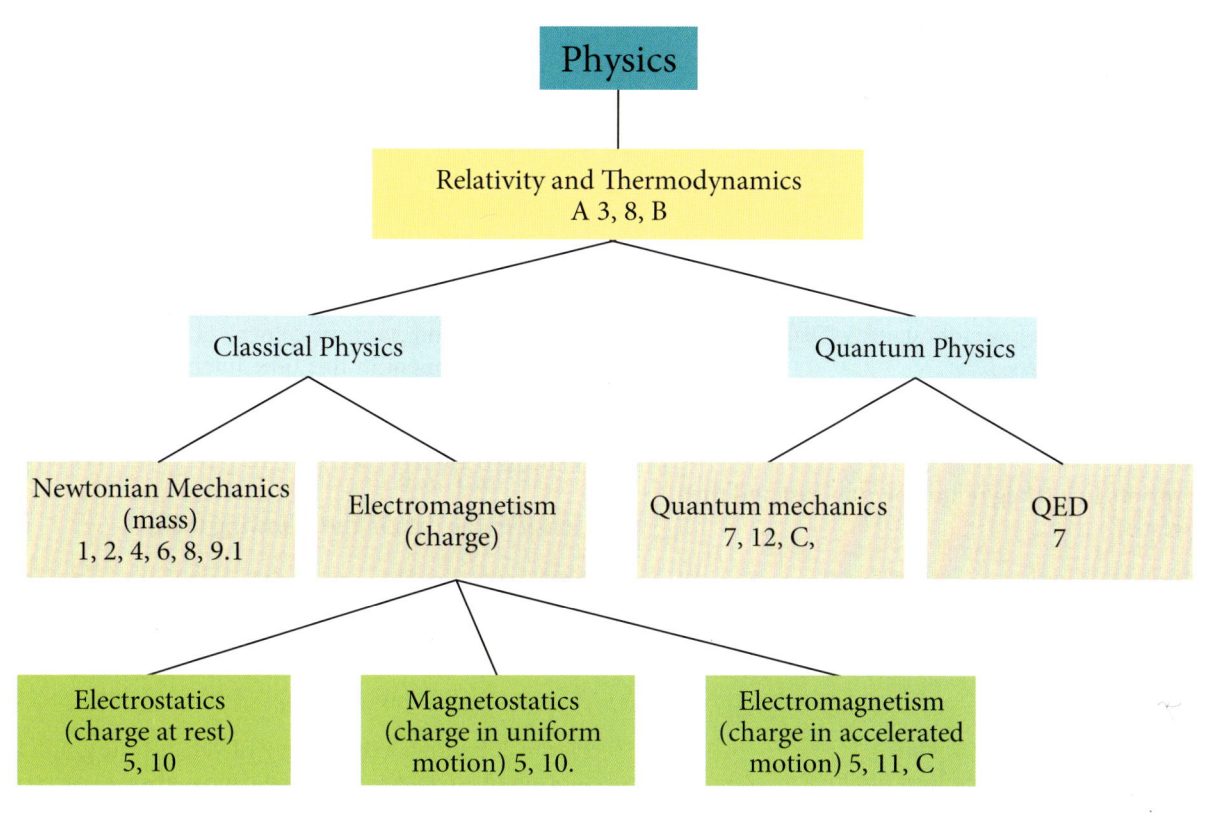

Figure 202 *The structure of this course*

At the present time we understand the two great pillars of Physics to be General Relativity (which describes space and time) and Quantum Physics (which describes everything else) and somewhere along the line Thermodynamics has to fit in as well. One of the great aims of physicists is to try and "unify" General relativity and Quantum Mechanics into a single theory.

In this part of the course we start our journey through Newtonian Mechanics, one of the great pillars of Classical Physics. The essential problem in Mechanics is this: if at any given instant in time we know the positions and velocities of all the particles that make up a particular system, can we predict the future position and velocities of all the particles?

This is the mechanics problem in its most general form. Specific examples are problems such as predicting solar eclipses, putting satellites into orbit, finding out how the positions of an oscillating object varies with time and finding out where a snooker ball ends up when it is struck by another snooker ball.

In 1687, Isaac Newton (1642-1727) published his Principia Mathematica in which he set out a method for solving these type of problems; hence the name 'Newtonian Mechanics'.

There are two parts to mechanics. Kinematics covers the concepts needed to describe motion. Dynamics deals with the effects forces have on motion.

2.1 Motion

NATURE OF SCIENCE:

Observations: The ideas of motion are fundamental to many areas of physics, providing a link to the consideration of forces and their implication. The kinematic equations for uniform acceleration were developed through careful observations of the natural world. (1.8)

© IBO 2014

Essential idea: Motion may be described and analysed by the use of graphs and equations.

Understandings:
- Distance and displacement
- Speed and velocity
- Acceleration
- Graphs describing motion
- Equations of motion for uniform acceleration
- Projectile motion
- Fluid resistance and terminal speed

CHAPTER 2

Distance and displacement

As already mentioned in Chapter 1, distance is a scalar quantity that has magnitude only and is measured from some reference point in metres.

Displacement is a measured distance in a given direction. It tells us not only the distance of an object from a particular reference point but also the direction from the reference point. So displacement is a vector quantity. Displacements are also measured in metres.

Example

An athlete sprints in a straight line for 50m east and then runs 40m west.

1. What was the distance that he ran?
2. What is his final displacement?

Solution

1. The distance run is 50 + 40 = 90m.
2. Let us make east positive and west negative.
   The final displacement is the sum of +50 m and -40 m
   = +10 m east.

In many situations we will often measure the displacement from the origin of a Cartesian co-ordinate system and the displacement of a particle can be simply given as the co-ordinates of a point (x, y, z). (Strictly speaking the displacement should be written as a matrix $[x, y, z]$ but that need not concern us here since in most instances in this chapter we will usually restrict ourselves to displacement in the x-direction only).

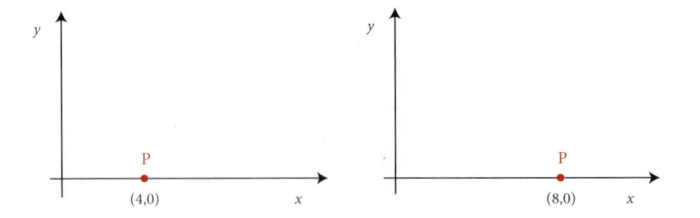

Figure 203   One method of calculating displacement

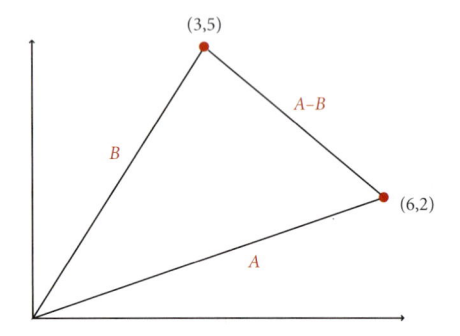

Figure 204   Another method of calculating displacement

Figures 203 and 204 are two examples of calculating displacements. In Figure 203 the particle P is at the point (4,0) at some instant and at a certain time later it is at a point (8,0). Its displacement in this time interval is, therefore, 4 units in the positive x direction.

In Figure 204, the particle is at the point (6, 2) at some instant and vector A represents its displacement from the origin. At some interval later the particle has moved to the point (3, 5) and its displacement is represented by vector **B**. Its displacement in this time interval is therefore **B** − **A**. Remember from Chapter 1 that if you subtract vectors, you reverse the arrow direction for **A**, and then you add to vector **B**, that is **B** + (-**A**).

Consider the data in the following table.

Displacement / m	0	10	16	18	20	32
Time / s	0	2	4	6	8	10

Figure 205 shows a displacement versus time graph for the motion of an object. Note that time is always placed on the x-axis. You can see that 6s after it was at rest the object has a displacement of 18m, and so on. We can also see whether the object is speeding up or slowing down because in the graph the object moves 10m in the first 2 seconds but only 6m in the second 2 seconds.

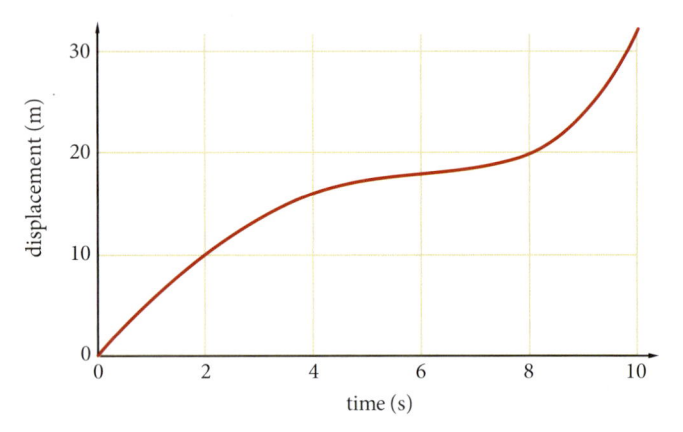

Figure 205   Displacement versus time graph 1

When an object moves backwards in some time period, then the displacement versus time graph will show this. Consider another set of data for an object in motion.

Displacement / m	0	5	6	7	10	13
Time / s	0.38	1	2	3	4	4.5

34

MECHANICS

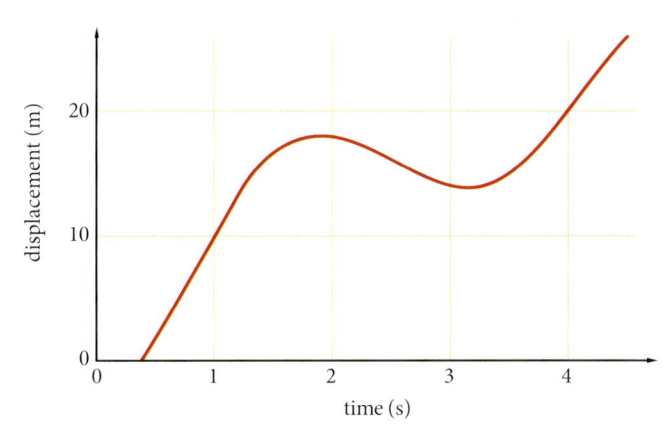

Figure 206   Displacement versus time graph 2

Figure 206 shows the object moving backwards when the graph has a negative slope between the t = 2s and t = 3s.

Speed and velocity

Speed tells us the rate at which a moving object covers distance with respect to time and it is measured in metres / second (m s^{-1}). Hence we have:

$$speed = \frac{distance}{time}$$

Velocity is speed in a given direction. It is therefore a vector quantity. Again, it is measured in metres / second (m s^{-1}).

To plot a course, an airline pilot needs to know not only the speed of the wind but from which direction it is blowing i.e. the wind velocity must be known.

Example

In still water a motorboat has a maximum speed of 5 m s^{-1}. The boat sets off to cross from one bank of a river to the other. The river flows with a speed of 2 m s^{-1} and the motorboat engine is set to maximum. Calculate the velocity of the motorboat.

Solution

The solution to this problem involves simple vector addition.

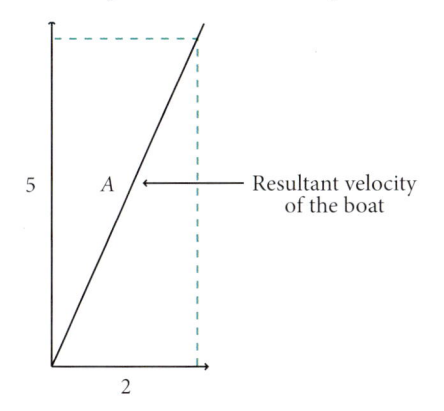

With reference to the above diagram the magnitude of the resultant velocity of the boat is

$$\sqrt{5^2 + 2^2} = 5.4 \text{ m s}^{-1}.$$

The direction can be measured relative to the original direction of the boat and is given by

the angle $\alpha = \tan^{-1}\left(\frac{2}{5}\right) = 21.8° \, (= 22°)$.

If you travel by car between two towns a distance of 200 km apart and your journey time is 4 hours your average speed would be 50 km hr^{-1}.

Average speed = total distance travelled / total time taken

Clearly this does not mean that every time that you look at the speedometer it would be reading 50 km h^{-1}. This may be possible on an uncongested road. However, it is likely that there will be congestion, traffic lights and stops for rest. This raises the question of what does the reading on the speedometer actually mean? On a very simplistic level it tells us that if we maintain this speed then in one hour we will cover a distance of 50 km. At another level it tells us that at that particular instant the car has an instantaneous speed of 50 km h^{-1}.

Speed and average speed are useful for indicating how fast an object is moving but they do not tell us the direction of the motion. To determine the speed we need to rather use the concept of average velocity. Suppose that an object has initial position x_0 at time t_0. A short time later, suppose it is at its final position x at time t. Then

$$\text{average velocity} = \frac{\text{total displacement}}{\text{total time taken}} = \frac{x - x_0}{t - t_0}$$

Motion is said to be **uniform** if equal displacements occur in equal periods of time.

Suppose you were on a trip and your average velocity was 80 kmh^{-1}. This just gives us an average value but it does not provide any information about the velocity at any instant of time during your trip. This is called the **instantaneous velocity** and it is an indicator of how fast an object moves and its direction for each instant of time (the magnitude of the instantaneous velocity is called the **instantaneous speed**).

An experiment with free fall can help us understand this. Figure 207 shows a ball being dropped between two photo-gates that are connected to an electronic counter.

35

CHAPTER 2

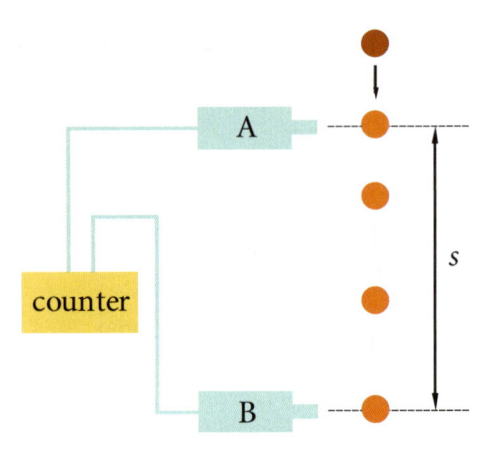

Figure 207   An experiment in free fall

The ball is dropped from a point somewhere above the light gate A. The two light gates A and B connected to the counter will record the time t that it takes the ball to fall the distance s.

The **average speed** of the ball as it falls between A and B is just $\frac{s}{t}$.

The ball is of course accelerating as it falls so its speed is changing. Now imagine the light gate B closer to A and repeat the experiment. We would obtain a different value for the average velocity. As we repeat the experiment several times, each time moving B closer to A, we will find that the values of the average speed obtained each time will be approaching some limiting value. This limiting value is actually the instantaneous speed of the ball as it passes A. When the distance between A and B becomes very small (as does the corresponding time of fall) then this distance divided by the time will very nearly be equal to the instantaneous speed at A.

If we let the small distance equal Δs and the time of fall equal Δt then the **average speed** v_{av} over this distance is

$$v_{av} = \frac{\Delta s}{\Delta t}$$

and the **instantaneous speed** v is given by

$$v = \lim_{\Delta t \to 0} \frac{\Delta s}{\Delta t}$$

or $v = \frac{\Delta s}{\Delta t}$ as $\Delta t \to 0$

(For those of you who do calculus you will recognise this as the derivative $\frac{ds}{dt}$ as $\Delta t \to$ zero.)

If we are dealing with velocities then we must write the above equation in vector form. The magnitude of instantaneous velocity is the instantaneous speed of the object at the instant measured and the direction of the velocity is the direction in which the object is moving at that instant.

Acceleration

Acceleration refers to the rate of change of velocity in a given direction. (Change in velocity ÷ time taken).

Average acceleration can be found by dividing the change in velocity of an object by the total time taken for this change to occur.

$$\text{average acceleration} = \frac{\text{change in velocity}}{\text{time taken}}$$

The change in velocity is found by subtracting the initial velocity **u** from the final velocity **v**. If a is the average acceleration and t is the time taken then

$$a = \frac{(v - u)}{t}$$
$$= \frac{\Delta V}{\Delta t}$$

The concept of instantaneous acceleration follows accordingly as

$$\boldsymbol{a} = \frac{\Delta v}{\Delta t} \text{ as } \Delta t \to 0$$

Where Δv is the change in velocity in time Δt.

In the SI system the unit is metres per second per second. i.e. the change in velocity measured in m s^{-1} every second. We write this as m s^{-2}. Since we define acceleration in terms of velocity it is therefore a vector quantity.

If the acceleration of an object is positive then we understand its rate of change of velocity to be positive and it could mean that the speed of the body is increasing. A body that is slowing down will have a negative acceleration. However, do not think of acceleration as always referring to a 'slowing up' or 'getting faster' process because if a car for example goes round a bend in the road at constant speed, it is accelerating. Why? Because the direction of the car is changing and therefore its velocity is changing. If its velocity is changing then it must have acceleration. This is sometimes difficult for people to grasp when they first meet the physics definition of acceleration because in everyday usage acceleration refers to something getting faster.

As we shall see later on in this chapter it is very important to keep in mind the vector nature of both velocity and acceleration.

Example

An object has an initial velocity of 30 ms^{-1} east. What is its acceleration if:

(a) it takes 2 seconds to reach 60 ms^{-1} east
(b) it takes 5 seconds to reach 20 ms^{-1} west?

MECHANICS

Solution

(a) $a = \dfrac{(v-u)}{t}$

$a = \dfrac{(60-30)}{2}$

$= 15 \text{ ms}^{-2}$

(b) $a = -\dfrac{20-30}{5}$

$= -10 \text{ ms}^{-2}$ i.e. 10 ms^{-2} west

Graphs describing motion

A very useful way to describe the motion of an object is to sketch a graph of its kinematics properties. This is best illustrated by means of some examples.

A distance-time graph will help us to understand the concept of instantaneous speed (or velocity).

Consider the distance-time graph of an object moving with constant speed as shown in Figure 208. In this situation equal distances are covered in equal times and clearly the speed is equal to the gradient of the graph – in this case, 10 m s^{-1}. The average speed is equal to the instantaneous speed at all points.

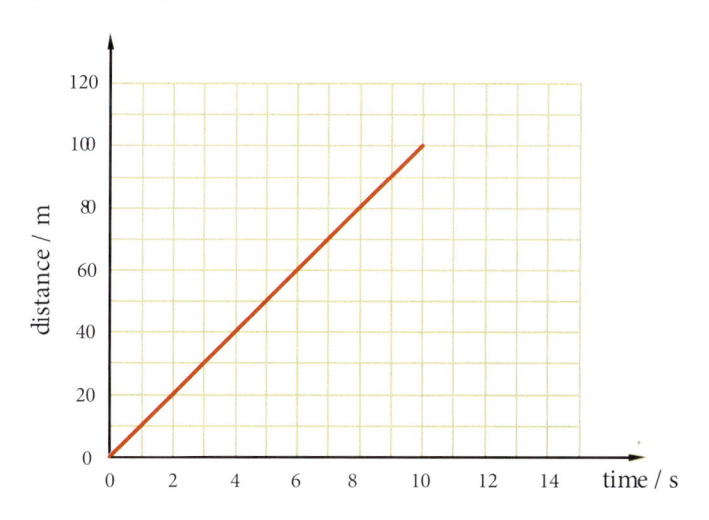

Figure 208   *Distance-time graph*

If the velocity is not constant we can still find average speeds and instantaneous speeds from displacement–time graphs. To demonstrate this, let us look at the distance–time graph for a falling ball this is shown in Figure 209.

From the graph we see that the time it takes the ball to fall 1.0 m is 0.4 s. The average speed over this distance is therefore 2.5 m s^{-1}. (Remember that speed is the magnitude of velocity).

To find say the instantaneous speed at 1.0 m we find the gradient of the curve at this point.

To do this we draw the tangent to the point as shown. From the tangent that is drawn, we see that the slope of the line is 1.8 (= Δs) divided by 0.4 (= Δt) = 4.5 m s^{-1}.

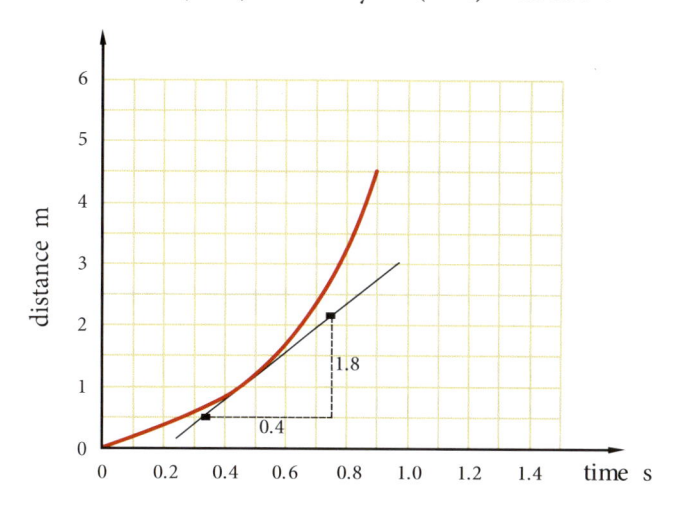

Figure 209   *Distance-time for a falling ball*

This is actually what we mean when we write an instantaneous speed as

$$v_P = \dfrac{\Delta s}{\Delta t} \text{ as } \Delta t \to 0$$

v_p is the instantaneous speed at the point P and

$\dfrac{\Delta s}{\Delta t}$ as $\Delta t \to 0$ is the gradient of the displacement–time

graph at the point P.

(Those students who are familiar with calculus will recognise this process as differentiation. The equation for the above

graph is $s = kt^2$ such that the derivative $\dfrac{ds}{dt} = 2kt$ is the

gradient at any time t and hence the instantaneous speed).

When sketching or plotting a displacement-time graph we have to bear in mind that displacement is a vector quantity. Consider for example, the situation of an object that leaves point A, travels with uniform speed in a straight line to point B, returns to point A at the same constant speed and passes through point A to a point C. If we ignore the accelerations at A and B and regard the point A as the zero reference point, then a sketch of the displacement-time graph will look like that shown in Figure 210.

CHAPTER 2

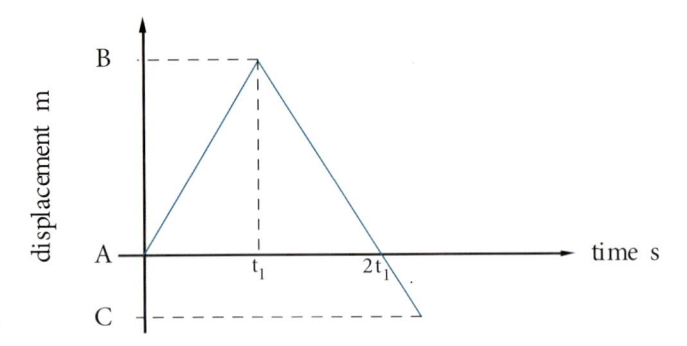

Figure 210 Displacement-time Graph

Velocity (speed) – time graphs

The graph in Figure 211 shows how the velocity of the falling ball varies with time.

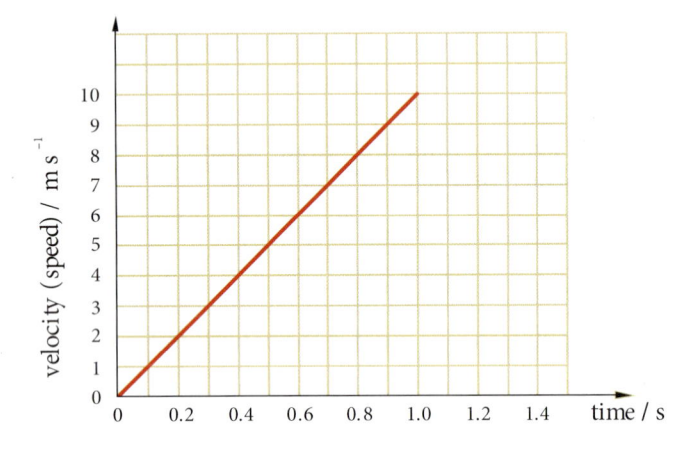

Figure 211 Velocity-time graph for a falling ball

The acceleration is the $\frac{\text{change of velocity}}{\text{time}}$ and in this case this is equal to the gradient of the straight line and is equal to 10 m s^{-2}. This is a situation of constant acceleration but even when the acceleration is not constant the acceleration at any instant is equal to the gradient of the velocity–time graph at that instant.

Earlier we saw that we defined instantaneous acceleration as

$$a = \frac{\Delta v}{\Delta t} \text{ as } \Delta t \to 0$$

Students familiar with calculus will recognise that acceleration is the derivative

$$a = \frac{dv}{dt}$$

which can also be written as

$$a = \frac{d}{dt}\left(\frac{ds}{dt}\right) = \frac{d^2 s}{dt^2}$$

We can also determine distances from speed–time graphs. This is easily demonstrated with the graph of Figure 212 which shows the speed time graph for constant speed.

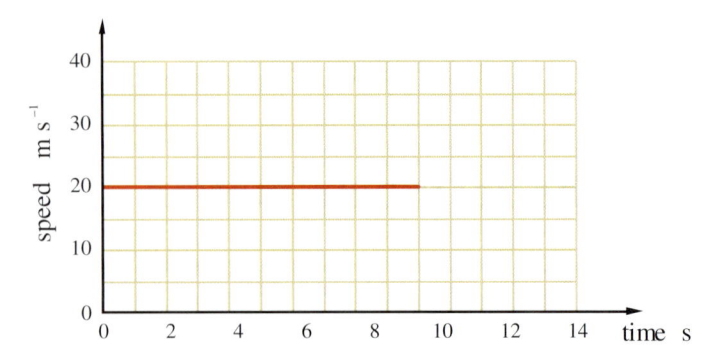

Figure 212 Speed-time graph to determine distance

The distance travelled is just speed × time. So at a constant speed of 20 m s^{-1} after 10 s the object will have travelled 200 m. This is of course equal to the area under the line between $t = 0$ and $t = 10$ s.

If the velocity is not constant then the area under a velocity–time graph will also be equal to the displacement. So, for the falling ball, we see from the velocity time graph, Figure 215, that the distance travelled after 1.0 s is equal to the area of the triangle of base 1.0 s and height 10 m s^{-1} equals = ½ × 1.0 s × 10 m s^{-1} = 5.0 m

Example

A train accelerates uniformly from rest to reach a speed of 45 m s-1 in a time of 3.0 min. It then travels at this speed for a further 4.0 min at which time the brakes are applied. It comes to rest with constant acceleration in a further 2.0 min.

Draw the velocity-time graph for the journey and from the graph calculate

(a) the magnitude of the acceleration between 0 and 3.0 min

(b) the magnitude of the acceleration after the brakes are applied

(c) the total distance travelled by the train

38

MECHANICS

The graph is shown plotted below

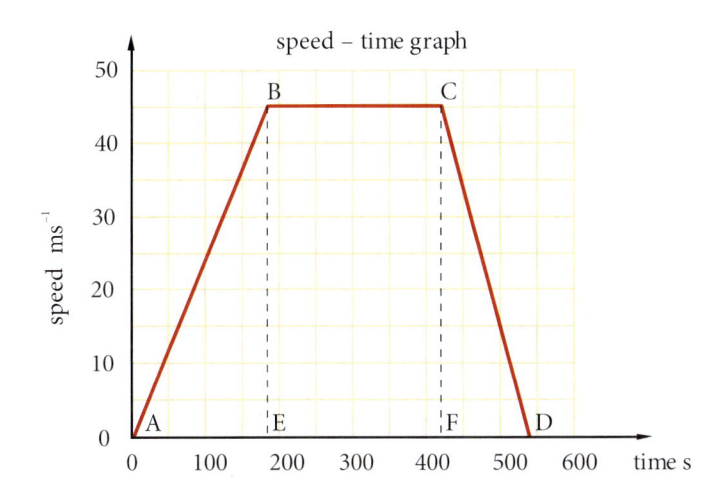

speed – time graph

(a) the acceleration of the train in the first 3 minutes is the gradient of the line AB.

Therefore, we have, $a = \dfrac{45}{180} = +0.25 \text{ m s}^{-2}$

(b) the acceleration of the train after the brakes are applied is the gradient of the line

$CD = -\dfrac{45}{120} = -0.38 \text{ m s}^{-2}$

(c) the distance travelled by the train is the total area under the graph.

Total area

= area of triangle ABE + area BCFE + area of triangle CDF

= 17550 ≅ 18000 m

= 18 km.

Acceleration-time graphs

If the acceleration of an object varies it is quite tricky to calculate the velocity of the object after a given time. However, we can use an acceleration-time graph to solve the problem. Just as the area under a velocity-time graph is the distance travelled then the area under an acceleration-time graph is the speed achieved. For the falling ball the acceleration is constant with a value of 10 m s^{-2}. A plot of acceleration against time will yield a straight line parallel to the time axis, Figure 213. If we wish for example to find the speed 3.0 s after the ball is dropped, then this is just the area under the graph:

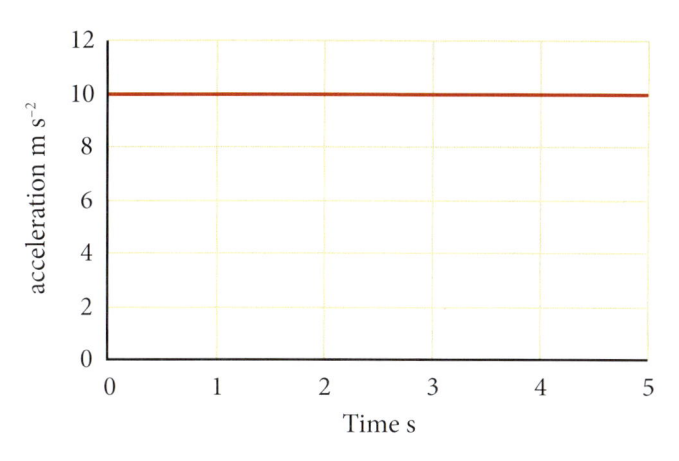

Figure 213   Acceleration - time graph

We could of course have found the speed directly from the definition of uniform acceleration i.e.

speed = acceleration × time

Example

The acceleration of an object increases uniformly at a rate of 3.0 m s^{-2} every second. If the object starts from rest, calculate its speed after 10 s.

Solution

A plot of the acceleration-time graph for this situation is shown in the graph below.

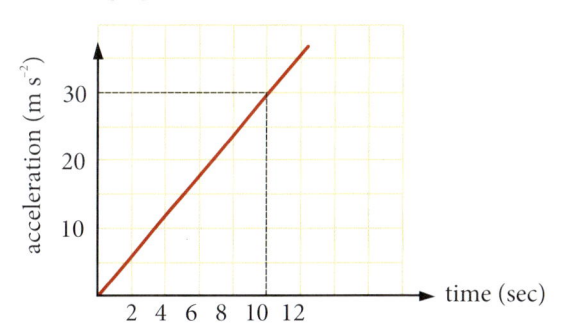

The speed attained by the object after 10 s is the area under the line.

Therefore, speed = area enclosed by acceleration–time graph

$= \frac{1}{2} \times 10 \times 30 = 150 \text{ m s}^{-2}$

To summarise :
- the gradient of a displacement-time graph is equal to the velocity (speed)
- the gradient of a velocity (speed)-time graph is equal to the acceleration
- the area under a velocity (speed)-time graph is equal to the distance.
- the area under an acceleration-time graph is the change in velocity.

39

CHAPTER 2

Consider the motion of a hard rubber ball that is dropped onto a hard surface. The ball accelerates uniformly under gravity. On each impact with the surface it will lose some energy such that after several bounces it will come to rest. After each successive bounce it will leave the ground with reduced speed and reach a reduced maximum height. On the way down to the surface it speeds up and on leaving the surface it slows down.

The graph in Figure 214 shows how we can best represent how the velocity changes with time i.e., the velocity–time graph.

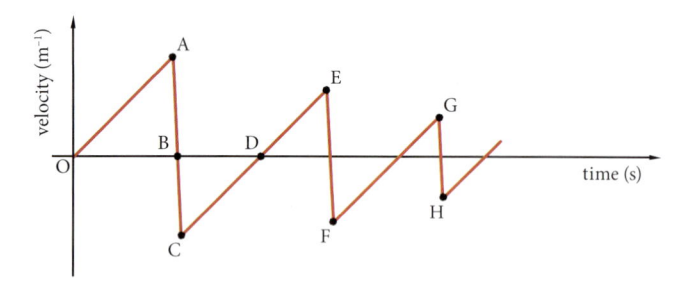

Figure 214   Changes of velocity with time

The ball leaves the hand at point O and accelerates uniformly until it hits the ground at A. At A it undergoes a large acceleration during which its velocity changes from positive to negative (being zero at B). The change in velocity between B and C is less than the change in velocity between A and B since the rebound velocity is lower than the impact velocity. The ball accelerates from C to D at which point it is at its maximum height and its velocity is zero. Notice that even though its velocity is zero its acceleration is not. The ball now falls back to the surface and hits the surface at point E. Neglecting air resistance the velocity of the ball at points C, and E will be the same. The process now repeats.

The lines OA, CE and FG are parallel and the gradient of these lines is the acceleration of free fall.g.

The lines AC, EF and GH are also parallel and the gradient of these lines is equal to the acceleration of the ball whilst it is in contact with the surface. The lines should not be vertical as this would mean that the acceleration would be infinite.

The acceleration of the ball at points such as A, C and E is not equal to g.

Exercise 2.1

1. The diagram shows a velocity-time graph for a truck

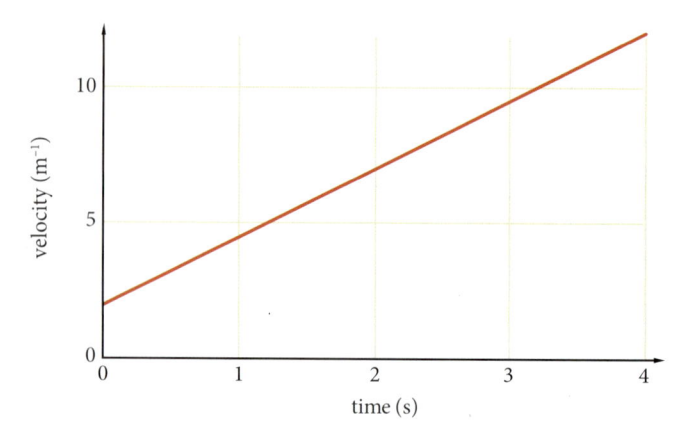

   What is the distance travelled in the first 4 seconds?

   A. 2.5 m   B. 3.0 m   C. 20 m   D. 28 m

2. A racing car travels round a circular track of radius 100 m as shown in the diagram below:

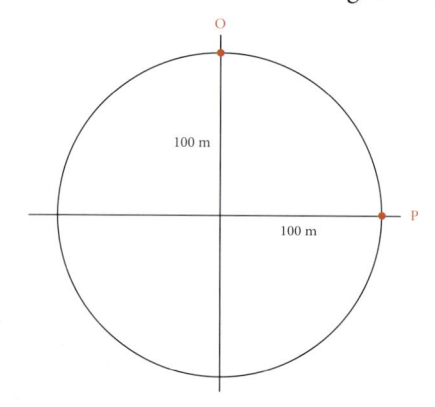

   The car starts at O. When it has travelled to P its displacement as measured from O is:

   A   100 m due east
   B   100 m due west
   C $100\sqrt{2}$ m south east
   D $100\sqrt{2}$ m south west

3. A self-propelled airship has a velocity of 40.0 ms^{-1} north-east. How long will it take to travel 280 km north-east?

4. A person runs 16 m south in 2.0 s and then 12 m east in 3.0 s.

   (a) What is the distance travelled?
   (b) What is the displacement?
   (c) What was the average speed of the person?
   (d) What is the average displacement of the person?

40

5. An electric car moving at 20.0 kmh^{-1} accelerates uniformly to a velocity of 30.0 kmh^{-1} in 20.0 s. Determine its acceleration.

6. A stone is dropped from the top of a canyon. When it hits the water in the river at the bottom it has a velocity of 73.79 ms^{-1}. If it takes 7.53 s to reach the water:

    (a) Determine the value of the acceleration of the stone.
    (b) Determine the average velocity of the stone.
    (c) What was the height of the canyon?

7. The following velocity graph shows a lift travelling from the ground floor to the 4th floor of a building.

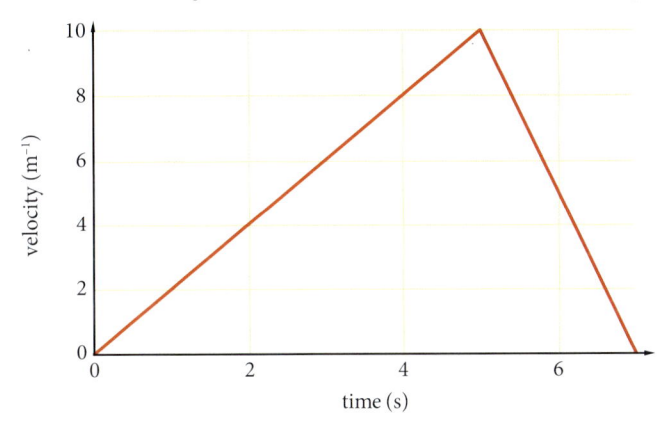

    (a) Describe the motion of the lift in as much detail as possible.
    (b) Find the total distance travelled by the lift.

8. A 'glider' bounces backwards and forwards between the 'buffers' of a linear air-track. Neglecting friction, which one of the graphs below best represents how the

    (a) velocity
    (b) acceleration
    (c) displacement and
    (d) speed

    of the glider varies with time?

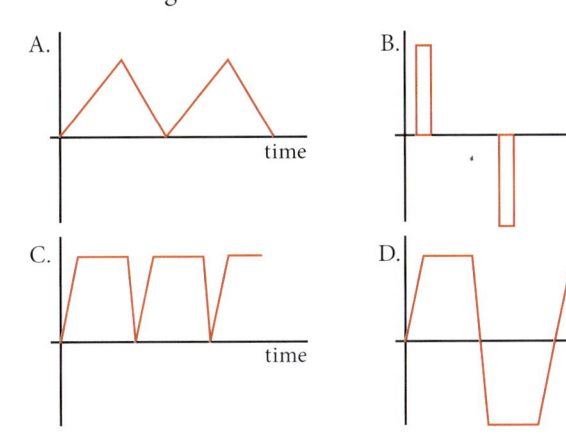

    A.   B.   C.   D.

9. The graph below shows the idealised velocity-time graph for a car pulling away from one set of traffic lights until it is stopped by the next set.

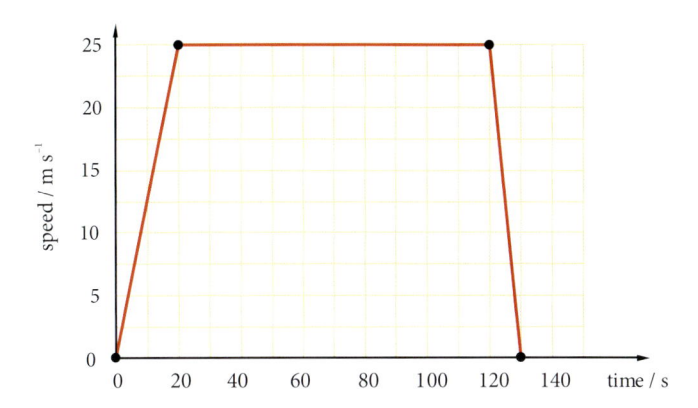

    Calculate the:
    i. acceleration of the car between 0 and 20 s and between 120 and 130 s
    ii. total distance travelled by the car during braking
    iii. total distance between the traffic lights.

10. Sketch the distance-time graph for the car in question 9.

Relative motion

Suppose that you are standing on a railway station platform and a train passes you travelling at 30 m s^{-1}. A passenger on the train is walking along the corridor in the direction of travel of the train at a speed of 1 m s^{-1}. From your point of view the speed of the person is 31 m s^{-1}. In another situation a car A is travelling at a speed of 25 m s^{-1} and overtakes a car B that is travelling in the same direction at a speed of 20 m s^{-1}. The speed of car A relative to car B is 5 m s^{-1}. Clearly the determination of speed (and therefore velocity and acceleration) depends on what it is measured relative to.

Generally speaking, if the speed of a particle A relative to an assigned point or reference frame O is V_A and the speed of a particle B relative to the same point is V_B, then the *velocity of A relative to B is the vector difference $V_A - V_B$* (See Topic 1.3)

The example that follows illustrates how relative velocity is very important in situations such as plotting a correct course for an aircraft or for a boat or ship.

Example

The Figure below shows the two banks of a river. A ferryboat operates between the two points P and Q that are directly opposite each other.

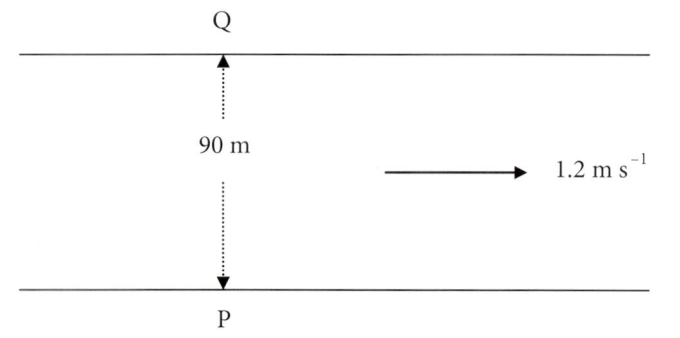

```
           Q
           ┆
           ┆
           ┆
         90 m        ──────▶  1.2 m s⁻¹
           ┆
           ┆
           ┆
           P
```

The speed of flow of the river relative to the riverbanks is 1.2 m s⁻¹ in the direction shown. The speed of the ferry boat in still water is 1.8 m s⁻¹ in a direction perpendicular to the river banks. The distance between P and Q is 90 m.

Clearly, if the ferryboat sets off from P directly towards Q, it will not land at Q.

It is left as an exercise for you to show that the speed of the ferryboat relative to the speed of the water is 2.2 m s⁻¹ and that it will land at a point 60 m downstream of Q. You should also show that, in order to land at Q, the ferry boat should leave P heading upstream at an angle of 56° to the riverbank and that the time taken to cross to Q is about 67 s.

For an aircraft, the rate of water flow becomes the wind speed and for a ship at sea it becomes a combination of wind, tide and current speeds.

The idea of relative measurement has far reaching consequences as will become clear to those of you who choose to study the Relativity (Option D or in Option H.)

Equations of motion for uniform acceleration

Throughout most of the course we will deal with constant accelerations and you will not be expected to solve problems that involve non-uniform accelerations. Vector notation will be used only when deemed necessary.

For objects moving with constant acceleration (in a straight line) there is a set of four very useful equations that relate displacement, velocity, acceleration and time (in some books they are called the *suvat* equations) and we will now derive these.

The equations only relate magnitudes and therefore we will write the symbols in the equations in scalar form.

Let s = the distance travelled in time t

u = the speed at time $t = 0$, the *initial* speed

v = the speed after time t, the *final* speed

a = acceleration

t = the time for which the body accelerates

The first equation is just the definition of acceleration.

Average acceleration = change in velocity / time taken

$$a = \frac{(v - u)}{t}$$

By making **v** the subject of the equation, the first suvat equation becomes:

$$v = u + at$$

Now:

$$\text{Average speed} = \frac{\text{total distance travelled}}{\text{time taken}}$$

The average speed for **v** and **u** would be $\frac{v + u}{2}$

Therefore:

$$\frac{(v + u)}{2} = \frac{s}{t}$$

So, we can use the first equation to replace the **v** in the second equation:

$$\frac{(u + at + u)}{2} = \frac{2u}{2} + \frac{at}{2} = \frac{2u}{2} + \frac{at}{2} = \frac{s}{t}$$

Therefore:

$$s = ut + \frac{1}{2}at^2$$

We can make **t** the subject of the first equation to get:

$$t = \frac{v - u}{a}$$

Then we can substitute in the second equation to get:

$$\frac{v + u}{2} = \frac{s}{v} - \frac{u}{a}$$

$$\frac{(v + u)(v - u)}{2a} = s \quad \text{[Remember that } (v+u)(v-u) = v^2 - u^2\text{]}$$

$$v^2 = u^2 + 2as$$

This is the fourth equation for the set.

MECHANICS

Acceleration of free fall

Kinematics tells us that:

$$s = ut + \tfrac{1}{2} gt^2$$

If the initial velocity u is zero, then a graph of the displacement s plotted against the time squared t^2 will be a straight line passing through the origin. The slope of the graph will equal half of the acceleration due to gravity, ½ g. If the initial velocity is not equal to zero, then the line of best fit will cut the y-intercept at some point.

There are a number of experimental designs with different timing devices that can be used. The simplest case is metal grooved inclined plane in which a marble rolls.

In this case,

$$s = ut + \tfrac{1}{2} g \sin\theta\, t^2$$

Another way is to take a strobe photograph of a falling object against the background of a measuring stick or even a video camera that normally has 25 frames per second.

Still another way is to use an accelerometer (ticker-tape timer) where a vibrating pointer places dots on a piece of tape every 0.02 seconds as shown in Figure 215.

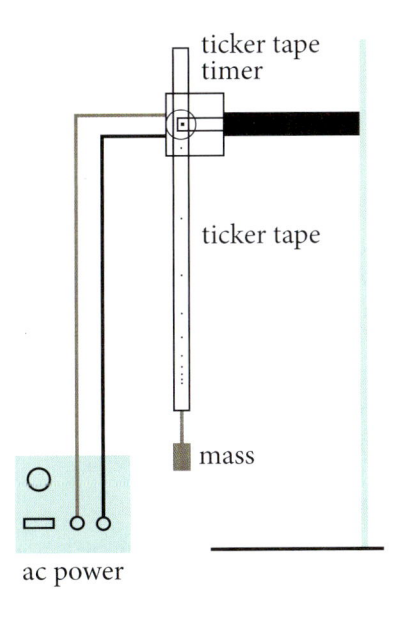

Figure 215   A ticker-tape timer

Another method is to drop a steel ball from an electromagnet using a two-way (SPDT) switch as shown in Figure 216. By switching the switch to cause the steel ball to fall, the timer is switched on. As the ball hits the trap door, the electric circuit becomes incomplete and the timer is switched off.

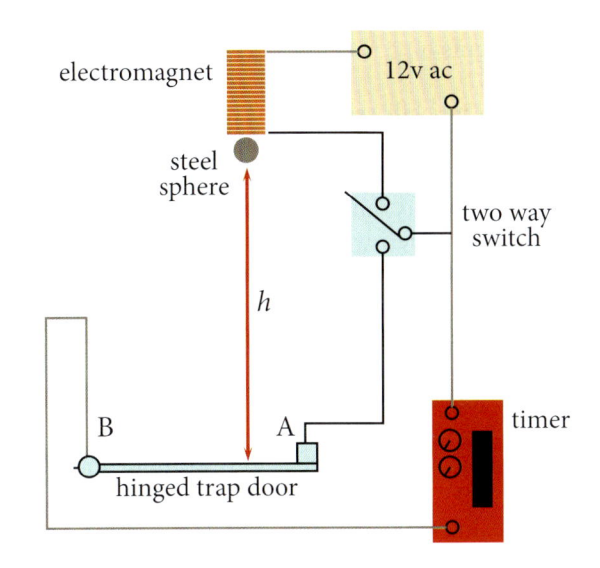

Figure 216   The trap door method

The following exercise demonstrates an alternative way of analysing data obtained using strobe photographic techniques. This method is based on the equations of uniform motion.

Figure 217 shows the results of an experiment in which the strobe photograph of a falling ball has been analysed. The strobe takes 20 pictures a second. The time between each picture is therefore 0.05 s. The distance column is the measured distance of each successive photograph of the ball from the origin. The error in the distance has been estimated from parallax error in reading from the scale against which the photographs have been taken and also in locating the centre of the ball in each photograph.

You are to plot a graph of s against t^2 and from the graph find a value of g. You should include error bars on the graph and use these to calculate the error in the value of g that you have determined.

time t/s ±0.01/s	distance s/cm ±0.4 cm
0	0
0.05	1.2
0.10	4.8
0.15	10.9
0.20	19.4
0.25	30.3
0.30	43.7
0.35	59.4
0.40	77.6
0.45	98.2
0.50	121.2
0.55	146.7
0.60	174.6

Figure 217   Data For Free Fall

Some comments on g

In Chapter 6 we will see that the value of g varies with position and with height above the Earth's surface and also in the absence of air resistance, the acceleration of free falling objects is independent of their mass. This was first noted by Galileo who is reputed to have timed the duration of fall for different objects dropped from the top of the leaning Tower of Pisa. The fact that the acceleration of free fall is independent of an object's mass has far reaching significance in Physics and is discussed in more detail in the Relativity Option.

If you carry out an experiment to measure g and obtain a value say of 9.4 m s^{-2}, then make sure that you calculate your error using the correct method. For example, do not assume that the value of g is 9.8 m s^{-2} and hence compute your error in measurement as ± 4%. You do not know what the value of g is at your location and that is why you are measuring it. One correct way to calculate the error is using maximum and minimum lines of best fit to calculate the greatest residual from the line of best fit.

Exercise 2.2

1. A stone is dropped down a well and a splash is heard 2.4 s later. Determine the distance from the top of the well to the surface of the water?

2. A girl stands on the edge of a vertical cliff and she throws a stone vertically upwards. The stone eventually lands in the sea below her. The stone leaves her hand with a speed of 15 m s^{-1} and the height of the cliff is 25 m.

   Calculate

   i. the maximum height reached by the stone.
   ii. the time to reach the maximum height.
   iii. the speed with which the stone hits the sea.
   iv. the time from leaving the girl's hand that it takes the stone to hit the sea.

3. A sprinter starts off down a track at a speed of 10 m s-1. At the same time a cyclist also starts off down the track. The cyclist accelerates to a top speed of 20 m s^{-1} in 4.0 s. Ignoring the acceleration of the sprinter, determine the distance from the start that the cyclist will pass the sprinter.

4. A boy drops a ball out of his apartment window and it hits the ground 2.0 s after it leaves his hand. Determine the height of the window sill.

5. An object has an initial velocity of 15 m s^{-1} east and a final velocity of 15 m s^{-1} west after 6 seconds.

   (a) Determine its acceleration.
   (b) What is its displacement after 6 seconds?

6. A girl throws a stone vertically upwards. The stone leaves her hand with a speed of 15.0 m s^{-1}.

   Determine:

   (i) the maximum height reached by the stone and
   (ii) how long the time it takes to return to the ground after leaving her hand?

7. A motorbike is moving in a straight line between four unevenly spaced trees ABCD with constant acceleration. The distance between tree A and B is 11 m, and the distance between tree B and C is 28 m. It takes 1 second to travel between A and B, and 2 seconds to travel between B and C. If the distance between tree C and D is 36 m, determine:

   (a) the initial velocity of the motorbike
   (b) the acceleration of the bike
   (c) the time taken to travel tree C to tree D.

8. A jet airliner leaves Shanghai at 22:00 hours and flies due south at a constant speed of 100.0 km h-1. At midnight, a second airliner travelling at 250.0 km h^{-1} follows the same flight path at a higher altitude.

   (a) At what time will the second airliner overtake the first?
   (b) At what place will the second airliner overtake the first?

9. A bullet is uniformly accelerated from rest down the barrel of a rifle of length 1 metre, and it leaves the muzzle of the rifle at 3.0 x 10^2 m s^{-1}.

   Calculate

   (a) the acceleration of the bullet
   (b) the time the bullet was in the barrel of the rifle.

10. A glider is projected up a long frictionless linear air track inclined at 60^0 to the horizontal with a velocity of 60.0 m s^{-1}.

    Determine:

    (a) the distance the glider moves up the track
    (b) the time the glider takes to move up the track

Projectile motion

In this section we shall look at the motion of a projectile that is launched horizontally from a point above the surface of the Earth.

In the Figure 218 a projectile is fired horizontally from a cliff of height h with an initial horizontal velocity v_h.

Our problem is effectively to find where it will land, with what velocity and the time of flight. We shall assume that we can ignore air resistance and that the acceleration due to gravity g is constant.

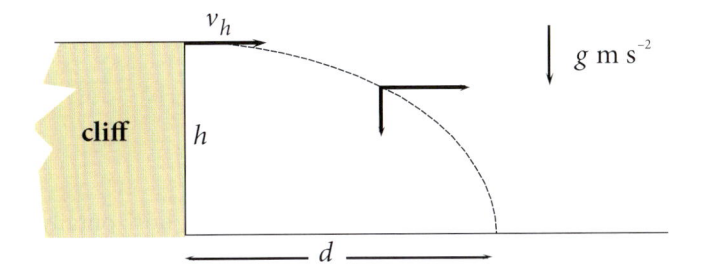

Figure 218   The path of a horizontal projectile

Since there is no force acting in the horizontal direction the horizontal velocity will remain unchanged throughout the flight of the particle. However, the vertical acceleration of the projectile will be equal to g.

We can find the time of flight t by finding the time it takes the particle to fall a height h.

To start with, we consider only the vertical motion of the object:

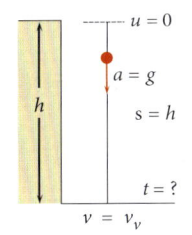

Figure 219   The vertical motion of the object

Time of flight

This is calculated from the definition of acceleration

i.e., using $v = u + at$, we have that

$$v_v = 0 + gt \Rightarrow t = \frac{v_v}{g}$$

where v_v is the vertical velocity with which the object strikes the ground.

To find v_v we use the equation $v^2 = u^2 + 2as$, so that as the initial vertical velocity (u) is zero, the acceleration $a = g$ and $s = h$.

Then
$$v_v^2 = 0^2 + 2gh \Rightarrow v_v = \sqrt{2gh}$$

From which we have that

$$t = \frac{\sqrt{2gh}}{g} = \sqrt{\frac{2h}{g}}$$

and so, the time of flight is given by

$$t = \sqrt{\frac{2h}{g}}$$

Since the horizontal velocity is constant, the horizontal distance d that the particle travels before striking the ground is $v_h \times t$. (i.e., using $s = ut + \frac{1}{2}at^2 = ut$, where in the horizontal direction we have that $a = 0$ and $u = v_h =$ constant)

This gives
$$d = v_v \times \sqrt{\frac{2h}{g}}$$

This is the general solution to the problem and it is not expected that you should remember the formula for this general result. You should always work from first principles with such problems.

An interesting point to note is that, since there is no horizontal acceleration, then if you were to drop a projectile from the top of the cliff vertically down, at the moment that the other projectile is fired horizontally, then both would reach the ground at the same time. This is illustrated by the copy of a multi-flash photograph, as shown in Figure 220.

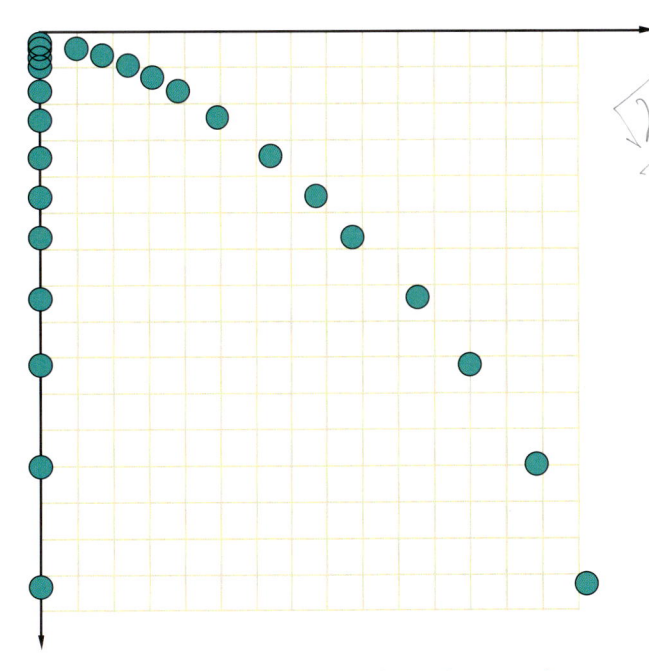

Figure 220   Using multi-flash photography

This is irrespective of the speed with which the particle is fired horizontally. The greater the horizontal speed, the further this projectile will travel from the base of the cliff. It is also possible to show that the path of the particle is parabolic.

To find the velocity with which the particle strikes the ground we must remember that velocity is a vector quantity. So, using Pythagoras' theorem at the point of impact (to take into account both the vertical component of velocity and the horizontal component of velocity) we have that the velocity has a magnitude of

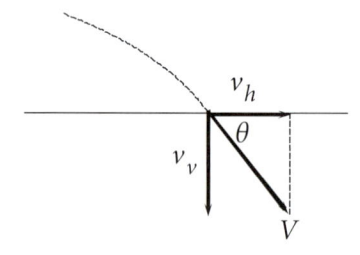

$$|V| = \sqrt{v_v^2 + v_h^2}$$

and the direction will be given by finding

$$\theta = \arctan\left(\frac{v_v}{v_h}\right)$$

where the angle is quoted relative to the horizontal. If the angle is to be given relative to the vertical then we evaluate

$$(90 - \theta)°$$

or

$$\theta = \arctan\left(\frac{v_h}{v_v}\right)$$

Notice that at impact the velocity vector is tangential to the path of motion. As a matter of fact, the velocity vector is always tangential to the path of motion and is made up of the horizontal and vertical components of the velocities of the object.

Projectiles launched at an angle to the horizontal

Consider the problem of a projectile that is launched from the surface of the Earth and at an angle to the surface of the Earth. Ignore air resistance and assume that g is constant. In Figure 221 the particle is launched with velocity v at angle θ to the surface.

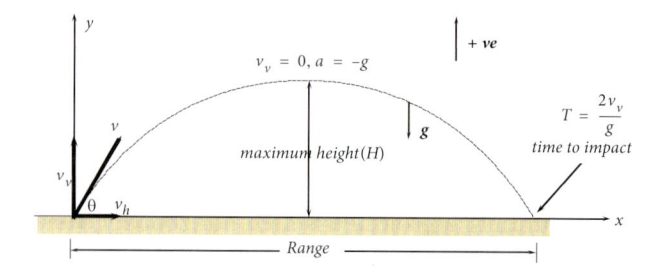

Figure 221   Projectile launched at an angle

The vertical component of the velocity, v_v, is

$$v_v = v\sin\theta$$

The horizontal component of the velocity, v_h, is

$$v_h = v\cos\theta$$

As in the case of the projectile launched horizontally, there is no acceleration in the horizontal direction and the acceleration in the vertical direction is g.

If we refer the motion of the projectile to a Cartesian co-ordinate system, then after a time t, the horizontal distance travelled will be given by

$$x = v_h t = (v\cos\theta)t$$

and the vertical distance can be found by using the equation

$$s = ut + \frac{1}{2}at^2$$

so that

$$y = v_v t + \frac{1}{2}(-g)t^2 \Rightarrow y = (v\sin\theta)t - \frac{1}{2}gt^2$$

If we now substitute for

$$t = \frac{x}{v\cos\theta}$$

into this equation we get

$$y = (v\sin\theta) \times \frac{x}{v\cos\theta} - \frac{1}{2}g\left(\frac{x}{v\cos\theta}\right)^2$$

$$= x \times \frac{\sin\theta}{\cos\theta} - \frac{1}{2}g\left(\frac{x}{v}\right)^2 \times \left(\frac{1}{\cos\theta}\right)^2$$

$$= x\tan\theta - \frac{1}{2}g\left(\frac{x}{v}\right)^2 \sec^2\theta$$

This is the general equation of the motion of the projectile that relates the vertical and horizontal distances. This equation is plotted below for a projectile that is launched with an initial speed of 20 m s⁻¹ at 60° to the horizontal. The path followed by the projectile is a parabola.

MECHANICS

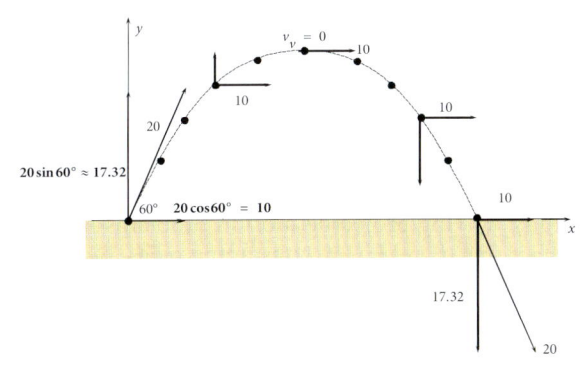

Figure 222 The parabolic path

The maximum height H that the projectile reaches can be found from the equation

$$v^2 = u^2 + 2as$$

where u is the initial vertical component of the velocity and v the final (vertical component) of the velocity at the highest point, where at this point, the vertical component is zero. So that,

$$0^2 = (v\sin\theta)^2 + 2 \times -g \times H \Rightarrow H = \frac{v^2 \sin^2\theta}{2g}$$

If we use the figures in the example above ($v = 20$ m s^{-1}, $\theta = 60°$) with g = 10 m s^{-2} then we see that $H = \frac{(20\sin 60°)^2}{2 \times 10} = 15$

i.e., the object reaches a maximum height of 15 m.

The time T to reach the maximum height is found using $v = u + at$, such $v = 0$,

$u = v\sin\theta$ and $a = -g$, to give

$0 = v\sin\theta - gT \rightarrow gT = v\sin\theta$

Hence, $T = \frac{v\sin\theta}{g}$

For the example above the value of T is 1.73 s. This means (using symmetry) that the projectile will strike the ground 3.46 s after the launch. The horizontal range R is given by $R = (v\cos\theta) \times 2T$ which for the example gives $R = 34.6$ m.

(We could also find the time for the projectile to strike the ground by putting $y = 0$ in the equation

$$y = (v\sin\theta)t - \frac{1}{2}gt^2$$

Although we have established a general solution, essentially solving projectile problems, remember that the horizontal velocity does not change and that when using the equations of uniform motion you must use the component values of the respective velocities.

You may like to determine the range of a projectile by an alternative method. The range is equal to:

$$\text{Range} = v_x t = v\cos\theta \, t$$

The time to reach the maximum height of a projectile is obtained from:

$$0 = v_y - gt = v\sin\theta - gt$$

Therefore:

$$t = \frac{v\sin\theta}{g}$$

So the total time of flight $= \frac{2v\sin\theta}{g}$

Substituting this value of t into the range equation we get:

$$\text{Range} = v\cos\theta \cdot \frac{2v\sin\theta}{g}$$

But $2\sin\theta \cdot \cos\theta = \sin 2\theta$

Therefore:

$$R = \frac{v^2 \sin 2\theta}{g}$$

This tells us that the maximum range occurs when $\sin 2\theta = 1$ that occurs when:

$\sin 2\theta = 90°$ $\theta = 45°$

Example

A particle is fired horizontally with a speed of 25 ms^{-1} from the top of a vertical cliff of height 80 m. Determine

(a) the time of flight

(b) The distance from the base of the cliff where it strikes the ground

(c) the velocity with which it strikes the ground

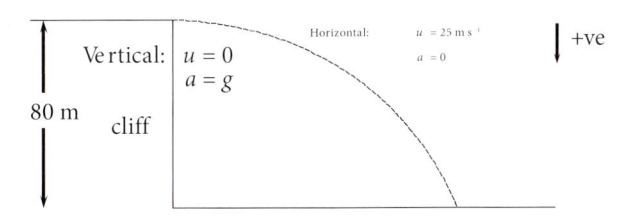

47

CHAPTER 2

Solution

(a) The vertical velocity with which it strikes the ground can be found using the equation

$$v^2 = u^2 + 2as, \text{ with } u = 0, a = g \text{ and } s = 80 (= h).$$

This then gives $v_v = \sqrt{2gh} = \sqrt{2 \times 10 \times 80}$

$= 40$

That is, the vertical velocity at impact is 40 m s^{-1}.

The time to strike the ground can be found using $v = u + at$, with $u = 0$, $a = g$ and $v = v_v$. So that,

$$t = \frac{v_v}{g} = \frac{40}{10} = 4.$$

That is, 4 seconds.

(b) The distance travelled from the base of the cliff using

$$s = ut + \frac{1}{2}at^2, \text{ with } u = 25,$$

$a = 0$ and $t = 4$ is given by

$s = 25 \times 4 = 100$.

That is, the range is 100 m.

(c) The velocity with which it strikes the ground is given by the resultant of the vertical and horizontal velocities as shown.

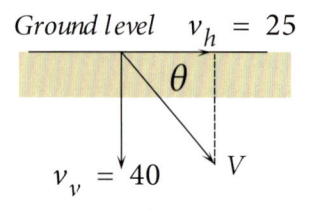

The magnitude of this velocity is

$\sqrt{40^2 + 25^2} = 47$ m s^{-1}

and it makes an angle to the horizontal of

$$\theta = \arctan\left(\frac{40}{25}\right)$$

$= 58°$ (or to the vertical of 32°).

Conservation of energy and projectile problems

In some situations the use of conservation of energy can be a much simpler method than using the kinematics equations. Solving projectile motion problems makes use of the fact that $E_k + E_p$ = constant at every point in the object's flight (assuming no loss of energy due to friction).

In Figure 223, using the conservation of energy principle we have that the

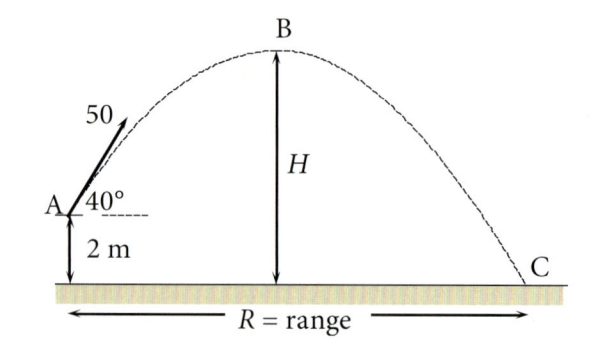

Total energy at A = Total energy at B = Total energy at C

i.e.,

$$mv_A^2 = \frac{1}{2}mv_B^2 + mgH = \frac{1}{2}mv_C^2 + mgh$$

Notice that at A, the potential energy is set at zero ($h = 0$).

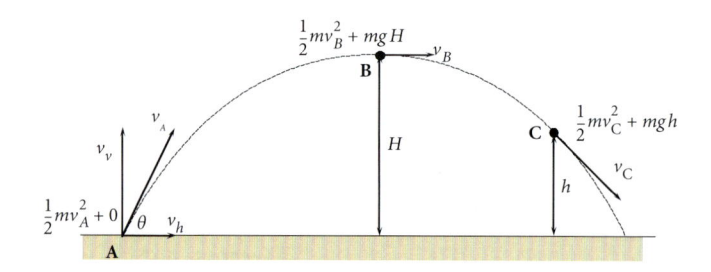

Figure 223 Energy problem

Example

A ball is projected at 50 ms^{-1} at an angle of 40° above the horizontal. The ball is released 2.00 m above ground level. Taking g = 10 m s^{-2}, determine

(a) the maximum height reached by the ball

(b) the speed of the ball as it hits the ground

Solution

48

a. The total energy at A is given by

$$E_k + E_p = \frac{1}{2}m(50.0)^2 + mg \times 2.00$$
$$= 1250\,m + 20\,m$$
$$= 1270\,m \approx 1300\,m$$

Next, to find the total energy at B we need to first determine the speed at B, which is given by the horizontal component of the speed at A.

Horizontal component: $50.0 \cos 40° = 38.3$ m s^{-1}.

Therefore, we have that

$$E_k + E_p = \frac{1}{2}m(38.3)^2 + mg \times H$$
$$y = 15t - 5t^2$$
$$= 733.53\,m + 10\,mH$$

Equating, we have

$$1270\,m = 733.53\,m + 10\,mH$$
$$\Leftrightarrow 1270 = 733.53 + 10\,H$$
$$\Leftrightarrow H = 53.6$$

That is, the maximum height reached is 53.6 m.

b. At C, the total energy is given by

$$E_k + E_p = \frac{1}{2}mv_C^2 + mg \times 0 = \frac{1}{2}mv_C^2$$

Using the total energy at A, $E_k + E_p = 1270\,m$

Equating, we have that

$$1270\,m = \frac{1}{2}mv_C^2 \Rightarrow v_C^2 = 2540$$
$$\therefore v_C = \sqrt{2540} = 50.4$$

That is, the ball hits the ground with a speed of 50.4 m s^{-1}.

Exercise 2.3

1. A projectile is fired from the edge of a vertical cliff with a speed of 30 m s^{-1} at an angle of 30° to the horizontal. The height of the cliff above the surface of the sea is 100 m.

    (a) If $g = 10$ m s^{-2} and air resistance is ignored show that at any time t after the launch the vertical displacement y of the projectile as measured from the top of the cliff is given by:
    $$y = 15t - 5t^2$$

    Hence show that the projectile will hit the surface of the sea about 6 s after it is launched.

    (b) Suggest the significance of the negative value of t that can be obtained in solving the equation?
    (c) Determine the maximum height reached by the projectile and the horizontal distance to where it strikes the sea as measured from the base of the cliff.

2. A man dives off a cliff 20 m high with zero initial vertical velocity and initial horizontal velocity of 5 m s^{-1}. If $g = 10$ m s^{-2}, determine:

    (a) the man's vertical velocity as he hits the water
    (b) the time it takes for him to hit the water
    (c) how far horizontally does he land from the cliff.

3. A projectile is launched horizontally at 50 m s^{-1}.

    Determine:

    (a) the horizontal and vertical components of its velocity after 1 s and 5 s.
    (b) the velocity of the projectile after 1 s and 5 s (Hint: use vectors)

4. A ball is thrown into the air with an initial velocity of 50 ms^{-1} at 37° to the horizontal. (let $g = 10$ m s^{-2})

    (a) Determine the total time the ball is in the air
    (b) Determine the total horizontal distance it travels
    (c) Sketch a graph vertical height versus time
    (d) Sketch the vertical distance versus the horizontal distance
    (e) Where does the ball land if the same ball is thrown off a 55 m cliff with the same velocity and angle of projection.

5. A goalkeeper kicks a soccer ball at an angle of 40.0° above the horizontal with an initial velocity of 14 m s^{-1}. Ignoring air resistance, calculate:

    (a) the maximum height reached by the soccer ball
    (b) the time of flight the soccer ball
    (c) the range of the soccer ball

6. A cannon is fired with an initial velocity $v_0 = 95$ m s^{-1} and an angle $\theta = 50°$. After 5 s it strikes the top of a hill. Determine:

    (a) the elevation of the hill above the point of firing
    (b) the range of the cannon.

7. A rocket from a rocket launcher with an initial speed of 40 m s^{-1} is projected downward at an angle of 30° to the horizontal from the top of a tower 170 m high. Ignoring air resistance, determine how long it takes to hit a target on the ground.

8. A tennis ball is thrown from ground level with an initial velocity 0f 20.0 m s^{-1} at an angle of 70° to the horizontal. Ignoring air resistance:

    (a) Determine the initial vertical and horizontal components of velocity respectively.
    (b) Determine the time the tennis ball is in the air.
    (c) Sketch 2 graphs on the same set of axes (including scales) to show the variation with time of the vertical and horizontal velocities
    (d) From the sketch, state how the maximum height reached by the ball, and the horizontal range of the ball may be determined from these graphs.

Fluid resistance and terminal speed

So far in this discussion of motion we have ignored the effects of air resistance. When an object moves through the air, it is subjected to a retarding force called the drag force or air resistance. The magnitude of the drag force depends on another of factors such as the shape of the object, the surface area of the object, the speed of the object and the nature of the fluid through which the object moves.

Cars, dolphins, sharks are streamlined to reduce drag. Larger parachutes will move slower to decrease air resistance. If an object moves at high speed, there is more friction and the object will reach its terminal speed more quickly. There is greater friction in water than in air due to the greater viscosity of the water. When a person dives into water, the diver fairly abruptly reaches a terminal velocity. The pressure increases with depth of the diver and there is an upthrust force that brings the diver to the surface.

However, For spherical objects moving at relatively low speeds, experiment shows that the retarding force due to air resistance is directly proportional to the speed of the object (provided that the density of the air stays constant). Effectively this means that as the object moves faster and faster, the drag force gets greater and greater until in fact it reaches a value equal to the value of the force accelerating the object. The weight force of the object downwards will be equal to the air resistance force upwards. We say the forces are balanced. When this occurs, the object will no longer accelerate and will move with constant speed. The object will not become stationary because Newton' First Law states that an object moving in a straight line will continue to move with constant motion provided there are no external forces. (We will discuss the effect of forces on the motion of objects in much more detail in the next section).

A parachute is designed to have a large drag force. As the person jumps from a plane, their gravitational potential energy is converted to kinetic and thermal energy as they move through the air fluid. They fall freely due to their weight force and this unbalanced force causes them to accelerate. Eventually, the air resistance force balances the weight force, and the parachutist reaches their terminal velocity, and they fall at a constant speed of about 50 ms^{-1}. When the parachute is opened there is a rapid negative acceleration and then they reach a new terminal velocity of about 8 m s^{-1}. A sketch graph of the speed versus time for the parachutist is shown in Figure 224.

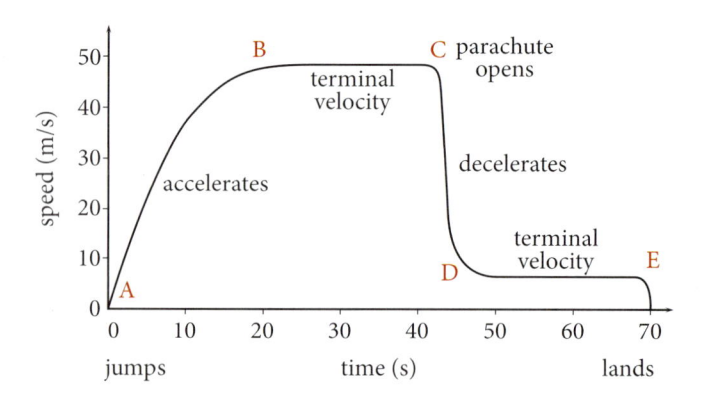

Figure 224   The effect on speed during a parachute jump

All objects such as a feather, a raindrop, a hailstone or a snowflake fall with their own distinct terminal velocity. If a stone is dropped from a balloon that is at a height of 5000 metres, then, if we ignore air resistance, the velocity with which it strikes the ground is about 320 m s^{-1} (about the speed of sound). Because of air resistance the actual speed is much less than this. For example if you fell out of the balloon, your terminal velocity would be about 60 m s^{-1}.

The effect of air resistance on projectile motion

We have seen that in the absence of air resistance, the path followed by a projectile is a parabola and that the path depends only on the initial speed and angle of projection. Of course, in the real world all projectiles are subject to air resistance. Fig 225 shows the free body force diagram for a projectile subject to air resistance.

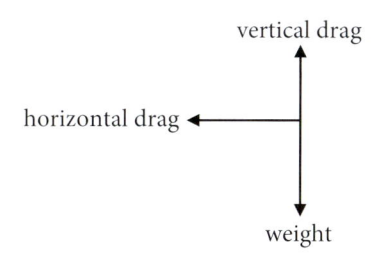

Figure 225   The effect of air resistance

Experiment shows that both the horizontal and vertical drag forces depend on the speed of the projectile. The effect of the horizontal drag will be to shorten the range of the projectile and the effect of the vertical drag will be to reduce the maximum height reached by the projectile. However, the presence or air resistance also means that the mass of the projectile will now affect the path followed by the projectile. Rather than follow a parabolic path, the path will be an "assymetric" parabola of shorter range as shown in Figure 226.

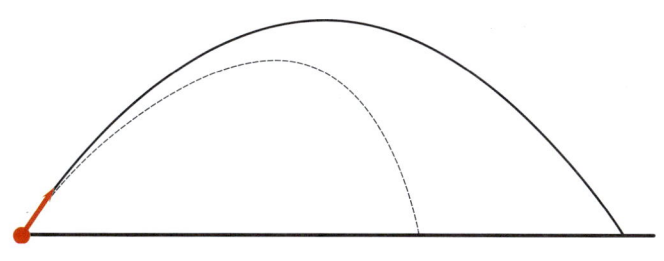

Figure 226   The effect of air resistance on a trajectory

In the absence of air resistance there is no acceleration in the horizontal direction and the acceleration in the vertical direction is g, the acceleration of free fall. With air resistance present, to find the horizontal (a_H) and vertical (a_V) accelerations we have to apply Newton's second law to both the directions. If we let the horizontal drag equal kv_H and the vertical drag equal Kv_V where k and K are constants and v_H and v_V are the horizontal and vertical speeds respectively at any instant, then we can write:

$$kv_H = ma_H \text{ and } mg - Kv_V = ma_V$$

From this we can now see why the mass affects the path since both aH and aV depend on height. *(For those of you doing HL maths, you will realise that the above equations can be written as differential equations but finding their solution is no easy matter.)* We have here another example of the Newton method for solving the general mechanics problem—know the forces acting at a particular instant and you can in principle predict the future behaviour of the system.

Notice that at impact the velocity vector is tangential to the path of motion. As a matter of fact, the velocity vector is always tangential to the path of motion and is made up of the horizontal and vertical components of the velocities of the object.

2.2 Forces

NATURE OF SCIENCE:

Using mathematics: Isaac Newton provided the basis for much of our understanding of forces and motion by formalizing the previous work of scientists through the application of mathematics by inventing calculus to assist with this. (2.4)

Intuition: The tale of the falling apple describes simply one of the many flashes of intuition that went into publication of Philosophiæ Naturalis Principia Mathematica in 1687. (1.5)

© IBO 2014

Essential idea: Classical physics requires a force to change a state of motion, as suggested by Newton in his laws of motion.

Understandings:
- Objects as point particles
- Free-body diagrams
- Translational equilibrium
- Newton's laws of motion
- Solid friction

Objects as point particles

The word *force* appears very often in everyday usage and the simple definition is that a force is a "push" or a "pull". But what actually does we mean by *force*? One thing is for certain, a physicist will not be able to tell you what a "force" is and in many respects the question has little meaning in physics. To a physicist a force is recognised by the effect or effects that it produces. A force is something that can cause an object to:

- deform i.e. change its shape
- speed up
- slow down
- change direction.

The last three of these can be summarised by stating that a force produces acceleration.

So if you were to see an object that is moving along in a straight line with constant speed suddenly change direction, you would know immediately that a force had acted on it even if you did not see anything tangible 'pushing' on the object. The fact that a free falling object accelerates means that a force is acting on it. This force is the force of gravity that the Earth exerts on the object and is the weight of the body.

Since a force can produce acceleration, it is clearly a vector quantity. Hence, if two or more forces act on a particle, to find the resultant acceleration we have to find the resultant force. We use the word *particle* here because if forces act on a body they can produce a deformation as well as acceleration. If they act on a particle that ideally has no physical dimensions then it can only produce acceleration.

If a force exists between two objects in physical contact with each other, then the force is a **contact force**. When forces are disconnected, they were called "action at a distance" forces by Newton. Michael Faraday later introduced the concept of "field" and the forces involved are called **field forces**. These lines of flux extend outwards in space from a mass or a charge or a magnet. For example, the Earth has a gravitational field that outwards from the Earth and the Moon interacts with the field of the Earth and allows the Moon to maintain its orbit.

The four fundamental forces of nature are all field forces. Although at first sight there seems to be a bewildering number of different types of force, pushes, pulls, friction, electrical, magnetic etc. physicists now recognise that all the different forces arise from just four fundamental forces.

The weakest of the four is the **gravitational force**. As we have seen, this is the force that gives rise to the weight of an object. However, as mentioned, we also understand this force to act between all particles in the Universe. A force that is about 10^{26} times stronger than gravity is the **weak nuclear force**. This is the interaction, which is responsible for certain aspects of the radioactive decay of nuclei. The **electromagnetic force** is some 10^{37} times stronger than gravity and this is the force that exists between particles as a consequence of the electrical charge that they carry. The strongest of all the forces is the **strong nuclear force**. This is some 1039 times stronger than gravity and it is the force that holds the protons and neutrons together in the nucleus.

All four of these interactions are discussed in more detail throughout this book. However, a simple way of looking at them is to think of the gravitational force as being the force that accounts for planetary motion and the way that galaxies are put together. The electromagnetic interaction is the force that accounts for the way in which the electrons are arranged in atoms and as such is the force that accounts for all chemical and biological processes. The strong interaction accounts for the nuclear structure of the atom and the weak interaction accounts for how the nucleus comes apart. Physicists would like to unify all these forces into just one force. That is they would like to find that all the interactions were just special cases of one fundamental interaction. There has been some success in unifying the weak and the electromagnetic interaction but that is the current situation.

We tend to use the words point and particle quite often in physics. We have a point source of a sound wave, a point source of light, a point image, and point charge (point particle with non-zero charge) and point mass (point particle with non-zero mass).

A **rigid body** is an idealised body that does not change its size or shape when forces are applied to it. Rigid objects that move without rotating are said to be in translational motion. When a body is in translational motion relative to a point of reference, all the particles within the body are moving in a similar manner and so any particle within the body could be used to represent the position of the body. This would not be true if the body was rotating because all the particles would move in different directions to a point of reference.

If you were at infinity, all objects of various shapes and sizes would appear to be like a point object. They would appear to be dimensionless with no spatial extension. So, a **point particle** is a good approximation for a rigid object with its mass positioned at its centre of mass. This rigid object is considered to be a system and everything outside the rigid object is called the **environment**. Forces due to the interaction of particles within the system are called **internal forces** and the forces due to objects in the environment that interact with the particles within the system are called **external forces**. This differentiation of the meaning of internal and external forces will soon become apparent when we discuss Newton's Laws of motion and momentum.

Free-Body Diagrams

We have seen that we can represent the forces acting on a point particle by lines with arrows, the lengths of which represented the relative magnitude of the forces. Such diagrams are a useful way to represent the forces acting on a body or particle and are called **free-body diagrams**.

Figure 227 shows an object of mass M that is suspended vertically by a thread of negligible mass. It is then pulled to one side by a force of magnitude F and held in the position shown.

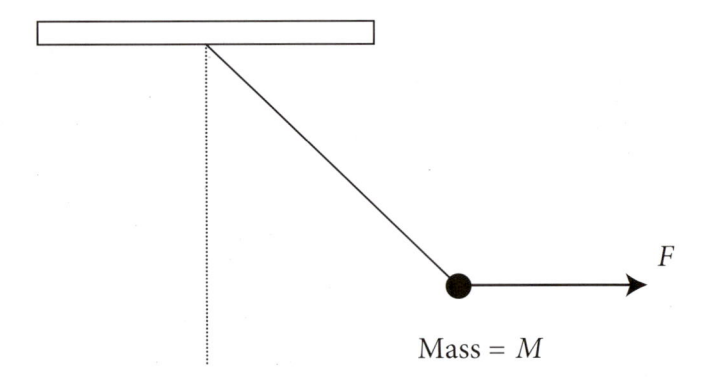

Figure 227   Forces acting on an object

The free-body diagram for the forces acting on the object in Figure 227 is shown in Figure 228

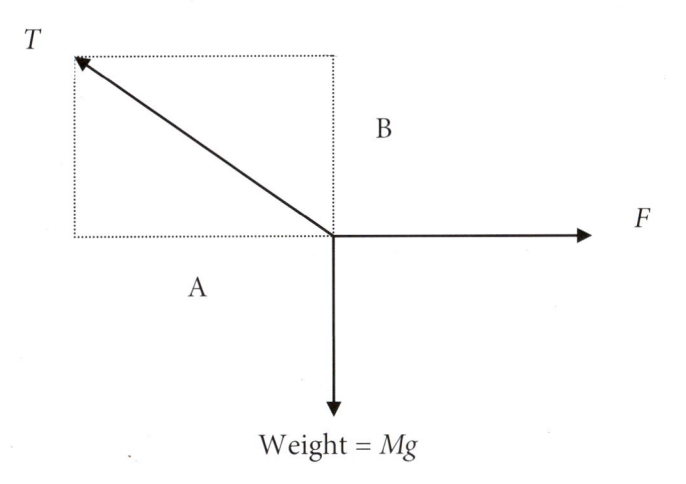

Figure 228 The Free-body Force Diagram for Fig. 227

The weight of the object is Mg and the magnitude of the tension in the thread is T.

The object is in equilibrium (see 2.2.6) and so the net force acting on it is zero. This means that the vertical components of the forces must be zero as must the horizontal components. Therefore, in the diagram the line A is equal in length to the arrow representing the force F and the line B is equal in length to the arrow representing the weight. The tension T, the "resultant" of A and B, is found by the using the dotted line constructions. When producing a free-body diagram. there is no need to show these constructions. However, as well as being in the appropriate directions, the lengths of the arrows representing the forces should be approximately proportional to the magnitudes of the forces. The following example is left as an exercise for you.

A more elaborate free-body force diagram of the forces acting on an aircraft passenger airliner that is flying horizontally with constant velocity is shown in Figure 229.

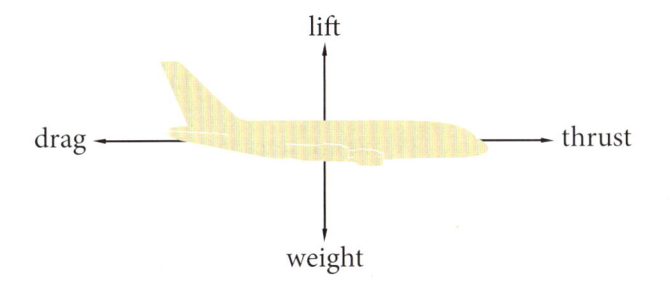

Figure 229 Free-body diagram for a constant speed passenger airliner.

An interesting and very important type of contact force arises in connection with springs. If you hold one end of a spring in one hand and pull the other end then clearly to extend the spring you have to exert a force on the spring as shown in Figure 230. If you don't hold one end of the spring, then when you pull, it will accelerate in the direction that you are pulling it. Holding one end and pulling the other produces a tension force in the spring.

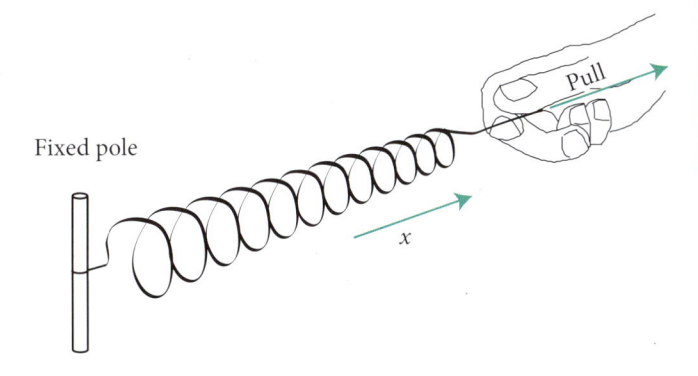

Figure 230 Forces on a spring

In Figure 231 the blue arrows show the tension forces set up in the spring. There is a force that opposes the pulling force and a force equal in magnitude to this force is also exerted by the spring on the fixed support.

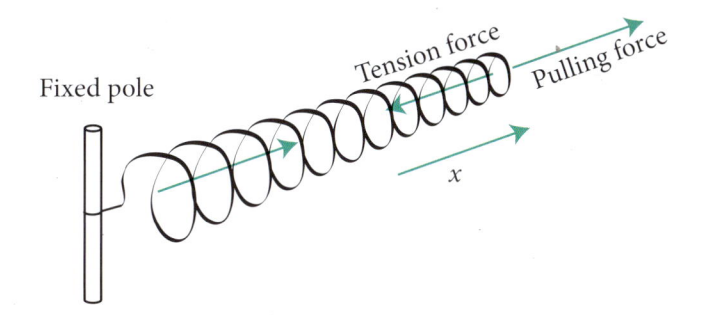

Figure 231 Tension forces in a spring

One thing that you will notice as you pull the spring is that the further you extend the spring the force that you need to apply becomes greater. You can investigate how the force required to extend the spring varies with the extension e of the spring by simply hanging masses of different values on the end of a vertically suspended spring. (The force of gravity measured in newtons that acts on a mass M can be found to a very good approximation by multiplying M by 10). The result for a typical spring in which the force F is plotted against the extension e is shown in Figure 232.

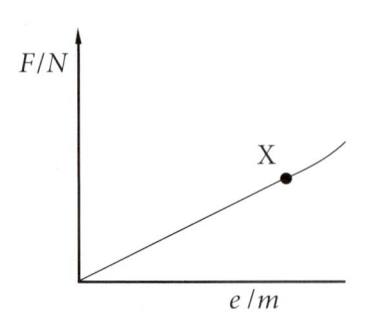

Figure 232 Graph of force against extension of a spring

Up to the point X the force F is directly proportional to e. beyond this point the proportionality is lost. If the point X is passed, the spring can become permanently deformed in such a way that when the weights are removed the spring will not go back to its original length. In the region of proportionality we can write

$$F = ke$$

where k is a constant whose value will depend on the particular spring. For this reason k is called the spring constant. This spring behaviour is a specific example of a more general rule known as Hooke's law of elasticity after the 17th century physicist Robert Hooke. For this reason the region of proportionality is often referred to as the 'Hookey region' or 'elastic region' and point X is called the 'elastic limit'. We can see that a spring can be calibrated to measure force and no doubt your physics laboratory has several so-called "newton metres". However, you will discover that these newton metres do not provide a particularly reliable method of measuring force.

Example

Two forces act on particle P as shown in the Figure below (N stands for 'newton' and is the SI unit of force as we shall see in the next section.)

Determine the magnitude of the net force acting in the horizontal direction and the magnitude of the net force acting in the vertical direction and hence determine the resultant force acting on P.

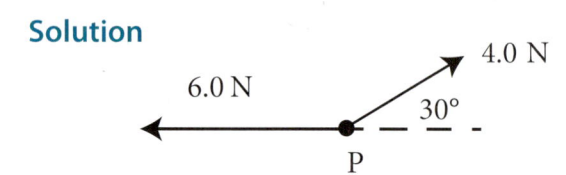

Solution

The component of the 4.0 N in the horizontal direction is 4.0 cos 30 = 3.5 N.

Hence the magnitude of the force in the horizontal direction is 2.5 N.

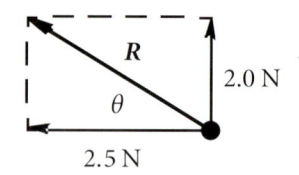

The component of the 4.0 N force in the vertical direction is 4.0 sin 30 = 2.0 N and this is the magnitude of the force in the vertical direction.

The Figure above shows the vector addition of the horizontal and vertical components.

The resultant R has a magnitude = 3.2 N and the angle θ = tan^{-1} (2.0/2.5) = 39°.

Translational equilibrium

There are essentially two types of equilibrium, static and dynamic. If an object is at rest, then it is in static equilibrium and, if it is moving with constant velocity, then it is in dynamic equilibrium. In both of these situations the resultant force on the object is zero and as such the object is said to be in **translational equilibrium**.

For an object to be in equilibrium two conditions need to be met:

- The sum of the forces must be equal to be zero
- The sum of the torques must be equal to zero

We will soon see from Newton's first law that the condition for both equilibriums is that the net force acting on the object is zero. We can express that mathematically as Σ F = 0. That is the vector sum of the forces acting on the object is zero.

Static equilibrium

Consider the simple case of a book resting on a table. Clearly gravity acts on the book and without the intervention of the table, the book would fall to the ground. The table therefore exerts an equal and opposite force on the book. This force we call the normal reaction. The forces acting on the book are shown in Figure 233.

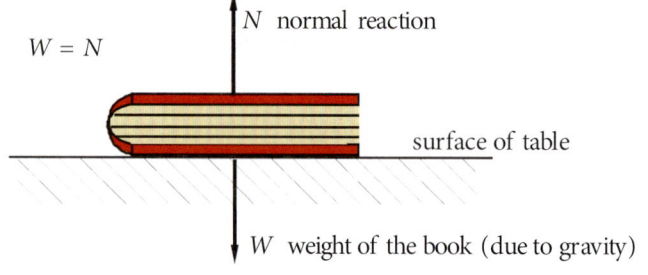

Figure 233 Static forces on a book

54

Dynamic equilibrium

Now consider the case where the book, or any other object, is pulled along the surface of the table with constant velocity. Gravity and the normal reaction are still acting but there is now a frictional force acting which is equal in magnitude but opposite in direction to the pulling force. The force diagram for this situation is shown in Figure 234.

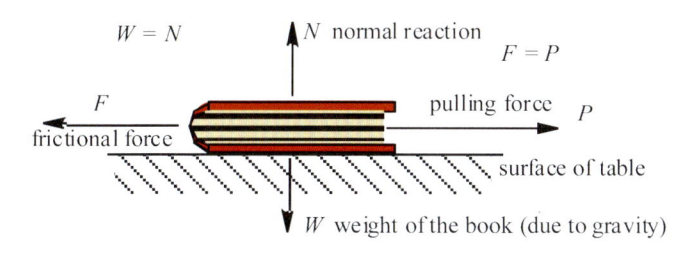

Figure 234 Dynamic forces on a book

Suppose that we were to pull the object with a force as applied to it as shown in Figure 235.

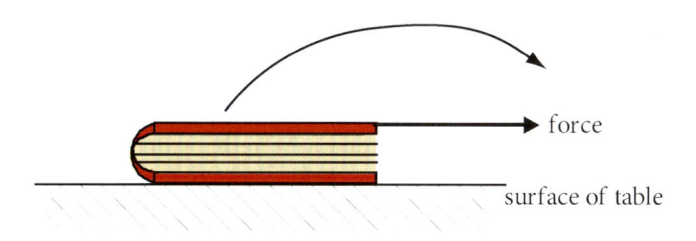

Figure 235 Forces on the book

In this situation, it is quite likely that the book will not move along the table but actually rotate in the direction shown by the arrow.

This again demonstrates that in fact we should apply the Newton laws of motion to particles. Unless the pulling force acts through what we call the centre of mass of the object then the pulling force and the frictional force can produce rotation. This is quite a subtle point. However, in many situations in this chapter we will refer to objects and bodies when strictly speaking we mean particles. There is in fact a branch of mechanics known as 'Rigid Body Mechanics' that specifically deals with the mechanics of extended bodies rather than particles. This is not covered in the IB course. However, we can still get quite a long way with mechanics by considering bodies to act as particles.

Newton's laws of motion

Before developing the ideas of Newtonian Mechanics, it is important to discuss the concepts of mass and weight. In everyday usage, mass and weight are often interchangeable; in physics there is a fundamental difference between the concepts.

So what is **mass**? Some textbooks will tell you that it is the 'quantity of matter' in a body. However, this is pretty meaningless for two reasons. It begs the question as to what is meant by 'matter' and also it gives no means of quantifying mass. In respect of the latter, if a physicist cannot measure something in the laboratory then that something does not belong to physics.

Mass is one of the fundamental properties of all matter and essentially measures a body's inertia, that is, a body's reluctance to change its state of motion. Moving bodies tend to keep on moving in the same direction and stationary bodies don't start moving by themselves. The more massive a body, the more reluctant it is to change its state of motion. It is quite possible to give a logical definition to mass that enables a physicist to quantify the concept. However, this definition does not tell you what mass is and why particles that make up bodies should have mass is one of the great unsolved mysteries of physics. The situation is further complicated by the fact that there are two types of mass. There is the mass we have just described, inertial mass and then there is the mass a body has that gives rise to the gravitational attraction between bodies. This is the gravitational mass of a body. We have already mentioned that experiment shows that the acceleration of free fall is independent of the mass of an object. This suggests that gravitational mass and inertial mass are in fact equivalent.

The equivalence of gravitational and inertial mass is one of the cornerstones of Einstein's theory of general relativity and is discussed in more detail in Option A. The SI unit of mass is the kilogram and the standard is the mass of a platinum alloy cylinder kept at the International Bureau of Weights and Measures at Sevres near Paris.

We shall see that when we talk about the **weight** of a body what we actually mean is the gravitational force that the Earth exerts on the body. So **weight is a force** and since the force of gravity varies from place to place and also with height above the Earth's surface, the weight of a body will also vary but it's inertial and gravitational mass remains constant.

Newton's First Law of motion

At the beginning of this chapter we stated that the general mechanics problem is, that given certain initial conditions of a system, to predict the future behaviour of the system. The method that Newton devised to solve this problem is encompassed in his celebrated three laws of motion that he published in his *Principia Mathematica* circa 1660. Essentially Newton tells us to find out the forces acting on the system. If we know these then we should in principle able to predict the future behaviour of the system.

Newton's First Law is essentially qualitative and is based on the work done by Galileo. Prior to Galileo's work on mechanics, the Aristotelian understanding of motion was the accepted view, that a constant force is needed to produce constant motion. This seems to fit in with every day experience; if you stop pushing something then it will stop moving, to keep it moving you have to keep pushing it. Galileo's brilliance was to recognise that the opposite is actually the case and his idea is summarised in the statement of Newton's First Law:

If a body is at rest it wants to remain at rest, and if the body is moving in a straight line with uniform motion it will continue to move with uniform motion unless acted upon by an external force.

Note that there essentially 2 parts to Newton's First Law.

The Aristotelian view does not take into account that when you push something another force is usually acting on the body that you are pushing, namely the force of friction. In some situations, as we have seen, the frictional force, acting on a moving object is a function of the object's velocity and in fact increases with velocity. Hence a greater engine power is required to move a car at high speed than at low speed. We shall return to this idea later.

If you eliminate friction and give an object a momentary push, it will continue moving in the direction of the push with constant velocity until it is acted upon by another force. This can be demonstrated to a certain degree using the linear air-track. It is to a limited degree since it is impossible to eliminate friction completely and the air track is not infinite in length. It is in this sense impossible to prove the first law with absolute certainty since sooner rather than later all objects will encounter a force of some kind or another.

In fact Arthur Eddington (1882–1944), is reputed to have quoted Newton's first law thus:

'Every object continues in a state of rest or uniform motion in a straight line in so far as it doesn't.'

By this he meant that nothing in the Universe is ever at rest and there is no such thing as straight-line motion.

However, we are inclined to believe that if a force does not act upon a body, then Galileo's description of its motion is correct.

It is sometimes difficult to discard the Aristotelian view of motion particularly in respect of objects that are subject to a momentary force. Consider the example shown in Figure 236:

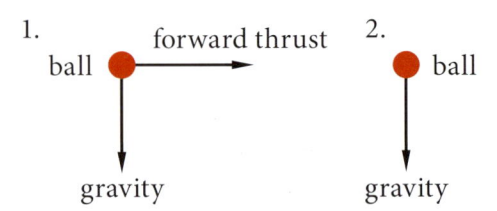

Figure 236   Aristotelian and Galilean forces on a ball

A girl throws a ball towards another girl standing some metres away from her, it is tempting to think that, as Aristotle did, there must be a forward thrust to keep the ball moving through the air as shown in 1. However, if air resistance is neglected, the only force acting on the ball is gravity as shown in 2.

Newton's second law of motion

The vector form of the equations relating to Newton's Second Law given in this section, show that the vectors act in the same straight line (for example, the force and acceleration in **F** = m**a**). In the IB Physics data booklet, all vector equations are given in scalar form i.e. they relate just the magnitudes of the quantities.

Newton's First Law as we have seen is qualitative in nature. With his second law we move into quantitative territory, for it is essentially the second law that enables us to solve general mechanics problem. A force produces acceleration when it acts on a particle and Newton argued that for a given particle the acceleration is:

- Directly proportional to the force acting and is in the same direction as the applied force.
- Inversely proportional to the inertial mass of the object (the greater the mass the less the acceleration).

We can therefore write that:

Acceleration $\propto \frac{\text{force}}{\text{mass}}$

$a \propto \frac{F}{m}$ or force $\propto$ mass $\times$ acceleration

$$F = ma$$

This is Newton's second law in its simplest form. There are however, many situations where the mass of the system does not remain constant e.g. a firework rocket, sand falling on to a conveyor belt etc. It is therefore helpful to express the law in a more general form.

We can express the acceleration a in terms of the rate of change of velocity i.e.

$$F = m \times \frac{\Delta v}{\Delta t}$$

We now define a quantity called the *linear momentum p* of the object as

$$p = mv$$

such that we can now write Newton's second law if the form

$$F = \frac{\Delta p}{\Delta t}$$

We shall discuss the concept of momentum in more detail in another section.

Although it is possible to verify Newton's Second Law directly by experiment its real validity is understood in terms of the experimentally verifiable results that it predicts. In a sense this law is the whole of Classical Mechanics and tells us that if we pay attention to the forces then we can find the acceleration and if we know the acceleration then we know the future behaviour of the particle. Newton in essence said that if you can find the force law governing a system, you would be able to predict its behaviour. Unfortunately, it is not always possible to know the force law. Newton himself gave one, his famous law of gravitation, which we will look at in Topic 6. However, in such situations such as the collision of two billiard balls we do not know the force acting nor do we know the force acting between the millions of molecules of a gas, solid or liquid. In situations such as these we have to find some other means of solving the problem and these we look at some of these methods shortly.

We can use Newton's Second Law to understand the equivalence between inertial and gravitational mass. A simple argument shows this to be so. If we assume that the gravitational force F_G that the Earth exerts on an object is proportional to the gravitational mass m_G of the object. We can write this as

$$F_G = K m_G$$

where K is a constant.

The acceleration g of the object is given by Newton's second law.

$$F_G = K m_G = m_I g$$

where m_I is the inertial mass of the object.

But experiment shows that g is a constant and has the same value for all objects. Hence, it follows from the above equation that $m_G = m_I$ with $g = K$.

If we have an independent definition of inertial mass, then we can use the second law to define a unit of force. The SI unit of force is the **newton** (N) and it is that force which produces an acceleration of 1 m s^{-2} in a mass of 1 kg. This is an **absolute** definition in that it does not depend on the properties of any material or any outside influence such as pressure and temperature.

The second law also enables us to quantify the relationship between mass and weight. As has been previously stated the acceleration of free fall is the same for all objects. If its value is g and an object has a mass m then from the second law we see that the gravitational force exerted on it by the Earth has a value mg. Hence mg is the weight of the object. If we take g to have a value of 10 m s^{-2} then a mass of 1 kg will have a weight of 10 N close to the surface of the Earth. On the Moon where the acceleration of gravity is about 1.7 m s^{-2} a mass of 1 kg will have a weight of about 1.7 N.

Newton's third law of motion

Students should understand that when two bodies A and B interact, the force that A exerts on B is equal and opposite to the force that B exerts on A.

As was suggested in the introduction, Newton's Second Law does not always work. It does not give the right answers when we apply it to atoms and molecules nor does it give the right answer when particle are moving with speed close to that of the speed of light. However his **third law of motion** as far as we know is valid across the whole of physics. It is really quite a remarkable piece of insight and has, as we shall see far reaching consequences. In fact, Ernst Mach a famous turn of the century physicist regarded the third law has Newton's greatest contribution to Mechanics.

Newton formulated this law based on an idea first put forward by Descartes. The law basically says that forces always appear in equal and opposite pairs. To state the law more formally:

> **When a force acts on a body, an equal and opposite force acts on another body somewhere in the universe.**

For example, when two bodies A and B collide, then whilst in contact, the force that A exerts on B is equal and opposite to the force that B exerts on A. It is important to remember that **forces always exist in pairs**.

In some textbooks and websites the law is stated as "to every action there is an equal and opposite reaction".

However, this can lead to some confusion. To see why, let us look at the situation of the book on the table in Figure 237. The book is in equilibrium so we conclude that the force that the book exerts on the table is balanced by a "reaction force", the force that the table exerts on the book. We can ask the question " what if the book is so heavy that it breaks the table"? Does this mean that Newton's third law no longer works? No, because this is not an example of Newton's third law. Let us look at the situation more carefully to see just how the third law is working.

There are four forces so there are two pairs. The weight of the book is the force **X** and this is the force that the earth exerts on the book. The force **Y** is the force that the book exerts on the Earth.

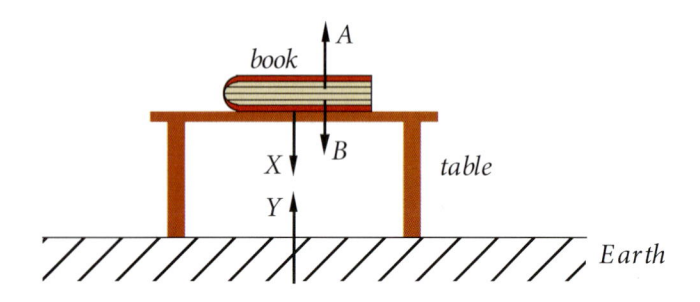

Figure 237   The four forces acting on a book resting on a table

The table is actually incidental to the action of these forces. The two forces are the equal and opposite pair referred to in the third law such that

X + Y = 0

If the table is not there, the book falls towards the Earth and the Earth falls towards the book. Both the book and the Earth will accelerate in accordance with Newton's second law. However, considering that the earth is some 10^{26} times more massive than the book, we do not observe the acceleration of the Earth.

The force *A* is the force that the table exerts on the book. The force *B* is the force that the book exerts on the table. These two forces are again an equal and opposite pair of forces referred to in the third law such that

A + B = 0

So what is the origin of these two forces? They actually arise from the interaction forces between the molecules of the book and the molecules of the table. The interaction force is complex but essentially between any two molecules there is either a force of repulsion or a force of attraction. Which force operates depends on the separation of the molecules and, in equilibrium, the two molecules will take up a separation at which the repulsion force balances the attraction force. If we "push" the molecules closer together the repulsion force becomes greater than the attraction force and if the "push" is released the molecules will move back to the equilibrium position. It is this repulsion force which stops you falling through the floor.

If we "pull" the molecules further apart then the attraction force becomes greater than the repulsion force and if we release the "pull", the molecules will move back to the equilibrium position. The attraction force is why you need to apply a force to stretch a spring and the repulsion force is why you need to apply a force to compress a spring.

The three situations are illustrated in Figure 238.

equilibrium:

The net force on each molecule is zero.

There is now a net force of repulsion on each molecule.

There is now a net force of attraction on each molecule.

Figure 238   Forces between molecules

The magnitude of both the attraction and repulsion forces depends on the separation of the molecules. The repulsion force increases very rapidly as the molecules get closer together and this is why it requires a much greater force to compress certain solids than it does to extend them. So we see that the table can still break without violating the third law. When the table breaks we just have a re-arrangement of its molecular structure.

When students first encounter the third law they sometime say things like ' if, when a force is exerted on an object there is an equal and opposite reaction, how can anything ever move? Equal and opposite forces must mean the net force on everything is always zero.' Think of the Earth and the book. When the table is not present, there is a net force on the book and it accelerates in accordance with Newton's second law. It is the net force on the system of the Earth and the book that is zero. Similarly if we think of a horse pulling a cart, the acceleration of the cart depends not on the forces that it might exert on something else, but on the forces that are exerted on it.

We have gone into the third law in some detail because it is an area of physics which is often misunderstood by students. Basically, just remember to be careful when identifying the system in which the equal and opposite forces appear. Also remember that when computing the acceleration of a body, it is the forces that act on the body that are considered, not the forces that it exerts on other bodies.

It is the third law, which enables physicists to give a logical definition of mass. This is something that you will not be expected to know for the examination and it is included for the inquisitive who are puzzled as to how two quantities, force and mass, can be defined from one equation, $F = m\,a$. Well, the answer is that they are not.

If a system consists of two isolated particles that exert equal and opposite forces on each other, then the ratio of their acceleration will be in the ratio of their masses. One of the particles can be considered to be a "standard mass" and the acceleration of other particles interacting with this standard can be measured in order to determine their mass.

In the rest of this section we will look at some examples of the application of the second law and give some exercises. Hopefully, this will help you gain familiarity with its use.

Example 1

The diagram below shows two masses connected by a non-stretch string. A 0.5 kg mass hangs over a light frictionless pulley and a 1.0 kg mass lies on a horizontal, frictionless surface. Taking g to be 10 ms^{-2}, calculate the acceleration of the system when released.

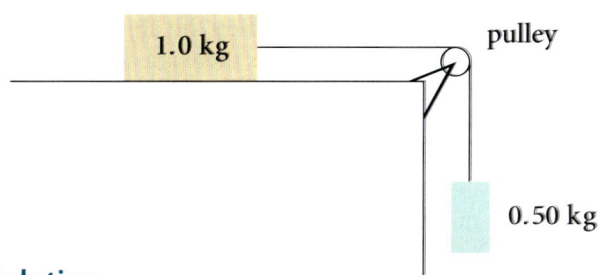

Solution

The force acting on the system is the weight of the hanging mass which is 0.50 kg = 5 N

Using Newton's second law $F = m\,a$
we have $5.0 = (1.5)\,a$
Hence $a = 3.3$

That is, the acceleration is 3.3 m s^{-2}.

Example 2

A person of mass 70 kg is strapped into the front seat of a car, which is travelling at a speed of 30 m s^{-1}. The car brakes and comes to rest after travelling a distance of 180 m. Estimate the average force exerted on the person during the braking process.

Solution

We can use the equation $v^2 = u^2 + 2as$ to calculate the magnitude of the average acceleration of the car.

In this instance we have that $u = 30$, $v = 0$ and $s = 180$, so that

$$0^2 = 30^2 + 2 \times 180 a \Rightarrow a = -\frac{900}{360}$$

$$= -2.5 \text{ m s}^{-2}$$

The negative sign indicates that the car is slowing down. Using Newton's second law $(F = ma)$ we have that $F = 70 \times -2.5 = -175$.

Hence the average force on the person, is 175 N.

Example 3

A light, inextensible string connects two masses of 5.00 kg and 3.00 kg, and they hang on opposite sides of a frictionless pulley as shown in the diagram. The masses are different. Therefore, the 5.00 kg mass will accelerate down and the 3.00 kg mass will accelerate up.

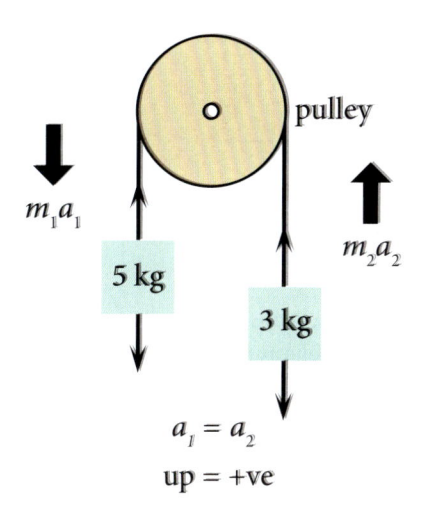

Calculate:

(a) the acceleration of the system
(b) the tension in the string.

Solution

The string is light so its weight can be ignored. The tension in the string is the same at all points along its length because the string does not stretch and the pulley is frictionless.

Let us say upwards is positive and downwards as negative.

Chapter 2

(a) For the 5.00 mass:
$ma = T - m_1 g$
$T = -5a + m_1 g$ Equation 1

For the 3.00 kg mass:
$ma = T - m_2 g$
$T = 3a + m_2 g$ Equation 2

Combining equation 1 and equation 2:

$-5a - m_1 g = 3a + m_2 g$

$-8a = m_2 g - m_1 g$

$-a = \frac{(3g - 5g)}{8} = \frac{-2g}{8} = \frac{-g}{4}$

$a = \frac{9.81}{4} = 2.453 = 2.45$ m s^{-2}

The acceleration of the system is 2.45 m s-2.

(d) From equation 1
$T = -5a + m_1 g = (-5 \times 2.45) + (5.00 \times 9.81) = 36.8$ N

From equation 2
$T = 3a + m_2 g = (3 \times 2.45) + (3.00 \times 9.81) = 36.8$ N

The tension in the string is 36.8 N.

Example 4

A person of mass 80.0 kg stands on a set of newton scales in an elevator. What is the reading on the newton scales when:

(a) the elevator moves up at a steady speed of 4 m s-1
(b) the elevator accelerates upward at 2.00 m s-2
(c) the elevator accelerates downward at 2.00 m s-2.

Solution

(a) The person is not accelerating. Therefore, the net force acting on him is zero. This means that the reaction force pushing upwards from the scales must equal his weight.
$W = mg = 80 \times 9.81 = 784.8$ N = **785 N**.

(b) The force pair acting on the person is the weight force (mg) and the normal reaction force (F). We will solve this problem two ways, letting the up direction be either positive or negative.

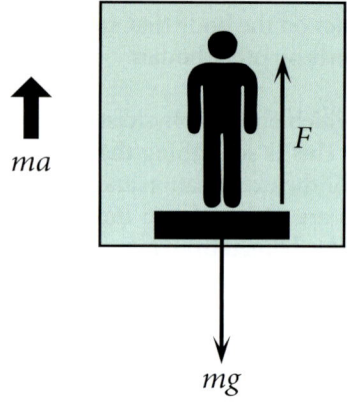

Let up be positive:
$ma = F - mg$
$80.0 \times 2.00 = F - 80.0 \times 9.81$
$1.60 \times 10^2 = F - 784.8$
$F = 944.8$ N = 945 N

Let up be negative:
$-ma = -F + mg$
$-(80 \times 2.00) = -F + (80 \times 9.81)$
$-1.60 \times 10^2 = -F + 784.8$
$-F = -944.8$ N = **945 N**

(c) Let up be positive
$-ma = +F - mg$
$F = mg - ma$
$F = 784.8 - 160$
$F = 625$ N

Let up be negative
$+ma = -F + mg$
$F = mg - ma$
$F = 625$ N

Solid Friction

When an object is in contact with some surface its weight force acting downwards is balanced by the normal reaction of the surface acting upwards. When the object moves or is about to move along the surface, there is a component of the force that is parallel to the surface called the frictional force. This frictional force acts to oppose the motion of the object.

There are a number of techniques that can be used to reduce friction. Polishing the object's surface reduces the microscopic roughness. Lubricating with oil or other fluids reduces the wear and tear of the object. Rolling rather than sliding is another way to reduce friction such as the ball bearings in a bike wheel. Hovercrafts work by separating the craft and the surface with a layer of air. Streamlining objects is yet another way to reduce friction.

At the same time friction also has its advantages. Friction allows us to walk and keeps moving objects in their required path. Friction between the tyres of a car and the road surface is absolutely necessary to steer the car.

Suppose you attached a newton balance to a block of wood in contact with a surface such as the bench top in a laboratory. You then pull on the newton balance but the object does not move because the component of the applied force parallel to the surface balances the static frictional force.

As you continue to pull on the newton balance, there will be a certain applied force registered when the block just begins to move. At this point the applied force is slightly greater than the maximum static frictional force.

Experimental evidence shows that the frictional force is independent of the actual area of the two surfaces in contact provided the surfaces in contact with each other are hard. The evidence suggests that the magnitude of the **frictional force is directly proportional to the magnitude of the normal reaction force.**

$F_f \propto FN$ or $F_f \propto R$

$F_f = \mu_s R$

where μ_s is called the **coefficient of static friction.**

When a force is then applied that causes the block of wood to slide along the surface, the magnitude of the force registered on the newton balance rapidly decreases and then becomes a steady value provided the block is moved with a uniform speed. The frictional force is then given by:

$F_f = \mu_k R$

where μ_k is called the **coefficient of kinetic friction.**

The coefficient of static friction will always be a larger value. As an approximate guide the coefficient of static friction between a polished wood block and polished wood surface is around 0.4 and the coefficient of kinetic friction for the same materials is 0.2. Both coefficients are dimensionless.

Example 5

A snowboard is at rest on some snow and the coefficient of static friction is $\mu_s = 0.42$. If the snowboard and the person have a combined mass of 55 kg, what will be the value of the horizontal force that needs to be applied to cause the snowboard just to move forward?

Solution

The normal reaction force R will be equal and opposite to the weight force = mg

=55 × 9.8 = 539N.

When the snowboard just moves, the maximum frictional force $F_f = \mu_s R$

= 226.38 = **226 N**.

Example 6

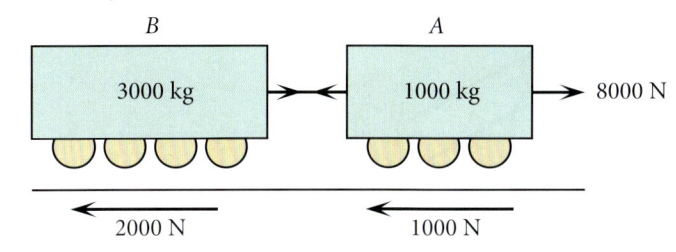

The figure shows a truck A of mass 1000kg pulling a trailer B of mass 3000kg. The frictional forces on A and B are 1000N and 2000N respectively. The truck exerts a horizontal force of 8000N. Calculate:

(a) the acceleration of the truck on the trailer.
(b) the tension in the tow bar connecting A and B.

Solution

(a) For B only   T - 2000 = 3000a   Equation 1
     For A only   8000 - 1000 - T = 1000a   Equation 2

Add equation 1 and equation 2

8000 - 1000 - 2000 = 4000a
5000 = 4000a
a = 1.25 ms⁻²

(b) From equation 1

T = 3000 × 1.25 + 2000 = **5750N**

Example 7

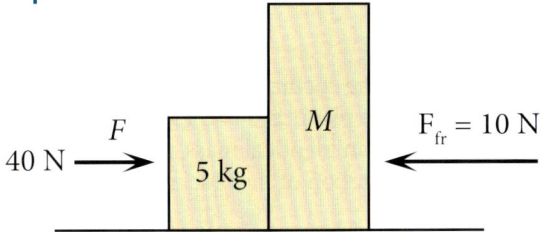

Two boxes, one of mass 5 kg and the other of unknown mass M are pushed along a rough floor with a force of 40N. If the frictional resistance force is 10N and the boxes accelerate at 2 ms⁻², what is the value of M?

Solution

Applying Newton's second Law:

$F - F_f = ma$
40 - 10 = (M + 5) × 2
30 = 2M + 10
M = **10 kg**

Example 8

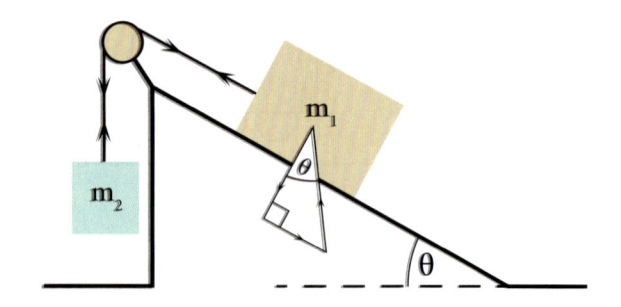

A block of mass m_1 lying on an inclined plane is connected to a mass m_2 by a massless cord passing over a pulley.

(a) Determine a general formula for the acceleration of the system in terms of m_1, m_2, θ and g.

The component of the weight of block m_1 parallel to the inclined plane will equal x

$$x = m_1 g \sin \theta$$

If $m_1 \sin \theta$ is greater than m_2 then the block will move down the plane. We will make this the positive direction:

$m_1 g \sin \theta - T = m_1 a$ Equation 1

$T - m_2 g = m_2 a$ Equation 2

Add equation 1 and equation 2

$m_1 g \sin \theta - m_2 g = m_1 a + m_2 a$

$$a = \frac{(m_1 \sin \theta - m_2)g}{(m_1 + m_2)}$$

(b) Suppose the coefficient of static friction between m_1 and the plane is $\mu_s = 0.15$ and that each block has an identical mass of 2.0 kg. Determine the acceleration of the blocks if $\theta = 30°$.

Solution

Now the magnitude of $m_1 g \sin \theta$ will be less than $m_2 g$. Therefore, mass 1 will move up the plane.

The component of the normal reaction force perpendicular to the plane = $m_1 g \cos \theta$

The frictional force on m_1 up the incline plane
$F_f = \mu_s R = \mu s\, m_1 g \cos \theta$

For mass 1:
$T - m_1 g \sin \theta - \mu s\, m_1 g \cos \theta = m_1 a$ Equation 1

For mass 2:
$-m_2 g + T = -m_2 a$ or $m_2 g - T = m_2 a$ Equation 2

Adding equations 1 and 2:

$(m_2 - m_1 \sin \theta - \mu_s m_1 \cos \theta)g = (m_1 + m_2) a$

$a = \dfrac{(m_2 - m_1 \sin \theta - \mu s\, m_1 \cos \theta) g}{(m_1 + m_2)}$

$a = [2 - 2 \sin 30 - (0.15 \times 2 \cos 30)] \times \dfrac{9.8}{4}$

$a = 1.81 = \mathbf{1.8\ ms^{-2}}$ **up the plane**.

Exercise 2.4

1. This exercise is designed to help you distinguish between the concepts of mass and weight. Here are six different hypothetical activities

    (a) You weigh yourself using bathroom scales.
    (b) You determine the mass of an object by using a chemical beam balance.
    (c) You determine the density of lead.
    (d) You drop a brick on your foot.
    (e) You trap your fingers in a car door.
    (f) You design a suspension bridge.

    Suppose that you were able to carry out these activities on the moon. Discuss how would the result of each activity compare with the result when carried out on Earth.

2. The graph below shows how the length of a spring varies with applied force.

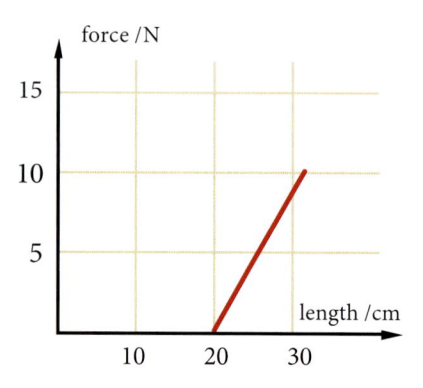

    i. State the value of the unstretched length of the spring.

    ii. Use data from the graph to plot another graph of force against extension and from this graph determine the spring constant.

3. A person stands on bathroom scales placed on the floor of a lift. When the lift is stationary the scales record a weight of 600 N. The person now presses the button for the 6th floor. During the journey to the 6th floor the scales read 680 N, then 600 N, then 500 N and finally 600 N. Explain these observations and calculate any accelerations that the lift might have during the journey.

4. Three forces act as shown on a particle of mass 0.500 kg as shown in this Figure. Calculate the acceleration of the particle.

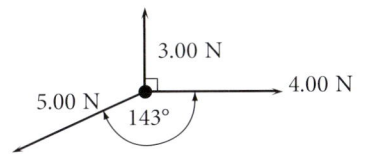

5. Sketch the distance-time, speed-time, velocity-time and acceleration-time for a free fall parachutist from the time that she leaves the aeroplane to the time that she lands on the ground. (Remember that she does not open a parachute until some time after jumping from the aeroplane).

6. An object is thrown through the air. Ignoring air resistance, draw a free body diagram of the forces acting on the ball whilst it is in flight.

7. An object of weight weighing 50 N is suspended vertically by two strings as shown in the following figure:

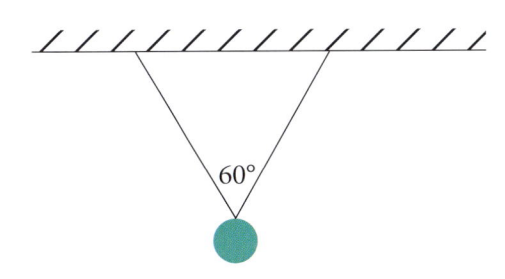

The strings are of the same length and the angle between them is 60°. Draw a free body diagram of the forces acting on the object. Calculate the tension in the strings.

8. When a person stands on bathroom scales the scale reads 60 kg. Suppose the person stands on the same scales when in an elevator (lift). The elevator accelerates upwards at 2.0 m s-2. Determine the new reading on the scale.

9. The diagram shows two blocks connected by a string that passes over a pulley.

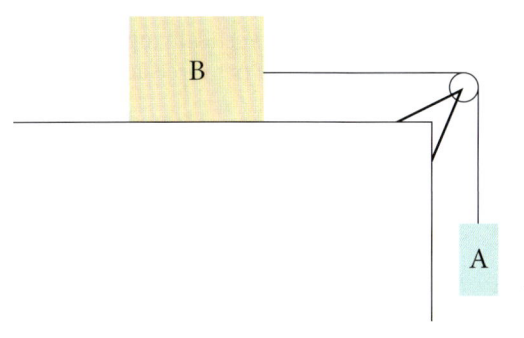

Block A has a mass of 2.0 kg and block B a mass of 4.0 kg and rests on a smooth table. Determine the acceleration of the two blocks?.

10. Here are four statements about a book resting on a table.

A. The book exerts a force on the table.
B. The table exerts a force on the book.
C. the book exerts a force on the Earth.
D. the Earth exerts a force on the book.

Which forces form a pair of forces as described by Newton's Third Law?

11. Which of the following units is dimensionally the same as the unit of acceleration?

A. Nm^{-1} C. Nkg^{-1}
B. Ns^{-1} D. $Nms^{-1}0$

12. An object is initially at rest. It is free to move without friction. It is acted upon by a force that varies with time as shown in the diagram.

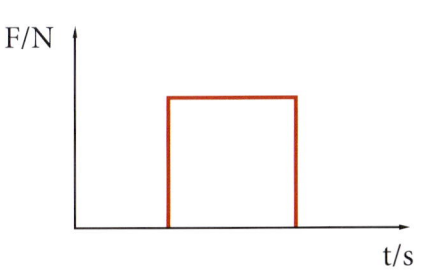

Which one of the following graphs best represents the speed of the object with time?

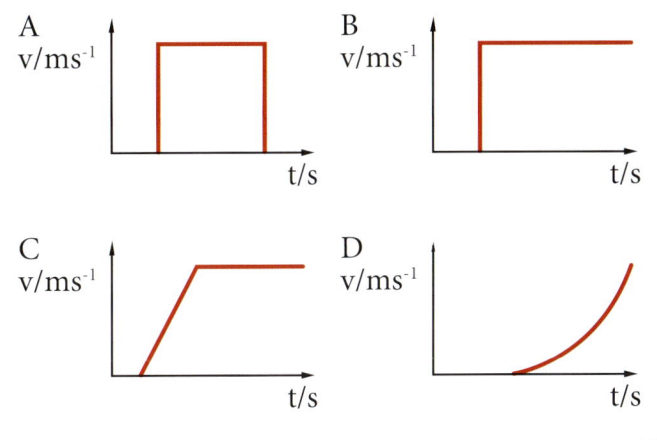

13. A helicopter ascends vertically upwards at constant velocity. The net force acting on the helicopter is equal to

    A. the weight of the helicopter
    B. the lift of the rotor on the helicopter
    C. the sum of the magnitude of the lift of the rotor and the magnitude of the weight of the helicopter
    D. zero

14. A horizontal force F is applied to the block of mass m which is in contact with a larger block of mass M, and accelerates both of them along a horizontal surface as shown in the diagram

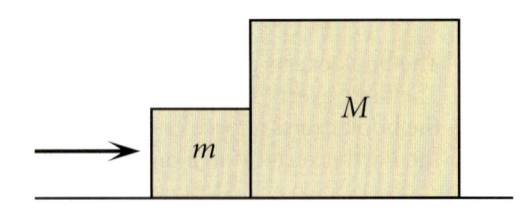

    Ignoring friction, the force on the block of mass M is:

    A. $M(M - m)^{-1} F$
    B. $M(M + m)^{-1} F$
    C. $m(M + m)^{-1} F$
    D. $m(M - m)^{-1} F$

15. Two constant forces of 3.0 N and 4.0 N act at right angles to each other on a point particle of mass 0.50 kg. The magnitude of the acceleration of the particle is

    A. 2.0 ms^{-2}
    B. 10 ms^{-2}
    C. 14 ms^{-2}
    D. 50 ms^{-2}

This information is for questions 16 and 17.

A tractor at an airport is pulling two loaded baggage carts A and B. The tractor exerts a force of 4000 N on them. Trolley A has a total mass of 200kg and it has a frictional force between the wheels and the tarmac of 2200 N. Trolley B has a total mass of 300kg and it has a frictional force between the wheels and the tarmac of 800 N.

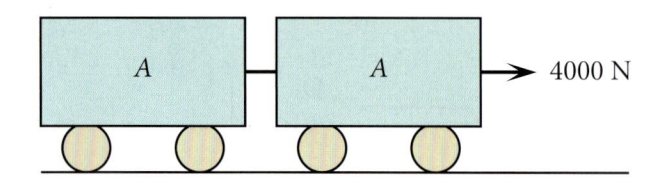

17. The acceleration of the baggage carts is

    A. 2 ms^{-2} C. 5 ms^{-2}
    B. 3.3 ms^{-2} D. 8 ms^{-2}

18. The tension in the coupling between A and B is

    A. 400 N   C. 1400 N
    B. 600 N   D. 2600 N

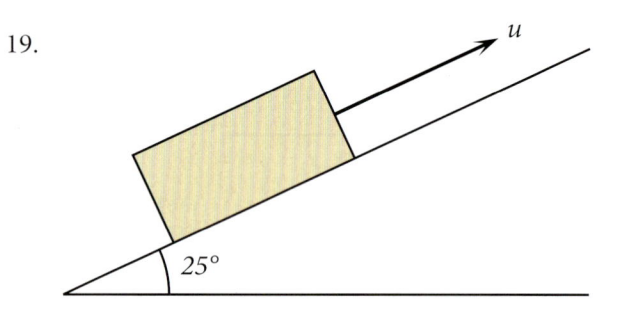

19.

    A 3 kg block is given an initial velocity, u, up a surface sloped at 25° to the horizontal. The surface between the block and the slope has a coefficient of kinetic friction μk = 0.30

    (a) Find the frictional force opposing the motion.
    (b) If the block travels 3.0 m up the slope before coming to rest, find the initial velocity, u, given to the block.

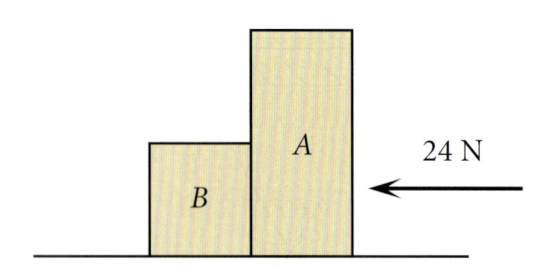

20.

    Two masses, A = 4 kg and B = 2 kg, at rest and in contact on a horizontal frictionless surface experience a force of 24 N for 10s. Find:

    (a) the acceleration of the system.
    (b) the force exerted by A on B.
    (c) the force exerted by B on A.

2.3 Work, energy and power

> **NATURE OF SCIENCE:**
>
> Theories: Many phenomena can be fundamentally understood through application of the theory of conservation of energy. Over time, scientists have utilized this theory both to explain natural phenomena and, more importantly, to predict the outcome of previously unknown interactions. The concept of energy has evolved as a result of recognition of the relationship between mass and energy. (2.2)
>
> © IBO 2014

Essential idea: The fundamental concept of energy lays the basis upon which much of science is built.

Understandings:
- Kinetic energy
- Gravitational potential energy
- Elastic potential energy
- Work done as energy transfer
- Power as rate of energy transfer
- Principle of conservation of energy
- Efficiency

Work done as energy transfer

The terms **work** and **energy** are used frequently in everyday speech. However, although it is possible to give an exact physical definition of 'work' it is not so easy to define precisely what is meant by 'energy'. Energy is defined as the *ability to do work*. We often use the term in the context of 'having enough energy to get something done'. So for the time being, let us be happy with this idea and also note that getting 'something done' usually involves a transfer of energy from one form to another. For example a car engine uses fuel in order to get the car moving and to keep it moving, you eat food in order to live, natural gas undergoes combustion in a power station to provide electrical energy and you can think of other examples.

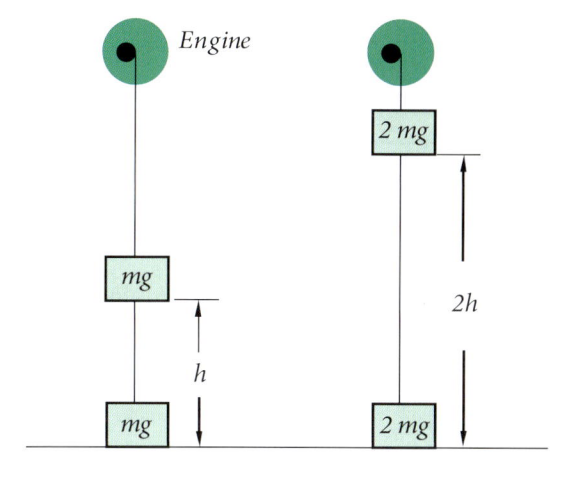

Figure 239 Fuel used to lift a weight

In Figure 239 we imagine a situation in which we are using fuel in an engine whose function is to lift a weight mg to a certain height.

In the first diagram the engine lifts the weight to a height h and in the second diagram the engine lifts a weight 2mg to a height 2h. In the first situation let us suppose that the engine requires 1 dm^3 of fuel to complete the task. Then in the second situation we would guess that the engine would require 4 dm^3 of fuel. The amount of fuel used is a measure of the energy that is transferred and to complete the tasks we say that the engine has done **work**. The amount of work that is done will be a measure of the energy that has been transferred in the performance of the tasks. Clearly the engine has done **work against a force**, in this situation, the force of gravity. If we double the force i.e. the weight, then the work done is doubled and if we double the distance through which the weight is moved then the work done is also doubled. If we double both together, then the work done is quadrupled. It seems that we can define work as:

Work = force × distance moved in the direction of the force

W = Fs

Work is only done in a scientific sense if the point of application of the force moves.

The SI unit of work is the **newton metre** and is called the **joule** named after the 19th Century physicist James Prescott Joule. Work is a scalar quantity so energy is also a scalar quantity.

Quite often, the force applied to an object is not in the same direction as the movement of the object. That is, the force is at angle to the direction of movement. Therefore, the **component of the force** in the direction of motion is used

Chapter 2

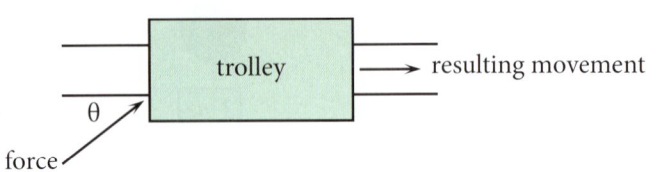

Figure 240 Work done at an angle

Consider a trolley being pushed on some rails as shown in Figure 240. The horizontal component of the force in the direction of the resulting motion is F cosθ. Therefore, the work done is:

W = Fs cosθ

When θ = 90°, cos θ = 0 and so W = 0. (This situation arises in uniform circular motion where the force acts only to change the direction of motion but not to do work as we will study in a later section).

Consider the situation in Figure 241 in which the engine lifts the weight up a slope to height h.

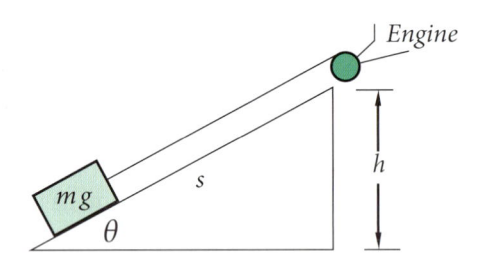

Figure 241 Using a slope

We shall assume that the surface of the slope is frictionless. The work W done by the engine using the above definition is:

$$W = force \times s$$

where s is the distance up the slope.

The force this time is the component of the weight down the slope. Hence,

$$W = mg \sin\theta \times s$$

But $\sin\theta = \dfrac{h}{s} \Rightarrow s = \dfrac{h}{\sin\theta}$,

meaning that $W = mg\sin\theta \times s = mg\sin\theta \times \dfrac{h}{\sin\theta}$

= mgh

That is,

$$W = mgh$$

This is just the amount of work that the engine would have performed if it had lifted the weight directly. Our definition of work therefore becomes

Work = magnitude of the force × displacement in the direction of the force.

Which is the same as

work = magnitude of the component of the force in the direction moved × the distance moved.

Kinetic energy

Any object that is moving has kinetic energy. Kinetic energy is energy due to motion. A moving object possesses the capacity to do work. A hammer, for example, by virtue of its motion can be used to do work in driving a nail into a piece of wood.

So how do we find out just how much work a moving object is capable of doing?

In Figure 242 below, a force F moves an object of mass m a distance d along a horizontal surface. There is no friction between the object and the surface.

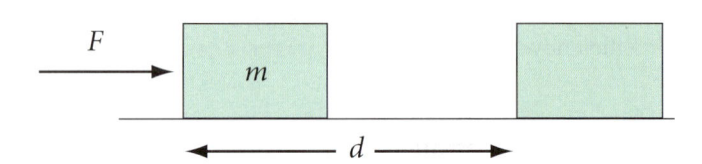

Figure 242 An example of kinetic energy

We will assume that the object starts from rest. The force F will accelerate the object in accordance with Newton's Second Law and the magnitude of the acceleration will be given by using the formula

F = ma, from which we obtain $a = \dfrac{F}{m}$

We can use the equation $v^2 = u^2 + 2as$ to find an expression of the speed of the object after it has moved a distance d (= s).

That is, $v^2 = (0)^2 + 2\left(\dfrac{F}{m}\right)d \Rightarrow v^2 = \dfrac{2Fd}{m}$

From which $Fd = \dfrac{1}{2}mv^2$

The work done (Fd) by the force F in moving the object a distance d in the direction of F is now, therefore, expressed in terms of the properties of the body and its motion.

The quantity $\dfrac{1}{2}mv^2$ is called the **kinetic energy** (KE) of the body and is the energy that a body possesses by virtue of its motion. The kinetic energy of a body essentially tells us how much work the body is capable of doing.

We denote the kinetic energy by E_k, so that

$$E_k = \frac{1}{2}mv^2$$

Potential energy

When an object is resting on top of a wall it has the potential to do work. If it falls off the wall onto a nail sticking out of a piece of wood then it could drive the nail further into the wood. Work is needed to lift the object on to the top of the wall and we can think of this work as being 'stored as **potential energy** in the object.

In general if an object of mass m is lifted to a height h above the surface of the earth then the work done is mgh and the object has a potential energy also equal to mgh (force = mg, displacement = h).

However, we have to be careful when using this equation since if h is too large then we can no longer regard g as being constant. Also, we must note that there is no zero of potential energy in this situation.

(In a later chapter we will see how we define the zero of potential energy). If the object rests upon the ground and we dig a hole, then the object has potential to do work when it falls down the hole. The object in fact has gravitational potential energy by virtue of its position in the Earth's gravitational field.

No matter where an object is placed in the Universe it will be attracted by the gravitational force of the Earth. We say that the Earth has an associated gravitational field. (Indeed all objects will have an associated gravitational field as we will discuss in the later chapters 6 and 9). At this stage you should appreciate that when an object is moved a distance Δh in the Earth's gravitational field and in the direction of the field, its change in potential energy ΔE_p is:

$$\Delta E_p = mg\Delta h$$

(provided that g is constant over the distance moved.)

To demonstrate the so-called principle of energy conservation we will solve a dynamics problem in two different ways, one using the principle of energy conservation and the other using Newton's laws and the kinematics equations.

Example

An object of mass 4.0 kg slides from rest without friction down an inclined plane. The plane makes an angle of 30° with the horizontal and the object starts from a vertical height of 0.50 m. Determine the speed of the object when it reaches the bottom of the plane.

Solution

The set-up is shown in Figure 243:

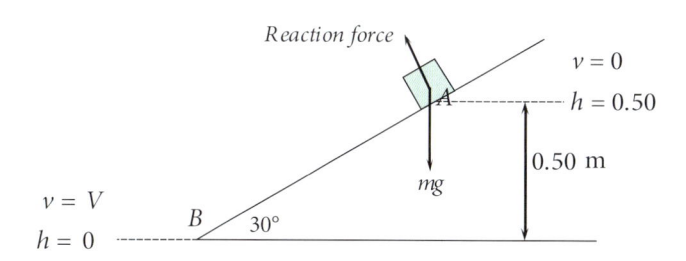

Figure 243

We set 'zero–level' at point B, the base of the incline, so that $h = 0$ and so that at point A, $h = 0.50$. At A, the object has no kinetic energy ($v = 0$) and at point B, the object has gained kinetic energy ($v = V$).

Method 1:   **Kinematic solution**

The force down the plane is given by $mg\sin\theta = 20$ N.

Using Newton's 2nd law ($F = ma$) gives the acceleration

$$a = \frac{20}{4.0} = 5.0 \text{ m s}^{-2}$$

(Note that we could have determined the acceleration by writing down the component of g down the plane)

Using $V^2 = u^2 + 2as$ with $u = 0$, and s (the distance down the plane)

$$s = \frac{0.50}{\sin\theta} = \frac{0.50}{\sin 30°} = 1.0$$

We have, $V^2 = 0^2 + 2 \times 5 \times 1 \Rightarrow V = \sqrt{10} = 3.2$ m s^{-1}

Method 2:   **Energy Principle**

As the object slides down the plane its potential energy becomes transformed into kinetic energy. If we assume that no energy is 'lost' we can write

change in PE = mgh = gain in KE = $\frac{1}{2}mv^2$

So that $\frac{1}{2}mv^2 = mgh \Rightarrow V^2 = 2gh$

$$\Rightarrow V = \sqrt{2gh}$$

67

Using the values of g and h, we have that V = 3.2 m s⁻¹.

That is, the object reaches a speed of 3.2 m s⁻¹. As the object slides down the plane its potential energy becomes transformed into kinetic energy. If we assume that no energy is 'lost' we can write

change in PE = mgh = gain in KE = 1/2 mv²

So that 1/2 mv² = mgh ⇒ V² = 2gh

⇒ V = √2gh

Using the values of g and h, we have that V = 3.2 m s⁻¹.

That is, the object reaches a speed of 3.2 m s⁻¹.

Note that the mass of the object does not come into the question, nor does the distance travelled down the plane. When using the energy principle we are only concerned with the initial and final conditions and not with what happens in between. If you go on to study physics in more depth you will find that this fact is of enormous importance.

The second solution involves making the assumption that potential energy is transformed into kinetic energy and that no energy is lost. This is the so-called 'energy principle', this means the energy is conserved.

Clearly in this example it is much quicker to use the energy principle. This is often the case with many problems and in fact with some problems the solution can only be achieved using energy considerations.

What happens if friction acts in the above example? Suppose a constant force of 16 N acts on the object as it slides down the plane. Now, even using the energy principle, we need the distance down the plane so we can calculate the work done against friction

This is 16 × 1.0 = 16 J. The work done **by** gravity i.e. the change in PE = 20 J

The total work done **on** the object is therefore

20 – 16 = 4.0 J.

Hence the speed is now given by $\frac{1}{2}mv^2$ = 4.0.

To give v = 1.4 m s⁻¹.

So in this problem, not all the work done has gone into accelerating the object. We say that the frictional force has dissipated energy. If we are to retain to the idea of energy conservation then we must "account" for this "lost" energy. It was the great triumph of some late eighteenth and early nineteenth physicists and engineers to recognise that this 'lost energy' is transformed into thermal energy. If you rub your finger along the top of a table you will definitely feel it getting warm. This is where the "lost" energy has gone. It has in fact been used to make the molecules of the table and the molecules of your finger vibrate more vigorously.

Another thing to notice is that this energy is "lost" in the sense that we can't get it back to do useful work. If there were no friction between the object and the surface of the plane then, when it reached the bottom of the plane, work could be done to take it up to the top of the plane and this cycle could go on indefinitely. (We could actually set up the arrangement such that the objects KE at the bottom of the plane could be used to get it back to the same height again). It is what we call a *reversible process*.

The presence of friction stops this. If the object is dragged back up the plane you won't get the energy back that has been "lost" due to friction, you will just "lose" more energy. This is an *irreversible process*. We can now start to glimpse why, even though it is impossible to destroy energy, it is possible to "run out" of "useful energy". Energy becomes as we say, *degraded*.

The general principle of energy conservation finds its formulation in the 'First Law of Thermodynamics' and the consequences of this law and the idea of energy degradation and its implication on World energy sources is discussed in much more detail in a later chapter.

Elastic potential energy

Clearly, when a spring is stretched or compressed, work is done. That a stretched spring or a compressed spring has the potential to do work is apparent from many everyday examples and it is left as an exercise for you to describe some of the examples.

Figure 244(a) below shows a plot of a constant force against displacement and Figure 244(b) shows a plot of the spring force against displacement.displacement.

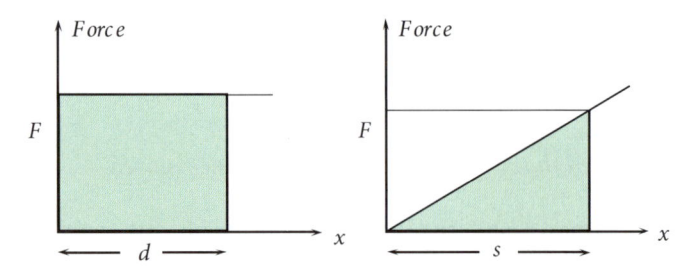

Figure 244 (a) and (b)   Force versus displacement graphs

The area under each of the graphs is clearly equal to the work done. In Figure 244(a) when the force F undergoes a displacement d the work done is Fd. In Figure 244(b) when the force F produces an extension s then work done is $\frac{1}{2}Fs$.

But in this case $F = ks$, hence the work done is

$$\frac{1}{2} \times (ks) \times s = \frac{1}{2}ks^2$$

The work that has been done is stored in the spring as elastic potential energy E_{elas}. The adjective potential in this context essentially means "hidden". Clearly

$$E_{elas} = \frac{1}{2}ks^2 \text{ or } E_p = \frac{1}{2}k\Delta x^2$$

We can extend this idea to find the work done by any non-constant force. If we know how the force depends on displacement then to find the work done by the force we just compute the area under the force-displacement graph.

Example

A force of 100 N pulls a box of weight 200 N along a smooth horizontal surface as shown in the Figure below.

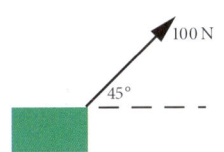

Caculate the work done by the force

(a) in moving the box a distance of 25 m along the horizontal

(b) against gravity.

Solution

(a) The component of the force along the direction of motion, i.e., the horizontal component, F_h, can be determined by using the fact that

$$\cos 45° = \frac{F_h}{100}$$

$$\Leftrightarrow F_h = 100 \cos 45°$$

$$= 71 \text{ N}$$

That is, the component of the force along the direction of motion is 71N. Therefore the work done

(F × s) = 71 × 25 = 1780

That is, the work done is 1780 J.

(b) There is no displacement by the force in the direction of gravity. Hence the work done by the force against gravity is zero.

Suppose that a constant frictional force of 50 N acts on the box. How much work is done against friction? Again, using the fact that W = force × distance, we have that the work done is simply 50 × 25 = 1250 N.

Notice that in the above example we use the expressions 'work done by' and 'work done against'. This occurs over and over again in physics. For example the engine lifting the weight does work against gravity. However, if the weight is allowed to fall, the work is done by gravity. Strictly speaking we have a sign convention for work. The convention is that

Work done **on** a system is **negative**.

Work done **by** a system is **positive**.

In the example above, the box can be identified as the system.

Although this convention is not always important in many mechanics situations, it is of great importance when we come to consider the relationship between heat and work and in the relation between field strength and potential

Power

Suppose that we have two machines A and B that are used in lifting objects. Machine A lifts an object of weight 100 N to a height of 5.0 m in a time of 10 s. When machine B is used to perform the same task it takes 0.1 s. Ones instinct tells us that machine B is more powerful than machine A. To quantify the concept of power we define power as

Power = the rate of doing work

i.e., $power = \dfrac{work}{time}$

The unit that is used to measure power is the joule per second that is called the watt, (W) after the 19th Century Scottish engineer James Watt. Again, power is a scalar quantity.

In the example above, of our two machines, A will have a power output of 50 W and machine B a power output of 5000 W.

In Figure 245 some sort of engine is used to pull an object at a constant speed along the horizontal.

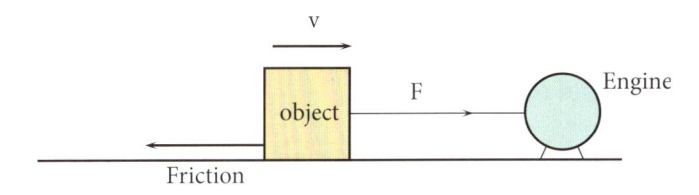

Figure 245   Force Against Friction

The pulling force F (which is the tension in the rope) produced by the engine will be equal to the frictional force between the object and the floor.

Suppose that the engine moves the object a distance Δs in time Δt. The work done against the frictional force (i.e., the work done by the engine) is:

$$\Delta W = F \Delta s$$

The power P developed by the engine is therefore

$$P = \frac{\Delta W}{\Delta t} = F \cdot \frac{\Delta s}{\Delta t}$$

But, $v = \frac{\Delta s}{\Delta t}$ hence, we have that

$P = Fv$

Example

A diesel locomotive is pulling a train with a maximum speed of 60 ms^{-1}. At this speed the power output of the engine is 3.0 MW. Calculate the tractive force exerted by the wheels on the track.

Solution

Using the formula $P = Fv$, we have that $F = \frac{P}{v}$.

Hence, $F = \frac{3\,000\,000}{60} = 50$ kN.

Our answer is at best an approximation since the situation is in fact much more complicated than at first glance. The train reaches a maximum speed because as its speed increases the frictional force due to air resistance also increases. Hence, at its maximum speed all the energy produced by the motors is used to overcome air resistance, energy lost by friction between wheels and track and friction between moving parts of the motors and connected parts.

Principle of conservation of energy

There are many different forms of energy and their transformations of which some examples are given here.

Thermal energy

This is essentially the kinetic energy of atoms and molecules. It is sometimes incorrectly referred to as 'heat'. The term heat actually refers to a transfer of energy between systems.

Chemical energy

This is energy that is associated with the electronic structure of atoms and is therefore associated with the electromagnetic force.

An example of this is combustion in which carbon combines with oxygen to release thermal energy, light energy and sound energy.

Nuclear energy

This is the energy that is associated with the nuclear structure of atoms and is therefore associated with the strong nuclear force.

An example of this is the splitting of nuclei of uranium by thermal neutrons to produce energy.

Electrical energy

This is energy that is usually associated with an electric current and is sometimes referred to incorrectly as *electricity*. For example the *thermal energy* from a chemical reaction (*chemical energy*) can be used to boil water and produce steam. The *kinetic energy* of the molecules of steam (*thermal energy*) can be used to rotate magnets and this rotation generates an electric current. The electric current transfers the energy to consumers where it is transformed into for example *thermal* and *light energy* (filament lamps) and *kinetic energy* (electric motors). We shall learn later that these different forms of energy all fall into the category of either potential or kinetic energy and are all associated with one or other of the fundamental forces.

Energy can be transformed from one form into another and as far as we know energy can never be created nor can it be destroyed. This is perhaps one of the most fundamental laws of nature and any new theories which might be proposed must always satisfy the principle of energy conservation. A simple example of the principle is, as we have seen, to be found in the transformation of gravitational potential energy into kinetic energy.

Efficiency

The **efficiency** of an engine is defined as follows

$$Eff = \frac{W_{OUT}}{W_{IN}} = \frac{P_{OUT}}{P_{IN}}$$

Where W_{OUT} is how much useful work the engine produces, and W_{IN} is how much work (energy) is delivered to the engine. The ratio of these two quantities is clearly the same as the ratio power output of the engine to its power output. To understand the idea of efficiency, we will look at the following example.

Example

An engine with a power output of 1.2 kW drags an object of weight 1000 N at a constant speed up an inclined plane that makes an angle of 30° with the horizontal. A constant frictional force of 300 N acts between the object and the plane and the object is dragged a distance of 8.0 m.

Determine the speed of the object and the efficiency of the engine.

Solution

We first need to draw a diagram to visualise the situation.

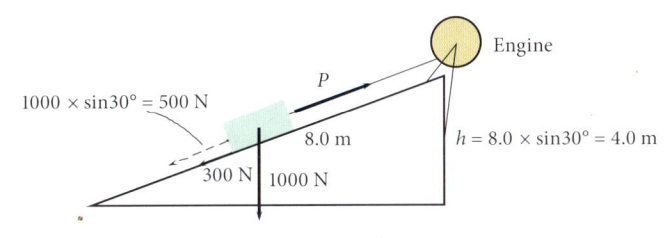

The component of the object's weight down the plane is $1000 \times sin30° = 500$ N.

The total force against which the machine does work is therefore 500 + 300 = 800 N.

Using $P = Fv$, we have $1200 = 800 \times v$

so that $v = 1.5$ m s^{-1}.

The machine lifts the object to a height h,

where $h = 8.0 \times sin30° = 4.0$ m.

The useful work that is done by the machine is therefore $1000 \times 4.0 = 4000$ J.

The actual work that is done by the machine is $800 \times 8.0 = 6400$ J.

The efficiency of the machine is the useful work done divided by the actual work done which is

$$\frac{4000}{6400} \times 100\% = 63\%.$$

We can also calculate the energy output per second of the fuel used by the machine. The machine has a useful power output of 1.2 kW and if it is 63% efficient then the fuel must produce energy at the rate of

$$\frac{1.2}{0.63} = 1.9 \text{ kJ s}^{-1}.$$

Exercise 2.5

1. A box weighing 1000 N needs to be put on a platform that is 6 m high. If it is pushed up a frictionless ramp that is 30.0 m long, the total work that needs to be done on the box is:

   A. 180. J
   B. 6.00 kJ
   C. 30.0 kJ
   D. 180. kJ

2. A spring with a spring constant of 10 Ncm^{-1} is stretched 10 cm from its equilibrium position. Its potential energy increased by:

   A. 1 J
   B. 5 x 10^{-1} J
   C. 500 J
   D. 5000 J

3. A block rests on a rough horizontal plane and a force P is applied to the block as shown in the diagram.

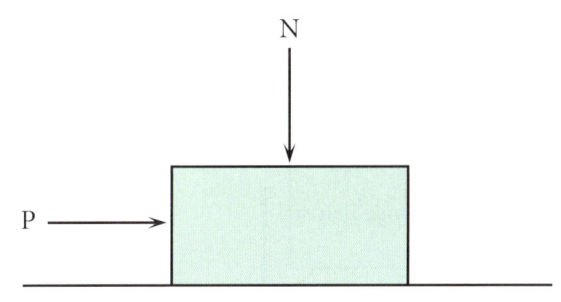

   The normal reaction between the plane and the block is N and the frictional force between the block and the plane is F_f. The coefficient of static friction between the block and the plane is μ_s and initially P is zero.

   As P is increased in value, which one of the following statements is true concerning the relationship between F, N and μs?

   A. F_f is always equal to μsN
   B. F_f is always greater than μsN
   C. F_f is always less than μsN
   D. F_f can be equal to μsN

5. A body starts from a point in a gravitational field, moves along a very long path, then returns to the point. Over the entire process, it:

   A. loses energy
   B. gains energy
   C. neither gains nor loses energy
   D. overcomes frictional forces

71

CHAPTER 2

6. Which one of the following pairs contains a scalar quantity and a vector quantity?

    A. displacement and velocity
    B. power and kinetic energy
    C. momentum and acceleration
    D. force and work

7. A man drags a sack of flour of mass 50 kg at constant speed up an inclined plane to a height of 5.0 m. The plane makes an angle of 30° with the horizontal and a constant frictional force of 200 N acts on the sack down the plane.

    Calculate the work that the man does against:

    (a) friction
    (b) gravity

8. A man drags a sack of flour of mass 100 kg at constant speed up an inclined plane to a height of 6.0 m. The plane makes an angle of 30° with the horizontal and a constant frictional force of 250 N acts on the sack down the plane. Determine the efficiency of the inclined plane.

9. The diagram shows a pile driver that is used to drive a metal bar (the pile) into the ground.

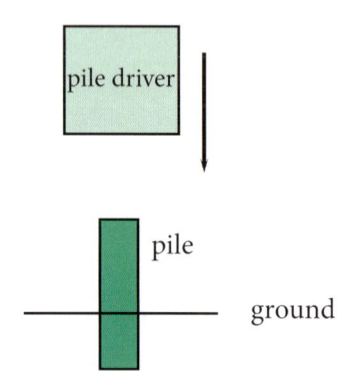

    A particular pile driver has a mass of 500 kg and it falls through a height of 2.5 m before striking the top of the pile. It stays in contact with the pile and drives it a distance of 0.40 m into the ground.

    Calculate the average force exerted by the ground on the pile by using

    i. energy considerations
    ii. the equations of uniform motion and Newton's Second Law. (assume that the mass of the pile driver is much greater than the mass of the pile.)

10. A man slides a box of mass 50 kg at constant speed up an inclined slope to a height of 2.0 m. The slope makes an angle of 30° with the horizontal and it takes him 4 s to reach the height of 2.0 m and a constant frictional force of 250 N acts on the block.

    Calculate

    i. the work the man does against friction
    ii. the work the man does against gravity
    iii. the efficiency of the "man-slope machine"
    iv. the power the man develops to push the block up the slope.

11. This question is about calculating the power output of a car engine. Here is some data about a car that travels along a level road at a speed of 25ms^{-1}.
    Fuel consumption = 0.20 litre km^{-1}
    Calorific value of the fuel = 5.0×10^6 J litre^{-1}
    Engine efficiency 50%

    Determine:

    i. the rate at which the engine consumes fuel
    ii. the rate at which the fuel supplies energy
    iii. the power output of the engine
    iv. the power used to overcome the frictional forces acting on the car
    v. the average frictional force acting on the car.

    Explain why:

    (i) the power supplied by the engine is not all used to overcome friction
    (ii) the fuel consumption increases as the speed of the car increases

12. A light aircraft with a mass of 800kg starts from rest on the tarmac and 4 minutes later it has an altitude of 300m and it is flying with a velocity of 80ms^{-1}. Calculate the work done by the engine during these 4 minutes.

13. A child on a swing is raised to a height of 1.20 m above the lowest point of its motion. If the swing is released with an initial speed of 0.750ms^{-1} calculate the maximum height the swing and the child will reach after half an oscillation. (Ignore friction and air resistance).

14. A gymnast bouncing on a trampoline vertically leaves the trampoline at a height of 1.25 m above the ground and reaches a maximum height of 5.10 m before coming back down. Ignoring air resistance, calculate the initial velocity with which the gymnast leaves the trampoline.

MECHANICS

15. An object with a mass of 5.0 kg moves horizontally on a frictionless surface with a speed of 12ms⁻¹. An unbalanced force gives it an acceleration of 2.0ms⁻² for 45 seconds.

    Determine:

    (a) the work done by the force.
    (b) the power developed by the force.

16. A student releases a block of mass M placed at the top of an inclined plane and measures the time it takes to move a certain distance down the plane as shown in the following diagram:

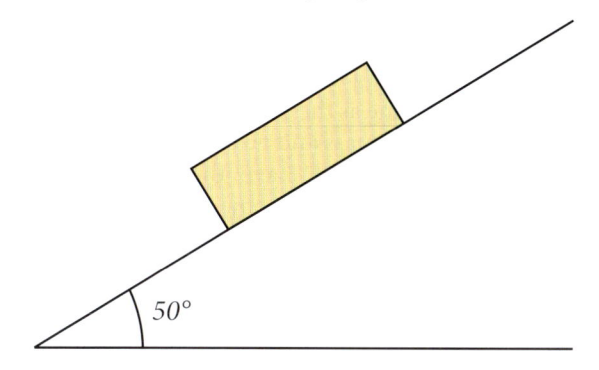

    (a) On a similar diagram, draw and name the forces acting on the block

    (b) If the block takes 1.80 s to travel a distance of 4.00 m down the plane, calculate:

        (i) the acceleration of the block down the plane.
        (ii) the component of the weight down the plane in terms of M.

    (c) If the coefficient of kinetic friction between the block and the incline is μ_k, what is the value of the frictional force in terms of μ_k and M.

    (d) Calculate the value of μ_k.

17. How much energy is stored in a 100.0 kg bell tower clock weight when it is wound up to a height of 20.0 m?

18. A 12 kg trolley can move on a smooth track ABCD as shown below..

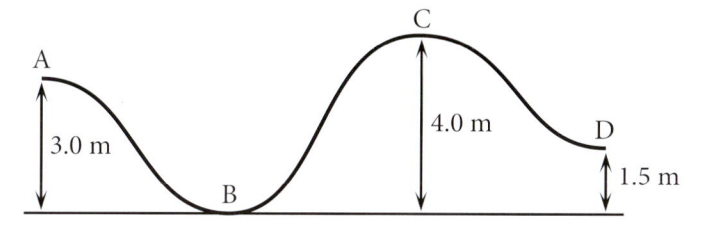

    At point A, the trolley has a speed of 5.0ms⁻¹.

    (a) Calculate the total energy of the trolley at point A.
    (b) Calculate the speed of the trolley at point C.
    (c) Calculate the kinetic energy of the trolley at point D.

19. Estimate the minimum take-off power of a grasshopper (cicada).

20. An elastic band of length 2d is attached to a horizontal board as shown in the diagram below.

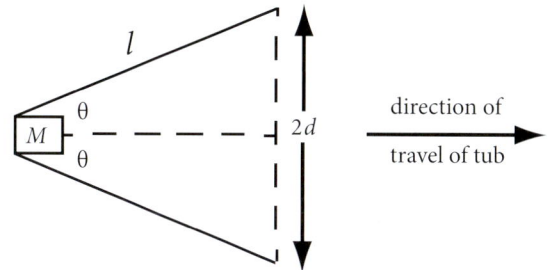

    A margarine tub has some weights attached to the bottom of the inside of the tub such that the total mass of the weights and the tub is M. The tub is placed at the centre of the band and is pulled back until the tub makes an angle θ with the band as shown. The tub is then released such that it is projected down the runway for a distance s before coming to rest. The problem is to deduce an expression for the speed with which the tub leaves the band. The force constant of the elastic band is k.

73

2.4 Momentum and impulse

NATURE OF SCIENCE:

The concept of momentum and the principle of momentum conservation can be used to analyse and predict the outcome of a wide range of physical interactions, from macroscopic motion to microscopic collisions. (1.9)

© IBO 2014

Understandings:
- Newton's second law expressed in terms of rate of change of momentum
- Impulse and force-time graphs
- Conservation of linear momentum
- Elastic collisions, inelastic collisions and explosions

Newton's second law expressed in terms of rate of change of momentum

Section 2.4 deals with interactions between objects. Interactions can be thought of as occurring when two or more objects exert forces on each other. Interactions may involve contact forces such as the collision of two pool/snooker/billiard balls, or as forces over a distance such as the lift off of a rocket. In the previous sections of this chapter we examined interactions between objects in terms of changes in displacement, velocity, acceleration or force. In this section we will concentrate on interactions in terms of unchanging quantities, that is, quantities that are the same before and after an interaction – a quantity that is conserved.

The famous French chemist Antoine Lavoisier (1743 – 1794) carried out many chemical reactions and from his findings he concluded that mass is conserved provided the system is isolated (nothing can enter or leave the reaction vessel/system). There are a number of conservation laws in physics such as the law of conservation of energy, the law of conservation of charge and the law of conservation of momentum, to mention but a few.

In his *Principia Mathematica* Newton introduces the idea of quantity of motion in which he brings together the inertia of a body and the speed with which it is moving. The two quantities mass and speed do seemed to be linked in the sense that taken together they have an effect on inertia. An ocean liner of mass several thousand tonnes moving at even a very low speed takes a lot of stopping. If it does not stop before coming into dock it will do an enormous amount of damage when it hits the dock. Similarly a bullet with a mass of several grams moving with very high speed also takes a lot of stopping and can do great damage in the process of coming to rest. We can consider momentum as a measure of how hard it is to stop an object.

We define **linear momentum p** as the product of an object's mass and linear velocity.

$$\boldsymbol{p} = m\boldsymbol{v}$$

So a bullet of mass 10 g (0.01 kg) moving with a speed of 400 ms^{-1} has a momentum of 4 kg ms.

Note that **momentum is a vector quantity** and that the common units is kg ms^{-1} or Ns. The direction of the momentum is the same as the direction of velocity. So, two objects with the same mass and velocity that are travelling in opposite directions will have the same magnitude of momentum. However, one object will have positive momentum and the other will have negative momentum.

In his Principia Newton states his Second Law of Motion as follows: "The rate of change of linear momentum of a particle is directly proportional to the impressed force acting upon it and takes place in the direction of the impressed force". In other words, **the time rate of change of momentum is directly proportional to the resultant force and acts in the direction of the force**. This is in fact the form in which you should remember Newton's second law of motion since the law in the form $\mathbf{F} = m\mathbf{a}$ is actually, as we have seen, a special case.

Consider a force $\boldsymbol{F}$ acting on an object of mass m for a time t so that it accelerates from an initial velocity $\boldsymbol{u}$ to a final velocity $\boldsymbol{v}$. From the definition of acceleration given earlier we found:

$$a = \frac{(v-u)}{t}$$

We have seen that Newton's second law can be put in the form:

$$\boldsymbol{F} = m\boldsymbol{a}$$

Therefore:

$$F = ma = \frac{m(v-u)}{t}$$

$$F = \frac{\text{change in momentum}}{\text{change in time}}$$

Also, in terms of the momentum change we have seen that we can express the second law as

$$\boldsymbol{F} = \frac{\Delta \boldsymbol{p}}{\Delta t}$$

Here $\Delta \boldsymbol{p}$ is the change in momentum produced by the force F in time Δt.

From the second law in the form $F = \dfrac{\Delta p}{\Delta t}$ we see that

$F \Delta t = \Delta p$

The term $F \Delta t$ is called the *impulse* of the force and it is a very useful concept in solving certain types of problem particularly in situations where the force acts for a short time such as kicking a football. We also see by expressing the second law in this way that an equivalent unit for momentum is Ns.

Impulse and force – time graphs

Consider a constant force F applied for a certain time Δt as shown in Figure 246.

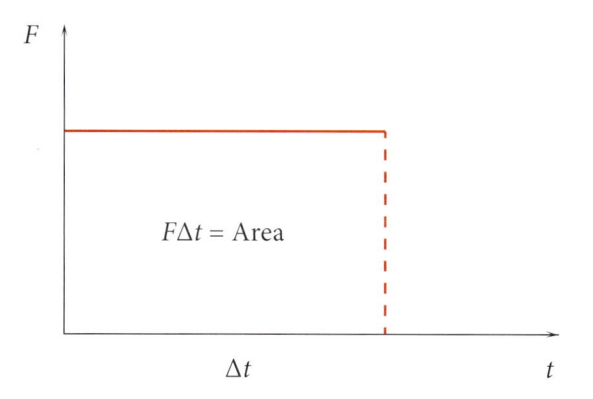

Figure 246 Force – time graph for a constant force.

$F \Delta t$ = the area under the curve = impulse = change in momentum

In real situations, the force is usually not constant as shown in Figure 247.

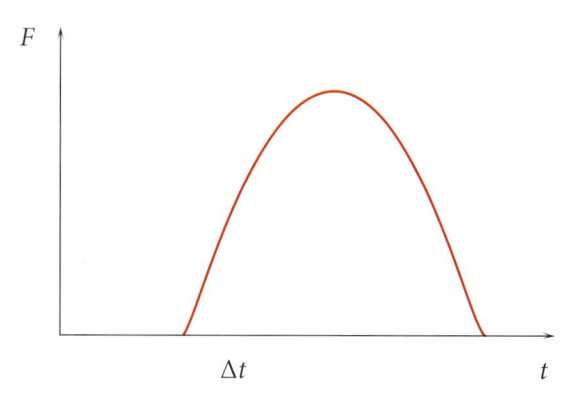

Figure 247 Force–time graph for a non-constant force.

The area under the curve equals the change in momentum or impulse. If the graph was on a graph paper background, you could estimate the area of the total number of grids enclosed by the curve.

Suppose that the force exerted on football when it is kicked varies with time as shown in Figure 248.

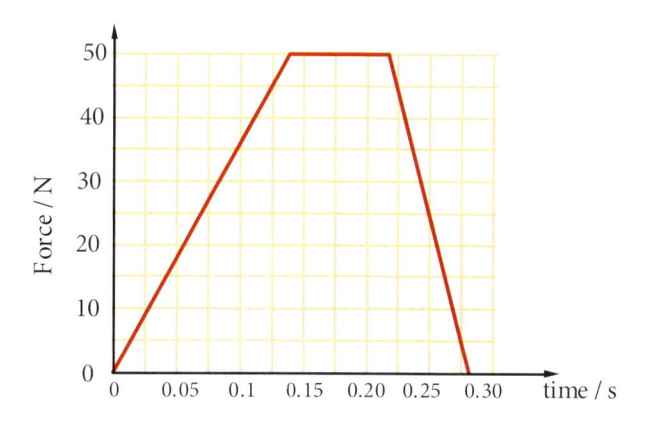

Figure 248 Idealised Force-time Graph

Since the area under the graph is equal to the impulse we can calculate the speed with which the football leaves the foot.

The area equals

$\dfrac{1}{2} \times (50 \times 0.14) + (0.14 \times 50) + \dfrac{1}{2} \times (50 \times 0.08) = 13.5$ N s

Suppose that the mass of the football is 0.40 kg then from

$F \Delta t = \Delta p = m \Delta v$, we have that $\Delta v = \dfrac{13.5}{0.40} = 34$.

That is, the change in velocity is 34 ms^{-1}.

In actuality one is much more likely to use the measurement of the speed of the football to estimate the *average* force that is exerted by the foot on the football. The time that the foot is in contact with the ball can be measured electronically and the speed of the football can be computed by measuring its time of flight.

Here is another example in which we use the ideas of impulse and the rate of change of momentum.

Example

Water is poured from a height of 0.50 m on to a top pan balance at the rate of 30 litres per minute. Estimate the reading on the scale of the balance.

Solution

We shall assume here that the water bounces off the top of the balance horizontally. Again we can calculate the speed with which the water hits the balance.

75

From the equation $v^2 = 2as$

we have $v^2 = 2 \times 10 \times 0.50 = 10$. So, $v = 3.2$ m s^{-1}.

The mass of water arriving at the balance per second is 0.50 kg s^{-1}.

The rate of change of momentum is therefore 0.50×32.

= 1.6 N.

If the balance is calibrated in grams the reading will therefore be about 160 g.

In any collision or explosion, a force is exerted for a period of time.

The longer the time of collision, the smaller will be the force exerted.

In all ball/bat sports such as golf, baseball and tennis games, your coach will tell you to follow through when making contact with the ball. Some factors such as your strength and radius of swing are set and you cannot change these. However, by following through, the contact time of the bat with the ball increases, so the impulse

$F \Delta t$ increases. The change in momentum increases, so the ball moves away with increased velocity.

Many of the safety design features in cars use impulse. Seat belts are designed to stretch a little to make the time longer in a collision. This also applies when air bags are released during collisions. The bumper bars on the back and front of a car are designed to crumple during a collision so that the force applied to the car is over a longer period of time. crash barriers between the opposite lanes on a highway are made of crumpling material. The crumpling means that on collision, a car will come to rest in a greater time. The change in momentum of the car would be the same as if the barrier were made of concrete but the F in FΔt is now smaller since Δt is larger.

When a skydiver hits the ground from a height, they bend their knees or roll upon landing. When you catch a heavy or hard object, you pull back your hands. Again, the longer the time of contact during a collision, the smaller will be the force exerted.

On the other hand, there are certain advantages to having the contact time of a shorter duration so the force becomes larger. Hammers, jackhammers and nail guns use this to their advantage.

Conservation of linear momentum

One of the most important aspects of Newton's laws is that the second and third law lead to the idea that in any interaction in a closed (isolated) system, linear momentum is always conserved, that is it stays constant. By a closed system, we mean one in **which no external forces act**. We can get an idea of this from the fact that the second law tells us that the net external force is equal to the rate of change of momentum. If there is no external force then there is no rate of change of momentum of the system as a whole and therefore the momentum does not change, that is, it remains constant.

Consider for example, two balls rolling towards each other and colliding with each other as shown in Figure 249.

Figure 249   Interacting particles

When two particles A and B collide, they exert a force **F** on each other for a short period of time **t**. Therefore, the momentum of each ball changes. when they are in contact, It follows from Newton's third law that the force that A exerts on B is equal and opposite to the force that B exerts on A. This means that the net force that is exerted on the system comprising A and B is zero. If the net force is zero then it follows that there is no change in momentum of the system because the changes in momentum are equal and opposite. So, the momentum **gained** by one ball is equal to the momentum **lost** by the other ball.

We can now write these qualitative statements in a mathematical way:

From Newton's third law:

$$\mathbf{F}_{BA} = -\mathbf{F}_{AB}$$

From Newton's second law:

$$m_A a_A = -m_B a_B$$

From the definition of acceleration:

$$\frac{m_A(v_A - u_A)}{\Delta t} = -\frac{m_B(v_B - u_B)}{\Delta t}$$

Since the time of interaction is the same for both balls:

$$m_A(v_A - u_A) = -m_B(v_B - u_B)$$

Rearranging the previous equation:

$$m_A u_A + m_B u_B = m_A v_A + m_B v_B$$

In collision processes we can therefore express the law of conservation of momentum as:

Momentum before collision = momentum after collision.

In collision processes we can therefore express the law of conservation of momentum as **momentum before collision = momentum after collision.** However, the more general statement of the law is:

If the total external force acting on a system is zero then the momentum of the system remains constant (is conserved).

Or simply that: **The momentum of a closed system is constant (conserved).**

However, the more general statement of the law is:

If the total external force acting on a system is zero then the momentum of the system remains constant (is conserved).

Or simply that:

The momentum of a closed system is constant (conserved).

We now look at a problem that cannot be solved by using the Second Law in the form $F = ma$.

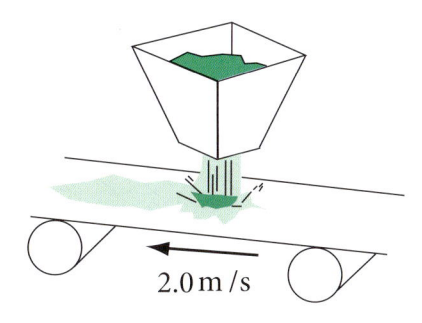

2.0 m/s

Figure 250   The conveyor belt problem

Suppose that sand is poured vertically at a constant rate of 400 kg s^{-1} on to a horizontal conveyor belt that is moving with constant speed of 2.0 m s^{-1} as shown in Figure 250.

We wish to find the minimum power required to keep the conveyor belt moving with constant speed. In every second the horizontal momentum of the sand changes by 400×2.0 kg m s^{-1}. This means that the *rate of change of momentum* of the sand is 800 kg m s^{-2}.

The force exerted on the conveyor belt by the sand is therefore 800 kg m s^{-2}. This force is the frictional force between the sand and the conveyor belt and it is this force which accelerates the sand to the speed of the conveyor belt. The power therefore to keep the conveyor belt moving at this speed is this force multiplied by the speed of the belt. i.e. the power (P) equals (Fv) = 1600 W.

We can also work out the rate of change of kinetic energy of the sand since the change in KE every second is 800 s^{-1}. This is quite interesting since we see that whatever the nature of the sand or the belt we always lose half the power in dissipation by the frictional force.

Example

Estimate the force exerted on a man who jumps off a wall of height 2.0 m and lands in soft earth. Explain why he would be likely to hurt himself if he landed on concrete.

Solution

We estimate the mass of the man as 70 kg and the time that he comes to rest on landing on the earth to be 2.0 s.

We can find the speed with which he hits the ground from $v^2 = 2gh$. i.e.

$v = 6.3$ m s^{-1}

His change in momentum on coming to rest is therefore 70×6.3 N s.

This is equal to the impulse $F\Delta t$. Hence we see that

$F = 70 \times \dfrac{6.3}{2.0} = 220$ N.

Therefore total force = 220 + 700 = 920 J.

Chapter 2

Admittedly this problem could have been solved by computing the acceleration of the man on landing and coming to rest from the equation $v = at$. However, this involves another step and is not as elegant a solution since it doesn't really get to the physics.

If he were to land on concrete then he would come to rest much more quickly. However, his change in momentum would be the same hence the F in the impulse $F\Delta t$ would be much greater.

Along with the law of conservation of energy, the law of conservation of momentum is of great importance in Physics. Although Newton's laws are found not to work when applied to atoms and molecules and are also modified by Relativity theory, the law of conservation of momentum still stands. If you were to invent a new theory, no matter how elegant the theory, if it violates conservation of energy and momentum then you can forget it.

The beauty of a law such as the conservation of momentum is that we are able to predict an outcome without knowing the intricacies of what actually is going on. When two billiard balls collide, the forces that act during collision are very complicated and we have no idea of their spatial and time dependence. However, because we know that they are equal and opposite we are able to predict the outcome of the collision.

Figures 251–254 below show some examples of collisions and their possible outcomes.

The outcome of a collision will depend on the mass of each particle, their initial velocities and also how much energy is lost in the collision. However, whatever the outcome, momentum will always be conserved. Any predicted outcome that that violates the conservation of linear momentum will not be accepted.

A very useful relationship exists between kinetic energy and momentum.

We have that $p = mv$ such that $p^2 = m^2v^2$

Hence substituting for v^2 we have

$$E_K = \frac{p^2}{2m} \quad \text{and} \quad \frac{p^2}{m} = mv^2 = 2E_K$$

Example

A railway truck, B, of mass 2000 kg is at rest on a horizontal track. Another truck, A, of the same mass moving with a speed of 5.0 m s^{-1} collides with the stationary truck and they link up and move off together.

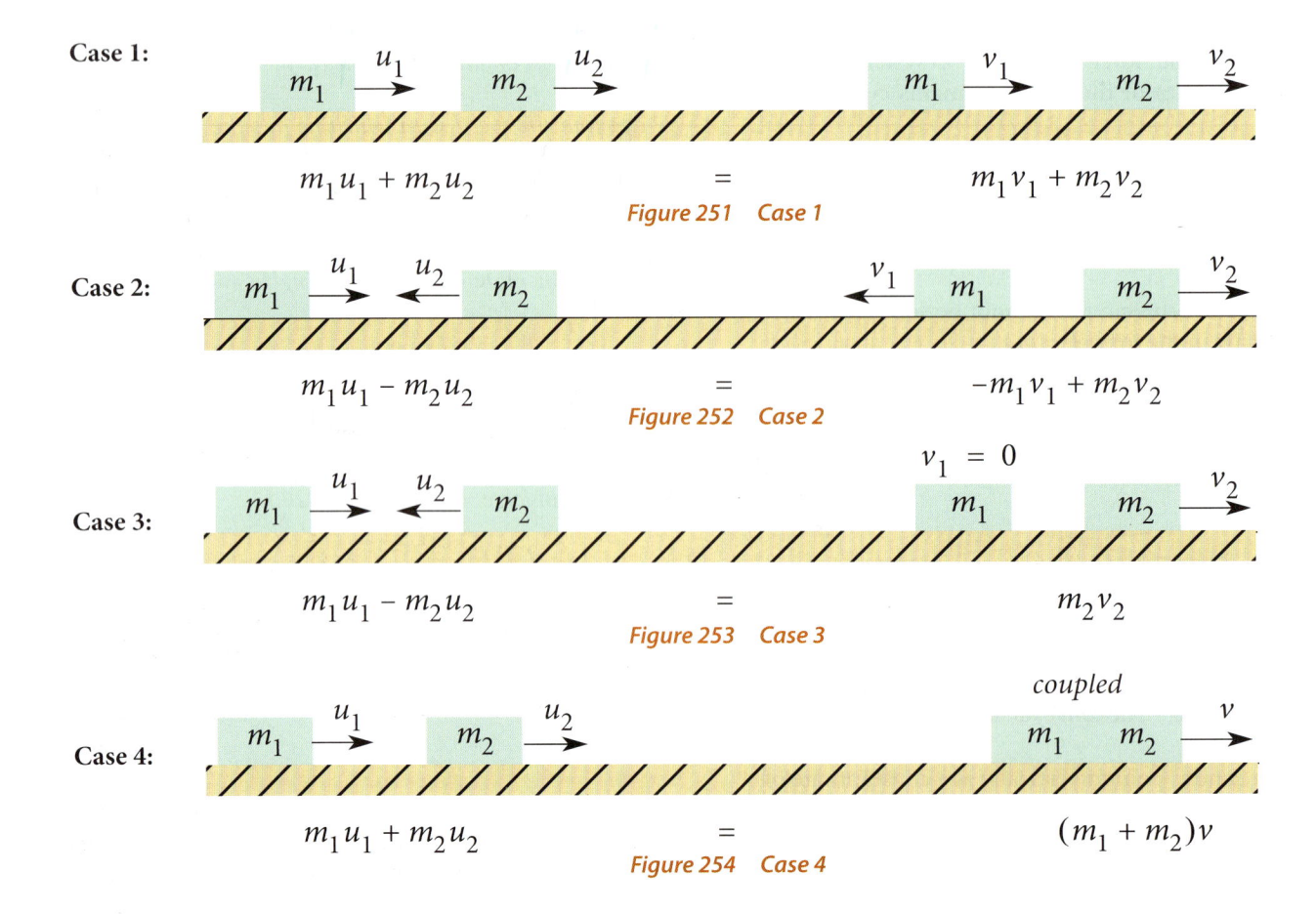

Case 1:

$m_1 u_1 + m_2 u_2 = m_1 v_1 + m_2 v_2$

Figure 251   Case 1

Case 2:

$m_1 u_1 - m_2 u_2 = -m_1 v_1 + m_2 v_2$

Figure 252   Case 2

Case 3:

$v_1 = 0$

$m_1 u_1 - m_2 u_2 = m_2 v_2$

Figure 253   Case 3

Case 4:

coupled

$m_1 u_1 + m_2 u_2 = (m_1 + m_2)v$

Figure 254   Case 4

Determine the speed with which the two trucks move off and also the loss of kinetic energy on collision.

Solution

It helps to start by drawing a diagram:

Before collision | *After collision*

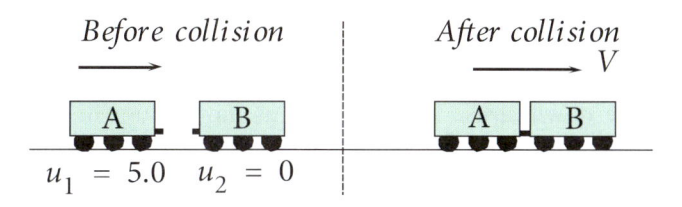

A B | A B
$u_1 = 5.0$ $u_2 = 0$ | V

Figure 255 Collision of two trucks

As the trucks couple after the collision, the conservation of momentum law states that:

$$m_1 u_1 + m_2 u_2 = (m_1 + m_2) V$$

where, $m_1 = m_2 = 2000$,

so that

$$2000 \times 5 + 0 = (2000 + 2000) V$$

$$\Rightarrow V = \frac{10000}{4000} = 2.5 \text{ m s}^{-1}.$$

The total KE before collision equals

$$\frac{1}{2} \times 2000 \times 5^2 + 0 = 25000 \text{ J}$$

The total KE after collision equals

$$\frac{1}{2} \times 4000 \times (2.5)^2 = 12500$$

The kinetic energy lost on collision is

$(25000 - 12500) = 12500$ J.

By lost energy we mean that the energy has been dissipated to the surroundings. Some of it will be converted into sound and most will heat up the coupling between the trucks.

Example

A billiard ball of mass 100 g strikes the cushion of a billiard table with a velocity of 10 m s^{-1} at an angle of 45° to the cushion. It rebounds at the same speed and angle to the cushion. What is the change in momentum of the billiard ball?

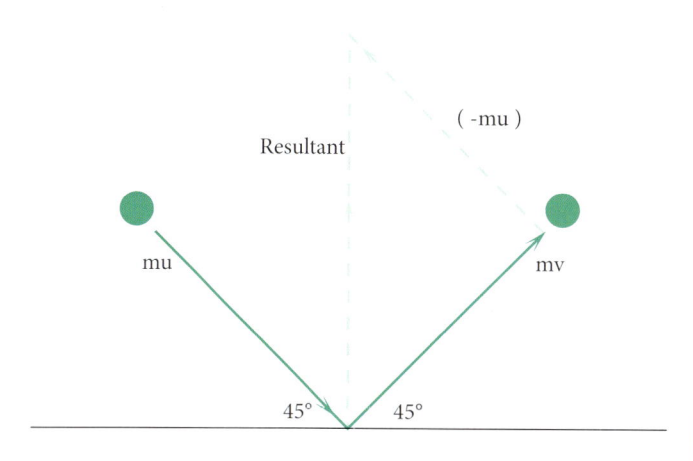

Solution

Change in momentum = final momentum – initial momentum = mv – mu

But momentum is a vector quantity. To subtract a vector, you reverse the direction of the arrow for mu, and then add the vectors to find the resultant vector. Using Pythagoras theorem:

Resultant vector = $\sqrt{(0.1 \times 10) + (0.1 \times 10)} = \sqrt{2}$

= 1.4 kg ms^{-1} at 90° away from the cushion

Elastic collisions, inelastic collisions and explosions

Collisions in which mechanical energy is not lost are called elastic collisions whereas collisions in which mechanical energy is lost are called inelastic collisions. If in a collision kinetic energy is conserved, then the collision is said to be elastic. If colliding objects coalesce (stick together) or are deformed, the collision is inelastic.

Perfectly elastic collisions occur when gas molecules collide with each other and with the sides of the container within which they are enclosed. If the collisions were not perfectly elastic, the gas molecules energy would be converted to other forms of energy and the gas molecules would be changed to liquid molecules. Collisions between sub-atomic particles are also perfectly elastic.

In the real (macroscopic) world mechanical energy is always lost during a collision. However, some collisions do approximate quite well to being elastic. The collision of two snooker (pool) balls is a very close approximation (some of the energy is converted into heat and sound), as is the collision between two steel ball bearings. An interesting situation arises when the balls are of the same mass and one is at rest before the collision and the collision takes place along a line joining their centres as shown Figure 256.

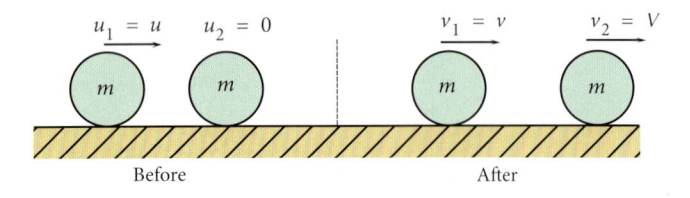

Figure 256   Rolling Balls

Suppose that the speed of the moving ball is u and that the respective speeds of the balls after collision are v and V. If we now apply the laws of momentum and energy conservation we have **conservation of momentum**:

$$mu = mv + mV$$

conservation of energy

$$\tfrac{1}{2}mu^2 = \tfrac{1}{2}mv^2 + \tfrac{1}{2}mV^2$$

From which we see that $u = v + V$ and $u^2 = v^2 + V^2$

The only solution to these equations is that $u = V$ and $v = 0$.

This means that the moving ball comes to rest after collision and the ball that was at rest moves off with the speed that the moving ball had before collision. This situation is demonstrated in that well known "toy", the Newton's Cradle as shown in Figure 257.

When solid spheres are suspended in a row (a), and the sphere on one end is pulled up and let go (b), its momentum will be transferred to the sphere on the other end (c). However, if two balls on one end are pulled up and let go, two balls at the other end lift up with velocities equal to the original two balls.

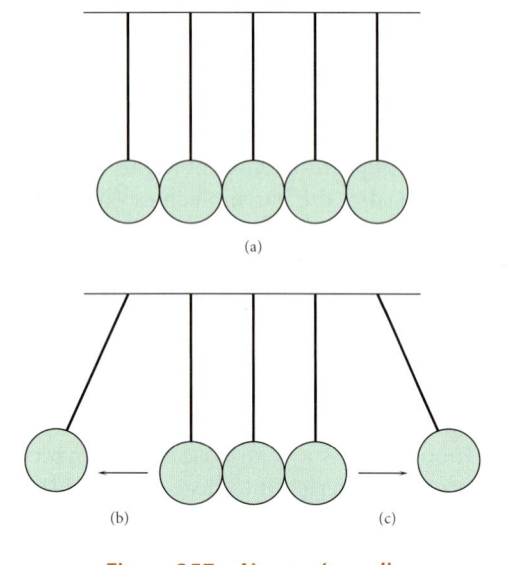

Figure 257   Newton's cradle

You may ask yourself the question why do two balls lift up (2m × v) and not one ball with twice the velocity (m × 2v), as this will still conserve momentum? To answer tis question, one would have to consider not only the law of conservation of momentum but also the law of conservation of kinetic energy for an elastic collision.

Explosions can be viewed as being the opposite of a collision because objects move apart rather than coming together. When a rocket is launched using a controlled explosion, the rocket moves upwards and the hot gases move downwards. The rocket gains momentum in one direction and the ejected hot gases gains equal momentum in the opposite direction.

Example

A rifle has a mass of 4 kg. A bullet of mass 10g is fired from the rifle at a horizontal velocity of 400 m s^{-1}. Calculate the recoil velocity of the gun.

Solution

A rifle firing a bullet is similar to a rocket ejecting fuel. Newton's third law tells us that (a) forces exist in pairs and (b) the force of the rifle on the bullet will be equal and opposite to force of the bullet on the rifle. The bullet goes forward (say in the positive direction) and the rifle "kicks" backward (the negative direction). We say there is a **recoil** velocity. The bullet gains momentum in one direction while the rifle recoils with its momentum being in the opposite direction.

total momentum before the blast = total momentum after the blast

$$0 = (0.01 \times 400) - (4 \times v)$$

$$4v = 4$$

$$v = 1 \text{ m s}^{-1} \text{ backwards.}$$

Exercise 2.6

1. A mass of sand m kg falls vertically onto a conveyor belt at a rate of m kg s^{-1} and the sand and the conveyor belt move with a constant speed of v m s^{-1}.

    In order to keep the belt moving at this constant speed, the horizontal force that must be exerted on the belt is:

    A. m **v**
    B. ½ m **v**
    C. m **v**^{2}
    D. ½ m **v**^{2}

2. A rocket accelerates vertically upwards by ejecting high-speed gases vertically downwards. The weight of the rocket is W and the magnitude of the thrust force the rocket exerts on the gases is T.

    The magnitude of the net force on the rocket would be

    A. W.
    B. T.
    C. T - W.
    D. T + W.   C

3. The momentum of a system is conserved if:

    A. the forces acting on the system are in equilibrium.
    B. no friction forces act within the system.
    C. no kinetic energy is lost or gained by the system.
    D. no external forces act on the system.   D

4. If an egg is dropped onto the floor, it is likely to break. However, when it is wrapped in a cloth, the chances of it breaking become less. This is because the cloth:

    A. reduces the momentum of the egg.
    B. increases the time for which the force of the ground acts on the egg.
    C. reduces the change of momentum of the egg.
    D. reduces the impulse acting on the egg.   B

5. If the velocity of a particle is changing, the rate of change of the momentum of the particle is equal to:

    A. the acceleration of the particle.
    B. the work done on the particle.
    C. the net force acting on the particle.
    D. the change in kinetic energy of the particle.

6. A bullet of mass 9.0 g leaves the barrel of a rifle with a speed of 8.0 ×10^2 m s^{-1}. The mass of the rifle is 1.8 kg. If the rifle is free to move, calculate the speed with which it recoils.

7. Sarah is standing on the horizontal surface of a frozen pond. She throws a ball of mass 250 g. The ball leaves her hand with a horizontal speed of 8.0 m s^{-1}. As a result, Sarah moves with an initial speed of 5.0 cm s^{-1}. Estimate Sarah's mass.

8. In a particular thunderstorm, the hailstones and raindrops have the same mass and terminal velocity. Explain, with reference to Newton's Second Law why the hailstones hurt more that raindrops when they hit you.

9. A football of mass 0.46 kg attains a speed of 7.7 m s^{-1} when kicked. The toe of the football boot with which it is kicked is in contact with the ball for 0.26 s.

    Calculate

    (i)   the impulse given to the ball
    (ii) the average force exerted on the ball.

10. Refer again to Figure 256. Suppose that after the collision, truck A and truck B do not link but instead truck A moves with a speed of 1 m s^{-1} and in the same direction as prior to the collision. Determine:

    (i)   the speed of truck B after collision
    (ii) the kinetic energy lost on collision

11. A man drags a sack of flour of mass 100 kg at constant speed up an inclined plane to a height of 6.0 m. The plane makes an angle of 30° with the horizontal and a constant frictional force of 250 N acts on the sack down the plane.

    Determine the efficiency of the inclined plane?

12. A car of mass 1000 kg is parked on a level road with its handbrake on. Another car of mass 1500 kg travelling at 10 m s^{-1} collides into the back of the stationary car. The two cars move together after collision in the same straight line. They travel 25m before finally coming to rest.

    Determine the average frictional force exerted on the cars as they come to rest.

13. When a golfer strikes a golf-ball it is in contact with the club head for about 1 ms and the ball leaves the club head with a speed of about 70 m s^{-1}. If the mass of the ball is 50 g estimate the maximum accelerating force exerted on the golf ball, stating any assumptions that you make.

14. A ball of mass 0.1 kg is dropped from a height of 2.0 m onto a hard surface. It rebounds to a height of 1.5 m and it is in contact with the surface for 0.05 s.

    Calculate the

    (i) speed with which it strikes the surface.
    (ii) speed with which it leaves the surface.
    (iii) change in momentum of the ball.
    (iv) impulse given to the ball on contact with the surface.
    (v) average force that the surface exerts on the ball.

15. A bullet of mass 0.02 kg is fired into a block of wood of mass 1.5 kg resting on a horizontal table. The block moves off with an initial speed of 8.0 m s^{-1}. Estimate the speed with which the bullet strikes the block.

16. The bullet in question 15 is fired from a rifle of mass 2.5 kg. Assuming that the bullet leaves the barrel of the rifle with the speed calculated above, find the recoil speed of the rifle if it is free to move. In reality the rifle is held and for a certain person the rifle recoils a distance of 0.12 m.

    Determine the average force that the person exerts on the rifle?

17. Two identical balls A and B with mass m are on a horizontal surface as shown in Figure 258. Ball B is at rest and Ball A is moving with a velocity v along a line joining the centres of the balls. During the collision of the balls, the magnitude of the force that ball A exerts on ball B is F_{AB} the magnitude of the force that ball B exerts on ball A is F_{BA}.

    Figure 258   Collision between balls

    (a) On a diagram similar to the one below, add labelled arrows to show the magnitude and direction of the forces during the collision.

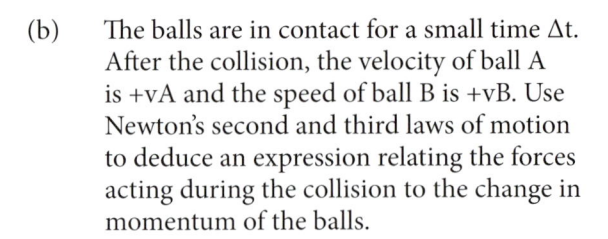

    (b) The balls are in contact for a small time Δt. After the collision, the velocity of ball A is +vA and the speed of ball B is +vB. Use Newton's second and third laws of motion to deduce an expression relating the forces acting during the collision to the change in momentum of the balls.

    (c) Deduce that if the kinetic energy is conserved in the collision, then after the collision, ball A will come to rest and ball B will move with speed v.

18. A satellite of mass 2.40 x 10² kg is moving in free space at a velocity of 6.00 x 10³ m s^{-1}. It collides with an unknown object that causes it to be deflected at right angles to its original path during a time of 0.500 s. Determine the impulse that acted on the satellite if it continued to move at 6.00 x 10³ m s^{-1} after deflection.

20. A ball of mass 5.0 g and velocity of 1.0 m s-1 strikes a glancing blow on a second ball of mass 1.0 x 10¹ g. After the collision, the first ball moves at 0.80 m s-1 at 45⁰ to its original direction. Assuming that the system is isolated, determine:

    (a) the magnitude of the second ball after the collision
    (b) the direction that the second ball was moving after the collision.

3. Thermal Physics

Contents

3.1 – Thermal concepts
3.2 – Modelling a gas

Essential Ideas

Thermal physics deftly demonstrates the links between the macroscopic measurements essential to many scientific models with the microscopic properties that underlie these models.

The properties of ideal gases allow scientists to make predictions of the behaviour of real gases. © IBO 2014

Chapter 3

3.1 Thermal Physics

NATURE OF SCIENCE:

Evidence through experimentation: Scientists from the 17th and 18th centuries were working without the knowledge of atomic structure and sometimes developed theories that were later found to be incorrect, such as phlogiston and perpetual motion capabilities. Our current understanding relies on statistical mechanics providing a basis for our use and understanding of energy transfer in science. (1.8)

© IBO 2014

Essential idea: Thermal physics deftly demonstrates the links between the macroscopic measurements essential to many scientific models with the microscopic properties that underlie these models.

Understandings:
- Molecular theory of solids, liquids and gases
- Temperature and absolute temperature
- Internal energy
- Specific heat capacity
- Phase change
- Specific latent heat

TOK   THERMAL CONCEPTS

In looking at the ways of knowing described in the Theory of knowledge guide, scientists could legitimately claim that science encompasses all these. Driven by emotion, imagination, faith, using sense perception, enhanced by technology and combined with reason and memory, it communicates through language, principally the universal language of mathematics. Science cannot suppose to be the truth as so many paradigm shifts have occurred over the centuries. It is hoped that the following historical perspective will help in showing how science has changed the way of reasoning in thermal physics.

Phlogiston/Caloric Theory

The concept of heat has been studied for many centuries. Aristotle (384 –322 B.C.) considered fire one of the five basic elements of the Universe. Over 2000 years ago, Greek philosophers believed that matter was made of "atomos", elemental atoms in rapid motion, and that the result of this rapid motion was heat. It was understood that heat flowed from hot bodies to colder ones, somewhat analogous to water or another fluid flowing from a higher to lower elevation. It is not surprising that the early theory of heat flow regarded heat as a type of fluid.

Around the time of Galileo Galilei (1564 –1642), this heat fluid was known as phlogiston – the soul of matter. Phlogiston was believed to have a negative mass, and, upon heating or cooling, the phlogiston was driven out or absorbed by an object.

Further refinements of the phlogiston theory were carried out by Antoine Lavoisier (1743–1794), and it became known as the caloric theory. Sir Isaac Newton (1642–1727) and other famous scientists supported the caloric theory. Calorists believed that a hot object had more caloric than a cold object. They explained expansion by saying that the caloric filled up the spaces between atoms pushing them apart. The total amount of caloric was unchanged when a hot and cold body came into contact.

However, the caloric theory did not adequately explain some phenomena involving heat. It was difficult to understand how the conservation of caloric fluid applied to friction and the expansion of liquids and gases. Some calorists' answer to the friction concept was that the latent heat was released which implies that a change of state was involved. Others argued that during friction the material is "damaged" and that it "bleeds" heat. No satisfactory answers were forthcoming.

Count Rumford

Much of the credit for dismantling the idea that heat was motion rather than substance or caloric goes to Benjamin Thompson (1753 –1814), also known as Count Rumford of Bavaria.

During the American Revolution, he was a Tory or loyalist in the disputes between Britain and its American colonies serving as a major in a company of militia. It is believed that he invented a cork flotation system for cannons while being transported by horses across rivers. He also designed a gun carriage that could be carried by three horses and could be assembled ready for firing in 75 seconds. He was knighted by King George III of England, and made a Count in 1791 by Theodor in his brief reign as elector of the Holy Roman Empire.

In 1793, Thompson left England ultimately to take up a post with the before mentioned Theodor, elector of Bavaria. He was appointed a major general in the Bavarian army. He designed fortifications and worked as an administrator in munitions. It was here that he observed that a large amount of heat was generated in the boring of cannons. He read the following extracts before the Royal Society of London in 1798.

'Being engaged, lately, in superintending the boring of cannon, in the workshops of the military arsenal at

Munich, I was struck with the very considerable degree of Heat which a brass gun acquires, in a short time, in being bored; and with the still more intense Heat (much greater than that of boiling water as I found by experiment) of the metallic chips separated from it by the borer.'

'From whence comes the Heat actually produced in the mechanical operation above mentioned? Is it furnished by the metallic chips which are separated by the borer from the solid mass of metal?"... "If this were the case, then, according to the modern doctrines of latent heat, and of caloric, the capacity for Heat of the parts of the metal, so reduced to chips, ought not only to be changed, but the change undergone by them should be sufficiently great to account for all the Heat produced.'

Count Rumford was saying that the metal chips should have undergone some alteration in their properties after the production of so much thermal energy. He noted that some cannon shavings were hot enough to glow, but he continued:

"But no such change had taken place; for I found, upon taking equal quantities, by weight, of these chips, and of thin slips of the same block of metal separated by means of a fine saw, and putting them, at the same temperature (that of boiling water), into equal quantities of cold water, the portion of water into which the chips were put was not, to all appearance, heated either less or more than the other portion, into which the slips of metal were put. From whence it is evident that the Heat produced [by boring the cannon] could not possibly be furnished at the expense of the latent Heat of the metallic chips."

Rumford further went on to explain that he had immersed cannons in water while they were being bored and noted the rate at which the temperature rose. His results showed that the cannon would have melted had it not been cooled. Rumford concluded that heat was not a caloric fluid in which caloric is conserved but rather a concept of motion. He argued that heat is generated when work is done, and that the heat will continue to be generated as long as work is done. He estimated a heat to work ratio within the order of magnitude accepted today.

However, many scientists of the time were not convinced because Rumford could not give a clear explanation of exactly what heat was in terms of the accepted model for matter at that time. It would take another half century before Joule supplied the accepted answers.

James Prescott Joule

James Prescott Joule (1818-1889) conducted a series of brilliant experiments between 1842 and 1870 that proved beyond doubt that heat was a type of energy – internal energy – of the particles of matter. The caloric theory lost popularity very quickly.

Joule was the son of a wealthy brewer in Manchester, UK. Because of his wealth, he never worked for a living. His experiments were performed in a laboratory that he built at his own expense while he was in his twenties. He became interested in ways to develop more efficient engines that were driving various components of the brewing process. Encouraged by the work of Count Rumford and others, he began to investigate whether mechanical work could produce heat.

Joule performed a variety of experiments and he refined and elaborated his apparatus and his techniques. In one of his first experiments, he used a falling weight to drive a small electric generator. The current produced heated a wire that was immersed in a definite mass of water, and the change in temperature was noted. He reasoned that the work done as the weight decreases its gravitational potential energy should be equivalent to the heat energy gained by the water. In another experiment he mounted a large container filled with air into a tub of water. When the air was compressed, the temperature of the gas increased. He measured the amount of work needed to compress the gas and the amount of heat energy given to the water as a result of compression.

Perhaps Joule's most famous experiment consisted of a paddlewheel mounted inside a cylinder of water that was driven by falling weights as shown in Figure 301. He wanted to see if one could raise the temperature of the water simply by turning the paddles. He repeated this experiment many times continually improving the apparatus and refining his analysis of the data. For example, he took great care to insulate the container so that no heat was lost to the surroundings, and he developed his own thermometer so that he could measure the temperature with a precision of a fraction of a degree.

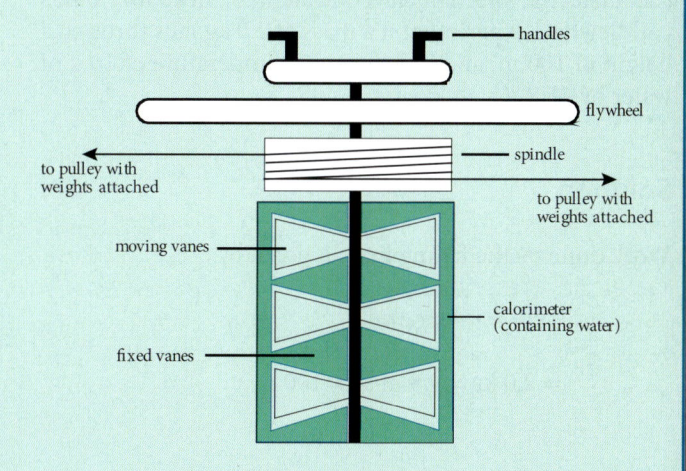

Figure 301 Schematic diagram of Joule's paddlewheel experiment.

Joule arranged the vanes of the paddlewheel so that they would not interfere with the particles of water set in motion. He didn't want to bruise or damage the water particles so that they might "bleed" heat.

Chapter 3

In 1849 he published his results in which he reported

"...the quantity of heat produced by friction of bodies, whether solid or liquid, is always proportional to the quantity of [energy] expended."

"...the quantity of heat capable of increasing the temperature of a pound of water by 1 ° Fahrenheit requires for its evolution the expenditure of a mechanical energy represented by the fall of 772 pound through the distance of one foot".

Joule found that about 4.2 joules of work would yield one calorie of heat or that the quantity of heat required to raise the temperature of one gram of water by 1 °C is one calorie.

A modern day value for the mechanical equivalent of heat is 4.18605 joules = 1 calorie.

The experiments proved beyond doubt that mechanical work can produce heat and as such no caloric fluid can be created or destroyed. Furthermore, Joule reasoned that the temperature increase must be related to the energy of the microscopic motions of the particles.

Finally, a paradigm shift in our way of reasoning had again proved that science is not the ultimate truth.

© IBO 2014

Example

Calculate the mechanical equivalent of heat for Joule's paddlewheel experiment if a mass of 2.0 kg falls through a height of 100 m, and increases the temperature of 10 g of water by 46.8 °C.

Solution

Work done by the falling mass is given by

$$W = E_p = mg\Delta h$$

$$= 2.0 \text{ kg} \times 9.8 \text{ m s}^{-2} \times 100 \text{ m}$$

$$= 1.96 \times 10^3 \text{ J}$$

Heat energy produced is given by

$$Q = m \times c \times \Delta T$$

$$= 10 \text{ g} \times 1 \text{ cal} \times 46.8 \text{ °C}$$

$$= 4.68 \times 10^2 \text{ calories}$$

Mechanical equivalent of heat is given by

$$\frac{W}{Q} = \frac{1.96 \times 10^3 \text{ J}}{4.68 \times 10^2 \text{ cal}}$$

$$= 4.19 \text{ J cal}^{-1}$$

The mechanical equivalent of heat for water is **4.2 J cal^{-1}**.

Molecular theory of solids, liquids and gases

An understanding of thermal energy is based upon a theory called the **moving particle theory** or **kinetic theory** (for gases) that uses models (Figure 302) to explain the structure and nature of matter. The basic assumptions of this moving particle theory relevant to thermal energy are:

- all matter is composed of extremely small particles
- all particles are in constant motion
- if particles collide with neighbouring particles, they conserve their kinetic energy
- a mutual attractive force exists between particles

solid   liquid   gas

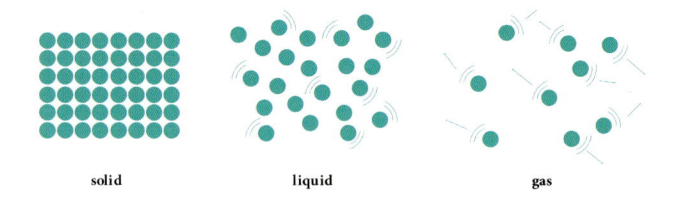

Figure 302   Arrangement of particles in solids, liquids and gases

An atom is the smallest neutral particle that represents an element as displayed in a periodic table of elements. Atoms contain protons, neutrons and electrons and an array of other sub-atomic particles. Atomic diameters are of the order of magnitude 10^{-10} m. Atoms can combine to form molecules of substances. In chemistry, the choice of the terms e.g. 'atoms, molecules, ions' are specific to elements and compounds. In physics, the word 'particle' is used to describe any of these specific chemistry terms at this stage of the course.

Evidence for the constant motion of particles can be gained from observation of what is known as **Brownian Motion**. If pollen grains from flowers are placed on water and observed under a microscope, the pollen grains undergo constant random zig-zag motion. The motion becomes more vigorous as the thermal energy is increased with heating. A Whitley Bay smoke cell uses smoke in air to achieve the same Brownian motion. In both cases, the motion is due to the smaller particles striking the larger particles and causing them to move.

Thermal Physics

The large number of particles in a volume of a solid, liquid or gas ensures that the number of particles moving in all directions with a certain velocity is constant over time. There would be no gaseous state if the particles were losing kinetic energy.

A mutual attractive force must exist between particles otherwise the particles of nature would not be combined as we know them. Further explanation of this assumption will be given later in this topic.

Matter is defined as anything that has mass and occupies space. There are four states of matter which are also called the four phases of matter – solids, liquids, gases and plasma. Most matter on Earth is in the form of solids, liquids and gases, but most matter in the Universe is in the plasma state. Liquids, gases and plasma are fluids. Plasma is made by heating gaseous atoms and molecules to a sufficient temperature to cause them to ionise. The resulting plasma consists then of some neutral particles but mostly positive ions and electrons or other negative ions. The Sun and other stars are mainly composed of plasma.

The remainder of this chapter will concentrate on the other three states of matter, and their behaviour will be explained in terms of their macroscopic and microscopic characteristics of which some are given in Figures 303.

A macroscopic property is one that can be observed. Physical properties such as melting point, boiling point, density, thermal conductivity, thermal expansion and electrical conductivity can be observed and measured.

Macroscopic properties are also the observable behaviours of that material such as shape, volume, compressibility, diffusion and comparative density.

The many macroscopic or physical properties of a substance can provide evidence for the nature and structure of that substance.

Characteristic	Solid	Liquid	Gas
Shape	Definite	Variable	Variable
Volume	Definite	Definite	Variable
Compressibility	Almost Incompressible	Very slightly Compressible	Highly Compressible
Diffusion	Small	Slow	Fast
Comparative Density	High	High	Low

Figure 303   Some macroscopic characteristics of solids, liquids and gases

The modern technique of X-ray diffraction that will be studied in detail in a later chapter has enabled scientists to determine the arrangement of particles in solids. The particles are closely packed and each particle is strongly bonded to its neighbour and is held fairly rigidly in a fixed position to give it definite shape in a crystalline lattice. Some patterns are disordered, as is the case for ceramics, rubber, plastics and glass. These substances are said to be amorphous. The particles have vibrational kinetic energy in their fixed positions and the force of attraction between the particles gives them potential energy.

In liquids the particles are still closely packed and the bonding between particles is still quite strong. However, they are not held as rigidly in position and the bonds can break and reform. This infers that the particles can slowly and randomly move relative to each other to produce variable shape and slow diffusion. Particles in a liquid have vibrational, rotational and some translational kinetic energy due to their higher mean speeds. The potential energy of the particles in a liquid is somewhat higher than for a solid because the spacing between the particles is large.

In gases the particles are widely spaced and the particles only interact significantly on collision or very close approach. Because of the rapid random zig-zag zigzag motion of the particles, a gas will become dispersed throughout any container into which it is placed. Diffusion (the spreading out from the point of release) can occur readily. Gases are compressible because the particles are widely spaced at a distance much greater than the size of the particles. The much higher mean speeds are due to an increased translational kinetic energy of the particles. Gases have a much higher potential energy than liquids because the particles are much further apart.

Temperature and absolute temperature

Everyone seems to have a feel for the concept of heat because there is so much talk in everyday conversation of how hot or how cold it is. We endure the seasons of the year wanting either to cool our bodies or heat up our surroundings. We are aware of how difficult it was for our ancestors to keep warm, and our present dwellings are designed and insulated to suit the climate. Our consumption of electrical and chemical energy for heating and other purposes is a continual concern. We have become aware that increased global warming could spell the end of the world, as we know it.

But what is the difference between heat and temperature in physics? They are definitely not the same physical quantity. If you fill a cup and a dish with hot water at the same temperature, and then place ice cubes into each, the dish full of hot water can melt more ice cubes than the cup of hot water even though the water in each was at the same temperature. The dish containing the larger amount of hot water has a greater mass of water as well as a greater amount of heat or thermal energy. A greater mass infers a greater number of water molecules and more thermal energy infers that these molecules would have a greater overall amount of energy. We cannot see the interaction of the water molecules at the microscopic level because we cannot see atoms. However, we can observe and monitor the temperature change by using a macroscopic temperature-measuring device.

Thermal energy is a measure of the kinetic and potential energy of the component particles of an object and is measured in joules. Heat is the thermal energy that is absorbed, given up or transferred from one object to another.

Temperature is a scalar quantity that gives an indication of the degree of hotness or coldness of a body. Alternatively, temperature is a macroscopic property that measures the average kinetic energy of particles on a defined scale such as the Celsius or Kelvin scales. The chosen scale determines the direction of thermal energy transfer between two bodies in contact from the body at higher temperature to that of lower temperature. Eventually, the two bodies will be in thermal equilibrium when they acquire the same temperature in an isolated system. It will be deduced later in this text that thermal energy cannot be transferred from a body at lower temperature to that of higher temperature.

There is no instrument that directly measures the amount of thermal energy a body gives off or absorbs. A property that varies with temperature is called a thermometric property. This property can be used to establish a temperature scale and construct a thermometer. Thermometers are made using the thermometric properties of a substance such as:

- the expansion of a column of liquid in a capillary tube (laboratory and clinical thermometers).

- the electrical resistance of a wire (resistance and thermistor thermometers).

- the difference in the rates of expansion of two metals in contact (bimetallic strips).

- the pressure of a gas at constant volume.

- the volume of a gas at constant pressure (gases expand by a greater amount and more evenly than liquids).

- the heating of two metal wires wound together (thermocouple thermometers rely on the two metals producing different currents).

- the colour of a solid heated to high temperatures (pyrometers).

A typical laboratory thermometer as shown in Figure 304 contains a liquid such as mercury or coloured alcohol. The expansion of alcohol is six times greater than mercury. Alcohol thermometers are safer and can be used at lower temperatures than mercury which turns to a solid below -38.9 °C. Its disadvantage is that it boils above 78.5 °C. To make the thermometer sensitive, it has a narrow bore tube and a large bulb. The bulb is made of thin glass so that heat can be transferred quickly between the bulb liquid and the material being observed. There is a vacuum above the thermometer liquid and it can move easily along the glass bore.

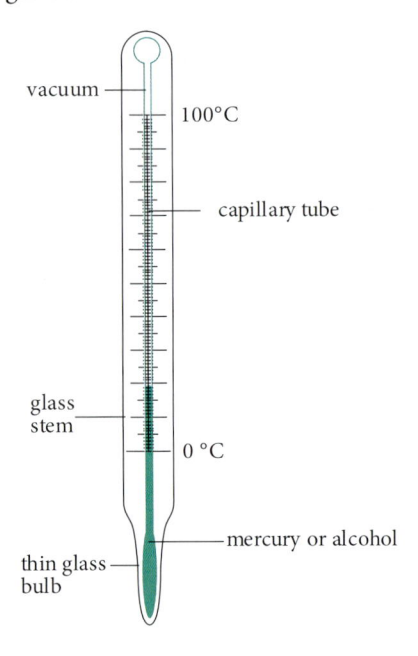

Figure 304   Typical laboratory thermometer

A clinical thermometer as shown in Figure 305 does not need the temperature range of a laboratory thermometer. It is designed so that the maximum temperature remains constant after the patient's temperature is taken. It has a small constriction to stop the mercury flowing back into the bulb. The mercury is then shaken back into the bulb after the temperature has been taken.

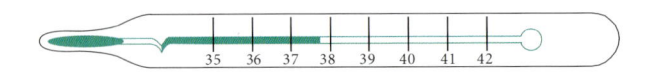

Figure 305   A Clinical Thermometer

In order to calibrate these thermometers, two fixed points are used to define the standard temperature interval. The ice point (the lower fixed point) marked at 0 °C is the temperature of pure ice at standard atmospheric pressure and is in thermal equilibrium with the liquid in the bulb. The steam point (the upper fixed point) marked at 100 °C is the temperature of steam at standard atmospheric pressure and is in thermal equilibrium with the liquid in the bulb. The scale between these values is marked with even spaces. The Celsius temperature scale named after the Swedish astronomer *Anders Celsius* (1701-1774) is constructed in such a manner.

Although thermometers constructed using thermometric properties are useful for everyday use, they are not accurate enough for scientific work. Unfortunately, two thermometers constructed using different thermometric properties do not necessarily agree with each other as they do not vary linearly over large temperature ranges. (They are of course in agreement at the lower and upper fixed points). For example, different thermometers will give different values for the boiling point of zinc (907 °C).

The standard fundamental temperature scale in the SI system is denoted by the symbol *T* and is measured in Kelvin, K. It is the thermodynamic temperature scale used in scientific measurement and it is a fundamental unit.

The lower fixed point is **absolute zero** and is assigned a value of 0 K. This is the point where molecular vibrations become a minimum – the molecules have minimum kinetic energy but molecular motion does not cease. The upper fixed point is the triple point of water. This is the temperature at which saturated water vapour, pure water and melting ice are all in equilibrium. For historical reasons, it is assigned a value of 273.16 K.

***T* in K = *T* in °C + 273.16**

Exercise 3.1

1. At room temperature, an iron rod feels cooler when held in the hand than wood held in the same hand. This is because:

    A. thermal energy tends to flow from the metal to the wood
    B. wood has a higher specific heat capacity than the iron rod
    C. wood has a lower specific heat capacity than the iron rod
    D. the iron rod conducts thermal energy better than the wood

2. Explain the difference between heat and temperature.

3. If you were travelling to Antarctica, deduce what would be the better thermometer to take – mercury or alcohol?

4. State one advantage and one disadvantage of a

    i. mercury in glass thermometer
    ii. constant volume thermometer.

5. The triple point of water is 273.16 K. Express this as a Celsius temperature.

6. Determine the ice point and the steam point of pure water on the Kelvin scale?

7. Define absolute zero.

8. If normal body temperature is 37.0 °C, what is it on the thermodynamic temperature scale?

Internal energy

Thermal energy refers to the **non-mechanical transfer** of energy between a system and its surroundings.

Thermal energy of a system is referred to as **internal energy**. Internal energy is the sum total of the potential energy and the random kinetic energy of the molecules of the substance making up a system. In order to apply the Law of conservation to thermal systems, one has to assume that a system has internal energy.

At the macroscopic level, it can be observed that molecules are moving. As already mentioned, when pollen (a fine powder produced by flowers) is sprinkled on the surface of water and the set-up is viewed under magnification, it can be seen that the pollen particles carry out zigzag motion called Brownian Motion. Their motion is caused by the kinetic energy of the water molecules.

Walking past a coffee shop you smell the aroma of the coffee due to diffusion caused by the kinetic energy of the air molecules allowing the aroma molecules to spread.

The expansion of solids, liquids and gases is a macroscopic property that allows us to understand that matter in a system has potential energy. When you heat a liquid it can be seen to expand as in a thermometer and this means that the potential energy of the system is increasing as the molecules move further apart. The compressibility of gases allows us to understand that the potential energy of the molecules is decreasing.

Although the internal energy in the examples above can never be absolutely determined, the change in internal energy can be observed.

The potential energy between particles is due to:

- the energy stored in bonds called bond energy
- intermolecular forces of attraction between particles.

Chapter 3

The bond energy is a form of chemical potential energy. It becomes significant in chemistry when a chemical reaction occurs, and bonds are broken and formed.

The intermolecular forces of attraction between particles is are due to the electromagnetic fundamental force since the gravitational force is too small to be of any significance.

Microscopic characteristics help to explain what is happening at the atomic level, and this part of the model will be interpreted further at a later stage. some of these characteristics are summarised in Figure 306.

Characteristic	Solid	Liquid	Gas
Kinetic energy	Vibrational	Vibrational Rotational Some translational	Mostly translational Higher rotational Higher vibrational
Potential energy	High	Higher	Highest
Mean molecular Separation (r_0)	r_0	r_0	$10 r_0$
Thermal energy of particles (ε)	$< \varepsilon/10$	$< \varepsilon > \varepsilon/10$	$> \varepsilon$
Molecules per m³	10^{28}	10^{28}	10^{25}

Figure 306 Some microscopic characteristics of solids, liquids and gases

Figure 307 indicates how the intermolecular electromagnetic force **F** between particles varies with the distance **r** between their centres.

At distances greater than r^0 (less than 2.5×10^{-10} m) attraction takes place, and at distances closer than r^0 the particles repel. At r^0 the particles are in equilibrium. Any displacement from the equilibrium position results in a simple harmonic oscillation of a particle or molecule.

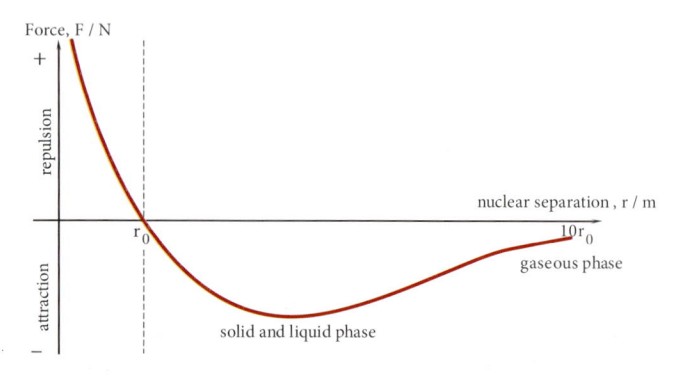

Figure 307 Force versus separation of particles

Figure 308 shows the relationship between the potential energy and the separation r of two molecules. At 0 K, the average separation of particles centres is r^0 and the overall force is zero. This is the point of minimum potential energy. Work will need to be done to move the particles apart and there will be an increase in potential energy.

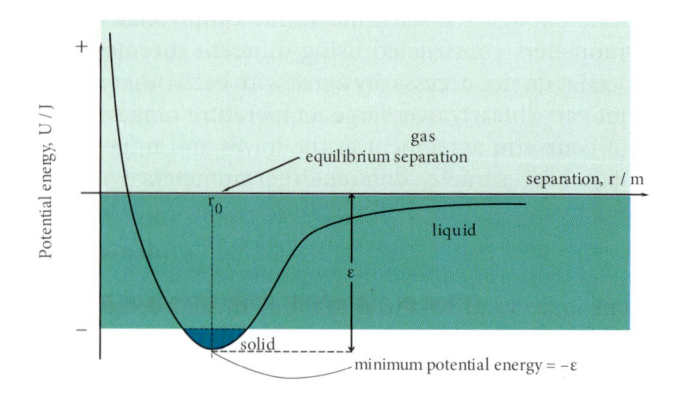

Figure 308 Potential energy versus separation of particles

$$\text{Work done} = \text{force} \times \text{distance} = F \times r = \text{change in potential energy}$$

$$\therefore F = \frac{\Delta E_p}{r}$$

In other words, the gradient of the potential energy curve at any point on the curve gives the force that must be applied to hold the molecules at that separation. We can classify the phases according to the sizes of the energy ε.

When less than $\frac{\varepsilon}{10}$, the vibrations occur about fixed positions and the particles are in the solid phase. When approximately equal to $\frac{\varepsilon}{10}$, the particles have sufficient energy to partly overcome the attractive forces and melting occurs.

When greater than $\frac{\varepsilon}{10}$, a liquid can form. When greater than ε, the particles have sufficient energy to leave the liquid and form a gas.

The **kinetic energy** is mainly due to the translational, rotational and vibrational motion of the particles as depicted in Figure 309.

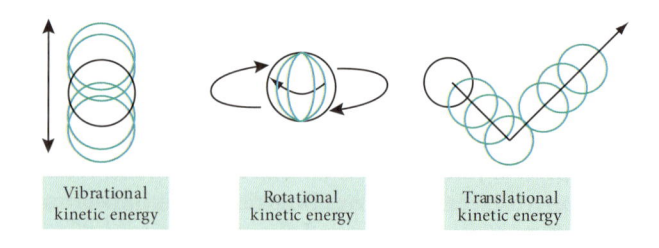

Figure 309 major particle motion and energy

90

THERMAL PHYSICS

Specific heat capacity

When different substances undergo the same temperature change they can store or release different amounts of thermal energy. They have different thermal (heat) capacities. If a substance has a high heat capacity it will take in the thermal energy at a slower rate than a substance with a low heat capacity because it needs more time to absorb a greater quantity of thermal energy. They also cool more slowly because they give out thermal energy at a slower rate.

We define the **thermal capacity** as the amount of energy required to change the temperature of a substance by one degree Kelvin.

We define the thermal (heat) capacity as,

$$\text{Thermal Capacity} = \frac{\Delta Q}{\Delta T} \; J\,K^{-1}$$

measured in J K^{-1} where:

ΔQ is the change in thermal energy in joules J

ΔT is the change in temperature in kelvin degrees K.

Water is used in car cooling systems and heating systems because of its high thermal capacity. A metal heat sink is used on the back of refrigerators because of its low thermal capacity.

Example

The thermal capacity of a sphere of lead is 3.2×10^3 JK^{-1}. Determine how much heat thermal energy can be released if the temperature changes from 61 °C to 25 °C.

Solution

$$\text{Heat capacity} = \frac{\Delta Q}{\Delta T} \Rightarrow \Delta Q = \text{Heat capacity} \times \Delta T$$

$= 3.2 \times 10^3 \, JK^{-1} \times (61 - 25)\,°C$

$= 3.2 \times 10^3 \, JK^{-1} \times 36\,°C$

$= 115200\,J$

$= 1.2 \times 10^5\,J$

- Note that a change in Kelvin temperature is the same as a change in Celsius temperature

Heat capacity does not take into account the fact that different masses of the same substance can absorb or release different amounts of thermal energy.

Consider three one kilogram blocks of aluminium, zinc and lead with the same sized base that have been heated to the same temperature of 80 °C. They are quickly placed on top of a large block of candle wax for a time period as shown in Figure 310.

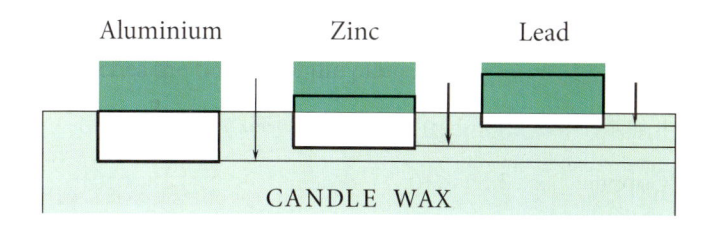

Figure 310 Front-on view of the metal blocks after a period of time.

The aluminium block melts the most wax and the lead melts the least. Therefore, the metals of the same mass give out different amounts of thermal energy in a certain time period. This can be explained from a microscopic viewpoint. The kilogram unit masses have different numbers of particles of different types and masses. The metal blocks were given the same amount of thermal energy when they were heated to 80 °C.

When the thermal energy gained by each metal is distributed amongst its particles, the average energy change of each particle will be different for each metal.

To obtain a characteristic value for the heat capacity of materials, equal masses of the materials must be considered. The physical property that includes the mass is called the specific heat capacity of a substance c.

Specific heat capacity or specific heat is the heat capacity per unit mass. It is defined as the quantity of thermal energy required to raise the temperature of a unit mass of a substance by one degree Kelvin.

$$c = \frac{\Delta Q}{m \Delta T}$$

measured in J kg^{-1} K^{-1}

$$\Delta Q = m \times c \times \Delta T$$

ΔQ = the change in thermal energy required to produce a temperature change in Joules, J.

m = mass of the material in grams (g) kilograms (kg)

ΔT = the temperature change in Kelvin, K.

Note that ΔT is always positive because heat always transfers from the higher temperature region to the lower temperature region.

Figure 311 shows the specific heat capacity for some common substances at room temperature (except ice)

91

CHAPTER 3

Substance	Specific heat J kg^{-1} K^{-1}	Substance	Specific heat J kg^{-1} K^{-1}
Lead	1.3×10^2	Iron	4.7×10^2
Mercury	1.4×10^2	Aluminium	9.1×10^2
Zinc	3.8×10^2	Sodium	1.23×10^3
Brass	3.8×10^2	Ice	2.1×10^3
Copper	3.85×10^2	Water	4.18×10^3

Figure 311 *Specific heat of some common substances*

Example

Determine how much thermal energy is released when 650 g of aluminium is cooled from 80 °C to 20 °C.

Solution

Using the fact that $\Delta Q = m.c.\Delta T$, we have,

$\Delta Q = 0.650 \, kg \times 9.1 \times 10^2 \, J \, kg^{-1} \, K^{-1} \times (80 - 20) \, K$

$= 3.549 \times 10^4 \, J$

$= 4 \times 10^4 \, J$

That is, $4 \times 10^4 \, J$ of heat is released.

Example

An active solar heater is used to heat 50 kg of water initially at a temperature of

12 °C. If the average rate that thermal energy is absorbed in a one hour period is 920 J min^{-1}, determine the equilibrium temperature after one hour

Solution

Quantity of heat absorbed in one hour = 920 J min^{-1} × 60 min = 5.52×10^4 J

Using the fact that $\Delta Q = m \, c \, \Delta T$, we have

$5.52 \times 10^4 \, J = 5.0 \times 10^1 \, kg \times 4.18 \times 10^3 \, J \, kg^{-1} \, K^{-1} \times (T_f - 12)K$

$5.52 \times 10^4 \, J = 2.09 \times 10^5 \, JK^{-1} \times (T_f - 12) \, °C$

$5.52 \times 10^4 \, J = 2.09 \times 10^5 \, T_f - 2.51 \times 10^6 \, J$

$2.09 \times 10^5 \, T_f = 5.52 \times 10^4 \, J + 2.51 \times 10^6 \, J$

$T_f = \dfrac{2.553 \times 10^6 \, J}{2.09 \times 10^5 \, (J \, K^{-1})}$

$T_f = 12.26 \, °C$

$\therefore T_f = 12 \, °C$

A calorimeter is a useful piece of equipment for investigations in Thermal Physics because it allows masses at different temperatures to be mixed with minimum energy loss to the surroundings. It is used for direct and indirect methods in determining the specific heat capacity of a substance. (The name of the instrument is derived from the Imperial unit, the calorie.)

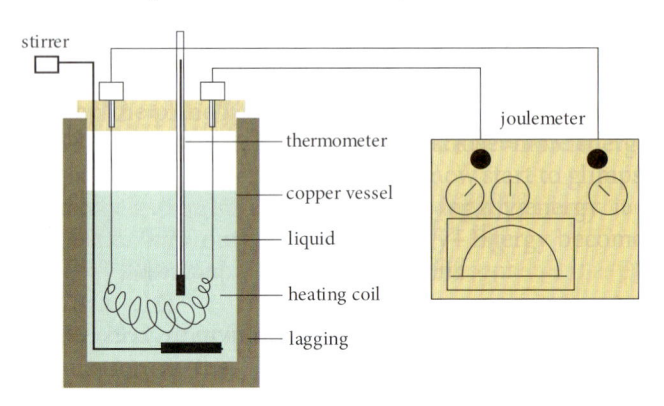

Figure 312 *Calorimeter being used to measure the heating effect of a current*

Figure 312 illustrates the use of a calorimeter to determine the specific heat capacity of a liquid, in this case water. The heating coil is used to convert electrical energy to thermal energy. The electrical energy can be measured by a joulemeter or by using a voltmeter/ammeter circuit. The duration of time of electrical input is noted.

The thermal energy gained by the calorimeter cup and the water is equal to the electrical energy lost to the calorimeter cup and water.

Electrical energy lost =

$V \times I \times t = [m \times c \times \Delta T]_{calorimeter \, cup} + [m \times c \times \Delta T]_{water}$

where V is the potential difference across the heating coil in volts V and I is the current in the amperes, A.

The specific heat capacity, c, of the calorimeter cup is obtained from published values. The other quantities are recorded and the specific heat capacity of the water is calculated.

In calorimeter investigations, heat losses to the surroundings need to be minimised. It is normal to polish the calorimeter cup to reduce loss of heat due to radiation. The calorimeter is also insulated with lagging materials such as wool or polystyrene to reduce heat loss due to conduction and convection.

After the power supply is switched off, the temperature should continue to rise for a period, and then level out for an infinite time. However, heat is lost to the surroundings, and the maximum temperature that could be achieved, in theory, is never reached. Instead appreciable cooling occurs. One method used to estimate the theoretical maximum temperature is to use a **cooling correction curve** as shown in Figure 313.

{Note that cooling correction is not required in the syllabus but is included for possible extended essays.}

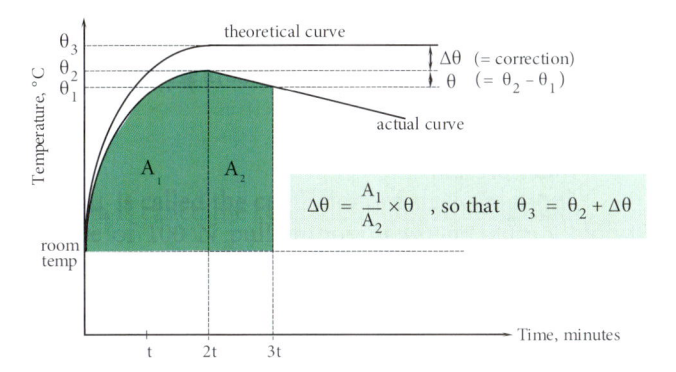

Figure 313   Graph of cooling correction.

A cooling correction is based on Newton's Law of Cooling. It states that the rate of loss of heat of a body is proportional to the difference in temperature between the body and its surroundings (excess temperature). A full explanation of this Law will not be given. If the power supply is switched off at time 2t minutes, then the temperature should continue to be recorded for a further t minutes. The correction to the temperature θ can be obtained from the graph as shown. The final temperature, θ_3, is then given as the final temperature of the thermometer plus the correction θ.

Another direct electrical method used to determine the specific heat capacity of a metal is shown in Figure 314. An immersion heater is placed into a metal block. The hole for the heater is lubricated with oil to allow even heat transmission. The electrical energy lost to the block is recorded for a given period of time and the specific heat of the metal is calculated. Cooling correction is more important in this case because the temperatures under which the investigation is carried out could be much higher than was the case when using a calorimeter.

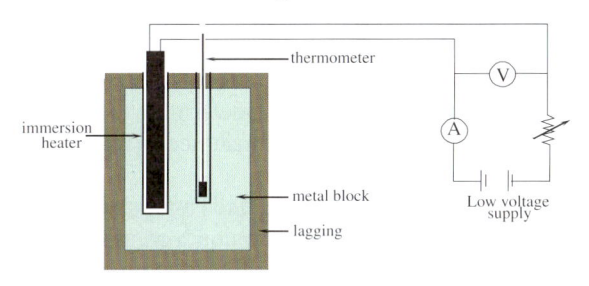

Figure 314   Electrical method using an immersion heater and a metal block

A common **indirect** method to determine the specific heat capacity of a solid and liquids is called the **method of mixtures**. In the case of a solid, a known mass of the solid is heated to a certain temperature, and then transferred to a known mass of liquid in a calorimeter whose specific heat capacity is known. The change in temperature is recorded and the specific heat of the solid is calculated from the results obtained. In the case of a liquid, a hot solid of known specific heat is transferred to a liquid of unknown specific heat capacity.

Example

A block of copper of mass 3.0 kg at a temperature of 90 °C is transferred to a calorimeter containing 2.00 kg of water at 20 °C. The mass of the copper calorimeter cup is 0.210 kg. Determine the final temperature of the water.

Solution

The thermal energy gained by the water and the calorimeter cup will be equal to the thermal energy lost by the copper.

That is, $[mc\Delta T]_{copper} = [mc\Delta T]_{calorimeter\ cup} + [mc\Delta T]_{cup}$

We also have that,

Thermal energy lost by the copper

$= (3.0\ kg)\ (3.85 \times 10^2\ J\ kg^{-1}\ K^{-1})\ (90.0 - Tf)\ K$

Thermal energy gained by the water

$= (2.0\ kg)\ (4.18 \times 10^3\ J\ kg^{-1}\ K^{-1})\ (Tf - 20.0)\ K$

Thermal energy gained by the cup

$= (0.21\ kg)\ (9.1 \times 10^2\ J\ kg^{-1}\ K^{-1})\ (Tf - 20.0)\ K$

$1.04 \times 10^5 - 1.155 \times 10^3\ Tf$

$= (8.36 \times 10^3\ Tf - 1.67 \times 10^5) + (1.91 \times 10^2\ Tf - 3.82 \times 10^3)$

That is,

$-9.71 \times 10^3\ Tf = -2.75 \times 10^5$

Giving

$Tf = 28.3\ °C$

The final temperature of the water is 28 °C

Exercise 3.2

1. The amount of thermal energy required to raise the temperature of 1.53×10^3 g of water from 15 K to 40 K is

    A. 1.6×10^7 J
    B. 1.6×10^5 J
    C. 4.4×10^7 J
    D. 4.4×10^5 J

2. The specific heat capacity of a metal block of mass m is determined by placing a heating coil in it, as shown in the following diagram.

    The block is electrically heated for time t and the maximum temperature change recorded is $\Delta\theta$. The constant ammeter and voltmeter readings during the heating are I and V respectively. The electrical energy supplied is equal to VIt.

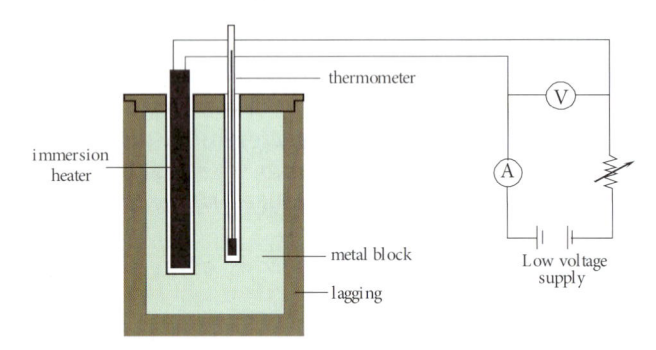

    The specific heat capacity is best calculated using which **one** of the following expressions?

    A. $c = \dfrac{m\Delta\theta}{VI}$
    B. $c = \dfrac{VI}{m\Delta\theta}$
    C. $c = \dfrac{VIt}{m\Delta\theta}$
    D. $c = \dfrac{m\Delta\theta}{VIt}$

3. 5.4×10^6 J of energy is required to heat a 28 kg mass of steel from 22 °C to 450 °C. Determine the specific heat capacity of the steel.

4. Liquid sodium is used as a coolant in some nuclear reactors. Describe the reason why liquid sodium is used in preference to water.

5. 6.00×10^2 kg of pyrex glass loses 8.70×10^6 J of thermal energy. If the temperature of the glass was initially 95.0 °C before cooling, calculate is its final temperature.

    (Take the specific heat capacity of pyrex glass to be 8.40×10^2 J kg^{-1} K^{-1})

6. A piece of wood placed in the Sun absorbs more thermal energy than a piece of shiny metal of the same mass. Explain why the wood feels cooler than the metal when you touch them.

7. A hot water vessel contains 3.0 dm^3 at 45 °C. Calculate the rate that the water is losing thermal energy (in joules per second) if it cools to 38 °C over an 8.0 h period.

8. Determine how many joules of energy are released when 870 g of aluminium is cooled from 155 °C to 20 °C.

9. If 2.93×10^6 J is used to raise the temperature of water from 288 K to 372 K, calculate the mass of water present.

10. 1.7 mJ of energy is required to cool a 15 kg mass of brass from 400 °C to 25 °C. Determine the specific heat capacity of brass.

11. A piece of iron is dropped from an aeroplane at a height of 1.2 km. If 75% of the kinetic energy of the iron is converted to thermal energy on impact with the ground, determine the rise in temperature.

12. If 115 g of water at 75.5 °C is mixed with 0.22 kg of water at 21 °C, determine the temperature of the resulting mixture.

13. Describe an experiment that would allow you to determine the specific heat capacity of a metal.

    (i) Sketch the apparatus.
    (ii) Describe what measurements need to be made and how they are obtained.
    (iii) State and explain the equation used to calculate the specific heat capacity of the metal.
    (iv) Describe 2 main sources of error that are likely to occur in the experiment.
    (v) Is the experimental value likely to be higher or lower than the theoretical value, if the experiment was carried out in a school laboratory? Explain your answer.

14. A heating fluid releases 4.2×10^7 Jkg^{-1} of heat as it undergoes combustion. If the fluid is used to heat 250 dm^3 of water from 15 °C to 71 °C, and the conversion is 65% efficient, determine the mass of the heating fluid that will be consumed in this process.

15. A large boulder of 125 kg falls off a cliff of height 122 m into a pool of water containing 120 kg of water. Determine the rise in temperature of the water. Assume that no water is lost in the entry of the boulder, and that all the heat goes to the water.

16. A thermally insulated container of water is dropped from a large height and collides inelastically with the ground. Determine the height from which it is dropped if the temperature of the water increases by 1.5 °C.

17. A piece of copper is dropped from a height of 225 m. If 75% of its kinetic energy is converted to heat energy on impact with the ground, calculate the rise in temperature of the copper. (Use the table of specific heat capacities to find the value for copper).

18. 5kg of lead shot is poured into a cylindrical cardboard tube 2.0 m long. The ends of the tube are sealed, and the tube is inverted 50 times. The temperature of the lead increases by 4.2 °C. If the specific heat of lead is 0.031 kcal kg^{-1} °C^{-1}, determine the number of work units in joules that are equivalent to the heat unit of 1 kilocalorie.

Phase change

A substance can undergo changes of state or phase changes at different temperatures. Pure substances (elements and compounds) have definite melting and boiling points that are characteristic of the particular pure substance being examined. For example, oxygen has a melting point of -218.8 °C and a boiling point of -183 °C at standard atmospheric pressure.

The heating curve for benzene is illustrated in Figure 315. A sample of benzene at 0 °C is heated in a closed container and the change in temperature is graphed as a function of time. The macroscopic behaviour of benzene can be described using the graph and the microscopic behaviour can be interpreted from the macroscopic behaviour.

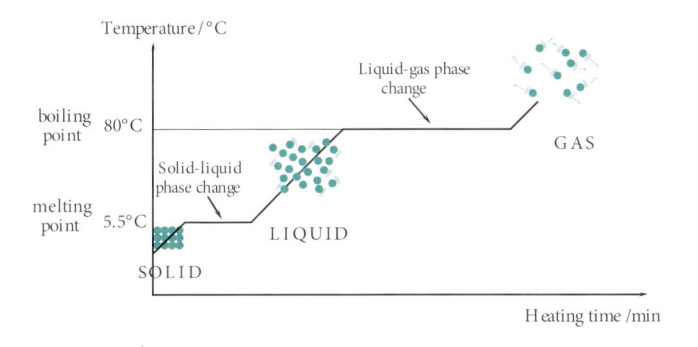

Figure 315   Heating curve for benzene.

When the solid benzene is heated the temperature begins to rise. When the temperature reaches 5.5 °C the benzene begins to melt. Although heating continues the temperature of the solid – liquid benzene mixture remains constant until **all** the benzene has melted. Once all the benzene has melted the temperature starts to rise until the liquid benzene begins to boil at a temperature of 80 °C. With continued heating the temperature remains constant until all the liquid benzene has been converted to the gaseous state. The temperature then continues to rise as the gas is in a closed container.

Molecular behaviour and phase changes

The moving particle theory can be used to explain the microscopic behaviour of these phase changes. Firstly, it is known that **temperature is related the average kinetic energy** of the particles in a system.

When solid benzene is heated, the particles of the solid vibrate at an increasing rate as the temperature is increased. The vibrational kinetic energy of the particles increases. At the melting point a temperature is reached at which the particles vibrate with sufficient thermal energy to break from their fixed positions and begin to slip over each other. The kinetic energy does not increase at 5.5 °C, and all the energy input is now in the form of potential energy.

As the solid continues to melt at 5.5 °C, more and more particles gain sufficient potential energy to overcome the forces between particles and over time all the solid particles change to a liquid at 5.5 °C. The potential energy of the system increases as the particles begin to move. As heating continues the temperature of the liquid rises due to an increase in the vibrational, rotational and part translational kinetic energy of the particles. At the boiling point a temperature is reached at which the particles gain sufficient energy to overcome the inter-particle forces present in the liquid benzene and escape into the gaseous state. Continued heating at the boiling point provides the potential energy needed for all the benzene molecules to be converted from a liquid to a gas. With further heating the temperature increases due to an increase in the kinetic energy of the gaseous molecules due to the larger translational motion.

Evaporation and boiling

When water is left in a container outside, exposed to the atmosphere, it will eventually evaporate. Mercury from broken thermometers has to be cleaned up immediately due to its harmful effects. Water has a boiling point of 100 °C and mercury has a boiling point of 357 °C. Yet they both evaporate at room temperature.

The process of evaporation is a change from the liquid state to the gaseous state that occurs at a temperature below the boiling point.

Chapter 3

The moving particle theory can be applied to understand the evaporation process. A substance at a particular temperature has a range of kinetic energies. So in a liquid at any particular instant, a small fraction of the molecules will have kinetic energies considerably greater then the average value. If these particles are near the surface of the liquid, they may have enough kinetic energy to overcome the attractive forces of neighbouring particles and escape from the liquid as a vapour. Now that the more energetic particles have escaped, the average kinetic energy of the remaining particles in the liquid has been lowered. Since temperature is proportional to the average kinetic energy of the particles, a lower kinetic energy implies a lower temperature, and this is the reason why the temperature of the liquid falls as evaporative cooling takes place. Another way of explaining the temperature drop is in terms of latent heat.

As a substance evaporates, it needs thermal energy input to replace its lost latent heat of vaporisation and this thermal energy can be obtained from the remaining liquid and its surroundings.

A substance that evaporates rapidly is said to be a volatile liquid. A liquid's volatility is controlled by a factor known as its equilibrium vapour pressure. There are forces that must be overcome before a particle can leave the surface of a liquid. Different liquids exert different vapour pressures that depend on the relative strengths of the intermolecular forces present in the liquids. Ether, chloroform and ethanol have relatively high vapour pressures.

The values in Figure 316 compare the vapour pressure of some liquids at 293 K.

Substance	Vapour pressure / kPa
Ether	58.9
Chloroform	19.3
Ethanol	5.8
Water	2.3
Mercury	0.0002

Figure 316   Some common vapour pressures

Although the vapour pressure of mercury is much lower than the other substances listed at room temperature, some evaporation does occur. Because of its extreme toxicity any mercury spill is treated seriously. Ether has a high vapour pressure. If a stream of air is blown through a sample of ether in a beaker that is placed on a thin film of water, the water will eventually turn to ice.

When overheating occurs in a human on hot days, the body starts to perspire. Evaporation of the perspiration results in a loss of thermal energy from the body so that body temperature can be controlled. Local anaesthetics with high vapour pressures are used to reduce pain on the skin. Thermal energy flows from the surrounding flesh causing its temperature to drop and thereby anaesthetises the area.

A liquid boils when the vapour pressure of the liquid equals the atmospheric pressure of its surroundings. As the boiling point is reached, tiny bubbles appear throughout the liquid. If the vapour pressure of the bubble is less than the atmospheric pressure the bubbles are crushed. However a point is reached when the pressures are equal. The bubble will then increase in size as it rises to the surface of the liquid.

Exercise 3.3

1. When smoke is strongly illuminated and viewed under a microscope it is possible to observe

    A. all particles moving in straight lines
    B. smoke particles moving randomly by air molecules
    C. smoke particles colliding with each other
    D. air molecules in random motion

2. The internal energy of a monatomic gas such as neon is mainly due to

    A. the potential energy holding the atoms in fixed positions
    B. the vibrational energy of the atoms
    C. the random translational energy of the atoms
    D. the rotational energy of the atoms

3. For a given mass of a certain liquid, the magnitude of the thermal energy transfer is the same for the following two processes

    A. freezing and sublimation
    B. melting and evaporation
    C. evaporation and condensation
    D. sublimation and condensation

4. Which of the following is a unit of thermal energy?

    A. watt
    B. the product of the newton and the metre
    C. the quotient of the watt and the second
    D. the product of the joule and the second

Base your answers to Questions 5 and 6 on the following graph. The graph shows the temperature of an unknown substance of mass 10.0 kg as heat is added at a constant rate of 6300 Jmin⁻¹. The substance is a solid at 0 °C.

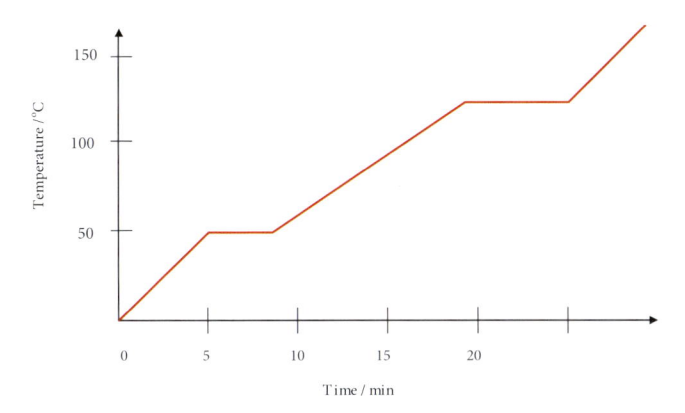

5. The internal potential energy of the unknown substance increases without any change in internal kinetic energy from the beginning of the:

   A. first minute to the end of the fourth minute
   B. seventh minute to the end of the seventeenth minute
   C. seventeenth minute to the end of the twenty first minute
   D. nineteenth to the end of the twenty fifth minute

6. The specific heat capacity of the substance when it is solid is:

   A. 63 Jkg⁻¹K⁻¹
   B. 105 Jkg⁻¹K⁻¹
   C. 126 Jkg⁻¹K⁻¹
   D. 504 Jkg⁻¹K⁻¹

7. Give five macroscopic and five microscopic characteristics of the liquid/gas in a butane lighter.

8. Describe the components of internal energy in each of the following situations

   (a) air at room temperature
   (b) a jar of honey
   (c) a melting ice cream.

9. Explain the difference between heat, thermal energy and temperature.

10. Does a block of ice contain any heat? Explain your answer fully.

11. Draw a fully labelled cooling curve for the situation when steam at 110 °C is converted to ice at −25 °C.

12. (a) Convert 63 °C to Kelvin
    (b) Convert 52 K to degrees Celsius

13. The temperatures of the same volume of air and water are raised by a small amount. Explain why a different amount of heat is required for each process.

14. If you increase the heat under a pot of water in which you are boiling potatoes, will the potatoes be cooked faster?

15. If you wanted to cool a bottle of soft drink at the beach, would you be better to wrap a wet towel around it or to put it into the seawater? Explain your answer.

16. Why is it important not to stand in a draught after vigorous exercise?

17. Describe and explain the process of evaporative cooling in terms of its microscopic properties.

18. A kettle made of stainless steel containing water is heated to a temperature of 95 °C. Describe the processes of thermal energy transfer that are occurring in the stainless steel kettle and the water.

Specific latent heat

The thermal energy which a particle absorbs in melting, evaporating or sublimating or gives out in freezing, condensing or sublimating is called **latent heat** because it does not produce a change in temperature. See Figure 317.

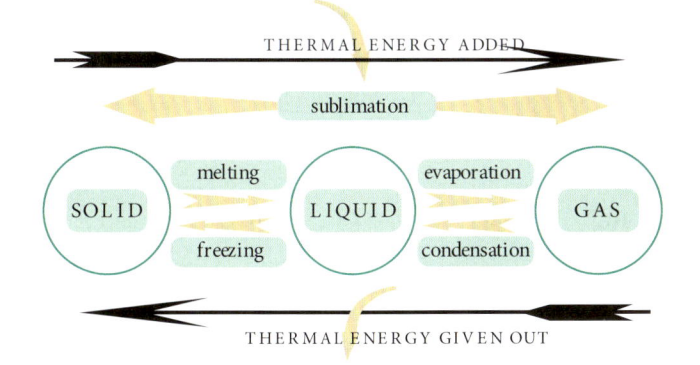

Figure 317   Macroscopic transformations between states of matter.

Sublimation is a change of phase directly from a solid to a gas or directly from a gas to a solid. Iodine and solid carbon dioxide are examples of substances that sublime.

97

CHAPTER 3

When thermal energy is absorbed/released by a body, the temperature may rise/fall, or it can remain constant. If the temperature remains constant then a phase change will occur as the thermal energy must either increase the potential energy of the particles as they move further apart or decrease the potential energy of the particles as they move closer together. If the temperature changes, then the energy must increase the average kinetic energy of the particles.

The quantity of heat required to change one kilogram of a substance from one phase to another is called the **latent heat of transformation**.

$\Delta Q = mL$

ΔQ is the quantity of heat absorbed or released during the phase change in J,

m is the mass of the substance in kg and

L is the latent heat of the substance in J kg^{-1}

L could be the latent heat of fusion L_f, the latent heat of vaporisation L_v or the latent heat of sublimation L_s. The latent heat of fusion of a substance is less than the latent heat of vaporisation or the latent heat of sublimation. More work has to be done to reorganise the particles as they increase their volume in vaporisation and sublimation than the work required to allow particles to move from their fixed position and slide over each other in fusion. Figure 318 lists the latent heat of some substances.

Substance	Melting point K	Latent heat of fusion 10⁵ J kg⁻¹	Boiling point K	Latent heat of Vaporisation 10⁵ J kg⁻¹
Oxygen	55	0.14	90	2.1
Ethanol	159	1.05	351	8.7
Lead	600	0.25	1893	7.3
Copper	1356	1.8	2573	73
Water	273	3.34	373	22.5

Figure 318   Some Latent Heat Values

Example 1

Calculate the heat energy required to evaporate 5.0 kg of ethanol at its boiling point.

Solution

Given that m = 5.0 kg and L_v = 8.7 × 10⁵ J kg^{-1}.

We then have,

= 4.35 × 10⁶ J = 4.4 × 10⁶ J

The heat energy required for the vaporisation is **4.4 × 10⁶ J**.

Example 2

Determine the heat energy released when 1.5 kg of gaseous water at 100 °C is placed in a freezer and converted to ice at -7 °C. The specific heat capacity of ice is 2.1 × 10³ J kg^{-1} K^{-1}.

Solution

The energy changes in this process can be represented as shown in Figure 322.

Using

$Q = mL_V + mc\Delta T_{WATER} + mL_f + mc\Delta T_{ICE}$

$= m [L_V + c\Delta T_{WATER} + L_f + c\Delta T_{ICE}]$

$= 1.5 [22.5 × 10^5 + (4180 × 100) + 3.34 × 10^5 + (2100 × 7)]$

$= 4.52 × 10^6$ J

That is, the energy released is **4.5 × 10⁶ J or 4.5 MJ**.

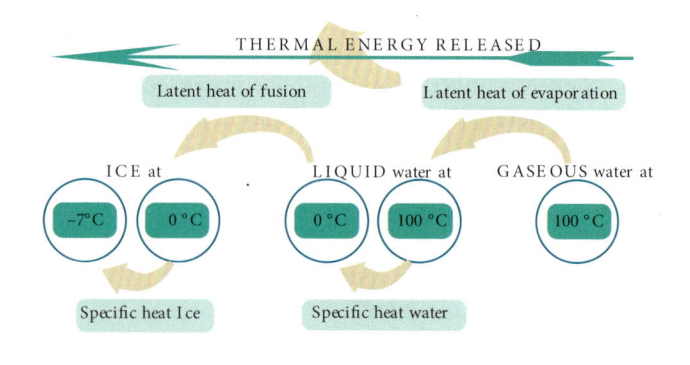

Figure 319   Energy released in steam-ice change.

The latent heat of vaporisation can be found using a self-jacketing vaporiser as shown in Figure 322. The liquid to be vaporised is heated electrically so that it boils at a steady rate. The vapour that is produced passes to the condenser through holes labelled H in the neck of the inner flask. Condensation occurs in the outer flask and the condenser.

THERMAL PHYSICS

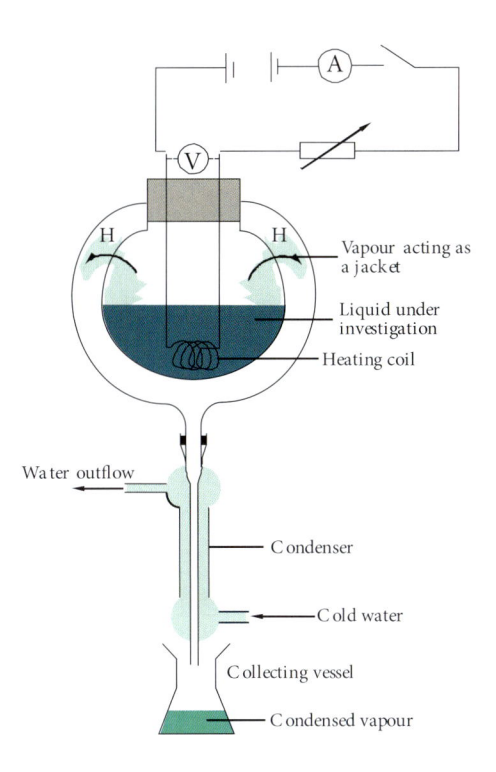

Figure 320 Latent heat of vaporisation apparatus.

Eventually, the temperature of all the parts of the apparatus becomes steady. When this steady state is reached, a container of known mass is placed under the condenser outlet for a measured time t, and the measured mass of the condensed vapour m is determined. The heater current I is measured with the ammeter 'A' and potential difference V is measured with a voltmeter 'V'. They are closely monitored and kept constant with a rheostat.

The electrical energy supplied is used to vaporise the liquid and some thermal energy H is lost to the surroundings.

Therefore:

$$V_1 I_1 t = m_1 L_V + H$$

In order to eliminate H from the relationship, the process is repeated using a different heater potential difference and current. The vapour is collected for the same time t The rate of vaporisation will be different but the heat lost to the surroundings will be the same as each part of the apparatus will be at the same temperature as it was with the initial rate vaporisation.

Therefore:

$$V_2 I_2 t = m_2 L_V + H$$

By subtracting the two equations:

$$(V_1 I_1 - V_2 I_2) t = (m_1 - m_2) L_V$$

From this equation, the value of the latent heat of vaporisation of the unknown substance can be determined.

Excercise 3.4

1. The specific latent heat of fusion of ice is the heat required to

   A. raise the temperature of ice from 0 °C to 10 °C
   B. change 1 dm³ of ice at 0 °C to water at 0 °C
   C. change 1kg of ice at 0 °C to water at 0 °C
   D. change the temperature of 1 kg by 10 °C

2. A substance changes from liquid to gas at its normal boiling temperature. What change, if any, occurs in the average kinetic energy and the average potential energy of its molecules?

	Average kinetic energy	Average potential energy
A.	constant	increases
B.	increases	constant
C.	increases	decreases
D.	constant	constant

3. Thermal energy is transferred to a mass of water in four steps. Which one of the four steps requires the most thermal energy?

   A. 5 °C to 20 °C
   B. 15 °C to 35 °C
   C. 75 °C to 90 °C
   D. 95 °C to 101 °C

4. Determine the amount of thermal energy that is required to melt 35 kg of ice at its melting point.

5. A 5.0×10^2 g aluminium block is heated to 350 °C. Determine the number of kilograms of ice at 0 °C that the aluminium block will melt as it cools.

6. Steam coming from a kettle will give you a nastier burn than boiling water. Explain why.

7. An immersion heater can supply heat at a rate of 5.2×10^2 J s^{-1}. Calculate the time that it will take to completely evaporate 1.25×10^{-1} kg of water initially at a temperature of 21 °C?

8. A 3.45 kg sample of iron is heated to a temperature of 295 °C and is then transferred to a 2.0 kg copper vessel containing 10.0 kg of a liquid at an initial temperature of 21.0 °C. If the final temperature of the mixture is 31.5 °C, determine the specific heat capacity of the liquid?

9. A mass of dry steam at 1.0×10^2 °C is blown over a 1.5 kg of ice at 0.0 °C in an isolated container. Calculate the mass of steam needed to convert the ice to water at 21.5 °C.

99

10. A freezer in a refrigerator takes 2.00 hours to convert 2.15 kg of water initially at 21.5 °C to just frozen ice. Calculate the rate at which the freezer absorbs heat.

11. Describe an experiment to determine the specific heat capacity of an unknown metal. Sketch the apparatus used and describe what measurements are made. State the main sources of error and explain how they can be minimised.

12. Calculate how much thermal energy is released when 1.2 kg of steam at 100 °C is condensed to water at the same temperature.
($L_v = 2.25 \times 10^6$ Jkg^{-1})

13. Determine how much energy is released when 1.5 kg of gaseous water at 100 °C is placed in a freezer and converted to ice at −7 °C. (the specific heat capacity of ice is 2.1×10^3 J kg^{-1} K^{-1}).

14. Describe an experiment that can be used to determine the latent heat of vaporisation of a liquid.

3.2 Modelling a gas

NATURE OF SCIENCE:

Collaboration: Scientists in the 19th century made valuable progress on the modern theories that form the basis of thermodynamics, making important links with other sciences, especially chemistry. The scientific method was in evidence with contrasting but complementary statements of some laws derived by different scientists. Empirical and theoretical thinking both have their place in science and this is evident in the comparison between the unattainable ideal gas and real gases. (4.1)

© IBO 2014

Essential idea: The properties of ideal gases allow scientists to make predictions of the behaviour of real gases.

Understandings:
- Pressure
- Equation of state for an ideal gas
- Kinetic model of an ideal gas
- Mole, molar mass and the Avogadro constant
- Differences between real and ideal gases

Pressure

Investigations into the behaviour of gases involve measurement of "states" of a system such as pressure, volume, temperature, density, mass and the number of moles of the particles present. Experiments use these macroscopic properties of a gas to formulate a number of gas laws.

In 1643 Torricelli found that the atmosphere could support a vertical column of mercury about 76 cm high and the first mercury barometer became the standard instrument for measuring pressure. The pressure unit 760 mm Hg (760 millimetres of mercury) represented standard atmospheric pressure. In 1646, Pascal found that the atmosphere could support a vertical column of water about 10.4 m high.

Pressure can be defined as the force exerted over an area.

$$\text{Pressure} = \frac{\text{Force}}{\text{Area}}$$

$$p = \frac{F}{A}$$

The SI unit of pressure is the pascal Pa.
1 atm = 1.01×10^5 Nm^{-2} = 101.3 kPa = 760 mmHg

THERMAL PHYSICS

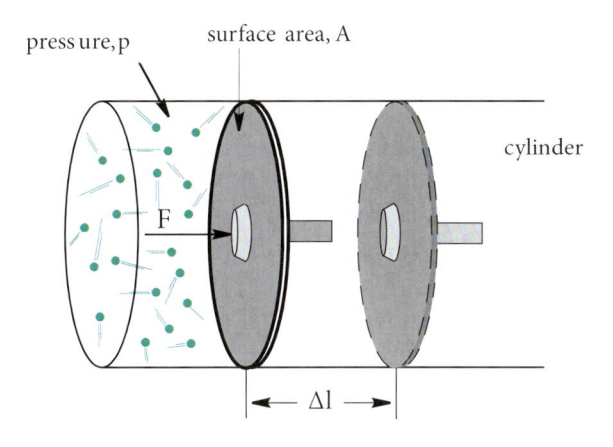

Figure 321 Expansion of a gas at constant pressure

The pressure inside the cylinder is caused by the gas particles colliding with the sides of the container and the surface of the cylinder. The pressure is not dependent on the gas particles colliding with each other. When work is done on the gas in the system, the cylinder will move through a distance, say Δ l. The gas expands and the volume becomes larger. The same number of collisions will occur as before but the pressure will become less as the force of the particles is distributed over a larger surface area.

Equation of state for an ideal gas

Boyle's Law

Robert Boyle (1627-1691) discussed that the pressure of a gas at constant temperature is proportional to its density. He also investigated how the pressure is related to the volume for a fixed mass of gas at constant temperature. Boyle's Law relates pressure and volume for a gas at fixed temperature.

Boyle's Law for gases states that the pressure of a fixed mass of gas is inversely proportional to its volume at constant temperature.

$$P \alpha \frac{1}{V} \Leftrightarrow PV = \text{constant}$$

When the conditions are changed, with the temperature still constant

$$P_1 V_1 = P_2 V_2$$

The readings of P and V must be taken slowly to maintain constant temperature because when air is compressed, it warms up slightly.

There are three common experimental designs, and a schematic diagram for each design is shown in Figure 322.

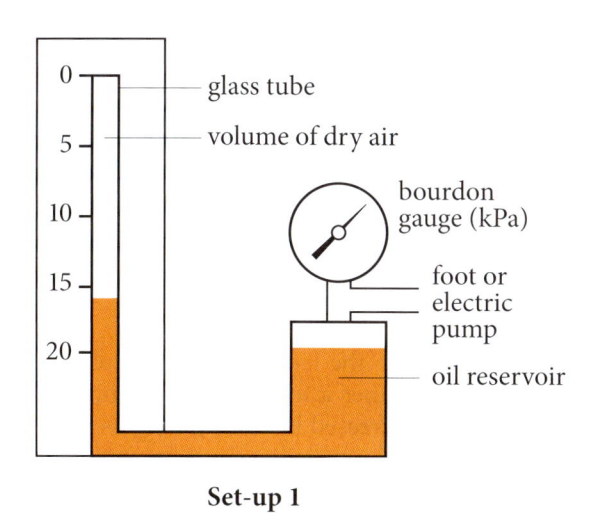

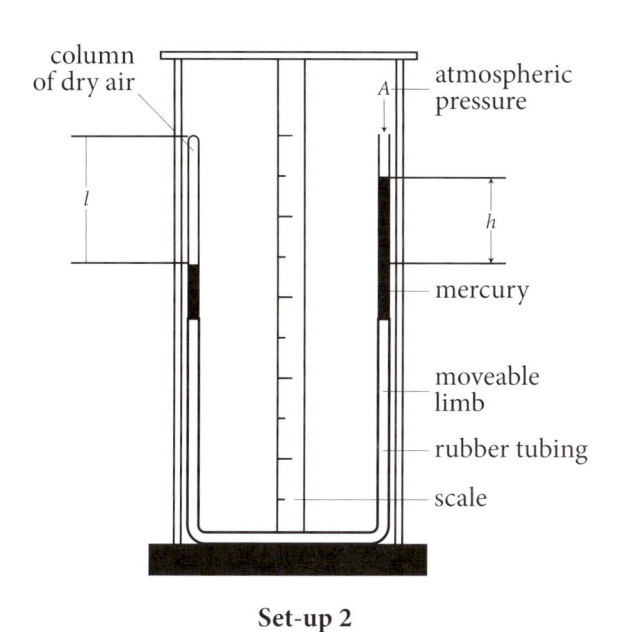

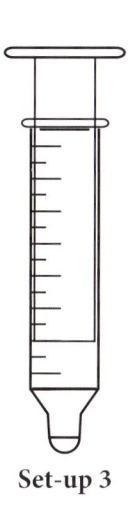

Set-up 1

Set-up 2

Set-up 3

Figure 322 Some common experimental techniques for studying Boyle's Law

101

CHAPTER 3

In set-up one, different pressures are applied to obtain different volumes of dry air.

In the second set-up, the length, l, is directly proportional to the volume, and the pressure is equal to atmospheric pressure ± the distance, h. Find various values of the air column length when different pressures are applied to it. The pressure is altered by raising and lowering the open-ended tube.

In the third set-up the length of the air column can be varied applying either a tension or a compression force. The pressure is increased by adding weights to the top of the plunger of the syringe.

When a pressure versus volume graph is drawn for the collected data a hyperbola shape is obtained, and when pressure is plotted against the reciprocal of volume a straight line (direct proportionality) is obtained. See Figure 323.

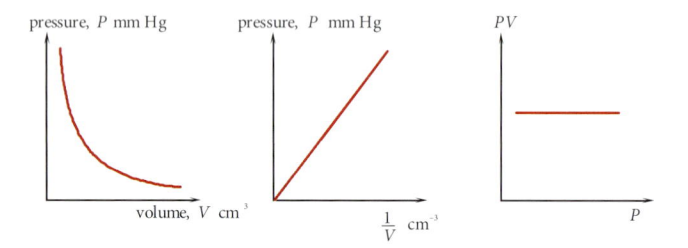

Figure 323 Pressure-volume graphs.

As already mentioned, the pressure that the molecules exert is due to their collisions with the sides of the container. When the volume of the container is decreased, the frequency of the particle collisions with the walls of the container increases. This means that there is a greater force in a smaller area leading to an increase in pressure. The pressure increase has nothing to do with the collisions of the particles with each other.

Charles' Law

In 1787 *Jacques Charles* (1746–1823) performed experiments to investigate how the volume of a gas changed with temperature. *Gay-Lussac* (1778–1850) published more accurate investigations in 1802.

A very simple apparatus to investigate Charles' Law is shown in Figure 324. A sample of dry air is trapped in a capillary tube by a bead of concentrated sulfuric acid (the acid absorbs moisture from the air). The capillary tube is heated in a water bath and the water is constantly stirred to ensure that the whole air column is at the same temperature.

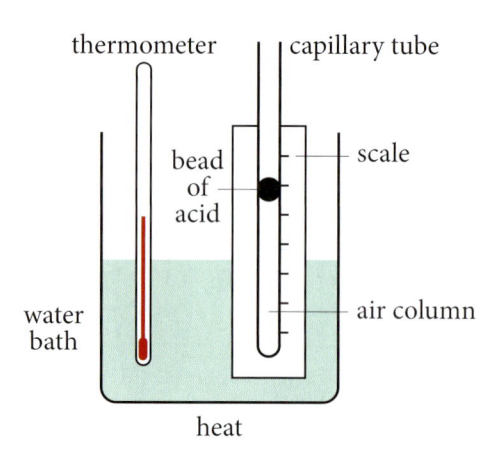

Figure 324 Apparatus for Charles' law.

The investigation should be carried out slowly to allow thermal energy to pass into or out of the thick glass walls of the capillary tube. When the volume and temperature measurements are plotted, a graph similar to Figure 325 is obtained.

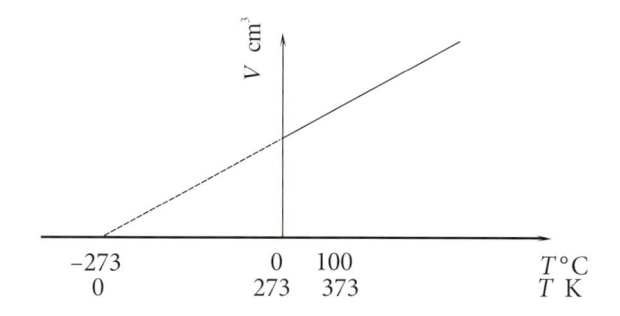

Figure 325 Variation of volume with temperature.

Note that from the extrapolation of the straight line that the volume of gases would be theoretically zero at −273 °C called absolute zero. The scale chosen is called the Kelvin scale K.

The Charles' (Gay-Lussac) Law of gases states that:
The volume of a fixed mass of gas at constant pressure is directly proportional to its absolute (Kelvin) temperature.

This can also be stated as:

The volume of a fixed mass of gas increases by 1/273 of its volume at 0 °C for every degree Celsius rise in temperature provided the pressure is constant.

$$V \alpha T \Rightarrow V = kt \text{ so that } \frac{V_1}{T_1} = k$$

Therefore,

$$\frac{V_1}{T_1} = \frac{V_2}{T_2}$$

As the temperature of a gas is increased, the average kinetic energy per molecule increases. The increase in velocity of the molecules leads to a greater rate of collisions, and each collision involves greater impulse. Hence the volume of the gas increases as the collisions with the sides of the container increase.

Thermal Physics

Experiments were similarly carried out to investigate the relationship between the pressure and temperature of a fixed mass of various gases.

The essential parts of the apparatus shown in Figure 326 are a metal sphere or round bottomed flask, and a Bourdon pressure gauge. The sphere/flask and bourdon gauge are connected by a short column of metal tubing/capillary tube to ensure that as little air as possible is at a different temperature from the main body of enclosed gas. The apparatus in Figure 326 allows the pressure of a fixed volume of gas to be determined as the gas is heated.

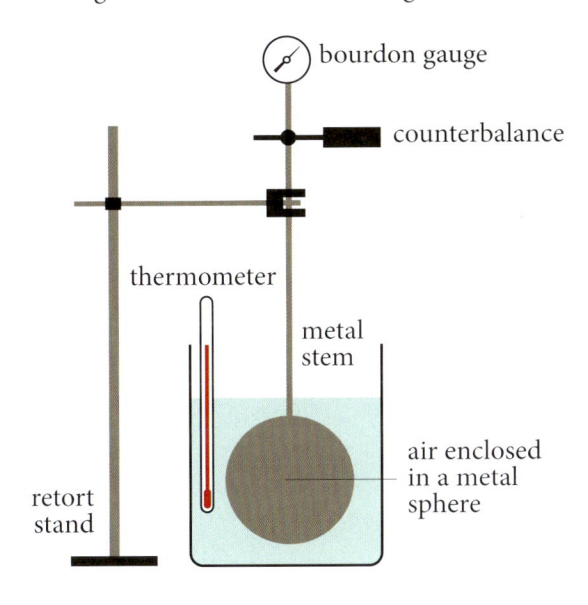

Figure 326 Pressure law apparatus.

The variation in pressure as the temperature is changed is measured and graphed. A typical graph is shown in Figure 327.

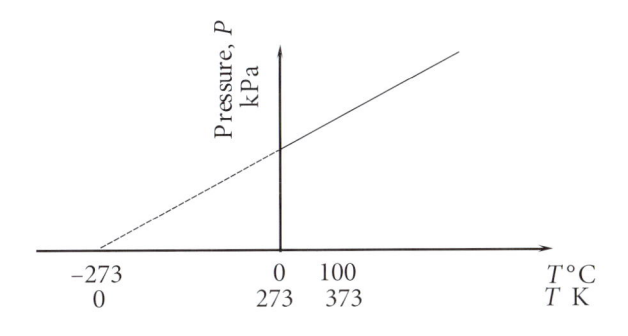

Figure 327 Variation of pressure with temperature.

The Pressure (Admonton) Law of Gases states that:

The pressure of a fixed mass of gas at constant volume is directly proportional to its absolute (Kelvin) temperature.

$$P \propto T \Leftrightarrow P = kT \therefore \frac{P_1}{T_1} = k$$

Therefore,

$$\frac{P_1}{T_1} = \frac{P_2}{T_2}$$

As the temperature of a gas is increased, the average kinetic energy per molecule increases. The increase in velocity of the molecules leads to a greater rate of collisions, and each collision involves greater impulse. Hence the pressure of the gas increases as the collisions with the sides of the container increase.

Combined Gas Law

The three gas Laws mentioned above can be combined into an overall equation to give:

$$\frac{P_1 V_1}{T_1} = \frac{P_2 V_2}{T_2}$$

Example

An ideal gas has a volume V at a pressure P and absolute temperature T. What would be the new pressure if the volume was halved and the temperature was doubled?

Solution

By using the combined gas equation:

$$\frac{P_1 V_1}{T_1} = \frac{P_2 V_2}{T_2}$$

And making P_2 the subject of the equation

$$P_2 = \frac{P_1 V_1}{T_1} \times \frac{T_2}{V_2} = \frac{PV}{T} \times \frac{2T}{\frac{1}{2}V}$$

$$P_2 = 4P$$

Mole, molar mass and the Avogadro constant

The mass of an atom is exceedingly small. For example, the mass of a fluorine atom is 3.16×10^{-23} g and the mass of the isotope carbon–12 is 1.99×10^{-23} g. Because the masses of atoms are so small, it is more convenient to describe the mass of an atom by comparing its mass with those of other atoms.

In 1961 the International Union of Pure and Applied Chemistry (IUPAC) defined the masses of atoms relative to carbon–12 that was assigned a value of 12.0000. Therefore, the relative atomic mass is defined as the mass of an atom when compared with 1/12 the mass of carbon–12 atom.

Just as the relative atomic mass is used to describe the masses of atoms, the relative molecular mass is used to

103

describe the masses of molecules. Therefore, the relative molecular mass is defined as the mass of a molecule when compared with 1/12 of the mass of a carbon–12 atom.

It is convenient to group things into quantities. For example, a box of diskettes, a ream of photocopy paper (500 sheets), a dozen eggs are common groupings. The SI fundamental unit for the amount of a substance is the **mole** (mol).

The mole is the amount of substance that contains as many elementary particles as there are in 0.012 kg of carbon–12.

Amadeo Avogadro (1776 – 1856) found that equal volumes of gases at the same temperature and pressure contained the same number of particles.

One mole of any gas contains the **Avogadro number** of particles N_A. It is now known that one mole of a gas occupies 22.4 dm³ at 0 °C and 101.3 kPa pressure (STP) and contains 6.01×10^{23} **particles**.

When using the mole, the atoms or molecules should be clearly stipulated. For example, one mole of copper atoms contains 6.02×10^{23} copper atoms. One mole of nitrogen molecules (N_2) contains 6.02×10^{23} of nitrogen molecules and 12.04×10^{23} nitrogen atoms.

The amount of substance (the moles) is related to the mass and the molar mass according to the following equation:

$$n = \frac{m}{M}$$

where

n = amount of a substance in mol,
m = the mass in g and
M = the molar mass in g mol⁻¹.

The molar mass can be obtained from the periodic table.

Example

1. Calculate the number of moles of oxygen molecules contained in 64.0 g of oxygen gas, O_2.

2. Calculate the number of oxygen molecules in part 1 of this example.

3. Determine the volume of oxygen gas that would be present at STP.

4. Calculate the mass in 0.75 mol of carbon dioxide gas.

Solution

1. $n = \frac{m}{M} = \frac{64.0 \, g}{(16.0 + 16.0) \, g \, mol^{-1}} = 2 \, mol$

2. The number of oxygen molecules

   $= 2N_A \times$ the number of molecules

   $= 6.02 \times 10^{23} \times 2$

   $= 1.024 \times 10^{24}$ molecules.

3. Volume

   $= 2 \, mol \times 22.4 \, dm^3$

   $= 44.8 \, dm^3$

4. $m = n M$

   $= 0.75 \, mol \times (12 + 16 + 16) \, g \, mol^{-1}$

   $= 33 \, g$

Exercise 3.5

1. The internal energy of a substance is equal to:

   A. the total potential energy stored in the bonds of a substance
   B. the potential and kinetic energy of molecules in a substance
   C. the energy stored in bonds and intermolecular forces of a substance
   D. the translational, rotational and vibrational motion of particles in the substance

2. The number of moles of sodium chloride (NaCl) in 100g of pure sodium chloride is
   (Mr of NaCl = 58.5 gmol-1)

   A. 5850 mol
   B. 0.585 mol
   C. 1.71 mol
   D. 41.5 mol

3. Two different objects with different temperatures are in thermal contact with one another. When the objects reach thermal equilibrium, the direction of transfer of thermal energy will be

   A. from the lower temperature object to the higher temperature object

B. half way between the temperatures of the two objects
C. from the higher temperature object to the lower temperature object
D. in many different directions

4. A sealed flask contains 16 g of oxygen (mass number 16) and also 8 g of hydrogen (mass number 1). The ratio of the number atoms of hydrogen to the number of atoms of oxygen is

   A. 16
   B. 8
   C. 4
   D. 2

5. The number of molecules present in 0.5 mol SO_3 is

   A. 3×10^{23}
   B. 6×10^{23}
   C. 12×10^{23}
   D. 24×10^{23}

6. The number of **atoms present in 0.5 mol SO_3** is

   A. 3×10^{23}
   B. 6×10^{23}
   C. 12×10^{23}
   D. 24×10^{23}

7. Calculate the approximate molar masses of each of the following molecules and compounds:

   (a) Cl_2
   (b) HCl
   (c) $CuSO_4$
   (d) Na_2CO_3
   (e) CH_4

8. Calculate the mass of the given amounts of each of the following substances:

   (a) 2.0 mole of iron, Fe
   (b) 0.2 mole of zinc, Zn
   (c) 2.5 mole of carbon dioxide, CO_2
   (d) 0.001 mole of sulfur dioxide, SO_2
   (e) 50 mole of benzene, C_6H_6

9. Calculate the amount of subtance (number of mole) in:

   (a) 100 g of copper
   (b) 5.0 g of oxygen molecules
   (c) 100 g of calcium carbonate, $CaCO_3$
   (d) 4.4 g of carbon dioxide
   (e) 13.88 g of lithium

10. A sample of aluminium sulfate $Al_2(SO_4)_3$ has a mass of 34.2 g. Calculate:

    (a) the number of alumimium ions Al^{3+} in the sample
    (b) the number of sulfate ions SO_4^{2-} in the sample

11. Classify the following as a macroscopic or microscopic property of a gas

    (a) volume
    (b) specific heat capacity
    (c) kinetic energy of a particle
    (d) pressure
    (e) temperature

Kinetic model of an ideal gas

An ideal gas is a theoretical gas that obeys the ideal gas equation exactly.

They obey the equation

PV = nRT

when there are no intermolecular forces between molecules at all pressures, volumes and temperatures.

Remember from Avogadro's hypothesis that one mole of any gas contains the Avogadro number of particles NA equal to 6.02×10^{23} particles. It also occupies 22.4 dm3 at 0 °C and 101.3 kPa pressure (STP).

The **internal energy of an ideal gas would be entirely kinetic energy** as there would be no intermolecular forces between the gaseous atoms. As temperature is related to the average kinetic energy of the atoms, the kinetic energy of the atoms would depend only on the temperature of the ideal gas.

From the combined gas laws, we determined that:

$$\frac{PV}{T} = k$$

or $PV = kT$

If the value of the universal gas constant is compared for different masses of different gases, it can be demonstrated that the constant depends not on the size of the atoms but rather on the number of particles present (the number of moles). Thus for n moles of any ideal gas:

$$\frac{PV}{nT} = R$$

or $PV = nRT$

105

CHAPTER 3

This is called the 'equation of state' of an ideal gas, where R is the universal gas constant and is equal to 8.31 J mol⁻¹ K⁻¹.

The equation of state of an ideal gas is determined from the gas laws and Avogadro's law.

n mol = N (total particles present) / N_A (Avogadro's number of particles)

Thus, the ideal gas equation becomes:

$$PV = \frac{N}{N_A} RT$$
$$= NkT$$

Where, $k = \frac{R}{N_A}$ = Boltzmann's constant = 1.38×10^{-23} J K⁻¹

Example 1

A weather balloon of volume 1.0 m³ contains helium at a pressure of 1.01×10^5 N m⁻² and a temperature of 35 °C. What is the mass of the helium in the balloon if one mole of helium has a mass of 4.003×10^{-3} kg?

Solution

Use the equation, $PV = nRT$, we have

$(1.01 \times 10^5 \, N \, m^{-2}) \times (1.0 \, m^3) = n \times (8.31 \, J \, mol^{-1} \, K^{-1}) \times (35 + 273 \, K)$

So that, $n = 39.46$ mol

Then, the mass of helium = $(39.46 \, mol) \times (4.003 \times 10^{-3} \, kg \, mol^{-1})$

= 0.158 kg.

The mass of helium in the balloon is 0.16 kg.

Example 2

An ideal gas has a density of 1.25 kg m⁻³ at STP. Determine the molar mass of the ideal gas.

Solution

Use the equation, $PV = nRT$, with V = molar mass/density. For 1 mole

$(1.01 \times 10^5 \, Nm^{-1}) \times \left(\frac{molar \, mass}{1.25 \, kg \, m^{-3}}\right) = 1 \times (8.31 \, J \, mol^{-1} \, K^{-1}) \times (0 + 273 \, K)$

Molar mass = $1 \times 8.31 \, J \, mol^{-1} \, K^{-1} \times 273 \, K \times \frac{1.25 \, kg \, m^{-3}}{1.01 \times 10^5 \, Nm^{-1}}$

$n = 39.46$ mol

= 28×10^{-3} kg mol⁻¹

The ideal gas is helium with a molar mas of 2.8×10^{-2} kg mol⁻¹.

Real gases conform to the gas laws under certain limited conditions but they can condense to liquids, and then solidify if the temperature is lowered. Furthermore, there are relatively small forces of attraction between particles of a real gas, and even this is not allowable for an ideal gas.

Most gases, at temperatures well above their boiling points and pressures that are not too high, behave like an ideal gas. In other words, real gases vary from ideal gas behaviour at high pressures and low temperatures.

When the moving particle theory is applied to gases it is generally called the kinetic theory of gases. The kinetic theory relates the macroscopic behaviour of an ideal gas to the behaviour of its molecules.

The assumptions or postulates of the moving particle theory are extended for an ideal gas to include:

- Gases consist of tiny particles called atoms (monatomic gases such as neon and argon) or molecules.
- The total number of molecules in any sample of a gas is extremely large.
- The molecules are in constant random motion.
- The range of the intermolecular forces is small compared to the average separation of the molecules.
- The size of the particles is relatively small compared with the distance between them.
- Collisions of short duration occur between molecules and the walls of the container and the collisions are perfectly elastic.
- No forces act between particles except when they collide, and hence particles move in straight lines.
- Between collisions the molecules obey Newton's Laws of motion.

Based on these postulates the view of an ideal gas is one of molecules moving in random straight-line paths at constant speeds until they collide with the sides of the container or with one another. Their paths over time are therefore zigzags. Because the gas molecules can move freely and are relatively far apart, they occupy the total volume of a container.

The large number of particles ensures that the number of particles moving in all directions is constant at any time.

THERMAL PHYSICS

Temperature is a measure of the average random kinetic energy of an ideal gas.

At the **microscopic** level, temperature is regarded as the measure of the **average kinetic energy per molecule** associated with its movements. For gases, it can be shown that the average kinetic energy is

$$\bar{E}_k = \frac{1}{2}m\bar{v}^2 = \frac{3}{2}kT \quad \text{where } k = \text{Boltzmann constant}$$

and is equal to 1.38 x 10-23 JK^{-1}

$$\therefore \bar{v}^2 \propto T$$

The term average kinetic energy is used because, at a particular temperature different particles have a wide range of velocities, especially when they are converted to a gas. This is to say that at any given temperature the average speed is definite but the velocities of particular molecules can change as a result of collision.

Figure 328 shows a series of graphs for the same gas at three different temperatures. In 1859 James Clerk Maxwell (1831-1879) and in 1861 Ludwig Boltzmann (1844-1906) developed the mathematics of the kinetic theory of gases. The curve is called a **Maxwell-Boltzmann speed distribution** and it is calculated using statistical mechanics. It shows the relationship between the relative number of particles N in a sample of gas and the speeds v that the particles have when the temperature is changed. ($T_3 > T_2 > T_1$)

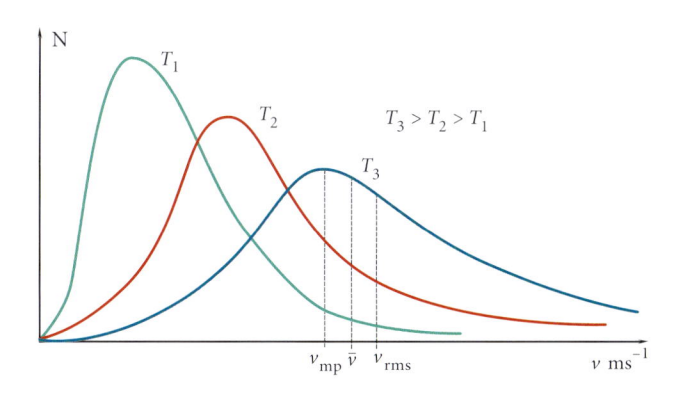

Figure 328 Maxwell-Boltzmann speed distribution for the same gas at different temperatures

The graphs do not show a normal distribution as the graphs are not bell-shaped. They are slightly skewed to the left. The minimum speed is zero at the left end of the graphs. At the right end they do not touch the x-axis because a small number of particles have very high speeds. The peak of each curve is at the most probable speed v_p a large number of particles in a sample of gas have their speeds in this region. When the mathematics of statistical mechanics is applied it is found that mean squared speed v_{av}^2 is higher than the most probable speed. Another quantity more often used is called the root mean square speed V_{rms} and it is equal to the square root of the mean squared speed.

$$v_{rms} = \sqrt{\bar{v}^2}$$

The root mean square is higher than the mean squared speed.

Other features of the graphs show that the higher the temperature, the more symmetric the curves becomes. The average speed of the particles increases and the peak is lowered and shifted to the right. The areas under the graphs only have significance when N is defined in a different way from above.

Figure 329 shows the distribution of the number of particles with a particular energy N against the kinetic energy of the particles E_k at a particular temperature. The shape of the kinetic energy distribution curve is similar to the speed distribution curve and the area under the curve gives the total energy of the gas.

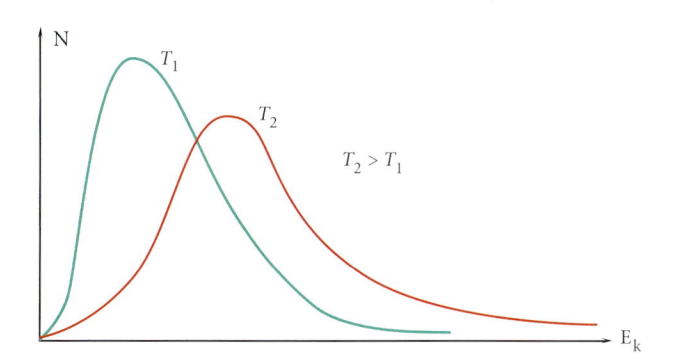

Figure 329 Distribution of kinetic energies for the same gas at different temperatures.

The average kinetic energy of the particles of all gases is the same. However, gases have different masses. Hydrogen molecules have about one-sixteenth the mass of oxygen molecules and therefore have higher speeds if the average kinetic energy of the hydrogen and the oxygen are the same.

Because the collisions are perfectly elastic there is no loss in kinetic energy as a result of the collisions.

107

Chapter 3

Mathematics of the kinetic theory

Consider a cube of length l. A particle of mass m is moving and has a component of velocity v_x. It collides with the side of the container, and rebounds with the same speed.

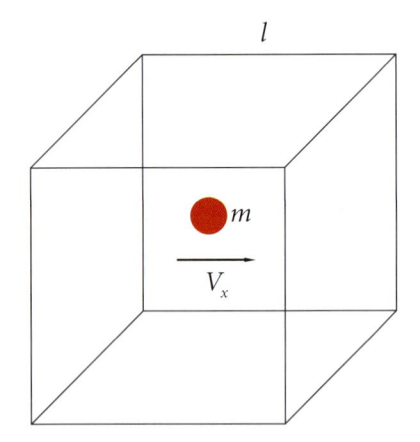

The change in momentum of the particle would be:

$$\Delta mv = -2 m v_x$$

The force of the wall on the particle would be:

$$F = \frac{\Delta p}{\Delta t} = \frac{-2 m v_x}{\Delta t}$$

From Newton's third law it follows that the force of the particle on the wall (F) is given by:

$$F = -F'$$

That is

$$F = +\frac{2m v_x}{\Delta t}$$

The molecule then travels a distance $2l$ before hitting the same end again.

Therefore, the average time between collisions $= \frac{2l}{v_x}$ and

$$\text{Average force} = \frac{\frac{2 m v_x}{2l}}{v_x} = \frac{m v_x^2}{l}$$

Therefore, the average pressure $= \frac{\text{average force}}{\text{area}} = \frac{\frac{m v_x^2}{l}}{A}$

$$= \frac{m v_x^2}{V}$$

Now if there are N particles with the same speed v. Their motion is random and an average of $\frac{1}{3}$ N particles would be moving in the x, y and z directions.

The pressure acting on the wall would be:

$$P = \frac{1}{3} N m \frac{v^2}{V}$$

$$PV = \frac{1}{3} N m v^2$$

In reality, the particles collide with each other, so they would not all have the same speed. Therefore we refer to the particles having an average speed squared $\bar{v}^2$.

Therefore:

$$PV = \frac{1}{3} Nm\bar{v}^2$$

This can also be written as:

$$PV = \frac{2}{3} N \times \frac{1}{2} m\bar{v}^2$$

Now ½ $m v^2$ is the average kinetic energy of a single particle. The product of the average kinetic energy and the number of particles N equals the total random kinetic energy, and this is equal to the internal energy U of the gas particles.

$$PV = \frac{2}{3} U$$

But

$$PV = N kT$$

Therefore:
where k = Boltzmann constant

Example

Avogadro's Law states that equal volumes of gases contain the same number of molecules provided they are at the same temperature and pressure. Show a proof to confirm Avogadro's Law.

Solution

$$PV = \frac{1}{3} N_1 m_1 \bar{v}_1$$

Since the pressures and volumes are the same:

$$\frac{1}{3} N_1 m_1 \bar{v}_1^2 = \frac{1}{3} N_2 m_2 \bar{v}_2^2$$

Now if the temperatures are the same the average kinetic energy of each gas will be the same:

$$\frac{1}{2} m\bar{v}^2 = \frac{1}{2} m_2 v_2^2$$

Therefore:
$$N_1 = N_2$$

Example

Calculate the average kinetic energy for an ideal gas at a temperature of 27 °C.

Solution

$E_k = \frac{1}{2}m\bar{v}^2 = \frac{3}{2}kRT$

$E_k = \frac{3}{2}kBT = \frac{3}{2}$ X and is equal to:

1.38 x 10^{-23} JK^{-1} X 300 K = **6.2 x 10^{-21} J**

Differences between real and ideal gases

An ideal gas is a theoretical gas that obeys the ideal gas equation exactly. Real gases conform to the gas laws under certain limited conditions but they can condense to liquids, then solidify if the temperature is lowered. Furthermore, there are relatively small forces of attraction between particles of a real gas, and even this is not allowable for an ideal gas.

A real gas obeys the gas laws at low pressure and at high temperature (above the temperature at which they liquefy), and we refer to such a gas as an ideal gas.
Most gases, at temperatures well above their boiling points and pressures that are not too high, behave like an ideal gas. In other words, real gases vary from ideal gas behaviour at high pressures and low temperatures.

Exercise 3.6

1. If the average translational kinetic energy EK at a temperature T of helium (molar mass 4 g mol^{-1}), then the average translational kinetic energy of neon (molar mass 20 g mol^{-1}) at the same temperature would be:

   A. 1/5 E_K C. √5 E_K
   B. 5 E_K D. E_K

2. A sample of gas is contained in a vessel at 20 °C at a pressure P. If the pressure of the gas is to be doubled and the volume remain constant, the gas has to be heated to:

   A. 40 °C C. 586 °C
   B. 293 °C D. 313 °C

3. Real gases behave most like ideal gases at

   A. low temperatures and high pressures
   B. high temperatures and low pressures
   C. low temperatures and low pressures
   D. high temperatures and high pressures

4. The Kelvin temperature of an ideal gas is a measure of:

   A. the average potential energy of the gas molecules
   B. the average speed of the gas molecules
   C. the average pressure of the gas molecules
   D. the average kinetic energy of the gas molecules

5. Two identical containers A and B contain an ideal gas under the different conditions as shown below.

Container A	Container A
N Moles	3N Moles
Temperature T	Temperature T/3
Pressure P_A	Pressure P_B

   The ratio $P_A : P_B$ would be:

   A. 3 C. 1
   B. 2 D. 2/3

6. The internal volume of a gas cylinder is 3.0 × 10^{-2} m^3. An ideal gas is pumped into the cylinder until the pressure is 15 MPa at a temperature of 25 °C.

   (a) Determine the number of moles of the gas in the cylinder
   (b) Determine the number of gas atoms in the cylinder
   (c) Determine the average volume occupied by one atom of the gas
   (d) Estimate the average separation of the gas atoms

7. A cylinder of an ideal gas with a volume of 0.2 m^3 and a temperature of 25 °C contains 1.202 × 10^{24} molecules. Determine the pressure in the cylinder.

8. (a) State what is meant by the term *ideal gas*.
   (b) In terms of the kinetic theory of gases, state what is meant by an *ideal gas*.
   (c) Explain why the internal energy of an ideal gas is kinetic energy only.

THERMAL PHYSICS

109

Chapter 3

This page is intentionally left blank

4. Waves

Contents

4.1 – Oscillations

4.2 – Travelling waves

4.3 – Wave characteristics

4.4 – Wave behaviour

4.5 – Standing waves

Essential Ideas

A study of oscillations underpins many areas of physics with simple harmonic motion (shm) a fundamental oscillation that appears in various natural phenomena.

There are many forms of waves available to be studied. A common characteristic of all travelling waves is that they carry energy, but generally the medium through which they travel will not be permanently disturbed.

Waves interact with media and each other in a number of ways that can be unexpected and useful.

When travelling waves meet they can superpose to form standing waves in which energy may not be transferred. © IBO 2014

Chapter 4

4.1 Oscillations

NATURE OF SCIENCE:

Models: Oscillations play a great part in our lives, from the tides to the motion of the swinging pendulum that once governed our perception of time. General principles govern this area of physics, from water waves in the deep ocean or the oscillations of a car suspension system. This introduction to the topic reminds us that not all oscillations are isochronous. However, the simple harmonic oscillator is of great importance to physicists because all periodic oscillations can be described through the mathematics of simple harmonic motion. (1.10)

© IBO 2014

Essential idea: A study of oscillations underpins many areas of physics with simple harmonic motion (shm) a fundamental oscillation that appears in various natural phenomena.

Understandings:
- Simple harmonic oscillations
- Time period, frequency, amplitude, displacement and phase difference
- Conditions for simple harmonic motion

TOK THE NATURE OF LIGHT

In the 17th century there were two opposing views expressed regarding the nature of light by *Sir Isaac Newton* (1642-1727), and *Christian Huygens* (1629-1695). Newton was of the opinion that light consisted of particles and Huygens was of the opinion that light consisted of waves. They both believed, however, that a medium – the ether – was necessary for the propagation of light.

In Galileo's book "Two New Sciences" published in 1638, he pointed out that the flash from an artillery gun was seen before the sound of the blast was heard. He concluded that the flash of light appeared instantaneously. However, he stated that we would not know whether it was instantaneous unless there was some accurate way to measure its speed. He suggested an experiment could be performed where a person with a lantern stood on one hill and an observer stood on another hill, a known distance away. By flashing the lantern on one hill, the observer on the distant hill could time how long it took the flash to reach him. This was highly unlikely to produce a result, and there is only anecdotal evidence to suggest that the experiment was actually performed.

In 1676, the Danish astronomer *Ole Römer* made the first recognised prediction of the speed of light. Through observation of Jupiter's moon Io, he noticed that the period between eclipses (when Io disappears behind Jupiter) of Io is about 42.5 hours. However, he observed that there was an irregularity in the times between successive eclipse periods as the Earth orbits the Sun. He reasoned that because of the Earth's motion about the Sun, it spends about six months of the year moving away from Jupiter, and the remaining time moving towards it. Therefore, if the eclipsing at the close distance and the far distance was compared, he estimated that the extra time for the light to reach Earth in the extreme was about 22 minutes. A short time later, the Dutch physicist, *Christian Huygens*, used Römer's value of 22 minutes and his estimate of the Earth's orbit to obtain a value equivalent in SI units of 2×10^8 m s^{-1} – about two-thirds the presently accepted value (we now know that the actual value is 16 minutes).

In 1849, the French physicist, *A.H.L. Fizeau* made the first non-astronomical measurement of the **speed of light**. He realised that the use of a rotating toothed wheel would make it possible to measure very short time intervals. He placed a light source on a hill in Paris and a rotating toothed wheel on another hill 8.63 km away as shown in Figure 1801.

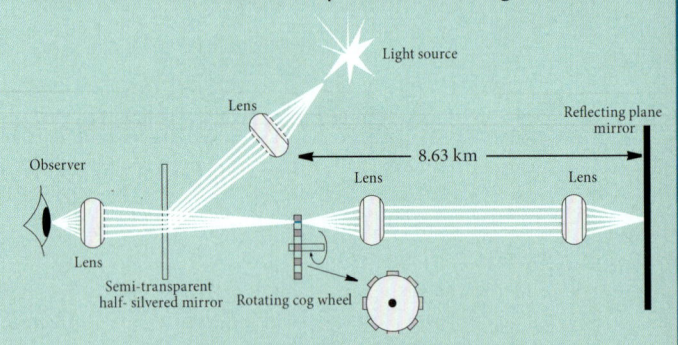

Figure 401   Fizeau's apparatus to determine the speed of light

Using a system of lenses and a half-silvered mirror, he was able to focus a beam of light onto a gap in the toothed wheel placed on one hill. The beam then travelled to the other hill where it was reflected off a plane mirror and returned to the rotating wheel. The light intensity of the returning light then passed through the transparent part of the half-silvered mirror where it was observed. At low speeds of rotation of the toothed wheel, he found that little light was visible because the path of the reflected light was obstructed by the teeth of the rotating wheel. He then increased the rotation of the wheel until the reflected light was visible. He reasoned that this would occur when the rotation speed was such that the reflected light passed through the next gap between teeth. The wheel had 720 teeth and the light was bright when the wheel rotated at 25.2 revolutions per second.

In 1860, *Jean Foucalt* improved Fizeau's method by replacing the toothed wheel with an eight-sided rotating mirror. He used the improved apparatus to measure the speed of light through air and water, and discovered that the speed of light in water was less than in air.

In the 1860s, *James Clerk Maxwell* (1831-1879) put Faraday's theory of electricity and magnetism into a mathematical form that could be generalised and applied to electric and magnetic fields in conductors, insulators and space free of matter. According to **Maxwell's theory**, a changing electric field in space produces a magnetic field, and a changing magnetic field in space produces an electric field. He predicted that by mutual induction, a sequence of time-varying and space-changing electric and magnetic fields would be propagated from an original source such as an oscillating charge. He proposed that these fluctuating, interlocked electric and magnetic fields would propagate through space in the form of an electromagnetic wave. He still believed that the ether was the medium for propagation. Following on from other earlier attempts to determine the speed of light, Maxwell proposed that the speed of electromagnetic waves was about 311 000 000 m s^{-1}.

In 1887 *Heinrich Hertz* (1857-1894) demonstrated experimental evidence for the existence of electromagnetic waves as proposed by Maxwell in 1865. Hertz's apparatus is shown schematically in Figure 402.

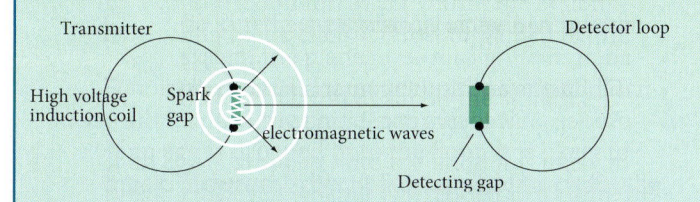

Figure 402   Hertz's apparatus

The induction coil produced a large potential difference between the gap in the source loop, and sparks were produced. When the detector loop was brought near the source loop, sparks were also noticed jumping across the air gap in the detector loop. Hertz hypothesised that the sparks in the source loop set up changing electric and magnetic fields that propagated as an electromagnetic wave, as postulated by Maxwell. These waves then aligned in the air gap of the detector loop setting up electric and magnetic fields. These induced a spark in the air gap.

Between the years of 1880 to 1930, the American physicist, *Albert. A. Michelson*, made a series of precise measurements to determine the speed of light. Using a rotating octagonal steel prism, he constructed the apparatus is shown in Figure 403.

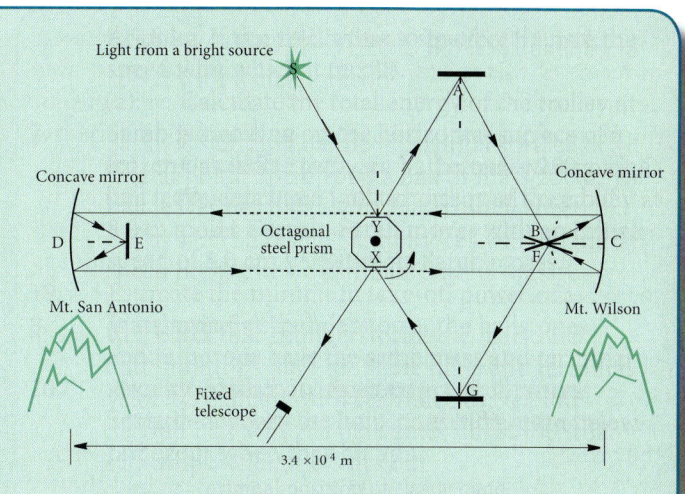

Figure 403   Michelson's method for determining the speed of light

His apparatus consisted of the main "set-up" on the right of Figure 403, placed on Mount Wilson in the USA, and a concave mirror to the left of the diagram placed on Mount San Antonio 3.4 × 10^4 m away.

When the octagonal steel prism is stationary, the light from the source S follows the path SYABCDEDCFGX, and an image can be seen through the fixed telescope. When the rotating prism is rotated slowly, the image disappears because the faces of the prism are not in a suitable position. However, if the prism is rotated so that it turns through one- eighth of a revolution in the time it takes light to travel from X to Y, then an image is seen in the fixed telescope. Michelson found that this occurred when the prism was rotated at 530 revolutions per second.

Michelson then corrected for the fact that the light was travelling through air rather than a vacuum, and obtained a value that is today considered to be accurate to one part in a hundred thousand.

Another method to determine the speed of light involves the use of the electrical constants ε_0 and μ_0 rather than the speed of light directly. According to Maxwell's electromagnetic theory, the speed of light is given by

$$c = \frac{1}{\sqrt{\varepsilon_0 \mu_0}}$$

where ε_0 is the permittivity of free space and is equal to 8.85 × 10^{-12} C^2 N^{-1} m^{-2}, and μ_0 is the permeability of free space and is equal to 4.25 × 10^{-7} T m A^{-1}.

When we substitute these values into the equation, the value obtained for the speed of light is 299 863 380.5 m s^{-1}. With modern electronic devices being used physicists are continuing to improve their methods of measurement. The most accurate measurements indicate that the speed of light in a vacuum is 299 792 458 m s^{-1} with an uncertainty of 1 m s^{-1}.

Chapter 4

The standard unit of length, the metre, is now defined in terms of this speed. In 1960, the standard metre was defined as the length equal to 1 650 763.73 wavelengths of the orange-red line of the isotope krypton-86 undergoing electrical discharge. Since 1983, the metre has been defined as "the metre is the length of path travelled by light in a vacuum during a time interval of 1 / 299 792 458 s". The speed of all electromagnetic waves is the same as the speed of light.

Further investigation showed that the electromagnetic waves exhibit the properties of reflection, refraction, interference, diffraction and polarisation. Furthermore, they travelled at the speed of light. Since these properties are also exhibited by light, Hertz's experiment had shown that light is a form of a transverse electromagnetic wave.

In 1905, *Albert Einstein* in his special theory of relativity proposed that the ether did not exist, and that the speed of light is constant in all frames of reference.

Introduction

What we learn in this chapter about oscillations in mechanical systems and of the waves that oscillating systems may set up, forms the basis for gaining an understanding of many other areas of physics.

A study of oscillations is important for many reasons, not least, safety in design. For example, oscillations may be instigated in a bridge as traffic passes over it. In a worse case scenario these oscillations can lead to structural damage in the bridge. Many types of machines (lathes, car engines etc.) are also subject to oscillations and again, depending on the nature of the oscillations, machines can be damaged.

Of course oscillating systems may also be very useful. The oscillations of a simple pendulum may be used as an accurate timing device and the oscillations set up in a quartz crystal may be used as an even more accurate timing device. In fact, oscillations are used to define the international time systems so that the world can be kept in synchronization.

If an oscillating body causes other particles with which it is in contact to oscillate, then the energy of the oscillating body may be propagated as a wave. An oscillating tuning fork, vibrating string and vibrating reed cause the air molecules with which they are in contact to oscillate thereby giving rise to a sound wave that we may hear as a musical note. As we shall see, oscillating systems and waves are intimately connected.

On a fundamental level, all atoms and molecules are in effect oscillating systems. An understanding of these oscillations is crucial to understanding both the microscopic and macroscopic properties of a substance. For example, the dependence of specific heat capacity on temperature (a topic well beyond the scope of an IB Physics Course) arises from studying atomic oscillations. Also, by analysing the oscillations of atoms and molecules, we gain an understanding of the interaction between matter and radiation. For example, the Greenhouse Effect is essentially due to the interaction between infrared radiation and gases such as carbon dioxide.

It must also be mentioned that the oscillations of electrically charged particles give rise to electromagnetic waves (light, radio waves, X-rays etc). This is examined in more detail in several other places in both the Core and AHL material.

We will look first at the oscillations in mechanical systems.

Simple harmonic oscillations

Harmonic motion is the motion that repeats the same cycle exactly in a given period of time. An example of simple harmonic motion would be a mass oscillating periodically on a spring.

In the introduction, we mentioned several oscillating systems. Here are a few more:

- a boat at anchor at sea
- the human vocal chords
- an oscillating cantilever
- the Earth's atmosphere after a large explosion
- models of icebergs oscillating vertically in the ocean.

Time period, frequency, amplitude, displacement and phase difference

In order to understand the terms associated with oscillating systems, let us look at the oscillation of a so-called simple pendulum. A simple pendulum consists of a weight (called the bob) that is suspended vertically by a light string or thread as shown in Figure 404. The thread from the point P suspends the bob X.

WAVES

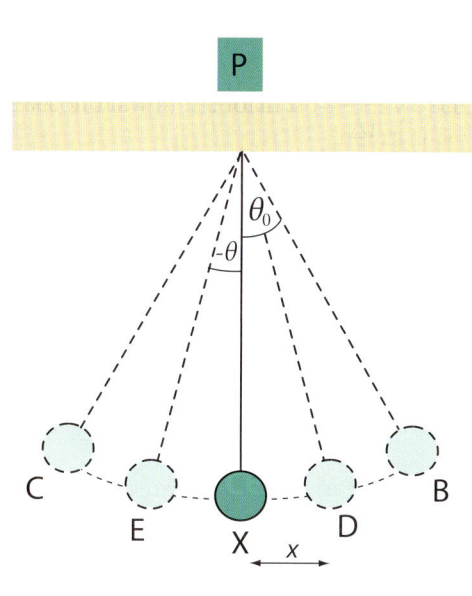

Figure 404 *A simple oscillating system*

To set the pendulum oscillating, the bob is pulled up to a position such as B where the angle XPB is θ_0. It is then released. The bob will now oscillate between the positions B and C.

Displacement (x, θ)

This refers to the distance that an oscillating system is from its equilibrium position at any particular instant during the oscillation. In the case of the simple pendulum the displacement is best measured as an angular displacement. For example, in Figure 404 when the bob is at the position D, the displacement is the angle XPD = θ and when at E, is the angle XPE = $-\theta$. The displacement when the bob is at X is $\theta = 0$.

Amplitude (x_0, θ_0)

This is the maximum displacement of an oscillating system from its equilibrium position. For the simple pendulum in Figure 404 this is clearly θ_0.

Period (*T*)

This is the time it takes an oscillating system to make one complete oscillation. For the simple pendulum in Figure 404, this is the time it takes to go from X to B, B to C and then back to X.

Frequency (*f*)

This is the number of complete oscillations made by the system in one second.

Relation between frequency and period

The time for one complete oscillation is the period *T*. Therefore the number of oscillations made in one second is $\frac{1}{T}$. The number of oscillations made in one second is also defined as the frequency *f*, hence:

$$f = \frac{1}{T}$$ **Equation 4.1**

Phase difference

Suppose we have for instance, two identical pendulums oscillating next to each other. If the displacements of the pendulums are the same at all instances of time, then we say that they are oscillating in phase. If on the other hand the maximum displacement of one of them is θ_0 when the maximum displacement of the other is $-\theta_0$, then we say that they are oscillating in anti-phase or that the **phase difference** between them is 180°. The reason for the specification in terms of angle will become clear in section 4.1.5. In general, the phase difference between two identical systems oscillating with the same frequency can have any value between 0 and 360° (or 0 to 2π radians). We shall see that the concept of phase difference is very important when discussing certain aspects of wave motion.

Radian measure

When dealing with angular displacements, it is often useful to measure the displacement in radians rather than in degrees. In Figure 405, the angle θ measured in radians is defined as the arc length AB (*l*) divided by the radius *r* of the circle i.e

$$\theta(\text{rad}) = \frac{l}{r}$$ **Equation 4.2**

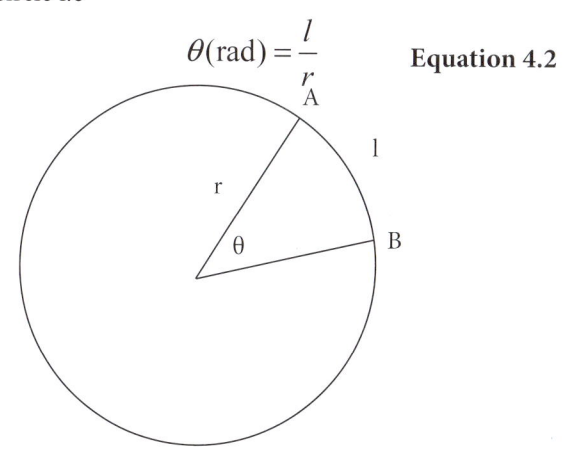

Figure 405 *The radian*

If $\theta = 180°$ then *l* is equal to half the circumference of the circle i.e. $l = \pi r$. Hence from equation 4.2, we have that

$$\theta(180^0) = \frac{\pi r}{r} = \pi \text{ (rad)}$$

Hence 1 radian (rad) is equal to 57.3°.

115

CHAPTER 4

Conditions for simple harmonic motion

Description of SHM

Suppose we were to attach a fine marking pen to the bottom of the bob of a simple pendulum and arrange for this pen to be in contact with a long sheet of white paper as shown in Figure 406.

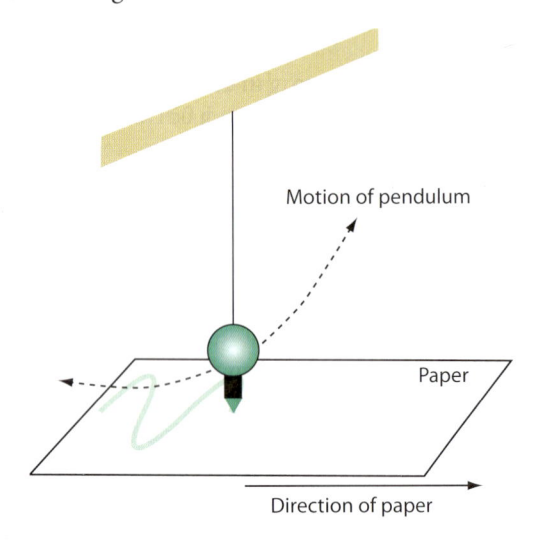

Figure 406　Arrangement for demonstrating SHM

As the pendulum oscillates, the paper is pulled at a constant speed in the direction shown. Figure 407 shows a particular example of what is traced on the paper by the marker pen.

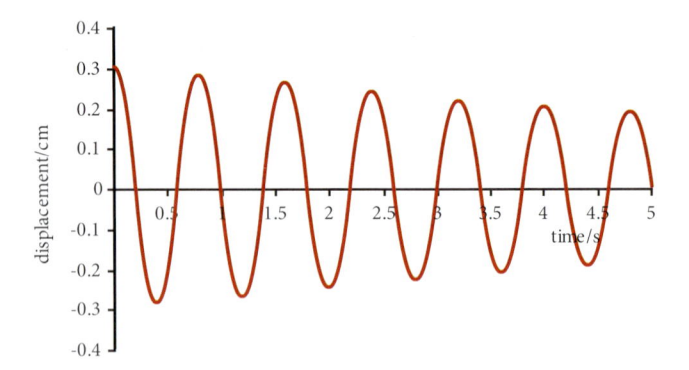

Figure 407　A sample trace

The displacement is measured directly from the trace and the time is calculated from the speed with which the paper is pulled.

There are several things to notice about the trace.

1. One complete oscillation is similar to a sine or cosine graph.

2. The period stays constant at about 0.8 s. (Oscillations in which the period is constant are called *isochronous*.)

3. The amplitude is decaying with time. This is because the pendulum is losing energy to the surroundings due to friction at the point of suspension and to air resistance.

Based on this sort of time trace, we can define a special type of oscillatory motion. Oscillators that are perfectly isochronous and whose amplitude does not change with time are called simple harmonic oscillators and their motion is referred to as **simple harmonic motion** (SHM). Clearly SHM does not exist in the real world as the oscillations of any vibrating system will eventually die out. Interestingly enough, the simple pendulum does not perform SHM for yet another reason, namely that the period is actually dependant on the amplitude. However, this only becomes noticeable as θ_0 exceeds about 40°.

Although SHM does not exist in the real world, many oscillatory systems approximate to this motion. Furthermore, as part of the scientific method, it makes good sense to analyse a simple situation before moving onto more complex situations.

Angular frequency (Angular Velocity) (ω)

A very useful quantity associated with oscillatory motion is angular frequency, ω. This is defined in terms of the linear frequency as

$$\omega = 2\pi f \qquad \text{Equation 4.3}$$

Using equation 4.1 we also have that

$$\omega = \frac{2\pi}{T} \qquad \text{Equation 4.4}$$

There is a connection between angular frequency and the angular speed of a particle moving in a circle with constant speed. The angular speed of the particle is defined as the number of radians through which the particle moves in one second. If the time for one complete revolution of the circle is T, then from equation 4.2 we have that

$$T = \frac{2\pi}{\omega} \text{ or } \omega = \frac{2\pi}{T}$$

There is actually a physical connection between angular speed and SHM in the respect that it can be shown that the projection of the particle onto any diameter moves with SHM. See Figure 408.

WAVES

Figure 408 Circular and harmonic motion

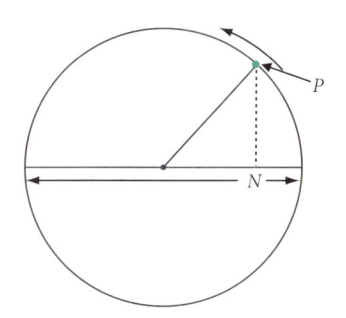

As the particle *P* moves round the circle, its projection *N* onto a diameter moves backwards and forwards along the diameter with SHM

Definition of SHM

If it were possible to remove all frictional forces acting on an oscillating pendulum, then the displacement – time graph for the motion would look like that in Figure 409. The amplitude does not decay with time. This therefore is a displacement-time graph for SHM

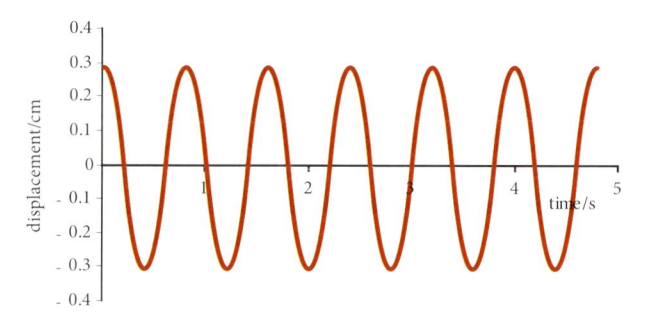

Figure 409 A displacement – time graph

Figure 410 shows the relationship between displacement, velocity and acceleration versus time for a certain harmonic oscillator.

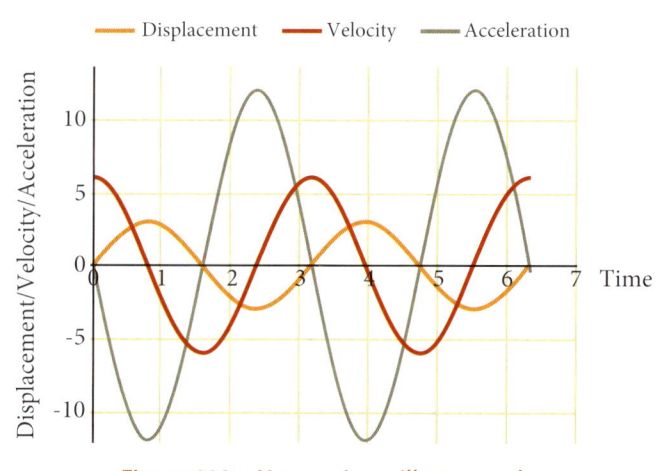

Figure 410 Harmonic oscillator graphs.

The three graphs depict trigonometric functions. If we examine the graphs from a qualitative point of view, one can see that when the displacement is a maximum, the velocity is zero and the acceleration is a maximum in the negative direction. Similarly, when the velocity is a maximum, the displacement from the equilibrium position is zero and the acceleration is zero.

It turns out that if the acceleration a of a system is:

- directly proportional to its displacement x from its equilibrium position

and

- directed towards the equilibrium position

then the system will execute SHM.

This is the formal definition of SHM.

We can express this definition mathematically as

$$a = -\text{const} \times x \qquad \textbf{Equation 4.5}$$

The negative sign indicates that the acceleration is directed towards the equilibrium position. Mathematical analysis shows that the constant is in fact equal to ω^2 where ω is the angular frequency (defined above) of the system. Hence equation 4.5 becomes

$$a = -\omega^2 x \qquad \textbf{Equation 4.6}$$

Figure 411 shows the relationship that exists between the acceleration and the displacement from the equilibrium position for a harmonic oscillator.

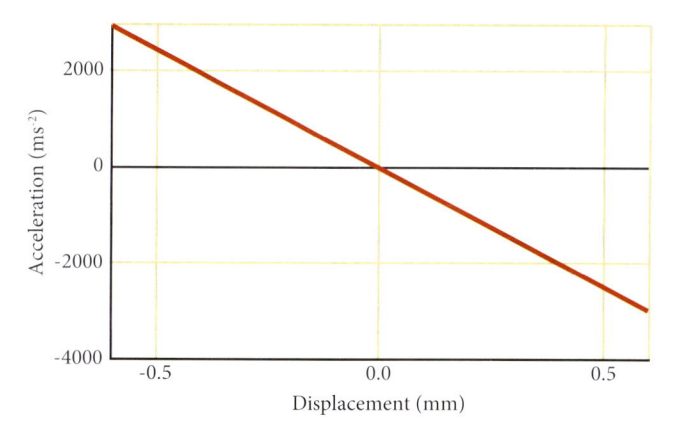

Figure 411 Acceleration versus displacement for a harmonic oscillator

Using the graph, one can determine the gradient of the graph is negative and equal to ω2.

If a system is performing SHM, then to produce the acceleration, a force must be acting on the system in the direction of the acceleration. From our definition of SHM, the magnitude of the force *F* is given by

$$F = -kx \qquad \textbf{Equation 4.7}$$

117

CHAPTER 4

where k is a constant and the negative sign indicates that the force is directed towards the equilibrium position of the system. (Do not confuse this constant k with the spring constant. However, when dealing with the oscillations of a mass on the end of a spring, k will be the spring constant.)

Solving problems involving $a = -\omega^2 x$

To understand how the equation $a = -\omega^2 x$ is used in particular situations involving SHM let us look at an example.

Example

1. A cylindrical piece of wood floats upright in water as shown in Figure 412.

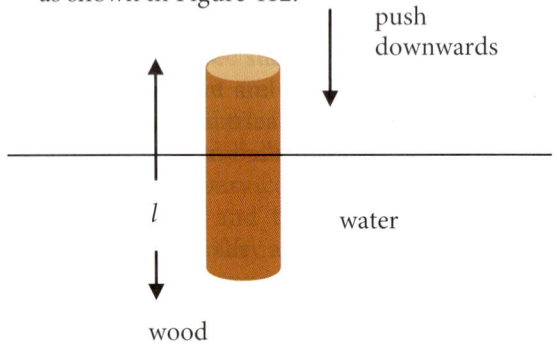

Figure 412 SHM of a floating piece of wood

The wood is pushed downwards and then released. The subsequent acceleration a of the wood is given by the expression

$$a = -\frac{\rho g}{\sigma l} x$$

where ρ = density of water, σ = density of the wood, l = length of wood, g = acceleration of free fall and x = displacement of the wood from its equilibrium position.

(a) Explain why the wood executes SHM.

Answer

The equation shows that the acceleration of the wood is proportional to its displacement from equilibrium and directed towards the equilibrium position.

(b) The length of the wood is 52 cm and it is pushed downwards a distance of 24 cm. Calculate the maximum acceleration of the wood.
($\rho = 1.0 \times 10^3$ kg m^{-3}, $\sigma = 8.4 \times 10^2$ kg m^{-3}, $g = 9.8$ m s^{-2}).

Answer

Maximum acceleration is when $x = A$ (the amplitude)

$= 2.4 \times 10^{-2}$ m

therefore maximum acceleration

$$= \frac{1.0 \times 10^3 \times 9.8 \times 24 \times 10^{-2}}{8.4 \times 10^2 \times 52 \times 10^{-2}}$$

$= 5.4$ m s^{-2}

(c) Calculate the period T of oscillation of the wood.

Answer

$$\omega^2 = \frac{a_{max}}{A} = \frac{5.4}{24 \times 10^{-2}} = 23 \text{ rad}^2$$

$\omega = 4.8$ rad

$T = \frac{2\pi}{\omega} = 1.3$ s

(d) State and explain in terms of the period T of oscillation of the wood, the first two instances when the acceleration is a maximum.

Answer

The amplitude will be a maximum when $t = 0$ and again when

$$t = \frac{T}{2}$$

So acceleration is a maximum at $t = 0$ and $t = 0.65$ s

Examples

The graph in Figure 413 shows the variation with time t of the displacement x of a system executing SHM.

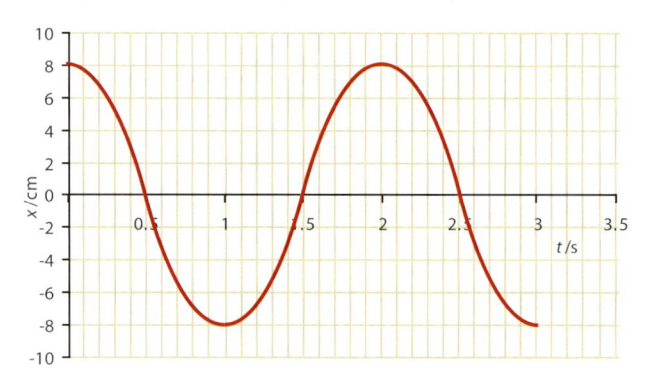

Figure 413 Displacement – time graph for SHM

118

Use the graph to determine the

(i) period of oscillation
(ii) amplitude of oscillation
(iii) maximum acceleration

Solutions

(i) 2.0 s (time for one cycle)
(ii) 8.0 cm
(iii) Using $a_{max} = \omega^2 x_0 = \pi^2 \times 8.0 = 79$ m s^{-2}

1. (a) Answer the same questions (a)(i) to (a)(iii) in the above example for the system oscillating with **SHM** as described by the graph in Figure 414 below.

   (b) Also state two values of t for when the magnitude of the velocity is a maximum and two values of t for when the magnitude of the acceleration is a maximum.

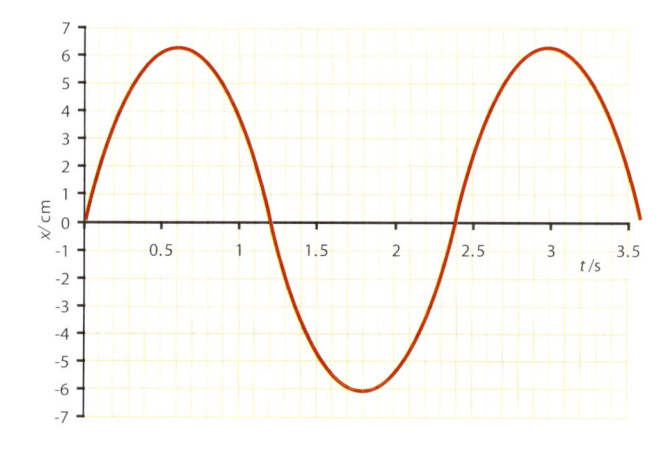

Figure 414

Kinetic and potential energy changes

We must now look at the energy changes involved in SHM. To do so, we will again concentrate on the harmonic oscillator. The mass is stationary at $x = +x_0$ (maximum extension) and also at $x = -x_0$ (maximum compression). At these two positions the energy of the system is all potential energy and is in fact the elastic potential energy stored in the spring. This is the total energy of the system E_T and clearly

$$E_T = \frac{1}{2}kx_0^2 \qquad \text{Equation 8}$$

That is, that for any system performing SHM, the energy of the system is proportional to the square of the amplitude. This is an important result and one that we shall return to when we discuss wave motion.

At $x = 0$ the spring is at its equilibrium extension and the magnitude of the velocity v of the oscillating mass is a maximum v_0. The energy is all kinetic and again is equal to E_T. We can see that this is indeed the case as the expression for the maximum kinetic energy E_{max} in terms of v_0 is

$$E_{max} = \frac{1}{2}mv_0^2 \qquad \text{Equation 9}$$

Clearly E_T and E_{max} are equal such that

$$E_T = \frac{1}{2}kx_0^2 = E_{max} = \frac{1}{2}mv_0^2$$

From which

$$v_0^2 = \frac{k}{m}x_0^2 \quad (\text{as } v_0 \geq 0)$$

Therefore

$$v_0 = \sqrt{\frac{k}{m}}\, x_0 = \omega x_0$$

which ties in with the velocity being equal to the gradient of the displacement-time graph (see 4.1.5). As the system oscillates there is a continual interchange between kinetic energy and potential energy such that the loss in kinetic energy equals the gain in potential energy and $E_T = E_K + E_P$.

The sketch graph in Figure 415 shows the variation with displacement x of E_K and E_P for one period.

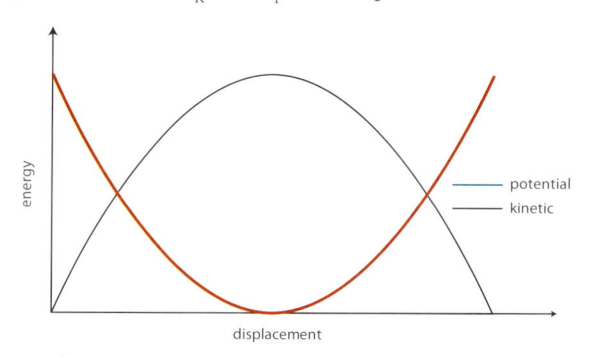

Figure 415 Energy and displacement

Let us now look at solving a problem involving energy in SHM

Example

The amplitude of oscillation of a mass suspended by a vertical spring is 8.0 cm. The spring constant of the spring is 74 N m^{-1}. Determine

(a) the total energy of the oscillator

(b) the potential and the kinetic energy of the oscillator at a displacement of 4.8 cm from equilibrium.

119

CHAPTER 4

Solution

(a) Spring constant $k = 74$ N m^{-1} and

$x_0 = 8.0 \times 10^{-2}$ m

$E_T = \frac{1}{2} kx_0^2$

$= \frac{1}{2} \times 74 \times 64 \times 10^{-4} = 0.24$ J

(b) At $x = 4.8$ cm

Therefore $E_p = \frac{1}{2} \times 74 \times (4.8 \times 10^{-2})^2$

$= 0.085 \times 10^{-4}$ J

$E_K = E_T - E_p = 0.24 - 0.085 = 0.16$ J

Exercise 4.2

In a simple atomic model of a solid, the atoms vibrate with a frequency of 2.0×10^{11} Hz. The amplitude of vibration of the atoms is 5.5×10^{-10} m and the mass of each atom is 4.8×10^{-26} kg. Calculate the total energy of the oscillations of an atom.

Damping

In this section, we look at oscillations of real systems. In section 4.1.3, we described an arrangement by which the oscillations of a pendulum could be transcribed onto paper. Refer to Figure 416.

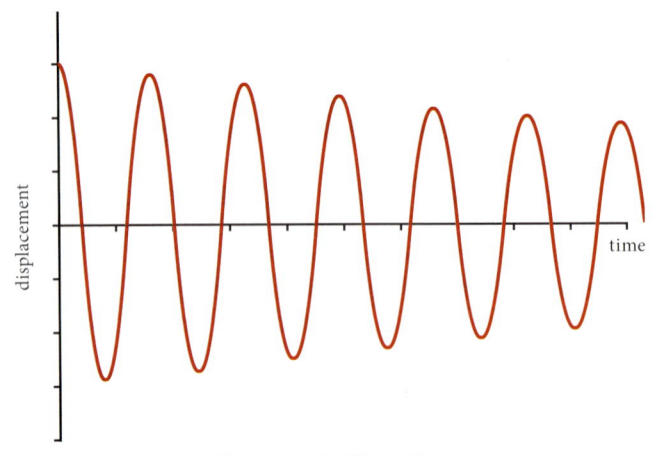

Figure 416 Damping

The amplitude of the oscillations gradually decreases with time, whereas for SHM, the amplitude stays at the same value forever. Clearly, the pendulum is losing energy as it oscillates. The reason for this is that dissipative forces are acting that oppose the motion of the pendulum. As mentioned earlier, these forces arise from air resistance and though friction at the support. Oscillations, for which the amplitude decreases with time, are called **damped oscillations.**

Examples of damped oscillations

All oscillating systems are subject to damping as it is impossible to completely remove friction. Because of this, oscillating systems are often classified by the degree of damping. The oscillations shown in Figure 416 relatively slow and the pendulum will make quite a few oscillations before finally coming to rest. Whereas the amplitude of the oscillations shown in Fig 417 decay very rapidly and the system quickly comes to rest. Such oscillations are said to be **heavily damped**.

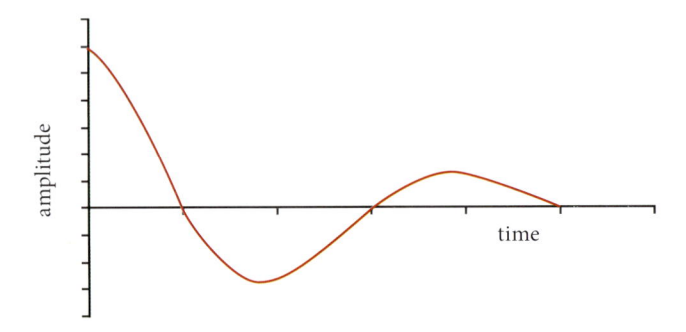

Figure 416   Heavily damped oscillations

Consider a harmonic oscillator in which the mass is pulled down and when released, and the mass comes to rest at its equilibrium position without oscillating. The friction forces acting are such that they prevent oscillations. However, suppose a very small reduction in the friction forces would result in heavily damped oscillation of the oscillator, then the oscillator is said to be **critically damped.**

The graph in Figure 417 shows this special case of damping known as **critical damping.**

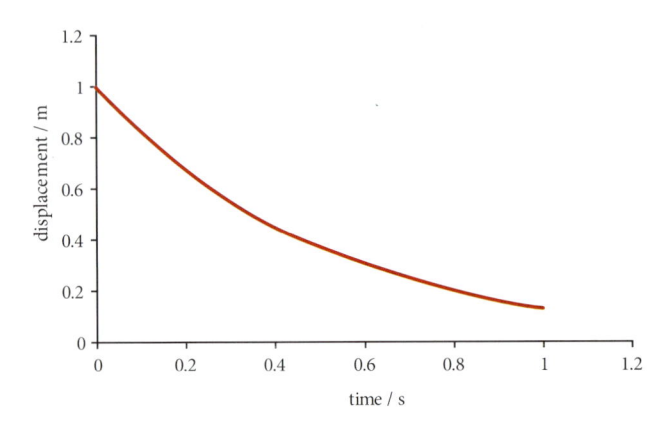

Figure 417   Critical damping

120

Although not in the IB syllabus, a useful way of classifying oscillating systems, is by a quantity known as the **quality factor** or *Q*-factor. The *Q*-factor does have a formal definition but it is approximately equal in value to the number of oscillations that occur before all the energy of the oscillator is dissipated. For example, a simple pendulum has a *Q*-factor of about 1000.

As mentioned in the introduction to this chapter, the oscillations (vibrations) made by certain oscillatory systems can produce undesirable and sometimes, dangerous effects. Critical damping plays an important role in these situations. For example, when a ball strikes the strings of a tennis racquet, it sets the racquet vibrating and these vibrations will cause the player to lose some control over his or her shot. For this reason, some players fix a "damper" to the springs. If placed on the strings in the correct position, this has the effect of producing critically damped oscillations and as a result the struck tennis racquet moves smoothly back to equilibrium. The same effect can be achieved by making sure that the ball strikes the strings at a point known as the 'sweet spot' of which there are two, one of which is known as the 'centre of percussion (COP)'. Cricket and baseball bats likewise have two sweet spots.

Another example is one that involves vibrations that may be set up in buildings when there is an earthquake. For this reason, in regions prone to earthquakes, the foundations of some buildings are fitted with damping mechanisms. These mechanisms insure any oscillations set up in the building are critically damped.

Exercise 4.3

Identify which of the following oscillatory systems are likely to be lightly damped and which are likely to be heavily damped.

1. Atoms in a solid
2. Car suspension
3. Guitar string
4. Harmonic oscillator under water
5. Quartz crystal
6. A cantilever that is not firmly clamped
7. Oil in a U-tube
8. Water in a U-tube

Natural frequency and forced oscillations

Consider a small child sitting on a swing. If you give the swing a single push, the swing will oscillate. With no further pushes, that is energy input, the oscillations of the swing will die out and the swing will eventually come to rest. This is an example of damped harmonic motion. The frequency of oscillation of the swing under these conditions is called the **natural frequency** of oscillation (vibration). So far in this topic, all the systems we have looked at have been systems oscillating at their natural frequency.

Suppose now when each time the swing returns to you, you give it another push. See Figure 418

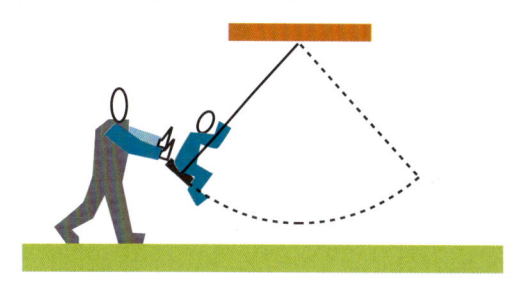

Figure 418   A forced oscillation

The amplitude of the swing will get larger and larger and if you are not careful your little brother or sister, or who ever the small child might be, will end up looping the loop.

The frequency with which you push the swing is exactly equal to the natural frequency of oscillation of the swing and importantly, is also in phase with the oscillations of the spring. Since you are actually forcing the swing to oscillate, the swing is said to be undergoing **forced oscillations.** In this situation the frequency of the so-called *driver* (in this case, you) is equal to the natural frequency of oscillation of the system that is being driven (in this case, the swing). If you just push the swing occasionally when it returns to you, then the swing is being forced at a different frequency to its natural frequency. In general, the variation of the amplitude of the oscillations of a driven system with time will depend on the

- frequency of the driving force
- frequency of natural oscillations
- amplitude of the driving force
- phase difference between driving force frequency and natural frequency
- amount of damping on the system

(There are many very good computer simulations available that enable you to explore the relation between forced and natural oscillations in detail.)

CHAPTER 4

Resonance

We have seen that when an **oscillatory system is driven at a frequency equal to its natural frequency**, the amplitude of oscillation is a maximum. This phenomenon is known as **resonance**. The frequency at which resonance occurs is often referred to as the resonant frequency.

In the introduction to this topic, we referred to resonance phenomena without actually mentioning the term resonance. For example we can now understand why oscillations in machinery can be destructive. If a piece of machinery has a natural frequency of oscillation and moving parts in the machine act as a driver of forced oscillations and have a frequency of oscillation equal to the natural frequency of oscillation of the machine, then the amplitude of vibration set up in the machine could be sufficient to cause damage.

Similarly, if a car is driven along a bumpy road, it is possible that the frequency, with which the car crosses the bumps, will just equal the natural frequency of oscillation of the chassis of the car. If this is the case then the result can be very uncomfortable.

We now also see why it is important that systems such as machines, car suspensions, suspension bridges and tall buildings are critically or heavily damped.

Resonance can also be very useful. For example, the current in a particular type of electrical circuit oscillates. However, the oscillating current quickly dies out because of resistance in the circuit. Such circuits have a resonant frequency and if driven by an alternating current supply, the amplitude of the current may become very large, particularly if the resistance of the circuit is small. Circuits such as this are referred to as *resonant circuits*. Television and radios have resonant circuits that can be tuned to oscillate electrically at different frequencies. In this way they can respond to the different frequencies of electromagnetic waves that are sent by the transmitting station as these waves now act as the driving frequency.

In the introduction we mentioned the use of quartz crystal as timing devices. If a crystal is set oscillating at its natural frequency, electric charge constantly builds up and dies away on it surface in time with the vibration of the crystal (This is known as the *piezoelectric effect*.). This makes it easy to maintain the oscillations using an alternating voltage supply as the driving frequency. The vibrations of the crystal are then used to maintain the frequency of oscillation in a resonant circuit. It is the oscillations in the resonant circuit that control the hands of an analogue watch or the display of a digital watch.

4.2 Travelling waves

NATURE OF SCIENCE:

Patterns, trends and discrepancies: Scientists have discovered common features of wave motion through careful observations of the natural world, looking for patterns, trends and discrepancies and asking further questions based on these findings. (3.1)

© IBO 2014

Essential idea: There are many forms of waves available to be studied. A common characteristic of all travelling waves is that they carry energy, but generally the medium through which they travel will not be permanently disturbed.

Understandings:
- Travelling waves
- Wavelength, frequency, period and wave speed
- Transverse and longitudinal waves
- The nature of electromagnetic waves
- The nature of sound waves

Perhaps one of the most familiar types of wave motion is a water wave. However, we can also set up waves in strings very easily. A simple demonstration is to take a length of rubber tubing. Hold one end of it and shake that end up and down. A wave will travel down the tube. If we give the end of the tube just one shake then we observe a pulse to travel down the tube. By this we can see that we can have either a continuous travelling wave or a travelling pulse. This is illustrated in Figure 419 in which we have taken an instantaneous snap shot of the tube.

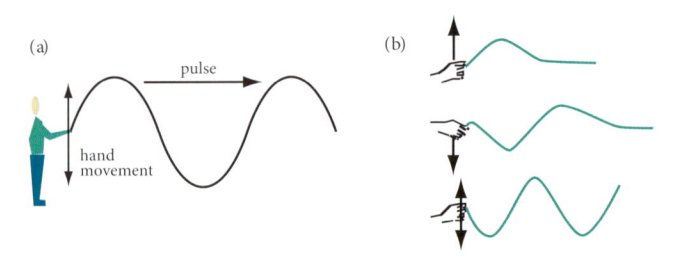

Figure 419   Simple waves

We can also set up another type of wave by using a slinky spring. In this demonstration we lay the spring along the floor. Hold one end of it and move our hand backwards and forwards in the direction of the spring. In this way we see a wave travelling down the spring as a series of compressions and expansions of the spring as illustrated in Figure 420.

We can also set up a pulse in the spring by moving our hand backwards and forwards just once in the direction of the spring.

WAVES

Of course we can set up a wave in the spring that is similar to the one we set up in the rubber tube. We shake the spring in a direction that is at right angles to the spring.

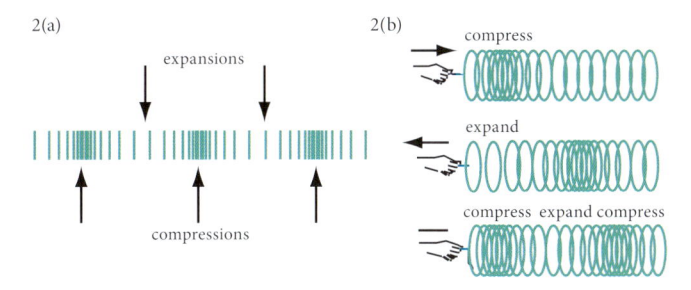

Figure 420 Slinky springs

A very important property associated with all waves is their so-called **periodicity**. Waves in fact are periodic both in time and space and this sometimes makes it difficult to appreciate what actually is going on in wave motion. For example, in our demonstration of a wave in a rubber tube we actually drew a diagram that froze time- an instantaneous snapshot of the whole string. Hence Figure 420 shows the periodicity of the wave in space.

The diagram is repeated as a sketch graph in Figure 421. The y-axis shows the displacement of the tube from its equilibrium position. The graph is a *displacement space* graph.

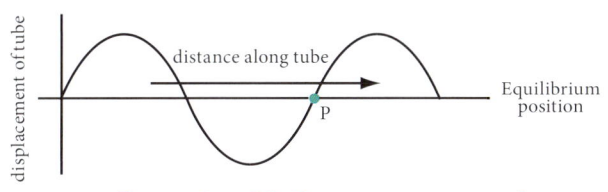

Figure 421 Displacement-space graph

We now look at one particle of the tube labelled P and "unfreeze" time. The diagram in Figure 422 shows how the position of P varies with time. This illustrates the periodicity of the wave in time. We recognise that the point P is actually oscillating with SHM.

The y-axis now shows the displacement of the point P from equilibrium. The graph is a *displacement-time* graph.

The space diagram and the time diagram are both identical in shape and if we mentally combine them we have the whole wave moving both in space and time.

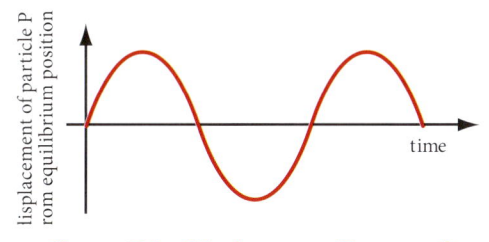

Figure 422 Displacement-Time graph

For the longitudinal wave in the slinky spring, the displacement-space graph actually shows the displacement of the individual turns of the spring from their equilibrium position as a function of distance along the spring. However, it could equally show how the density of turns of the spring varies with length along the spring. The displacement-time graph shows the displacement of one turn of the spring from its equilibrium positions as a function of time.

We have shown that waves fall into two distinct classes but as yet we haven't actually said what a wave is. Quite simply a wave is a means by which energy is transferred between two points in a medium without any net transfer of the medium itself. By a medium, we mean the substance or object in which the wave is travelling. The important thing is that when a wave travels in a medium, parts of the medium do not end up at different places. The energy of the source of the wave is carried to different parts of the medium by the wave. When for example you see a field of corn waving in the wind you actually see a wave carry across the field but none of the corn moves to another field. Similarly with the previous demonstrations, the tube and the spring do not end up in a different part of the laboratory. Water waves however, can be a bit disconcerting. Waves at sea do not transport water but the tides do. Similarly, a wave on a lake does not transport water but water can actually be blown along by the wind. However, if you set up a ripple tank you will see that water is not transported by the wave set up by the vibrating dipper.

Transverse and longitudinal waves

Transverse waves

In these types of wave, the **source** that produces the wave **vibrates at right angles** to the direction of energy propagation. It also means that the particles of the medium through which the wave travels vibrate at right angle to the direction of travel of the wave (direction of energy propagation).

Figure 420 illustrates an example of a transverse wave. Light is another example of a transverse wave although this a very special kind of wave.

An important property of transverse waves is that they cannot propagate through fluids (liquid or gases). This is one reason why light wave are special; they are transverse and yet can propagate through fluids and through a vacuum.

Longitudinal waves

In these types of wave, the **source** that produces the wave vibrates in the same direction as the direction of travel of the wave i.e. the direction in which the **energy carried by the wave is propagated**. It also means that the particles of

Chapter 4

the medium through which the wave travels vibrate in the same direction of travel of the wave (direction of energy propagation). The wave in the slinky spring in Figure 421 is a longitudinal wave, as are sound waves.

Wavelength, frequency, period and wave speed

Definitions associated with waves

Amplitude (*A, a*)

This is the **maximum displacement of a particle from its equilibrium position**. (It is also equal to the maximum displacement of the source that produces the wave).

Period (*T*)

This is the time that it takes a particle to make one complete oscillation. (It is also equal to the time for the source of the wave to make one complete oscillation).

Frequency (*f, v*)

This is the number of oscillations made per second by a particle. (It is also equal to the number of oscillations made per second by the source of the wave). The SI unit of frequency is the Hertz-Hz or s^{-1}. Clearly then, $f = \frac{1}{T}$

Wavelength (λ)

This is the **distance** along the medium between **two successive particles that have the same displacement**.

Wave speed (*v, c*)

This is the **speed with which energy is carried in the medium** by the wave. A very important fact is that wave speed depends only on the nature and properties of the medium.

You can demonstrate this by sending pulses along different types of rubber tubes or by sending pulses along a slinky in which you alter the distance between successive turns by stretching it.

(In some circumstances the wave speed is a function of wavelength, a phenomenon known as dispersion).

Figure 423 shows how the different terms and definitions associated with waves relate to both transverse and longitudinal waves. From these diagrams, we also see that the wavelength of a transverse wave is equal to the distance between successive crests and also between successive troughs. For a longitudinal wave the wavelength is equal to the distance between successive points of maximum compression and also between successive points of maximum rarefaction.

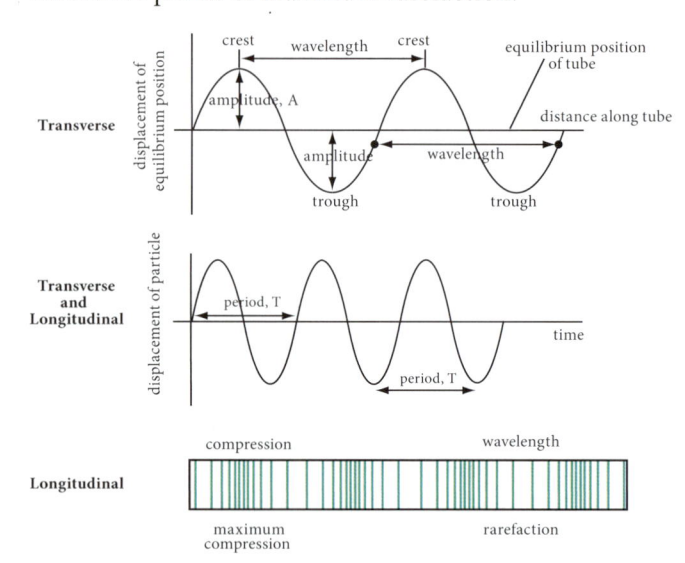

Figure 423 *How the definitions apply*

Figure 424 shows an instantaneous snapshot of a medium through which a wave is travelling. A particle of the medium is labelled 'P'.

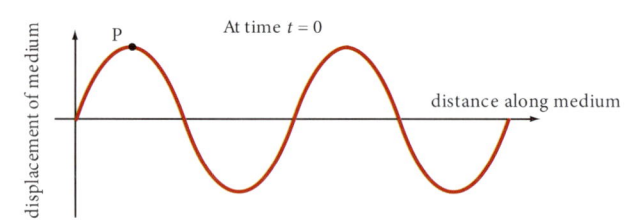

Figure 424 *Instantaneous snapshot of displacement of medium*

If we take another photograph half a period later then the particle 'P' will be in the position shown in Figure 425.

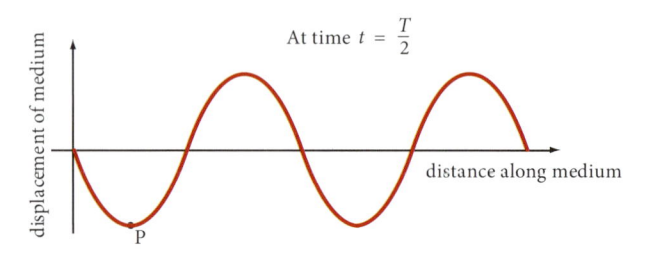

Figure 425 *Particle 'P' half a period later*

In this time the wave will have moved forward a distance of half a wavelength $\left(\frac{\lambda}{2}\right)$.

We have therefore that the speed v of the wave

$= \frac{\text{distance}}{\text{time}} = \frac{\lambda}{T}$, but $f = \frac{1}{T}$,

Hence we have

$$v = f\lambda$$

Example

Water waves of wavelength 5.0 cm are travelling with a speed of 1.0 ms^{-1}.

Calculate the frequency of the source producing the waves?

The waves travel into deeper water where their speed is now 2.0 m s^{-1}. Calculate the new wavelength λ_{new} of the waves.

Solution

Using $v = f\lambda$ we have that $1.0 = f \times .05$

and so $f = 20$ Hz.

In the deeper water using $v = f\lambda$ we now have that $2.0 = 20 \times \lambda_{new}$

and so $\lambda_{new} = 10$ cm.

The nature of electromagnetic waves

We mentioned above that light is a transverse wave but in fact light is just one example of a most important class of waves known as **electromagnetic waves**. They are called electromagnetic waves (em waves) because they actually consist of an electric field and a magnetic field oscillating at right angles to each other. Their existence was first predicted by *Clerk Maxwell* in 1864 and verified some 20 years later by *Heinrich Hertz* (hence Hz for the unit of frequency). This was one of the great unifications in Physics.

Electromagnetic waves are produced because of the accelerated motion of electric charges. When electrons are accelerated in aerials or within atoms, they produce changing electric fields. These changing electric fields generate changing magnetic fields in a plane perpendicular to the electric field plane. As this process continues, a self-propagating electromagnetic wave is produced. As the charge oscillates with simple harmonic motion, the strength of the electric and magnetic vectors vary with time, and produce sine curves (sinusoidal waves) perpendicular to each other and to the direction of the wave velocity v as shown in Figure 426. The waves are therefore periodic and transverse.

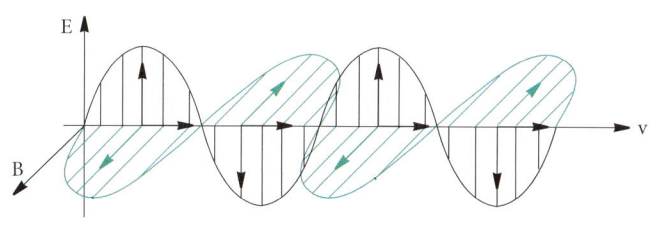

Figure 426   The nature of EM waves

All electromagnetic waves travel at the same speed in a vacuum. This speed is known as the *free space velocity of light* and is given the symbol c. It is a fundamental constant of nature and has a velocity of 2.998×10^8 m s^{-1} ($\sim 3 \times 10^8$ m s^{-1}). Electromagnetic waves can also, of course, propagate in matter. However, when they travel in different media such as glass, they are refracted and therefore travel at a different speed.

An electromagnetic wave also carries momentum and energy, and the energy factor will be expanded on shortly.

The so-called spectrum of em waves is vast. Suppose in a thought experiment, we were to have a charged metal sphere and were able to set it oscillating at different frequencies. When vibrating with a frequency of 10^3 Hz, the oscillating charged sphere would be a source of long wave radio signals,

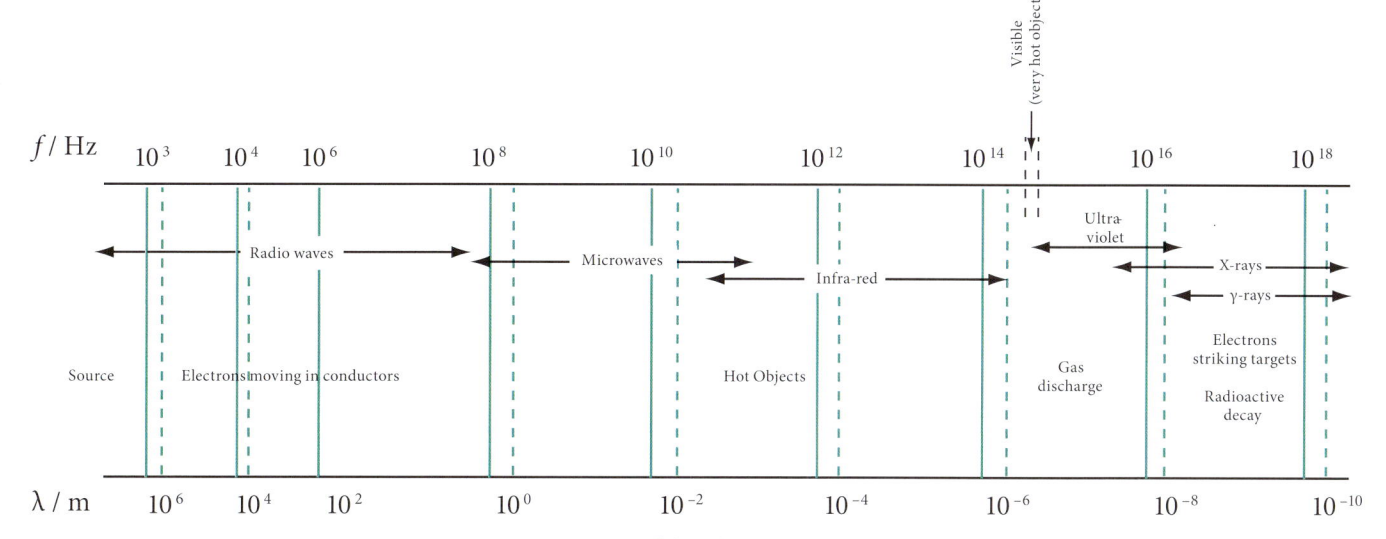

Figure 427   Regions of the electromagnetic spectrum

125

at 10^9 Hz a source of television signals, at 10^{15} Hz a source of visible light and at 10^{18} Hz, it would emit X-rays. Of course this is just a thought experiment and we should identify the actual sources of the different regions of the em spectrum. These are shown in Figure 427 (previous page).

Recall from Chapter 4 the wave terminology concerning wave speed, wavelength, frequency and period. Because speed equals the time rate of change of distance, in electromagnetic wave terms:

$$c = \frac{\text{wavelength}}{\text{period}} = \frac{\lambda}{T}$$

but, $T = 1/f$ $\therefore c = f \times \lambda$

Regions of the electromagnetic spectrum

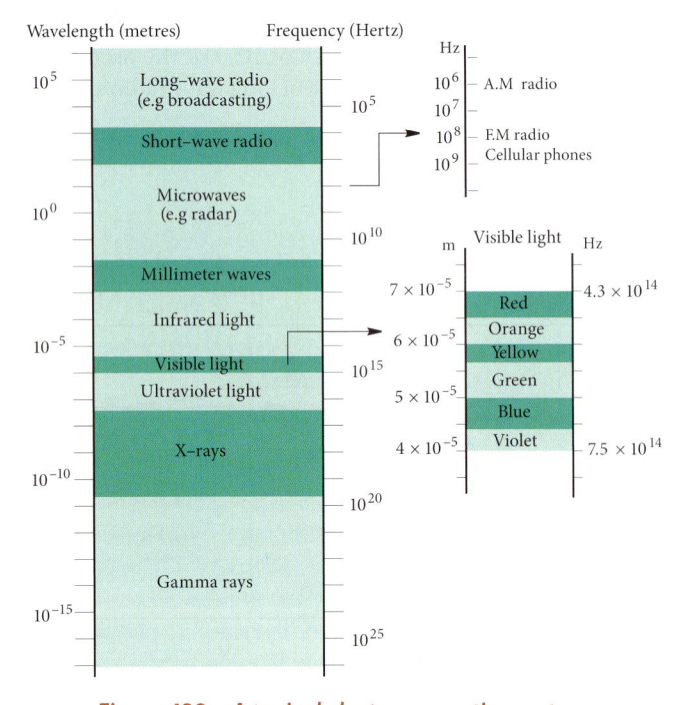

Figure 428 shows another typical electromagnetic spectrum. You will see the **EM spectrum** depicted in different textbooks in different ways. Sometimes it is shown horizontally and other times vertically. Sometimes the higher energy end is to the left and at other times it is reversed. The spectrum is only a representation of the different radiation bands and therefore there is no convention as to how it should be represented.

Visible light as detected by the human eye is only a small part of the range of electromagnetic waves. The wavelength of visible light ranges from 400 nm for the shorter wavelength violet near the ultra-violet region of the spectrum to 700 nm for the longer wavelength red near the infrared region.

Note that there are no sharp divisions or rapid changes of properties between the various regions, rather a gradual merging into each other. The names used for each region are for convenience only and the classification scheme has developed due to the origin of their manner of production. For example, the ranges of gamma radiation and X-radiation overlap. X-rays could be produced with wavelengths similar to gamma radiation emitted by radioactive substances, but they are called X-rays because they are produced when electrons hit a metal target.

You should become familiar with the order of magnitude of the frequencies (and wavelengths) of the different regions. Figure 429 indicates the orders of magnitude of the wavelengths for the regions of the spectrum. If you become familiar with these you can determine the order of magnitude of the frequencies using the equation $f = c / \lambda$. Note that the range of gamma rays extends beyond the range of instrumentation that is presently available.

Because the speed of electromagnetic waves is constant ($\sim 3 \times 10^8$ m s^{-1}), as the wavelength gets longer (increases), the frequency decreases. Similarly, if the wavelength decreases, the frequency increases. Therefore, there is a range of wavelength values (and frequency values) that electromagnetic waves can have. The entire possible range is called the electromagnetic spectrum. A range of wavelengths from about 10^8 m to 10^{-17} m corresponding to a frequency range of 1 Hz to 10^{25} Hz have been studied by scientists. Refer to Figure 430.

Although electromagnetic waves can be produced and detected in different ways, all waves in the electromagnetic spectrum behave as predicted by Maxwell's theory of electromagnetic waves.

Wave type	Order of magnitude of the wavelength (m)
Radio	$10^4 - 10^{-2}$
Microwave	$10^{-2} - 10^{-4}$
Infrared	$10^{-4} - 10^{-6}$
Visible	$7 \times 10^{-7} - 3 \times 10^{-7}$
Ultra-violet	$3 \times 10^{-7} - 10^{-9}$
X-rays	$10^{-9} - 10^{-11}$
Gamma rays	$10^{-11} - 10^{-13}$

Figure 429   Wavelengths and wave types

Electromagnetic radiation has particular properties in the different regions of the EM spectrum, and they are generated and detected in different manners. The regions will now be discussed in some detail.

Figure 428   A typical electromagnetic spectrum

Radio waves

Radio waves have the longest wavelengths, and lie in the frequency range from 30 Hz to greater than 3000 MHz. They have specialised uses in radio communication including AM (amplitude modulated) and FM (frequency modulated) radio, television, CB radio, radio microphones and scanning devices in MRI (magnetic resonance imaging). Because of their many uses governments must regulate the bandwidth that can be used in communication devices in order to avoid congestion of the airwaves. The internationally agreed frequency bands used for carrier waves, and their uses are given in Figure 430.

Radio waves are generated by an electric circuit called an oscillator and are radiated from an aerial. A tuned oscillatory electric circuit that is part of a radio/television receiver detects the radio waves. As they do not penetrate solid materials, radio waves are relatively easily reflected off surfaces and this makes them ideal for communication technology.

Band Frequency		Frequency	Wavelength (m)	Use
Extremely high EHF Super high	SHF	300–3GHz	<0.1	Space satellite link
Ultra-high	UHF	3–0.3GHz	1–0.1	Television
Very high	VHF	300–30MHz	10–1	FM radio
High	HF	30–3MHz	100–10 short wave	AM radio
Medium	MF	3–0.3MHz	1000–100 medium wave	AM radio
Low Very low	LF VLF	300–3kHz	>100 000-1000	Defence use

Figure 430   Radio frequencies

Long and medium wavelength radio waves easily diffract around obstacles such as small mountains and buildings, and they can be reflected by the earth's ionosphere. Therefore, there does not have to be a direct line of sight between the antenna and the receiver and they can be broadcasted over large distances provided the transmitter used is powerful.

Television and FM broadcasting stations have wavelengths from 1–10 m and they are not easily diffracted around objects. Therefore, coaxial cables or relay stations are necessary to transmit signals between points more than 80 km apart, even if there is a direct line free of obstacles between them.

Waves in the UHF, SHF and EHF bands are not reflected off the ionosphere in the upper atmosphere but rather pass into space. For this reason, SHF and EHF bands are used for outer space and satellite communications. These outer bands are overlapping in the microwave region of the EM spectrum.

Microwaves

Microwaves have many applications – including mobile phone, satellite communication, radar (radio detection and ranging) and cooking. They are also used in the analysis of fine details concerning atomic and molecular structure. Microwaves are the main carriers of communication between repeater stations. They use 'line-of-sight' technology where relay stations are placed in high positions 50 km apart.

They are produced by special electronic semi-conductor devices called Gunn diodes, or by vacuum tube devices such as klystrons and magnetrons. Point contact diodes, thermistor bolometers and valve circuits detect them.

Radar consists of short pulses of microwaves. They are used to detect the speed of vehicles by police, and to find distances to aeroplanes and ships. It is a microwave system that guides large aircraft into airports.

Microwaves can interact with matter and this is the basis of the microwave oven. Water readily absorbs radiation (energy) with a wavelength of 10 cm, and this absorbed energy causes the molecules to produce thermal energy due to their vibration. The heat is therefore generated in the substance itself rather than conducted in from the outside, and this allows food to be cooked rapidly. At short distances from the source, microwave radiation can damage living tissue.

Infra-red radiation

In heated bodies, the outer electrons in atoms and molecules give off electromagnetic waves with wavelengths shorter than 10^{-4} m due to a change in the rotational and vibrational kinetic energy of these particles. Because the radiation given off has a wavelength slightly longer than the red end of the visible spectrum, it is called **infrared radiation**. Infrared radiation allows us to receive warmth from the Sun and other heat sources. In fact most of the emitted radiation from any hot object is infrared.

Infrared radiation can be detected by our skin, by thermometers, thermistors, photoconductive cells, special photographic film. Special photographic film is used to identify heat sources such as human beings trying to hide from the scene of a crime or soldiers moving in a war situation. They can also identify environmental problems. Because this radiation is scattered by small particles in the atmosphere, they can be used in haze photography. It is also employed in the identification of the molecular structure of many organic compounds.

Visible light

As already mentioned, the receptors in the human eye are sensitive to electromagnetic radiation between about 400 nm to 700 nm, and radiation in this region is referred to as 'visible light'. Visible light is detected by stimulating nerve endings of the retina of the eye or by photographic film and photocells. The eye is most sensitive to the green and yellow parts of the visible spectrum. It can be generated by the re-arrangement of outer orbital electrons in atoms and molecules. These excited electrons emit light and other electromagnetic radiation of a certain frequency when they lose energy as happens in gas discharge tubes. Visible light can cause photochemical reactions in which radiant energy is converted into chemical energy.

Ultra-Violet radiation

Ultra-violet radiation produces EM waves between about 10^{-7} m to 10^{-9} m. The orbital electrons of atoms of the Sun generate it, and other instruments such as high-voltage discharge tubes and mercury vapour lamps. Like visible light, UV radiation can cause photochemical reactions in which radiant energy is converted into chemical energy as in the production of ozone in the atmosphere and the production of the dark pigment (melanin) that causes tanning in the skin. It also helps to produce vitamin D in our skin. However, too much UV radiation can cause melanoma cancers.

UV radiation has the ability to ionise atoms and this is the reason why ozone is produced in the atmosphere. This ozone is capable of killing bacteria and therefore it can be put to good use in the sterilisation of many objects.

The atoms of many elements emit UV radiations that are characteristic of those elements, and this quality allows many unknown substances to be identified. UV radiation can be detected by photography and the photo-electric effect. Furthermore, certain crystals fluoresce when they absorb UV radiation, and this is put to use in washing powders to make the 'whites look whiter'.

X-Radiation

X-radiation produces EM waves between about 10^{-8} m to 10^{-18} m. It can be generated by the rapid deceleration (stopping) or deflection of fast-moving electrons when they strike a metal target or other hard objects. It can also be generated by the sudden change in energy of innermost orbital electrons in atoms. The maximum frequency produced is determined by the energy with which the fast-moving electrons from the source strike a target. The accelerating voltage of an X-ray machine in turn determines this energy.

X-rays can penetrate different materials to different degrees. 'Hard' X-rays (near the gamma end) are more penetrating than 'soft' X-rays (near the ultra-violet end). Apart from the penetration increasing with increased frequency of the X-rays, the penetrating power also depends on the nature of the material being penetrated. The higher the atomic mass of a material, the less is the penetration. For example, X-rays can easily penetrate skin and flesh containing mainly the lighter atoms carbon, hydrogen and oxygen, but are less able to penetrate dense material such as bone containing the heavier element calcium. Because X-rays are detected by photography, the photographic plate placed beneath the body can be used to identify possible bone fractures.

X-rays are ideal for identifying flaws in metals. They are also used in CAT scans (computerised axial tomography). Because tissues absorb X-rays differently, when a body is scanned, the internal organs and tissues can be identified from the analysis of the images produced by the computer.

X-radiation can ionise gases and cause fluorescence. Because X-rays produce interference patterns when they interact with crystals in rocks and salts, the structure of these regular patterns of atoms and molecules can be determined by this process of X-ray diffraction.

X-rays can damage living cells and continued use and exposure to X-rays is discouraged. Radiologists who work in X-ray Departments always stand behind lead-lined walls when an X-ray is being taken. On the other hand, some types of diseased cells are damaged more easily than are healthy cells. Therefore, if X-rays are carefully controlled, they can be used to destroy cancerous cells, as is the case with the use of certain lasers in radiation therapy.

Gamma radiation

The **gamma radiation** region of the EM spectrum overlaps with the X-ray region and their use in cancer therapy overlaps with the last statement made about X-ray use in radiotherapy. These high frequency rays are highly penetrating and are produced by natural and artificial radioactive materials. As such, gamma rays will pass through metres of air and need large thicknesses of concrete or lead to absorb them in order to protect humans from danger. An ionisation chamber as found in a Geiger-Müller counter can detect gamma radiation.

WAVES

Exercise 4.4

1. Describe how are electromagnetic waves produced?

2. List five characteristics of electromagnetic waves.

3. A photon of red light has a wavelength of 700 nm. Calculate the:

   (a) photon's frequency
   (b) photon's energy.

4. The wavelength of the electromagnetic waves detected in the air loop in the Hertz experiment were about 1 m. In which region of the EM spectrum are they found?

5. Describe how long and medium radio waves pass around obstacles such as buildings and mountains.

6. Draw a table showing the order of magnitude of the frequencies for the different regions of the electromagnetic spectrum.

7. Describe how the frequency of X-rays is related to their penetration of matter.

8. Identify a possible source of the radiation in each region of the EM spectrum. Give one way of detecting each source.

9. Describe how the SI unit, the metre, is defined in terms of the speed of light.

10. (a) With the aid of a diagram, describe the dispersion of white light by a prism.
    (b) Explain how a factor of the prism causes the dispersion to occur.

11. (a) What does the term LASER stand for?
    (b) In simple terms, how is laser radiation produced?
    (c) Give one application of the use of lasers in each of technology, industry and medicine.

The nature of sound waves

Sound waves are an example of a longitudinal wave and as such the oscillating particles of the medium move in the same direction as the energy propagation as shown in Figure 431. A series of compressions and rarefactions are propagated and these correspond to the crests and troughs of a transverse wave.

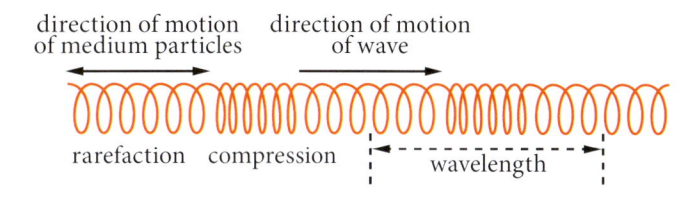

Figure 431 Longitudinal or compression wave

The compressions are areas of high pressure and the rarefactions are regions of lower pressure.

Sound waves can only be created and transmitted in a medium because the medium must contain particles for the disturbance to move from one place to another. Therefore, sound waves cannot travel in a vacuum. Sound waves travel fastest in the solid medium and slowest in the gaseous medium.

When the source of sound arrives at the ear, the eardrum vibrates at the same frequency as the source.

The human ear as shown in Figure 432 consists of 3 sections:

1. the external or outer ear
2. the middle ear
3. the internal or inner ear

The **outer ear** consists of the **pinna**, the external **auditory canal** and the **ear-drum** (tympanic membrane).

Sound waves reaching the ear are collected by the pinna – it has an important spatial focussing role in hearing. The sound waves are directed down the external auditory canal that is about 2.5 cm long and about 7 mm in diameter. It is closed at one end by the ear drum (tympanic membrane) that consists of a combination of radial and concentric fibres about 0.1 mm thick with an area of about 60 mm^2 that vibrate with a small amplitude.

The ear canal is like a closed organ pipe as the canal is a shaped tube enclosing a resonating column of air that vibrates with an optimum resonant frequency around 3 kHz.

The eardrum acts as an interface between the external and middle ear. As sound travels down the ear canal, air pressure waves from the sounds set up sympathetic vibrations in the taut membrane of the ear and pass these vibrations on to the middle ear structure.

129

CHAPTER 4

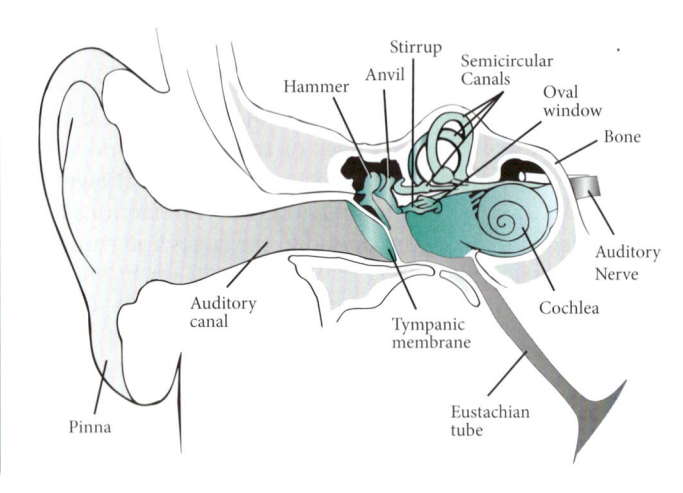

Figure 432   Typical human ear

The **middle ear** consists of a small (around 6 cm^3), irregular, air-filled cavity in which, suspended by ligaments, are the **ossicles** – a chain of three bones called the malleus, incus and stapes, more commonly known as the hammer, anvil and stirrup.

They act as a series of levers with a combined mechanical advantage of 1.3. Because of their combined inertia as a result of the ossicle orientation, size and attachments, they cannot vibrate at frequencies much greater than 20 kHz. The malleus is attached to the inner wall of the tympanic membrane and the flat end of the stapes comes up against the **oval window** (a membrane called fenestra ovalis).

The inner ear is a complicated bony chamber filled with fluid and embedded in the bone of the skull. It is divided into two parts:

The range of frequency audible to the human ear varies considerably from individual to individual, but the average range is from 20–20 000 Hz or 20 Hz to 20 kHz. The upper limit is around 20 kHz for a young, normal individual but it decreases with age.

Sound waves below 20 Hz are termed subsonic or infrasonic. Sound waves of frequencies greater than 20 kHz are called ultrasonic.

Although humans can hear up to 20 000 Hz, some animals can hear in the ultrasonic region such as dogs, dolphins and bats.

Sound waves can be reflected off solid surfaces to produce echoes. Ultrasonic echoes are used as a diagnostic tool in medicine. Whenever a sound wave travels from skin, muscle or bone for example, it is reflected back to give echoes. An electronic device to produce an image on a picture monitor can detect these signals. Echo sounding is also used in geology for oil exploration, in medicine for breaking up kidney stones, in dentistry for removing plaque from your teeth and in jewellery for cleaning metal surfaces.

The characteristics of a musical note are its pitch, its loudness and its quality, or timbre.

The **pitch** is determined by the note's frequency. This has already been defined as the number of vibrations that occur in one second measured in Hz or s^{-1}.

The quality, or **timbre**, of a note is very much determined by the source or instrument producing the note. It is determined by both the frequency and the relative amplitude of the note/s being produced. The timbre of noise is poor because it consists of a random mixture of unrelated frequencies. The timbre of a violin is good because the notes produced are mixtures of fundamental frequencies and harmonics.

The **loudness** of a note depends on the amplitude of the produced wave. The greater the amplitude, the greater will be the loudness because the wave carries more energy.

Just as the resonance of two vibrating tuning forks mounted in two "sound boards" is the selective reinforcement of the natural frequency of vibration of the "sound boards", so too the response of the ear to sound is essentially one of **resonance** due to the vibrations of the sound matching the **natural frequencies** of the vibration of parts of the ear.

The external auditory canal is like a closed pipe and exhibits slight resonance at approximately 3000 Hz. The middle ear displays a slight but broad resonance between about 700 Hz and 1400 Hz, and is greatest at about 1200 Hz. The cochlea of the inner ear displays excellent transmission between about 600 Hz and 6000 Hz.

The human perception of loudness at different frequencies is not constant and varies considerably over the audible frequency range as shown in the intensity-logarithmic frequency diagram of Figure 433. A logarithmic scale for frequency is preferred so that the wide range of audible frequencies can be examined on the same graph.

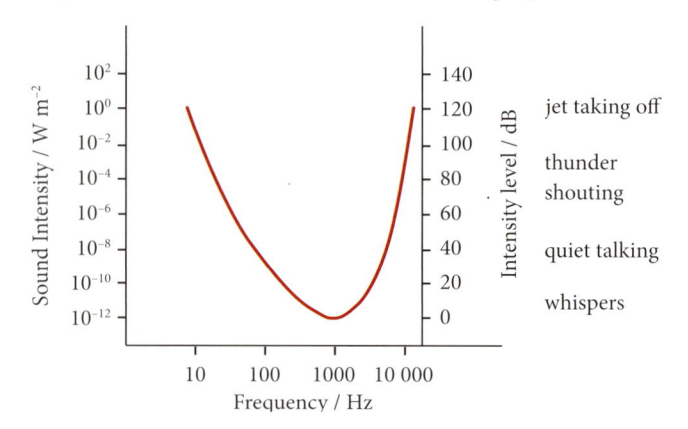

Figure 433   Frequency range of the average ear

The minimum detectable intensity for a given frequency is called the **threshold intensity of hearing** and it is the envelope of the curve in the above diagram. For example, the threshold of hearing between 2-3 kHz is 10^{-12} Wm^{-2} and at 90 Hz, the threshold of hearing is about 10^{-8} Wm^{-2}. From the graph, we can see that the intensity level of the sound increases as the range of frequencies that can be detected increases, up to a maximum intensity level of 100 dB where the audio range of 20 Hz to 20 kHz is reached. The ear is most sensitive to sounds at a frequency around 3 kHz. This is no accident as the the cochlear tube length is 2.5 cm and it acts like a closed pipe with a standing wave of ¼ λ. (this is approximate because it is more a travelling wave due to changing speed and wavelength). This would give us λ = 10cm. If the speed of sound is 330 ms^{-1}, then the frequency would equal 3000 Hz. Why do you think alarms and the human scream are around 3000 Hz?

The loudness of a note is a subjective quantity. It is not a measure of intensity level. Although loudness takes into account the logarithmic measure of the ear's response to intensity dB, it does not take into account the hearing of the listener or the frequency response of the listener's ear to different frequencies. Therefore, loudness is determined by the intensity and the frequency of a sound.

Just as an audible frequency is dependent on the sound level intensity, so too is loudness intensity. However, it is also dependent on the energy transfer mechanisms of the ear.

Figure 434 hereunder shows some apparatus that could be used for determining the speed of sound in air using the phenomenon of resonance. Waves produced by the tuning fork or the signal generator are propagated down the tube and are reflected at the water boundary. Standing waves are produced for certain water levels within the cylinder. These standing waves can be detected using resonance that occurs when the frequency of the source is equal to the natural frequency of the air column.

Resonance occurs when the standing or stationary waves produced have maximum amplitude. A node of high pressure and density is formed at the water boundary and an antinode of low pressure and density is formed a short distance, e, above the open end of the tube. The temperature of the air column will affect the speed of sound. The average speed of the air particles is greater at higher temperatures and therefore they can pass on compressions and rarefactions more quickly.

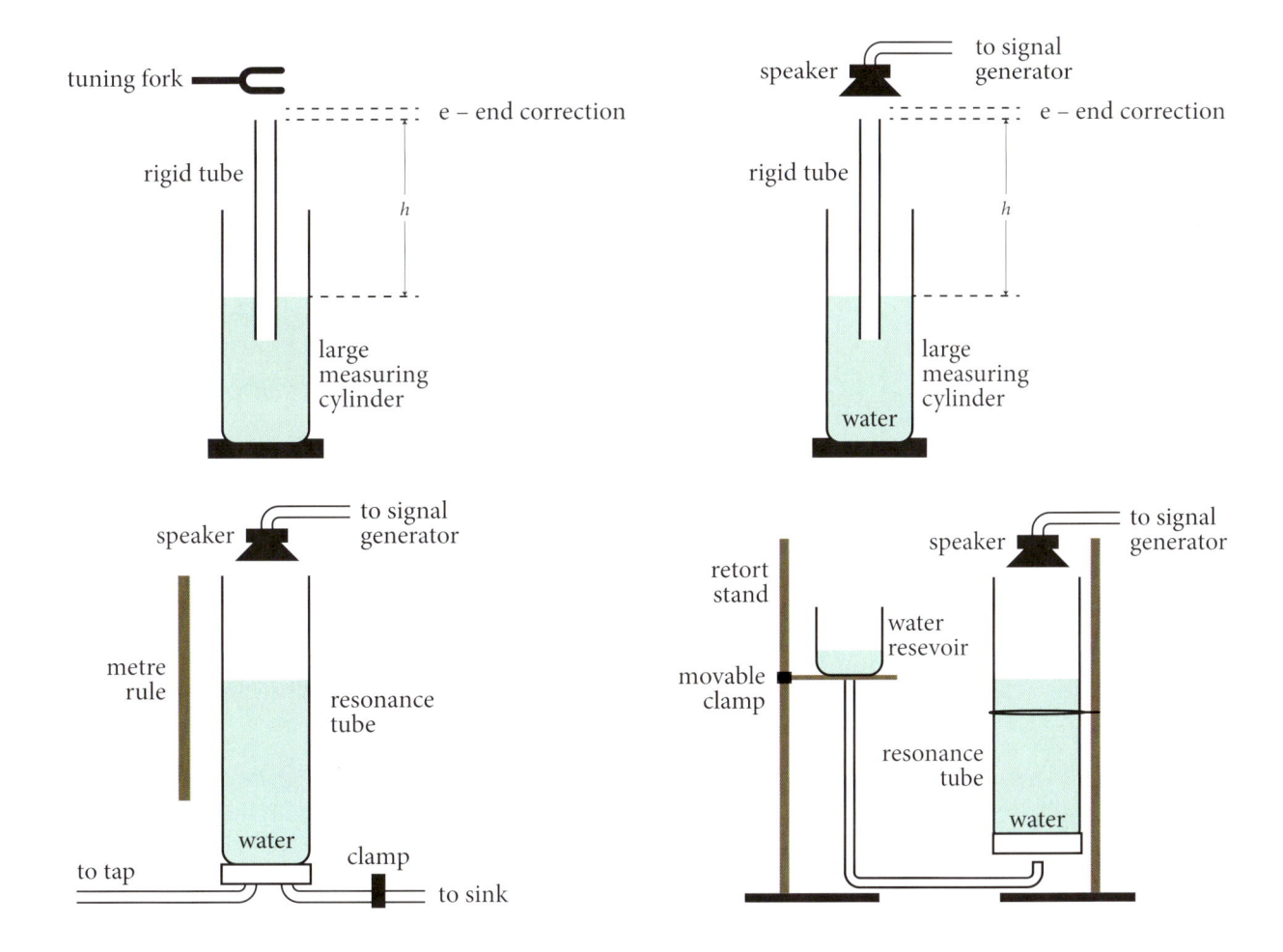

Figure 434   Resonance and the speed of sound investigations.

The longest wavelength producing resonance is equal to four times the length of the air column, L. This can also be stated as the length, L, of the air column is equal to one quarter of the wavelength plus the end correction, e.

$$\frac{\lambda}{4} = L + e$$

The speed of sound at a particular temperature is given by $v = f\lambda$

$$L + e = \frac{v}{4f} \quad \text{or} \quad L = \frac{v}{4f} - e$$

If the fundamental frequency, f, is determined using different tuning forks or signals from the signal generator for different lengths of the air column, a graph of L against 1/f produces a linear graph with a slope of v/4 and a y-intercept of –e.

Experimental measurements show that

$$\frac{v_1}{v_2} = \sqrt{\frac{T_1}{T_2}}$$

where $c = 331$ m s^{-1} at 273 K.

4.3 Wave characteristics

NATURE OF SCIENCE:

Imagination: It is speculated that polarization had been utilized by the Vikings through their use of Iceland Spar over 1300 years ago for navigation (prior to the introduction of the magnetic compass). Scientists across Europe in the 17th–19th centuries continued to contribute to wave theory by building on the theories and models proposed as our understanding developed. (1.4)

© IBO 2014

4.4 Wave behaviour

NATURE OF SCIENCE:

Competing theories: The conflicting work of Huygens and Newton on their theories of light and the related debate between Fresnel, Arago and Poisson are demonstrations of two theories that were valid yet flawed and incomplete. This is an historical example of the progress of science that led to the acceptance of the duality of the nature of light. (1.9)nd models proposed as our understanding developed. (1.4)

© IBO 2014

Essential idea: Waves interact with media and each other in a number of ways that can be unexpected and useful.

Understandings:
- Wave fronts and rays
- Reflection and refraction
- Snell's law, critical angle and total internal reflection
- Diffraction through a single-slit and around objects
- Interference patterns
- Double-slit interference
- Path difference
- Amplitude and intensity
- Superposition
- Polarization

Wave fronts and rays

Place a piece of white card beneath a ripple tank and shine a light on the ripples from above. The image that you get on the card will look something like that shown in Figure 435.

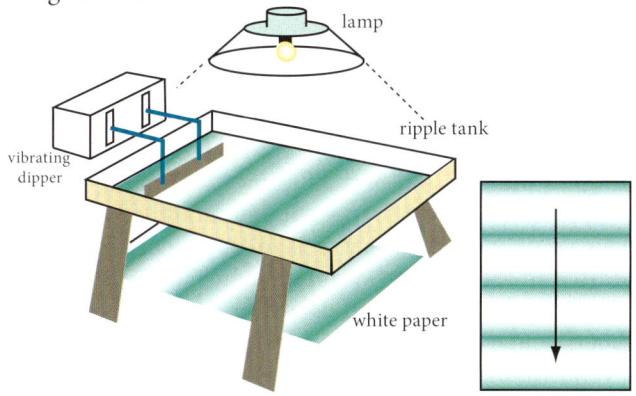

Figure 435   A ripple tank

Again this is a snapshot. However, by using a stroboscope as the source of illumination it is possible to "freeze" the waves.

Each bright area of illumination represents a trough or crest. (Light incident on the top or bottom of a crest will be transmitted through the water and not reflected by the surface).

Each bright line representing a crest can be thought of as a "wavefront" and this is a very good way of representing a travelling wave as shown in Figure 436.

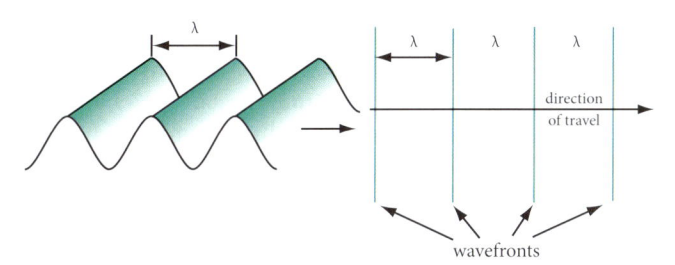

Figure 436   A wavefront

A **wavefront** is a line joining points in phase on a wave. If the wave is a light wave then the arrow that shows the direction of travel of the wave is none other than what we call a light ray. So a ray is a line drawn at right angles to the wavefront as shown in Figure 436.

Qualitative terms associated with waves

Crest This is a term coined from water waves and is used in connection with transverse waves. It refers to the maximum height of the wave i.e. the point at which the particles in the medium in which the waves travel, have maximum displacement.

Trough Again, a term coined from water waves and used in connection with transverse waves. It refers to the lowest point of the wave i.e. the point at which the particles in the medium in which the waves travel, have minimum displacement (strictly speaking, maximum negative displacement).

Compression This is a term used in connection with longitudinal wave and refers to the region where the particles of the medium are "bunched up".

Rarefaction Again, a term used in connection with longitudinal waves referring to the regions where the particles are "stretched out".

In a sound wave, the compression is the region in which the molecules of the air are pushed together (region of greatest pressure) and the rarefaction is the region where the molecules are well separated (region of least pressure). In a physiological sense, sound is the ear's response to the changes in pressure in a longitudinal wave in air. If the frequency of the wave is too low (≈ 10 Hz) then the ear will not detect the pressure changes nor will it if the frequency is too high (≈ 15 kHz).

A longitudinal wave in a slinky spring is analogous to a sound wave in that each turn of the spring represents an air molecule.

Reflection and refraction

Reflection of a single pulse

We now look to see what happens to a wave when it is incident on the boundary between two media. First of all, we shall look at a single pulse travelling along a string.

Figure 437 shows a pulse travelling in a string that is fixed to a rigid support, it is essentially the positive part of a sine curve.

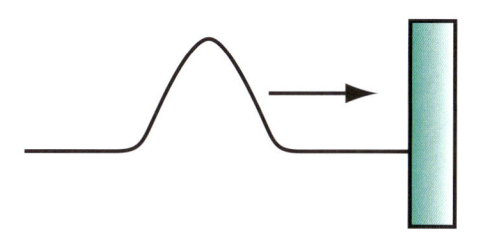

Figure 437   The incident pulse

When the pulse reaches the end of the string it is reflected. Figure 438 shows the reflected pulse, this pulse is the negative part of a sine curve. Some of the energy of the pulse will actually be absorbed at the support and as such, the amplitude of the reflected pulse will be less than that of the incident pulse.

133

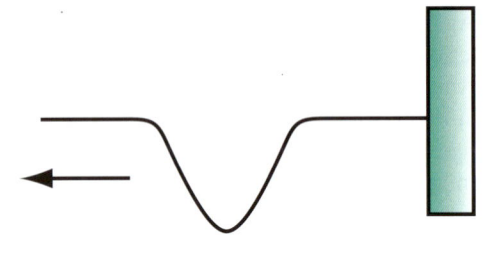

Figure 438   The reflected pulse

The main thing to note is that the pulse keeps its shape except that now it is inverted i.e. the pulse has undergone a 180° (π) phase change.

The reason for this is a little tricky to understand but it is essentially because the end of string that is fixed cannot move. As any part of the forward pulse reaches the fixed end the associated point on the string is moving upward and so if the fixed point is not to move, the point on the string that is moving upward must be cancelled out by a point moving downwards. This is an example of the so-called **principle of superposition** that we look at in a later section.

If the string is not attached to a support then a pulse is still reflected from the end of the string but this time there is no phase change and so the reflected pulse is not inverted.

Reflection of wavefronts

Reflection

We can use the idea of wavefronts to see what happens when a wave strikes a barrier. This is demonstrated with a ripple tank.

Figure 439 shows the incident wavefronts and the reflected wavefronts.

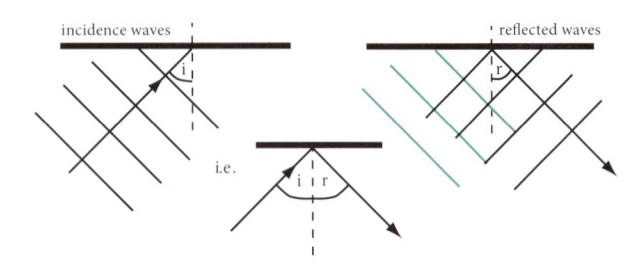

Figure 439   Incident and reflected wavefronts

By constructing the associated rays (see Figure 440), we see that the angle at which the waves are reflected from the barrier is equal to the angle at which they are incident on the barrier (the angles are measured to the normal to the barrier). That is $\angle i = \angle r$. All waves, including light, sound and water obey this rule, the so-called **law of reflection.**

We can use the idea of reflection to make a very simple measurement of the speed of sound. If you stand about 100 m away from a tall wall and clap your hands once, a short time later you will hear an echo of the clap. The sound pulse produced by your hand clap travels to the wall and is reflected back. The trick now is to clap your hands continuously until each clap is synchronous with the echo. You should then get a friend to time the number of claps that you make in say 30 seconds. Let us suppose that this is 50. The number of claps per second is therefore 0.6. This means that it takes the sound 0.6 seconds to travel to the wall and back. The speed of sound can then be approximated as 330 m s^{-1}. If you should get this result then you have done very well indeed since this is very nearly the value of the speed of sound in air at standard temperature and pressure. However, in the true spirit of experimental physics, you should repeat the experiment several times and at different distances and you should also assess the quantitative error in your result.

Refraction

We now look to see what happens to a wave when it is incident on the boundary between two media and passes from one medium to the other (**transmission**). As for the single pulse, some energy will be **absorbed** at the boundary. Also, as well as energy being transmitted by the wave, some energy will be **reflected** at the boundary. Here we will concentrate on the transmitted energy.

We have discussed previously the idea that the speed of a wave depends only on the nature and properties of the medium through which the wave travels. This gives rise to the phenomenon of **refraction**. That is the **change in direction** of travel of a wave resulting from a **change in speed** of the wave. This is easily demonstrated with a ripple tank by arranging two regions of different depth. To achieve this, a piece of flat perspex or glass is placed a short distance from the source of the waves as shown in Figure 440.

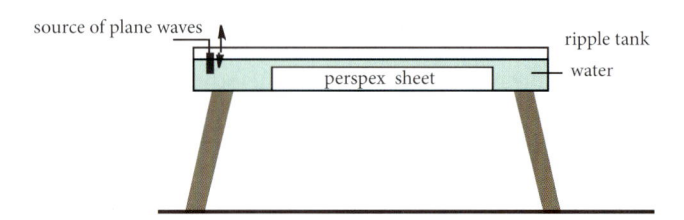

Figure 440   Ripple tank setup to demonstrate refraction

Figure 441, shows the result of a continuous plane wave going from deep to shallow water.

WAVES

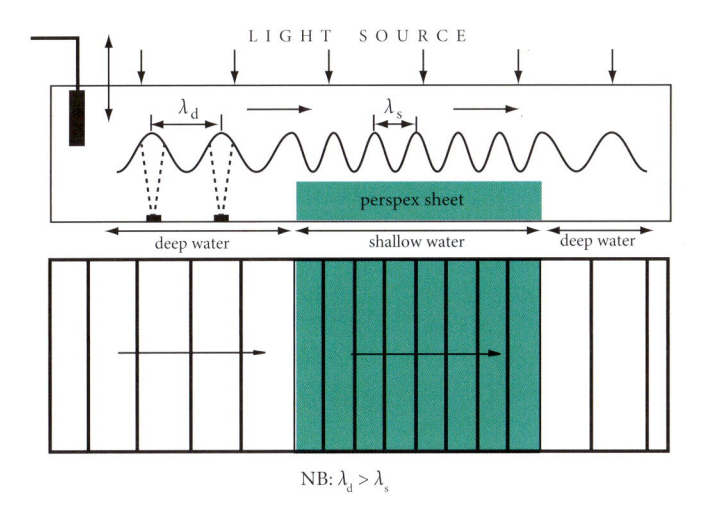

Figure 441 Wavefronts in water of different depth

In this diagram the wave fronts are parallel to the boundary between the two regions. The frequency of the waves does not alter so, as we have mentioned before, the wavelength in the shallow water will be smaller. If the speed of the waves in the deep water is v_d and the speed in the shallow water is v_s,

then, $f = \dfrac{v_d}{\lambda_d} = \dfrac{v_s}{\lambda_s}$

such that

$$\dfrac{v_d}{v_s} = \dfrac{\lambda_d}{\lambda_s}$$

In Figure 442 the wavefronts are now incident at an angle to the boundary between the deep and shallow water.

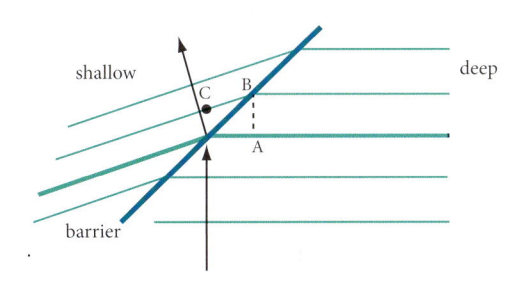

Figure 442 Waves incident at an angle

As well as the wavelength being smaller in the shallow water the direction of travel of the wavefronts also alters. We can understand this by looking at the wavefront drawn in bold. By the time that part A of this wavefront reaches the barrier at B the refracted wave originated from the barrier will have only reached C since it is travelling more slowly.

Snell's law, critical angle and total internal reflection

In 1621 the Dutch physicist *Willebrord Snell* discovered and published a very important rule in connection with the refraction of light.

Figure 443 shows a light ray travelling from one medium to another. The line AB (called the normal) is a line constructed such that it is at right angles to the surface between the two media. It is used as a reference to enable the of the angle of incidence, i, and the angle of refraction r.

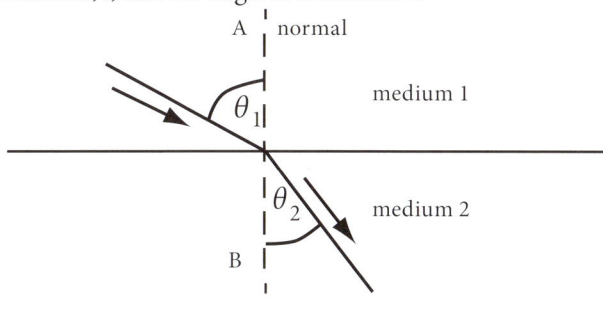

Figure 443 Refraction of light rays

Snell discovered that for any two media

$$\dfrac{\sin \theta_1}{\sin \theta_2} = \text{constant}$$

This is known as **Snell's law**. In fact it enables us to define a property of a given optical medium by measuring the angles when medium 1 is a vacuum. (In the school laboratory air will suffice.). The constant is then a property of medium 2 alone called its **refractive index** n.

We usually write

$$n = \dfrac{\sin i}{\sin r}$$

Figure 444 shows some refractive index values for some media.

Medium	Refractive index	Speed of light in medium ms^{-1}
Air	1.0	3×10^8
Water	1.33	2.25×10^8
Perspex	1.5	2×10^8
Glass	≈ 1.5	2×10^8
Diamond	2.4	1.2×10^8

Figure 444

135

CHAPTER 4

When Snell published his law it was essentially an empirical law and the argument as to the nature of light was still debatable.

Although we have described Snell's law for light rays, we must remember that a ray is a line that is perpendicular to the wavefronts of a wave. In this respect Snell's law is true for all waves.

In Figure 445 the wavefronts in medium 1 travel with speed v_1 and in medium 2 with speed v_2.

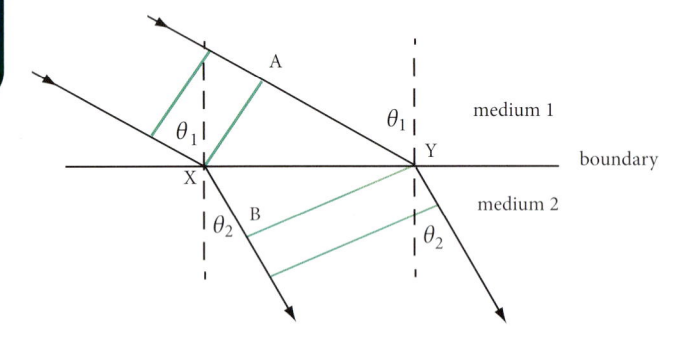

Figure 445   Snell's law

In the time that it takes point A on the wavefront XA to reach Y, point X will have travelled to point B. If we let this time be Δt then we have

$AY = v_1 \Delta t$ and $XB = v_2 \Delta t$

However from the geometry of the situation we have that

$AY = XY \sin\theta_1$ and $XB = XY \sin\theta_2$

From which

$$\frac{AY}{XB} = \frac{XY \sin\theta_1}{XY \sin\theta_2} = \frac{v_1 \Delta t}{v_2 \Delta t}$$

that is

$$\frac{\sin\theta_1}{\sin\theta_2} = \frac{v_1}{v_2} = \frac{n_1}{n_2}$$

That is, the constant in Snell's law is the ratio of the speed of light in medium 1 to that in medium 2. The result will of course be valid for all types of waves.

For light we have that the refractive index of a material is the ratio of the speed of light *in vacuo* (c) to that in the material (v). We can write this as

$$n = \frac{c}{v}$$

As mentioned above this result has been confirmed for light by a direct measurement of the speed of light in water and subsequently in other materials.

Example

The refractive index of a certain type of glass is 1.5. The speed of light in free space is 3.0×10^8 m s^{-1}. Calculate the speed of light in glass.

Solution

From Snell's law, the speed of light in glass can be found using $c = 3 \times 10^8$ m s^{-1}.

So, with $n = 1.5$ and , we have that $v = 2.0 \times 10^8$ m s^{-1}.

Critical angle and total internal reflection

If we consider, for example, light waves incident on the surface of a glass block in air, then some of the light will be **absorbed** at the surface, some **reflected** and some **transmitted**. However, let us just concentrate on the transmitted light. In Figure 446 (a), the direction of the incident and transmitted waves are represented by the rays labelled I and T. Figure 446 (b) shows the waves travelling from glass to air.

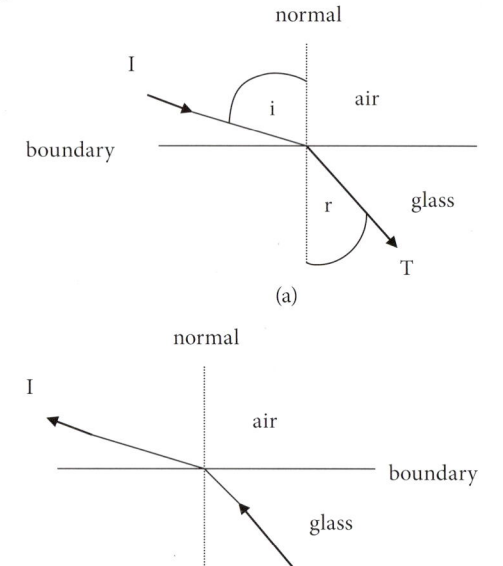

Figure 446   Refraction: (a) air to glass transmission (b) glass to air transmission

In Figure 446(b) the light clearly just follows the same path as in Figure 446(a) but in the opposite direction. However, whereas any angle of incidence between 0 and 90° is possible when the light is travelling from air to glass, this is not the case when travelling from glass to air. When angle $i = 90°$ in (a), this will correspond to the maximum possible angle of refraction when the light is travelling from glass to air as is shown in Figure 447.

136

WAVES

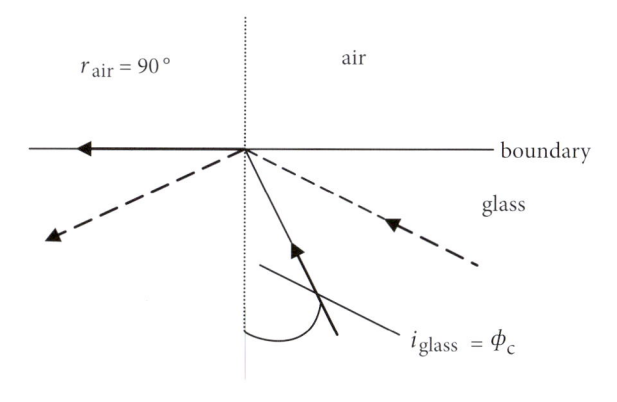

Figure 447 Total internal reflection

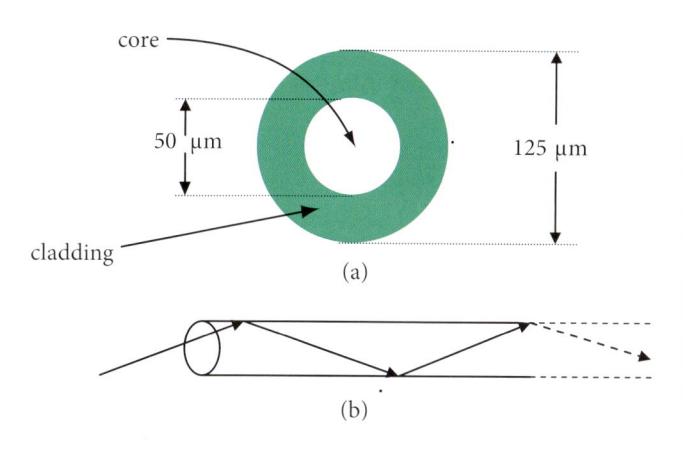

Figure 448 Optical fibre (a) structure (b) transmission of light by TIR

For light travelling from air to glass, we have from Snell's law that the refractive index of the glass n is given by

$$n = \frac{\sin i}{\sin r}$$

Therefore travelling from glass to air, we have

$$\frac{1}{n} = \frac{\sin i_{glass}}{\sin r_{air}}$$

For the maximum value of $r_{air} = 90°$ therefore

$$\frac{1}{n} = \frac{\sin i_{glass}}{1} = \sin \phi_c$$

where ϕ_c is the angle of incidence in the glass that corresponds to an angle of refraction in air of 90°.

The dotted rays in Figure 448 show what happens to light that is incident on the glass-air boundary at angles greater than ϕ_c; the light is **totally internally reflected**. The reflection is total since no light is transmitted into the air. The angle of incidence at which total reflection just occurs is the angle ϕ_c and for this reason ϕ_c is called the **critical angle**.

Example

The critical angle for a certain type of glass is 40.5°. Determine the refractive index of the glass.

Solution

The phenomenon of total internal reflection along with the principles of modulation and analogue to digital conversion form the basis for data transmission using optic fibres. In optic fibres, the carrier wave is light. There are essentially three types of optic fibres but we will only concern ourselves with the so-called step-index fibre. Figure 448 shows the essential structure of a single step fibre and of light transmission along the fibre by total internal reflection.

Figure 448 (a), shows that there is a two-layer structure of the fibre, the core and the cladding. The diameter of the core is constant, at approximately 50 to 60 μm and a typical refractive index index would be 1.440. The surface of the core is kept as smooth as possible. The outer layer, the cladding, is bonded at all points to the surface of the core and a typical refractive index would be 1.411. The cladding ensures that the refractive index of the outside of the core is always less than the inside. A layer of plastic that protects and strengthens the fibre usually surrounds the cladding.

The refractive index of a medium can be easily determined experimentally.

Aim:
To determine the refractive index of a semi-circular Perspex (Lucite) slab using Snell's Law.

Materials:
Ray box, semi-circular block, rectangular Perspex (Lucite) block, rotating platform, single slit, sheets of A4 paper and a pencil, protractor, graph paper.

Procedure:
1. Position a semi-circular perspex block on top of a piece of A4 paper at the centre of a rotating platform as shown in Figure 449.

2. Turn on the ray box and rotate the platform so that the incident ray is directed onto the straight side of the semi-circular block.

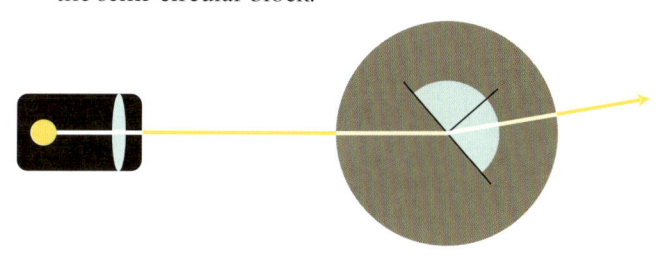

Figure 449

137

3. Measure and record the angle of incidence and the angle of refraction at this point.

4. Rotate the platform to obtain a number of angles of incidence and refraction over as wide a range as possible.

5. From the data determine the refractive index of perspex graphically.

Diffraction through a single-slit and around objects

When waves pass through a slit or any aperture, or pass the edge of a barrier, they always spread out to some extent into the region that is not directly in the path of the waves. This phenomenon is called **diffraction**. This is clearly demonstrated in a ripple tank.

Figure 450 (a) shows plane waves incident on a barrier in which there is a narrow slit, the width of which is similar in size to the wavelength of the incident waves. Figure 450 (b) shows plane waves passing the edge of a barrier.

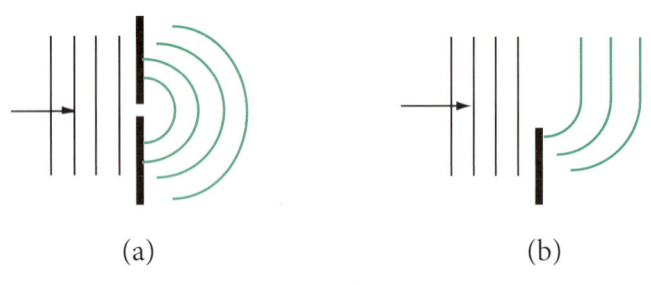

**Figure 450   (a) Diffraction at a slit
(b) Diffraction at an edge**

To understand this, we use an idea put forward by Christiaan Huygens (1629-1695). Huygens suggested that each point on any wavefront acts as a source of a secondary wave that produce waves with a wavelength equal to the wavelength associated with the wavefront. In Figure 451 we can see how this works in the case of plane wavefronts.

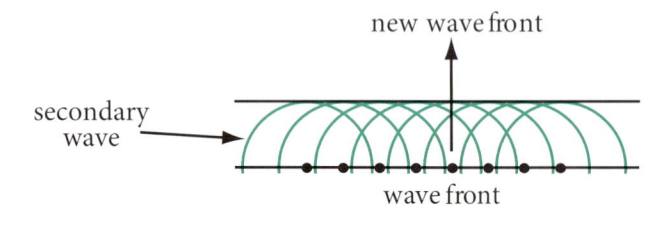

Figure 451   Planar waves

Each point on the wavefront is the source of a secondary wave and the new wavefront is found by linking together the effect of all the secondary waves. Since there are, in this case an infinite number of them, we end up with a wavefront parallel to the first wavefront. In the case of the plane waves travelling through the slit in Figure 451 (b), it is as if the slit becomes a secondary point source.

If we look at the effect of plane waves incident on a slit whose width is much larger than the incident wavelength as shown in Figure 452 then we see that diffraction effects are minimal. We can understand this from the fact that each point on the slit acts as a secondary source and we now have a situation where the waves from the secondary sources results in a wavefront that is nearly planar.

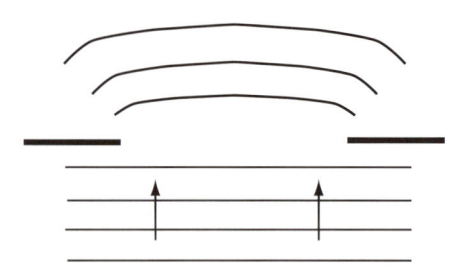

Figure 452   Diffraction at a large aperture

Diffraction effects at edges can also be understood on the basis of Huygens' suggestion. By sketching the wavefronts of secondary sources at points on a wavefront close to the edge, it easy to see that diffraction effects become more pronounced as the wavelength of the incident wave is increased.

We have seen that the ripple tank can be used to demonstrate the phenomenon of diffraction, i.e. the spreading out of a wave has it goes through an aperture or encounters an object.

Diffraction of sound waves can be demonstrated using the set up shown Figure 453.

The barrier would be expected to prevent the waves reaching points such as X. However, because of diffraction at the slit, the waves spread out in the region beyond the slit and the microphone will detect sound at points such as X.

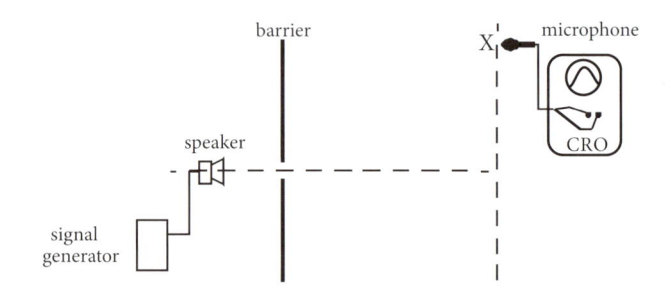

Figure 453   Diffraction of sound waves

You can also demonstrate the diffraction of light by shining laser light through a single narrow slit such that after passing through the slit, the light is incident on a screen. The effect of diffraction at the slit produces a pattern on the screen that consists of areas of illumination (bright fringes) separated by dark areas (dark fringes). In this situation, each point on the slit is acting as a secondary source and the pattern of light and dark fringes is a result of the interference (see next section) of the waves from these sources.

As mentioned above, diffraction effects at a slit really only becomes noticeable when the slit width is comparable to the wavelength. In this respect, if the laser is replaced by a point source, then as the slit is made wider, the diffraction pattern tends to disappear and the illumination on the screen becomes more like what one would expect if light consisted of rays rather than waves. Historically, the diffraction of light was strong evidence for believing that light did indeed consist of waves.

The diffraction of light can also be demonstrated by looking directly at a point source through a narrow slit. Unless the point source is monochromatic – a monochromatic source is one that emits light of a single colour (i.e. wavelength), you will see a series of different coloured fringes interspersed by dark fringes.

Exercise 4.5

Determine the critical angle for light travelling from air to water of refractive index 1.33

1. (a) Calculate the wavelengths of
    (i) FM radio waves of frequency 96 MHz
    (ii) long wave radio waves of frequency 200 kHz. (Speed of light in free space $c = 3.0 \times 10^8$ m s^{-1})

   (b) Use your answers to (i) and (ii) to explain why if your car is tuned to FM, it cuts out when you enter a tunnel but doesn't if you are tuned to long wave reception. (Hint: consider diffraction)

2. Suggest one reason why ships at sea use a very low frequency sound for their foghorn.

Superposition

We now look at what happens when two waves overlap. For example, in Figure 454 what will happen when the two pulses on a string travelling in opposite directions cross each other?

Figure 454   Two pulses approaching each other

There is a very important principle in physics that applies not only to waves but to other situations as well. This is the **principle of superposition** – when two or more waves overlap, the resultant wave is the vector sum of the displacements of the waves. What it effectively tells us is that if you want to find out the effect of two separate causes then you need to add the effects of each separate cause. In the example of the two pulses, to find the net displacement at any point we add the displacement of pulse 1 at that point to the **displacement** of pulse 2 at the point. We show two examples of this in Figure 454 below as the pulses move across each other.

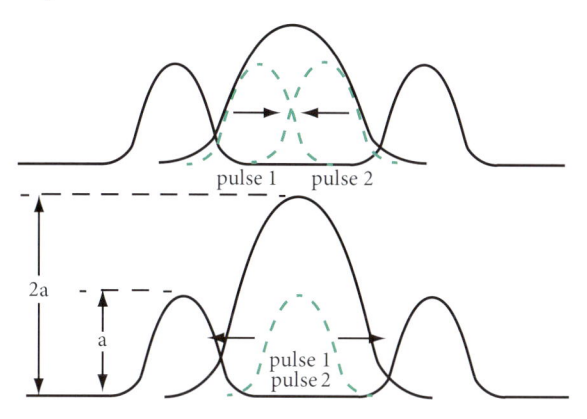

Figure 455   (a) Constructive interference

1. In the Figure 455(a) the pulses do not fully overlap and in the second diagram they do. In the second diagram we have we what call full **constructive interference**. The two pulses add to give a single pulse of twice the amplitude of each separate pulse.

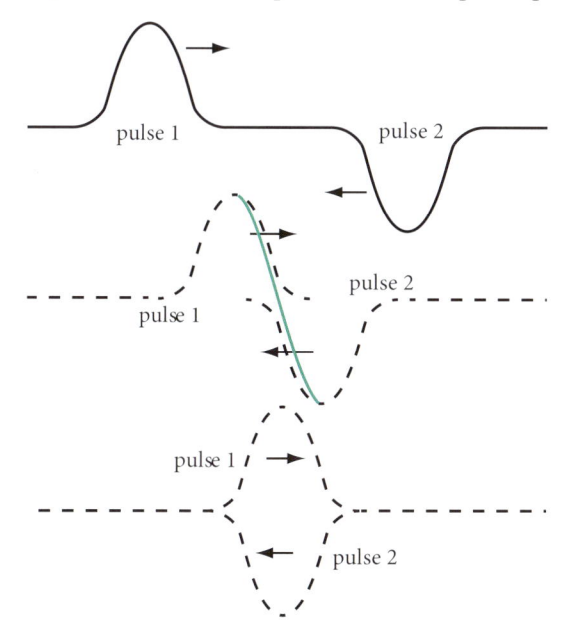

Figure 456 (b)   Destructive interference

2. We now consider two pulses as shown in Figure 456 (b).

139

CHAPTER 4

When these two pulses completely overlap the net displacement of the string will be zero. We now have what we call complete destructive interference.

Path difference and phase difference

The schematic diagram in Figure 457 shows part of the interference pattern that we get from two point sources, S_1 and S_2. The left hand side is a reflection of the pattern on the right hand side. The dashed line shows where the crests from S_1 meet the crests from S_2, creating a *double crest*, i.e., constructive interference.

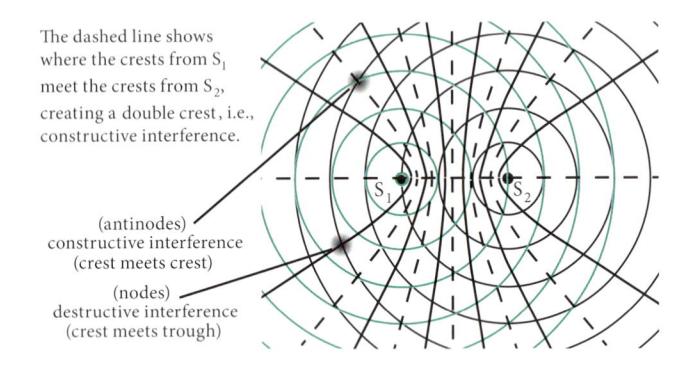

Figure 457   Two source interference

The sources are two dippers connected to a bar that is vibrated by an electric motor. The dippers just touch the surface of the water in a ripple tank. The sources therefore have the same frequency. They also are in phase. By this we mean that when a crest is created by one dipper a crest is created by the other dipper. Sources that have the same frequency and that are in phase are called **coherent sources**.

The bold lines in the diagram show the places where a trough of one wave meets the crest of the other wave producing complete destructive interference. The water at these points will not be displaced and such points are called nodal points or **nodes**. Points of maximum constructive interference are called **antinodes**. The bold lines are therefore called nodal lines.

The overall pattern produced by the interfering waves is called an **interference pattern**.

To emphasise the idea of phase the diagrams in Figure 458 show snapshots of two waves in phase and two waves that are π out of phase.

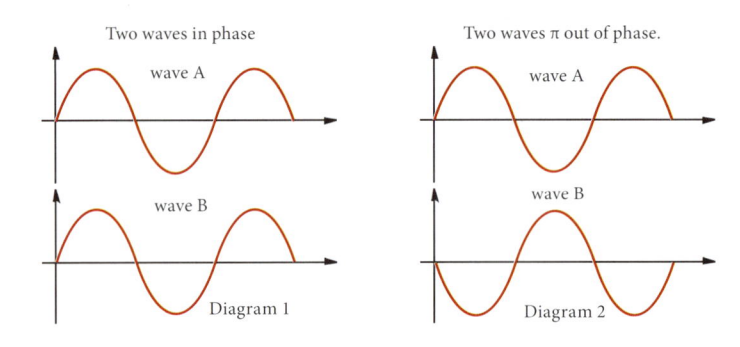

Figure 458   Phase difference

The idea of π out of phase comes from the idea that if the space displacement of wave A in diagram 2 is represented by $y = A\sin\theta$, then the space displacement of wave B is $y = A\sin(\theta + \pi)$. At the instant shown the waves in diagram 1 will reinforce and produce constructive interference whereas the waves in diagram 2 will produce destructive interference.

Let us now look at the interference between wave sources from two points in a little more detail.

In Figure 459 we want to know what will be the condition for there to be a point of maximum or minimum interference at P.

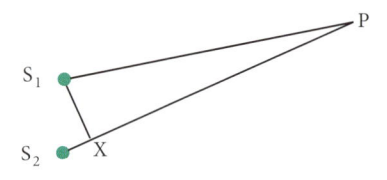

Figure 459   Interference

S_1 and S_2 are two coherent point sources. For there to be a maximum at P, a trough must meet a trough or a crest must meet a crest. The waves from the sources will have travelled a different distance to P and therefore will not be necessarily in phase when they reach P. However, if a crest meets a crest (or trough a trough) at P then they will be in phase. They will be in phase only if the difference in the distances travelled by the two waves is an integral number of wavelengths. The interference of waves is discussed in further detail in Option G (Chapter 18).

This means that path difference

$$S_2P - S_1P = S_2X = n\lambda, \; n = 0, 1, 2, \ldots$$

The waves will be out of phase if the difference in the distance travelled, is an odd number of half-wavelengths.

So for a minimum we have path difference =

$$S_2P - S_1P = S_2X = \left(n + \tfrac{1}{2}\right)\lambda, \; n = 0, 1, 2, \ldots$$

Amplitude and intensity

The energy that a wave transports per unit time across unit area of the medium through which it is travelling is called the intensity (I). From our knowledge of SHM we know that the energy of the oscillating system is proportional to the square of the amplitude. Hence for a wave of amplitude A, we have that

$I \propto A^2$

The ear and the brain in combination produce a very sensitive instrument that can detect properties such as pitch and loudness. Pitch is a measure of frequency. Loudness is a measure of the power carried in a longitudinal sound wave. Sound waves travel in a three-dimensional medium such as air (gas) or bone (solid) but these travelling sound waves form two-dimensional fronts that are perpendicular to the direction of the wave propagation. Therefore, it is more appropriate to define the transport of sound energy as the average power per unit area rather than as the total power in the wave because of the two-dimensional wavefronts that are detected in the air-brain combination.

The average power per unit area of a sound wave that is incident perpendicular to the direction of propagation is called the sound intensity. The units of sound intensity are watts per square metre, W m⁻².

As the sound intensity spreads out from its source, the intensity I is reduced as the inverse square of the distance d from the source. Therefore,

For example, doubling the distance from a sound source would cause the sound intensity to become one-quarter that at the original distance.

In a sound wave, the medium particles, such as air molecules, vibrate in simple harmonic motion parallel to the direction of wave propagation. Just as the energy in simple harmonic motion is proportional to the square of the amplitude (A) and the square of the frequency, so too sound intensity is directly proportional to the square of the amplitude and the square of the frequency of the particle vibrations.

$I \propto A^2$

The ear-brain combination can accommodate a range of sound intensities. The just detectable sound intensity known as the threshold of hearing I_0 is taken to be 10^{-12} W m⁻². The sound intensity that produces a sensation of pain called the pain threshold is taken to be 1 W m⁻².

A sound wave can also be considered as a pressure wave because the particle vibrations cause harmonic variations in the density of the medium. The pressure variations corresponding to the hearing threshold and the pain threshold are 3×10^{-5} Pa to 30 Pa respectively superimposed on atmospheric pressure of about 101 kPa.

Because of the extreme range of sound intensities to which the ear is sensitive - a factor of 10^{12} - (curiously about the same factor as the range of light intensities detectable by the human eye), and because loudness (a very subjective quantity) varies with intensity in a non-linear manner, a logarithmic scale is used to describe the intensity level of a sound wave. The intensity level of sound, IL, is defined as:

$$b = \log_{10}\left(\frac{I}{I_0}\right)$$

where I is the intensity corresponding to the level b and I_0 is the threshold intensity or threshold of hearing taken as 10^{-12} W m⁻². b is measured in bels B, named after Alexander Graham Bell, one of the inventors of the telephone. Because the bel is a large unit, it is more convenient to use the decibel dB (one-tenth of a bel).

$$b = 10\log_{10}\left(\frac{I}{I_0}\right) \text{ dB}$$

Using this scale, the threshold of hearing is:

$$b = 10\log_{10}\left(\frac{I_0}{I_0}\right) = 10\log_{10} 1 = 0 \text{ dB}$$

The pain threshold is:

$$b = 10\log_{10}\left(\frac{10^0}{10^{-12}}\right) = 10\log_{10} 10^{12} = 120 \text{ dB}$$

Example

The sound intensity at a distance of 20 m from a fire alarm is 5.0×10^{-3} W m⁻². Calculate the sound intensity at a distance of 50 m.

Solution

Given that $I \propto \frac{1}{d^2} \Rightarrow Id^2 = k$ (a constant).

Therefore,

$I_1 d_1^2 = I_2 d_2^2$.

So that $(5.0 \times 10^{-3} \text{ W m}^{-2})(20 \text{ m})^2 = I_2 \times (50 \text{ m})^2$

$\therefore I_2 = \dfrac{(5.0 \times 10^{-3} \text{ W m}^{-2})(20 \text{ m})^2}{(50 \text{ m})^2}$

$= 8.0 \times 10^{-4}$ W m⁻²

Chapter 4

Polarization

This section will specifically address the **polarization** of light. However, the discussion is equally valid for all electromagnetic waves.

As discussed previously, electromagnetic waves consist of oscillating electric and magnetic fields. The electric field vector is perpendicular to the magnetic field vector. Since em waves are transverse, the oscillations of the fields are confined to the plane of the wavefront. If we consider just the electric field a vector, then the angle of vector within this plane can take any value between zero and 360°. In general, this angle is continually changing as the wave advances. However, we can simplify the situation by resolving the vector into a horizontal component and a vertical component. In Figure 460, the direction of travel of the em wave is out of the paper and the electric field vector is shown resolved into the components E_V and E_H.

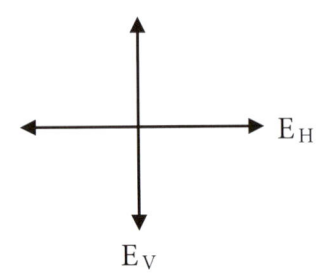

Figure 460   The electric field components of light

Light in which the plane of vibration of the electric vector is continually changing is said to be *unpolarized*.

When light passes through natural crystals of tourmaline and of calcite, and through certain synthetic materials, only the E_V or E_H (depending on one's viewpoint) is transmitted, a process called **preferential absorption**. Because of this, the emergent light is said to be plane polarized.

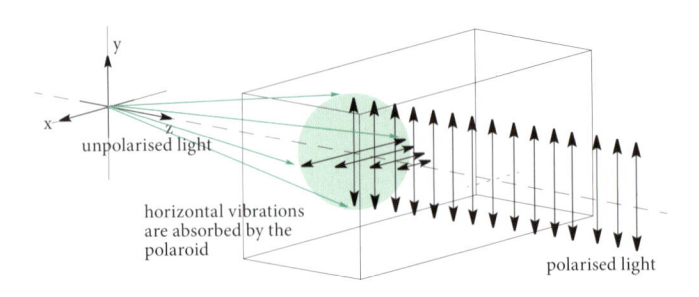

Figure 461   Polarization by preferential absorption

Figure 461 shows how unpolarized light upon entering and leaving a sheet of a synthetic material called a **polaroid** is polarized. Through the process of preferential absorption, the vibrations that are parallel to the transmission plane of the polaroid (the E_H components) are removed and the light emerges polarized in the vertical plane. If the polaroid is rotated through 90° the E_V components of the unpolarized light are removed and the E_H components are transmitted.

Polarization by reflection and Brewster's law

Apart from light being partially or totally polarized by transmission through certain materials, it can also be partially or totally polarized by reflection. In 1808, E.L. Malus (1775-1812) showed that light reflected from a horizontal sheet of glass is polarized. This is the reason that polarized sunglasses cut down glare from the surface of water.

Figure 462 shows a ray of unpolarized light being reflected and transmitted (refracted) at the surface of water. The dots and arrows show the components of the electric field vector.

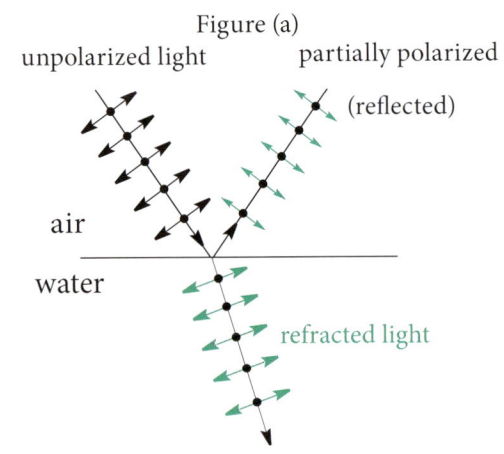

Figure 462   Partial polarization of reflected light at the surface of water

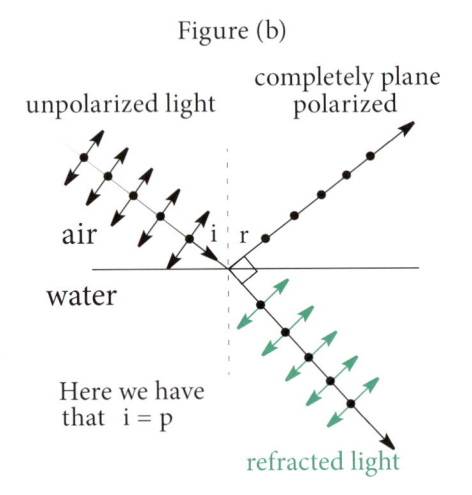

Figure 463   Complete polarization of reflected light at the surface of water

In Figure 463 the unpolarized beam strikes the water at a certain angle of incidence and is partially polarized on reflection (*the refracted beam is also partially polarized but to a much lesser extent than the reflected light*).

In Figure 463, at a particular angle where the reflected ray is perpendicular to the refracted ray, the reflected ray is **completely plane-polarized**. The angle to the normal at which this occurs is called the **Brewster angle** or the 'polarising angle' after its discoverer *David Brewster*, a Scottish physicist (1781-1868).

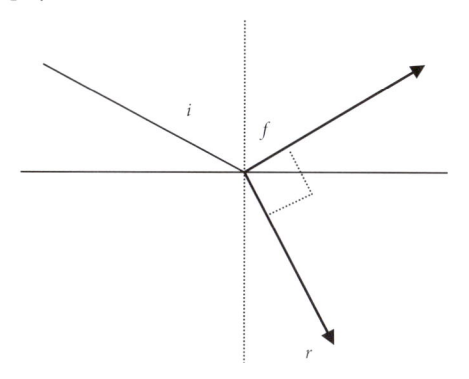

Figure 464   Illustrating Brewster's angle

In Figure 464 the light is incident at the Brewster angle ϕ. From the diagram we see that the refracted angle $r = (90 - \phi)$ such that $\sin r = \sin(90 - \phi) = \cos \phi$. The incident angle $i = \phi$

From the definition of refractive index n there is therefore

$$n = \frac{\sin i}{\sin r} = \frac{\sin \phi}{\cos \phi}$$

or $n = \tan \phi$

This is known as **Brewster's law**.

For water of refractive index 1.3, the Brewster angle $\phi = \tan^{-1}(1.3) = 52°$. This means if you look at the surface of water at an angle of 38° to the surface, the light reflected from the surface to your eyes will be plane polarized. If you are wearing Polaroid sunglasses then the only light entering your eyes will be light originating from below the surface of the water. This phenomenon is put to good use by anglers.

We have assumed in deriving Brewster's law that the incident light is in air. If the incident light is in a medium of refractive index n_1 and is incident on the surface of a medium if refractive index n_2, then Brewster's Law becomes

$$\tan \phi = \frac{n_1}{n_2}$$

Since the refractive index is different for different wavelengths (Topic G.1.3), the Brewster angle will vary with wavelength. However for substances such as glass and water, the angle does not change very much over the visible spectrum as the example below shows.

The refractive index for crown glass for red light of wavelength 660 nm is 1.52 and for violet light of wavelength 480 nm is 1.54. Calculate the difference in the Brewster angle for these two wavelengths.

$\phi_{red} = \tan^{-1}(1.52) = 56.7°$

$\phi_{blue} = \tan^{-1}(1.54) = 57.0°$

therefore $\Delta \phi = 0.30°$

When light is incident on a surface, the electric field vector of the light sets the electrons in the surface into oscillation. The radiation from these oscillating electrons is the origin of the reflected light. If the reflected light is observed at 90° to the refracted light, only the vibrations of the electric field vector that are perpendicular to the plane of incidence will be in the reflected beam. This is because the components of electric field in the plane of incidence cannot have a component at 90° to this plane. Hence at the Brewster angle, the reflected light is plane polarized. Clearly, longitudinal waves cannot be polarized. However, remember that all other electromagnetic waves such as radio and microwaves can be polarized.

Polarizers and analysers

In Figure 465 unpolarized light is incident on a polaroid sheet (called the polarizer) and after transmission, is incident on another polaroid (called the analyser) whose axis is at 90° to the polarizer.

After passing through the polarizer, the intensity of light incident on this ideal polarizer is halved by preferential absorption and after passing through the analyser, is reduced to zero.

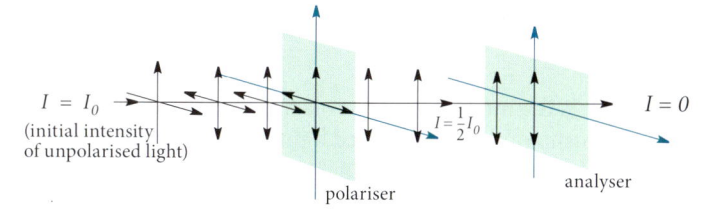

Figure 465   Crossed polaroids

The analyser can be rotated from 0° to 90° to reduce the emerging light intensity from a maximum to zero. When the intensity is zero the polarizer and analyser are said to be 'crossed'.

Polaroid is a material composed of sheets of nitro-cellulose containing crystals of iodosulfate arranged as long molecules. These long molecules reflect and preferentially absorb electric vector components along their length. As mentioned above, this material is used in sunglasses to reduce glare.

Chapter 4

Malus' law

In the situation shown in Figure 1129, the intensity of the light is incident on the polarizer is I_0 and on the analyser is ½ I_0. The intensity of the light after transmission through the analyser is zero. Let us now consider the case where the transmission axis of the analyser is inclined at an angle θ to the direction of the field vector of the light incident on the analyser. The electric field vector of the light incident on the analyser may be resolved into a component parallel to the transmission plane of the polarizer and one at right angles to it as shown in Figure 466. In this Figure the amplitude of the light is considered to be proportional to the amplitude of the electric field.

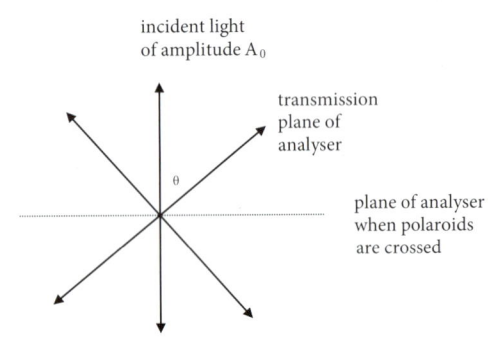

Figure 466

Clearly the component that is at right angles to the transmission plane of the analyser will not be transmitted. The amplitude A_1 of the component parallel to the transmission plane is given by

$$A_1 = A_0 \cos\theta$$

But the intensity of the light is proportional the square of the amplitude, therefore we can write that

$$I = A_0^2 \cos^2\theta = I_0 \cos^2\theta$$

where I_0 is the intensity of the polarized light (which is half the intensity of the unpolarized light) and I the intensity of the transmitted polarized light. We have then that

$$I = I_0 \cos^2\theta$$

This is **Malus' law** named after *Etienne Malus* (1775-1812) who discovered, by accident, polarization of light by reflection.

Optically active substances

Substances that are able to rotate the plane of polarization of light are said to be optically active. Such substances include solutions of so-called chiral molecules such as sucrose (sugar). Chiral molecules of a substance may be identical in all respects except that some of the molecules "left-handed" and some are "right-handed". By this we mean that a right-handed molecule cannot be superimposed on the mirror image of a left-handed molecule, just as your own right hand and the mirror image of your left hand cannot be superimposed. All amino acids, the building blocks of life occur in both left-handed and right handed forms. However, all life on Earth is made exclusively of left-handed amino acids and no one knows why. Substances whose molecules rotate the plane of polarisation exclusively clockwise usually have their name prefixed by "dextro" and those whose molecules rotate the plane exclusively anticlockwise, with the prefix "levto".

Solids with rotated crystal planes such as quartz and calcite also rotate the plane of polarization of light. In the Figure 467 a solution of sugar is contained in an instrument called a **polarimeter**. A polarimeter is essentially a tube that is bounded at either end with a polaroid.

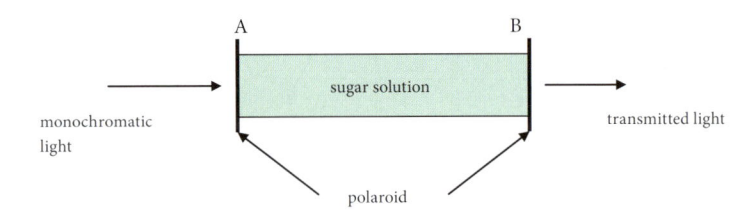

Figure 467   The principle of a polarimeter

Monochromatic light is polarized by the polaroid A and after passing through the sugar solution, the plane of polarization has been rotated by the sugar molecules. The number of sugar molecules present determines the degree of rotation and the length of the path traversed by the polarized light. The polaroids A and B are initially crossed such when no solution is present, no light is transmitted by B. When a solution of sugar is added, light will now be transmitted by polaroid B since the solution has rotated the plane of polarization. The polaroid B may be rotated and is also provided with an angular scale. By rotating B until no light is transmitted, the degree of rotation of the plane of polarisation can be measured. If a polarimeter of length l with a concentration of the sugar solution C produces an angle of rotation α, then we can define a specific angle of rotation α_S from

$$\alpha_S = \frac{\alpha}{lC}$$

The angle α_S is often referred to as the optical activity. Because it also depends on the temperature of the solution and wavelength of light, measurements of optical activity are usually standardised at 20 °C and 589 nm (the wavelength of the D-line in the line emission spectrum of sodium). The phenomenon of optical activity is used in the sugar industry to measure the concentrations of syrups and it is also being developed to measure blood sugar levels in people with diabetes. Chemists also use it to identify the presence of certain substances present in solutions.

Optical activity and stress

Many kinds of glass and transparent plastics become optically active when put under stress. The effect of the stress is to alter the speed of the two components of polarized light. A **strain viewer** consists of two polaroids with a material under strain placed between them as shown in Figure 468.

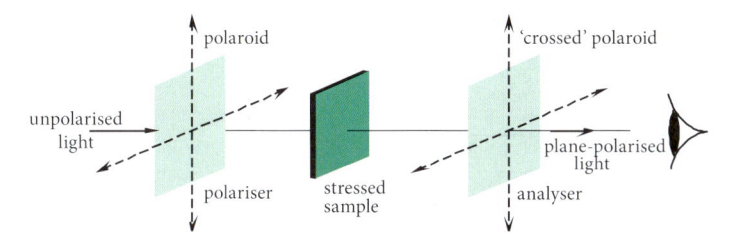

Figure 468   Stress and polarization

When the stress is increased, the concentration of the optical activity increases. A series of light and dark coloured bands are seen that can be analysed. Plastic models of objects are made to determine any places that might cause mechanical breakdown.

Liquid crystal displays (LCD)

The molecules of solids keep a fixed orientation and a fixed position whereas molecules of a liquid continually change their orientation and position. Liquid crystals are substances that are not quite solid and not quite liquid in the respect that, although the molecules keep a fixed orientation, the molecules do change their positions. The molecules of one particular type of LC called a **nematic liquid crystal**, essentially have the shape of a twisted helix and as such, can rotate the plane of polarization of polarized light. If an electric field is applied across the crystal, then the amount of "twist" of the crystal is altered and hence the degree of polarization of incident will also be altered.

A liquid crystal display consists essentially of two crossed polaroids between which is placed a LC. Without the LC, light polarized by the polarizer will be blocked by the analyser. However, with the LC present, the plane of polarisation is rotated and light will now be transmitted by the analyser. This light can again be blocked by applying an electric field to the LC.

Figure 469, shows the basic set-up of the different layers necessary to produce a liquid crystal display.

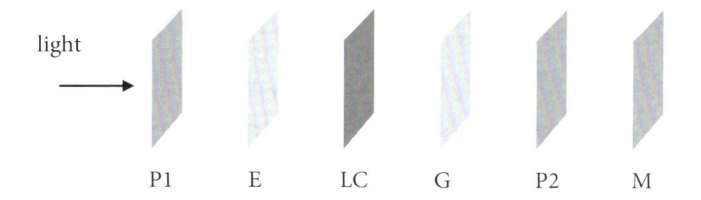

Figure 469   The principle of a liquid crystal display (LCD)

Incident light is polarized by the polarizer P1. E is an indium-tin oxide electrode and G is a piece of glass upon which there is another electrode in the shape of the final display (e.g. a letter or a number). P2 is the analyser and M is a mirror. In the absence of an electric field, the LC rotates the plane of polarization such that the light is transmitted by P2 and reflected back to the observer by the mirror M. When a potential difference is applied to the electrodes, the resulting field untwists the nematic crystal such that the plane of polarization of the light is no longer rotated. However, only those parts of the LC outlined by the shape on the electrode will be affected such that the field of view will now contain a black area corresponding to the shape of the electrode on G. By varying the strength of the applied field and hence the degree of twist of the LC, the displayed shape can be varied from different shades of grey through to black.

Whole pictures can be displayed by breaking the picture down into small areas called picture elements or **pixels** and using a LC for each pixel.

Example

Unpolarized light of intensity 6.0 W m^{-2} is incident on a polarizer. The light transmitted by the polarizer is then incident on an analyser. The angle between the transmission axes of the polarizer and analyser is 60°. Calculate the intensity of the light transmitted by the analyser.

Solution

Using the formula for intensity,

we have $I = I_0 \cos^2 \theta$

$I = (3.0 \text{ Wm}^{-2}) \times \cos^2 60°$

$= 0.75 \text{ W m}^{-2}$

Chapter 4

Exercise 4.5

1. Which of the following correctly describes the direction of a ray drawn relative to a wavefront for longitudinal and transverse waves?

	longitudinal	transverse
A.	parallel	parallel
B.	perpendicular	perpendicular
C.	parallel	perpendicular
D.	perpendicular	parallel

2. If the refractive indices of water and a particular glass are 1.33 and 1.56 respectively, then total internal reflection at an interface between them

   A. can never occur.
   B. may occur only when the light goes from glass to water.
   C. may occur only when the light goes from water to glass.
   D. may occur for light going either from water to glass or from glass to water.

3. The diagram below shows incident wavefronts approaching the boundary between two different media. The speed of the wavefronts in medium 2 is greater than the speed in medium 2.

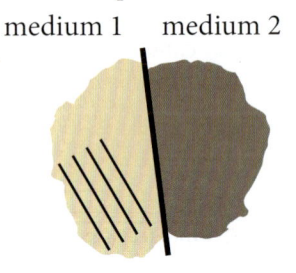

   Which one of the following sketches correctly shows the refracted wavefronts?

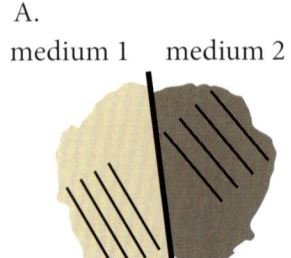

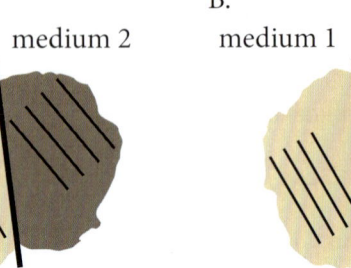

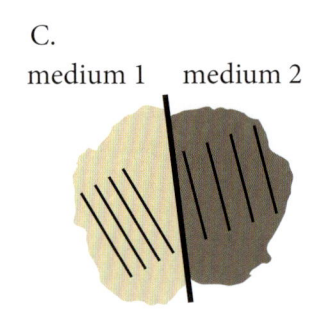

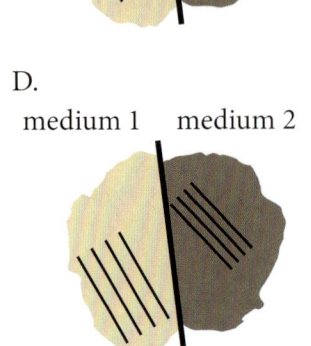

4. The phenomena of interference can be demonstrated using sound. This is evidence that

   A. sound is a longitudinal wave
   B. sound is electromagnetic in nature
   C. sound has wave characteristics
   D. sound behaves as a stream of particles

5. The figure below shows two pulses travelling in the shown direction along a string towards a fixed boundary.

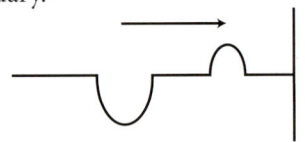

   After the two pulses have both been reflected the string shape could appear as

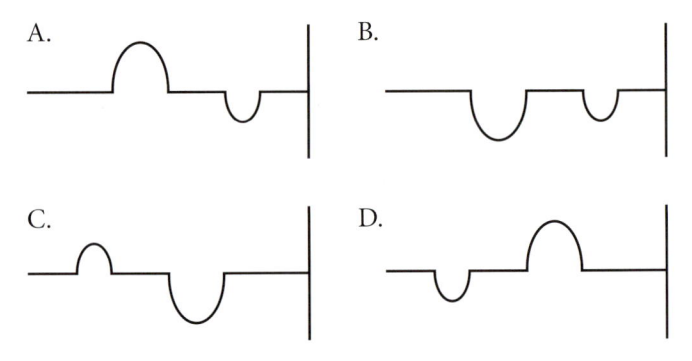

6. (a) Calculate the wavelengths of

   (i) FM radio waves of frequency 96 MHz

   (ii) long wave radio waves of frequency 200 kHz. (Speed of light in free space $c = 3.0 \times 10^8$ m s^{-1})

   (b) Use your answers to (i) and (ii) to explain why if your car is tuned to FM, it cuts out when you enter a tunnel but doesn't if you are tuned to long wave reception. (Hint: consider diffraction).

7. Suggest one reason why ships at sea use a very low frequency sound for their foghorn.

8. Sophia sounds two tuning forks A and B together and places them close to her ear.

   The graph in Figure 470 shows the variation with time t of the air pressure close to her ear over a short period of time.

146

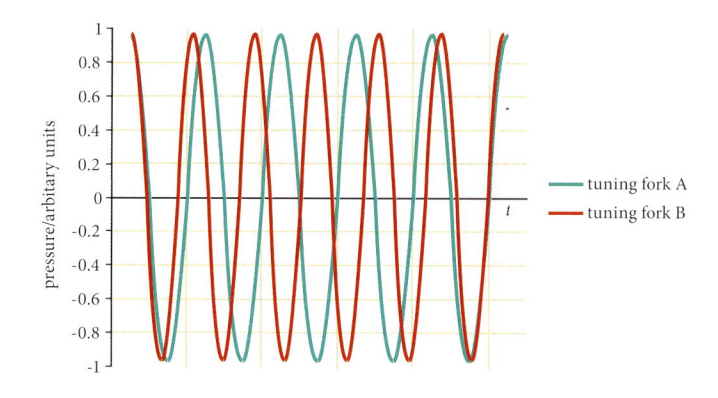

Figure 470   Tuning fork graph

On a separate piece of graph paper, use the principle of superposition to construct a graph that shows the variation with time t of the resultant pressure close to Sophia's ear.

Use your graph to suggest the nature of the sound that Sophia will hear up until the time that the vibrations of the tuning forks die out.

9. A ray of yellow light goes from air into water with a refractive index of 1.33 at an angle of incidence of 34.0°. Determine the angle of refraction of the ray.

10. Determine the refractive index of a certain type of glass in which the velocity of light is 2.40×10^8 ms^{-1}.

11. Determine the velocity of light in benzene that has a refractive index of 1.501.

12. Determine the critical angle for light travelling from water with a refractive of index 1.33 to air.

13. A ray of orange light passes from water to crown glass that has a refractive index of 1.5. The ray is refracted at an angle of 410 in the crown glass. Determine the original angle of incidence.

14. A ray of light passes from air into diamond with a refractive index of 2.42 at an angle of 410. Determine the angle through which the wave is deviated from its original angle of incidence.

15. Light reflected from the surface of a material in air at an angle of 56.5° to the normal is completely polarized. Calculate the angle of refraction of the glass.

4.5 Standing Waves

NATURE OF SCIENCE:

Common reasoning process: From the time of Pythagoras onwards the connections between the formation of standing waves on strings and in pipes have been modelled mathematically and linked to the observations of the oscillating systems. In the case of sound in air and light, the system can be visualized in order to recognize the underlying processes occurring in the standing waves. (1.6)

© IBO 2014

Essential idea: When travelling waves meet they can superpose to form standing waves in which energy may not be transferred.

Understandings:
- The nature of standing waves
- Boundary conditions
- Nodes and antinodes

The nature of standing waves

A very interesting situation arises when considering a travelling wave that is reflected and the reflected wave interferes with the forward moving wave. It can be demonstrated either with a rubber tube, stretched string or a slinky spring attached to a rigid support. In the diagram the tube is attached at A and you set up a wave by moving your hand back and forth at B. If you get the frequency of your hand movement just right then the tube appears to take the shape as shown in Figure 471. When you move your hand faster you can get the tube to take the shape shown in Figure 472. All points other than A and B on the tube are of course oscillating but the wave is not moving forward. You can actually get the wave to appear to stand still by illuminating it with a strobe light that flashes at the same frequency of vibration as your hand. The fact that the wave is not progressing is the reason why such waves are called standing or **stationary waves**.

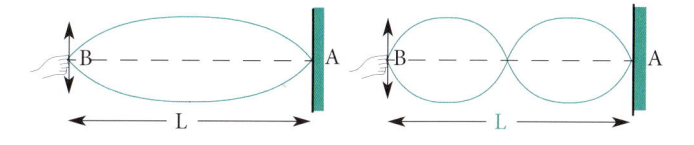

Figures 471 and 472

There are several things to note about standing waves. The fact the wave is not moving forward means that no energy is being propagated. If you increase the amplitude with which you shake your hand, this increase in energy input to the wave will result in a greater maximum displacement of the tube.

CHAPTER 4

It can also be seen that points along the wave oscillate with different amplitudes. In this respect, the amplitude of a standing wave clearly varies along its length.

To illustrate this, Figure 473 shows a standing wave set up in a string of length 30 cm at a particular instant in time. Figure 474 shows the variation with time t of the amplitude x_0 of the string at a point 13 cm along the string.

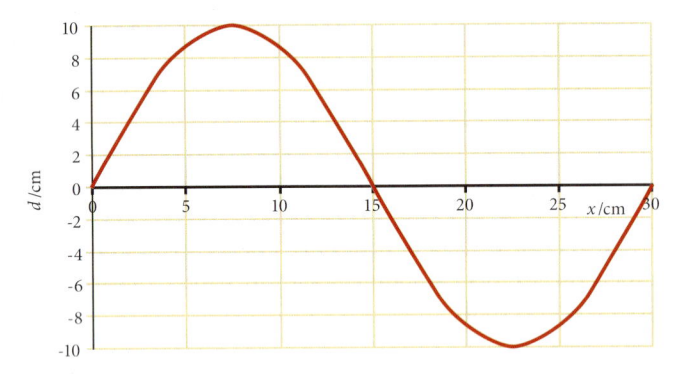

displacement d of string vs distance x along string

Figure 473   A standing wave

amplitude x_0 vs time t

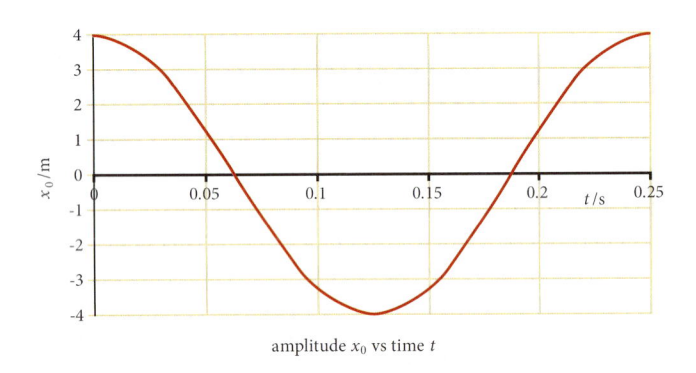

Figure 474   Variation of amplitude with time

Since energy is not propagated by a standing wave, it doesn't really make a lot of sense to talk about the speed of a standing wave.

This speed is the speed of a travelling wave in the string. As we have seen, the speed of a travelling wave is determined by the nature and properties of the material through which it travels. For the string therefore, the speed of the travelling wave in the string determines the frequency with which you have to oscillate the string to produce the standing wave. (From Figure 473 we see that for this situation the frequency of oscillation of the string is f = 1/T = 1/0.25 = 4.0 Hz)

Also, since at any one time all the particles in a standing wave are either moving "up" or moving "down", it follows that all the particles are either in phase or in anti-phase with each other.

Nodes and Antinodes

A standing wave arises from the interference of two waves moving in opposite directions. To understand this, let us look at the situation of a standing wave in a string or tube as described above. Initially when you start moving the free end of the tube up and down, a wave travels along the tube. When it reaches the fixed end, it is reflected and, as described in Topic 4, the reflected wave is π out of phase with the forward (incident) wave. The forward wave and reflected wave interfere and the resultant displacement of the tube is found from the principle of superposition. This is illustrated in Figure 475 which shows, at a particular incident of time, the displacement of the tube due to the incident wave, the displacement of the tube due to the reflected wave and the resultant displacement due to the interference of the two waves.

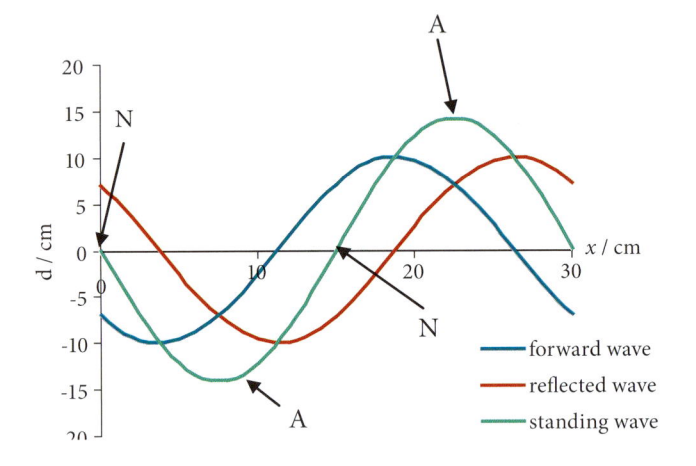

Figure 475   Nodes and anti-nodes

If at the fixed end, due to the incident wave, the tube is moving upwards, then due to the reflected wave it will be moving downwards such that the net displacement is always zero. Similarly, at the mid-point of the string, the displacements of the tube due to each wave are always in anti-phase. Hence the net displacement at this point is always zero. Points on a standing wave at which the displacement is always zero, are called **nodes** or nodal points. These are labelled N in Figure 474.

The points at which a standing wave reach maximum displacement are called **antinodes**. In Figure 474, the antinodes are at points one quarter and three-quarters the length of the tube (7.5 cm and 22.5 cm) and are labelled A. The amplitude of the forward wave and the reflected wave is 10 cm, hence the maximum displacement at an antinode in the tube is 20 cm. The maximum displacement at the antinodes will occur at the times when the forward and reflected waves are in phase.

Figure 476 shows an instance in time when the interference of the forward and reflected wave produce an the overall displacement of zero in the standing wave.

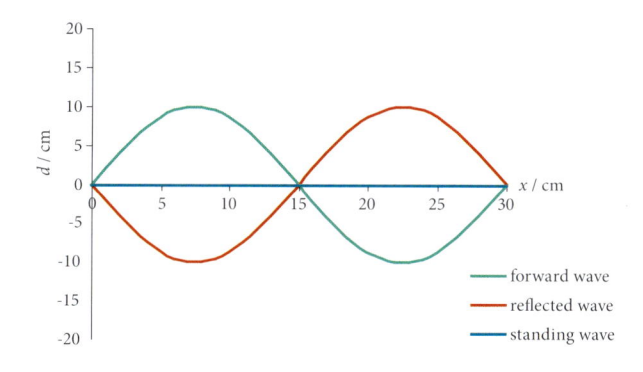

Figure 476 *Forward and reflected waves*

Sketch the shape of the forward and reflected wave in a string at an instant in time that results in the antinodes of the standing wave having maximum displacement.

Boundary conditions

If you take a wire and stretch it between two points and pluck it then you will actually set up standing waves in the wire as described above. The number of standing waves that you set up will actually be infinite.

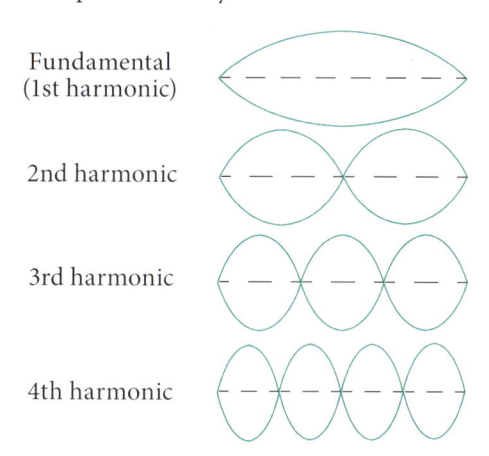

Figure 477 *The first four modes of vibration*

Figure 477 shows the first four modes of vibration i.e. standing waves in the string The modes of vibration are called **harmonics**. The first harmonic is called the **fundamental**. This is the dominant vibration and will in fact be the one that the ear will hear above all the others. This is what enables you to sing in tune with the note emitted by the plucked string.

If we were to vibrate the stretched string at a frequency equal to the fundamental or to one of its harmonics rather than just pluck it, then we set up a standing wave as earlier described. We have used the phenomenon of resonance to produce a single standing wave. In the case of the stretched string it has an infinite number of natural frequencies of oscillations, each corresponding to a standing wave. Hence, when plucked, we obtain an infinite number of harmonics.

Figure 477 shows part of what is called a **harmonic series**. Different fundamentals can be obtained by pinching the string along its length and then plucking it or by altering its tension. In a violin for example the four strings are of the same length, but under different tensions so that each produces a different fundamental. Holding the string down at different places and then bowing it obtain the notes of the harmonic series associated with each fundamental.

The harmonics essentially affect the quality of the note that you hear. The presence of harmonics is one reason why different types of musical instruments sounding a note of the same frequency actually sound different. It is not the only reason that musical instruments have different sound qualities. An "A" string on a guitar sounds different from the "A" string of a violin, because they are also produced in different ways and the sound box of each instrument is very different in construction. The actual construction of the violin for example distinguishes the quality of the notes produced by a Stradivarius from those produced by a plastic replica.

Figure 477 enables us to derive a relationship between the wavelengths of the harmonics and length of the string. If the length of the string is L then clearly, the wavelength of the fundamental (first harmonic) is $\lambda = 2L$, for the second harmonic $\lambda = L$ and for the third harmonic $\lambda = \frac{2L}{3}$. From this sequence, we see that the wavelength λ_n of the nth harmonic is given by

$$\lambda_n = \frac{2L}{n}$$

Resonance and standing waves also play their part in the production of sound from pipes. If you take a pipe that is open at one end and blow across the top it will produce a sound. By blowing faster you can produce a different sound. In this situation you have used resonance to set up a standing wave in the pipe. It is now the air molecules inside the pipe that are set vibrating. The sound wave that you create at the open-end travels to the bottom of the pipe, is reflected back and then again reflected when it reaches the open end. The waves interfere to produce a standing wave. However, when waves are reflected from an "open" boundary they do not undergo a phase change so that there is always an antinode at the open end.

The fundamental and the first three harmonics for a pipe open at one end are shown Figure 478.

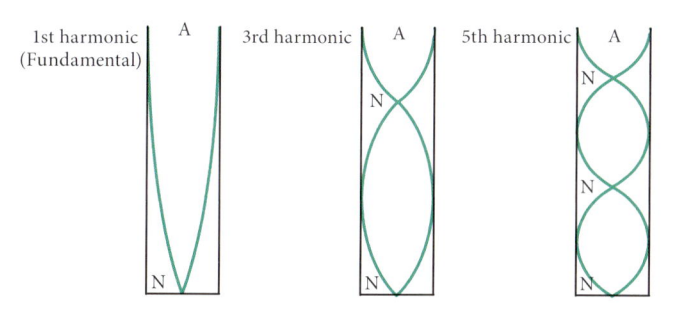

Figure 478 *The harmonics of a pipe open at one end*

CHAPTER 4

A pipe that is open at both ends and that has the same dimensions, as the previous pipe, will produce a different fundamental note and a different harmonic series. This is shown in Figure 479.

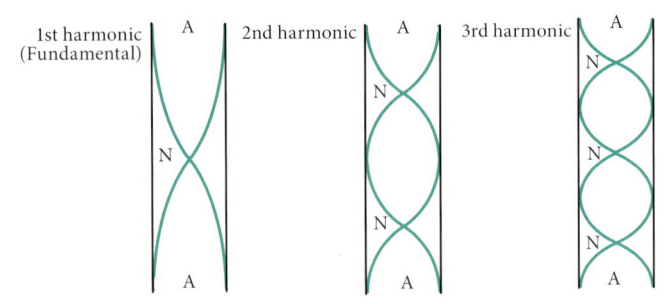

Figure 479 The harmonics of a pipe open at both ends

We see that, whereas a pipe open at both ends produces all the odd and even harmonics of the fundamental $\lambda = 2L$, ($\lambda_n = \frac{2L}{n}$), a pipe closed at one end can produce only the odd harmonics of the fundamental $\lambda = 4L$. With open and close pipes we are essentially looking at the way in which organs, brass instruments and woodwind instruments produce musical sounds.

The Figure 480 summarises the differences between travelling and standing waves.

Property	Travelling wave	Standing wave
Energy propagation	Propagated	Not propagated
Amplitude	Single amplitude	Variable amplitude
Phase difference	All phase differences between 0 and 2π	Only 0, 2π and π phase difference

Figure 480 Comparing travelling and standing waves

Since travelling waves have a single amplitude, it follows that there are no nodal or anti-nodal points in a travelling wave.

Example

The speed v of a wave travelling in a string is given by the expression

$$v = \sqrt{\frac{T}{\mu}}$$

where T is the tension in the string and μ is the mass per unit length of the string.

(a) Deduce an expression for the frequency f of the fundamental in a string of length L.

(b) Use your answer to (a) to estimate the tension in the 'A' string of a violin (frequency = 440 Hz).

Solution

(a) The wavelength of the fundamental is 2L

Using $f = \dfrac{v}{\lambda}$ we have $f = \dfrac{1}{2L}\sqrt{\dfrac{T}{\mu}}$

(b) As this is an estimate, we are not looking for an exact value but we do need to make some sensible estimates of L and μ.

$L = 0.5$ m say and $\mu = 2.0 \times 10^{-3}$ kg m^{-1}
i.e. about 2 g per metre.

From $f = \dfrac{1}{2L}\sqrt{\dfrac{T}{\mu}}$ we have that $T = 4L^2\mu f^2 =$

$4 \times 0.25 \times 2.0 \times 10^{-3} \times (440)^2 \approx 400$ N

Excercise 4.6

1. For a closed organ pipe of length L, the fundamental frequency is proportional to

   A. $\sqrt{L}$ C. L
   B. $1/\sqrt{L}$ D. 2L

2. An organ pipe is open at both ends and the fundamental note produced has a frequency of 300Hz. The frequency of the next harmonic would be

   A. 600 Hz C. 900 Hz
   B. 800 Hz D. 1200 Hz

3. An organ pipe is closed at one end and produces a fundamental note of frequency 128 Hz.

   (a) Calculate
   (i) the frequencies of the next two harmonics in the harmonic series of the pipe.
   (ii) the frequencies of the corresponding harmonics for an open pipe whose fundamental is 128 Hz
   (iii) the ratio of the length of the closed pipe to that of the open pipe.

   (b) Suggest why organ pipes that emit notes at the lower end of the organ's frequency range are usually open pipes.

4. Calculate the shortest length of a closed pipe that will resonate in air at 0 0C with a tuning fork with a frequency 320 Hz.

5. Using the same conditions as question 2 but now with an open pipe, calculate the shortest length of an open pipe that will resonate.

150

5. Electricity and magnetism

Contents

5.1 – Electric fields
5.2 – Heating effect of electric currents
5.3 – Electric cells
5.4 – Magnetic effects of electric currents

Essential Ideas

When charges move an electric current is created.

One of the earliest uses for electricity was to produce light and heat. This technology continues to have a major impact on the lives of people around the world.

Electric cells allow us to store energy in a chemical form.

The effect scientists call magnetism arises when one charge moves in the vicinity of another moving charge. © IBO 2014

CHAPTER 5

5.1 Electric fields

NATURE OF SCIENCE:

Modelling: Electrical theory demonstrates the scientific thought involved in the development of a microscopic model (behaviour of charge carriers) from macroscopic observation. The historical development and refinement of these scientific ideas when the microscopic properties were unknown and unobservable is testament to the deep thinking shown by the scientists of the time. (1.10)

© IBO 2014

Essential idea: A study of oscillations underpins many areas of physics When charges move an electric current is created.

Understandings:
- Charge
- Electric field
- Coulomb's law
- Electric current
- Direct current (dc)
- Potential difference

Charge

There are 2 types of electric charge – positive charge and negative charge.

It was not until the late 1890s through the work of *J.J. Thomson* that the true nature of electrons was discovered through experiments with cathode ray tubes. With this exploration of atoms and quantum mechanics in the 1900s, the electrical properties of matter were understood.

We now know that

- charge is conserved
- charge is quantised
- the force between two point charges varies as the inverse square law of the distance between the two charges.

These properties will be outlined further in a later section of this chapter.

We will begin with the study of stationary electric charges-**electrostatics**.

The present quantum mechanical model of the atom suggests that the positively charged protons and the neutral neutrons are held tightly in the nucleus by the short-range strong nuclear force, and that the negatively charged electrons are held in energy levels around the nucleus by the electromagnetic force. The electrons in a material are relatively free to move, and some electrons, given the right conditions, can move from one material to another. The materials can become electrically charged. The materials can either have an excess of electrons or a deficiency of electrons.

- substances with an excess of electrons are negatively charged.
- substances with a deficiency of electrons are positively charged.

When a perspex (lucite) rod is rubbed with a piece of silk, the perspex rod becomes positively charged and the silk becomes negatively charged as demonstrated in Figure 501.

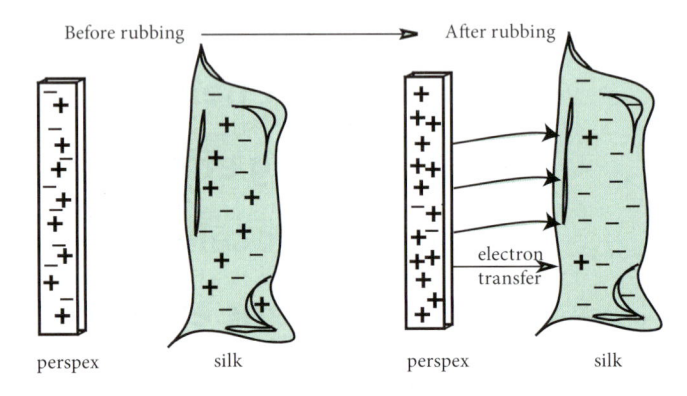

Figure 501   Charging a perspex rod rubbed with silk.

The perspex rod and the silk are initially electrically neutral as each material has the same number of positive and negative charges. The action of friction allows the less tightly held electrons of the perspex rod to be transferred to the silk. With time, the excess electrons on the silk will leak off its surface.

It has been further found that

- ebonite (a certain black material) rubbed with fur becomes negatively charged.
- polythene rubbed with a woollen cloth becomes negatively charged.
- cellulose acetate rubbed with a woollen cloth becomes positively charged.

Figure 502 shows a charged perspex rod placed on a pivot balance. If a charged perspex strip is brought near the perspex rod, the two repel each other causing the rod to rotate. The opposite can be observed with a perspex rod/ebonite strip situation.

By the use of this simple observation, it is possible to conclude that

- Like charges repel each other
- Unlike charges attract each other

ELECTRICITY AND MAGNETISM

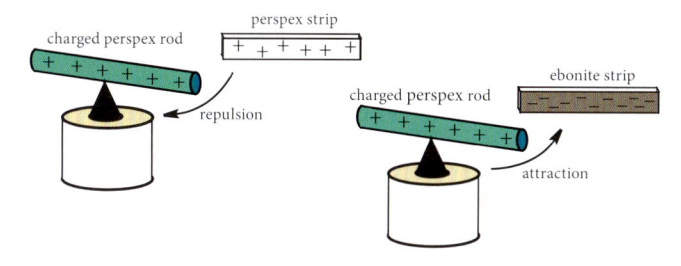

Figure 502   Attraction and repulsion

When charging objects by friction, charge is not created but rather redistributed on the two surfaces. This can be stated according to the **Law of conservation of electric charge** that states that in a closed system, the amount of charge is constant. Examine Figure 502 – there is a total of 16 positive charges and a total of 16 negative charges on the perspex rod and the silk before rubbing. After the process of " electrification " by friction, the charge is redistributed but the same number of positive and negative charges exists. In other words, charge is conserved.

Metals consist of positive ions surrounded by a 'sea' of delocalised electrons (electrons that are not all attached to a specific atom). Hence there are many electrons available for conduction. Solutions that conduct (electrolytes) contain electrons and positive and negative ions that are free to move. Therefore, conductors have a low electrical resistance.

If a **conductor** is held in the hand, any excess of electron charge that forms on the conductor will be transferred to the earth through the body of the person holding the conductor. Conversely, any deficiency of electron charge that forms on the conductor will be transferred from the earth through the body of the person holding the conductor. It is said that the conductor is earthed.

In an **insulator**, the electrons are held tightly by the atomic nuclei and are not as free to move through a material. They can accumulate on the surface of the insulator but they are not conducting.

According to the **energy band theory** that is used to explain the properties of conductors, semiconductors (such as germanium and silicon), and insulators, the valence or outer-shell electrons are held in the valence band that is full or partially filled with electrons. When there are many atoms in close proximity (as there is with all materials), there also exists an upper energy band known as the conduction band. The conduction band is empty. A forbidden energy gap exists between the valence and conduction bands.

For conductors such as metals, the valence and conduction bands overlap. However, in insulators, the energy gap between the valence band and the conduction band is large. Therefore, electrons cannot move across the forbidden energy gap. Insulators thus have a high electrical resistance and when a perspex or other insulating material is held, the electrons remain on the surface of the insulator and are not able to be conducted through the person.

The charge on an insulator will remain for a short period of time until it leaks off the surface or is discharged.

Coulomb's law

The French physicist, *Charles Augustin Coulomb* (1738-1806), using a torsion balance of his own invention, confirmed the existence of an inverse square law of electric charge. A typical Coulomb torsion balance model is shown in Figure 503.

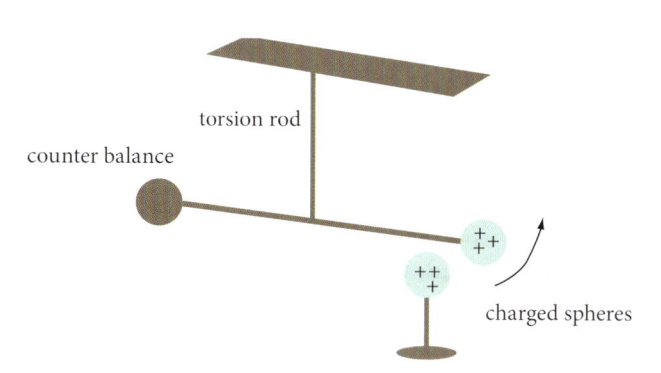

Figure 503   A Coulomb torsion balance

It consists of two conducting spheres fixed on an insulated rod that is suspended by a thin wire fibre connected to a suspension head. The whole apparatus is enclosed in a container to make sure that air currents do not disturb the degree of twist of the thin fibre when a test charge is lowered through a small opening in the apparatus component containing the spheres. The twist on the wire can be calibrated from the twist produced by small known forces and is read off the scale on the suspension head. Using the apparatus, Coulomb could determine the relationship that exists between the magnitude of two charges and force, and the distance and force.

The quantitative relationship that exists between electric point charges separated by a distance was first stated by Coulomb in 1785. Using the torsion balance in Figure 503 he measured the quantity of charge, the distance separating the 'point' charges and the force acting on the charged spheres.

On the basis of his experiments, he concluded that

1. the force F between two point charges q_1 and q_2 was directly proportional to the product of the two point charges.

$$F \propto q_1 \times q_2$$

2. the force between the two point charges was inversely proportional to the square of the distance between them r^2.

153

$$F \propto \frac{1}{r^2}$$

Therefore, it follows that

$$F \propto \frac{q_1 \times q_2}{r^2}$$

3. $F = \dfrac{kq_1 q_2}{r^2}$

When F is measured in newtons (N), q_1 and q_2 in coulombs (C), and r in metres (m), the quantitative statement of Coulomb's Law can be expressed mathematically as

$$F = \frac{q_1 q_2}{4\pi\varepsilon_0 r^2}$$

where $k = 1/4\pi\varepsilon_0$ is the constant of proportionality. Its value is 9.0×10^9 N m^2 C^{-2}. The part of the constant ε_0 is called the **permittivity constant** of free space. On its own, it has a value of 8.9×10^{-12} N m^2 C^{-2}, and this value applies if the experiment is carried out in air or in a vacuum. If the experiment is carried out in another medium, the value of ε_0 will need to be substituted with another value.

The **coulomb** is a large unit of charge, and it is more common to measure in micro-coulombs μC and nano-coulombs nC. One coulomb is the charge on 6.25×10^{18} electrons or protons. Therefore, the charge on a single proton or on a single electron is 1.60×10^{-19} C. The coulomb will be defined at a later stage as it is based upon the definition of electric current.

Note that it is not necessary to include the sign of the charge when carrying out the calculations. Simply use the magnitude of the point charge and then draw a diagram with the signs of each point charge shown, this will also indicate whether the force is attractive or repulsive.

Electric field

There is a similarity between the expressions for the gravitational force between two point masses and the coulombic force between two point charges in a vacuum. They both obey a product/inverse square law. There is no gravitational analogue of electrical permittivity as the gravitational force does not depend on the medium in which they are situated. Furthermore, the gravitational force, unlike the coulombic force, is always a force of attraction. But overall, the similarities outweigh the differences.

Michael Faraday (1791-1867) reasoned that just as Newton had developed the gravitational field concept, so too the concept of electric field could be an analogy of this. Faraday argued that an electric field is a region of influence around a point charge or group of point charges. If an electric field is present at a particular place, when a second point charge is placed near it, it feels a force. The stronger the field, the stronger the electric force will be. The electric field has direction because the force on the charged particle has direction. The direction of the electric field is defined as the direction of the force it causes to act on a small positive test charge.

Faraday introduced the concept of lines of force or lines of electric flux to show the direction that an isolated charge would follow if placed in the field. The field around a positive point charge is shown in Figure 504.

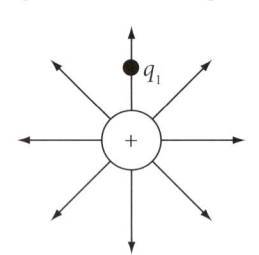

Figure 504   Electric field around a positive point charge

Lines of force or lines of electric flux are imaginary, and they are simply used to assist in an understanding of the nature of electric fields. They obey the following rules:

1. they start on a positive charge and end on a negative charge.

2. they meet electrostatically charged conductors at right angles.

3. they never cross over one another.

4. their density is an indication of the strength of the electric field.

5. there is no electric field in a hollow conductor.

6. the electric field is uniform between two oppositely charged parallel conducting plates.

The **electric field strength** or electric field intensity at any point in space, E is equal to the force per unit charge exerted on a positive test charge.

$$E = \frac{F}{q}$$

or

$$F = E \times q$$

Electric field strength is a vector quantity and it is measured in N C^{-1} or V m^{-1}.

The electric field due to an isolated point charge can be determined using **Coulomb's Law** and the definition of electric field strength. In Figure 505, if the point charges q_1 and q_2 are both positive and point charge q_2 is held stationary, then q_1 would experience a force to the right. If the point charge q_1 was negative it would experience a force to the left.

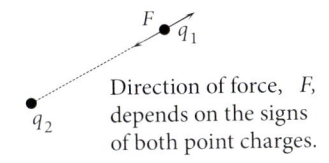

Figure 505   The direction of the force depends on both point charges

Since the forces are the same, that is, $F = k\,q_1 q_2 / r^2$ and $F = Eq_1$, we then have that

$$\frac{kq_1 q_2}{r^2} = E q_1$$

Therefore, the electric field due to a point charge q_2, is given by

$$E = \frac{kq_2}{r^2}$$

or

$$E = \frac{q_2}{4\pi\varepsilon_0 r^2}$$

If q_2 is positive, the direction of the electric field is radially outwards from q_2 as shown in Figure 506. If q_1 is negative, then the direction of the electric field is radially inwards (towards q_2).

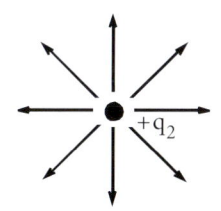

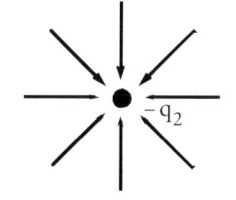

E field due +ve charge   E field due to –ve charge

Figure 506   Electric fields around point charges

Lines of electric flux for various point charge configurations are shown in Figure 507. Although it is not obvious in the diagrams, when a positive point test charge is moved further from the point charges, the radial lines of electric flux become further spaced apart.

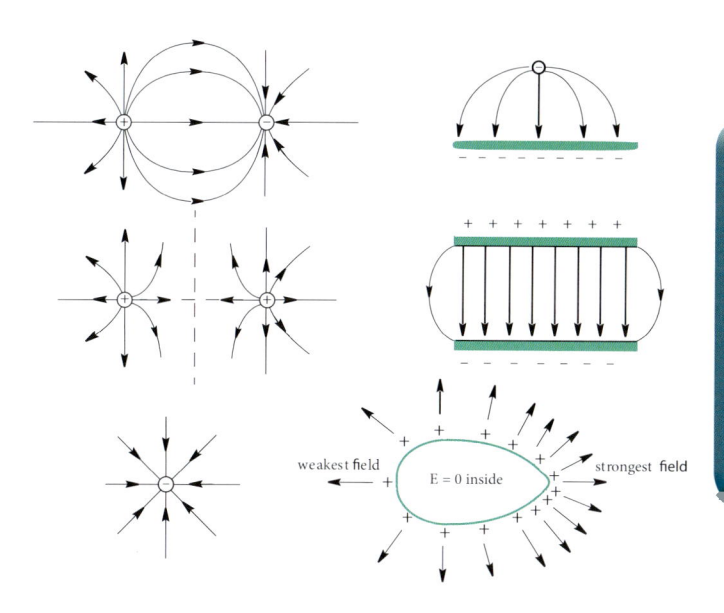

Figure 507   Electric fields for some charge distributions

For the oppositely charged parallel plates, the electric field is approximately uniform meaning that the electric field strength is the same at all points within the plates. Note the edge effect where the electric field lines are now radial at the ends of the plates and thus the electric field strength changes.

Example

Calculate the force acting between two point charges of +10.0 μC and -5.0 μC separated by a distance of 10.0 cm in a vacuum.

Solution

Using $F = k\,q_1 q_2 / r^2$, $q_1 = 10\,\mu C = 10 \times 10^{-6}$ C and $q_2 = (-5)$ μC $= -5 \times 10^{-6}$ C. We have,

$$F = \frac{9.0 \times 10^9 \text{ N m}^2\text{C}^{-2} \times (10 \times 10^{-6} \text{ C}) \times (-5 \times 10^{-6} \text{ C})}{(0.1 \text{ m})^2} = -45 \text{ N}$$

As the answer is negative, it implies that the force is attractive.

That is, there is a force of attraction of 45 N.

Example

The force between two point charges is 20.0 N. If one charge is doubled, the other charge tripled and the distance between them is halved, calculate the resultant force between them.

Solution

Let the original charges be Q_1 and Q_2, and their separation be R, so that the force between these two charges is given by

155

CHAPTER 5

$$F_1 = \frac{kQ_1Q_2}{R^2}$$

Now, let the new charges be q_1 and q_2 and their separation be r, so that $q_1 = 2Q_1$,

$q_2 = 3Q_2$, and $r = \frac{1}{2} R$.

The force between q_1 and q_2 is $F = \frac{kq_1q_2}{r^2}$

Substituting the respective values, we have

$$F = \frac{kq_1q_2}{r^2} = \frac{k(2Q_1)(3Q_2)}{\left(\frac{1}{2}R\right)^2}$$

$$= \frac{6kQ_1Q_2}{0.25\,R^2}$$

$$= 24\frac{kQ_1Q_2}{R^2}$$

$$= 24F_1$$

This means that the force is 24 times larger than it was originally. (i.e., 24 × 20.0 N)

The resultant force is 480 N.

Example

Charges of +1C are located at the corners of a 45° right-angled triangle as shown in the Figure below.

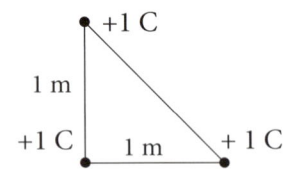

Determine the resultant force on the charge located at the right angle.

Solution

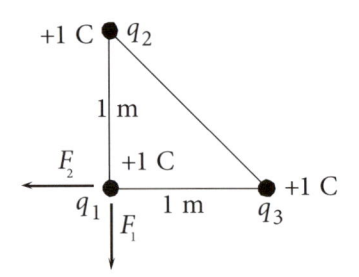

The rough vector diagram could be shown as above:

If several point charges are present, the net force on any one of them will be the vector sum of the forces due to each of the others as shown in the figure above. Since the three point charges are positive, then there will be repulsion on the bottom point charge due to each of the two point charges.

The force on the point charge on the right angle due to the top point charge is calculated as:

$$F_1 = \frac{kq_1q_2}{r^2} = \frac{9.0 \times 10^9\ \text{N m}^2\text{C}^{-2} \times (1\ \text{C}) \times (1\ \text{C})}{(1\,\text{m})^2}$$

$$= 9 \times 10^9\ N$$

The force on the point charge on the right angle due to the right point charge is calculated as

$$F_2 = \frac{kq_1q_3}{r^2} = \frac{9.0 \times 10^9\ \text{N m}^2\text{C}^{-2} \times (1\ \text{C}) \times (1\ \text{C})}{(1\,\text{m})^2}$$

$$= 9 \times 10^9\ N$$

The resultant force is given by the vector addition of the two forces that can be obtained by Pythagorean theorem as in the Figure below.

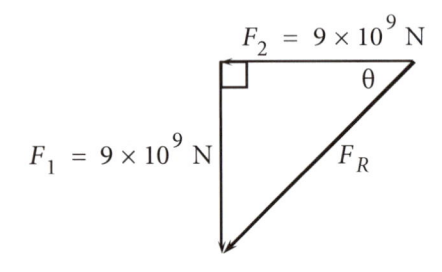

$$(F_R)^2 = (F_1)^2 + (F_2)^2$$

$$= (9 \times 10^9)^2 + (9 \times 10^9)^2$$

$$= 2 \times (9 \times 10^9)^2$$

$$F = 12.7 \times 10^9\ N$$

The direction of the resultant force can be calculated using trigonometry:

$$\tan\theta = \frac{\text{opposite}}{\text{adjacent}} = \frac{9 \times 10^9}{9 \times 10^9} = 1 \therefore \theta = 45°$$

The resultant force is 1.3×10^{10} N in a direction of 45° from the horizontal and downwards towards the right.

ELECTRICITY AND MAGNETISM

Example

Now consider the same problem as the previous one, but this time, the charges are set up as shown in the figure below.

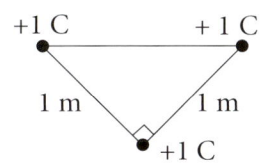

```
+1 C ————————— +1 C
  \      1 m    /
 1 m\         /1 m
     \  +1 C /
```

Determine the resultant force on the charge located at the right angle?

Solution

The rough vector diagram could be shown as follows:

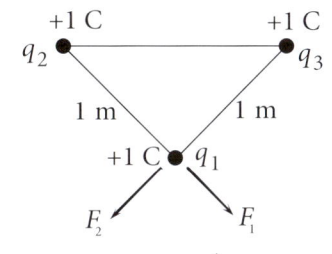

```
+1 C q₂ ————— +1 C q₃
       1 m  1 m
        +1 C q₁
       F₂ / \ F₁
```

The force on the point charge on the right angle due to the two top point charges is still calculated as before.

i.e., $F_1 = \dfrac{kq_1q_2}{r^2} = \dfrac{9.0 \times 10^9 \times (1) \times (1)}{(1m)^2}$

$= 9.0 \times 10^9$

$F_2 = \dfrac{kq_1q_3}{r^2} = \dfrac{9.0 \times 10^9 \times (1) \times (1)}{(1m)^2}$

$= 9.0 \times 10^9$

However, this time the resultant force is shown in the figure below Again, using Pythagoras' theorem, we have that

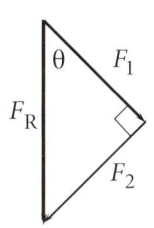

$(F_R)^2 = (F_1)^2 + (F_2)^2$

$= (9 \times 10^9)^2 + (9 \times 10^9)^2$

$= 2 \times (9 \times 10^9)^2$

$= 12.7 \times 10^9 \text{ N}$

The direction of the resultant force can be calculated using trigonometry:

$\tan\theta = \dfrac{\text{opposite}}{\text{adjacent}} = \dfrac{9 \times 10^9}{9 \times 10^9} = 1 \therefore \theta = 45°$

The resultant force is 1.3×10^{10} N in a vertical downwards direction.

Example

A point charge of 25 µC experiences a force of 1.0×10^{-4} N. Calculate the electric field strength producing this force.

Solution

Using the formula, $E = F / q$, we have that

$E = \dfrac{1.0 \times 10^{-4} \text{ N}}{2.5 \times 10^{-5} \text{ C}} = 4.0 \text{ NC}^{-1}$

The electric field strength is 4.0 N C⁻¹ (in the direction of the force).

Example

Calculate the electric field strength 1.5 cm from a point charge of 1.00×10^2 pC in a vacuum.

Solution

Using the formula, $E = q / 4\pi\varepsilon_0 r^2$

$E = (9 \times 10^9 \, Nm^2C^{-2}) \times (1.00 \times 10^{-10} \, C) \div (1.5 \times 10^{-2} \, m)^2$

$= 4.0 \times 10^3 \text{ N C}^{-1}$

The electric field strength is 4.0×10^3 N C⁻¹ (radially outwards).

CHAPTER 5

Example

The field at a particular place due to more than one point charge is the vector sum of the fields caused by each point charge on its own.

Calculate the electric field strength at X due to the charges shown in the following Figure.

Solution

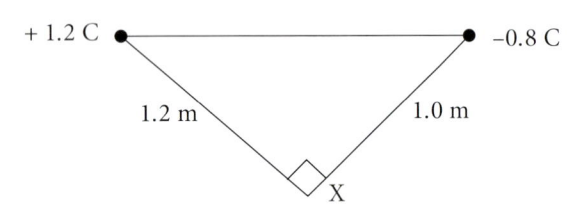

```
+1.2 C •————————————————• −0.8 C
         \                /
          \ 1.2 m    1.0 m /
           \              /
            \            /
             \⌐_____/
                  X
```

E due to −0.80 C point charge is given by kq/r^2

$= (9 \times 10^9 \, Nm^2C^{-2}) \times (-0.8 \, C) \div (1.0 \, m)^2 = -7.2 \times 10^9 \, NC^{-1}$

i.e., the E field, E_1, has a magnitude of $7.2 \times 10^9 \, NC^{-1}$ (radially inwards). The approximate direction is north-east.

E due to +1.2 C point charge is given by kq/r^2

$= (9 \times 10^9 \, Nm^2C^{-2}) \times (+1.2 \, C) \div (1.2 \, m)^2 = 7.5 \times 10^9 \, NC^{-1}$

i.e., the E field, E_2, has a magnitude of $7.5 \times 10^9 \, NC^{-1}$ (radially outwards).

The approximate direction is south-east

The Figure below shows the vectors of the two electric fields, and chooses an angle to measure as a reference for the direction of the resultant electric field. (Another reference angle could be chosen).

Final vector diagram

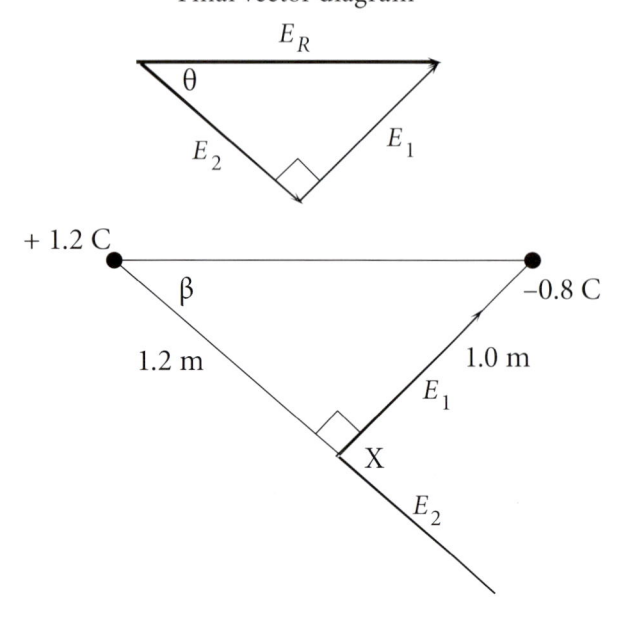

From Pythagoras theorem, we have
$E_R^2 = (7.2 \times 10^9)^2 + (7.5 \times 10^9)^2$

$\tan\theta = \dfrac{\text{opposite}}{\text{adjacent}} = \dfrac{7.2 \times 10^9}{7.5 \times 10^9} = 0.96, \therefore \theta = 43°50'$

so that $E_R = 10.4 \times 10^9 \, N \, C^{-1}$

Next, we have that

i.e. $\theta = 44°$ and $\beta = \tan^{-1}(1/1.2) = 40°$

Resultant field at X is $1.0 \times 10^{10} \, NC^{-1}$ at 1° below a horizontal line drawn through X.

Exercise 5.1

1. A metal sphere with an excess of 11 electrons touches an identical metal sphere with an excess of 15 electrons. After the spheres touch, the number of excess electrons on the second sphere is

    A. 26
    B. 2
    C. 13
    D. 4

2. When hair is combed with a plastic comb, the hair becomes positively charged because the comb

    A   transfers electrons to the hair.
    B. transfers protons to the hair.
    C. removes protons from the hair.
    D. removes electrons from the hair.

3. Which of the diagrams below describe the greatest coulombic force?

    A   q ———l——— 2q

    B   2q ———l——— 3q

    C   2q ———2l——— 3q

    D   2q ———2l——— 6q

158

ELECTRICITY AND MAGNETISM

4. The diagram below represents two charged parallel plates.

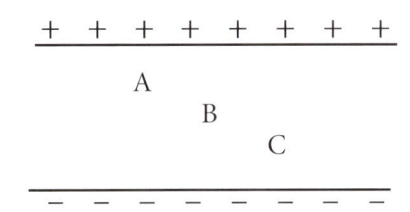

   What is the strength of the electric field relative to locations A, B and C?

   A. greater at A than at B
   B. greater at C than at A
   C. greater at B than at C
   D. the same at A, B and C

5. A negatively charged sphere of negligible mass is moving horizontally in an easterly direction. It enters two long parallel plates carrying opposite charges. Which one of the following figures best shows the path followed by the sphere?

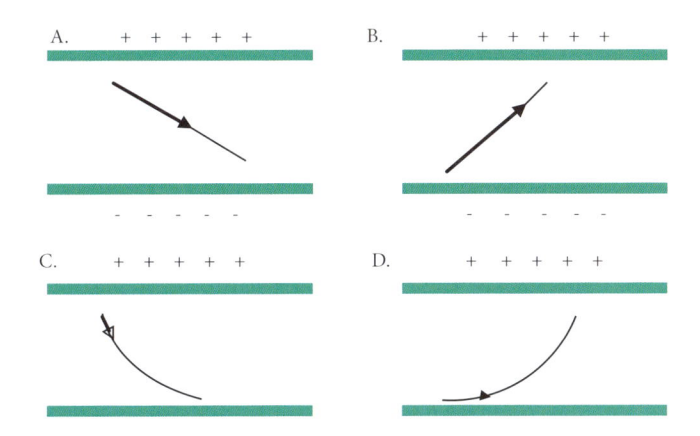

6. Electric field strength may be defined as

   A. the force per unit point charge
   B. the force exerted on a test point charge
   C. the force per unit charge exerted on a positive test charge
   D. the force per unit positive charge

7. Two conducting spheres of charge Q_1 and Q_2 whose centres are separated by a distance d attract each other with an electrostatic force F. If the charge on each sphere is halved and their separation is reduced to one-quarter of its original value, the new force of attraction is given by:

   A. F
   B. 4F
   C. 8F
   D. 64F

8. Three charges are placed at equal distances from each other as shown

   The arrow that best represents the resultant force on X is

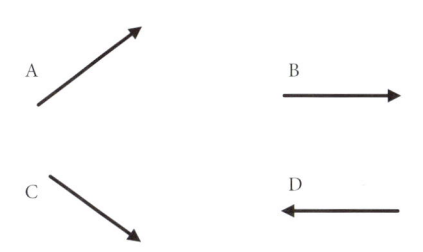

9. If the magnitude of the charge on each of two negatively charged objects is halved, the electrostatic force between the objects will:

   A. remain the same
   B. decrease to one-half
   C. decrease to one-quarter
   D. decrease to one-sixteenth

10. Which of the following is a vector quantity?

    A. potential difference
    B. electric field intensity
    C. electric charge
    D. electric power

11. An electrostatic force F exists between two points with a separation of d metres. Which graph best represents the relationship between F and d?

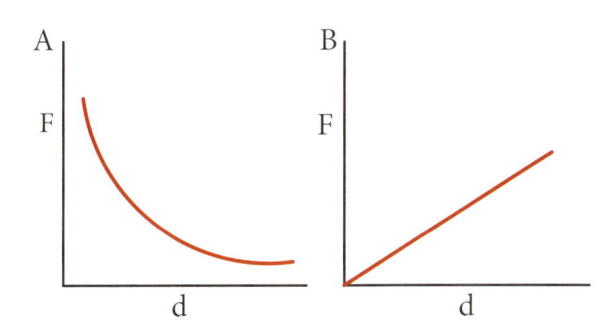

CHAPTER 5

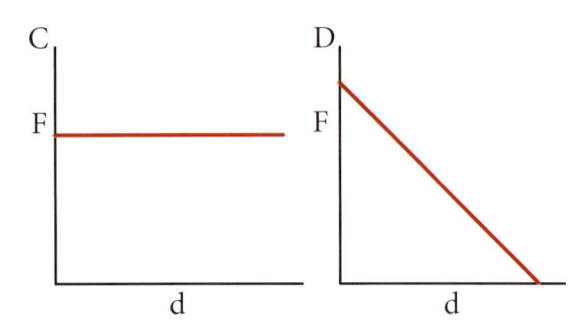

12. In terms of electrostatic induction, explain why road petrol tankers have a length of chain attached to the rear of the truck that touches the road.

13. Explain why charging the nozzle of a spray painting device will use less paint.

14. Why should spare petrol for cars be carried in a metal rather than a plastic container?

15. Explain why it is not wise to play golf during a thunderstorm.

16. Explain why the inside of a car is a safe place during a thunderstorm.

17. Calculate is the charge on 4.0×10^{20} protons?

18. Two charges of –6.00 μC and +8.00 μC attract each other with a force of 3.0×10^3 N in a vacuum. Calculate the distance between the particles?

19. Calculate the electric field strength at a point 2.4 m from a point charge of 5.7 μC in air.

20. Deduce the electric field strength at a point midway between charges of $+7.2 \times 10^{-6}$ C and -3.4×10^{-6} C that are 2.0 m apart in air.

21. Describe how the electric field strength at a point is similar to, and different from the gravitational field strength at a point.

22. Sketch the electric field around two negatively charged point charges separated by distance d if one point charge has twice the charge of the other point charge.

23. Two point charges placed 2.5×10^{-1} m apart in paraffin oil, carry charges of +7.00 pC and +9.00 pC. Calculate the force on each point charge. ($\varepsilon = 4.18 \times 10^{-11}$ $C^{-1}N^{-1}m^2$).

24. Three identical 2.00×10^{-5} C point charges are placed at the corners of an equilateral triangle of sides 1.0 m. The triangle has one apex C pointing up the page and 2 base angles A and B. Deduce the magnitude and direction of the force acting at A.

25. What is the force acting between two point charges of +10.0μC and –5.0μC when separated by a distance of 10.0 cm in a vacuum?

26. Point charges of +1C are located at the corners of a 45^0 right-angled triangle as shown in the diagram.

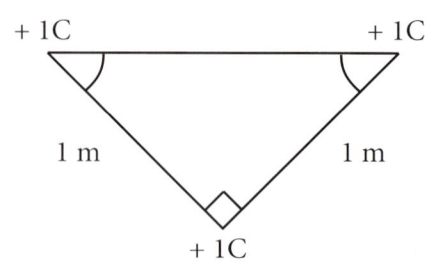

What is the resultant force on the charge located at the right angle?

27. Determine the electric field strength at a point 3.0 m from a point charge of 5.7 mC in air.

28. Describe what is meant by the term zero potential.

29. Determine how many electron-volts of energy are gained by an electron when it is accelerated through 250 V in a television tube. Determine many joules of energy does it gain.

30. Two point charges, X and Y, are separated by a distance of 100 cm as shown in the following diagram:

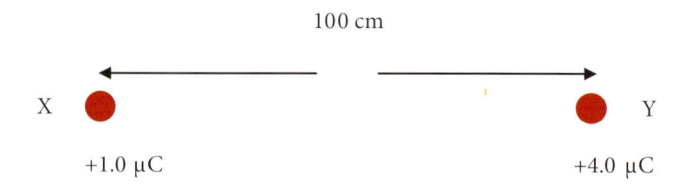

Deduce at what point the magnitude of the electric field is equal to zero.

Electricity and Magnetism

Electric current

An **electric current** is a movement of electric charge that can occur in solids, liquids and gases. A steady current can be maintained when there is a drift of charge-carriers between two points of different electric potential. The charges responsible for the drift are:

Solids
electrons in metals and graphite, and holes in semi-conductors.

Liquids
positive and negative ions in molten and aqueous electrolytes.

Gases
electrons and positive ions formed by electrons stripped from gaseous molecules by large potential differences.

At this point, we will define 'the electric current as the rate at which charge flows past a given cross-section'.

$$I = \frac{\Delta q}{\Delta t}$$

It makes sense that for an electric current to flow, there must be a complete circuit for it to flow through. The unit of current from the above equation is the coulomb per second C s^{-1} and this unit is called the **ampere** (A).

The ampere or 'amp' is a rather large unit. Current is often expressed in milliamps (mA) and microamps (μA) or even nanoamps and picoamps. Some relevant currents for situations are given in Figure 508.

A computer chip	$10^{-12} - 10^{-6}$ A
Electron beam of a television	10^{-3} A
Current dangerous to a human	$10^{-2} - 10^{-1}$ A
Household light bulb	0.5 A
Car starter motor	200 A
Lightning	10^{4} A

Figure 508   Some typical current situations.

When a current flows in the same direction around an electric circuit, the current is said to be a direct current (dc). Dry cell and wet cell batteries supply dc. When the direction of the current changes with time it is said to be an alternating current (ac). Household electrical (non-electronic) appliances are ac.

Electric current is a fundamental quantity in physics, and it is the primary quantity on which electrodynamics is based. Its SI unit is the ampere (A). The ampere is defined in terms of 'the force per unit length between parallel current-carrying conductors'. This definition is all that is required for this course. This definition can be extended further.

When two long parallel current-carrying wires are placed near each other, the force exerted by each will be a force of attraction or repulsion depending upon the direction of the current in each wire. The force is attractive if the two currents flow in the same direction, and the force is repulsive if the two currents flow in opposite directions.

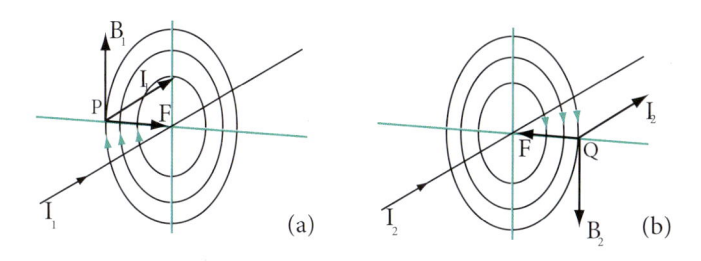

Figure 509 (a) and (b)   Force between two parallel current-carrying wires.

Figure 509 shows currents I1 and I2 flowing in the same direction in two wires. In part a, at point P on the right wire, the magnetic field B1 is upwards (right-hand screw rule). Then using the right-hand palm rule, the direction of the force F is in an easterly direction. Similar analysis of the diagram in part b reveals that the force is in a westerly direction. Both forces are inwards, and the wires will attract each other. Check that you obtain the same result as the diagrams.

The vector diagrams and lines of force or lines of magnetic flux for currents in the same direction and in opposite directions are shown in Figures 509 (a) and (b) respectively.

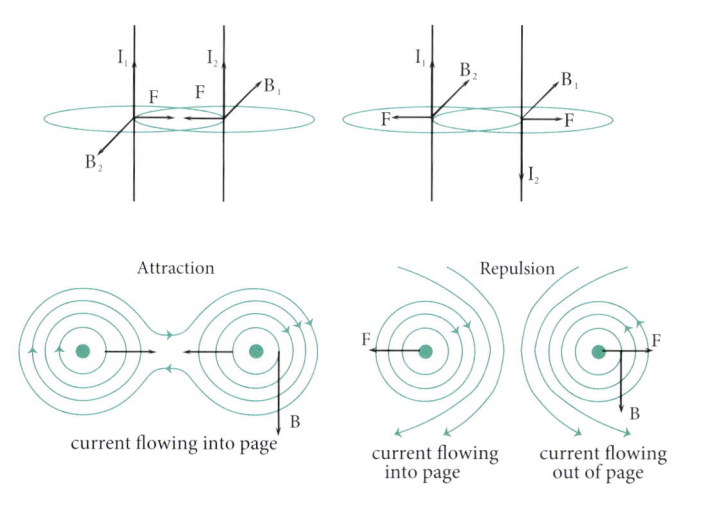

Figure 510   Forces and fields between currents.

The French physicist *Andre Marie Ampère* showed that the quantitative relationship for the force **F** per unit length l between two parallel wires carrying currents I_1 and I_2 separated by a distance r in a vacuum was given by

$$\frac{F}{l} \propto I_1 \times I_2 \times \frac{1}{r} \Leftrightarrow \frac{F}{l} = kI_1 \times I_2 \times \frac{1}{r}$$

161

Therefore,

$$F = \frac{\mu_0}{2\pi} \times I_1 \times I_2 \times \frac{l}{r} = \frac{\mu_0 I_1 I_2 l}{2\pi r}$$

The constant μ_0 is called the **permeability of free space**, is defined to equal

$\mu_0 = 4\pi \times 10^{-7}$ T m A^{-1}

This last equation now allows us to finally define the fundamental unit the ampere.

Thus one ampere is defined as that current flowing in each of two indefinitely-long parallel wires of negligible cross-sectional area separated by a distance of one metre in a vacuum that results in a force of exactly 2×10^{-7} N per metre of length of each wire.

Example

Calculate the current flowing through a hair drier if it takes 2.40×10^3 C of charge to dry a person's hair in 4.0 minutes.

Solution

To determine the current, given that there exists a charge of 2.40×10^3 C and that it is flowing for a period of 4.0 minutes (= 240 seconds).

This gives, $I = \dfrac{\Delta q}{\Delta t} = \dfrac{2.40 \times 10^3}{240} = 10$ A

The current flowing is 1.0×10^1 A.

Current is a scalar quantity but it is useful to indicate the direction of flow of current.

Before the electron was discovered, the direction of the charge-carriers was already defined by scientists and engineers to be from positive to negative.

Benjamin Franklin stated that an excess of fluid produced one kind of electric charge which he termed "positive", and a lack of the same fluid produced the other type of electric charge which he called "negative". Franklin's designation of (+) being an excess of electric charge and of (−) being a deficiency of electric charge was unfortunate.

When batteries and generators, the first source of continuous current, were developed in the 1800s, it was assumed that electric current represented the flow of positive charge as defined by Franklin. This is partly the reason that electric fields are defined in terms of a positive test charge, and the lines of electric flux being explained as going outwards from positive charges.

It is now known that in a metal an electric current is a flow of electrons from **negative to positive**.

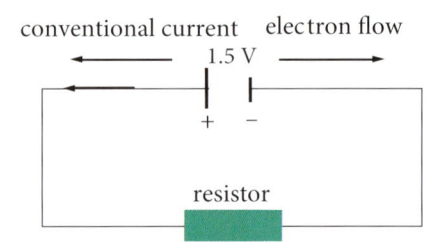

Figure 511   Conventional and electron currents

Unfortunately, this convention has been kept. Figure 511 shows a simple circuit diagram of a 1.5 V dry cell connected to a resistor. When drawing and interpreting circuit diagrams just remember that **conventional current flows from the positive to negative terminal unless you are specifically asked for the correct electron flow that flows from the negative to the positive terminal.**

Direct current (dc)

When a current flows in the same direction around an electric circuit, the current is said to be a direct current (dc). Dry cell and wet cell batteries supply dc. When the direction of the current changes with time it is said to be an alternating current (ac). Household electrical (non-electronic) appliances are ac.

Figure 512 illustrates a typical model of a metal. Metals are good electrical conductors because of the mobility of electrons.

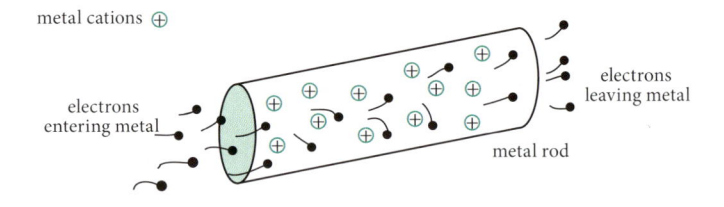

Figure 512   Model of electrical conduction in metals.

Scientists have gathered evidence to suggest that the electrons in outer shells of metals move about freely, within a three-dimensional metal lattice of positively charged metal ions. Thus, metal structure consists of positive ions in a "sea" of delocalised electrons.

However, most of the electrons are tightly bound to the positive metal ions and they are not free to move about to conduct an electric current. Only one or two electrons per atom are loosely bound and thus free to move. For example, in a copper conductor there are about 1029 free electrons per cubic metre.

ELECTRICITY AND MAGNETISM

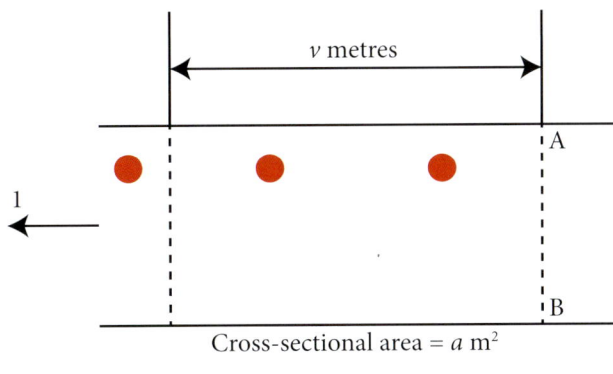

Figure 513

Figure 513 shows a conductor of cross-sectional area a m² carrying a conventional current I. The three electrons shown have a drift velocity of v ms-1. The two vertical lines are v metres apart. Let us assume that in a second the electrons in the box are able to cross the line AB, but the third electron does not have the required velocity v in that second.

If there are n electrons that can cross the line in one second, we can find the total charge by multiplying n electrons by the charge on each electron q and thus the current. (The charge on an electron is 1.6 x 10-19 C). So

$I = nq$

The volume of wire within the shown area is equal to the cross-sectional area A multiplied by the v metres which equals vA.

Therefore, the total current passing the line AB in one second will be:

$I = nAvq$

Positive ions and free electrons have internal energy that depends on the temperature of the metal at that point in time, and the delocalised electrons move about at enormous speeds of about 10^6 m s^{-1}, colliding with the positive ions in the metal's lattice. However, as much as the speed is high, there is no net movement of charge unless the conductor is connected to a source of potential difference (or voltage).

When a dry cell or battery is connected across the ends of a metal wire, an electric field is produced in the wire. The electrons drift with a **drift velocity** to the high positive potential end as shown in Figure 514. Electrons entering at one end of the metal cause a similar number of electrons to be displaced from the other end, and the metal conducts. Even though they are accelerated along their path, it is estimated that the drift velocity is only a small fraction of a metre each second (about 10^{-4} m s^{-1}).

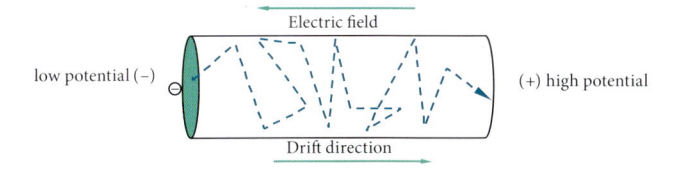

Figure 514 Drift velocity of electrons.

An electric field is created when a **potential difference** is supplied to a metal wire. The average speed and thus the average kinetic energy of the electrons increases. When they collide with positive ions in the lattice, they give up some of their energy to them. After this event, they are again accelerated because of the electric field until the next collision occurs. At each stage, these collisions generate heat that causes the temperature of the metal to increase. We say that a current produces a heating effect.

Potential difference

In an electric circuit, charge is moved from one point in the circuit to another and in this process electrical energy supplied by a dry cell or power supply is converted to other forms of energy. For example, a filament lamp converts electrical energy into heat and light energy. As charges move in the circuit, they gain energy supplied by the energy source.

For a charge to move from one point to another in a circuit, there has to be a potential difference between the two points because an amount of electrical energy is changed to other forms of energy as the charge moves from one point to another such as from one side of a filament lamp to the other.

In Chapter 2, we learnt that when a force acts on a mass and moves it in the direction of the force, then work is done on the mass. Work is a scalar quantity and is defined as the force F multiplied by the distance s moved in the direction of the force. Mathematically, this was stated as:

$$W = Fs \text{ or } W = Fs\cos\theta$$

Doing work on an object changes its energy.

There are a number of different ways to define **electric potential difference**:

1. The electric potential difference between two points in a conductor carrying a current is defined as 'the power dissipated in a load per unit current in moving from one point to another'. V = P / I

2. The electrical potential difference between two points in a conductor carrying a current is defined as 'the amount of electrical energy that is changed to other

163

CHAPTER 5

forms of energy when an amount of charge moves from one point to another, where *P* is the power dissipated in watts (W) and *I* is the current in amps (A)'.

The concept of electric fields will be explored more fully in Chapter 6. Here is a third way to define potential difference in terms of electric potential.

3. 'The electrical potential difference between two points in a conductor carrying a current is defined as the work done per unit charge in moving a positive charge from a point at higher potential to a point at lower potential.'

$$\Delta V = \frac{\Delta E_p}{q} = \frac{\Delta W}{q}$$

where *q* is the amount of charge measured in coulombs C. *(If work is done in moving an electron, it would therefore move from a point at lower potential to a point at higher potential)*. So as charge *q* moves through a potential difference ΔV, it does work equal to $\Delta V q$ and this work is equal to the energy given to the load component in the circuit. Therefore, potential difference is a measure of the energy released by an electric charge or group of charges.

The accepted unit of electric potential difference is the volt, V. It follows that it can also be measured in JC^{-1}. Just as work is a scalar quantity, so too electrical potential difference is a scalar quantity. When we use the term voltage, we are loosely talking about electric potential difference. If the potential difference between the terminals of a dry cell is 1.5V then the energy used in carrying a positive charge of +q from one terminal to the other is 1.5 J C^{-1}.

The work done in transferring a charge from two points does not depend on the path along which the charge is taken because potential difference is constant between two points.

Suppose there is charge moving in a closed circuit as shown in Figure 515. When the charge reaches the first load resistor at A, power is dissipated in the load resistor and the electrical energy is converted to heat energy. There is a potential (energy) difference between points A and B because work has been done on the charge flowing through the resistor and there will be an energy difference per unit charge through the resistor.

When it gets to the second load, again energy is dissipated and there is a potential difference between either side of the resistor. If the electric potential energy dissipated is the same in both of the loads, the energy difference per unit charge will be the same amount. We say the potential difference across each load resistor would be 3 volts.

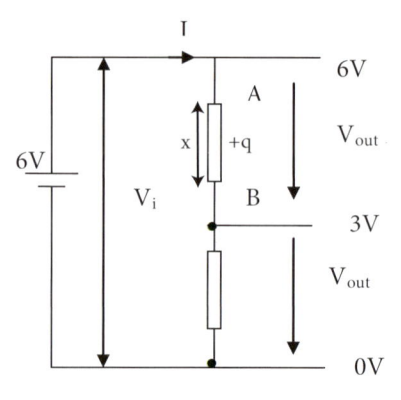

Figure 515 *Charge moving in a uniform electric field*

The Electron-volt (eV)

In atomic and nuclear physics, we often determine energy by multiplying charge times voltage ($W = qV$). Because the charge on an electron and the charge on the proton are used so frequently, a useful unit of energy is the electron-volt (eV).

One **electron-volt** (1 eV) is defined as the energy acquired by an electron as a result of moving through a potential difference of one volt.

Since $W = q\Delta V$, and the magnitude of charge on the electron (or a proton) is 1.6×10^{-19} C, then it follows that

$$W = (1.6 \times 10^{-19} \text{ C}) \times (1 \text{ V}) = 1.6 \times 10^{-19} \text{ J}$$

So that, 1 eV = 1.6×10^{-19} J

An electron that accelerates through a potential difference of 1000 V will lose 1000 eV of potential energy and will thus gain 1000 eV (1 keV) of kinetic energy.

The electron-volt is not an SI unit. In atomic physics, we find that energy is typically a few electron-volts. In nuclear physics the energy is more likely to be measured in MeV (10^6 eV), and in particle physics it is typically in GeV (10^9 eV).

Example 1

Calculate the work done in moving 10.0 μC through a potential difference of 150 V

Solution

Using the formula $\Delta W = q\Delta V$ where the potential difference, $\Delta V = 150$ V, we have $W = F \times x = Eq \times x$

$\Delta W = q\Delta V = (1.00 \times 10^{-5} \text{ C}) \times (150 \text{ V})$

$= 1.5 \times 10^{-3}$ J

The work done is 1.5×10^{-3} **J**.

Example 2

A 2.0 μC charge acquires 2.0×10^{-4} J of kinetic energy when it is accelerated by an electric field between two points. Calculate the potential difference between the 2 points.

Solution

Using the formula $\dfrac{\Delta W}{q} = \Delta V = \dfrac{2.0 \times 10^{-4} \text{ J}}{2.0 \times 10^{-6} \text{ C}}$

$= 100$ V

The potential difference is 1.0×10^2 **V**.

Example 3

The work done by an external force to move a -8.0 μC charge from point a to point b is 8.0×10^{-3} J. If the charge initially at rest had 4.0×10^{-3} J of kinetic energy at point b, calculate the potential difference between a and b.

Solution

Using the formula $\Delta V = \dfrac{W}{q} = \dfrac{8.0 \times 10^{-3} \text{ J}}{-8.0 \times 10^{-6} \text{ C}}$

$= -1.0 \times 10^2$ V

The work done to have a gain in kinetic energy

$= \dfrac{4.0 \times 10^{-3} \text{ J}}{8.0 \times 10^{-6} \text{ C}}$

$= 5.0 \times 10^1$ V.

$\Delta V = 1.0 \times 10^2 - 5.0 \times 10^1$ V

The potential difference is 5.0×10^1 **V**.

5.2 Heating effect of electric currents

> **NATURE OF SCIENCE:**
>
> Peer review: Although Ohm and Barlow published their findings on the nature of electric current around the same time, little credence was given to Ohm. Barlow's incorrect law was not initially criticized or investigated further. This is a reflection of the nature of academia of the time with physics in Germany being largely non-mathematical and Barlow held in high respect in England. It indicates the need for the publication and peer review of research findings in recognized scientific journals. (4.4)
>
> © IBO 2014

Essential idea: One of the earliest uses for electricity was to produce light and heat. This technology continues to have a major impact on the lives of people around the world.

Understandings:
- Circuit diagrams
- Kirchhoff's circuit laws
- Heating effect of current and its consequences
- Resistance expressed as
- Ohm's law
- Resistivity
- Power dissipation

Circuit diagrams

Electricity is transmitted to household localities using alternating current ac as the voltage can be stepped-down to the local supply voltage before entering houses. Some household appliances such as lights, stoves, water heaters, toasters, irons operate using ac. However, electronic devices such as stereos, computers and televisions operate using direct current dc. The power supply is stepped–down to say 12 V by a transformer inside the appliance, and a diode is inserted (diodes are devices that will only allow current to flow in one direction). Direct current is produced.

An electric circuit has three essential components

1. A source of emf.

2. A conducting pathway obtained by conducting wires or some alternative.

3. A load to consume energy such as a filament globe, other resistors and electronic components.

Chapter 5

A current will always take the easiest path in the circuit. If for some reason the current finds a way back to the source without passing through the essential components, a short-circuit occurs. Fuses are used in appliances to stop damage due to short-circuits.

Resistance

Electrical resistance, R, is a measure of how easily charge flows in a material. Conductors, semiconductors and insulators differ in their resistance to current flow. An ohmic material of significant resistance placed in an electric circuit to control current or potential difference is called a **resistor**.

The **electrical resistance** of a piece of material is defined by the ratio of the potential difference across the material to the current that flows through it.

$$R = \frac{V}{I}$$

The units of resistance are volts per ampere (VA^{-1}). However, a separate SI unit called the ohm Ω is defined as the resistance through which a current of 1 A flows when a potential difference of 1 V is applied.

Since the term resistance refers to a small resistance, it is common for a resistor to have kilo ohm (kΩ) and mega ohm (MΩ) values.

Factors affecting resistance

The **resistance** of a conducting wire depends on four main factors:

- length
- cross-sectional area
- resistivity
- temperature

It can be shown that when the temperature is kept constant

$$R = \rho \cdot \frac{l}{A}$$

where R is the resistance in Ω, ρ is the resistivity in Ω m, l is the length of the conductor in m, and A is the cross-sectional area of the conductor in m^2.

As the length of a conductor increases, the resistance increases proportionally

$$R \propto l$$

It seems logical that as the length increases so too does the resistance to the flow of charge across the conductor.

As the cross-sectional area of a conductor increases, the resistance decreases proportionally.

$$R \propto \frac{1}{A}$$

With increased cross-sectional area, there is a greater surface through which the charge can drift.

The **resistivity** is specific to the type of material being used as a resistance and is affected by the nature of the de-localised electrons and the positive ions within the material. The resistivity of a material is the resistance across opposite faces of a cube with sides of one metre. It can be found that by plotting a graph of R versus (l / A), the slope of the linear graph is the resistivity ρ measured in Ω m.

Figure 516 gives the resistivities of various materials at 20° C. The first three values show that although silver has a low resistivity, it is expensive to use in electrical circuits. Copper is the preferred metal although aluminum is commonly used in electricity transmission cables because of its lower density. Constantan and nichrome are commonly used in wire form, and are termed high resistance wires. Nichrome is used as the heating element in many toasters and electric radiators. The semiconductors behave in a special manner, and pyrex glass is an obvious insulator.

Material	Resistivity Ω m
Silver	1.6×10^{-8}
Copper	1.7×10^{-8}
Aluminium	2.8×10^{-8}
Tungsten	5.6×10^{-8}
Constantan (alloy of copper and nickel)	49×10^{-8}
Nichrome (alloy of nickel, iron & chromium)	100×10^{-8}
Graphite	$(3 - 60) \times 10^{-5}$
Silicon	$0.1 - 60$
Germanium	$(1 - 500) \times 10^{-3}$
Pyrex glass	10^{12}

Figure 516   Resistivities for certain materials at 20°C

Example

Determine the resistance of a piece of copper wire that is 10.0 m long and 1.2 mm in diameter?

Solution

The resistance, R, is given by the formula $R = \rho l / A$, where $A = \pi r^2$.

This means that

$R = (1.7 \times 10^{-8}\ \Omega m)\ (10.0\ m) / \pi (6.0 \times 10^{-4})^2\ m^2 = 0.150\ \Omega$.

The resistance of the copper wire is **0.15 Ω**.

The resistance of a material increases with temperature because of the thermal agitation of the atoms it contains, and this impedes the movement of electrons that make up the current.

The increase in resistance can be shown as

$$R_f = R_0(1 + \alpha t)$$

where R_0 equals the resistance at some reference temperature say 0 °C, R_f is the resistance at some temperature, t °C, above the reference temperature, and α is the temperature coefficient for the material being used.

One interesting phenomenon of the effect of temperature on resistance is superconductivity. In 1911, H. Kammerlingh Onnes found that mercury loses all its resistance abruptly at a critical temperature of 4.1 K. When a material attains zero resistance at some critical temperature, it is called a superconductor. The possibility of having a material that has an induced electric current that lasts forever has become a topic for research physicists. Just think of the energy saving if the perfect superconductor is found that can give zero resistance at room temperature.

Ohm's law

The German physicist, *Georg Simon Ohm* (1787-1854), studied the resistance of different materials systematically. In 1826, he published his findings for many materials including metals. He stated that:

Provided the physical conditions such as temperature are kept constant, the resistance is constant over a wide range of applied potential differences, and therefore the potential difference is directly proportional to the current flowing.

$$V \propto I \Leftrightarrow \frac{V}{I} = \text{constant}$$

This is known as **Ohm's Law**.

There are two relevant statements to be made here. Firstly, Ohm's Law is not really a law but rather an empirical statement of how materials behave. Many materials are non-ohmic and this law is only applicable to ohmic conductors. Secondly, the law should not be written as R = V / I as this statement defines resistance.

The formula is commonly written as:

$$V = IR$$

where *V* is the potential difference across the resistor (in volts V), *I* is the current in the resistor (in amperes A) and *R* is the resistance (in ohms Ω). When written in this form *R* is understood to be independent of *V*.

As the current moves through the resistance of a device, it loses electric potential energy. The potential energy of a charge is less upon leaving the resistor than it was upon entering. We say that there is a potential drop across the device.

A typical apparatus used in the confirmation of Ohm's Law is shown using a circuit diagram in Figure 517.

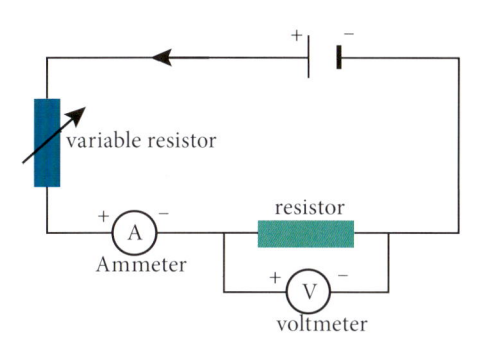

Figure 517   Typical Ohm's law apparatus

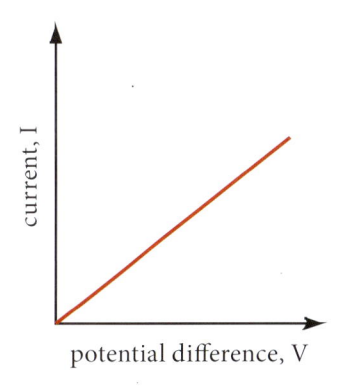

Figure 518   Typical Ohm's law graph.

A graph of current versus the potential difference is a straight line (see Figure 518). Devices that obey the linear relationship of the graph are said to be 'ohmic devices' or 'ohmic conductors'. There are very few devices that are ohmic although some metals can be if there is no temperature increase due to the heating effect of the current. However, many useful devices obey the law at least over a reasonable range. Remember that a resistor is any device with a potential difference and is not only restricted to the typical resistor used in the laboratory. It could be any useful electrical device – a filament lamp or globe, a diode, a thermistor. Figure 519 (overleaf) shows some typical V versus I graphs for some conductors. When a device does not obey Ohm's Law, it is said to be non-ohmic. The filament lamp is non-ohmic because Ohm's Law requires that the temperature remains constant, and this is not the case for a globe operating.

CHAPTER 5

Example

An iron draws 6.0 A of current when operating in a country with a mains supply of 120 V. Calculate the resistance of the iron?

Solution

We can transpose the formula V = IR to obtain R = V / I.

So that

$$R = \frac{120 \text{ V}}{6.0 \text{ A}} = 20 \text{ }\Omega$$

The resistance of the iron is 2.0×10^1 Ω.

Power dissipation

Electric power is the rate at which energy is supplied to or used by a device. It is measured in J s^{-1} called watts W.

When a steady current is flowing through a load such as a resistor, it dissipates energy in it. This energy is equal to the potential energy lost by the charge as it moves through the potential difference that exists between the terminals of the load.

If a vacuum cleaner has a **power rating** of 500 W, it means it is converting electrical energy to mechanical, sound and heat energy at the rate of 500 J s^{-1}. A 60 W light globe converts electrical energy to light and heat energy at the rate of 60 J s^{-1}.

Some common power ratings of household appliances are given in Table 520.

Appliance	Power rating
Blow heater	2 kW
Kettle	1.5 kW
Toaster	1.2 kW
Iron	850 W
Vacuum cleaner	1.2 kW
Television	250 W

Table 520   Some possible power ratings of household appliances

KEY: a ohmic conductor   b. gas discharge tube

c. thermionic diode   d. junction diode

e. filament lamp   f. thermistor

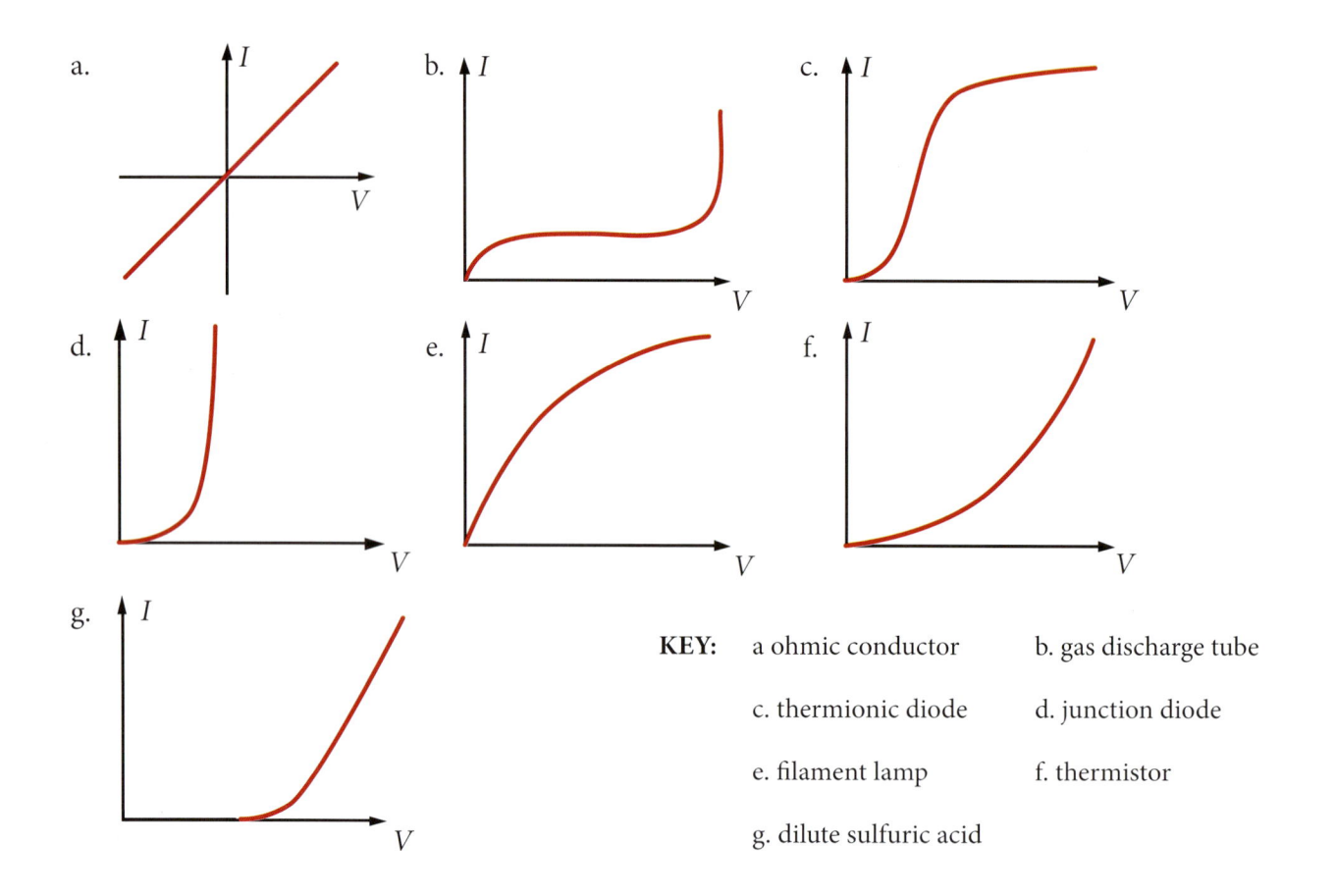

g. dilute sulfuric acid

Figure 519   Current/voltage relationships for some devices.

168

ELECTRICITY AND MAGNETISM

Deriving expressions for determining power

We start with the basic definition of power, $P = W / t$, then, we introduce the different expressions that we have already developed.

Using the result, $W = qV$ and $q = It \Leftrightarrow t = q / I$, we have

$$P = \frac{qV}{\left(\frac{q}{I}\right)} = IV$$

The same result could have been obtained using the fact that the electrical energy,

$$W = qV = ItV$$

so that

$$P = \frac{W}{t} = \frac{ItV}{t} = IV$$

However, we also have that $V = IR$, so that

$$P = I \times IR = I^2 R$$

We also could have used $I = V / R$, giving the result,

$$P = IV = \frac{V}{R} \times V = \frac{V^2}{R}$$

Summary

$$P = \frac{W}{t} = I^2 R = \frac{V^2}{R} = VI$$

The commercial unit of electrical energy is the **kilowatt-hour** (kW h). It is the energy consumed when 1 kW of power is used for one hour. The consumer has to pay a certain cost per kilowatt-hour say 14 cents per kW h.

Heating effect of current and its consequences

James Joule investigated the heating effect of a current in 1841. He was able to demonstrate that by supplying electrical energy to a high resistance coil of wire this energy could be converted to thermal energy.

$$V \times I \times t = m \times c \times \Delta T$$

Example

An electrical appliance is rated as 2.5 kW, 240 V.

(a) Determine the current needed for it to operate.

(b) Calculate the energy it would consume in 2.0 hours.

Solution

(a) Given that $P = 2.5 \times 10^3$ and $V = 240$ V, we use the formula, $P = IV$. Now,

$$P = IV \Rightarrow I = \frac{P}{V} = \frac{2.5 \times 10^3}{240} = 10.4$$

the current drawn is $\mathbf{1.0 \times 10^1}$ **A**.

(b) Next, we use the formula $W = VIt$, so that

$$W = (240 \text{ V}) \times (10.4 \text{ A}) \times (7.2 \times 10^3 \text{ s})$$

$$= 1.8 \times 10^7 \text{ J}$$

The energy consumed is $\mathbf{1.8 \times 10^7}$ **J**.

Example

A 2.5 kW blow heater is used for eight hours. Calculate the cost if electricity is sold at 12 cents per kilowatt-hour?

Solution

Using the fact that energy consumed (E) = power × time, we have

$$E = (2.5 \text{ kW}) \times (8 \text{ h}) = 20 \text{ kW h}$$

Therefore, the Cost = (20 kW h) × $0.12 per kW h = $2.40

The cost to run the heater is two dollars forty cents ($2.40).

Example

A 1.2 kW electric water heater that is made of aluminium is used to heat 2.5 L of water from 25 °C to the boil. If the mass of aluminium water heater is 350 g and all the electrical energy is converted to heat energy, determine the time in minutes to bring the water to the boil.

(Specific heat capacity of aluminium is 9.1×10^2 J kg^{-1}K^{-1})

Solution

A 1.2 kW heater delivers 1200 joules each second. 2.5 L = 2.5 kg

The electrical energy is used to heat the heater and the water from 25 °C to 100 °C.

169

$Q = mc\Delta T_{heater} + mc\Delta T_{water}$

$= (0.35 \text{ kg} \times 9.1 \times 10^2 \text{ J kg}^{-1} \text{K}^{-1} \times 75 \text{ °C}) +$
$(2.5 \text{kg} \times 4180 \text{ J kg}^{-1} \text{K}^{-1} \times 75 \text{ °C})$

$= 8.08 \times 10^5 \text{ J}$

Now if the heater delivers 1200 J per second, then to consume 8.08×10^5 J will take:

$\dfrac{8.08 \times 10^5 \text{ J}}{1200 \text{ J}}$

$= \dfrac{673 \text{ s}}{60}$

$= 11.2 \text{ min} \approx 11 \text{ min}$

Exercise 5.2

1. When a current is flowing through a wire attached to a dry cell

    A. positive charges flow from negative to positive terminal
    B. positive charges flow from positive to negative terminal
    C. negative charges flow from negative to positive terminal
    D. negative charges flow from positive to negative terminal

2. The total charge passing the same point in a conductor during 2.0 μs of an electric current of 7.5 mA is

    A. 15 nC
    B. 15 mC
    C. 15 μC
    D. 1.5×10^4 μC

3. An electric heater raises the temperature of a measured quantity of water. 6.00×10^3 J of energy is absorbed by the water in 30.0 seconds. What is the minimum power rating of the heater?

    A. 5.00×10^2 W
    B. 2.00×10^2 W
    C. 2.00×10^3 W
    D. 1.80×10^5 W

4. If the power developed in an electric circuit is doubled, the energy used in one second is

    A. halved
    B. doubled
    C. quartered
    D. quadrupled

5. The electron volt is **defined** as

    A. the energy acquired by an electron when it passes through a potential difference of 1.0 V.
    B. the voltage of an electron.
    C. a fraction of the ionisation of an electron.
    D. unit of energy exactly equal to 1.6×10^{-19} J.

6. The definition of the unit of current, the ampere, is based on

    A. the charge per unit time delivered by an emf of 1.0 V.
    B. the force per unit length on parallel current-carrying wires.
    C. the force per unit length on a conductor in a magnetic field
    D. the charge passing a point per unit time.

7. Identify the charge-carriers in

    a. a length of copper wire.
    b. an aqueous solution of sodium chloride.
    c. the atmosphere during a lightning storm.

8. The speed with which electrons move through a copper wire is typically 10^{-4} m s^{-1}.

    a. Explain why is it that the electrons cannot travel faster in the conductor?
    b. Explain why the electron drift produces heat?

9. Calculate the resistance of a wire if 0.5 V across it causes a current of 2.5 A to flow.

10. Calculate current flow through a 20 MΩ resistor connected across a 100 kV power supply.

11. A thin copper wire 200 cm in length has a 9 V dry cell connected between its ends. Determine the voltage drop that occurs along 30 cm of this wire.

12. Determine the length of tungsten wire with a diameter of 1.0 mm that is used to make a 20.0 Ω resistor.

13. A nichrome wire has a diameter of 0.40 mm. Calculate the length of this wire needed to carry a current of 30 mA when there is a potential difference of 12 V across it.

14. Explain in terms of atomic and electron movement, why resistance increases with temperature.

15. Determine how many coulombs there are when 2.0 A flows for 2.0 hours?

16. Distinguish the difference between an ohmic and a non-ohmic material.

17. Copy out and complete the table

Appliance	Power (Watt)	p.d (Volt)	Current (Ampere)	Fuse rating needed (3,5,10,13 A)
Digital clock	4	240		
Television	200	240		
Hair dryer		110	5	
Iron		230	4	
Kettle		240	10	

18. Calculate the cost of heating the water to wash the dishes if the sink is 48 cm long, 25 cm wide and the water is 25 cm high. The tap water is at 14 °C and the final temperature before washing up is 62 °C. Power is sold to the consumer at 14 cents per kilowatt-hour.

19. The element of an electric jug has a resistance of 60 Ω and draws a current of 3.0 A. Determine by how much the temperature of 5.0 kg of water will rise if it is on for 6 minutes.

20. Calculate the cost to heat 200 kg of water from 12°C to its boiling point if power costs 14 cents per kilowatt-hour.

21. Determine the work done in moving a charge of 10.0 nC through a potential difference of 1.50×10^2 V?

22. An electron in an electron gun of a picture tube is accelerated by a potential 2.5×10^3 V. Calculate the kinetic energy gained by the electron in electron-volts.

Kirchhoff's circuit laws

In the mid-nineteenth century, G.R. Kirchoff (1824-1887) stated two simple rules using the laws of conservation of energy and charge to help in the analysis of direct current circuits. These rules are

Kirchoff's current law – junction rule

'The sum of the currents flowing into a point in a circuit equals the sum of the currents flowing out at that point'.

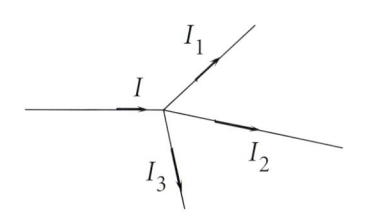

Figure 521   Kirchoff's current law.

In Figure 521, we have that $I = I_1 + I_2 + I_3 + ...$

This is based on the Law of Conservation of charge.

Kirchoff's voltage law – loop rule

'In a closed loop, the sum of the emfs equals the sum of the potential drops'.

In the Figure 522, we have that $V = V_1 + V_2 + V_3 + ...$

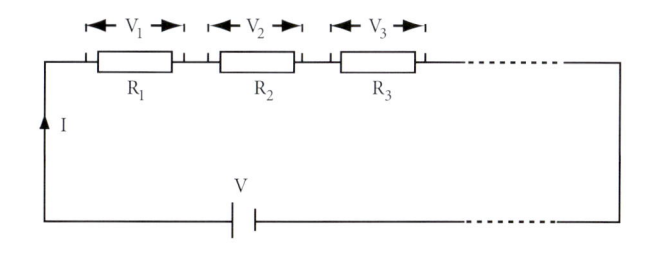

Figure 522   Kirchoff's voltage law.

This is a statement of the law of conservation of energy. 'Energy supplied equals the energy released in this closed path'.

Series circuits

In a series circuit with one cell:

1. All the components have only one current pathway.

2. All components have the same current through them.

3. The sum of the potential drop across each component is equal to the emf of the cell.

CHAPTER 5

From Kirchoff's laws, we have that

$I = I_1 = I_2 = I_3 = \ldots$ and $V = V_1 + V_2 + V_3 + \ldots$

Then, from Ohm's law

$$R = \frac{V}{I} = \frac{V_1 + V_2 + V_3 + \ldots}{I} = \frac{V_1}{I} + \frac{V_2}{I} + \frac{V_3}{I} + \ldots$$

But, the terms in this sum are equal to $R_1, R_2, R_3, \ldots$ (i.e., $R_1 = V_1 / I, R_2 = V_2 / I \ldots$).

So that,

$$R_{eff} = R_1 + R_2 + R_3 + \ldots$$

The total or **effective resistance** R_{eff} of a series circuit is equal to the sum of the separate resistances.

Example

From the diagram given in the Figure below of a potential divider, calculate

(a) the effective resistance of the circuit
(b) the current flowing
(c) the potential difference across each resistor

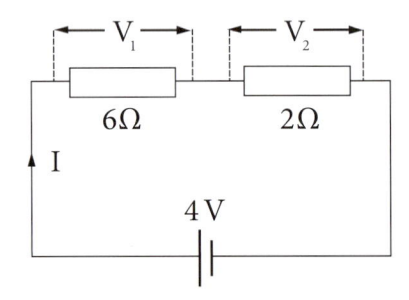

Solution

(a) Using our rule (from above), we have that the effective resistance is given by

$$R_{eff} = R_1 + R_2 = 6\,\Omega + 2\,\Omega = 8\,\Omega$$

That is, the effective resistance is 8 Ω.

(b) We can determine the current by making use of the formula, $I = V / R_{eff}$.

So that in this instance, we have that

$I = 4V / 8\Omega = 0.5$ A

The current flowing is 0.5 A.

(c) The potential drop across each resistor, R_i, is given by, $V_i = IR_i$.

So that for the 6 ohm resistor, the potential difference (or potential drop) is given by $V_1 = IR_1 = 0.5 \times 6 = 3$ V.

Similarly, for the 2 ohm resistor, the potential difference (or potential drop) is given by $V_2 = IR_2 = 0.5 \times 2 = 1$ V.

The potential differences in the 6 Ω and the 2 Ω resistors are 3 V and 1 V respectively.

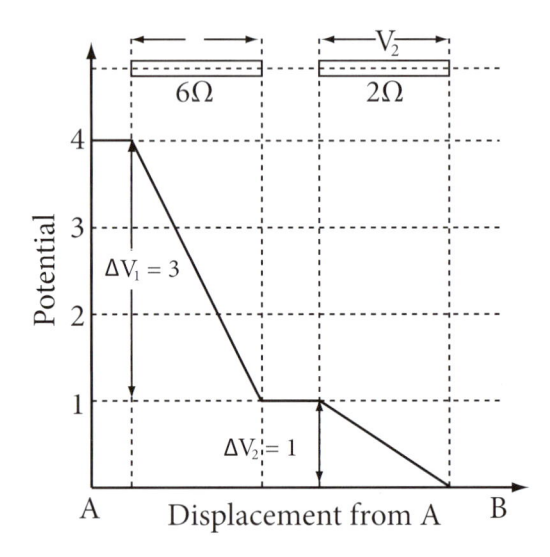

Figure 523   Potential – displacement graph.

Figure 523 shows the corresponding potential/displacement graph for the circuit in the example above.

Notice how we refer to the potential difference, ΔV_1 as opposed to simply V_1.

Parallel circuits

In the parallel circuit in Figure 524

There is more than one current pathway.

1. All components have the same potential difference across them.

2. The sum of the currents flowing into any point is equal to the sum of the currents flowing out at that point.

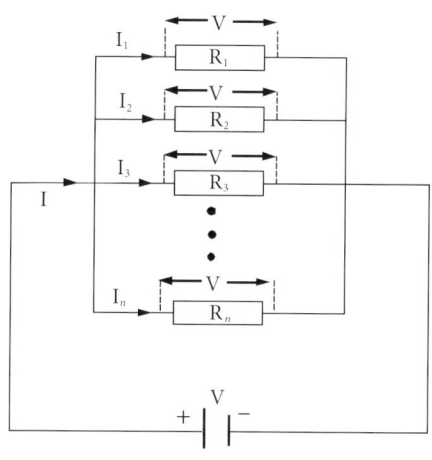

Figure 524 A parallel circuit

From Kirchoff's laws, we have that

$$I = I_1 + I_2 + I_3 + \ldots$$

and

$$V = V_1 = V_2 = V_3 = \ldots$$

From Ohm's law, we have $V = IR \Leftrightarrow V/R = I$ or $1/R = I/V$

Meaning that,

$$\frac{1}{R_{eff}} = \frac{I_1 + I_2 + I_3 + \ldots}{V} = \frac{I_1}{V} + \frac{I_2}{V} + \frac{I_3}{V} + \ldots$$

However, $V = V_1 = V_2 = V_3 = \ldots$, so that

$$\frac{1}{R_{eff}} = \frac{I_1}{V_1} + \frac{I_2}{V_2} + \frac{I_3}{V_3} + \ldots = \frac{1}{R_1} + \frac{1}{R_2} + \frac{1}{R_3} + \ldots$$

The effective resistance R_{eff} of a parallel circuit is less than the sum of the separate resistances.

Example

From the diagram given in the Figure below, calculate

(a) the effective resistance of the circuit,
(a) the current flowing in the main circuit,
(a) the current in each resistor.

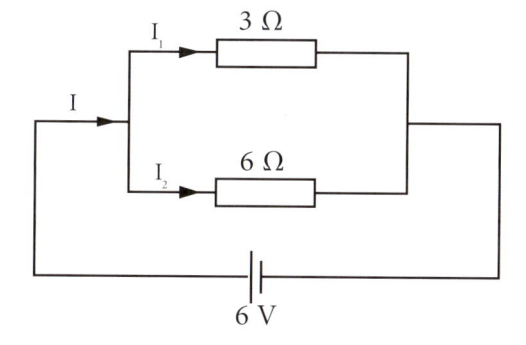

Solution

(a) Using $1/R_{eff} = 1/R_1 + 1/R_2$, we have,

$$\frac{1}{R_{eff}} = \frac{1}{3} + \frac{1}{6} = \frac{3}{6} \Rightarrow R_{eff} = 2$$

The effective resistance is 2 Ω.

(b) We can now determine the current flowing through the system:

$$V = IR_{eff} \Rightarrow I = \frac{V}{R_{eff}}$$

$$I = \frac{6 \text{ V}}{2 \text{ Ω}} = 3 \text{ A}$$

The current flowing in the main circuit is 3 A.

(c) Using $V_i = I_i R_i$, we have:

For the 3 ohm resistor; $I = 6/3 = 2A$

For the 6 ohm resistor; $I = 6/6 = 1A$

The currents in the 3 Ω and the 6 Ω resistors are 2 A and 1 A respectively.

Example

From the diagram given in the figure (below), calculate

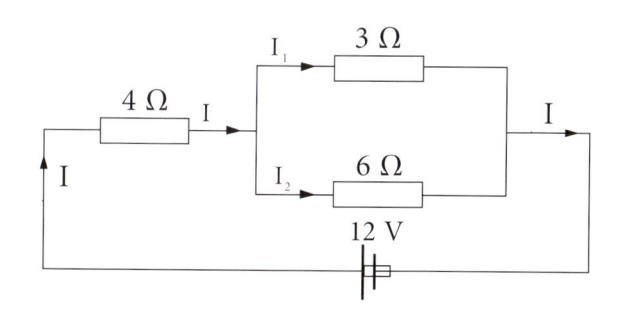

(a) the total resistance of the circuit
(b) the current flowing in the main circuit
(c) the current in each resistor

Solution

(a) For the parallel circuit we have,

$$\frac{1}{R_{eff}} = \frac{1}{3} + \frac{1}{6} = \frac{3}{6} \Rightarrow R_{eff} = 2$$

So that, $R_{eff} = 2Ω$

CHAPTER 5

Thus the total resistance (of the parallel and series arms), R_T, is given by

$R_T = 2\,\Omega + 4\,\Omega$
$\quad = 6\,\Omega$

That is, the effective resistance is 6 Ω.

(b) To determine the current, once again we use, $V = IR$, so that,

$$I = \frac{V}{R_T} = \frac{12}{6} = 2\text{ A}$$

The current flowing in the main circuit is 2 A.

(c) The current in the 4 Ω resistor is 2 A.
The potential difference across the 4 Ω resistor = IR = (2 A) . (4 Ω) = 8 V
The potential difference across the parallel network = 12 V – 8 V
 = 4 V

The current in the network resistors is given by,

$$I = I_1 + I_2 = \frac{V}{R_1} + \frac{V}{R_2}$$

$$I = \frac{4}{3}\text{ A} + \frac{4}{6}\text{ A} = 1.33\text{ A} + 0.67\text{ A}$$

The current in the 4 Ω resistor is **2 A**. The currents in the 3 Ω and the 6 Ω resistors are **1.33 A and 0.67 A respectively**.

joined wires	wires crossing (not joined)	cell
battery	**lamp**	**a.c. supply**
switch	**ammeter**	**voltmeter**
galvanometer	**resistor**	**variable resistor**
potentiometer	**heating element**	**fuse**
transformer	**oscilloscope**	**diode** / **cathode**

Table 525   Common circuit symbols

174

The circuit symbols shown in Table 525 are used for drawing circuit diagrams in this course. This standard set of symbols is also provided in the data booklet that is used during examination sessions. You should use these symbols in place of any other that you may have learnt.

Galvanometers are used to detect electric currents. They use a property of electromagnetism – a coil with a current flowing in it experiences a force when placed in a magnetic field. Most non-digital ammeters and voltmeters consist of a moving coil galvanometer connected to resistors. Digital meters are becoming more common, and the digital multimeter can act as an ammeter, a voltmeter or an ohmmeter.

A voltmeter

1. is always connected across a device (in parallel).

2. has a very high resistance so that it takes very little current from the device whose potential difference is being measured.

3. has a high resistor (a multiplier) connected in series with a galvanometer.

4. an ideal voltmeter would have infinite resistance with no current passing through it and no energy would be dissipated in it.

Example

A galvanometer has a resistance of $1.0 \times 10^3 \, \Omega$ (mainly due to the resistance of the coil), and gives a full-scale deflection (fsd) for 1.0 mA of current. Calculate the size of a multiplier resistor.

Solution

The Figure below demonstrates a possible set up.

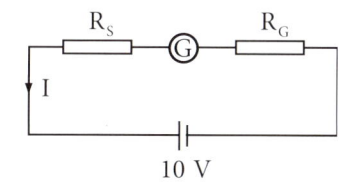

Because the resistors R_g and R_s are in series then the potential difference across the resistors will be 10 volts. Therefore, $I(R_g + R_s) = 10$

$0.001 \times (1000 + R_s) = 10$ $0.001 R_s = 9$

$R_s = 9000 \, \Omega$

The multiplier resistor = $\mathbf{9.0 \times 10^3 \, \Omega}$

An ammeter

1. is always connected in series with a circuit.

2. has a very low resistance compared with the resistance of the circuit so that it will not significantly alter the current flowing in the circuit.

3. has a low resistor (a shunt) connected in parallel with a galvanometer.

4. would ideally have no resistance with no potential difference across it and no energy would be dissipated in it.

Example

A galvanometer has a resistance of $1.0 \times 10^3 \, \Omega$ (mainly due to the resistace of the coil), and gives a full-scale deflection (fsd) for 1.0 mA of current. Calculate the size of a shunt resistor that would be needed to convert this to a meter with a fsd of 5.0 A.

Solution

The Figure below demonstrates a possible set up.

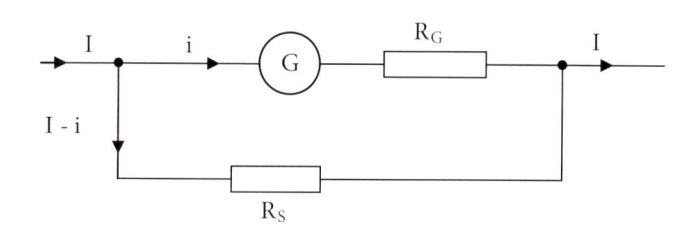

Because the resistors R_g and R_s are in parallel then the potential difference across the resistors is the same. Therefore, $iR_g = (I - i) R_s$.

$R_s = (i / I - i) R_g = [10^{-3} \, A / (5 - 0.001) \, A] \times 1000 \, \Omega = 0.20 \, \Omega$

The shunt resistor = $0.20 \, \Omega$

A potential divider

In electronic systems, it is often necessary to obtain smaller voltages from larger voltages for the various electronic circuits. A **potential divider** is a device that produces the required voltage for a component from a larger voltage. It consists of a series of resistors or a rheostat (variable resistor) connected in series in a circuit. A simple voltage divider is shown in Figure 526.

CHAPTER 5

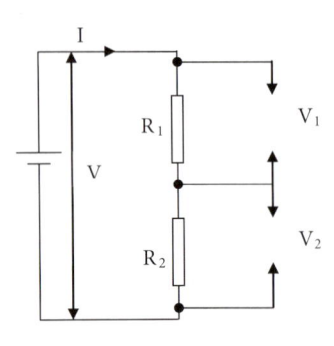

Figure 526 A simple voltage divider

Using Ohm's Law, $V_1 = IR_1$ and $V = I(R_1 + R_2)$.

$$\frac{V_1}{V} = \frac{IR_1}{I(R_1 + R_2)}.$$ Therefore,

$$V_1 = \frac{R_1}{(R_1 + R_2)} \times V$$

This is known as the potential divider equation.

Example

In the potential divider shown below, calculate:

(a) the total current in the circuit

(b) the potential difference across each resistor

(c) the voltmeter reading if it was connected between terminals 2 and 6.

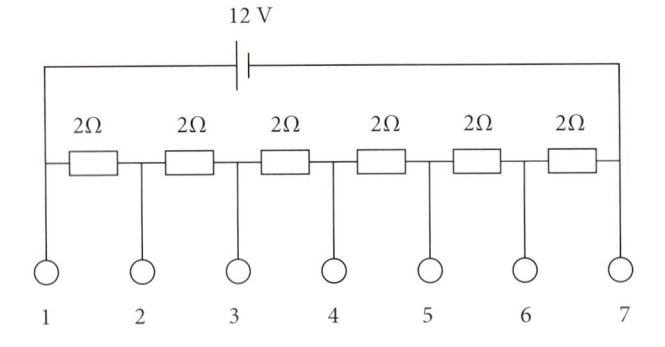

Solution

(a) The total resistance = 12 Ω. From Ohm's law

$$V = IR \text{ and } I = \frac{V}{R}$$

$$I = \frac{12 \text{ V}}{12 \text{ Ω}} = 1 \text{ A}$$

(b) The 12 V is equally shared by each 2 Ω resistor. Therefore, the potential difference across each resistor = 2 V. Alternatively, $V = IR = 1 \text{ A} \times 2 \text{ Ω} = 2 \text{ V}$

(c) Between terminals 2 and 6 there are 4 resistors each of 2 Ω. Therefore, the potential difference the terminals
= 4 × 2 V = 8 V.

Because resistance is directly proportional to the length of a resistor, a variable resistor also known as a potentiometer or colloquially as a "pot" can also be used to determine the potential difference across an output transducer (device for converting energy from one form to another) such as a filament lamp in Figure 527.

If the pointer was at A then the potential difference would be zero as there is no power dissipated per unit current in the potentiometer (the load), and there would be no output voltage. However, if the pointer is moved up to two-thirds the length of the potentiometer as in the figure, then the output voltage across the filament lamp would be ⅔ × 6V = 4V.

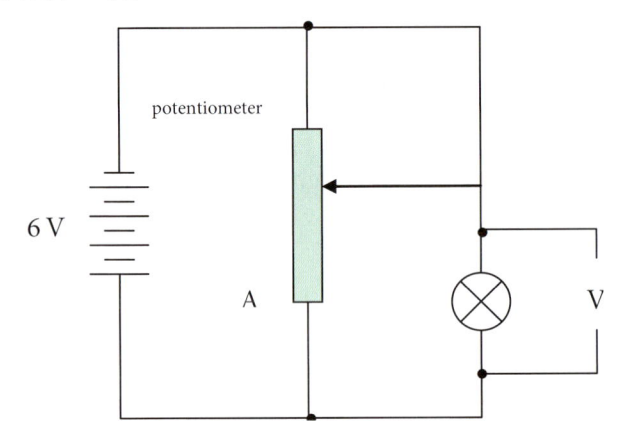

Figure 527 Laboratory potentiometer

Pots have a rotating wheel mounted in plastic and they are commonly used as volume and tone controls in sound systems. They can be made from wire, metal oxides or carbon compounds.

Sensors in potential divider circuits

A number of sensors (input transducers) that rely on a change in resistance can be used in conjunction with potential dividers to allow for the transfer of energy from one form to another. Three such sensors are:

- the light dependent resistor (LDR)
- the negative temperature coefficient thermistor (NTC)
- strain gauges.

176

ELECTRICITY AND MAGNETISM

A **light dependent resistor** (LDR) is a photo-conductive cell whose resistance changes with the intensity of the incident light. Typically, it contains a grid of interlocking electrodes made of gold deposited on glass over which is deposited a layer of the semiconductor, cadmium sulfide. Its range of resistance is from over 10 MΩ in the dark to about 100 Ω in sunlight. A simple LDR and its circuit symbol are shown in Figure 528.

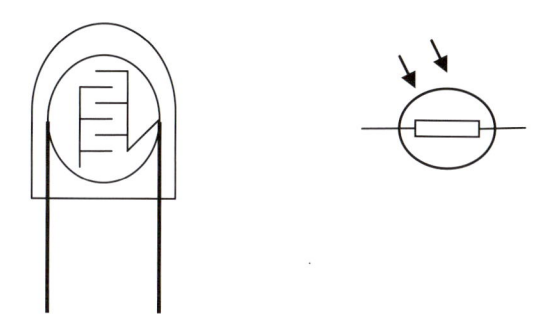

Figure 528 A light dependent resistor

There are only two ways to construct a voltage divider with an LDR sensor – with the LDR at the top, or with the LDR at the bottom as shown in Figure 529.

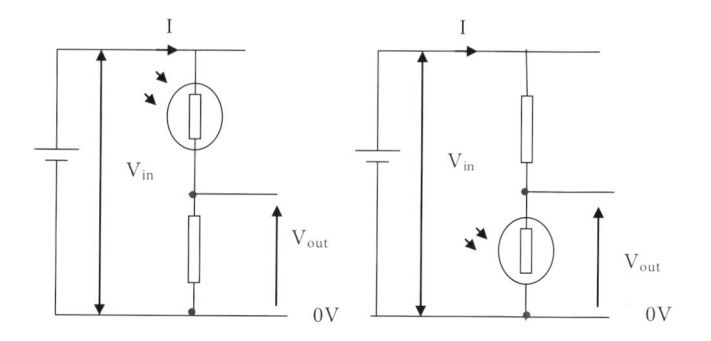

Figure 529 (a) and (b) two ways to construct an L.D.R.

Light dependent resistors have many uses in electronic circuits including smoke detectors, burglar alarms, camera light meters, camera aperture controls in automatic cameras and controls for switching streetlights off and on.

Resistors that change resistance with temperature are called **thermistors** (derived from thermal resistors). They are made from ceramic materials containing a semiconductor the main types being bead and rod thermistors. The NTC (negative temperature coefficient) thermistor contains a mixture of iron, nickel and cobalt oxides with small amounts of other substances. They may have a positive (PTC) or negative (NTC) temperature coefficient according to the equation:

$$R_f = R_0(1 + \alpha t)$$

where R_0 equals the resistance at some reference temperature say 0 °C, R_f is the resistance at some temperature, t °C, above the reference temperature, and α is the temperature coefficient for the material being used.

The circuit symbol for a thermistor is shown in Figure 530.

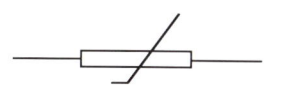

Figure 530 The circuit symbol for a thermistor

For a NTC thermistor, the resistance decreases when the temperature rises and therefore they can pass more current. This current could be used to to operate a galvonometer with a scale calibrated in degrees as used in electronic thermometers or car coolant system gauges. Thermistors are also used in data-logging temperature probes but the analogue signal has to be converted to digital signal using an analogue to digital converter (ADC).

With normal resistors the resistance becomes higher when the temperature increases and they therefore have a small positive temperature coefficient. Figures 531 and 532 demonstrate how the resistance changes with temperature for both types of thermistors.

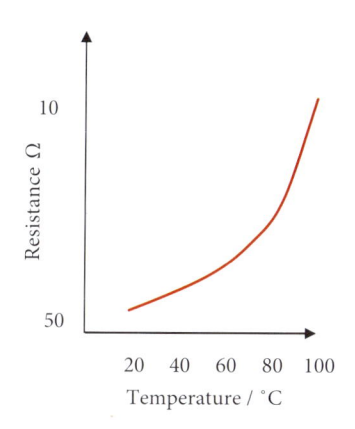

Figure 531 A PTC thermistor

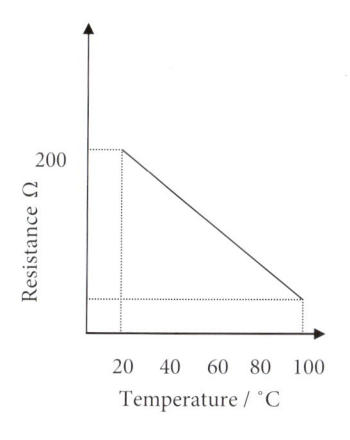

Figure 532 An NTC thermistor

An electronic thermometer can be made using an NTC thermistor as shown in Figure 533.

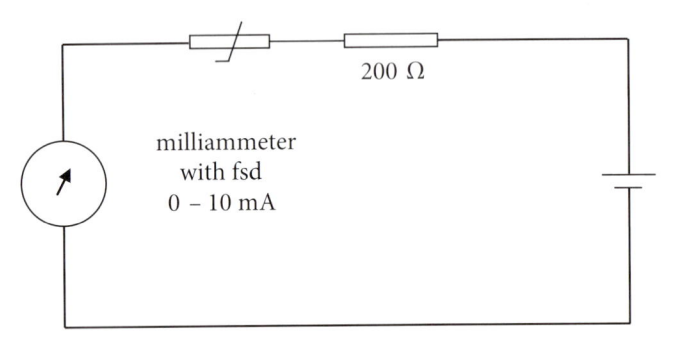

Figure 533   An electronic thermometer

When a metal conducting wire is put under vertical strain, it will become longer and thinner and as a result its resistance will increase. An **electrical strain gauge** is a device that employs this principle. It can be used to obtain information about the size and distribution of strains in structures such as metal bridges and aircraft to name but two. A simple gauge consists of very fine parallel threads of a continuous metal alloy wire cemented to a thin piece of paper that are hooked up to a resistance measuring device with thick connecting wires as shown in Figure 534. When it is securely attached on the metal to be tested, it will experience the same strain as the test metal and as this happens the strain gauge wire become longer and thinner and as such the resistance increases.

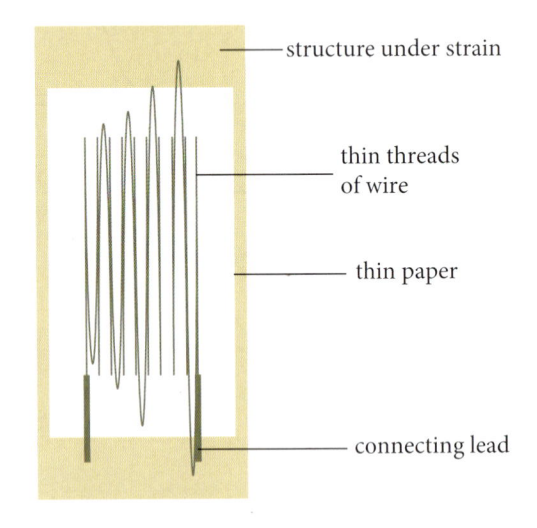

Figure 534   A resistance measuring device

5.3 Electric cells

> **NATURE OF SCIENCE:**
>
> Long-term risks: Scientists need to balance the research into electric cells that can store energy with greater energy density to provide longer device lifetimes with the long-term risks associated with the disposal of the chemicals involved when batteries are discarded. (4.8)
>
> © IBO 2014

Essential idea: Electric cells allow us to store energy in a chemical form.

Understandings:
- Cells
- Internal resistance
- Secondary cells
- Terminal potential difference
- Electromotive force (emf)

Cells

There are many practical applications and devices based on the principles of static electricity. Electrostatic devices were able to produce a small flow of charge at a high potential. However, once the electric cell was discovered, a continuous flow of charge for a long period of time could be obtained at low potential. This meant that the rate of flow of charge could be controlled in electrical circuits.

In the 1780s Luigi Galvani, a professor of Anatomy at the University of Bologna in Italy published a book on the contraction of frog's leg muscle when they were subjected to an electrical impulse supplied by a static electricity machine. Furthermore, he connected a copper wire to the freshly dissected spinal marrow of the frog's leg and an iron wire to the frog's foot, and when the wires were connected, the leg contracted or twitched. He believed that the source of the electric charge was either the muscle or the nerve.

A fellow countryman, *Alessandro Volta*, who was working at the University of Pavia about 200 km from Bologna was skeptical of Galvani's finding. He reasoned that the source of the electrical charge was not from the frog's muscle or nerve but rather from the two different metals that were touched together. He also reasoned that moisture was needed at the metal contact point or at the frog muscle, and that the contraction was a detecting device that measured the "electromotive force".

In order to support his thoughts, he constructed a series of silver and zinc disks piled on top of each other with

Electricity and Magnetism

each disc being separated with a piece of paper soaked in concentrated salt solution – a voltaic pile. When he touched the pile with his fingers, he felt a small electric shock. Further experiments found that different metals in combination produced different "electromotive force" and that graphite could also be used in place of one of the metals. He also used different solutions of salts and acids to see whether he obtained similar results. A diagram of a voltaic pile is shown in Figure 535.

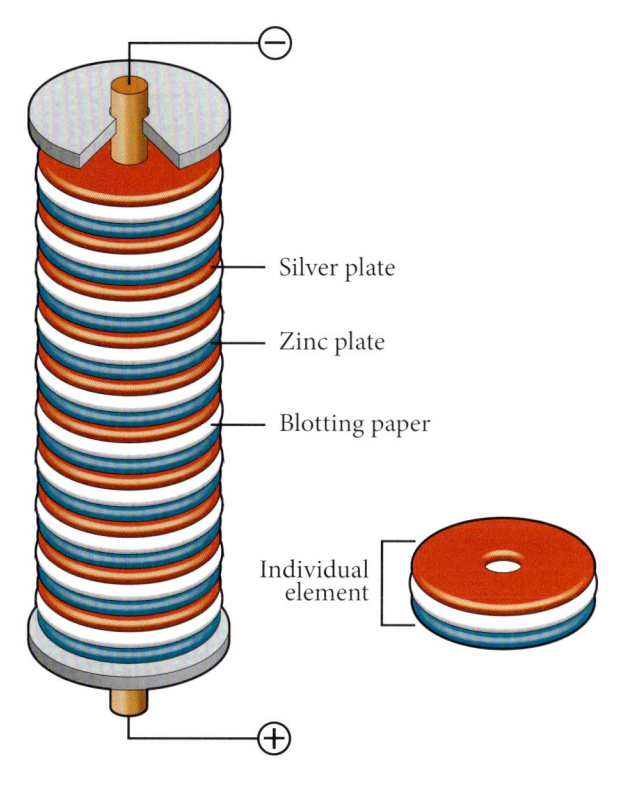

Figure 535   A voltaic pile

The voltaic pile method was very messy as the weight of the discs would squeeze out the electrolyte between the discs and the pile would malfunction. He refined his pile to produce a series of cells that made up a battery that he called "a crown of cups". It was so named because it consisted of a circle of glass containers each containing a series of copper and zinc electrodes dipping in sulfuric acid as shown in Figure 536.

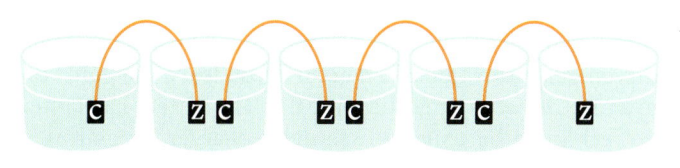

Figure 536   Volta's crown of cups

Volta published his results in the year 1800, and his findings revolutionized the future of electrodynamics.

Electric cells convert chemical energy into electrical energy, and they are often called electrochemical cells. In modern times there is a great variety of cells and batteries used in flashlights, remote controls, digital watches and the storage battery in a car to name but a few.

Primary cells

The most common one in the home is the dry cell as shown in Figure 537.

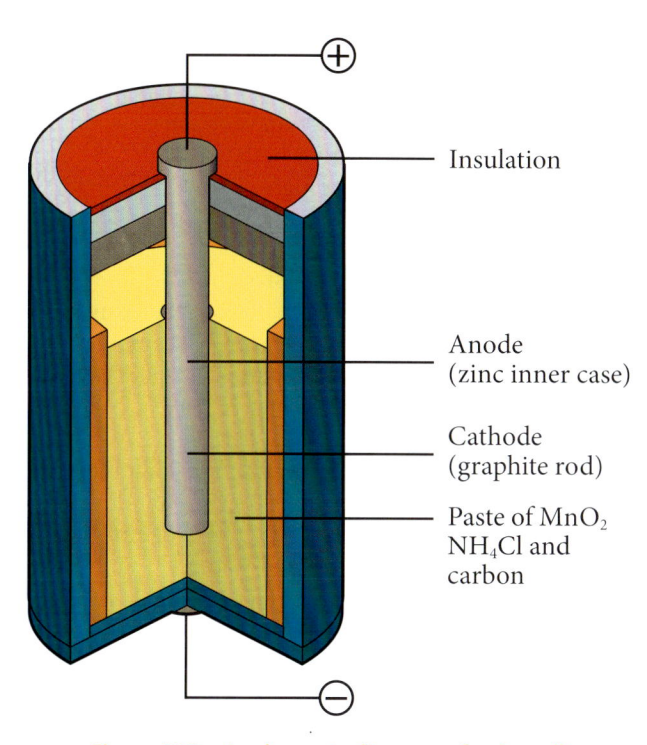

Figure 537   A schematic diagram of a dry cell.

It consists of a zinc-casing **electrode** (anode) with a graphite rod electrode in the centre surrounded by a paste of manganese (IV) oxide (cathode). The zinc metal is oxidized to Mn2+ ions and the manganese (IV) oxide is reduced to manganese (III) oxide. The manganese (IV) oxide also contains ammonium chloride and zinc chloride that act as the **electrolyte** for the cell. This type of dry cell has a maximum emf of 1.48 volts. When a number of dry cells are connected in series we call it a **battery**, although today in common talk a single cell is also called a battery. When used as a battery in the true sense, the total emf is the addition of the emf from each cell.

A dry cell is called a **primary cell** because once the total energy is discharged it cannot be recharged and it has to be disposed of. Some commonly used primary cells are listed in Figure 538.

179

Chapter 5

Cell	Nominal emf	Properties	Common use
Zinc - graphite	1.5V	Cheap, popular, best for low currents, emf falls as current increases.	Flashlight, toys Remote controls
Alkaline-manganese	1.5V	Longer use time, medium price, better for high currents.	Calculators Remote controls
Mercury	1.4V	Expensive, large capacity for size, constant emf.	Hearing aids, cameras, watches
Weston standard	1.0186V at RTP	emf contant at low currents	Laboratory standard of emf

Figure 538 *Types of primary cells*

An alkaline version of the dry cell has superior performance to the acid version describe above. A powdered zinc anode is used and the electrolyte used is potassium hydroxide instead of the ammonium chloride.

Internal resistance

A cell or battery is not a source of constant current because the current varies according to the resistance of a circuit. For example, if you try to turn on the starter motor of a car while using other electrical components of a car such as the heater or headlights, the heater temporarily cuts out and the headlights dim. The starter motor draws a large current around 100A and the voltage drops because the chemical reactions occurring in the cell or battery cannot maintain a constant charge fast enough to maintain the rated emf. Thus, the cell or battery itself has some resistance to the flow of charge. This type of resistance is called **internal resistance** and is usually designated with the symbol r.

Therefore, the terminal voltage of a typical electric cell loses its initial value rapidly in the beginning of its life. It then stabilises and becomes a fairly constant value for most of its "shelf" life. Then as the chemicals are consumed, it rapidly decreases to zero as the cell is completely discharged.

Secondary cells

In contrast to primary cells, secondary cells can be recharged and they can maintain a high continual flow of current for long periods of time. Most importantly, passing a current through the cell in the opposite direction to its discharge can recharge them. They are also called storage cells or accumulators and the two most common types are the lead-acid battery and the nickel-cadmium cell.

A schematic lead-acid battery is shown in Figure 539.

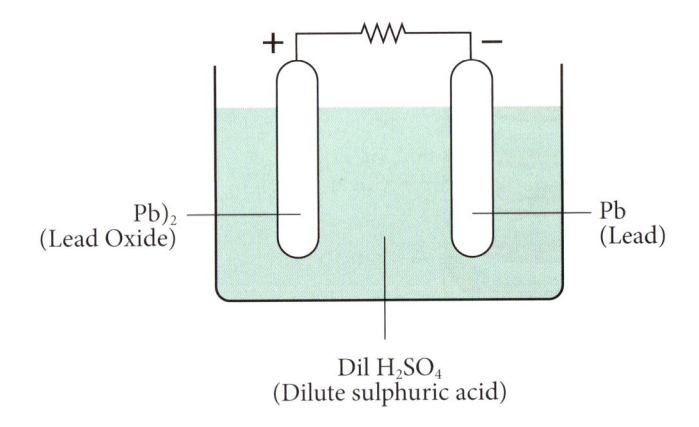

Figure 539 *A typical lead-acid battery used in motor vehicles*

The 12V storage battery consists of six cells connected in series with each cell having an emf of 2V. The cells are contained in a heavy-duty plastic external casing. The lead electrodes in each cell consist of a bank of lead plates to increase the surface area of the electrode material. One negative (anode) bank is filled with spongy metallic lead and one positive (cathode) bank is filled with brown lead dioxide. The electrolyte is a fairly concentrated solution of sulfuric acid (4.5 M).

When discharging, the electrodes become coated with lead sulfate and the density of the sulfuric acid decreases. When the battery is recharging, connecting the terminals to another electrical source of higher voltage, and reversing the direction of the electric current through the circuit reverse the electrode reactions.

Recharging occurs during driving time by using a motor-driven generator (an alternator). If you leave the lights on in a car, the battery may become flat. The battery can either be recharged by using a external recharger, or getting a "jump start" from the battery of another car.

The rechargeable cells used in digital cameras, cell phones and other cordless devices are often "ni-cad" cells.

180

Electricity and Magnetism

Electromotive force

In the previous section, potential difference was defined in terms of the amount of work that has to be done per unit charge to move a positively-charged body in an electric field.

$$\Delta V = \frac{\Delta W}{\Delta q} \quad \text{or} \quad \Delta W = (\Delta q)\Delta V$$

From the present definition of electric current, we have,

$$I = \frac{\Delta q}{\Delta t} \quad \text{or} \quad \Delta q = I\Delta t$$

By substituting for q in the top equation, we end up with

$$\Delta W = I \times \Delta t \times \Delta V \quad \text{or} \quad \Delta V = \frac{\Delta W}{I\Delta t}$$

Therefore it can be stated that

'Potential difference in external circuits is the power, (P), dissipated (released) per unit current'

Either definition for potential difference is acceptable. However, the above-mentioned ties voltage and current together.

Note: It is common place to use the expression $V = W / It$ as opposed to $\Delta V = \Delta W / I\Delta t$ and $q = It$ as opposed to $q = I\Delta t$. That is, to leave out the 'change in' (or 'Δ') notation when solving problems.

The terms emf, potential difference and voltage are commonly interchanged in talking about electricity. The term electromotive force is a misnomer as it is not a force at all but rather a potential energy difference. Again, there are historical reasons for the use of the term in the first place as already mentioned.

In the true sense, **electromotive force (emf) is the work per unit charge made available by an electrical source.** For the simple circuit in Figure 511, the emf of the dry cell is 1.5 V. However, when a voltmeter is used to test the potential difference across the resistor the reading on the voltmeter is found to be less than 1.5 V. This is due to the cell not being an 'ideal dry cell' because it has some internal resistance.

For the moment, let us say that emf is the energy **supplied** per unit charge across the terminals when it is not producing current in the circuit, and potential difference is the energy **released** (dissipated) per unit charge.

Example

A battery supplies 15.0 J of energy to 4.00 C of charge passing through it. Determine the emf of the battery

Solution

Using the formula, $V = W / q$, we have, $V = 15.0J / 4.00 C$

= 3.75 V

The emf of the battery is **3.75 V**.

In order to maintain the flow of charge through a conductor, an energy source must provide energy to the charge carriers. The major sources of emf are:

1. Electromagnetic

When a coil of wire is rotated in a magnetic field, an induced current is produced. Power stations use generators to produce a current.

2. Chemical

Oxidation-reduction reactions transfer electrons between chemicals. Dry cells, fuel cells and batteries are examples.

3. Photoelectric effect

Electrons are emitted from certain metal surfaces when high frequency light is shone on their surfaces. These photocells are used in watches, clocks, automatic doors.

4. Piezoelectric effect

Certain crystals can produce a charge on one side when placed under stress. If one side of the crystal is charged and the other side is uncharged, a potential difference exists across the crystal. This is used in crystal microphones.

5. Thermoelectric effect

When two pieces of certain metals are wound together and one end is heated while the other end is cooled, a current is produced. Thermocouples can be used as temperature-measuring devices.

Terminal voltage, emf and internal resistance

When electrons flow around a circuit, they gain potential energy in the cell and then lose the energy in the resistors. In a previous section, it was stated that the emf is the energy **supplied** per unit charge, and the potential difference is the energy **released** (dissipated) per unit charge in a circuit.

CHAPTER 5

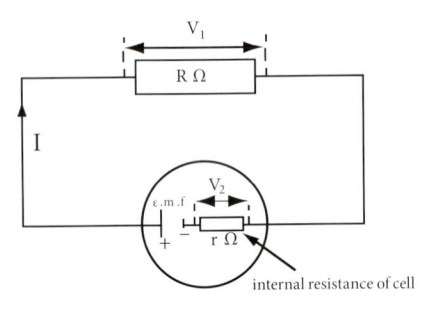

Figure 540 Internal resistance circuit.

In the circuit in Figure 540, the total energy supplied is determined by the value of the emf. The total energy released (per unit charge) is equal to the potential difference across resistor R plus the potential difference of the **internal resistance** r.

$$\text{emf} = V_1 + V_2 = IR + Ir = I(R + r)$$

IR is known as the terminal voltage.

That is,

$$IR = \text{emf} - Ir$$

All sources of emf have some internal resistance and it is a factor in determining how useful a source of emf is.

A dry cell being a primary cell releases its energy and becomes "dead". When it is just about inoperable, the build–up of the products of the relevant oxidation-reduction reaction increases the internal resistance beyond its normal value. The emf available does not decrease from giving the maximum number of joules per coulomb of charge. However, when much of the emf is used up in overcoming the potential difference across the internal resistor, the terminal voltage will drop to zero. Batteries are generally regarded as being a secondary cell because they can be recharged many times.

If a series of potential difference readings measured with a voltmeter across a variable resistor R are taken concurrently with the current flowing in the circuit measured with an ammeter for each variable resistance, and a graph of voltage versus current is plotted, from the relationship

$$\text{Terminal voltage} = V = \text{emf} - Ir$$

The y–intercept of the graph will be the value of the emf, and the negative gradient is the internal resistance.

Example

A dry cell has an internal resistance of 1.50 Ω. A resistor of 12.0 Ω is connected in series with the dry cell. If the potential difference across the 12.0 Ω resistor is 1.20 V, calculate the emf of the cell.

Solution

We first need to determine the current flowing through the system. This is done by using the formula, V = IR from where we obtain, I = V / R.

Therefore, we have that

$$I = \frac{V}{R} = \frac{1.20 \text{ V}}{12.0 \text{ Ω}} = 0.100 \text{ A}$$

Next, we use the formula, emf = I(R + r), where the internal resistance, r = 1.5 Ω.

That is, emf = I(R + r) = 0.100 A × (12.0 Ω + 1.50 Ω)

$$= 1.35 \text{ V}$$

The emf of the cell is **1.35 V**.

Exercise 5.3

1. Three identical resistors of 3 Ω are connected in parallel in a circuit. The effective resistance would be

    A.   1 Ω
    B.   3 Ω
    C.   6 Ω
    D.   9 Ω

2. In a television tube the picture is produced in a fluorescent material at the front of the picture tube by

    A.   an electrical discharge
    B.   a beam of positive ions
    C.   an electrolytic deposition of metal atoms
    D.   a stream of electrons

3. A 100 W light globe gives out a brighter light than a 60 W globe mainly because

    A.   a larger potential difference is used to run it
    B.   its resistance wire filament is longer
    C.   more electric current flows through it
    D.   it has a higher amount of inert gas in it

4. In the circuit below, a heater with resistance R is connected in series with a 48 V supply and a resistor S.

ELECTRICITY AND MAGNETISM

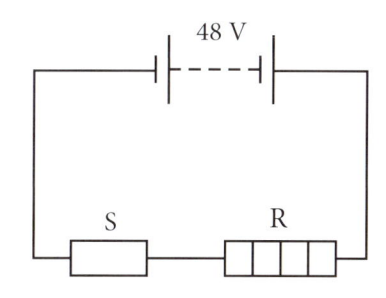

If the potential difference across the heater is to be maintained at 12 V, the resistance of the resistor S should be

A. R / 2
B. R / 4
C. R / 3
D. 3R

5. The diagrams below show circuits X, Y and Z of three resistors, each resistor having the same resistance.

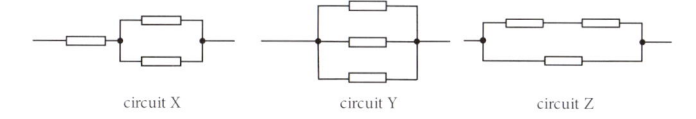

circuit X   circuit Y   circuit Z

Which **one** of the following shows the resistances of the circuits in increasing order of magnitude?

	lowest	→	highest
A.	Y	Z	X
B.	Z	X	Y
C.	X	Y	Z
D.	Z	Y	X

6. The graph below shows the current/voltage characteristics of a filament lamp.

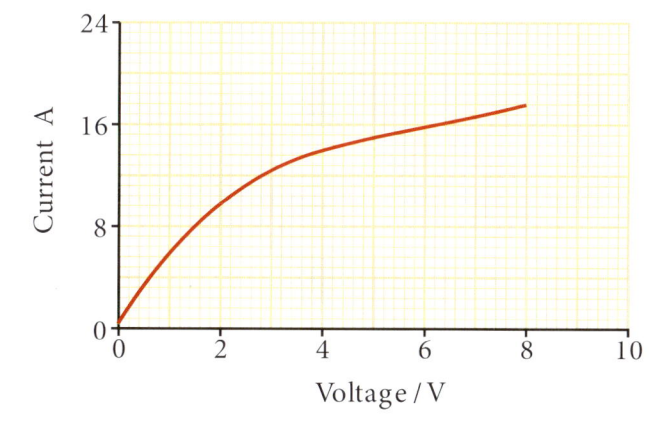

The resistance of the filament at 3.0 V is

A. 0.25 Ω
B. 250 Ω
C. 4000 Ω
D. 8000 Ω

7. Metal X has half the resistivity of metal Y and length three times that of Y. If both X and Y have the same surface area, the ratio of their resistance R_X / R_Y is:

A. 3 : 4
B. 2 : 3
C. 1.5 : 1
D. 2 : 9

8. Three identical resistors of 3 Ω are connected in parallel in a circuit. The effective resistance would be

A. 1 Ω
B. 3 Ω
C. 6 Ω
D. 9 Ω

9. If 18 J of work must be done to move 2.0 C of charge from point A to point B in an electric field, the potential difference between points A and B is:

A. 0.1 V
B. 9.0 V
C. 12 V
D. 20 V

10. The fundamental SI unit – the ampere – is defined in terms of:

A. potential difference and resistance
B. the time rate of change of charge in a circuit
C. the force acting between two current carrying wires
D. the product of charge and time

11. A 100 W light globe gives out a brighter light than a 60 W globe mainly because

A. a larger potential difference is used to run it
B. its resistance wire filament is longer
C. more electric current flows through it
D. it has a higher amount of inert gas in it

12. Three identical lamps L_1, L_2 and L_3 are connected as shown in the following diagram.

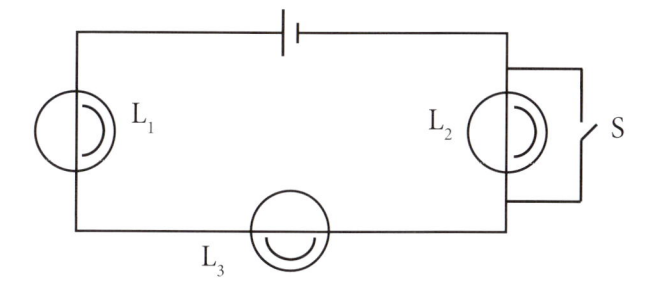

When switch S is closed

183

CHAPTER 5

A. L₁ and L₃ brighten and L₂ goes out
B. all three lamps glow with the same brightness
C. L₂ brightens and L₁ and L₃ remain unchanged
D. L₁ and L₃ go dimmer and L₂ goes out

13. Two resistors are connected in parallel and have the currents I₁ and I₂ as shown in the diagram

   If the effective resistance of the circuit is R then

   A. R = (R₁ + R₂) / R₁R₂
   B. I₁/R₁ = I₂ / R₂
   C. R = R₁R₂ / R₁ + R₂
   D. R = I (R₁ + R₂) / R₁R₂

14. The following circuit was set up to determine the internal resistance r of a dry cell. The load resistor was varied from 100Ω to 150Ω, and the current in the circuit was measured using an ammeter. The two respective values of the current are given in the circuits.

   The internal resistance r of the dry cell is

   A. 3.0 Ω
   B. 7.1 Ω
   C. 58.8Ω

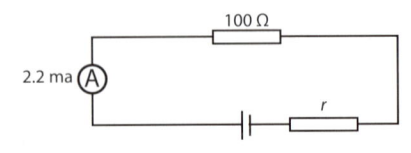

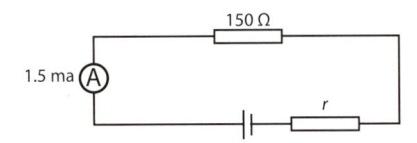

   The internal resistance r of the dry cell is

   A. 3.0 Ω
   B. 7.1 Ω
   C. 58.8Ω
   D. 250 Ω

15. This question concerns the electric circuit below

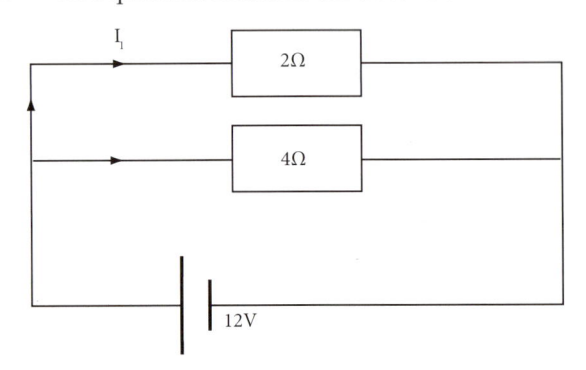

   The cell voltage is 12 V. The value of current I₁ is

   A. 2 A
   B. 4 A
   C. 6 A
   D. 9 A

16. Consider the circuit below that contains a 15 V battery with zero internal resistance.

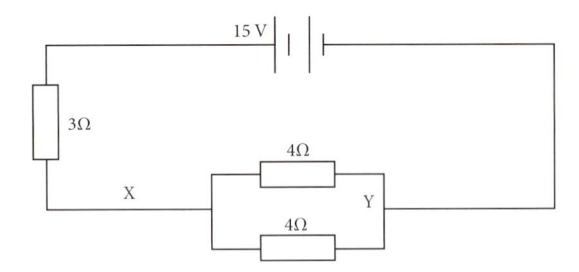

   A voltmeter connected between the points X and Y should read:

   A. 0 V
   B. 3 V
   C. 6 V
   D. 9 V

17. Determine the equivalent resistance when 12 Ω, 6 Ω and 4 Ω are placed in

   (a) series
   (b) parallel

18. Calculate the work done in moving a 12.0 μC through a p.d of 240 V.

19. The diagram shows resistances joined in a compound circuit.

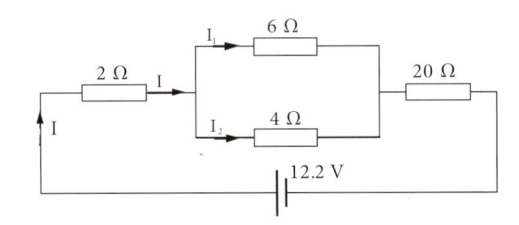

   (a) Determine the total resistance of the circuit.
   (b) Calculate current flows through the 2.0 Ω resistor.
   (c) Deduce the potential difference across the 20.0 Ω resistor.
   (d) Determine is the potential difference across the 6.0 Ω resistor.
   (e) Calculate is the current through the 4.0 Ω resistor.

184

ELECTRICITY AND MAGNETISM

20. A photo-electric cell draws a current of 0.12 A when driving a small load of resistance 2 Ω. If the emf of the cell is 0.8 V, determine the internal resistance of it under these conditions.

21. In terms of emf and internal resistance, explain why is it possible to re-charge a nickel-cadmium cell while normal dry cells have to be discarded once they are flat.

22. When a dry cell is connected to a circuit with a load resistor of 4.0 Ω, there is a terminal voltage of 1.3 V. When the load resistor is changed to 12 Ω, the terminal voltage is found to be 1.45 V.

    Calculate

    (a) the emf of the cell.
    (b) the internal resistance of the cell.

23. Calculate the current flowing through a hair drier if it takes 2.40×10^3 C of charge to dry a person's hair in 3.0 minutes.

24. An iron draws 6.0 A of current when operating in a country with a mains-supply of 240 V. Calculate the resistance of the iron.

25. An electrical appliance is rated as 2.5 kW, 240 V

    (a) Calculate the current it needs to draw in order to operate.
    (b) Determine how much energy would be consumed in 2 hours.

26. A 2.5 kW blow heater is used for eight hours. Calculate the cost of running the blow heater if the electricity is sold at 15 cents per kilowatt-hour.

27. The circuit below refers to the following questions:

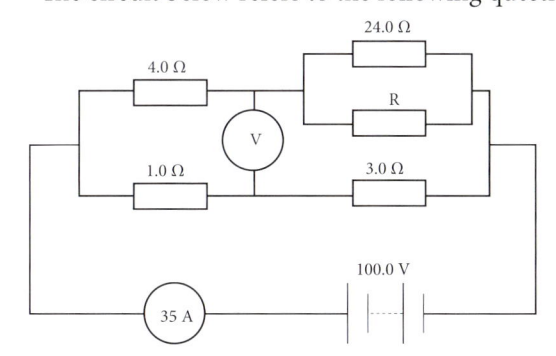

    (i) Determine the current flowing through R, and the value of resistor R.
    (ii) Deduce the reading on the voltmeter V.

28. (i) Describe the meaning of the "12 V" on a 12 volt car battery.
    (ii) A 14 V car battery drops to 12 V when supplying a current of 5.0 A. Determine the internal resistance of the battery.

29. A circuit was set up to investigate the relationship between the current *I* through a resistor and the magnitude of the resistance *R* while a constant electromotive force was supplied by a dry cell. The results of the investigation are given in the following table:

R ± 0.5Ω	I ± 0.1 A	
2.0	5.0	
6.0	1.7	
12	0.83	
16	0.63	
18	0.56	

    (a) Complete the last column for the inverse of the current giving the correct unit.
    (b) Plot a graph of R against 1 / I
    (c) Describe the relationship that exists between the resistance and the current.
    (d) Determine the electromotive force of the dry cell

30. Starting from the laws of conservation of energy and conservation of charge, derive a formula for calculating the effective resistance of two resistors in parallel.

31. The diagram shows a typical circuit.

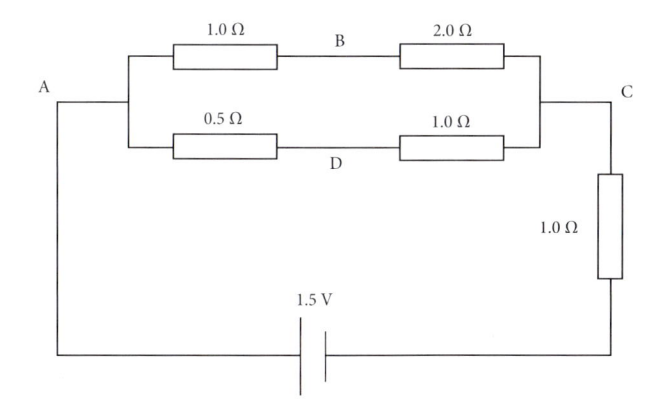

    (a) Determine the effective resistance of the whole circuit.
    (b) Determine the currents flowing in each network resistor.
    (c) Determine the potential differences V_{AB} and V_{AD}.
    (d) Determine the potential difference between B and D.

185

32. Determine the resistance of the LDR in the diagram below if a current of 4.5 mA is flowing in the circuit.

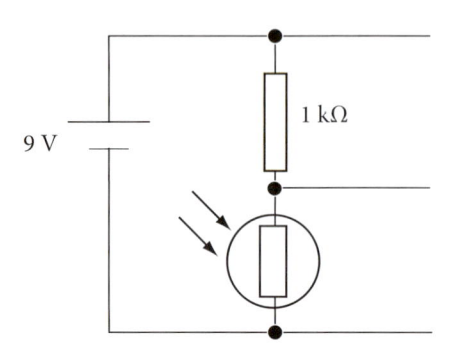

5.4 Magnetic effects of electric currents

NATURE OF SCIENCE:

Models and visualization: Magnetic field lines provide a powerful visualization of a magnetic field. Historically, the field lines helped scientists and engineers to understand a link that begins with the influence of one moving charge on another and leads onto relativity. (1.10)

© IBO 2014

Essential idea: The effect scientists call magnetism arises when one charge moves in the vicinity of another moving charge.

Understandings:
- Magnetic fields
- Magnetic force

Magnetic fields

Just as scientists use the concept of gravitational fields and electric fields to explain certain phenomena, so too the concept of magnetic fields and lines of magnetic flux are used to explain the nature of magnetism. (There are some differences as will be discussed later).

A **magnetic field** is said to exist at a point if a compass needle placed there experiences a force. The appearance of a magnetic field can be obtained with the use of iron filings or plotting compasses. The direction of the field is given by the direction that the compass needles point.

Figure 541 demonstrates the use of iron filings and plotting compasses to detect the magnetic fields of a bar magnet and bar magnets used in combination. The compass needles shown surrounding the bar magnet all point along the lines of magnetic flux of the magnet. The magnetic field produced for the two like poles have no magnetic field at some point P. If there are no lines of magnetic flux, there is no magnetic field.

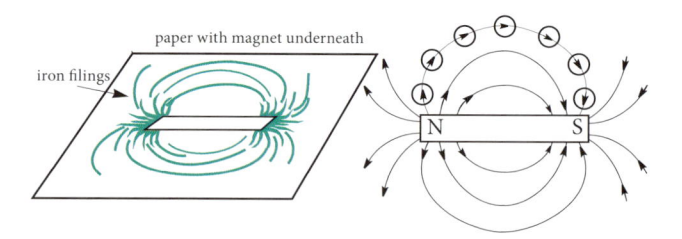

Figure 541   Magnetic field patterns of bar magnets

ELECTRICITY AND MAGNETISM

The Danish physicist, *Hans Christian Oersted* (1777-1851), in 1819, showed conclusively that there existed a relationship between electricity and magnetism. He placed a magnetic needle on a freely rotating pivot point beneath and parallel to a conducting wire. He aligned the compass needle and wire so that it lay along the earth's magnetic north-south orientation. When no current was flowing in the wire, there was no deflection in the needle. However, when the current was switched on, the needle swung to an east-west direction almost perpendicular to the wire. When he reversed the direction of the current, the needle swung in the opposite direction. This is shown in Figure 542.

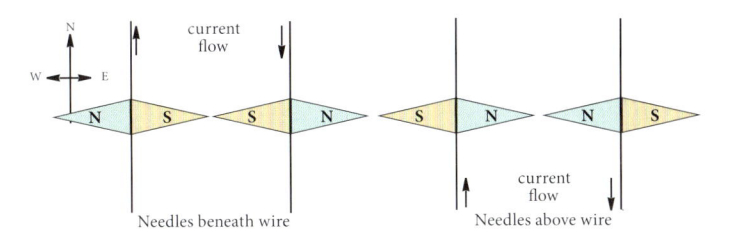

Figure 542   Oersted's experiment.

Up to this stage, all forces were believed to act along a line joining the sources such as the force between two masses, the force between two charges or the force between two magnetic poles. With Oersted's findings, the force did not act along the line joining the forces but rather it acted perpendicular to the line of action. On closer examination and analysis, it was determined that the conducting wire produced its own magnetic field. The magnetic needle, upon interaction with the conducting wire's magnetic field, turns so that it is tangentially (not radially) perpendicular to the wire. **Therefore, the magnetic field produced by the conducting wire produces a circular magnetic field.**

A force is also experienced when a moving charge or a beam of moving charges is placed in a magnetic field. This is what happens in a television set. We will look more closely at moving charges in the next chapter on Atomic and Nuclear physics.

A charged particle can be accelerated by an electric field or by a magnetic field because it experiences a force when influenced by either or both fields. Electric fields can accelerate both moving and stationary charged particles. However, a charge must be moving for a magnetic field to influence it. Furthermore, it cannot be moving parallel to the magnetic field to experience a force.

As stated, stationary charges experience no force in a magnetic field. However, if charged particles move in a magnetic field, a force is exerted on them. This force causes them to deflect if they are not confined in a conductor.

Magnetic field patterns due to currents

The **magnetic field** of a straight current-carrying wire can be investigated using plotting compasses or iron filings sprinkled around the wire as demonstrated with plotting compasses in Figure 543. The lines of force (magnetic flux) around the wire can be seen to be a series of concentric circles that are drawn further apart as the disance away from the wire increases.

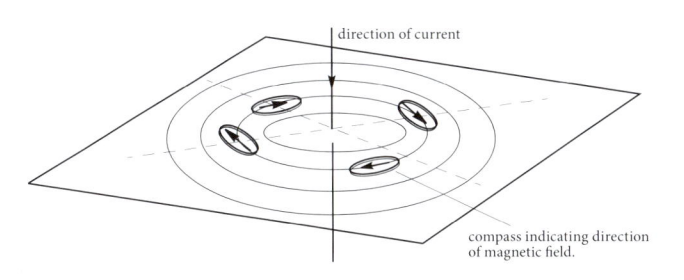

Figure 543   Magnetic field around wire carrying a current

The direction of the magnetic field for a straight conducting wire can be obtained using the "right-hand grip rule" demonstrated in Figure 544.

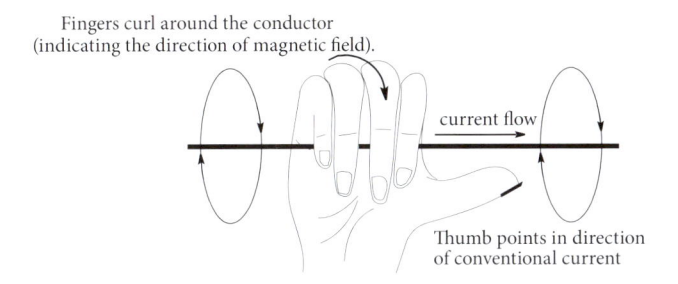

Figure 544   The right hand grip rule

When the thumb of the right hand points in the direction of the **conventional current,** the fingers curl in the direction of the magnetic field.

A more convenient two-dimensional representation of currents and magnetic fields is often used as shown in Figure 545. A cross (×) indicates that the current is into the page and a dot (.) indicates a current flow out of the page.

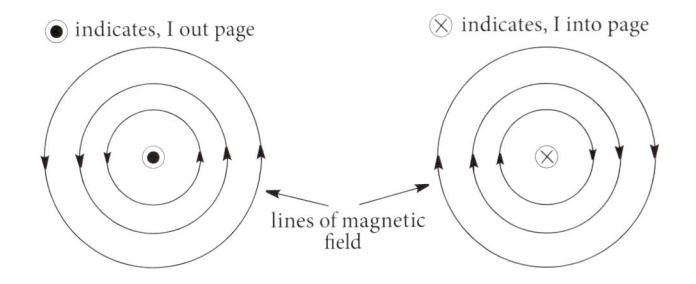

Figure 545   The magnetic field around a conductor

187

The magnetic field due to current in a flat coil (single loop) is shown in Figure 546. Note that the lines of magnetic flux are dense in the middle of the coil and towards the left and right current-carrying wires. This is similar to what happens with a U-shaped magnet (horseshoe magnet). The strength of the magnetic field increases in a coil.

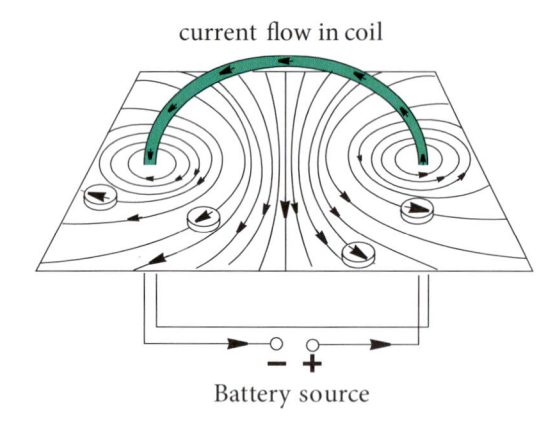

Figure 546   The magnetic field due to current in a single loop

A useful method for determining the polarity of the flat coil is shown in Figure 547.

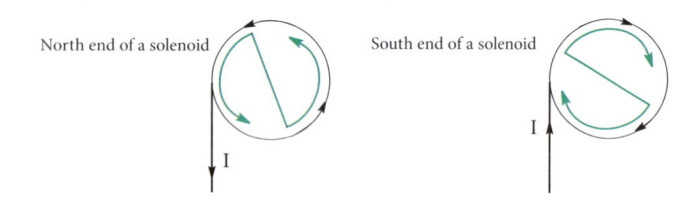

Figure 547   The polarity of a flat coil

If the conventional current is moving anti-clockwise, that end of the loop is a north pole. In this case the left side of the loop is a north pole. If the current flows clockwise as on the right side of the loop, it is a south pole.

A solenoid consists of many coils of a single long wire, and when a current flows in it, a magnetic field similar to a bar magnet is produced. By using plotting compasses as shown in Figure 548 the direction of the magnetic field can be determined.

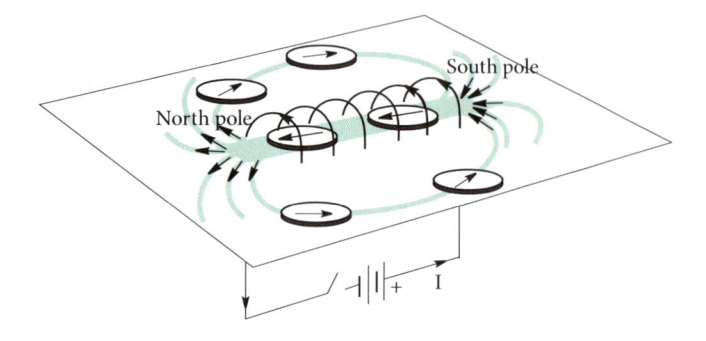

Figure 548   Magnetic field of a solenoid

The field inside the coil is very strong and uniform and this makes solenoids useful devices in science and technology.

The polarity of each end of the solenoid can be determined using the same method as shown in Figure 547. If the conventional current when **viewed head-on** is moving anti-clockwise, that end of the solenoid is a north pole. If the current flows clockwise, when viewed head-on it is a south pole. The north and south poles are shown in Figure 548. Check them for yourself.

The strength of the magnetic field inside a solenoid can be increased by:

1. Increasing the current flowing.

2. Increasing the number of coils per unit length.

3. Inserting a soft iron core in the coil.

When a soft-iron core is inserted into a solenoid and the current is switched on, an electromagnet is produced. If the current is switched off, the solenoid loses its magnetic properties. We say it is a temporary magnet in this case. However, electromagnets can be left on for long periods of time and most magnets in science and industry are of this sort. Electromagnets have many practical uses in scrap metal yards, in electric bells, in particle accelerators and maglev trains.

A relay is an electromagnet switch using a small current to switch on a larger current. This can be employed to switch on motors or electronic components commonly used in security systems.

Exercise 5.4

1. Explain why steel ships tend to become magnetised during the shipbuilding construction.

2. Magnets are often fitted to the doors of refrigerators to keep them closed. Use the concept of magnetic induction to explain this practical application.

3. Draw a diagram to show the magnetic field pattern round two magnets with their unlike poles close together where the strength of the field of one magnet is twice the strength of the other field.

4. If a solenoid is viewed from one end, and the current travels in an anti-clockwise direction, what is the polarity of that end?

ELECTRICITY AND MAGNETISM

Force on a moving current

Suppose a long straight current-carrying wire is hung perpendicular to the direction of the magnetic field between the poles of a U-shaped magnet, as shown in Figure 549. If a conventional current is then allowed to flow in the wire in a downwards direction, the wire experiences a force and it tends to want to be catapulted out of the magnet. This is known as the **motor effect** and this effect is put to practical use in electric motors.

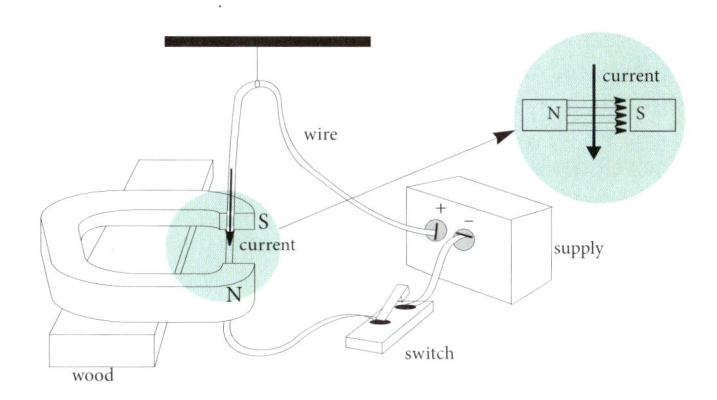

Figure 549   Conductor in a magnetic field

The reason for the movement is due to the interaction of the two magnetic fields - that of the magnet and the magnetic field produced by the current-carrying wire. Figure 550 shows the resultant magnetic field in this case. If the current was reversed, then the wire would be catapulted inwards.

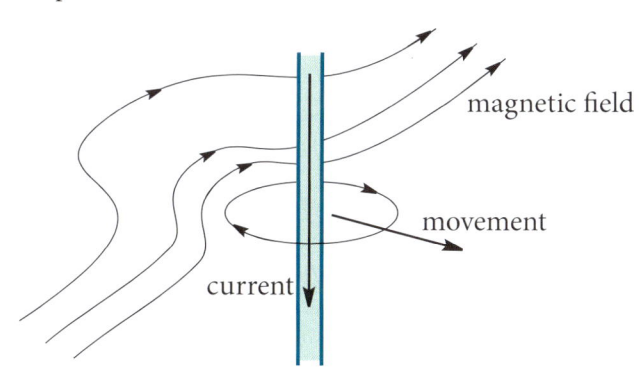

Figure 550   The motor effect

The direction of the force experienced by moving currents in a magnetic field can be determined by the vector addition of the two fields. However, an easier way for determining the direction of the force is to use a **right-hand palm rule** or **Fleming's left-hand rule**. There are a variety of hand rules used and it very much depends on the textbook you use as to what rules will be given. It is really up to you to use the hand rule that you prefer. Figure 551 shows three "hand rules" commonly used.

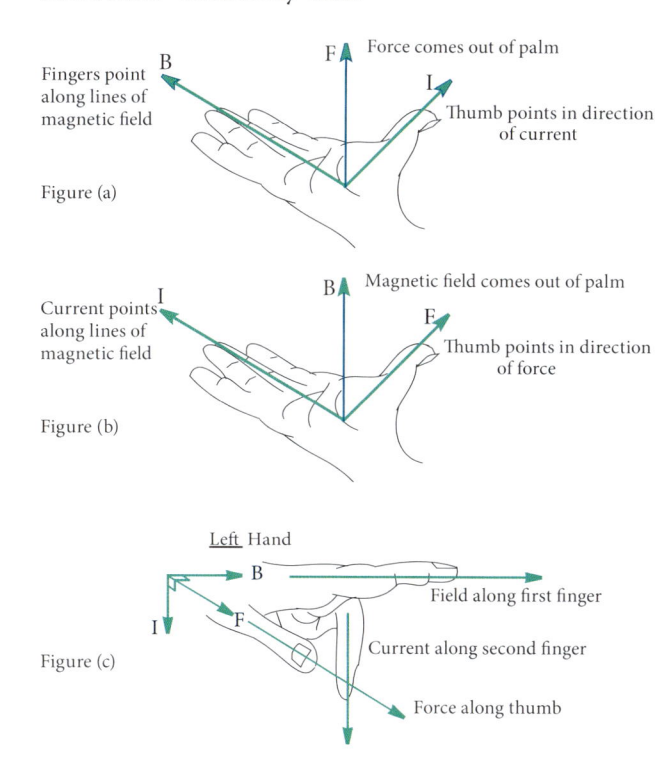

Figure 551   Hand rules used to show the direction of force.

In Figure 551 (a), if the fingers of your right hand point in the direction of the magnetic field B, and your thumb points in the direction of the conventional current, then your palm points in the direction of the force. This rule is called the right-hand palm rule. An alternative to this is shown in Figure 551 (b). (*This is the rule preferrd by the author because the fingers give a sense of flow of conventional current and the palm points north-south like a bar magnet and the thumb is the direction of movement or force*).

Figure 551 (c) is Fleming's left hand rule. The **f**irst **f**inger gives the direction of the magnetic **f**ield, the se**c**ond finger gives the direction of conventional **c**urrent, and the thu**m**b gives the **m**ovement or force direction

Try these rules for the examples in Figures 549 and 550 to see which one you prefer.

Note that these rules are for conventional current and not true electron flow. If electron flow is to be determined, apply a rule of choice and find the force for conventional flow, say north – then state your answer as the opposite direction, in this case south.

189

CHAPTER 5

Force on a moving charge

As stated, stationary charges experience no force in a magnetic field. However, if charged particles move in a magnetic field, a force is exerted on them. This force causes them to deflect if they are not confined in a conductor.

The same hand rules can be used to determine the direction of the force experience on a charge. However, remember that they apply to moving **positive** charge. If an electron is moving, as is more commonly the case, remember to apply the rule and then reverse the force direction.

When the positive charges move as shown in Figure 552, they will experience an upward force. Check this for yourself using one of the hand rules.

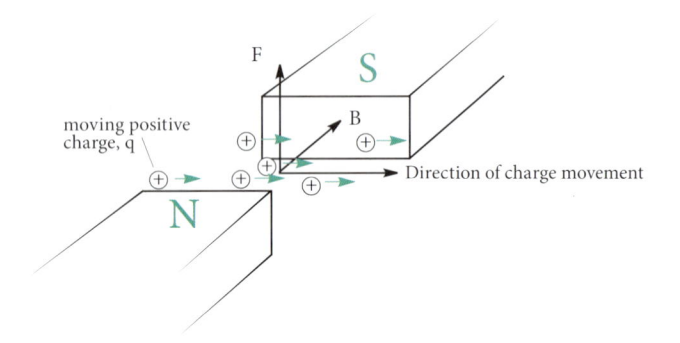

Figure 552   Force on a moving charge

Magnitude of moving currents and charges

When an electric current flows in a conductor, and the conductor is placed in a magnetic field, the force on the conductor is due to the individual forces on each of the individual charges in the conductor. The magnitude of the magnetic force F is found to be directly proportional to:

1. the strength of the magnetic field **B** measured in teslas (T)

2. the current flowing in the wire I measured in amperes (A)

3. the length of the conductor in the magnetic field l measured in metres (m).

So that
$$F = IlB$$

This force is greatest when the magnetic field is perpendicular to the conductor. Sometimes the wire in the magnetic field is at an angle θ to the magnetic field. In this case

$$F = IlB \sin \theta$$

Therefore, as θ decreases, so too does the force. When $\theta = 0°$ the current in the conductor is moving parallel to the magnetic field and no force on the conductor occurs.

The force experienced can be increased if the number of turns of wire carrying the current is increased. In this case the force is given by $F = IlBn$ where n is the number of turns of wire. When there are a number of turns of wire suspended between a magnetic field, the device is commonly called a wire toroid.

In order to determine the magnitude of the force experienced by a single point charge q, we will follow through the following derivation.

The velocity of the particle is given by

$$v = \frac{\text{displacement}}{\text{time}} = \frac{s}{t}$$

If the length of the charge carrier is l, then

$$v = \frac{l}{t} \Leftrightarrow l = vt$$

But, $q = It \Leftrightarrow I = q/t$ and $F = IlB$

By substitution, the force on an individual charge is given by

$$F = \frac{q}{t} \times (vt) \times B$$

That is,
$$F = qvB$$

If a charged particle enters a uniform magnetic field at an angle other than 90°, the force it experiences is given by

$$F = qvB \sin\theta$$

When $\theta = 90°$, and the magnetic field is uniform, the particle will undergo uniform circular motion as the force it experiences is at right angles to its motion. The radius of its circular motion is given by:

$$qvB = \frac{mv^2}{r} \Leftrightarrow r = \frac{mv}{qB}$$

When the particle enters the field at an angle other than a right angle, it will follow a helical path.

Example

A wire that is carrying a current of 3.50 A east has 2.00 m of its length in a uniform magnetic field of magnetic flux density of 5.00×10^{-7} T directed vertically downwards into the paper. Determine the magnitude and direction of the force it experiences.

ELECTRICITY AND MAGNETISM

Solution

Using the formula for the force on a wire in a magnetic field, we have:

$F = IlB = (3.5 \text{ A}) \times (2.00 \text{ m}) \times (5.00 \times 10^{-7} \text{ T}) = 3.50 \times 10^{-6} \text{ N}$.

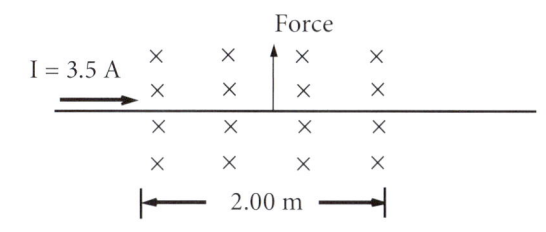

```
                    Force
         ×   ×   ↑×   ×
I = 3.5 A
──────→  ×   ×    ×   ×
         ×   ×    ×   ×
         ×   ×    ×   ×
         |←—— 2.00 m ——→|
```

The force on the conductor is 3.50×10^{-6} N north.

Example

An electron is moving with a speed of 3.0×10^5 m s^{-1} in a direction that is at right angles to a uniform magnetic field of 3.0×10^{-3} T. Calculate

a. the force exerted on the electron.

b. the radius of the path of the electron.

Solution

a. Using the formula, $F = qv \times B$, we have

$F = (1.6 \times 10^{-19} \text{ C}) \times (3.0 \times 10^5 \text{ ms}^{-1}) \times (3.0 \times 10^{-3} \text{ T})$

$= 1.44 \times 10^{-16}$ N

The force exerted on the electron is 1.4×10^{-16} N at right angles to the magnetic field and the path of its motion.

b. The force on each charge is given by either of the formulas, $F = qv \times B$ and $F = mv^2/r$.

Equating these two expressions we can determine the radius of the path:

$\frac{mv^2}{r} = qvB \Leftrightarrow r = \frac{mv^2}{qvB} \therefore r = \frac{mv}{qB}$

That is, $r = [(9.11 \times 10^{-31} \text{ kg}) \times (3.0 \times 10^5 \text{ ms}^{-1})] \div$

$[(1.6 \times 10^{-19} \text{ C}) \times (3.0 \times 10^{-3} \text{ T})]$

$= 5.69 \times 10^{-4}$

The radius of the path is 5.7×10^{-4} m.

Exercise 5.5

1. A suitable unit of magnetic field strength is

    A. A N^{-1} m^{-1}
    B. kg s^{-2} A^{-1}
    C. A m N^{-1}
    D. kg A s^2

2. An electron enters a uniform magnetic field that is at right angles to its original direction of movement. The path of the electron is:

    A. an arc of a circle.
    B. helical.
    C. part of a parabola.
    D. a straight line.

3. Two long straight wires with currents flowing in opposite directions experience a force because:

    A. the current in both wires increases
    B. the current in both wires decreases
    C. the current in the wires produces an attraction
    D. the current in the wires produces a repulsion

4. Determine in which direction the wire moves in the diagram shown.

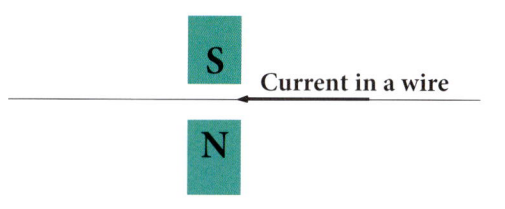

    ```
              ┌───┐
              │ S │
              └───┘      ← Current in a wire
    ──────────────────────────
              ┌───┐
              │ N │
              └───┘
    ```

    A. outwards
    B. inwards
    C. it does not move
    D. sideways

5. An electron passes through a uniform magnetic field of 0.050 T at right angles to the direction of the field at a velocity of 2.5×10^6 ms^{-1}. The magnitude of the force on the electron in newtons is:

    A. 2.0×10^{-14}
    B. 4.0×10^{-14}
    C. 8.0×10^{-14}
    D. zero

6. Two parallel wires carry currents I of equal magnitude in opposite directions as shown in the diagram

Chapter 5

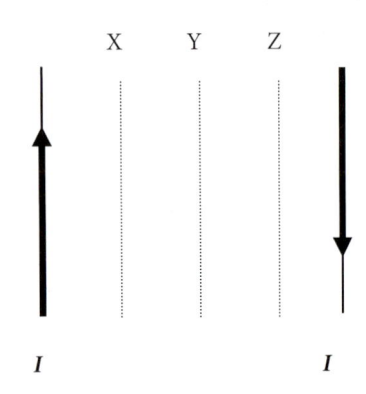

The line along which the magnetic fields cancel is

A. X
B. Y
C. Z
D. the magnetic fields do not cancel

7. A beam of protons enter a uniform magnetic field directed into the page as shown

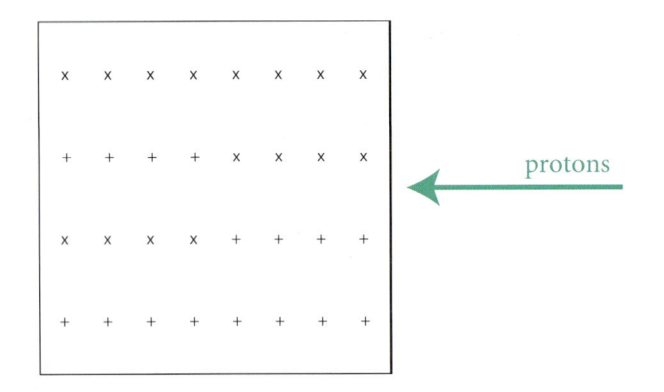

The protons will experience a force that pushes them

A. into the page
B. out of the page
C. upwards
D. downwards

8. Below is a schematic diagram of a coil connected to a battery.

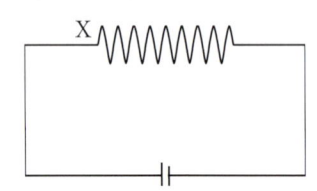

When an electric current flows in the circuit, the end of the coil labelled X will be:

A. a south pole
B. a north pole
C. either a north or a south pole
D. neither a north or a south pole

9. An ion carrying a charge of 3.2×10^{-19} C enters a field of magnetic flux density of 1.5 T with a velocity of 2.5×10^5 m s^{-1} perpendicular to the field. Calculate the force on the ion.

10. A straight wire of length 50 cm carries a current of 50 A. The wire is at right angles to a magnetic field of 0.3 T. Calculate the force on the wire.

11. A straight wire of length 1.4 m carries a current of 2.5 A. If the wire is in a direction of 25° to a magnetic field of 0.7 T, calculate the force on the wire.

12. A beam of electrons enters a pair of crossed electric and magnetic fields in which the electric field strength of 3.0×10^4 V m^{-1} and magnetic flux density of 1.0×10^{-2} T. If the beam is not deflected from its path by the fields, what must be the speed of the electrons?

13. An electron in one of the electron guns of a television picture tube is accelerated by a potential difference of 1.2×10^4 V. It is then deflected by a magnetic field of 6.0×10^{-4} T. Determine

   i. the velocity of the electron when it enters the magnetic field.
   ii. the radius of curvature of the electron while it is in the magnetic field.

14. A point charge of – 15 C is moving due north at 1.0×10^3 ms^{-1} enters a uniform magnetic field of 1.2×10^{-4} T directed into the page. Determine the magnitude and direction of the force on the charge.

15. A vertical wire 50 cm long carries a current of 1.5 A from the north to the south. It experiences a force of 0.2 N.

(a) Determine the magnitude of the magnetic field

(b) Determine how the force could be increased to 2 N.

6. Circular motion and gravitation

Contents

6.1 – Circular motion

6.2 – Newton's law of gravitation

Essential Ideas

A force applied perpendicular to its displacement can result in circular motion.

The Newtonian idea of gravitational force acting between two spherical bodies and the laws of mechanics create a model that can be used to calculate the motion of planets. © IBO 2014

CHAPTER 6

6.1 Circular Motion

> **NATURE OF SCIENCE:**
>
> Observable universe: Observations and subsequent deductions led to the realization that the force must act radially inwards in all cases of circular motion. (1.1)
>
> © IBO 2014

TOK INTRODUCTION

There are few things more spectacular than viewing the night sky on a clear night and our ancient ancestors were no less sensitive to beauty than we are. In fact, there were no brightly lit cities and polluting industries in the past and the heavenly bodies were like their giant movie theatre. It is likely that they had a greater appreciation of the heavens than we do.

We are all born with a strong curiosity of our environment and this same curiosity was bestowed upon our ancestors. They did not have scientific instruments like we do and they needed to understand the change of seasons and how to navigate the seas and oceans to a far greater degree than we do. They relied on the observations of the Sun, the moon and the stars to help them predict the events within the course of the year.

Our ancestors knew that the Sun, moon and the stars rise in the east and set in he west each day and at sunset, the brighter stars form regular patterns in the sky. These constellations rise at different time during the year but they remained fixed relative to the background stars.

However, around this fixed background there are five planets (from Greek meaning "wanderer' that can be seen with the unaided eye to move relative to the background - Mercury, Venus, Mars, Jupiter and Saturn.

When Christopher Columbus crossed the Atlantic ocean in 1492, some of his sailors were scared that they would fall off the edge of the earth. Yet the ancient Greeks had concluded that the earth was spherical. In fact, Eratosthenes (275 -194 BC) had calculated the circumference of the earth by studying the length of shadows.

There were many great mathematicians and philosophers from many countries who set us on our incredible journey of insight and observations -Plato, Eudoxus, Aristotle, Ptolemy, Aristarchus to name but a few. But it was not until the times of Nicholas Copernicus (1473 – 1543) that there began a paradigm shift away from a geocentric (earth-centred) model to a heliocentric (Sun-centred) model of our Solar system.

It was finally the works of the great astronomers and scientists such as Tycho Brahe (1546 -1601), Johannes Kepler (1571 – 1630), Galileo Galilei (1564 – 1642) and Isaac Newton (1642 – 1727) that allowed us to more fully understand uniform circular motion and the Laws of gravitational motion.

Essential idea: A force applied perpendicular to its displacement can result in circular motion.

Understandings:
- Period, frequency, angular displacement and angular velocity
- Centripetal force
- Centripetal acceleration

Period, frequency, angular displacement and angular velocity

A common two-dimensional motion is that of circular motion. An object is carrying out uniform circular motion if it is:

- moving in a circular path and if it
- sweeping out equal angles in equal periods of time.

Examples of uniform motion include the moon revolving around the earth, the earth revoving around the Sun (to a good approximation), wheels rotating on cars and bicycles, car going around banked corners, particles in centrifuges and charged particles in cyclotrons to mention some.

In order to describe uniform circular motion we need to be familiar with both degrees and radians. A radian (rad) is the angle subtended at the centre of a circle by an arc that is equal in length to the radius of a circle. See Figure 601.

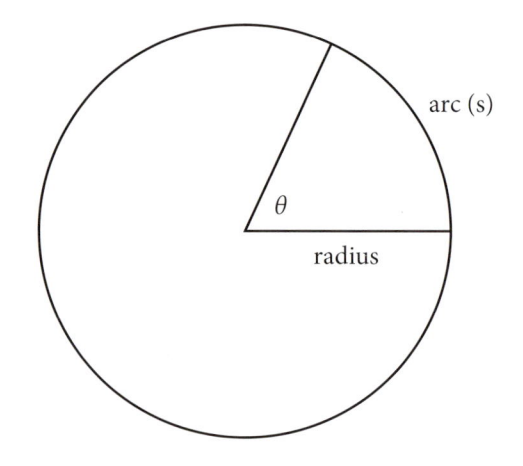

Figure 601   Defining the radian

194

CIRCULAR MOTION AND GRAVITATION

To find the number or fraction of radians in a given angle θ, one can divide the arc s subtending that angle by the radius

$$\theta = \frac{arc}{radius} = \frac{s}{r}$$

For example, a right angle (90°) is subtended at the centre of a circle by an arc that is one quarter of the circumference of a circle So to calculate the degrees in radians we simply do the following:

$$\theta = \frac{arc}{radius} = \frac{s}{r}$$
$$= \frac{1}{4} \times \frac{circumference}{radius}$$
$$= \frac{1}{4} \times \frac{2\pi r}{r}$$
$$= \frac{\pi}{2} \text{ radians}$$

To convert degrees to radians, divide by 360, and multiply by 2 π. To convert radians to degrees, divide by 2 π, and multiply by 360.

Therefore, 1 rad = $\frac{1}{2\pi} \times 360° \approx 57.30°$

We also need to familiarise ourselves with the quantity called angular displacement θ. **Angular displacement** θ of an object is the angle between an arbitrary reference line and the line joining the observer and the object.

Since the object is moving in a circular path, its angular displacement may change. **Angular velocity** ω is the time rate of change of angular displacement measured in radians per second (rad s).

$$\omega = \frac{\theta}{t}$$

We now have two ways of describing position – displacement and angular displacement, and two ways of describing the time rate of change of position – velocity and angular velocity. Let us consider a bicycle wheel spinning at a steady speed of 5 m s⁻¹. While the wheel's speed remains constant at all times, it is constantly changing direction, and thus its velocity is changing.

The instantaneous velocity of an object moving with uniform circular motion will be its linear velocity. At any moment, the instantaneous velocity with have magnitude equal to its speed, and direction equal to the tangent to the circle at that point in time. So what is the relationship that exists between linear velocity and angular velocity?

From the definition of the radian we have:

$$\theta = \frac{arc}{radius} = \frac{s}{r}$$

Therefore,

$$s = r\theta$$

So

$$\frac{s}{t} = \frac{r\theta}{t} \quad \text{and } \omega = \frac{\theta}{t}$$

Therefore,

$$v = \omega\, r$$

We know that the period T of motion is the time taken for one complete revolution. Since on revolution is equal to 2π radians, then the angular velocity is given by:

$$\omega = \frac{2\pi}{T} \quad \text{and } T = \frac{2\pi}{\omega}$$

We also know that the number of revolutions per second is the frequency f. Therefore,

$$f = \frac{\omega}{2\pi}$$

We have mentioned earlier in this chapter that when a car goes round a circular bend in the road moving with constant speed it must be accelerating. Acceleration is a vector and is defined as the rate of change of velocity. Velocity is also a vector and so it follows that if the direction of motion of an object is changing then even though its speed might be constant, its velocity is changing and hence it is accelerating. A force is needed to accelerate an object and we might ask what is the origin of the force that causes the car to go round the bend in the road. Do not fall into the trap of thinking that this force comes from the engine. This would imply that if you are on a bicycle then you can only go round a corner if you keep pedalling. To understand the origin of the force just think what happens if the road is icy. Yes, it is this friction between the tyres and the road that produces the force. If there is no friction then you cannot negotiate a bend in the road.

Let us think of the example where you whirl an object tied to a string about your head with constant speed. Clearly the force that produces the circular motion in this case in a horizontal plane is the tension in the string. If the string were to snap then the object would fly off at a tangent to the circle. This is the direction of the velocity vector of the object. The tension in the string acts at right angles to this vector and this is the prerequisite for an object to move in a circle with constant speed. If a force acts at right angles to the direction of motion of an object then there is no component of force in the direction of motion and therefore no acceleration in the direction of motion. If the force is constant then the direction of the path that the object follows will change by equal amounts in equal time intervals hence the overall path of motion must be a circle. Figure 602 shows the relation between the direction of the velocity vector and the force acting on a particle P moving with constant speed in a circular path.

Chapter 6

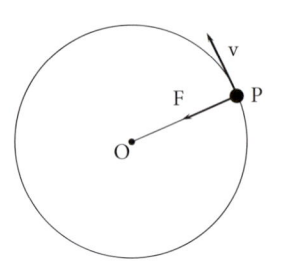

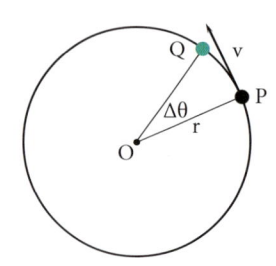

Figure 602 Centripetal Force

The force causing the circular motion is called the **centripetal force** and this force causes the particle to accelerate towards the centre of the circle and this acceleration is called the **centripetal acceleration**. However, be careful to realise that the centripetal acceleration is always at right angles to the velocity of the particle. If the speed of the particle is reduced then it will spiral towards the centre of the circle, accelerating rapidly as it does so. This is in effect what happens as an orbiting satellite encounters the Earth's atmosphere.

People sometimes talk of a **centrifugal force** in connection with circular motion. They say something along the lines that when a car goes round a bend in the road you feel a force throwing you outwards and this force is the centrifugal force. But there is no such force. All that is happening is that you are moving in accordance with Newton's laws. Before the car entered the bend you were moving in a straight line and you still want to keep moving in a straight line. Fortunately the force exerted on you by a side of the car as you push up against it stops you moving in a straight line. Take the side away and you will continue moving in a straight line as the car turns the bend.

Centripetal acceleration

In this section we shall derive an expression of the centripetal acceleration of a particle moving with uniform speed v in a circle of radius r. You will not be expected to derive this relation in an IB examination; it is given here for completeness.

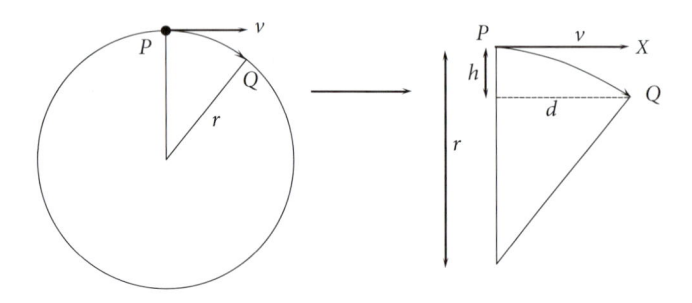

Figure 603 Centripetal Acceleration

In Figure 603 suppose that the particle moves from P to Q in time Δt

In the absence of a centripetal force the particle would reach the point X in this time. The force therefore effectively causes the particle to "fall" a distance h. For intersecting chords of a circle we have in this situation

$$d^2 + (r-h)^2 = r^2 \Leftrightarrow d^2 = 2rh - h^2$$

Now suppose that we consider a very small time interval then h^2 will be very small compared to $2rh$. Hence we can write

$$2rh = d^2$$

However d is the horizontal distance travelled in time Δt such that $d = v\Delta t$. Hence

$$h = \frac{d^2}{2r} = \frac{(v\Delta t)^2}{2r}$$

However h is the distance "fallen" in time Δt

So using $s = \frac{1}{2}at^2$ $\qquad h = \frac{1}{2}a(\Delta t)^2$

From which we have

$$\frac{(v\Delta t)^2}{2r} = \frac{1}{2}a(\Delta t)^2 \Rightarrow \frac{v^2(\Delta t)^2}{2r} = \frac{1}{2}a(\Delta t)^2$$

So that,

$$a = \frac{v^2}{r}$$

Since

$$v = \frac{d}{t} = \frac{2\pi r}{T}$$

It follows that

$$a = \frac{v^2}{r} = \frac{(2\pi r)^2}{rT^2}$$

$$a = \frac{4\pi^2 r}{T^2}$$

Furthermore,

$$v = \omega r$$

So

$$a = \frac{v^2}{r} = \omega^2 r$$

Example

A geo-stationary satellite is one that orbits Earth in 24 hours. The orbital radius of the satellite is 4.2×10^7 m. Calculate the acceleration of the satellite and state the direction of the acceleration. (24 hours = 8.64×10^4 s)

Solution

Let r = the orbital radius and T the period. The speed v of the satellite is then given by

$$v = \frac{d}{t} = \frac{2\pi r}{T}$$

the acceleration a is therefore given by

$$a = \frac{v^2}{r} = \frac{4\pi^2 r}{T^2} = \frac{40 \times 4.2 \times 10^7}{(8.64)^2 \times 10^8} = 0.23 \ m\ s^{-2}$$

The acceleration will be directed towards the centre of Earth.

Centripetal force

We can find the expression for the centripetal force F simply by using Newton's 2nd law. i.e. $F = ma$, so that, with $a = \frac{v^2}{r}$ we have:

$$F = \frac{mv^2}{r}$$

where m is the mass of the particle.

It is important to realise that the above equation is effectively Newton's Second Law as applied to particles moving in a circle. In order for the particle to move in a circle a force must act at right angles to the velocity vector of the particle and the speed of the particle must remain constant. This means that the force must also remain constant

The effect of the centripetal force is to produce an acceleration towards the centre of the circle. The magnitude of the particle's linear velocity and the magnitude of the force acting on it will determine the circular path that a particular particle describes.

Below are some examples of forces which provide centripetal forces:

1. **Gravitational force**. This is discussed in detail in the next section.

2. **Frictional forces** between the wheels of a vehicle and the ground.

3. **Magnetic forces**. This force is discussed in Topic 5.3

Then of course we have the example of objects constrained to move in circles by strings or wires attached to the object.

Example

A model airplane of mass 0.25 kg has a control wire of length 10.0 m attached to it. Whilst held in the hand of the controller it flies in a horizontal circle with a speed of 20 m s⁻¹. Calculate the tension in the wire.

Solution

The tension in the wire provides the centripetal force and is equal to $\frac{mv^2}{r}$ which in this situation is equal to

$$\frac{0.25 \times 400}{10} = 10\ N.$$

An interesting situation arises when we have circular motion in a vertical plane. Consider a situation in which you attach a length of string to an object of mass m and then whirl it in a vertical circle of radius r. If the speed of the object is v at its lowest point then the tension in the string at this point will be $mg + \frac{mv^2}{r}$ and at the highest point the tension will be $mg - \frac{mv^2}{r}$.

We now see why this is the case.

Lowest point:

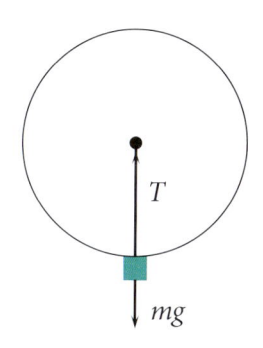

Figure 604   Lowest Point

The resultant force on mass, m / kg, throughout its motion in a circle (as long as the speed is constant) is always $\frac{mv^2}{r}$.

Taking the positive direction to be towards the centre of the circle, at its lowest point, the resultant force is provided by the expression $T - mg$, so that

$$\frac{mv^2}{r} = T - mg$$

That is, $T = \frac{mv^2}{r} + mg$

CHAPTER 6

Highest point:

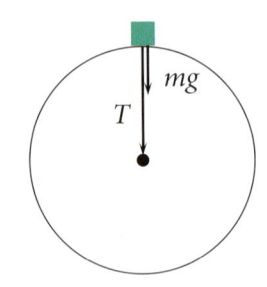

Figure 605 Highest Point

Again, we have that the resultant force on mass, m / kg, throughout its motion in a circle (as long as the speed is constant) is always $\frac{mv^2}{r}$.

Taking the positive direction to be towards the centre of the circle, at its highest point, the resultant force is provided by the expression T + mg, so that

$\frac{mv^2}{r} = T + mg$, i.e., $T = \frac{mv^2}{r} - mg$

Example

A "wall of death" motorcyclist rides his motorcycle in a vertical circle of radius 20 m. Calculate the minimum speed that he must have at the top of the circle in order to complete the loop.

Solution

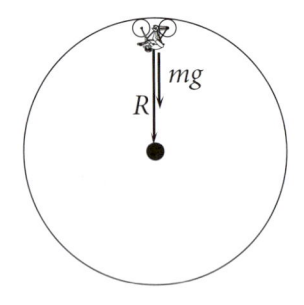

Let R be the reaction force on the bike, then we need to use the expression $\frac{mv^2}{r} = R + mg$ (when the bike is at its highest point).

However, the bike must always make contact with the track, that is, we must have that R ≥ 0.

Now, re-arranging the expression we have that

$R = \frac{mv^2}{r} - mg$, however, as R ≥ 0, we have that

$\frac{mv^2}{r} - mg \geq 0 \Rightarrow \frac{mv^2}{r} \geq mg$. So, the minimum speed will be given by $g = \frac{v^2}{r}$ since at any lower speed mg will be greater than $\frac{mv^2}{r}$ and the motor bike will leave the track.

So in this case the speed will be 14 m s^{-1}.

It is also worthwhile noting that in circular motion with constant speed, there is no change in kinetic energy. This is because the speed, v, is constant and so the expression for the kinetic energy, ½ mv², is also always constant (anywhere along its motion). Another way to look at this is that since the force acts at right angles to the particle then no work is done on the particle by the force.

Exercise 6.1

1. What length of arc would subtend an angle of 2.1 radians at the centre of a circle of radius 4.0 m?

2. Convert the following angles to radians

    (a) 45°   (b) 270°
    (c) 35°

3. An electron circles 6.6 x 10^{14} times per second with a speed of 3.0 x 10^7 m s^{-1}. Determine:

    (a) the radius of its path
    (b) the acceleration it experiences.

4. A person ties a mass of 2.00 kg to a length of rope and then swings it about his head in a horizontal circle of radius 1.50 m so that it completes 30 revolutions per minute. Determine:

    (a) the angular velocity of the object
    (b) the linear velocity of the object
    (c) the centripetal acceleration
    (d) the tension in the string

5. A 2.0 kg bucket is whirled by hand in a vertical circle of radius 1.00 m at a constant speed of 6.0 m s^{-1}. Calculate the tension in the arm

    (a) at the top, and
    (b) at the bottom of the motion.

6. An object is travelling at a constant speed of 40 ms^{-1} in a circular path of radius 80 m. Calculate the acceleration of the object?

7. A body of mass 5.0 kg, lying on a horizontal smooth table attached to an inextensible string of length 0.35 m, while the other end of the string is

fixed to the table. The mass is whirled at a constant speed of 2.0 m s⁻¹. Calculate

(a) the centripetal acceleration.
(b) the tension in the string.
(c) the period of motion.

8. The radius of the path of an object in uniform circular motion is doubled. The centripetal force needed if its speed remains the same is

A. half as great as before.
B. the same as before.
C. twice as great as before.
D. four times as great as before.

9. A car rounds a curve of radius 70 m at a speed of 12 m s⁻¹ on a level road. Calculate its centripetal acceleration?

10. A 500 g sphere is hung from an inextensible string 1.25 m long and swung around to form a 'conical pendulum'. The sphere moves in a circular horizontal path of radius 0.75 m. Determine the tension in the string.

11. Determine the maximum (constant) speed at which a car can safely round a circular curve of radius 50 m on a horizontal road if the coefficient of static friction (μ) between the tyres and the road is 0.7. (Use $g = 10$ m s⁻²). (HINT: if the normal reaction is N, the relationship is $F = \mu N$)

12. A sphere of mass m, attached to an inextensible string as shown in the diagram is released from rest at an angle θ with the vertical. When the sphere passes through its lowest point, show that the tension in the string is given by $mg(3 - 2\cos\theta)$

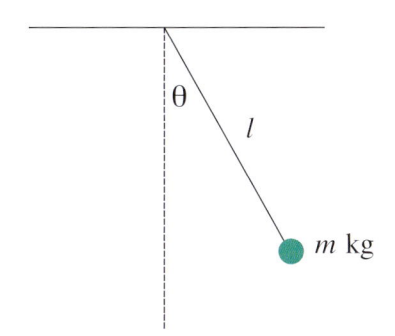

13. A 3 kg mass attached to a string 6 m long is to be swung in a circle at a constant speed making one complete revolution in 1.25 s. Determine the value that the breaking strain that the string must not exceed if the string is not to break when the circular motion is in

(a) a horizontal plane?
(b) a vertical plane?

14. A mass, m kg, is released from point A, down a smooth inclined plane and once it reaches point B, it completes the circular motion, via the smooth circular track B to C to D and then back through B, which is connected to the end of the incline and has a radius a / m.

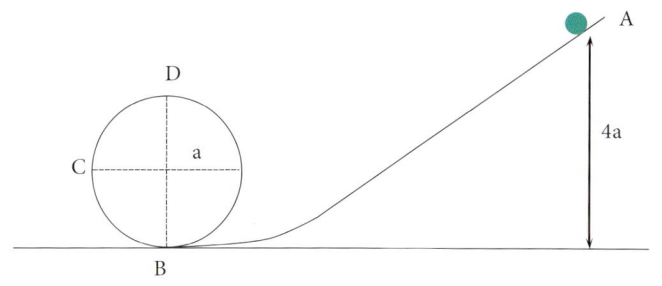

Determine the normal reaction of the track on the mass at (a) D. (b) C. (c) B.

6.2 Newton's law of gravitation

> **NATURE OF SCIENCE:**
>
> Laws: Newton's law of gravitation and the laws of mechanics are the foundation for deterministic classical physics. These can be used to make predictions but do not explain why the observed phenomena exist. (2.4)
>
> © IBO 2014

Essential idea: The Newtonian idea of gravitational force acting between two spherical bodies and the laws of mechanics create a model that can be used to calculate the motion of planets.

Understandings:
- Newton's law of gravitation
- Gravitational field strength

Newton's universal law of gravitation

In the *Principia* Newton stated his Universal **Law of Gravitation** as follows:

'Every material particle in the Universe attracts every other material particle with a force that is directly proportional to the product of the masses of the particles and that is inversely proportional to the square of the distance between them'.

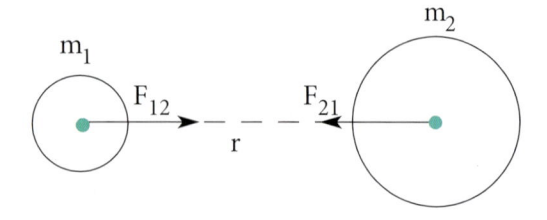

Figure 606   Newton's law of gravitation

We can write this law mathematically as:

$$F_{12} = \frac{Gm_1 m_2}{r^2} \hat{a}_{12} = -F_{21}$$

F_{12} is the force that particle 1 exerts on particle 2 and F_{21} is the force that particle 2 exerts on particle 1.

$\hat{a}_{12}$ is a unit vector directed along the line joining the particles.

m_1 and m_2 are the masses of the two particles respectively and r is their separation.

G is a constant known as the *Universal Gravitational Constant* and its accepted present day value is 6.67×10^{-11} N m^{-2} kg^{-2}.

There are several things to note about this equation. The forces between the particles obey Newton's third law as discussed in Section 2.7. That is, the forces are equal and opposite. The mass of the particles is in fact their gravitational mass as discussed in Section 2.3.3.

Every particle in the Universe, according to Newton, obeys this law and this is why the law is known as a 'universal' law. This is the first time in the history of physics that we come across the idea of the universal application of a physical law. It is now an accepted fact that if a physical law is indeed to be a law and not just a rule then it must be universal. Newton was also very careful to specify the word particle. Clearly any two objects will attract each other because of the attraction between the respective particles of each object. However, this will be a very complicated force and will depend on the respective shapes of the bodies. Do not be fooled into thinking that for objects we need only specify the distance r as the distance between their respective centres of mass. If this were the case it would be impossible to peel an orange. The centre of mass of the orange is at its centre and the centre of the mass of the peel is also at this point. If we think that r in the Newton Law refers to the distance between the centres of mass of objects the distance between the two centres of mass is zero. The force therefore between the peel and the orange is infinite.

You will almost invariably in the IB course and elsewhere, come across the law written in its scalar form as

$$F = \frac{Gm_1 m_2}{r^2}$$

However, do not forget its vector nature nor that it is a force law between particles and not objects or "masses".

Gravitational field strength

The Law of Universal Gravitation is an inverse square law and in this sense is very similar to the Coulomb force law discussed earlier. In fact if you replace m with q in the Newton law and G with $\frac{1}{4\pi\varepsilon_0}$ then we have the Coulomb law (except of course we can have negative charge but as far as we know there is no negative mass). So all that follows in the rest of this section is very similar to Coulomb's Law.

Circular Motion And Gravitation

Any particle in the Universe exerts a gravitational force on any other particle in the Universe. In this sense we can think of the effect that a particle P for example produces on other particles without even knowing the location of P. We can think of a "field of influence radiating out" from P. This influence we call the **Gravitational Field** and by introducing this concept we are essentially moving our attention from the source of the field to the effect that the source produces.

In Figure 607 a particle of mass m is placed at point X somewhere in the Universe.

Figure 607   Force and acceleration

The particle is observed to accelerate in the direction shown. We deduce that this acceleration is due to a gravitational field at X. We do not know the source of the field but that at this stage does not matter. We are only concerned with the effect of the field. If the mass of P is small then it will not effect the field at X with its own field. We define the gravitational field strength I at X in terms of the force that is exerted on P as follows

$$I = \frac{F}{m}$$

That is 'the gravitational field strength at a point is the force exerted per unit mass on a particle of small mass placed at that point'.

From Newton's 2nd law $F = ma$ we see that the field strength is actually equal to the acceleration of the particle. The gravitational field strength is often given the symbol 'g' So we can express the magnitude of the field strength in either N kg^{-1} or m s^{-2}. However, if we are dealing explicitly with field strengths then we tend to use the unit N kg^{-1}.

Figure 608(a) shows the "field pattern" for an isolated particle of mass M.

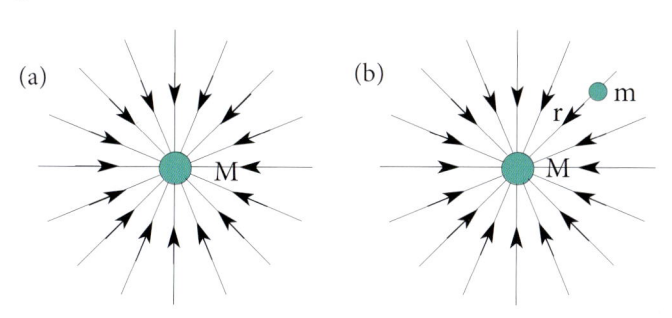

Figure 608 (a) and (b)

Clearly this is only a representation since the field due to M acts at all points in space.

Suppose we now place a particle of mass m a distance r from M as shown in Figure 608 (b).

The gravitational law gives the magnitude of the force that M exerts on m

$$F = \frac{GMm}{r^2}$$

So the magnitude of the gravitational field strength $I = \frac{F}{m}$ is given by

$$I = \frac{GM}{r^2}$$

If we wish to find the field strength at a point due to two or more point masses, then we use vector addition. (This is another example of the general principle of superposition - see 4.5.5). In Figure 609 the magnitude of the field strength produced by the point mass M_1 at point P is I_1 and that of point mass M_2 is I_2.

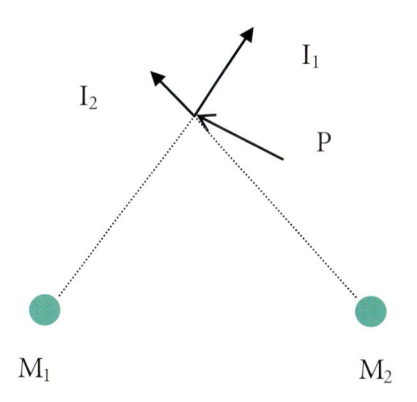

Figure 609   Vector addition

If I_1 and I_2 are at right angles to each other, the resultant magnitude of the field strength I at P is given by:

$$I = \sqrt{I_1^2 + I_2^2}$$

If the particle of mass M is replaced with a sphere of mass M and radius R, as shown in Figure 610, then rely on the fact that the sphere behaves as a point mass situated at its centre, the field strength at the surface of the sphere will be given by

$$I = \frac{GM}{R^2}$$

Figure 610   Field strength

201

CHAPTER 6

If the sphere is the Earth then

$$I = \frac{GM_e}{R_e^2}$$

But the field strength is equal to the acceleration that is produced on a mass hence the acceleration of free fall at the surface of the Earth, g_0, is given by

$$g_0 = \frac{GM_e}{R_e^2}$$

This actually means that whenever you determine the acceleration of free fall g_0 at any point on the Earth you are in fact measuring the gravitational field strength at that point. It can also be seen now why the value of g varies with height above the surface of the Earth. Since at a height of h above the surface of the Earth the field strength, g, is given by

$$g = \frac{GM_e}{(R_e + h)^2}$$

It can be shown that if we have a hollow sphere then the field strength at all points within the sphere is zero. This fact can be used to deduce an expression for the field strength at points inside the Earth.

It is left as an exercise to demonstrate (*if desired since this is not in the syllabus*) that if ρ is the mean density of the Earth then at a point distance r from the centre of the Earth the value of g is given by

$$g = \tfrac{4}{3}\pi r\, G\rho$$

Newton's theory of gravity can be used to explain planetary motion. For a planet orbiting the Sun, we can equate the force of gravity to the centripetal force acting on the planet.

$$F = \frac{GM_s m_p}{r^2} = m_p \omega^2 r$$

$$\frac{GM_s}{\omega^2} = r^3$$

From circular motion we have

$$T = \frac{2\pi}{\omega} \quad \text{Hence } \omega = \frac{2\pi}{T}$$

$$\omega^2 = \frac{4\pi^2}{T^2}$$

$$\frac{GM_s T^2}{4\pi^2} = r^3 \quad \text{Hence } T^2 = \frac{4\pi^2}{GM_s} r^3$$

Since $\frac{4\pi^2}{GM_s}$ will yield a constant value, we can see that

$$T^2 \alpha\, r^3$$

This is a derivation of Kepler's third law of planetary motion.

Example 1

Take the value of g_0 = 10 N kg^{-1} and the mean radius of the Earth to be 6.4×10^6 m to estimate a value for the mass of the Earth.

Solution

We have $g_0 = \dfrac{GM_e}{R_e^2}$, therefore

$$M_e = \frac{g_0 R_e^2}{G} = \frac{10 \times (6.4 \times 10^6)^2}{6.7 \times 10^{-11}}$$

$$\approx 6 \times 10^{24} \text{ kg}.$$

Example 2

Assuming the Earth and Moon to be isolated from all other masses, use the following data to estimate the mass of the Moon.

mass of Earth = 6.0×10^{24} kg

distance between centre of Earth and centre of Moon = 3.8×10^8 m

distance from centre of Earth at which gravitational field is zero = 3.42×10^8 m

Solution

Since $\dfrac{M_e}{(3.42 \times 10^8)^2} = \dfrac{M_e}{(3.8 \times 10^8 - 3.42 \times 10^8)^2}$

$M_m = 7.4 \times 10^{22}$ kg

For a satellite s travelling in circular motion around a planet p, the gravitational force of attraction between the planet and the satellite provides the required centripetal force. That is:

$$F_C = F_G$$

Therefore

$$\frac{m_s v^2}{r} = \frac{G m_s M_p}{r^2}$$

$$v^2 = \frac{G M_p}{r}$$

Example 3

At what speed must a satellite be moving so that remains in a circular orbit of 1.6×10^3 km above the earth's surface given that the mass of the earth is 6.0×10^{24} kg and the radius of the earth as 6.4×10^6 m.

Solution

$$v^2 = \frac{G M_E}{r}$$

$$= \frac{6.67 \times 10^{-11} \times 6.0 \times 10^{24}}{0.16 \times 10^6 + 6.4 \times 10^6}$$

$$= 5.00 \times 10^7$$

Therefore:

$v = 7.1 \times 10^3$ m s^{-1}

Example 4

A geostationary weather satellite is positioned above a particular point on the equator. It moves with uniform circular motion about the centre of the earth. Take the radius of the earth as being 6400 km.

(a) Determine the period of the of the weather satellite

(b) Determine the orbital radius and the orbital speed of the satellite

(c) Calculate the height of the satellite above the earth's surface.

Solution

(a) The period = 24 hours because it is geostationary = 8.64 x 10⁴ s.

(b) (b) $T^2 = \frac{4 \pi^2}{G M_E} r^3$

$(8.64 \times 10^4)^2 = \frac{4 \pi^2 r^3}{6.7 \times 10^{11} \times 6.0 \times 10^{24}}$

r = orbital radius = 4.236×10^7 m

Since $T = \frac{2\pi}{\omega}$ and $v = r\omega$

$v = \frac{2\pi r}{T} = \frac{2\pi \times 4.24 \times 10^7}{8.64 \times 10^4}$

= orbital speed = 3.1×10^3 ms^{-1}

(c) height above earth's surface = (42 360 – 6400) km
= 36 000 km.

Exercise 6.2

1. If the circumference of the earth is 4.02 x 10⁷ m, calculate

   (a) the radius of the earth
   (b) the mass of the earth
   (c) the average density of the earth.

2. What is the force of gravity between two masses of 15 kg and 5 kg separated by a distance of 2 m?

3. A satellite is orbiting the earth in a circular orbit at a distance of 1.28 x 10⁷ m above the surface of the earth. Determine

   (a) the acceleration due to gravity on the satellite
   (b) the weight of a 65 kg astronaut in the satellite
   (c) Explain why the astronaut feels "weightless".

4. Two objects of m1 and m2 respectively are separated by a distance d. They exert a force F on each other. If the masses halved and the distance between them is tripled, the new force would be

   A. F / 8   C. F / 36
   B. F / 12   D. F / 64

5. Determine the gravitational field strength on Jupiter given that its radius is 7.18 x 10⁷ m.

6. Determine the height above the earth where a satellite would have half the force of gravity that it has at sea level, given the radius of the earth is 6.38 x 10⁶ m.

7. Two objects have masses of 7.5 × 10⁴ kg and 5.0 × 10⁴, and the gravitational force between them is 1.0 × 10⁻⁸ N. Determine the distance between the two objects.

8. Io, one of the moons of Jupiter, has a period of rotation of 1.53 × 105 s when moving in an approximately circular orbit of 4.20 × 108 m. Use the provided information to determine the mass of Jupiter.

9. The earth is 150 000 000 km from the Sun, and takes 355.25 days to complete its orbit. Determine the mass of the Sun.

10. A moon of Saturn period of 1.00 x 10⁶ s, and it moves in a circular orbit with a adius of 1.00 x 10⁹ m. Calculate

    (a) the orbital speed of the moon of Saturn
    (b) the mass of Saturn.

Chapter 6

This page is intentionally left blank

7. Atomic, nuclear and particle physics

Contents

7.1 – Discrete energy and radioactivity
7.2 – Nuclear reactions
7.3 – The structure of matter

Essential Ideas

In the microscopic world energy is discrete.

Energy can be released in nuclear decays and reactions as a result of the relationship between mass and energy.

It is believed that all the matter around us is made up of fundamental particles called quarks and leptons. It is known that matter has a hierarchical structure with quarks making up nucleons, nucleons making up nuclei, nuclei and electrons making up atoms and atoms making up molecules. In this hierarchical structure, the smallest scale is seen for quarks and leptons (10−18m). © IBO 2014

Chapter 7

TOK INTRODUCTION

The word "atom" is derived from the Greek word "atomos" and it first appeared in the writings of Greek philosophers after the time of Democritus (born c. 460 BC). Although these philosophers speculated about the nature of matter, no scientific experimentations in this area have been found. Aristotle (c.340 BC) was skeptical of the atomic theory of Democritus and proposed an alternative theory based upon the four "elements" – earth, wind, fire and water.

Ideas developed slowly over hundreds of years and it wasn't until the eighteenth century when Antoine Lavoisier (1743 – 1794) proposed the Law of conservation of mass – the sum of the masses of reactants in a chemical reaction equals the sum of the masses of the products. In this period, many of the elements, including most of the atmospheric gases were identified.

In 1801, the English chemist John Dalton proposed his "atomic theory of matter":

- Matter is composed of small sold, indivisible atoms.
- Elements contain only one type of atom.
- Compounds contain more than one type of atom.

Michael Faraday (1791 – 1867) carried out experiments concerning the electrolysis of solutions. He suggested that small but definite amount of electric charge may be associated with each atom.

In 1855, the German physicist and glass blower Heinrich Geissler (1814 – 1879) invented a powerful vacuum pump, and developed the low-pressure discharge tube that came to be known as the Geissler tube. Before long, Geissler's friend, Julius Plücker, made a fascinating discovery using the new vacuum tubes. He sealed a wire into either end of a Geissler tube, attaching each wire to a metal plate inside the tube as shown in Figure 701. Outside the tube, the wires were attached to an induction coil with a potential difference of 10 000 V. Plücker was surprised to find that electricity passed through the low-pressure gas in the tube. Equally surprising was that as the pressure inside the tube was reduced to around one-hundredth that of atmospheric pressure, a mauve glow filled the tube. Plücker and others were intrigued by the glow produced by this early version of the cathode ray tube, but no further research was done for another twenty years.

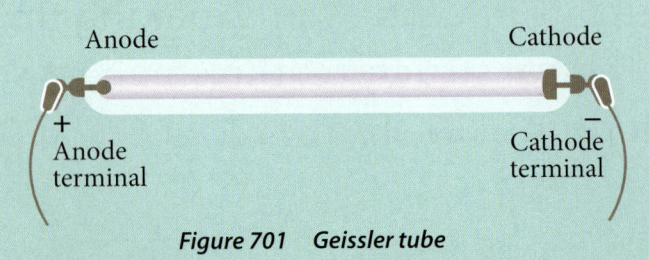

Figure 701   Geissler tube

Two decades later, Sir William Crookes (1832 - 1919) redesigned the Geissler tube for detailed studies of the phenomena that had come to be known as cathode rays as shown in Figure 702. There is a metal anode in the shape of a Maltese cross. A sharp light shadow of the positive electrode is cast on the wall of the tube opposite to the cathode.

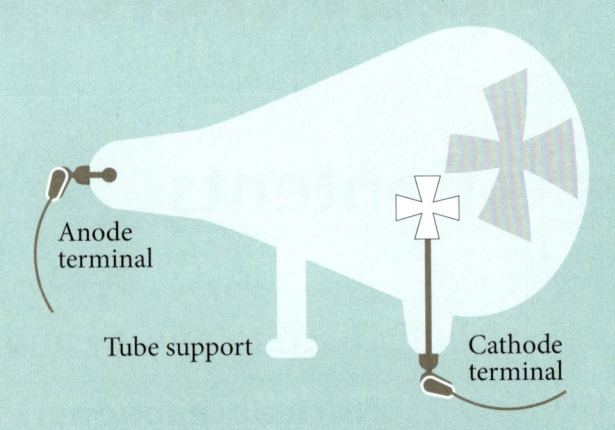

Figure 702   Crookes cathode ray

Cathode rays were so named because experiments had shown that whatever was causing the greenish glow in the Geissler tubes seemed to originate at the cathode, or negative plate. At the same time, the discharge appeared to be attracted to the anode, or positive plate.

Using various own Crookes' tubes, Crookes was able to determine several properties of cathode rays that were always the same, no matter what material made up the cathode.

Those properties are:

1. Cathode rays always travel in straight lines in the absence of a magnetic field.

2. In the presence of a magnetic field, the rays are deflected in the same way that magnetic fields deflect negatively charged particles as shown in Figure 703. This suggests the rays are negatively charged.

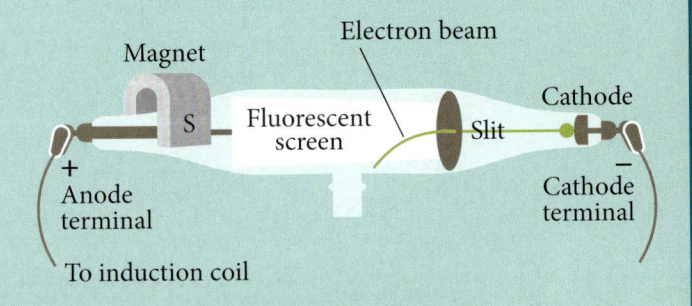

Figure 703   Deflection of cathode rays with a magnetic field.

ATOMIC, NUCLEAR AND PARTICLE PHYSICS

(Crookes also believed that cathode rays would be deflected in an electric field as shown in Figure 704 but he was not successful in demonstrating this).

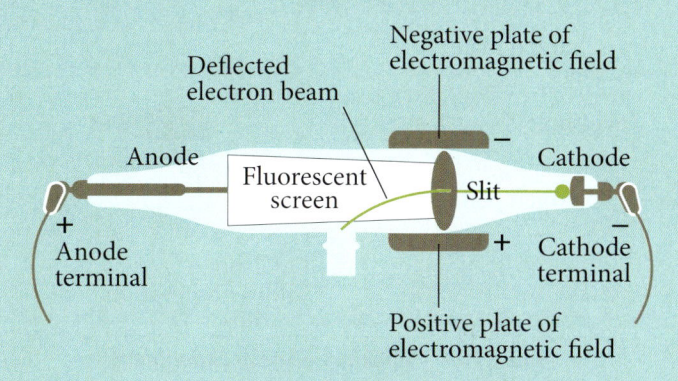

Figure 704 Deflection of cathode rays with an electric field.

3. The rays produce some chemical reactions similar to the reactions produced by light. This property led some scientists to conclude the rays must be a form of light.

4. The rays are able to turn tiny mica vaned paddle wheels, which suggests they have a particle nature because a force $F = ma$ is needed to do work on the paddle wheel. See figure 705.

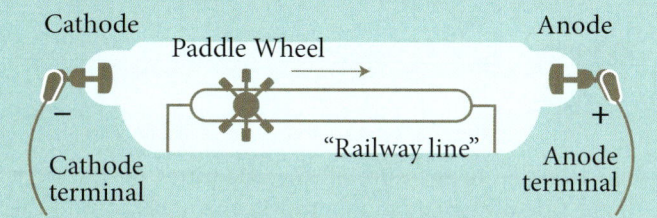

Figure 705 Paddle wheel cathode ray tube.

The fascinating properties of cathode rays began a long debate amongst scientists over the true nature of these rays. Were they made up of light waves, as some believed, or were they some kind of new particle? Were they charged, and if so, was the charge indeed negative? It would be twenty-five years before a definitive answer was obtained.

In 1897, the British physicist Sir Joseph John (J. J.) Thomson (1856 – 1940) did a series of experiments on cathode rays that left no doubt as to their nature.

By the time Thomson began his experiments on cathode rays, it was well known that the rays could be deflected by both electric and magnetic fields. Thomson reasoned that if the rays were indeed negatively charged particles, he could make predictions as to their behaviour in either a magnetic or an electric field. Furthermore, he reasoned that he could set up a magnetic field and an electric field such that the forces on the cathode rays due to each field would cancel each other out. In doing so, he could make a quantitative measurement of the charge to mass ratio of the particles thought to make up the cathode rays.

The force on a charged particle passing through a magnetic field is given by the equation

$$F_{magnetic} = evB$$

where B is the magnitude of the field.

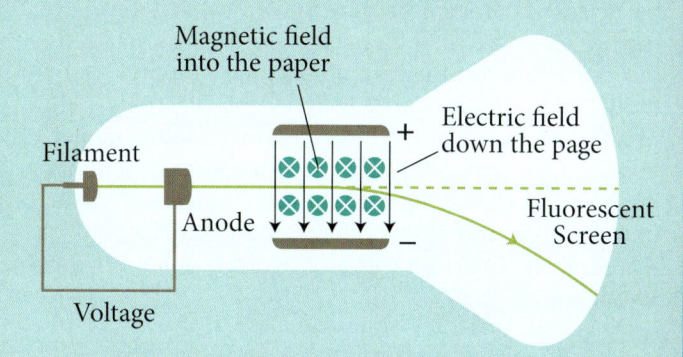

Figure 706 Crossed electric and magnetic fields

If the charged particle travels in a line perpendicular to the magnetic field, the path of the particle becomes a curve as shown in Figure 706. Since particles can only travel in a circular path when acted upon by a centripetal force, we must have that

$$F_{magnetic} = F_{centripetal}$$

Therefore

$$evB = \frac{mv^2}{r}$$

where r is the radius of the circular path.

Solving for the ratio of the charge to its mass, i.e., we have

$$\frac{e}{m} = \frac{v}{Br}$$

The velocity v can be found by applying an electric field simultaneously, so that the electric force and magnetic force are equal to one another. This results in the charged particle travelling in a straight line, and the velocity of the particle is found as follows:

$$F_{electric} = F_{magnetic}$$

$$eE = evB$$

Solving for v,

$$v = \frac{E}{B}$$

207

Chapter 7

Substituting the relevant equations, we have

$$\frac{e}{m} = \frac{E}{2B^2 r}$$

Using the equation above, Thomson was able to calculate the charge to mass ratio of cathode rays, which today is accepted to be 1.76×10^{11} C kg^{-1}. But simply being able to experimentally verify the charge to mass ratio of cathode rays was not enough for Thomson. He had his own ideas about the very nature of cathode rays. He suggested that cathode rays were composed of tiny negatively charged particles, which soon became known as electrons. He further suggested these electrons were not ions, or atoms, but rather a constituent of atoms, and he developed a model of the atom which included his electrons. His model became known as the "plum pudding model", because he envisioned the atom as a sphere of uniform positive charge, with the electrons embedded in it, rather like the raisins in a plum pudding. For his "discovery" of the electron, Thomson was awarded the Nobel Prize in Physics in 1906.

Between 1911 and 1913, Ernest Rutherford, a New Zealander working at Cambridge University in England, developed a model of the atom that in essence, was not replaced until 1926. In this model, the electrons orbit a positively charged nucleus. In a neutral atom, the number of orbital electrons is the same as the number of protons in the nucleus. The simplest atom, that of the element hydrogen, consists of a single electron orbiting a single proton. The centripetal force keeping the electron is orbit is provided by the Coulomb attraction between proton and electron, the proton being positively charged and the electron negatively charged. The magnitude of the charge on the electron is equal to the magnitude of the charge on the proton.

In 1910, the American physicist, Robert Millikan made the first precise determination of the charge on an electron as -1.602×10^{-19} C (The current value is $-1.60217733 \times 10^{-19}$ C).

He used the apparatus similar to Figure 707. Oil from an atomiser was charged by friction when sprayed above the top plate as electrons were stripped off the oil drops. Some of the positively charged oil drops entered the chamber through the tiny hole. They appeared as specks of light when viewed with a microscope. The two plates were connected to a variable voltage supply.

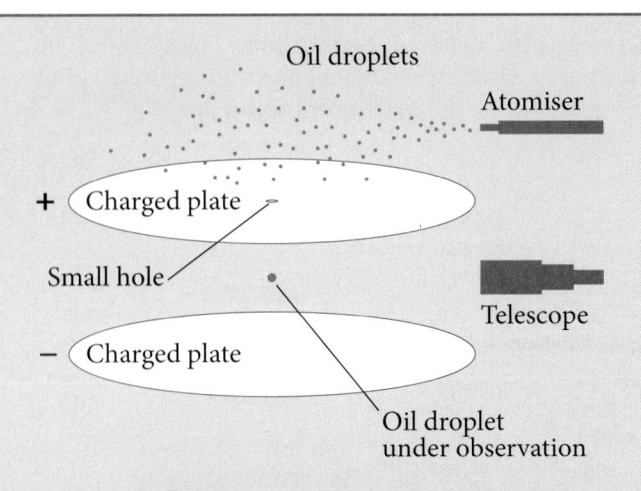

Figure 707   Millikan's oil drop experiment

He adjusted the voltage so that the electric force between the plates balanced the weight force of the droplet. Although the oil drops moved with constant velocity, we will consider them to be at rest.

$$mg = q\mathbf{E} \quad \text{or} \quad q = \frac{mg}{\mathbf{E}}$$

For a parallel plate capacitor, it can be shown that

$$E = \frac{V}{d}$$

Where V is the applied voltage and d is the distance between the plates.

By combining these two expressions we obtain

$$q = \frac{mgd}{V}$$

He was able to measure all the quantities on the right hand side of the equation and was then able to find the charge on the oil drop. In fact, he found that the charge on the oil drops could be written as an integer of the charge – the charge was quantised.

From the value of the charge to mass ratio found by Thomson, Millikan's determination of e enabled the mass of the electron to be determined. The current value for the electron mass me is $9.10938188 \times 10^{-31}$ kg. The current value for mass of the proton m_p is $1.67262158 \times 10^{-27}$ kg.

ATOMIC, NUCLEAR AND PARTICLE PHYSICS

7.1 Discrete energy and radioactivity

NATURE OF SCIENCE:

Accidental discovery: Radioactivity was discovered by accident when Becquerel developed photographic film that had accidentally been exposed to radiation from radioactive rocks. The marks on the photographic film seen by Becquerel probably would not lead to anything further for most people. What Becquerel did was to correlate the presence of the marks with the presence of the radioactive rocks and investigate the situation further. (1.4)

© IBO 2014

Essential idea: In the microscopic world energy is discrete.

Understandings:

- Discrete energy and discrete energy levels
- Transitions between energy levels
- Radioactive decay
- Fundamental forces and their properties
- Alpha particles, beta particles and gamma rays
- Half-life
- Absorption characteristics of decay particles
- Isotopes
- Background radiation

Discrete energy and discrete energy level

If a gas such as hydrogen is subjected to an electrical discharge, its spectrum is not continuous. Only certain wavelengths are present in a line emission spectrum as shown in Figure 708 for hydrogen and helium.

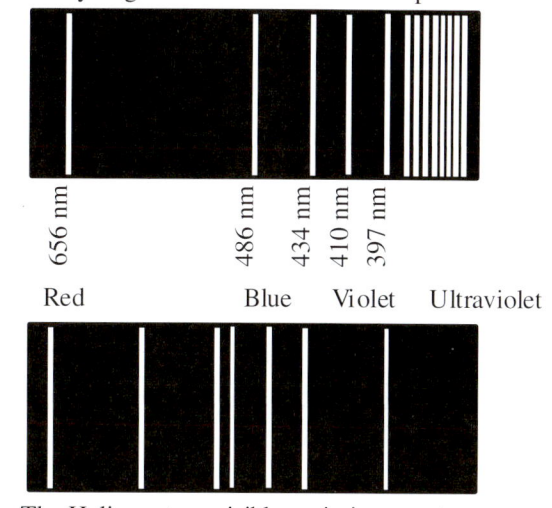

Figure 708 Emission spectra of hydrogen and helium

If a sufficiently high potential difference is applied between the ends of a **glass discharge tube** that is evacuated apart from the presence of a small amount of hydrogen vapour, the tube will glow a mauve colour. To study the radiation emitted by the tube, the emitted radiation could be passed through a slit and then through a dispersive medium such as a prism or a diffraction grating as shown in Figure 709.

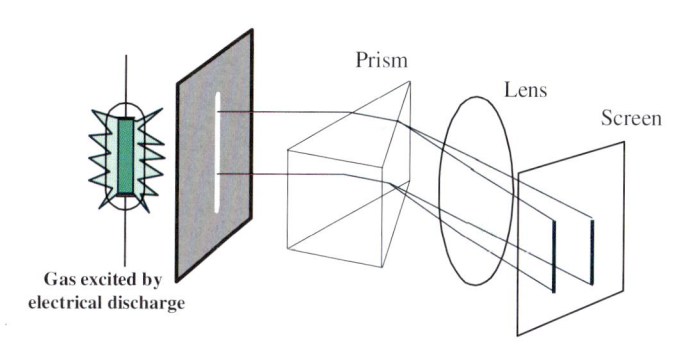

Figure 709 Electrical discharge tube

The prism splits the radiation into its component wavelengths. If the light emerging from the prism is brought to a focus, an image of the slit will be formed for each wavelength present in the radiation. Whereas the radiation from an incandescent solid (e.g the filament of a lit lamp) produces a continuous spectrum of colours, the hydrogen source produces a line spectrum. Each line in this spectrum is an image of the slit and in the visible region of the electromagnetic spectrum. Hydrogen gives rise to three distinct lines in the visible region – red, blue and violet. (There are other lines in other electromagnetic regions).

The study of line spectra is of great interest as it is found that all the elements in the gaseous phase give rise to a line spectrum that is characteristic of the particular element. In fact elements can be identified by their characteristic spectrum and is one way that astronomers are able to determine the elements present in the surface of a star (see Option E D). Also, the spectrum of an element provides clues as to the atomic structure of the atoms of the element.

The Scottish physicist Thomas Melvill (1726-1753) was the first to study the light emitted by various gases. He used a flame as a heat source, and passed the light emitted through a prism. Melvill discovered that the pattern produced by light from heated gases is very different from the continuous rainbow pattern produced when sunlight passes through a prism. The new type of spectrum consisted of a series of bright lines separated by dark gaps. This spectrum became known as a line spectrum. Melvill also noted the line spectrum produced by a particular gas was always the same. In other words, the spectrum was characteristic of the type of gas, a kind of "fingerprint" of the element or compound. This was a very important finding as it opened the door to further studies, and ultimately led scientists to a greater understanding of the atom.

Spectra can be categorised as either emission or absorption spectra. An emission spectrum is, as the name suggests, a spectrum of light emitted by an element. It appears as a series of bright lines, with dark gaps between the lines where no light is emitted. An absorption spectrum is just the opposite, consisting of a bright, continuous spectrum covering the full range of visible colours, with dark lines where the element literally absorbs light. The dark lines on an absorption spectrum will fall in exactly the same position as the bright lines on an emission spectrum for a given element, such as neon or sodium. For example, the emission spectrum of sodium shows a pair of characteristic bright lines in the yellow region of the visible spectrum. An absorption spectrum will show two dark lines in the same position as shown in Figure 710.

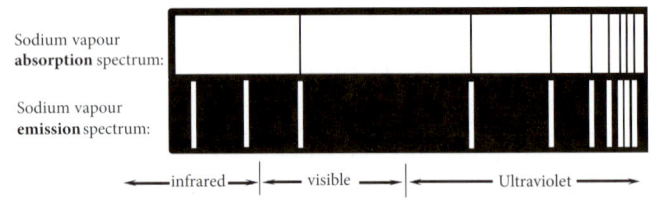

Figure 710   Absorption and emission spectra of sodium vapour

If the radiation from a filament lamp passes through a slit and then a tube containing unexcited mercury hydrogen vapour and is then focussed after passing through a prism, the resulting spectrum is continuous but is crossed with dark lines. These lines correspond exactly to the lines in the emission spectrum of hydrogen. To understand this, suppose the difference in energy between the lowest energy level and the next highest level in mercury atoms is E, then to make the transition between these levels, an electron must absorb a photon of energy E. There are many such photons of this energy present in the radiation from the filament. On absorbing one of these photons, the electron will move to the higher level but then almost immediately fall back to the lower level and in doing so, will emit a photon also of energy E. However, the direction in which this photon is emitted will not necessarily be in the direction of the incident radiation. The result of this absorption and re-emission is therefore, a sharp drop in intensity in the incident radiation that has a wavelength determined by the photon energy E. The phenomenon of absorption spectra is of great importance in the study of molecular structure since excitation of molecules will often cause them to dissociate before they reach excitation energies.

Melvill's early experiments created a new field of science called spectroscopy, which is the study of emission and absorption spectra. Spectroscopists make use of atomic spectra to identify unknown elements in mixtures, as well as to study the origin of spectra, the atom.

Once it had been established that an atomic spectrum is characteristic of the type of element producing the spectrum, scientists looked in vain for a pattern in the lines. In 1885, one pattern was finally established when the Swiss school teacher, Johann Jakob Balmer (1825–1898) published a paper in which he used a simple mathematical formula to describe the visible spectrum of hydrogen. His formula took the form

$$\lambda = b \left(\frac{n^2}{n^2 - 2^2} \right)$$

where λ is the wavelength, b is a constant equal to 3.645×10^{-5} cm, and n is a whole number. The pattern of lines in the visible region of the hydrogen spectrum came to be known as the Balmer series.

Balmer suggested there might be other lines that were not in the visible spectrum and that his formula could be used to predict where such other lines could be found. His original formula was soon modified by the Swedish spectroscopist, Johannes Rydberg (1854 – 1919). The new equation, called the Rydberg formula, took the form

$$\frac{1}{\lambda} = R \left(\frac{1}{2^2} - \frac{1}{n^2} \right)$$

where R was now a constant, called the Rydberg constant, equal to 1.096×10^7 m^{-1}. In both Rydberg's and Balmer's formulae, setting n equal to 3, 4, and 5 gave the wavelength of the first three lines of the Balmer series.

It wasn't long before additional series of lines were found using Rydberg's formula, just as Balmer had predicted. One set of lines was found in the ultraviolet region of the spectrum, named the Lyman series after discoverer, Theodore Lyman. Another set was found in the infrared region, named the Paschen series after discoverer, Friedrich Paschen. However, it still wasn't understood why the formula worked. That explanation came with the help of the work of the German physicist, Max Planck.

ATOMIC, NUCLEAR AND PARTICLE PHYSICS

Planck proposed that energy associated with vibrating molecules could be considered to occur in discrete amounts or **quanta**.

In 1905, based on the work of Max Planck, Einstein proposed that light is made of small packets of energy called photons. Each photon has an energy E given by:

E = hf

where h is a constant known as the Planck constant and has a value

6.6×10^{-34} J s.

Transitions between energy levels

The photon model of light suggests an atomic model that accounts for the existence of the line spectra of the elements. If it is assumed that the electrons in atoms can only have certain discrete energies or, looking at it another way, can only occupy certain allowed energy levels within an atom, then when an electron moves from one energy level to a lower energy level, it emits a photon whose energy is equal to the difference in the energy of the two levels. The situation is somewhat analogous to a ball bouncing down a flight of stairs; instead of the ball losing energy continuously when rolling down the banisters, it loses it in discrete amounts.

In 1913, the young Danish physicist, Niels Bohr (1885 – 1962) had recently completed his PhD and was working in the Manchester laboratory of Ernest Rutherford. Naturally, Bohr was fascinated by Rutherford's planetary model of the atom, and joined in the search for an explanation of why electrons are not drawn into the nucleus by electrostatic forces of attraction to the protons of the nucleus. He used the work of Balmer, Rydberg, and Planck to develop a brilliant theory of the atom that put forward some basic postulates:

1. The electron travels in circular orbits around the positively charged nucleus, but only certain orbits are allowed. The electron will not radiate energy as long as it is in one of these orbits.

2. If an electron falls from one orbit, or energy level, to another, it loses energy in the form of a photon of light. The energy of the photon equals the difference between the energies of the two orbits. The frequency of light emitted can be calculated using Planck's formula, $E = hf$. These calculated frequencies correspond to the lines found on an emission spectrum of the atom.

3. A hydrogen atom can absorb only those photons of light which will cause the electron to jump from a lower energy level to a higher energy level. Thus the dark lines found on an absorption spectrum correspond to energy absorbed as an electron transition from a lower to a higher energy level. Energy levels are also called **quantum levels**. Figure 711 shows the possible transitions between energy levels for hydrogen.

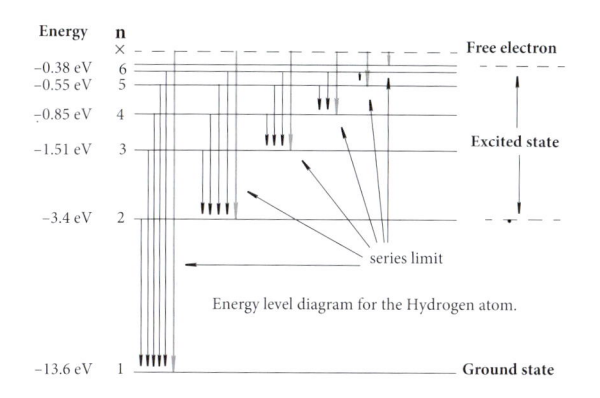

Figure 711   Transitions between energy levels for hydrogen.

Bohr's postulates provided a beautiful model of the hydrogen atom. The model was successful because it explained the hydrogen spectrum, particularly why hydrogen atoms can emit and absorb only certain frequencies of light. It provided a reasonable explanation as to why electrons do not fall into the nucleus. Like any successful model, it also made predictions, in this case of additional spectral series, that were later found to be correct. For his revolutionary view of the atom, Bohr was awarded the 1922 Nobel Prize in Physics, at the relatively young age of 37.

In spite of its success in explaining the hydrogen atom, Bohr's theory also had some difficulties. It failed to predict accurately the spectra for atoms that had two or more electrons. It could not predict or explain the relative brightness of spectral lines. It was also unable to account for two or more closely spaced spectral lines, such as the pair found in the yellow region of the visible spectrum of sodium. It was also unable to explain the bonding of atoms in molecules or in solids or liquids. Finally, Bohr was unable to provide any experimental evidence to support his postulates, which blended classical and quantum mechanics without any justification.

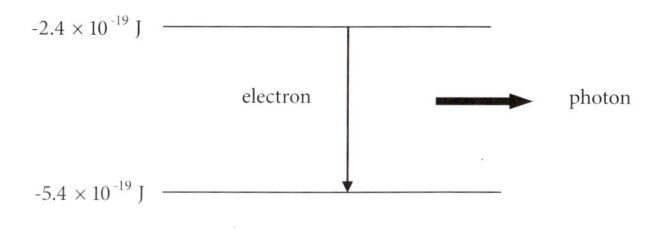

Figure 712

211

Figure 712 shows two of the allowed energy levels of atomic hydrogen.

When the electron makes the transition as shown in Figure 712, it emits a photon whose frequency is given by the Planck formula i.e

$$f = \frac{E}{h} = \frac{3.0 \times 10^{-19}}{6.6 \times 10^{-34}} = 4.5 \times 10^{14} \text{ Hz}$$

The associated wavelength is given by

$$\lambda = \frac{c}{f} = \frac{3 \times 10^8}{4.5 \times 10^{14}} = 6.6 \times 10^{-7} \text{ m}$$

Alternatively, we can show:

$$E = \frac{hc}{\lambda} \quad \text{or} \quad \lambda = \frac{hc}{E}$$

$$\lambda = \frac{6.6 \times 10^{-34} \times 3 \times 10^8}{3 \times 10^{-19}} = 6.6 \times 10^{-7} \text{ m}.$$

This is in fact the measured value of the wavelength of the red line in the visible spectrum of atomic hydrogen.

In most situations, the electrons in an atom will occupy the lowest possible energy states. Electrons will only move to higher energy levels if they obtain energy from somewhere such as when the element is heated or, as mentioned above, when subjected to an electrical discharge. When the electrons are in their lowest allowed levels, the atom is said to be 'unexcited' and when electrons are in higher energy levels, the atom is said to be 'excited'. To move from a lower level to a higher energy level, an electron must absorb an amount of energy exactly equal to the difference in the energy between the levels.

Many elements have forms that are chemically identical but each form has a different mass. These forms are called **isotopes**. The existence of isotopes explains why the chemical masses of elements do not have integral values since a sample of an element will consist of a mixture of isotopes present in the sample in different proportions.

The explanation for the existence of isotopes did not come until 1932 when *James Chadwick* an English physicist, isolated an uncharged particle that has a mass very nearly the same as the proton mass. Since 1920, both *Rutherford* and *Chadwick* had believed that an electrically neutral particle existed. An uncharged particle will not interact with the electric fields of the nuclei of matter through which it is passing and will therefore have considerable penetrating ability. In 1930 *Walther Bothe* and *Herbert Becker* found that when beryllium is bombarded with α-particles a very penetrating radiation was produced. It is this radiation that *Chadwick* showed to consist of identical uncharged particles. These particles he called **neutrons** and the current value of the neutron mass m_n is $1.67262158 \times 10^{-27}$ kg.

The neutron explains the existence of isotopes in the respect that, a nucleus is regarded as being made up of protons and neutrons. The nuclei of the different isotopes of an element have the same number of protons but have different numbers of neutrons. For example, there are three stable isotopes of oxygen; each nucleus has eight protons but the nuclei of the three isotopes have eight, nine and ten neutrons respectively.

In the study of particle physics, the proton and neutron are regarded as different charge states of the same particle called the **nucleon**.

The three isotopes of oxygen mentioned above are expressed symbolically as $^{16}_{8}O$, $^{17}_{8}O$ and $^{18}_{8}O$ respectively. The "O" is the chemical symbol for oxygen, the subscript is the number of protons in the nucleus of an atom of the isotope and the superscript, the number of protons plus neutrons. Each symbol therefore refers to a single nucleus and when expressed in this form, the nucleus is referred to as a **nuclide**. In general, any nuclide X is expressed as $^{A}_{Z}X$

A is the **nucleon number**

Z is the **proton number**

We also define the **neutron number** N as $N = A - Z$.

It should be mentioned that it is the proton number that identifies a particular element and hence the electronic configuration of the atoms of the element. And it is the electronic configuration that determines the chemical properties of the element and also many of the element's physical properties such as electrical conductivity and tensile strength.

Fundamental forces and their properties

Most people observe thousands of forces every day, but in fact there are only four types of forces responsible for all these observations: the gravitational force, the electromagnetic force, the strong nuclear force and the weak interaction force as shown in Figure 713.

Type	Relative Strength (approx.)	Field Particle
Gravitational	1	Graviton
Weak nuclear	10^{32}	W± and Z0
Electromagnetic	10^{36}	Photon
Strong nuclear	10^{38}	Gluons

*Relative strength is approx. for 2 protons in nucleus.

Figure 713   Fundamental forces

ATOMIC, NUCLEAR AND PARTICLE PHYSICS

Of these four fundamental forces, the weakest is the gravitational. Considering that gravitational forces hold the solar system and galaxies together, imagine how strong the other forces must be! The gravitational force is only attractive acting between any two bodies that have mass, and it has an infinite range, although its magnitude is inversely related to the square of the distance between the two bodies. This means the strength of the gravitational force diminishes rapidly as the distance between two bodies increases. Still, it is gravitational forces that hold the moon in orbit around the earth, and the earth in orbit around the sun, suggesting the range of these forces is quite extraordinary.

The electromagnetic force is the one responsible for most of our every day interactions. For example, when you push a door open, it is the interactions between the electrons in your hand and the electrons in the door that cause the door to move. The electromagnetic force acts between charged bodies, and can be attractive or repulsive. It is stronger than the gravitational force, but weaker than the strong nuclear force. Again, the electromagnetic force can act over great distances, but, like the gravitational force, the strength of this force varies inversely with the square of the distance between the charged objects.

Unlike the gravitational and electromagnetic forces, the remaining forces are confined to the nucleus of atoms, so they only operate on protons and neutrons. Protons are positively charged, which means they repel one another with electromagnetic forces that become stronger the closer the protons are to one another. Since the nucleus is of the order of 10^{-12} m in diameter, if the only force present was the electromagnetic force, the nucleus would be expected to disintegrate. However, stable nuclei do stay together, suggesting the presence of a force even stronger than the electromagnetic force. That force is called the strong nuclear force, and is believed to be an attractive force that exists between all nucleons, regardless of their charge.

Shortly after the discovery of the neutron, Hideki Yukawa, a Japanese physicist, postulated a strong force of attraction between nucleons that overcomes the Coulomb repulsion between protons. The existence of the force postulated by Yukawa is now well established and is known as the strong nuclear interaction. The force is independent of whether the particles involved are protons or neutrons and at nucleon separations of about 1.3 fm, the force is some 100 times stronger than the Coulomb force between protons. At separations greater than 1.3 fm, the force falls rapidly to zero. At smaller separations the force is strongly repulsive thereby keeping the nucleons at an average separation of about 1.3 fm. (1 femtometre = 10^{-15} m)

While there is still no real mathematical description of the strong nuclear force, such as those that exist for the electromagnetic and gravitational forces, some of its characteristics are known. The force also requires maintaining a delicate balance between the number of protons and neutrons in the nucleus. A difference of as little as one neutron more or less in a nucleus can lead to instability, which is essentially a reduction in the strength of the strong nuclear force. In the absence of this force, nuclei can and do disintegrate in the process called radioactive decay.

The fourth and last fundamental force is also found only in the nucleus and is called the weak interaction force. Evidence of the existence of this force appears in beta decay, in which a neutron is thought to disintegrate into a proton and an electron. The range of this force is of the order of 10^{-17} m, which means it only operates within certain elementary particles. Otherwise, not much is known about this particular force.

Exercise 7.1

1. Atoms that have the same atomic number but different mass numbers are called

    A. nucleons
    B. ions
    C. isotopes
    D. elements

2. The mass number of an isotope is equal to

    A. the number of protons.
    B. the number of neutrons.
    C. the number of electrons.
    D. the number of nucleons.

3. Which type of light in the electromagnetic spectrum has the highest energy photons?

    A. ultraviolet
    B. visible blue
    C. visible red
    D. infrared

4. An electron makes a transition from the $n=3$ energy level to the $n=2$ energy level. What happens?

    A. Energy is emitted in the form of light.
    B. Energy is absorbed in the form of light.
    C. Energy is neither emitted or absorbed.
    D. This kind of transition is impossible because electrons must remain in assigned orbits.

5. The ground state of a hydrogen atom is when

    A. its electron has escaped.
    B. the atom is in its lowest energy state.
    C. the atom is in its highest energy state.
    D. its electron has been absorbed into the nucleus.

6. Which of the following transitions will result in light of the highest frequency being emitted?

   A. $n = 1$ to $n = 2$
   B. $n = 2$ to $n = 1$
   C. $n = 3$ to $n = 1$
   D. $n = 1$ to $n = 3$

7. What energy would be associated with a photon of red light with a frequency of 3.85×10^{14} Hz?

8. Determine the frequency of a quantum of light that has energy of 1.0 J.

9. A neutron travelling at 1.98×10^4 m s^{-1} has the same energy as light with a frequency of 5.0×10^{14} Hz. Determine the mass of the neutron.

10. A metal requires energy of at least 4.0 eV to emit an electron from its surface. Determine the longest wavelength of light that will emit electrons.

11. Determine the frequency of the light emitted when electrons fall from an energy level of 6.68×10^{-19} J to an energy level of 2.46×10^{-19} J.

12. The first three energy levels od an atom are at 1.2375 eV, 3.0375 eV and 5.4063 eV. An electron transition results in the release of a photon of frequency of 5.72×10^{14} Hz. Determine the energy levels between which the electron moved.

13. Suppose an electron can exist in the ground state and three other energy levels. How many lines in the emission spectrum can be predicted?

14. An atom has the following quantum levels: ground state, 3.36×10^{-19} J, 4.96×10^{-19} J, 5.76×10^{-19} J, 5.92×10^{-19} J and 8.16×10^{-19} J.

   (a) What is the minimum energy required to remove an electron from the atom?
   (b) The atoms are bombarded with electrons with energies of 5.28×10^{-19} J. Determine the possible energies of the photons emitted.

Alpha particles, beta particles and gamma rays

Shortly after Becquerel's discovery two physicists working in France, Pierre and Marie Curie isolated two other radioactive elements, polonium and radium each of which is several million times more active than uranium. An extremely good account of the Curies' heroic experiments can be found in Madame Curie's biography written by her daughter, Irene. At first it was thought that the radiation emitted by radioactive elements were of the same nature as the X-rays that had been discovered the previous year. However, in 1897, Rutherford found that two types of radiation occurred in radioactivity, some of the rays being much more penetrating than the others. He called the less penetrating rays alpha (α) rays and the more penetrating ones beta (β) rays.

In 1900 Paul Villiard, also French, detected a third type radiation which was even more penetrating than β-rays. Naturally enough he called this third type gamma (γ) radiation.

Madame Curie deduced from their absorption properties that α-rays consisted of material particles and Rutherford showed that these particles carried a positive charge equal to about twice the electron charge but that they were very much more massive than electrons. Then in 1909, in conjunction with Royds, Rutherford identified α-particles as helium nuclei.

Alpha particles – α

Alpha rays are the least penetrating, and can be stopped by a layer of skin or a sheet of paper. However, the ionising power of alpha rays is much greater than the other two types of ray, meaning they have a much greater ability to knock electrons out of orbit in atomic collisions. The inverse relationship between penetrating and ionising powers made perfect sense once Rutherford was able to show that alpha rays were actually made up of particles, that is, helium nuclei. The large size of these nuclei allows them to be stopped very quickly, but the sudden expending of their energy increases their ionising power dramatically. Rutherford identified alpha particles as helium nuclei by trapping them in a modified cathode ray tube, and comparing their spectrum with that of helium.

Beta particles – β

These were determined to have a greater penetrating power than alpha particles, able to penetrate up to one to two centimetres of water or human flesh. They can however be stopped by a sheet of aluminium a few millimetres thick. The greater penetrating power of these rays that were made up of electrons means they have less ionising power than alpha particles. Essentially, the rays penetrate farther, meaning their energy is spent over a greater distance resulting in a reduced ability to ionise atoms.

ATOMIC, NUCLEAR AND PARTICLE PHYSICS

In 1928 the English physicist Paul Dirac predicted the existence of a positively charged electron and in 1932, this particle was found by the American physicist Carl Anderson, to be present in cosmic radiation. Then in 1934, the Curie's daughter Irene, along with her husband Frederic Joliot, discovered the positively charged electron, now called the positron (e+, β+), to be present in certain radioactive decay.

Gamma rays – γ

These have the greatest penetrating power of the three types of radiation, able to pass through the human body, a metre of concrete and many centimetres of lead. Consequently, these electromagnetic waves have the lowest ionising power of the three types of radiation. However, the fact that γ-rays are still a form of ionising radiation means they are still potentially harmful, and their great penetrating power means very thick radiation shielding is necessary to protect people who work with, or are exposed to, these rays.

Figure 714 summarises some of the properties of alpha and beta particles and gamma radiation.

One of the easiest ways to distinguish between alpha, beta and gamma radiations is to observe their interaction with a magnetic field as shown in Figure 715. Alpha particles are helium nuclei, so they are positively charged, while beta particles are negatively charged.electrons. Therefore, a magnetic field will deflect these two types of particles in opposite directions, an effect that can be very easily demonstrated using a photographic plate detector. Gamma rays, being electromagnetic waves, have no charge associated with them, and so will pass through a magnetic field undeflected.

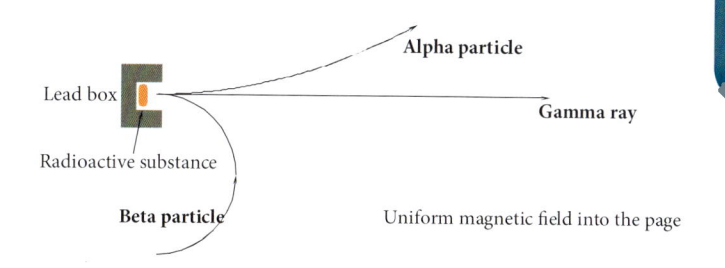

Figure 715   Deflection of alpha and beta particles

Type	Description	Charge, q, (in e units) and mass, m (in nucleons, u)	Energy, E, and speed, v.	Penetration (range)
alpha (α)	Helium nucleus	$q = +2e$ (i.e., charge of 2 protons) $m = 4u$	type: kinetic $E = \frac{1}{2}mv^2$ $v \sim 0.05c$	Stopped by a sheet of paper (a few centimetres)
beta (β)	fast moving electron	$q = -e$ (i.e., charge of an electron) $m = \frac{1}{1850}u$	type: kinetic $E = \frac{1}{2}mv^2$ $v = (0-99\%)$ of c	Stopped by a few millimetres of Al (a few metres)
gamma (γ)	electromagnetic radiation (with short wave-length, λ, and high frequency, f)	$q = 0$ $m = 0$	type: photon $(E = hf)$ $v = c$	many centimetres of lead (no maximum range)

Figure 714   Summary of properties (& observations)

CHAPTER 7

Radioactive decay

Certain elements emit radiation spontaneously, i.e. without any external excitation; this phenomenon is called natural **radioactive decay**.

The discovery of radioactivity in 1896 is credited to the French physicist Antoine Henri Becquerel (1852 – 1908). He had been inspired by the discovery two months previously of X-rays, and set about to determine the connection between X-rays and fluorescence or phosphorescence. Becquerel placed several salt samples, including the double sulphate of uranium and potassium on a photographic plate that was doubly wrapped in black paper to keep out light. He then exposed the salt samples to bright sunlight for several hours. He discovered the photographic plate showed no exposure due to sunlight, but did show a silhouette of the uranium salt, indicating the presence of a penetrating radiation. Becquerel did further research on the uranium salt and discovered that the rays emitted by this salt were **not affected by changes in temperature, exposure to light, or any other physical or chemical changes**. This indicated that the radioactive properties of uranium were due to the nucleus of the atom and not to the electronic structure. He also discovered the rays were able to ionise gas molecules, so they became known as ionising radiation.

α-decay

A nucleus of a radioactive element that emits an α-particle must transform into a nucleus of another element. The nucleus of the so-called 'parent' element loses two neutrons and two protons. Therefore the nucleon number (A) changes by 4 and the proton number (Z) by 2. The nucleus formed by this decay is called the 'daughter nucleus'. We may express such a nuclear decay by the nuclear reaction equation

$$^{A}_{Z}X \rightarrow {}^{A-4}_{Z-2}Y + {}^{4}_{2}He$$
(parent)   (daughter)   (α-particle)

For example the isotope uranium–238 is radioactive and decays by emitting α radiation to form the isotope thorium-234, the nuclear reaction equation being:

$$^{238}_{92}U \rightarrow {}^{234}_{90}Th + {}^{4}_{2}He$$

Or simply

$$^{238}_{92}U \rightarrow {}^{234}_{90}Th + \alpha$$

Clearly in any nuclear radioactive decay equation, the nucleon number and proton number of the left-hand side of the equation must equal the nucleon number and proton number of the right-hand side of the equation.

β⁻ and β⁺ decay (HL only)

In 1950 it was discovered that a free neutron decays into a proton and an electron after an average life of about 17 minutes. Another particle, the **antineutrino** is also emitted in this decay. In order to conserve momentum and energy in β decay, the Italian physicist *Enrico Fermi* postulated the existence of the neutrino in 1934. . This particle is extraordinarily difficult to detect. It has zero rest mass so travels at the speed of light; it is uncharged and rarely reacts with matter (Millions upon millions of neutrinos pass through the human body every second). However, in 1956, the neutrino was finally detected. It turns out though, that in β-decay in order to conserve other quantities, it must be an antineutrino and not a neutrino that is involved. (*Conservation laws along with particles and their antiparticles are discussed in detail in section 7.3*).

The decay equation of a free neutron is

$$^{1}_{0}n \rightarrow {}^{1}_{1}p + {}^{0}_{-1}e + \bar{\nu}$$

Or simply

$$n^0 \rightarrow p^+ + e^- + \bar{\nu}$$

$\bar{\nu}$ is the symbol for the antineutrino.

The origin of β- particle is the decay of a neutron within a nucleus into a proton. The nucleon number of a daughter nucleus of an element formed by β- decay will therefore remain the same as the nucleon number of the parent nucleus. However, its proton number will increase by 1. Hence we can write in general that

$$^{A}_{Z}X \rightarrow {}^{A}_{Z+1}Y + {}^{0}_{-1}\beta + \bar{\nu}$$
(parent) (daughter) (electron) (antineutrino)

For example, a nucleus of the isotope thorium-234 formed by the decay of uranium-238, undergoes β- decay to form a nucleus of the isotope protactinium-234. The nuclear reaction equation for this decay is

$$^{234}_{90}Th \rightarrow {}^{234}_{91}Pa + {}^{0}_{-1}e + \bar{\nu}$$

Or simply

$$^{234}_{90}Th \rightarrow {}^{234}_{91}Pa + e^- + \bar{\nu}$$

The origin of β+ particles is from the decay of a proton within a nucleus into a neutron. The decay equation of the proton is

$$^{1}_{1}p \rightarrow {}^{1}_{0}n + {}^{0}_{1}e + \nu$$

Or simply

$$p^+ \rightarrow n^0 + e^+ + \nu$$

ν is the symbol for the neutrino.

216

For example, a nucleus of the isotope ruthenium-90 decays to a nucleus of the isotope technetium-90. The nuclear reaction equation for this decay is

$$^{90}_{44}Ru \rightarrow ^{90}_{43}Tc + ^{0}_{+1}e + \nu$$

Or simply

$$^{90}_{44}Ru \rightarrow ^{90}_{43}Tc + e^+ + \nu$$

Unlike free neutrons, free protons are stable (although current theory suggests that they have an average life of 10^{30} years.). This probably explains why most of the observable matter in the universe is hydrogen.

γ radiation

The source of γ radiation in radioactive decay arises from the fact that the nucleus, just like the atom, possesses energy levels. In α and β decay, the parent nuclide often decays to an excited state of the daughter nuclide. The daughter nuclide then drops to its ground state by emitting a photon. Nuclear energy levels are of the order of MeV hence the high energy of the emitted photon. (Nuclear energy levels are discussed in more detail in topic 12.2).

Absorption characteristics of decay particles

As α-particles travel through air, they readily interact with the air molecules by "grabbing" two electrons from a molecule and so becoming a neutral helium atom. This leaves the air molecule electrically charged and the air through which the α-particle are passing is said to be 'ionized'. If the air is situated between two electrodes, the positive ions created by the passage of the α-particles will migrate to the negative electrode and an ionization current can be detected.

β-particles and γ radiation may also cause ionization but in this situation the particles of the radiation actually remove electrons from the air molecules by collision with the molecules, thereby creating an electron-ion pair.

The ionising power of radiation has been used to develop a number of detection methods over the past one hundred years. One of the first detectors used was the electroscope. A charged electroscope can be discharged in the presence of ions of the opposite charge. The rate at which an electroscope is discharged is indicative of the number of ions in the air near the electroscope. Marie Curie used the electroscope to study the radioactivity of uranium ore. She found that different ores discharged the electroscope at different rates, with higher rates being associated with greater radioactivity. It was these differences in radioactivity that ultimately led to her discovery of the highly radioactive element, radium.

The ionisation chamber is another detector that uses the ionising power of radiation. The chamber contains fixed electrodes, which attract electrons and ions produced by the passage through the chamber of high-speed particles or rays. When the electrodes detect ions or electrons, a circuit is activated and a pulse is sent to a recording device such as a light.

The Geiger-Müller detector was named for its inventors, Hans Geiger (1882–1945) and his student W. Müller (See Figure 716). It operates on the same general principle as the ionisation chamber, except that in a Geiger-Müller detector, the negative electrode is a hollow cylinder, while the positive electrode is a thin wire down the centre of the cylinder. Both electrodes are placed in a glass tube filled with low-pressure gas, and ionisation of the gas is used to detect the passage of radiation.

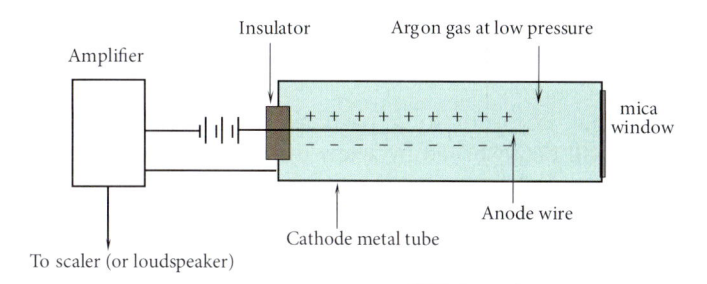

Figure 716   Geiger-Müller tube

Dosimetry is the study of radiation. Recall that radiation can be transmitted in the form of electromagnetic waves or as energetic particles, and that, when an atom absorbs sufficient energy, it can cause the release of electrons and the formation of positive ions. When radiation causes ions to form it is called ionising radiation. Ionising radiation is produced by X-rays, CAT, radioactive tracers and radiopharmaceuticals, as well as by many other natural and artificial means.

When ionising radiation penetrates living cells at the surface or within the body, it may transfer its energy to atoms and molecules through a series of random collisions. The most acute damage is caused when a large functioning molecule such as DNA is ionised leading to changes or mutations in its chemical structure. If the DNA is damaged it can cause premature cell death, prevention or delay of cell division, or permanent genetic modification. If genetic modification occurs, the mutated genes pass the information on to daughter cells. If genetic modification occurs in sperm or egg cells, the mutated genes may be passed on to offspring.

Since the body is 65% water by weight, most of the radiation energy is absorbed by the water content. This energy can produce ions (H+, OH-, H_3O+) and electrically neutral free radicals of water. These ions and free radicals can cause chemical reactions with other chemical constituents of the cell. For example, OH- ions and OH free radicals that form the strong oxidising reagent hydrogen peroxide (H_2O_2) that can interfere with the carbon-carbon double bonds within the DNA molecule, causing rupture of the double helical strands. Free radicals may also cause damage to enzymes that are required for the metabolism of the cell or they can affect the membranes that are vital for the transport of materials within the cell.

Ionising radiation appears to affect different cells in different ways. Cells of the reproductive organs are very radiation–sensitive and sterility is a common outcome after radiation exposure. Bone and nerve cells are relatively radiation-resistant. However, radiation of bone marrow leads to a rapid depletion of stem cells that can then induce anaemia or even leukemia.

Exposure to radiation results in a range of symptoms including skin burns, radiation sickness (nausea, vomiting, diarrhoea, loss of hair, loss of taste, fever) cancer, leukemia and death.

External exposure to α particles is fairly harmless as they will be absorbed by a few microns of skin. Internal exposure after ingestion is very damaging as the α-particles are very ionising and they can interact with body fluids and gases. β-particles are more penetrating but because of their irregular paths upon entering body tissue they are considered to have low ionising ability. X-rays and γ-radiation have high ionising ability (but not as high as ingested α-particles), and are a common cause of disruption to normal cellular metabolism and function. Simple cells are more sensitive than highly complex cells.

Cells that divide and multiply rapidly are more sensitive than those that replicate slowly.

Background radiation

All living things are being exposed to cosmic radiation from the Sun and space, and terrestrial radiation from the lithosphere (uranium and thorium series and isotopes of radon) It is important that we have measures in place to monitor both this background and artificial radiation produced in medicine and the nuclear industry. In 1928, the body now known as the International Commission on Radiological Protection (ICRP) was set up to make recommendations as to the maximum amounts of radiation that people could safely receive.

Radiation dosimetry deals with the measurement of the absorbed dose or dose rate resulting from the interaction of ionising radiation with matter.

The effect of all the radiations is essentially the same and the probability of genetic damage increases with increasing intensity and exposure to radiation. Because of this it is vitally important that we are exposed to the minimum of radiation. The effects of the atom bombs dropped on Nagasaki and Hiroshima are still with us and this should be a salutary lesson to all Governments.

The effects of radiation on the whole body have been studied in great detail. The radiation energy absorption by the body of about 0.25 J kg^{-1} causes changes in the blood and may lead to leukaemia; 0.8 –1.0 J kg^{-1} gives rise to severe illness (radiation sickness) with the chance of recovery within about 6 months, whereas about

5 J kg^{-1} is fatal. For this reason, people working in hospital X-ray departments, radioisotope laboratories, outer space and nuclear power stations take great precautions against exposure to radiation.

On the plus side, the controlled use of the radiations associated with radioactivity is of great benefit in the treatment of cancerous tumours.

Half-life

Figure 717 shows the variation with proton number (Z) of the neutron number (N) (number of neutrons) of the naturally occurring isotopes. The straight line is the plot of the points given by $N = Z$.

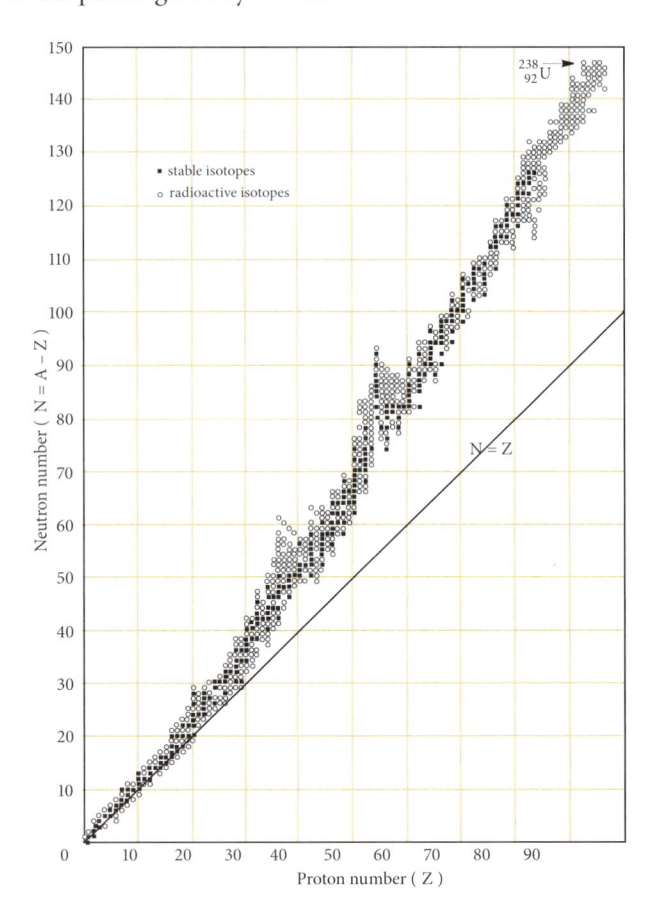

Figure 717 *The variation of neutron number with proton number*

From Figure 717 we see that for nuclei of the elements with Z less than about 20, the number of protons is equal to the number of neutrons. Above Z = 20, there is an excess of neutrons over protons and this excess increases with increasing Z. We can understand this from the fact that as Z increases so does the electrostatic force of repulsion between protons. To balance this, the number of neutrons is increased thereby increasing the strong nuclear force of attraction between the nucleons. When a proton is added to a nucleus, it will exert roughly the same force of repulsion on the other protons in the nucleus. This is because it is very nearly the same distance from each of the other protons. However, the strong nuclear force is very short range and is only really effective between adjacent neighbours. So as the size of the nucleus increases, proportionally more and more neutrons must be added. Each time protons and neutrons are added, they have to go into a higher energy state and eventually a nuclear size is reached at which the nucleus becomes unstable (a bit like piling bricks on top of one another) and the nucleus tries to reach a more stable state by emitting a nuclear sub-group consisting of two protons and two neutrons i.e. a helium nucleus (α particle).

Consider now a nucleus of the isotope $^{65}_{28}$Ni. This nucleus is unstable because the neutron excess is too great, each neutron added having to go into a higher energy state. To become stable, one of the neutrons will change into a proton by emitting an electron i.e. a β⁻-particle. On the other hand, a nucleus of the isotope $^{54}_{25}$Mn does not contain enough neutrons to be stable. To become stable, a proton changes into a neutron by emitting a positron i.e. a β⁺ particle.

We have no means of knowing when a particular atom in a sample of radioactive element will undergo radioactive decay. You could sit watching an atom of uranium-238 and it might undergo decay in the next few seconds. On the other hand, you might wait 10⁹ years or even longer, before it decayed. However, in any sample of a radioactive element there will be a very large number of atoms. It is therefore possible to predict the probability of decay of a particular atom. What we can say is that the more atoms in a sample, the more are likely to decay in a given time. **The rate of decay of atoms at a given instant is therefore proportional to the number of radioactive atoms of the element in the sample at that instant**. As time goes by, the number of atoms of the element in the sample will decrease and so therefore will the rate of decay or activity as it is called. Of course, the number of atoms in the sample will not change since, when an atom of the element decays, it decays into an atom of another element.

Exponential decay

There are many examples in nature where the rate of change at a particular instant of a quantity is proportional to the quantity at that instant. A very good example of this is the volume rate of flow of water from the hole in a bottom of the can. Here the volume rate is proportional to the volume of water in the can at any instant. Rates of change such as this, all possess a very important property, namely that the quantity halves in value in equal increments of time. For example, if the quantity Q in question has a value of 120 at time zero and a value of 60, 20 seconds later, then it will have a value of 30 a further 20 seconds later and a value of 15 another 20 seconds later. If the quantity Q is plotted against time t, we get the graph shown in Figure 718.

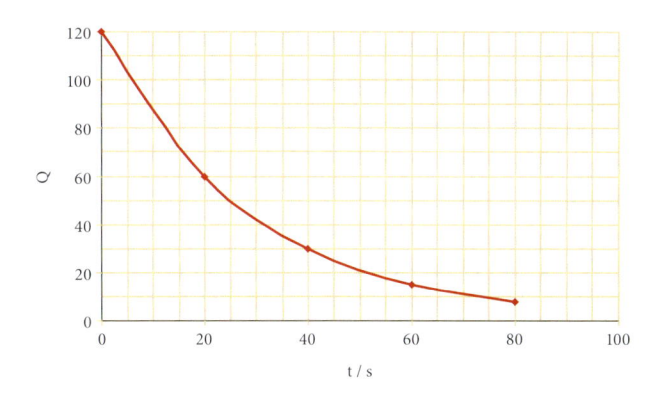

Figure 718   Exponential decay

This type of decay is called an exponential decay. The time it takes for the quantity to reach half its initial value is called the half-life. Clearly the half-life is independent of the initial value of the quantity and depends only on the physical nature to which the quantity refers. For instance, in the case of water flowing from a can, the half-life will depend on the size of the hole in the can and we might expect it to depend on the temperature of the water and the amount and type of impurities in the water. For radioactive elements, the half-life depends only on the particular element and nothing else.

A decay curve can be either a plot of the number of parent nuclei in a given sample over time, or the rate of decay in decays per second over time as shown in Figre 719. Both plots yield the same types of exponential curve, and both can be used to determine the half-life of the isotope whose decay is being plotted. On the first type of curve, **the amount of time it takes for the initial number of parent nuclei to decline by half** is the half-life. On the second plot, **the amount of time it takes for the activity of the sample to drop decline by half** is the half-life.

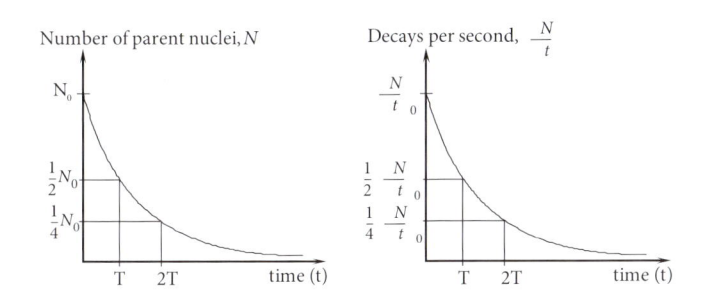

Figure 719   Exponential decay curves

CHAPTER 7

For the case of a radioactive element its half-life is therefore defined as the time it takes the activity of a sample of the element to halve in value or the time it takes for half the atoms in the sample of element to decay.

The number of atoms that decay in unit time i.e. the activity of the sample, has the SI unit the becquerel (Bq).

We also note from the graph, that theoretically it takes an infinite amount of time for the activity of a sample of a radioactive isotope to fall to zero. In this respect, we cannot ask the question "for how long does the activity of a sample last?"

Example

A freshly prepared sample of the isotope iodine-131 has an initial activity of 2.0×10^5 Bq. After 40 days the activity of the sample is 6.3×10^3 Bq.

Estimate the half life of iodine-131.

By plotting a suitable graph, estimate the activity of the sample after 12 days.

Solution

If we keep halving the activity 2.0×10^5, we get 1.0×10^5, 0.5×10^5, 0.25×10^5, 0.125×10^5, $0.0625 \times 10^5 (\approx 6.3 \times 10^3)$. So 5 half-lives = 40 days. Hence 1 half-life = 8 days.

Another way of looking at this is to note that the activity of a sample after n half-lives is

$\dfrac{A_0}{2^n}$ where A_0 is the initial activity. For this situation we have $2^n = \dfrac{2.0 \times 10^5}{6.3 \times 10^3} = 32$, giving

$n = 5$

The data points for a graph showing the variation with time of the activity are shown below:

time/days	activity/Bq × 10⁵
0	2
8	1
16	0.5
24	0.25

So the graph is as follows

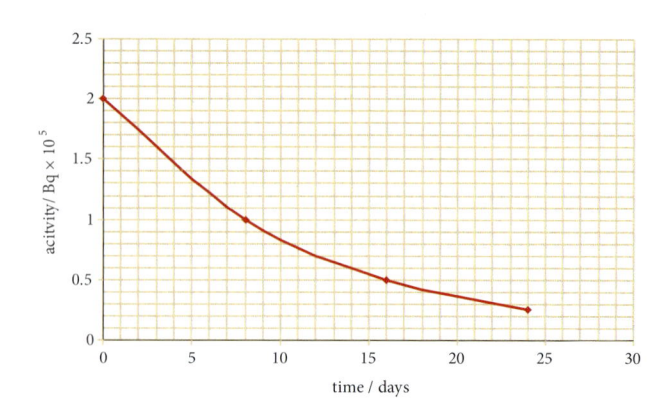

From which we see that after 12 days the activity is 7.0×10^4 Bq.

Exercise 7.2

1. A radioactive substance has a half-life of 4 days, and an initial mass of 10 kg. How much of the original isotope will remain after 12 days?

   A. 10 kg
   B. 5 kg
   C. 2.5 kg
   D. 1.25 kg

2. A sample of phosphorous–90 (half-life = 14 days) has a mass of 7.5 g. A few weeks later, only 1.88 g of the original isotope remains. How many days have gone by?

   A. 14 days
   B. 28 days
   C. 42 days
   D. 56 days

3. The half-life of a radioactive nuclide can be altered by

   A. heating the nuclide.
   B. exerting enormous pressure on the nuclide.
   C. chemical reaction of the nuclide with another element.
   D. none of the above.

4. The strong nuclear force is

   A. a relatively weak force.
   B. a repulsive force.
   C. an attractive force.
   D. a force which acts over great distances.

5. When a nucleus undergoes beta– decay,

   A. a neutrino is always emitted.
   B. an antineutrino is always emitted.
   C. a gamma ray is always emitted.
   D. a helium nucleus is always emitted.

6. When a nucleus undergoes alpha decay,

   A. only the atomic number changes.
   B. only the mass number changes.
   C. both the mass and atomic numbers change.
   D. neither the mass nor atomic number changes as only energy is emitted.

7. Which type of radioactive decay does not alter the mass or atomic numbers of the parent nucleus?

   A. α
   B. β
   C. electron capture
   D. γ

8. In the thorium series, $^{232}_{90}$Th is transformed to $^{228}_{90}$Th

   Which of the following series of decays could be responsible for this transformation?

   A. α, α, β
   B. β, α, γ
   C. α, β, β
   D. β, β, γ

9. A sample of a radioactive isotope has 1/8 its original activity after 48 hours. What is the half-life of the isotope? 16 hours

10. Radon–222 is a radioactive gas with a half-life of 3.8 days. It emits alpha particles when it decays. Explain why this gas is considered a serious hazard when found in home basements.

11. The initial activity of a sample of a radioactive isotope decreases by a factor of after 90 hours. Calculate the half-life of the isotope.

2. The graph below shows the variation with time t of the activity A of a sample of the isotope xenon-114. Use the graph to determine the half-life of xenon-114.

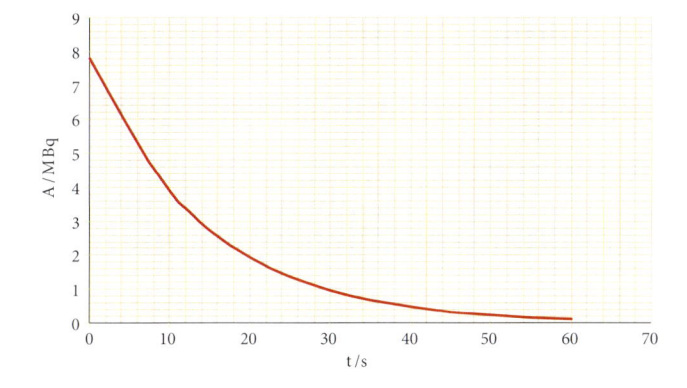

ATOMIC, NUCLEAR AND PARTICLE PHYSICS

7.2 Nuclear reactions

NATURE OF SCIENCE:

Patterns, trends and discrepancies: Graphs of binding energy per nucleon and of neutron number versus proton number reveal unmistakable patterns. This allows scientists to make predictions of isotope characteristics based on these graphs. (3.1)

© IBO 2014

Essential idea: Energy can be released in nuclear decays and reactions as a result of the relationship between mass and energy.

Understandings:

- The unified atomic mass unit
- Mass defect and nuclear binding energy
- Nuclear fission and nuclear fusion

In nuclear physics we are concerned with the interaction of different nuclei and therefore an explicit knowledge of individual isotopic masses is of fundamental importance. For this reason the old scale of atomic masses based on expressing the atomic mass of oxygen as 16, is of little use, oxygen alone having three isotopes. The atomic mass scale was therefore introduced based on the **atomic mass unit.** This unit is defined as $\frac{1}{12}$ the mass of an atom of carbon-12 or, to put it another way, a carbon atom has a mass of exactly 12 u. We know that 12 g of carbon (1 mole) has 6.02×10^{23} nuclei. Therefore 1 u is equivalent to

$\frac{1}{12}$ th of $\frac{12}{6.02 \times 10^{23}}$ = 1.661×10^{-24} g = 1.661×10^{-27} kg.

1 u = 1.66 10^{-27} kg

The following are some important quantities in unified mass units,

mass of an electron, m_e = 0.000549 u

mass of a proton, m_p = 1.007277 u

mass of a neutron, m_n = 1.008665 u

mass of one hydrogen atom, m_H = 1.007825 u

One consequence of *Einstein's Special Theory of Relativity* is that in order for the conservation of momentum to be conserved for observers in relative motion, the observer who considers him or herself to be at rest, will observe that the mass of an object in the moving reference frame increases with the relative speed of the frames. One of the frames of reference might be a moving object. For example, if we measure the mass of an electron moving relative to us, the mass will be measured as being greater than the mass of the electron when it is at rest relative to us, the **rest-mass**. However, an observer sitting on the electron will measure the rest mass. This leads to the famous Einstein equation $E_{tot} = mc^2$ where E_{tot} (often just written as E) is the total energy of the body, m is the measured mass and c is the speed of light in a vacuum. The total energy E_{tot} consists of two parts, 'the rest-mass energy' and the kinetic energy E_K of the object. So in general it is written:

$$E_{tot} = m_0 c^2 + E_K$$

Where m_0 is the rest-mass of the object.

Nuclear reactions are often only concerned with the rest mass.

Essentially, the Einstein equation tells us that energy and mass are interchangeable. If for example, it were possible to convert 1 kg of matter completely to energy we would get 9×10^{16} J of energy ($1 \times c^2$). Looking at it another way, if a coal-fired power station produces say 9 TJ of energy a day, then if you were to measure the mass of the coal used per day and then measure the mass of all the ash and fumes produced per day, you would find that the two masses would differ by about 0.1 g ($\frac{9 \times 10^{12}}{c^2}$).

We have seen that it is usual to express particle energies in electron volts rather than joules. Using the Einstein relation, we can express particle mass in derived energy units.

For example, the atomic mass unit = 1.661×10^{-27} kg

which is equivalent to $1.661 \times 10^{-27} \times (2.998 \times 10^8)^2$ J

or $\frac{1.661 \times 10^{-27} \times (2.998 \times 10^8)^2}{1.602 \times 10^{-19}}$ eV = 931.5 MeV

However, bearing in mind that energy and mass cannot have the same units and that $m = \frac{E}{c^2}$, we have that 1 atomic mass unit = 931.5 MeV c^{-2}.

Figure 720 shows the different units used for the rest mass of the electron, proton and neutron.

(All values, apart from values in atomic mass unit, are quoted to four significant digits).

particle	Rest-mass		
	kg	u	MeV c^{-2}
electron (m_e)	9.109×10^{-31}	0.00054858	0.5110
proton (m_p)	1.673×10^{-27}	1.007276	938.2
neutron (m_n)	1.675×10^{-27}	1.008665	939.6

Figure 720   Particle rest-mass (table)

Mass defect and nuclear binding energy

Mass defect

Let us now examine a nuclear reaction using the idea of mass-energy conversion. For example, consider the decay of a nucleus of radium-226 into a nucleus of radon-222.

The reaction equation is

$$^{226}_{88}\text{Ra} \rightarrow {}^{222}_{86}\text{Rn} + {}^{4}_{2}\text{He}$$

The rest masses of the nuclei are as follows

$^{226}_{88}\text{Ra}$ = 226.0254 u

$^{222}_{86}\text{Rn}$ = 222.0175 u

$^{4}_{2}\text{He}$ = 4.0026 u

The right-hand side of the reaction equation differs in mass from the left-hand side by +0.0053 u. This mass deficiency, or **mass defect** as it is usually referred to, just as in the coal-fired power station mentioned above, represents the energy released in the reaction. If ignoring any recoil energy of the radon nucleus, then this energy is the kinetic energy of the α-particle emitted in the decay.

Using the conversion of units, we see that 0.0053 u has a mass of 4.956 MeV c-2. This means that the kinetic energy of the α-particle is 4.956 MeV.

The fact that the mass defect is positive indicates that energy is released in the reaction and the reaction will take place spontaneously.

If the mass defect is negative in a reaction then this means that energy must be supplied for the reaction to take place. For example, let us postulate the following reaction:

$$^{23}_{11}\text{Na} \rightarrow {}^{19}_{9}\text{F} + {}^{4}_{2}\text{He} + Q$$

where Q is the mass defect.

222

ATOMIC, NUCLEAR AND PARTICLE PHYSICS

If factoring in the rest masses the equation becomes

22.9897 u → 18.9984 u + 4.0026 u + Q

This gives Q = - 0.0113 u = -10.4 MeV c⁻².

In other words for such a reaction to take place 10.4 MeV of energy must be supplied. $^{23}_{11}Na$ is therefore not radioactive but a stable nuclide.

Binding energy

A very important quantity associated with a nuclear reaction is the nuclear binding energy. To understand this concept, suppose we add up the individual masses of the individual nucleons that comprise the helium nucleus, then we find that this sum does not equal the mass of the nucleus as a whole. This is shown below

$2m_p + 2 m_n → {}^{4}_{2}He + Q$

$(2 \times 938.2 + 2 \times 939.6)$ MeV c⁻² → 3728 MeV c⁻² + Q

To give Q = 28.00 MeV c⁻².

This effectively means that when a helium nucleus is assembled from nucleons, 28 MeV of energy is released. Or looking at it another way, 28 MeV of energy is required to separate the nucleus into its individual nucleons since if we postulate, as we did above for the decay of $^{23}_{11}Na$, this reaction

${}^{4}_{2}He → 2m_p + 2 m_n + Q$

then Q = - 28 MeV c⁻².

The definition of **nuclear binding energy** is therefore either **the energy required to separate the nucleus into it individual nucleons** or **the energy that would be released in assembling a nucleus from its individual nucleons**. Since the potential energy of a nucleus is *less* than the potential energy of its separate nucleons, some texts take the binding energy to be a **negative** quantity. In this book, however, we will regard it to be a **positive** quantity on the basis that the greater the energy required to separate a nucleus into its nucleons, the greater the difference between the potential energy of the nucleus and its individual nucleons.

Binding energy per nucleon

Rather than just referring to the binding energy of the nuclei of different isotopes, it is much more important to consider the binding energy per nucleon. The addition of each nucleon to a nucleus increases the total binding energy of the nucleus by about 8 MeV. However, the increase is not linear and if we plot the binding energy per nucleon against nucleon number N, the graph (Figure 721) shows some very interesting features.

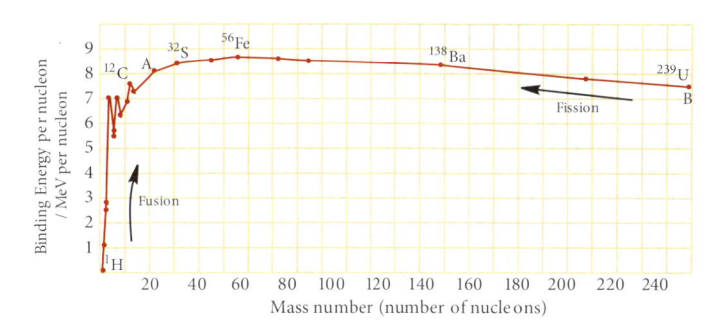

Figure 721 Binding energy per nucleon

The most stable nuclei are those with the greatest binding energy per nucleon as this means that more energy is required to separate the nucleus into its constituent nucleons. For example, much less energy is required per nucleon to "take apart" a nucleus of uranium-235 than a nucleus of helium-4. The most stable element is iron (Fe) as this has the greatest binding energy per nucleon.

Example

1. Calculate the kinetic energy in MeV of the tritium plus the helium nucleus in the following nuclear reaction.

    $${}^{6}_{3}Li + {}^{1}_{0}n → {}^{3}_{1}H + {}^{4}_{2}He$$

    mass of ${}^{6}_{3}Li$ = 6.015126 u

    mass of ${}^{3}_{1}H$ = 3.016030 u

    mass of ${}^{4}_{2}He$ = 4.002604 u

    neutron mass = 1.008665 u

Solution

Adding the masses of the left-hand and right-hand side of reaction equation gives:

Q = 0.005157 u. Using 1 u = 935.1 MeV, then Q = 4.822 MeV

223

Nuclear fission and nuclear fusion

Fission

Nuclear reactions produce very much more energy per particle than do chemical reactions. For example, the oxidization of one carbon atom produces about 4 eV of energy whereas the decay of a uranium atom produces about 4 MeV. However, natural radioactive isotopes do not occur in sufficient quantity to be a practical source of energy. It was not until the discovery of nuclear fission that the possibility of nuclear reactions as a cheap and abundant source of energy became possible. In 1934 Fermi discovered that when uranium was bombarded with neutrons, radioactive products were produced.

Recall that nuclear reactions are reactions that result in changes to the nuclei and occur whenever a parent nuclide gains, loses or changes one of its nucleons (a nuclear particle such as a proton or neutron). The products of the **transmutation** are daughter nuclides and elementary particles. One way to cause a nuclear transmutation is using a nuclear fission reaction. In this transmutation, a stable parent nuclide is bombarded with **slow or thermal** (with energy ≈ 1 eV or speeds of 1.2×10^3 ms^{-1} or 1.2 km s^{-1}) neutrons in what is called **neutron activation**.

Figure 722 shows a schematic representation of the nuclear fission of U–235. An intermediate unstable nucleus U–236 that fissions rapidly then produces two daughter nuclides (M_1 a smaller- and M_2 a larger fragment), fast neutrons with speeds near 2×10^7 m s^{-1} (20 000 km s^{-1}), elementary particles such as neutrinos (ν) and β–particles, and γ–radiation.

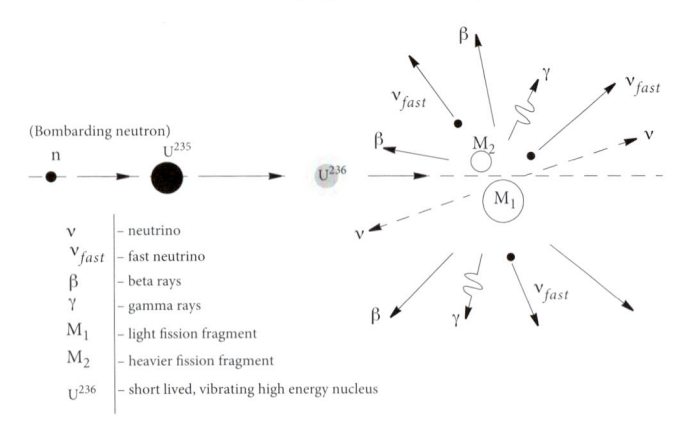

Figure 722   Nuclear fission of U–235

The process of nuclear fission and the manner of how a neutron fired at a nuclide can produce smaller fragments is explained using the **liquid drop model** as shown in Figure 723.

An analogy to help explain this could be a drop of water when it breaks up as it falls from a dripping tap. The water droplet can be seen to be stretching and wobbling before it breaks apart into smaller fragments. The counter forces of surface tension and intermolecular forces between the water molecules are holding the water droplet intact before it breaks up.

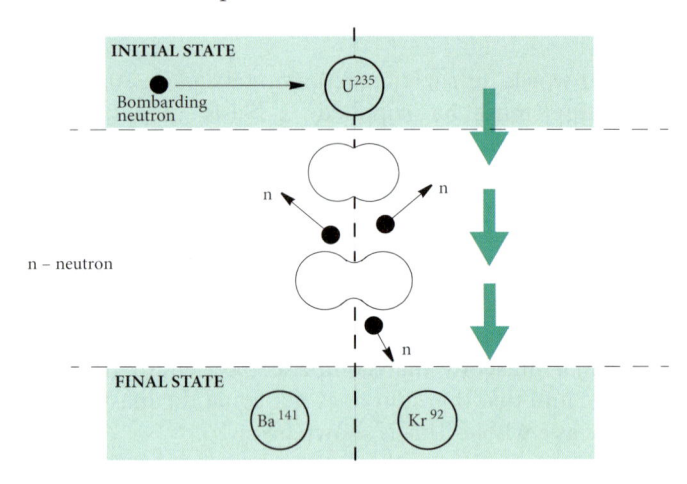

Figure 723   The liquid drop model

In a similar manner, the balance of electrostatic forces between protons and the short- range strong nuclear force holds the nucleus together. If a fast neutron was fired at the nucleus, it would pass straight through the nucleus leaving it unchanged. However, with a slow neutron absorbed, the unstable isotope U–236 is momentarily formed causing the nucleus to deform due to the balance of forces being disturbed. The nucleus oscillates and the electrostatic repulsive forces between the protons dominate causing the nucleus to break up into two smaller fragments. The long-ranged coulombic forces of repulsion between the protons now dominate over the short-ranged nuclear forces of attraction.

In any nuclear reaction, certain Conservation Laws govern the reaction:

1. momentum is conserved (relativistically)

2. total charge is conserved (charge of products = charge of reactants)

3. the number of nucleons remain constant

4. mass-energy is conserved ($\Delta E = mc^2$)

Since the parent nuclide is stationary, the kinetic energy before the reaction is due to the kinetic energy of the fired neutron. However, the kinetic energy after the reaction is greater than the kinetic energy of the initial neutron – the mass defect Δm has been changed into energy ΔE.

The smallest possible amount of fissionable material that will sustain a chain reaction is called the **critical mass**. The critical mass is determined when one of the neutrons released by fission will cause another fission. Apart from the amount of fissionable material, the shape of the material is also important. The preferred fuel for thermal reactors is solid uranium pellets contained in cylindrical fuel rods made of a zirconium alloy. This alloy is capable of withstanding high temperatures without distorting and becoming stuck in the fuel rod cavities. Not all the neutrons produced will leak from the reactor core and this is taken into consideration when determining the critical mass. Typically in a small thermal reactor, the critical mass (contained as pellets in the fuel rods) is a few kilograms in mass. There are about 150 fuel assemblies (see Figure 724) containing about 60 fuel rods each placed into the core of the reactor. In the *Chernobyl* nuclear accident of 1986, it was determined that the ziconium alloy cladding of the fuel rods overheated and led to the release of the fission products causing a meltdown in the reactor core.

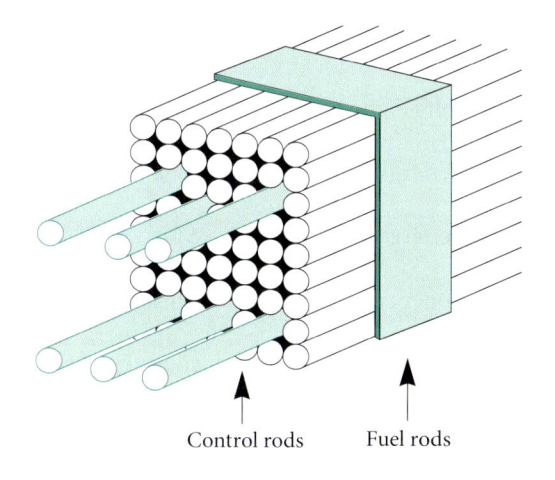

Figure 724   Fuel rod assembly

In 1938, two German scientists, *Otto Hahn* and *Fritz Strassmann* following on from the works of the Italian scientist, *Enrico Fermi*, found that when uranium-235 was bombarded with neutrons, two lighter fragment daughter nuclides and other rare Earth elements were formed. Two of the many possible nuclear transmutation reactions are

$$^{235}_{92}U + ^{1}_{0}n \rightarrow ^{148}_{57}La + ^{85}_{35}Br + 3^{1}_{0}n$$

$$^{235}_{92}U + ^{1}_{0}n \rightarrow ^{139}_{56}Ba + ^{95}_{36}Br + 2^{1}_{0}n$$

In other words, not every nucleus of U–235 produces the same two daughter nuclides although some daughter fragments are more probable than others. In general we can write:

$$^{235}_{92}U + ^{1}_{0}n \rightarrow A + B + x \cdot ^{1}_{0}n$$

where x is a positive integer.

Fission is a process where a large parent nucleus produces two smaller daughter nuclides that have a higher binding energy per nucleon.

The implications of this research would have far-reaching consequences in the history of our civilisation. In 1939, *Albert Einstein* wrote to the then US president *Franklin Roosevelt* discussing the military applications of fission research. Within a short time, The Manhattan Project, the precursor for the design of the bombs that destroyed *Hiroshima* and *Nagasaki* in Japan, had begun. Today, the threat of military use of nuclear fission is still real and the United Nations is concerned that some countries may have nuclear weapon ambition. However, the fission process has valid uses in medical research and power generation.

When fission takes place, additional neutrons (2 or 3) are released which leads to the possibility of producing a self-sustaining chain reaction as demonstrated in Figure 725.

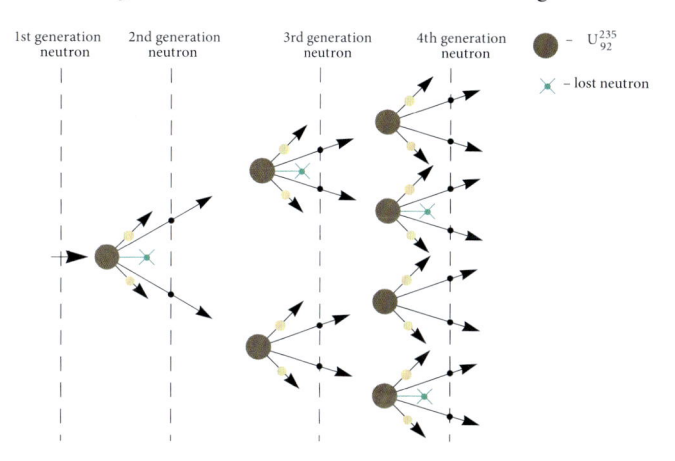

Figure 725   The principle of a chain reaction

If there are too few neutrons the nuclear chain reaction will cease and the reactor will shut down. If there are too many neutrons a runaway reaction will cause an explosion. The reaction will become uncontrolled as in a typical atomic bomb where enough fissionable material is in a sufficiently small space. In a nuclear reactor, if a chain reaction were to occur, the large amounts of energy would cause the fuel to melt and set fire to the reactor in what is called a meltdown.

Not all uranium atoms will undergo fission in a controlled manner. Natural uranium obtained from oxides of uranium ores such as pitchblende consists of 99.3% of the uranium–238 isotope, 0.7% of the uranium–235 isotope and 0.006% uranium–234.

U–238 and U–234 are not fissionable with thermal neutrons. Only U–235 is readily fissionable with thermal neutrons (neutrons that have energy comparable to gas particles at normal temperatures). Most commercial reactors are **thermal reactors**.

Most thermal reactor fuel is a mixture of fissionable and fertile material but it is enriched with U–235 to increase the probability of fission occurring. The common enrichment procedure involves the formation of uranium hexafluoride gas followed by the separation of the U–235 and U–238 isotopes in a gas centrifuge.

Which reaction takes place is dependant on the energy of the bombarding neutron.

A typical fission reaction might be

$$^{238}_{92}U + ^{1}_{0}n \rightarrow ^{90}_{38}Sr + ^{146}_{54}Xe + 3^{1}_{0}n$$

Given the following data, it is left as an exercise for you to show that energy released in this reaction is about 160 MeV:

mass of ^{238}U = 238.050788 u

mass of ^{90}Sr = 89.907737 u

mass of ^{146}Xe = 145.947750 u

The energy released appears in the form of kinetic energy of the fission nuclei and neutrons.

The three neutrons produced is the key to using fission as a sustainable energy source. Both the strontium isotope and xenon isotope produced are radioactive. Strontium-90 has a half-life of about 30 years and therein lies the main problem (as well as the large amounts of γ-radiation also produced) with nuclear fission as a sustainable energy source – the fact that the fission nuclei are radioactive often with relatively long half-lives.

$$^{235}_{92}U + ^{1}_{0}n \rightarrow ^{103}_{38}Sr + ^{131}_{54}I + 2^{1}_{0}n$$

Fusion

Energy can also be obtained from nuclear reactions by arranging for two nuclei to "fuse" together as we alluded to when we discussed nuclear binding energy above. To produce **nuclear fusion** very high temperatures and pressures are needed so that nuclei can overcome the coulombic repulsion force between them and thereby come under the influence of the strong nuclear force. A typical nuclear reaction might be

$$^{2}_{1}H + ^{3}_{1}H \rightarrow ^{4}_{2}He + ^{1}_{0}n$$

In this reaction a nucleus of deuterium combines with a nucleus of tritium to form a nucleus of helium and a free neutron.

Given the following data it is left as an exercise for you to show that energy released in this reaction is about 18 MeV:

mass of ^{2}H = 2.014102 u

mass of ^{3}H = 3.016049 u

mass of ^{4}He = 4.002604 u

The energy released appears in the form of kinetic energy of the helium nucleus and neutron.

The advantage that fusion has compared to fission as a source of sustainable energy is that no radioactive elements are produced. The disadvantage is obtaining and maintaining the high temperature and pressure needed to initiate fusion.

The graph in Figure 721 of binding energy per nucleon versus nucleon number shows that the nuclides with a nucleon number of about 60 are the most stable. This helps us to understand why the high nucleon number nuclides may undergo fission and the low nucleon number nuclides may under go fusion- they are trying to "reach" the nuclide that is most stable. For example consider the fission reaction

$$^{238}_{92}U + ^{1}_{0}n \rightarrow ^{90}_{38}Sr + ^{146}_{54}Xe + 3^{1}_{0}n$$

From the graph (Figure 711) we have that

total binding energy of ^{238}U = 7.6 × 238 = 1800 MeV

total binding energy of ^{90}Sr = 8.7 × 90 = 780 MeV

total binding energy of ^{146}Xe = 8.2 × 146 = 1200 MeV

Hence the sum of the total binding energies of the fission nuclei is greater than the total binding energy of the uranium-238 nucleus. Effectively the system has become more stable by losing energy.

Similarly for the fusion reaction $^{2}_{1}H + ^{3}_{1}H \rightarrow ^{4}_{2}He + ^{1}_{0}n$ the total binding energy of the helium nucleus is greater than the sum of binding energies of the tritium and deuterium nuclei. So, again as for fission, the system has effectively become more stable by losing energy.

Our Sun is an enormous factory in which hydrogen is converted into helium. At some time in its early life, due to gravitational collapse of the hydrogen making up the Sun, the pressure and temperature of the interior became high enough to initiate fusion of the hydrogen. Once started, the fusion will continue until all the hydrogen is used up, probably in about 10^{10} years from now.

One of the suggested fusion cycles that may take place in the Sun is

$$p^{+} + p^{+} \rightarrow ^{2}_{1}H + e^{+}$$

$$^{2}_{1}H + p^{+} \rightarrow ^{3}_{2}He$$

$$^{3}_{2}He + ^{3}_{2}He \rightarrow ^{4}_{2}He + 2p^{+}$$

For the complete cycle the first two reaction must occur twice and the final result is one helium nucleus, two positrons, two protons and two neutrinos. The protons are available for further fusion.

In stars that are much more massive than our Sun, as they age fusion of elements with higher atomic numbers takes place until finally iron is reached and no further fusion can take place as seen from the binding energy graph. This evolution of stars is discussed in detail in Option E.

The type of problems you might be expected to solve associated with fission and fusion have been looked at above. However, here are a two more.

Exercise 7.3

1. The mass of a nucleus is

    A. always greater than the mass of its constituent particles.
    B. always less than the mass of its constituent particles.
    C. always equal to the mass of its constituent particles.
    D. can be any of the above depending on how stable the nucleus is.

2. The mass defect refers to

    A. the difference between the mass of a nucleus and the mass of its constituent particles.
    B. the difference between the mass of the neutrons and the mass of the protons in a nucleus.
    C. the difference between the mass of a nucleus and the mass of the entire atom.
    D. the mass gained when an atom undergoes spontaneous radioactive decay.

3. Given the mass of $^{15}_{7}N$

    as 15.000108 u, the binding energy for nitrogen–15 is

    A. 0.123987 u B. 115 MeV
    C. 121 MeV   D. 6.25 MeV

4. Calculate the energy required to separate a nucleus of lithium-6 into its constituent nucleons. Hence find the binding energy per nucleon of lithium-6.

5. Calculate the binding energy per nucleon of an α-particle

6. Deduce, whether the following reaction may take place spontaneously.

    $^{212}_{83}Bi \rightarrow {}^{208}_{81}Tl + \alpha$

    mass of $^{212}_{83}Bi$ = 211.99127 u

    mass of $^{208}_{81}Tl$ = 207.98201 u

7. Determine the number x of neutrons produced and calculate the energy released in the following fission reaction

    $^{235}_{92}U + {}^{1}_{0}n \rightarrow {}^{144}_{56}Ba + {}^{90}_{36}Kr + x{}^{1}_{0}n$

    mass of ^{235}U = 235.043929 u

    mass of ^{146}Ba = 143.922952 u

    mass of ^{90}Kr = 89.919516 u x = 2,

8. Show that in the fusion cycle given in 7.3.10, the energy released is about 30 MeV.

9. (a) How many nucleons are there in lithium-6, carbon-12 and iron-56?

    (b) The mass defects of lithium-6 is 0.4431 u, carbon-12 is 0.9888 u and iron-56 is 0.52819 u. Calculate the binding energy and binding energy per nucleon for each atom.

    (c) Which of the isotopes would be the most stable?

CHAPTER 7

TOK THE STUDY OF PARTICLES

The study of particles has been fundamental to our understanding of the composition and the behaviour of matter and as such particle physics can be called the foundation stone on which all other branches of science are based. The exploration of the atom through the findings of *Henri Becquerel, Marie Curie, James Clerk Maxwell, Heinrick Geissler, Julius Plücker, William Crookes, Heinrick Hertz, J.J. Thomson, Robert Millikan, Ernest Rutherford, Hans Geiger, Ernest Marsden, Niels Bohr, James Chadwick, Max Planck, C.T.R. Wilson, Albert Einstein, Louis de Broglie, Erwin Schrödinger, Werner Heisenberg, Wilhelm Röntgen, Enrico Fermi, Paul Dirac, Wolfgang Pauli, Hideki Yukawa, Richard Feynmann, Murray Gell-Mann, Carlo Rubbia* and *Stephen Hawking* to mention but a few, have revolutionised the world over the last 120 years. Their research has led to the development of nuclear power stations (and unfortunately nuclear bombs), radioisotopes and medical imaging techniques, the chemical and petrochemical industries, lasers, electronics and computers, and, has revitalised the sciences of astrophysics and cosmology.

So when and how were atoms created? There is strong evidence to suggest that the elementary particles that make up atoms were created within the first seconds of the Big Bang – when time and space began from a "singularity". Atoms are the very complex end-products of this event as the elementary particles combined over the following few thousands of years. In order to understand the origin of matter, it is necessary to try and re-create the intense heat conditions of the **Big Bang** in the laboratory. Particle accelerators and their detectors try to imitate the original conditions in order to find these elementary particles and to develop a model of the nature of matter and energy. This is the branch of Physics that has become known as particle physics.

Particle physics had its beginnings in the 1920s when the nature of high-energy particles from outer space known as **cosmic rays** were studied at high altitudes. It was found that more particles existed other than protons, neutrons and electrons. The leptons – positron and muon, the baryons – Lamda, Sigma plus and the Xi minus, and the mesons – Kaon zero, Kaon plus and Kaon minus were discovered in cosmic radiation using cloud chamber and emulsion detectors.

Further particles were discovered in the 1950s using cloud chambers, bubble chambers, scintillation counters as detectors. More particles were discovered in nuclear reactions. It was soon realised that if more particles were to be found, then reactant particles would need to accelerated to high speeds in order to produce high-energy product particles of large mass or to resolve product particles of small size. By the beginning of the 1930s, the van der Graaf accelerator (named after *Robert van der Graaf*) was developed to accelerate positive charges with very high potential differences to give the charges kinetic energy up to 30 MeV.

In Figure 726, a voltage of 50 kV is applied to a pointed conductor at the bottom so that electrons are pulled off the moving belt insulator and the positive charges produced are moved to the top of the belt where they are transferred to the dome. Since there is no electric field inside a hollow conductor, the charges move to the outside of the dome because electrons are taken to earth by the belt. The conducting dome is hollow so as to allow a large charge build-up on its outside. The van der Graaf charge generator is connected to an evacuated accelerator tube containing hydrogen or helium ions. These are repelled by the high positive voltage and are accelerated to an earthed target.

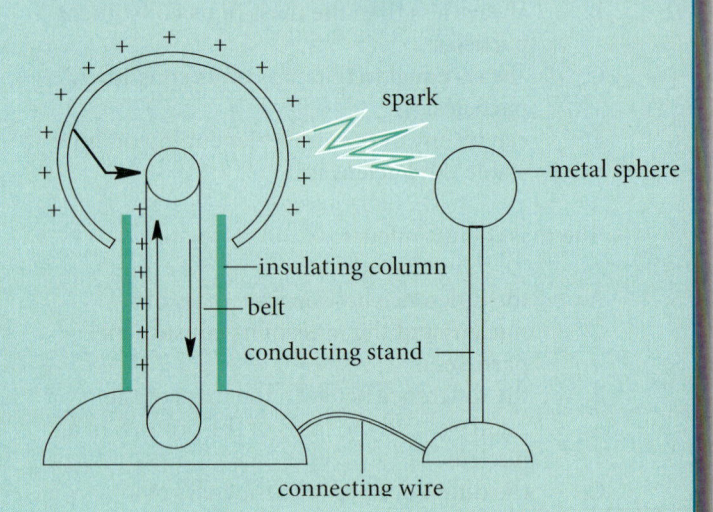

Figure 726 A van de Graaf accelerator

When these ions are crashed into a target material, new particles of different masses and sizes can be produced. There are currently several hundred composite and elementary particles with corresponding antiparticles that have been identified.

In 1930, *Ernest Lawrence* developed a small cyclotron, and between the 1980s and 1990s, we saw the developments of the Stanford Linear Accelerator Centre (SLAC, California) electron-positron linear collider (3 km long), the Geneva CERN (Conseil Européen pour la Recherche Nucléaire) large electron-positron circular collider (27 km circumference) and the Fermilab tevatron (6.4 km circumference near Chicago) which collides protons and anti-protons inside electronic detectors. CERN presently has 2 accelerators that are being continually upgraded. These accelerators and others in Germany, Japan, Russia and China will be studied fully in Section J2.

Particle phycisists have discovered hundreds of new varieties of particles with many weird names such as

ATOMIC, NUCLEAR AND PARTICLE PHYSICS

quarks, neutrinos, gauge bosons and muons with each having its own colour charge. Particle physicists talk about the "flavor" of some elementary particles. The up, down and strange quarks were originally called vanilla, chocolate and strawberry and thus the term flavor has stuck. *Murray Gell-Mann* was quite a comedian and when he came up with the term "quark" to describe a group of elementary particles, scientists asked him for the origins of the word. During various lectures he was known to call them quirks, quorks but the word quark stuck and it has its origins in James Joyce's novel "Finnegan's Wake" - …. "Three quarks for Muster Mark". Such is the menagerie of terms that some have coined the term "sub-atomic zoo" to describe the variety of particles.

As has been mentioned a number of times already in this textbook, there are 4 fundamental forces: the strong force, the electromagnetic force, the weak force and gravity, and all particles are governed by these forces. In the past 50 years, particle theorists have organised what has been found by particle experimenters into a theory that may explain a standard model of elementary particles and unite the forces into a Grand Unified Theory (GUT). Maxwell was able to show in the 1860s that the electric force and the magnetic force could be unified into a single electromagnetic force and his theory has been further refined into quantum electrodynamics (QED). In the 1960s, electromagnetism and the weak force were unified into the **electroweak** theory which predicted the existence of the Z boson particle as one of the exchange particles that mediate the weak force. Particle experimenters found the Z boson and particle theorists were found to be correct in their prediction. In 1973, the discovery of asymptotic freedom established **quantum chromodynamics** QCD as the correct theory to explain the nature of the strong force. The calculations done by *Gross*, *Wilczek* and *Politzer* showed that quarks were held together very strongly at distances that are comparable to the size of a proton, and this explained the concept of quark confinement. Now the search is on to find the link between the strong force, the electromagnetic, the weak force and gravity.

These major breakthroughs in particle physics have spurred on the developments in astrophysics and cosmology that hope to answer the questions concerning the origin and evolution of the universe - why does it have its shape and form and will it reach a point where it will stop expanding. Perhaps one of the best outcomes of particle physics is that it has brought scientists from many nationalities into a collaborative working environment. Accelerators are very expensive to construct and operate, and the CERN accelerator that passes underground into Switzerland and France is funded by 19 European countries and employs over 100 research physicists, over 800 applied physicists and engineers, over 1000 technicians and 1000+ office and administration staff and craftsmen.

7.3 The structure of matter

NATURE OF SCIENCE:

Predictions: Our present understanding of matter is called the standard model, consisting of six quarks and six leptons. Quarks were postulated on a completely mathematical basis in order to explain patterns in properties of particles. (1.9)

Collaboration: It was much later that large scale collaborative experimentation led to the discovery of the predicted fundamental particles. (4.3)

© IBO 2014

Essential idea: It is believed that all the matter around us is made up of fundamental particles called quarks and leptons. It is known that matter has a hierarchical structure with quarks making up nucleons, nucleons making up nuclei, nuclei and electrons making up atoms and atoms making up molecules. In this hierarchical structure, the smallest scale is seen for quarks and leptons (10^{-18} m).

Understandings:
- Quarks, leptons and their antiparticles
- Hadrons, baryons and mesons
- The conservation laws of charge, baryon number, lepton number and strangeness
- The nature and range of the strong nuclear force, weak nuclear force and electromagnetic force
- Exchange particles
- Feynman diagrams
- Confinement
- The Higgs boson

Rutherford and the Geiger Marsden experiment

The work of *J. J. Thomson* had indicated that the atom consists of a mixture of heavy positive particles, protons, and light negative particles, electrons. *Rutherford* devised an experiment to find out how the particles of the atom might be 'mixed' and the experiment was carried out by *Hans Geiger* and *Ernest Marsden*. Certain elements are unstable and disintegrate spontaneously. Such an element is radium and one of the products of the so-called **radioactive decay** of a radium atom, is a helium nucleus. Helium nuclei emitted as the result of radioactive decay are called alpha particles (α). When α-particles strike a fluorescent screen, a pinpoint flash of light is seen.

CHAPTER 7

Rutherford studied how α-particles were absorbed by matter and found that thin sheets of metal readily absorbed them. However, he found that they were able to penetrate gold foil which, due to the malleable nature of gold, can be made very thin.

Figure 727 illustrates the principle of the experiment carried out by *Geiger* and *Marsden*.

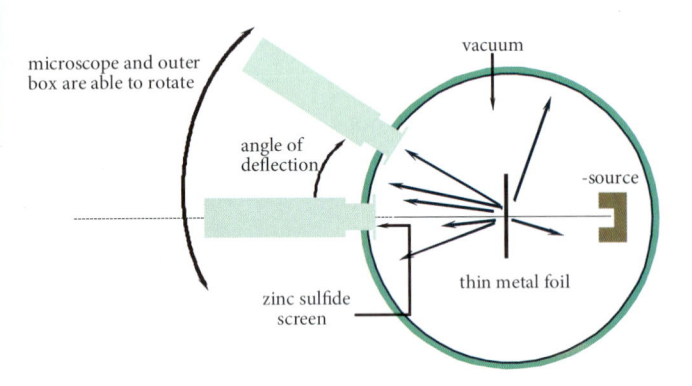

Figure 727   Geiger and Marsden's experiment

A piece of radium is placed in a lead casket such that a narrow beam of particles emerge from the tunnel in the casket The particle beam is incident on a piece of gold-foil behind which is placed a fluorescent screen. The whole apparatus is sealed in a vacuum.

The result is rather surprising. Most of the α-**particles** go straight through, but a few are scattered through quite large angle whilst some are even turned back on themselves. *Rutherford* was led to the conclusion that most of the gold foil was empty space. However, to account for the scattering, some of the particles must encounter a relative massive object that deflects them from their path. Imagine firing a stream of bullets at a bale of hay in which is embedded a few stones. Most of the bullets go straight through but, the ones which strike a stone, will ricochet at varying angles depending on how they hit the stone. *Rutherford* suggested that the atom consisted of a positively charged centre (the stones) about which there was a mist of electrons (the straw). He neglected the interaction of the α-particles with the electrons because of the latter's tiny mass and diffuse distribution. The significant reaction was between the massive positively charged centre of the gold atoms and the incoming α-particles. *Rutherford* was in fact quoted as saying "It was quite the most incredible event that has ever happened to me in my life. It was almost as incredible as if you had fired a 15–inch shell at a piece of tissue paper and it came back and hit you."

Figure 728 illustrates the paths that might be followed as the α-particles pass through the gold foil.

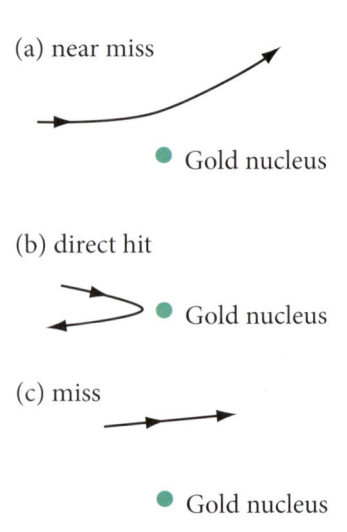

Figure 728   The paths of alpha particles through gold foil

It is the results of the Geiger-Marsden experiment that led to the atomic model outlined in 7.1.1 above. Detailed analysis of their results indicated that the gold atoms had a nuclear diameter of the order of 10^{-14} m, which meant that the radius of a proton is the order of 10^{-15} m. Work contemporary to that of *Geiger* and *Marsden*, using X-rays, had shown the atomic radius to be of the order of 10^{-10} m. To give some understanding to the meaning of the expression 'an atom is mainly empty space', consider the nucleus of the hydrogen atom to be the size of a tennis ball, then the radial orbit of the electron would be about 2 km.

The atomic **planetary model** proposed by *Rutherford* has certain problems associated with it. In particular, Maxwell's theory of electromagnetic (em) radiation demonstrated that accelerated electric charges emit em radiation. It is difficult therefore to see how the Rutherford atom could be stable since the orbiting electrons would be losing their energy by emitting radiation and as such, would soon spiral into the nucleus. As the orbits collapse, the acceleration of the electrons should change continuously. Therefore, the frequency of the emitted em-radiation should also change continuously. Hence, any radiation given out in this process should constitute a continuous spectrum in which all wavelengths are present. However, it is obvious that this is not the case.

Furthermore, how could the nucleus be stable; what was there to stop the protons flying apart from each other due to Coulomb repulsion?

The model was also unable to account for that the fact that many elements exhibited a range of atomic masses. For example, J.J. Thomson had shown that neon is made up of atoms some of whose atomic mass is 20 and others whose atomic mass is 22.

Rutherford recognised that his new model had these

ATOMIC, NUCLEAR AND PARTICLE PHYSICS

limitations and he actually made some predictions about an improved model of the atom, even hypothesising the existence of the neutron. However, it was *Niels Bohr* who in 1914 proposed a model that went some way to answering the electron radiation conundrum.

Quarks, leptons and their antiparticles

Particles are called **elementary particles** if they have no internal structure, that is, they are not made out of any smaller constituents. The elementary particles are the leptons and their antiparticles, quarks and their antiparticles, and exchange particles (gauge bosons).

1. Leptons

Leptons are elementary particles that are influenced by the weak force. Leptons are particles that can travel on their own, meaning that they are not trapped inside larger particles. Six distinct types called 'flavors' have been identified along with their antiparticles.

First generation ordinary matter included in this category are the electron with a size of less than 10^{-18} m, with its antiparticle the positron, and the neutrino, with its antiparticle the antineutrino. Electrons have a negative charge, while positrons have a positive charge. Neutrinos are neutral in charge. Leptons interact via the weak nuclear force, but not the strong, as well as the gravitational force, and where a lepton is charged, the electromagnetic force.

There are other leptons which are believed to have existed in the early moments of the Big Bang and may be found in cosmic rays and particle accelerators. They are the **second generation** muon and the muon neutrino and their antiparticles; that are heavier than the electron, and the **third generation** tau and tau neutrino (found at the Stanford collider but not found in nature) and their respective antiparticles, that are even heavier still than the muon.

Leptons have an electric charge of -1, +1 or 0 as can be seen in Figure 729. The top Roman numerals give the generation of each lepton and the bottom number in the table gives the rest mass of each lepton relative to the rest mass of a proton mp.

In spite of the different masses of each flavor of leptons, they have identical spin of a ½ and the same angular momentum.

The muon is unstable and decays into an electron, an electron antineutrino and a muon neutrino on average every 2.2 μs through the weak force interaction. The tau lepton decays on average every 3×10^{-13} s.

Leptons		
I	II	III
electron -1 e^- 0.005	muon -1 μ^- 0.1	tau -1 τ^- 1.9
electron neutrino 0 ν_e ~ 0	muon neutrino 0 ν_μ ~ 0	tau neutrino 0 ν_τ < 0
Antileptons		
I	II	III
positron +1 e^+ 0.005	antimuon+1 $\bar{u}$ 0.1	antitau +1 $\bar{\tau}-$ 1.9
electron antineutrino 0 $\bar{\nu}_{e+}$ ~ 0	muon antineutrino 0 $\bar{\nu}_\mu$ ~ 0	tau antineutrino 0 $\bar{\nu}_\tau$ < 0

Figure 729   Standard model for the leptons

Leptons are found in many environments. The electron is the charge-carrier in conductors and semi-conductors. Electron antineutrinos are found in the beta decay of a neutron into a proton. The remaining flavors are found in nuclear reactors, particle accelerators and cosmic rays.

2. Quarks

Quarks with a size of less than 10^{-18} m can never be found in isolation as they are trapped inside other composite particles called **hadrons** of which the proton, the neutron and mesons are examples.

Figure 730 shows the generation, names, symbols, charges and rest mass (bottom left) relative to the rest mass of the proton of the flavors of the quarks.

Quarks		
I	II	III
UP $u + 2/3$ 1/3	CHARM $c + 2/3$ 1.7	TOP $t +2/3$ 186
DOWN $d - 1/3$ 1/3	STRANGE $s - 1/3$ 0.5	BOTTOM $b - 1/3$ 4.9
Antiquarks		
I	II	III
ANTI UP $\bar{u} - 2/3$ 1/3	ANTI CHARM $\bar{c} - 2/3$ 1.7	ANTI TOP $\bar{t} - 2/3$ 186
ANTI DOWN $\bar{d} + 1/3$ 1/3	ANTI STRANGE $\bar{s} + 1/3$ 0.5	ANTI BOTTOM $\bar{b} + 1/3$ 4.9

Figure 730   Standard model for the quarks

All six 'flavors' of quarks have been identified. Quarks can experience weak interactions that can change them from one flavor to another.

Antiparticles

All particles have **antiparticles** which are identical to the particle in mass and half-integral spin but are opposite in charge to their corresponding particle. In 1928, Paul Dirac, while trying to combine special relativity and electromagnetic theories mathematically, came to the conclusion that particles with the same mass but opposite charge might exist somewhere in the Universe. In 1932, Carl Anderson found that cosmic radiation travelling through a bubble chamber during pair production split up into an electron path and a path with the same mass as an electron but in an opposite direction. Thus the positron had been discovered. In Figure 731, incoming gamma radiation (from the bottom) produces two circular tracks of a positron (to the left) and an electron (to the right), and a secondary electron-positron track.

Figure 731 Electron-positron pair creation

Further evidence came with the discovery of the antineutrino during beta-minus decay. It was not until 1955 that the antiproton was discovered. In particle accelerators, beams of electrons and positrons or protons and antiprotons are accelerated in opposite directions and are then collided with each other in detectors. These collisions produce enormous amounts of energy and the products of the collisions are studied. Today, most antiparticles have been identified in this way. The question still remains as to where has all the antimatter gone that must have been created during the Big Bang.

Although antiparticles have the same mass as their particle pair, they have opposite charge, lepton number, baryon number and strangeness. Some electrically neutral bosons and mesons are their own antiparticle.

Hadrons, baryons and mesons

Again, hadrons are composite of smaller particles (quarks) and are influenced by the strong force. Hadrons are classified as:

(a) Mesons

Mesons are intermediate mass particles that are made up of a quark – antiquark combination. Examples are the Pions π^+, π^- and π^0, the Kaons K^+, K^- and K^0, J/PSI, J/ψ and Eta η^0. Mesons are **bosons** that can mediate the strong nuclear force and this will be discussed further in a later section. Like the first and second generation leptons, mesons only exist for a short time and they are thus very unstable.

(b) Baryons

Baryons are three quark combinations and are called fermions. They are the heavyweights amongst particles that make up matter, including the proton and the neutron.

Other baryons include Lamda Λ^0, Sigma Σ^+, Σ^0 and Σ^-, Cascade Ξ^0 and Ξ^- and Omega Ω^- particles to name but a few.

Hadrons are not elementary particles because they are composed of quarks. Mesons consist of a quark and an antiquark. Baryons have three quarks. A proton has 2 up and 1 down quarks - uud, and the neutron has 2 down and 1 up quarks – ddu. Hadrons interact predominantly via the strong nuclear force, although they can also interact via the other forces.

In 1963, the American physicist Murray Gell-Mann (born in 1929), and the Israeli physicist Yuval Ne'eman independently developed a new classification system for the hundreds of elementary particles known at that time. Gell-Mann called his new theory the eightfold way.

Gell-Mann realised that if baryons and mesons were organised according to their charge, strangeness, and spin (angular momentum), simple patterns emerged. He used a new quantity called hypercharge, which was the sum of the baryon and strangeness numbers, as well as the spin number, to place each particle on a graph. What emerged were hexagon-shaped patterns, with six particles forming the hexagon and two particles at the centre of each hexagon, hence the name eightfold way. Subsequent work lead to a model of triangle shaped patterns consisting of ten particles.

The true test of any theory is its ability to not only explain known phenomena, but also to make predictions of new phenomena. When Gell-Mann's diagrams were initially set up, all of the particles on them were known, except one. That one particle was called omega minus ($\Omega-$). Using his diagrams, Gell-Mann was able to predict the strangeness, isotopic spin, charge, and mass of the new particle, which gave physicists more than enough information to begin searching for the particle. It took only seven months for the existence of the predicted particle to be verified. Murray Gell-Mann received the 1969 Nobel Prize in Physics for his new classification scheme.

Gell-Mann's work did not end with the discovery of the omega minus particle. His classification scheme suggested that the hadrons known in 1964 were not actually fundamental particles, but that these baryons and mesons were actually composed of even more fundamental particles.

The new fundamental particles were given the whimsical name quarks, a name which was apparently taken from a line in the James Joyce novel, Finnegan's Wake. The line reads "Three quarks for Muster Mark."

To build all the hadrons known in 1964, only three quarks were necessary. Gell-Mann soon determined the characteristics such as mass, charge and spin that these quarks would need. It is interesting to note that these quarks need to have fractions of the charge on an electron in order to work, a suggestion that was quite revolutionary in the wake of nearly sixty years of evidence that the electron was the smallest unit of charge.

The three quarks initially proposed were named Up (symbol u), Down (symbol d), and Strange (symbol s). Up and down have charges of +2/3e and -1/3e respectively, and zero strangeness, while strange has a charge of -1/3, but a strangeness value of –1. Baryons are built by grouping quarks in threes. For example, the proton is duu, while the neutron is ddu.

Note that these groupings support the law of conservation of charge.

Once the idea of quarks had been proposed, scientists began to hunt for evidence of their existence. Physicists at the Stanford Linear Accelerator Centre used electron beams to bombard liquid hydrogen (protons), and soon discovered tiny, point-like charge concentrations inside the proton. These charge points were determined to have the predicted charges, verifying that quarks do indeed exist.

Scientists then proceeded to try to make quarks, using colliding beams of electrons and positrons, such as the ones used in the LEP Collider at the European Laboratory for Particle Physics. In 1974, a new quark was discovered that had a charge of +2/3e. The new quark was named charm (symbol c) because of its "magical" ability to solve certain theoretical problems. Another new quark, named bottom or beauty (symbol b), was discovered in 1977. Bottom carries a charge of -1/3e. Because quarks must occur in pairs, a sixth quark was proposed, with the suggested names top or truth (symbol t) and a predicted charge of +2/3e, this was confirmed in 1995.

The conservation laws of charge, baryon number, lepton number and strangeness

Mass

All the classes of particles have distinct masses. Let us now introduce some common classes of particles in Figure 732 and Figure 733 showing their distinct masses, charge, spin and their life-spans. Because the kilogram is a large unit in which to measure mass, the preferred unit for mass in particle physics is the electron-volt/c^2.

However, one electron-volt is too small and as such we usually talk in mega and giga electron-volts. The rest energy E^0 of a particle can be defined as the energy associated with its rest mass m^0. The Theory of Relativity demonstrated that mass and energy are equivalent as given by the equation $E^0 = m_0 c^2$, so energy can be measured in electron-volt and rest mass can be measured in eV/c^2. It is often convenient to assume that the c^2 is therefore mass and just talk of a relative particle mass being measured in MeV or GeV.

Name	Mass / c^2	Charge	Spin	Lifespan
Leptons				
electron e⁻	0.511 MeV	-1	½	stable
positron e⁺	0.511 MeV	+1	½	stable
muon and antimuon µ⁺µ⁻	105.6 MeV	⁻₊1	½	2 × 10⁻⁶ s
tau and antitau τ⁻τ⁺	1.784 GeV	⁻₊1	½	3 × 10⁻¹³ s
electron neutrino/antineutrino $\gamma_e \bar{\gamma}_e$	< 50 eV	0	½	stable
muon neutrino/antineutrino $\gamma_\mu \bar{\gamma}_\mu$	< 0.5 MeV	0	½	stable
tau neutrino/antineutrino $\gamma_\tau \bar{\gamma}_\tau$	< 50 MeV	0	½	stable

CHAPTER 7

Name	Mass / c²	Charge	Spin	Lifespan
Quarks				
up and anti-up u ū	~ 5 MeV	$+2/3$ / $-2/3$	½	stable
down and anti-down d d̄	~ 10 MeV	$-1/3$ / $+1/3$	½	variable
strange and anti-strange s s̄	~ 100 MeV	$-1/3$ / $+1/3$	½	variable
charm and anti-charm c c̄	~ 1.5 GeV	$+2/3$ / $-2/3$	½	variable
bottom and anti-bottom b b̄	~ 4.7 GeV	$-1/3$ / $+1/3$	½	variable
top and anti-top t t̄	> 30 GeV	$-1/3$ / $+1/3$	½	variable
Gauge Bosons				
photon	0	0	1	stable
W-plus and W-minus	81 GeV	$+1$ / -1	1	10^{-25} s
Z	93 GeV	0	1	10^{-25} s
gluon	0	0	1	stable

Figure 732 Properties of leptons, quarks and exchange bosons

Name	Mass / c²	Charge	Spin	Lifespan
Mesons				
pion (pi-zero) π^0	135 MeV	0	0	0.8×10^{-16} s
pion (pi-plus) π^+ (pi-minus) π^-	140 MeV	$+1$ / -1	0	2.6×10^{-8} s
kaon (K-zero) K^0	498 MeV	0	0	short 0.9×10^{-10} s / long 5×10^{-8} s
kaon (K-plus) K^+ (K-minus) K^-	494 MeV	$+1$ / -1	0	1.2×10^{-8} s
J/psi	3.1 GeV	0	1	0.8×10^{-20} s
Baryons				
proton p	938.8 MeV	+1	½	stable
antiproton p̄	938.8 MeV	-1	½	stable
neutron n / anti-neutron n̄	939.6 MeV	0	½	in nuclei: stable
lamda Λ / anti-lamda $\bar{\Lambda}$	1.115 GeV	0	½	2.6×10^{-10} s
sigma (sigma-zero) Σ^0	1.192 GeV	0	½	6×10^{-20} s
sigma (sigma-plus) Σ^+	1.189 GeV	+1	½	0.8×10^{-10} s
sigma (sigma-minus) Σ^-	1.197 GeV	-1	½	1.5×10^{-10} s
xi cascade (xi-minus) Ξ^-	1.321 GeV	-1	½	1.6×10^{-10} s
xi cascade (xi-zero) Ξ^0	1.315 GeV	0	½	3×10^{-10} s
omega minus Ω^-	1.672 GeV	-1	3/2	0.8×10^{-10} s

Figure 733 Properties of some mesons and baryons

Quantum numbers

The **Quantum Mechanics** model of the atom was proposed in 1925 and 1926 to overcome some of the inadequacies of the Bohr model of the atom. Remember that the Bohr model had its limitations because although it could account for the wavelengths of light absorbed and emitted for the hydrogen atom, it could not be applied to any other atom. Furthermore, it could not account for the hyperfine lines that existed due the tiny splitting of energy levels.

While quantum mechanics retained some aspects of the Bohr model, it predicted that the electrons were spread out in space in electron clouds of negative charge as a result of the wave nature of the electrons. These electron clouds can have various three-dimensional shapes such as spheres, dumbbells and donuts. The different states in which an electron can exist are determined by four **quantum numbers**:

- the principal quantum number n where $n = 1, 2, 3…$
- the orbital quantum number l where $l = 0$ to n^{-1}
- the magnetic quantum number ml where $m_l = -l$ to $+l$
- the spin quantum number m_s where $m_s = +½$ or $-½$

Without going into too much depth here, the principal quantum number is similar to the n number of energy levels in the Bohr model, and it applies not only to electrons but to all leptons and baryons. These Baryon particles are called fermions (particles that have a half-integer spin). It defines the distinct energy levels or shells in which fermions can be located. The maximum number of fermions that the principal energy level can accommodate is $2n^2$. So, when n = 1, there can be two fermions, and when n=2, there can be 8 fermions and so on. Fermions move in orbitals which that can be depicted as a three-dimensional probability region. For each value of n, there are n^2 orbitals. So, for the third principal energy, level, n=3, and therefore there will be 9 orbitals.

Just as leptons can exist in lepton energy levels, so too can other elementary particles and their composites – the hadrons. Therefore, each elementary particle or a composite of elementary particles can be specified in terms of its mass and various quantum numbers.

Conservation of energy

In particle physics, the total energy of the particles before a reaction or decay must be equal to the total energy of the particles after the decay. The law of conservation of energy manifests itself in 2 forms in particle physics:

- The kinetic energy which is dependant on the velocity of the particles
- Their mass energy is given by $E = mc^2$.

It can be seen that the greater the mass of a particle, the greater is the mass energy that a particle has. It has been shown that if a particle decays into other particles then the mass of a decaying particle has to be greater or equal to the mass of the products. Let us say that particle X decays and forms Y and Z.

X → Y + Z

However, if energy is to be conserved, the kinetic energy of particles Y and Z must be taken into account:

X → Y + E_{KY} + Z + E_{KZ}

Therefore, the mass of X > the mass of Y + the mass of Z

On the other hand, a reaction between 2 particles in their initial state can have less mass than the total mass of the products because initial particles can introduce energy into the reaction so that energy is conserved.

Conservation of charge (Q)

Electric charge is given the symbol Q. As noted previously, antiparticles are identical to their corresponding particle in mass, but differ in electric charge. For example, the positron is the antiparticle which corresponds to the electron, so it has a mass of 9.11×10^{-31} kg, but an electric charge of '+1'. But charge is not the only difference between particles and antiparticles.

Example

Determine whether the following reactions can occur due to charge conservation:

(a) $e^- + p \to \nu + n$

(b) $\pi^- + p \to \Sigma^- + K^+$

(c) $p + n \to p + n + \pi^+$

(d) $e^+ + e^- \to \mu^+ + \nu_\mu$

Solution

(a) $e^- + p \to \nu + n$
      Q -1 + 1 = 0 + 0 charge is conserved.

(b) $\pi^- + p \to \Sigma^- + K^+$
      Q -1 + 1 = -1 + 1 charge is conserved.

(c) $p + n \to p + n + \pi^+$
      Q 1 + 0 ≠ 1 + 0 + 1 charge is not conserved.

(d) $e^+ + e^- \to \mu^+ + \nu_\mu$
      Q 1 + -1 ≠ 1 + 0 charge is not conserved.

Lepton number is also not conserved as we will see later.

Conservation of baryon number (B)

The **baryon number** is given the symbol B. It was introduced as a conservation law when it was found that reactions that conserved electric charge were not taking place.

For example, consider the following reaction:

$$p + p \to p + \pi^+$$

Q 1 + 1 = 1 + 1   charge is conserved but the reaction does not occur.

If we apply a couple of simple rules then we can decide whether baryon number is conseved. The rules are:

- the total number of baryons must remain constant
- all baryons are assigned a baryon number of 1 (p, n, Λ, Σ, Ξ)
- all non-baryons (leptons and mesons) are assigned a baryon number of 0 (π, K, e, μ, τ).
- an antiparticle has the opposite baryon number (-1) from its particle.

If we take the original reaction and look at the baryon numbers we have:

$$p + p \to p + \pi^+$$

Q 1 + 1 = 1 + 1

B 1 + 1 ≠ 1 + 0

The reaction does not occur because the baryon number is not conserved.

Now we will look at an earlier equation:

$$\pi^- + p \to \Sigma^- + K^+$$

Q -1 + 1 = -1 + 1   charge is conserved.

B 0 + 1 = 1 + 0   baryon number is conserved.

The reaction does occur because electric charge and the baryon number is conserved.

Finally, let us take an example that examines the 3 conservation laws introduced so far:

$$n \to \pi^+ + \pi^-$$

energy 1 = 1/7 + 1/7 mass is greater in the initial particle

Q 0 = 1 - 1   charge is conserved

B 1 ≠ 0 + 0   baryon number is not conserved.
               The reaction does not occur.

235

CHAPTER 7

Conservation of lepton number (L)

The **lepton number** is given the symbol L. As already mentioned, leptons carry the same electric charge and react via the weak and electromagnetic forces but not the strong force. They have partner neutrinos and they are split into 3 generations - L_e, L_μ and L_τ.

Neutrinos must accompany their partner leptons. But how does one know if a neutrino or an antineutrino is involved in a reaction or a decay. Basically, the rules are:

If the lepton and neutrino are on the same side of an equation

- electrons, negative muons and negative tau must be accompanied by an antineutrino
- positrons, positive muons and positive tau must be accompanied by an neutrino.

If the lepton and neutrino are on the opposite side of an equation

- electrons, negative muons and negative tau must be accompanied by an neutrino
- positrons, positive muons and positive tau must be accompanied by an antineutrino.

Again, this can be understood in terms of a new conservation rule – the conservation of lepton number. The rules for lepton number conservation are:

- the total number in each generation must always remain the same
- the electron and electron-neutrino are assigned a lepton electron number of 1
- the negative muon and muon-neutrino are assigned a lepton muon number of 1
- the negative tau and tau-neutrino are assigned a lepton tau number of 1
- all other particles are assigned a lepton number of 0
- an antiparticle has the opposite lepton number (-1) from its particle.

For example, consider the following reaction:

$$\nu_e + \mu_e \rightarrow e^- + \nu_\mu$$

L_e $1 + 0 = 1 + 0$ *first generation lepton number is conserved*

L_μ $0 + 1 = 0 + 1$ *second generation lepton number is conserved.*

Therefore the reaction can take place.

Now let us look at a second equation:

$$n + \nu_e \rightarrow p + \mu^-$$

L_e $0 + 1 \neq 0 + 0$ first generation lepton number is not conserved

L_μ $0 + 0 \neq 0 + 1$ second generation lepton number is not conserved.

Therefore the reaction cannot take place.

Conservation of strangeness (S)

Another property which differs for particles and their corresponding antiparticles is **strangeness**. Strangeness (symbol S) is a quantum number introduced in the early 1950s to explain the production and decay behaviour of some newly found particles, namely the kaon (symbol K^0) and the lambda hyperon (symbol Λ^0). These particles are always produced in pairs, even though conservation of mass-energy laws would allow production of single particles. Also, they are unstable even though the particles have much longer lifetimes than would be expected, leading scientists to believe they were **produced via the strong force but decay via the weak interaction force**. Strangeness was found to be conserved in strong nuclear interactions, but not in weak ones. Antiparticles of those particles that exhibit strange behaviour are assigned strangeness numbers that are opposite to their corresponding particle's strangeness number. So strangeness needs to be conserved in all strong and electromagnetic interactions but it does not need to be conserved in weak interactions.

Let us look at the following example

$$\pi^- + p \rightarrow \pi^- + \Sigma^+$$

Q $-1 + 1 = -1 + 1$ charge is conserved

B $0 + 1 = 0 + 1$ baryon number is conserved.

However, this reaction does not take place.

Here are some rules for strangeness:

- the strangeness of leptons is zero
- protons, neutrons and pions are assigned a strangeness of 0
- K^+ and K^0 mesons are assigned a strangeness of +1
- K^-, and Λ and Σ baryons are assigned a strangeness of -1
- Ξ baryons are assigned a baryon number of -2
- Ω baryons are assigned a baryon number of -3
- all antiparticles have the opposite strangeness to their particles.

Atomic, Nuclear and Particle Physics

So let us assign strangeness to the above example:

$$\pi^- + p \rightarrow \pi^- + \Sigma^+$$

Q -1 + 1 = -1 + 1 charge is conserved.

B 0 + 1 = 0 + 1   baryon number is conserved.

S 0 + 0 ≠ 0 + 1 strangeness is not conserved.

Therefore, the reaction does not occur.

Now let us examine a further nuclear reaction.

$$p + p \rightarrow p + p + K^+ + K^-$$

Q 1 + 1 = 1 + 1 + 1 + (-1) charge is conserved.

B 1 + 1 = 1 + 1 + 0 + 0 baryon number is conserved and the number of baryons is the same.

S 0 + 0 = 0 + 0 + 1 + (-1) strangeness is conserved.

Therefore, this reaction can occur.

Now let us examine the quark content of the particles that exhibit strangeness. The K$^+$ meson consist of an up quark and an anti-strange quark (strangeness of +1), and K$^-$ consists of a strange quark and an anti-up quark (strangeness of -1). Another example is the Λ baryon that contains uds quarks and has strangeness of -1. Finally, the Ξ$^-$ contains dss quarks and is assigned strangeness of -2. Therefore, it can be seen the number of the strangeness is negative for each strange quark present and positive for each anti-quark present.

The nature and range of the strong nuclear force, weak nuclear force and electromagnetic force

As often mentioned, there are four fundamental interactions and some of their properties are shown in Figure 734. As can be seen the fundamental interactions are:

- Gravitational
- Weak
- Electromagnetic
- Strong

Since the early 1970s the electromagnetic and weak interactions have been shown to be two aspects of the same interaction, the electroweak interaction.

The electromagnetic force is the cause for basic collisions between charged particles such as:

$$p + p \rightarrow p + p$$

The strong force is mainly responsible for reactions between hadrons. For example,

$$p + p \rightarrow p + p + \pi^0$$

The weak force is the only force in lepton reactions that produce neutrinos as they are electrically neutral and so will not be affected by the electromagnetic force. For example,

$$\nu_e + \mu^- \rightarrow e^- + \nu_\mu$$

Interaction	Gravitational	Weak	Electromagnetic	Strong	
				Fundamental	Composite
		Electroweak			
It acts on:	Mass-energy	Flavour	Electric charge	Colour charge	
On what particles:	All	Quarks Leptons	Electrically charged particles	Quarks Gluons	Hadrons
Exchange particle is:	Graviton?	W$^+$ W$^-$ Z^0	γ	Gluons	Mesons
Relative strength	10^{-38}	10^{-5}	10^{-2}	1	Not applicable
Range	∞	~10^{-18} m	∞	~10^{-15} m	

Figure 734   Fundamental interactions and some properties

CHAPTER 7

Exchange particles

Exchange particles – gauge bosons are elementary particles that transmit the forces of nature. Figure 735 lists the 6 exchange particles and some of their properties.

Force	Exchange particle	Rest mass m_p	Charge	Spin
strong	Gluons g	0	0	1
weak	W^+	89	+1	1
	W^-	89	-1	1
	Z^0	99	0	1
	Higgs boson (hypothetical)	> 86	0	1
electromagnetic	photons	0	0	1
gravity	graviton	0	0	2

Figure 735   Standard model for the exchange particles

Recall that when an electron transition occurs from a high energy level to a lower energy level, the energy difference is emitted in the form of a quantum of electromagnetic energy called a photon. The other exchange particles are similarly associated with quanta of energy. However, when we talk of a classical force we define it in terms of the rate of change in momentum. When looking at subatomic particles, quantum mechanics and relativity replace classical Newtonian mechanics and a more sophisticated notion of force is required that is described in terms of an **interaction.** It has become evident that the meaning of force transmission if a particular interaction is to occur is related to the energy and momentum that is carried by a quantum of the force field.

However, one thing that is obvious when looking at the standard model for exchange particles is that the weak force exchange particles have mass whereas the others do not. Why do they have mass nearly 100 times the mass of the proton while the other exchange particles have zero mass?

The graviton is the exchange particle for the gravitational force. It is an inverse square force with an infinite range that affects all particles and acts on all mass/energy and it has a rest mass of zero.

The W^+, W^- and Z^0 are the exchange particles involved in the weak nuclear interaction. They were predicted in 1979 and then measured at the CERN particle accelerator in 1982 by Carlo Rubbia. According to the electroweak unification theory, the weak and electromagnetic forces should have the same strength but the experimental data showed that this was not the case. However, under certain conditions a high strength force can have the appearance of a weak strength force provided the exchange particles have a large mass, as does the W and Z particles.

The photon is the exchange particle that is responsible for the electromagnetic force and its energy is given by the Planck relationship E = hf = hc / λ. The electromagnetic force is transmitted almost instantaneously, since photons travel at the speed of light.

In 1935, based on the earlier theory of electromagnetic forces, the Japanese physicist Hideki Yukawa (1907–1981) suggested the interchange of certain particles could also explain the strong nuclear force. These particles, members of the meson family, came to be known as pions (short for pi mesons). Yukawa suggested that the composite nucleons continuously exchange mesons with nearby nucleons without being altered themselves. Of course, this interaction must take place in a short enough time frame so as not to violate the law of conservation of mass-energy. Yukawa suggested the mesons mediate the strong nuclear force in the same way photons mediate the electromagnetic force and that the strong nuclear force is responsible for holding nucleons together in the nucleus. Yukawa's theory started a search for his predicted particle, and in 1947, pions were discovered in cosmic rays by Cecil Frank Powell (1905 – 1969). In 1949, Yukawa received the Nobel Prize in Physics for his prediction of the existence of pions. One year later, Powell also received the Nobel prize for his discovery of mesons, and developing the method that enabled their discovery.

The strong force due to gluons only occurs within hadrons. The force that holds the nucleus together is caused by "leakage" from the gluon exchange. For example, in interactions between protons and neutrons there is an exchange of pions. Gluons are the exchange particles that are responsible for quark colour. Just as the positive and negative charges are associated with the electromagnetic force, a three-colour charge is associated with quarks and gluons that bind the quarks together. The linking between quarks and antiquarks is done by gluon clumps called "glueballs". Colour will be dicussed in detail in section J3.

Feynman diagrams

The mediation of exchange forces can be easily represented using **Feynman diagrams**, so named for their inventor, the American physicist *Richard Feynman* (1918–1988). They were developed by Feynman as a graphical tool to examine the conservation laws that govern particle interactions according to quantum electrodynamic theory.

A typical Feynman diagram shown in Figure 736 depicts how 2 electrons approach each other, exchange a virtual photon and as a result repel each other. Be careful here – they do not depict the track of a particle or how a particle conserves energy and momentum in the classical sense because the Heisenberg uncertainty principle tells us of the uncertainty in position and momentum, so the path of the particle is not known in much detail. Each point at which lines come together is called a **vertex**. Lines with arrows represent particles and the direction of their travel while wavy lines represent the virtual particle.

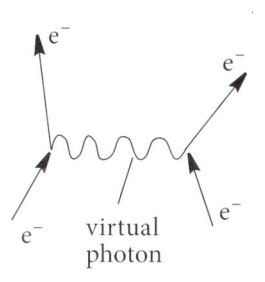

Figure 736 Feynman diagram for the electromagnetic force between 2 electrons

They are space time diagrams with the vertical direction representing time and the horizontal direction representing space, in this case *ct*. At each vertex, the conservation laws of charge, lepton number and baryon number must be obeyed. Feynman diagrams are a shorthand method for studying the probability for particle interactions.

Some of the interactions using Feynman diagrams are shown in Figure 736.

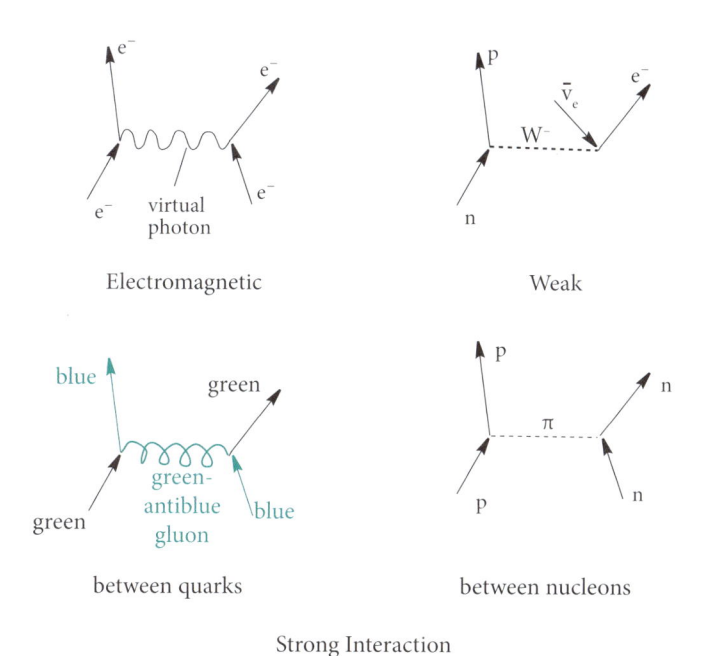

Figure 737 Some interactions using Feynman diagrams

Early in the last century, scientists began to search for an explanation of the four fundamental forces, particularly how they can act over a distance. One of the mechanisms suggested was the exchange of virtual particles. A **virtual particle** is a particle that cannot be observed during an interaction. A virtual photon is said to be the 'carrier' of the electromagnetic force. An example of this would be the exchange of virtual photons between charged particles as a mechanism for the repulsions and attractions that make up the electromagnetic force.

The analogy that is often used for repulsion of charged particles is to imagine the particles are like two people sitting on big frictionless frisbees on an icy pond. If one person throws a heavy object to the second person, the first person will recoil in the opposite direction to the object flight and the second person will recoil in the same direction as the object flight.

A similar analogy for attraction could be of children exchanging pillows. Imagine two children standing facing one another, each holding a pillow. The children can exchange their pillows in one of two ways. They can grab them from one another, at the same time pulling on each other, an action analogous to an attractive force. The other possibility is that they can throw them at one another, at the same time stepping back to catch the pillows, which is analogous to a repulsive force. If charged particles such as electrons exchanged photons in the same manner, the attractive and repulsive electromagnetic forces could be explained quite simply. Photons emitted by one electron cause it to recoil, as it transfers momentum and energy to the other electron. Then the second electron undergoes the same process almost immediately. The closer two charges are, the more energetic the virtual photons exchanged, while the further away two charges are, the less energetic their virtual photons. Because the exchange must be very rapid, the photons exchanged are called virtual photons, suggesting they are not observable. These virtual photons are said to carry the electromagnetic force, or in other words, to mediate the force. Figure 738 shows the Feynman diagram for the force of attraction between an electron and a positron.

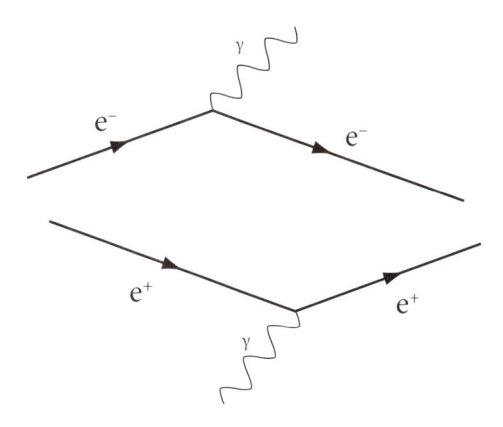

Figure 738 Force between an electron and positron

239

Chapter 7

When matter (such as an electron) collides with its corresponding antimatter (such as a positron), both particles are annihilated, and 2 gamma rays with the same energy but with a direction at 180° to each other are produced. This is called **pair annihilation**. The direction of the gamma rays produced is in accordance with the law of conservation of momentum and the electron-positron annihilation gives energy equal to $E = mc^2$. This is depicted in Figure 739.

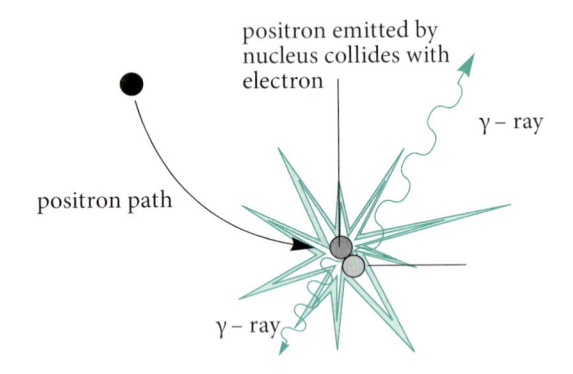

Figure 739 Electron-positron annihilation

When particles and antiparticles collide, they can annihilate one another, releasing energy in the form of gamma rays. The gamma rays result from conversion of matter to energy, according to Einstein's famous equation, $E = mc^2$. For example, an electron and positron annihilating one another at rest release an amount of energy equal to:

$E = mc^2 = (2 \times 9.11 \times 10^{-31}$ kg$) \times (3 \times 10^8$ ms$^{-1})^2$

$= 1.64 \times 10^{-13}$ J $= 1.03$ MeV.

The energy released is higher if the particles annihilate in a collision where one or both contribute kinetic energy to the process.

Particle–antiparticle pairs can also be produced when a gamma ray with sufficient energy passes close by a nucleus. The process is the reverse of annihilation and is called **pair production**. The law of conservation of mass-energy requires particles and antiparticles to be produced in pairs.

Figure 740 gives the Feynman diagrams for pair annihilation and pair production.

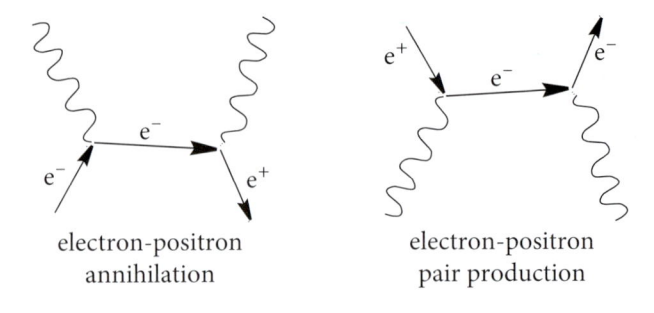

Figure 740 Feynman diagrams for electron-positron annihilation and production

Note that the backward arrow is an antiparticle, in this case a positron. Also remember the space-time concept of the Feynman diagram with time progressing upwards. Even though the arrow of the antiparticle is downwards, the antiparticle is still progressing upwards in time.

Providing sufficient energy is available, particles other than photons can be produced.

A number of particle processes have already been introduced in the preceding sections. Let us look at some more examples as shown in Figure 741.

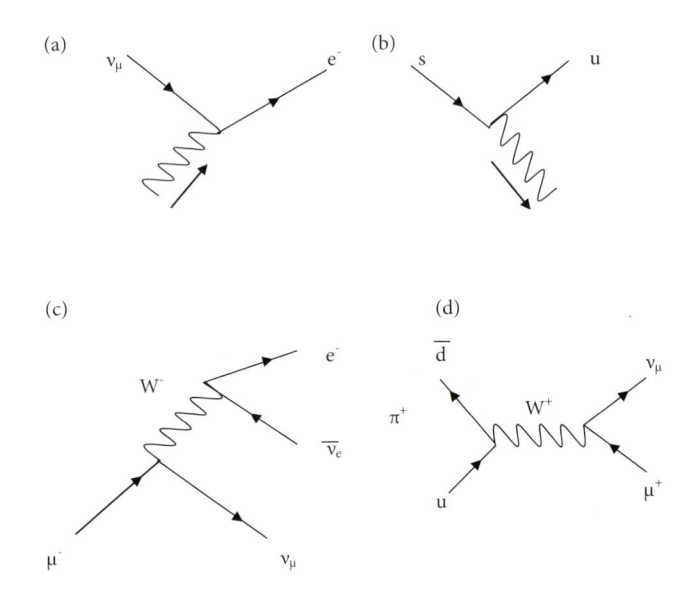

Figure 741 Some examples of Feynman diagrams

In (a), a muon neutrino interacts with a photon exchange particle to become an electron. In (b), a strange quark emits an exchange particle and becomes an up quark. This is an example of a flavour change as it transforms into a member of another generation. In (c), a negative muon emits a W⁻ particle and becomes a muon neutrino. The W⁻ particle changes to particle-antiparticle pair in the form of an electron and an electron antineutrino. In (d), a positive pion decays into a positive muon and a muon neutrino. The up quark and the antidown quark annihilate to produce a W⁺ particle. Note the backward direction of the antidown quark. The W⁺ then decays into a positive muon and a muon neutrino.

More complicated interactions can be demonstrated. For example, the electromagnetic interaction leads to photon–photon scattering (that is, scattering of light by light). The particles in the loop are electrons or positrons and this interaction is shown in Figure 742.

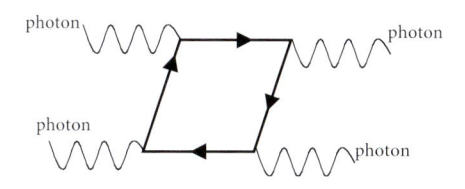

Figure 742   Electrons or positrons are in the loop

Quark confinement

Colour-charged particles such as quarks and gluons can never be found in isolation. They always exist in groups. In fact it is found that the strong (colour) interaction increases when you try to separate quarks. With the tremendous energy of modern particle accelerators, it was once thought that a quark may one day be found individually. However, all that we get in collisions is more 3-coloured baryons, and, colour and anti-colour mesons. A particle with four quarks has never been found. So, only combinations that are colour-neutral can be found. The property that quarks are always found in groups that are colourless is called **quark confinement**.

Electrons can be used in collisions to indirectly identify protons inside the nucleus, and we now know that at high energies the de Broglie wavelength of an electron is small enough to resolve particles inside the proton.

However, before the late 1960s, particle accelerators could not produce the energies needed to probe inside protons. However, when the SLAC linac came on line, electrons could be accelerated up to 20 GeV. Low energy electrons tend to be scattered away by protons. However, if the electron has sufficient energy, it can probe deep inside the proton. The collision is inelastic because the proton is disrupted and produces new particles. In effect, the electrons can diffract off quarks inside the protons causing one quark to move away from the other two, and this shattering produces a stream of hadrons.

These and later experiments have provided evidence for the existence of quarks, gluons and colour.

In 1973, *David Gross* and *Frank Wilczek* working at Princeton university and *David Politzer* working independently at Harvard university found that as quarks inside a meson or baryon get closer together, the force of attraction between them gets weaker so as to asymptotically approach zero for very close confinement. On the other hand, if the quarks move apart, the force of attraction becomes stronger. This discovery is known as **asymptotic freedom** and it established quantum chromodynamics QCD as the correct theory to explain the nature of the strong force. The calculations done by *Gross*, *Wilczek* and *Politzer* showed that quarks are held together very strongly at distances that are comparable to the size of a proton, and this explained the concept of quark confinement, that was previously mentioned in section J3.

Quark confinement can be visualized by using what is known as the "bag model" as shown in Figure 743.

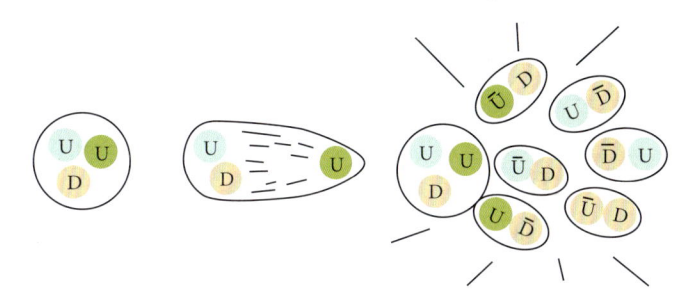

Figure 743   The bag model of quark confinement

In a normal proton, the quarks are close together and are free to move within the proton "bag". However, when you supply energy of the order of a GeV per femto distance, (10^{-15} m), the bag stretches like a balloon. The energy needed to remove a quark is much larger than that to produce a quark-antiquark pair. So instead of removing the quark, you just get a shower of mesons produced.

The Higgs boson

The Higgs boson is an elementary particle whose existence was first proposed by Peter Higgs in 1964. It is sometimes referred to as the 'God particle', a name disliked by many scientists, including Peter Higgs.

Up until July 4 2012, scientists had gathered a lot of evidence about the structure and constituents of the atom but it was not known how particles get their masses. Furthermore, particle theorists wondered why the W and Z bosons have large masses rather than being massless like the photon. Peter Higgs proposed that particles can acquire mass as a result of interactions with a hypothetical extra electroweak force field called the Higgs field. Higgs reasoned that if we start out with a particle H that has mass but no other conservation characteristics and bring it close to another particle, say a proton, then H can interact with the proton because there is a force between them. If H and the proton interact, then H must be a boson.

Chapter 7

When particles are created and annihilated in accelerators, particles are said to arise from 'fields' that are spread out in space and time. By using quantum mechanics mathematics, Higgs found that if H was in its lowest energy state of a field – empty space – the field would not be zero. Therefore, the Higgs particle (boson) that interacts with other particles can gain mass as a result of the interaction.

The hadron collider was commissioned at the end of 2007 at CERN in search of this elusive boson. It was important to find it because it plays an important role in the unification of different forces. If it was not found, then particle physics would have to go back to the drawing board, and a new theory would have to be proposed to replace the Grand Unifying Theory.

By March 2013, the particle had been proven to behave, interact and decay in many of the ways predicted by the Standard Model.

Exercise 7.4

1. Describe the difference between an elementary particle and a composite particle.

2. Neutrons and protons are classified as hadrons, whereas electrons are classified as leptons. Describe the concept underlying this type of classification.

3. (a) Describe the properties of the neutrino.
   (b) Explain the reasons that led Enrico Fermi to predict the existence of the neutrino.

4. The neutrino belongs to the same family as the

   A. neutron.
   B. proton.
   C. electron.
   D. baryon.

5. Hadrons differ from the other two families in that

   A. they mainly interact via the electromagnetic force.
   B. they mainly interact via the gravitational force.
   C. they mainly interact via the strong nuclear force.
   D. they mainly interact via the weak interaction force.

6. State the name and give the charge of each of the symbols given, and classify the particles as either quarks, baryons, mesons, leptons or gauge bosons:

   1. e^+ 2. d   3. π^+ 4. ν_e 5. Λ 6. Σ^+
   7. τ^+ 8. Ξ^0 9. K^- 10. g   11. Ω^- 12. γ
   13. μ^+ 14. Z   15. ν_μ 16. $\bar{u}$ 17. τ^- 18. c

7. Determine whether the following reactions can occur:

   (a) $e^- + p \rightarrow n + n$
   (b) $\pi^- + p \rightarrow \Sigma^- + K^+$
   (c) $p + n \rightarrow p + n + \pi^+$
   (d) $K^0 \rightarrow \pi^0 + \pi^0 + \pi^+$

8. Antiparticles differ from their corresponding particle in their
   A. charge.
   B. rest mass.
   C. family.
   D. rest energy.

9. Define the terms *antiparticle* and *antimass*.

10. Describe how baryons differ from mesons.

11. Of the proton, neutron, and electron, state which would

    a. travel in a straight line through a magnetic field.
    b. travel in a curved path in a magnetic field.
    c. travel in a curved path of greatest radius in a magnetic field.

12. Describe the quarks that make up a proton and a neutron in terms of the type of quark and their overall charge.

13. (a) Describe what is meant by pair annihilation.
    (b) Calculate the energy of each photon produced when an electron and positron, initially at rest, are annihilated.
    (c) Describe the direction of travel of the photons.
    (d) Determine the threshold photon frequency for electron-positron pair production.

14. Pair production results when

    A. particles and antiparticles annihilate one another.
    B. particles of sufficient energy pass close to a nucleus.
    C. gamma rays of sufficient energy pass close to a nucleus.
    D. gamma rays and particles annihilate one another.

242

15. Calculate the energy released when a proton and antiproton annihilate one another

   a. at rest.
   b. in a collision where each has kinetic energy of 25 MeV (assume no energy is lost).

16. If a γ-ray is to produce a neutron-antineutron pair, determine the minimum energy, in MeV, that it must have.

17. Explain how photon exchange mediates the force between two electrons.

18. Predict the Feynman diagram particle processes as shown in the figure below.

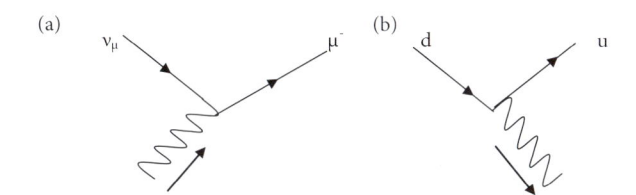

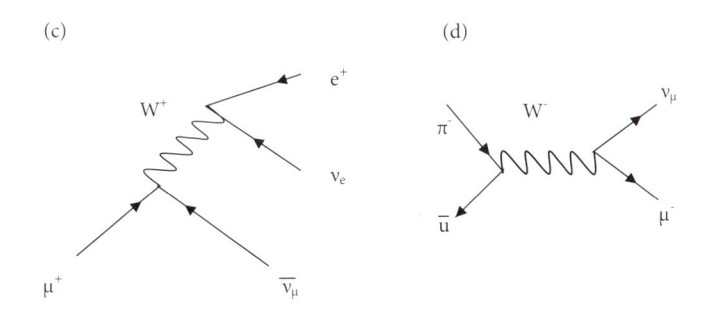

    (a) ν_μ , μ^-
    (b) d , u
    (c) W^+ , e^+ , ν_e , μ^+ , $\overline{\nu_\mu}$
    (d) π^- , W^- , ν_μ , $\overline{u}$, μ^-

19. The weak force is

    (a) the only force affecting neutrons
    (b) responsible for radioactive decay
    (c) the only force affecting protons
    (d) responsible for stability of the nucleus

20. Strangeness must be conserved in:

    (a) weak interactions
    (b) electromagnetic interactions only
    (c) strong interactions interactions only
    (d) both strong and electromagnetic interactions.

21. State the name of the force carrier in the Feynman diagram shown in the Figure below. Explain why you have chosen this force carrier.

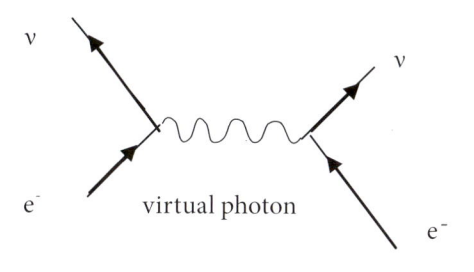

22. No particles of fractional charge have been confirmed thus far. Does this mean that quarks do not exist? Explain.

23. Explain why is it impossible thus far to detect single quarks?

24. Determine whether the following nuclear reactions will occur.

    (a) $p + p \rightarrow p + K^+ + \Lambda^0$
    (b) $n + \nu_e \rightarrow \mu^- + p$
    (c) $p + n \rightarrow$ energy
    (d) $\mu^+ + e^- \rightarrow$ energy
    (e) $p + \overline{\nu_e} \rightarrow e^+ + \Lambda^0 + K^0$
    (f) $p + \overline{\nu}\mu \rightarrow \mu^+ + n$
    (g) $p + \nu_e \rightarrow e^- + \Sigma^+ + K^+$
    (h) $K^+ \rightarrow \pi^+ + \pi^0$
    (i) $\Lambda^0 \rightarrow p + K^-$

Chapter 7

This page is intentionally left blank

8. Energy Production

Contents

8.1 – Energy sources

8.2 – Thermal energy transfer

Essential Ideas

The constant need for new energy sources implies decisions that may have a serious effect on the environment. The finite quantity of fossil fuels and their implication in global warming has led to the development of alternative sources of energy. This continues to be an area of rapidly changing technological innovation.

For simplified modelling purposes the Earth can be treated as a black-body radiator and the atmosphere treated as a grey-body. © IBO 2014

Chapter 8

8.1 Energy sources

NATURE OF SCIENCE:

Risks and problem solving: Since early times mankind understood the vital role of harnessing energy and large scale production of electricity has impacted all levels of society. Processes where energy is transformed require holistic approaches that involve many areas of knowledge. Research and development of alternative energy sources has lacked support in some countries for economic and political reasons. Scientists, however, have continued to collaborate and share new technologies that can reduce our dependence on non-renewable energy sources. (4.8)

© IBO 2014

Essential idea: The constant need for new energy sources implies decisions that may have a serious effect on the environment. The finite quantity of fossil fuels and their implication in global warming has led to the development of alternative sources of energy. This continues to be an area of rapidly changing technological innovation.

Understandings:
- Specific energy and energy density of fuel sources
- Sankey diagrams
- Primary energy sources
- Electricity as a secondary and versatile form of energy
- Renewable and non-renewable energy sources

Introduction

In Chapter 2, energy was defined as the capacity to do work. The various forms of energy can be classified as:

- Mechanical (kinetic and potential)
- Heat
- Radiant (electromagnetic)
- Chemical (potential)
- Sound
- Electrical/magnetic
- Nuclear

Mechanical energy includes both kinetic and potential energy. Friction is mechanical energy as it is caused by kinetic energy and potential energy of a body as a force is applied through a distance.

Heat energy is the energy a body possesses because of its internal energy due to the motion of the particles it contains.

Radiant energy is the source of all life on the Earth and is the greatest potential energy resource available for the future. When plants absorb radiant energy in the form of light, it is converted into stored chemical energy. This energy is available as food or biomass. Biomass is used to produce fuels. It is electromagnetic in nature and is possessed by all components of the electromagnetic spectrum (γ–rays, X–rays, UV radiation, visible light, IR radiation, microwaves and radio waves). Infra-red radiation falling on a body is converted into thermal energy. Solar energy is a form of radiant energy.

Chemical energy is the energy locked up in fuels and other chemicals. The energy obtained by the combustion of fuels represents a major source of energy in current use. All food that we eat is a store of chemical energy. It can be considered to be latent potential energy that a body possesses.

Sound energy is produced by longitudinal waves that have an organised and periodic pattern that causes the vibration of the particles in the same direction as the transfer of energy.

Electrical energy is the energy carried by moving charges and these moving charges produce a magnetic field. Through the process of electromagnetic induction, electricity has become the greatest form of energy used by man in everyday life.

Nuclear energy is the potential binding energy released during a nuclear reaction when mass is converted to thermal and perhaps light energy. Nuclear fission reactors are starting to gain a renewed acceptance and nuclear fusion has potential for the future when the technology becomes available.

Sankey diagrams

When energy is transferred from one form to other forms, the energy before the transformation is equal to the energy after (*Law of conservation of energy*). However, some of the energy after the transformation may be in a less useful form. We say that the energy has been 'degraded'. For example, in a simple battery operated flashlight, an energy input of 100 units of chemical potential energy will give a 10 unit output of light energy and the light energy is enhanced by placing a curved mirror behind the lamp to concentrate the light into a beam. The other 90 units of output is used in heating up the filament of the light bulb and in heating the battery and the surroundings. These 90 units of energy output have become degraded. The thermal energy that is transferred to the surroundings, the filament and the battery is no longer available to perform useful work.

The Second Law of Thermodynamics in one form states that engines are theoretically inefficient users of energy.

ENERGY PRODUCTION

	Mechanical	Electric	Radiant	Chemical	Thermal
Mechanical		99% (electric generator)			100% (brakes)
Gravitational potential	85% (water turbine)	90% (hydro-electricity)			
Electric	93% (electric motor)		40% (gas laser)	72% (wet cell battery)	100% (heating coil)
Radiant	55% (wind power)	27% (solar cell)		0.6% (photosynthesis)	100% (solar furnace)
Chemical	45% (animal muscle)	10% (dry cell battery)	15% (chemical laser)		88% (steam furnace)
Thermal	52% (steam turbine)	7% (thermocouple)	5% (fluorescent tube)		

Figure 801 Efficiency of some energy conversion devices table

The **efficiency** of an energy conversion process is a ratio of the useful energy output to the total energy input usually expressed as a percentage. In practice, the efficiency is even lower than this theoretical value. Figure 801 gives examples of the efficiency attainable by some devices in their energy conversion process.

Chemical energy and electrical energy are considered to be high-grade energy because they can be converted to other forms of energy. However, there is a gradual degradation of the high-grade energy to low-grade energy in the operation of machines as the entropy (amount of disorder in a system) increases. It has become increasingly more important that man explores the renewable energy sources so that the energy demands of the future can be met with new high-grade energy.

One useful way of showing the energy degradation is by using energy transfer diagrams. For a certain flashlight, the energy transfer can be represented as shown in the Sankey diagram in Figure 802.

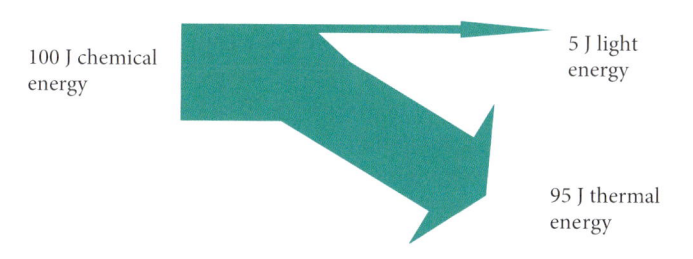

Figure 802 Sankey diagram for a torch

In a Sankey diagram, the thickness of each arrow gives an indication of the scale of each energy transformation. The total energy before the energy transfer is equal to the total energy after the transfer otherwise the Law of conservation of energy would be violated. The problem is that once the thermal energy is transferred to the surroundings, it cannot be used to do useful work. Scientists are becoming more aware of this waste and there are many innovations being made in building designs to use some of this energy for heating purposes.

The efficiency of this simple flashlight is 5%. The efficiency of any system can be determined by using the relationship:

$$\text{Efficiency} = \left(\frac{\text{useful energy output}}{\text{total energy input}}\right) \times 100\%$$

Power stations rely on thermal energy, gravitational potential energy or wind power to supply the kinetic energy to rotate a turbine. The turbine contains blades that are made to rotate by the force of water, gas, steam or wind. As the turbine rotates, it turns the shaft of a generator. Rotating coils in a magnetic field can produce the electrical energy.

Typically, fossil fuel power stations have a higher efficiency than nuclear power stations because current technology permits a higher temperature of 650 K versus the 570 K of a nuclear power plant.

The principle mechanisms involved in the production of electrical power with fossil fuels can be demonstrated by looking at the energy conversions in a coal-fired power station using a Sankey energy transfer diagram as shown in Figure 803.

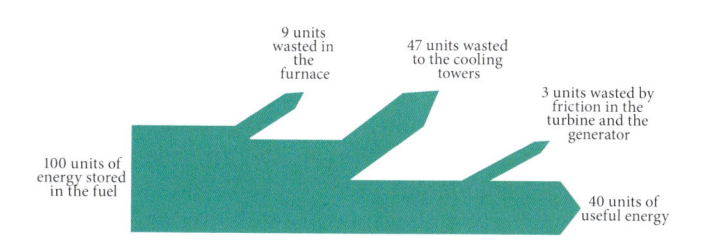

Figure 803 Sankey diagram for a coal-fired power station

Another useful energy transfer diagram is shown in Figure 804. The rectangles contain the different forms of energy, the circles show the conversion process, and the arrows show energy changes and energy outputs. The linked forms of energy can be said to form an energy chain.

247

Chapter 8

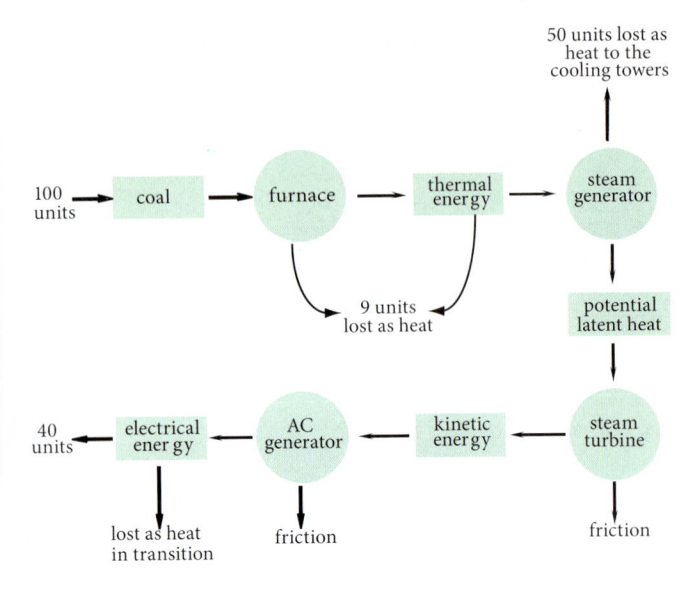

Figure 804 Energy flow diagram for a coal-fired power station

Heat energy is produced by the combustion of coal in a furnace. Liquid water absorbs the heat energy in a heat exchanger under pressure, and it is turned into steam. The steam contains latent potential energy as it has been converted from a liquid to a gas. Steam under pressure is capable of doing mechanical work to supply the rotational kinetic energy to turn the steam turbines. The turbine is coupled to the generator that produces electrical energy.

Energy is lost to the surroundings at many stages. For example, if 100 units of energy are supplied from the primary energy source then only 40 units of useful energy is available. The majority of the useful energy is lost to water in the cooling towers as heat is evolved in the condensation component of the heat exchanger cycle. Other forms of energy loss are shown on the arrowed parts of the diagram.

An oil-fired power station has a similar energy flow and efficiency to a coal-fired power station. However, a natural gas-fired power station is more efficient as they use combined cycle gas turbines (CCGT). A jet engine is used in place of the turbine to turn the generator. Natural gas is used to power the jet engine and the exhaust fumes from the jet engine are used to produce steam which turns the generator. These power stations can be up to 55% efficient.

The production of the majority of electrical power involves the combustion of coal, natural gas and oil or the fission of uranium-235. It depends on the energy sources available to countries. For example, most coal-fired power stations are found close to the coal source. Countries like Japan rely heavily on the importation of coal and natural gas as there are no reserves of fossil fuels available. In mountainous areas, hydro-electricity is common as water stored in dams can be used to rotate turbines.

By referring to the section on electromagnetic induction in Chapter 12 it can be deduced that a changing magnetic flux produces an induced eletromotive force (e.m.f). The rotating turbines contain coils of a conducting wire. When the coils are rotated in a magnetic field, the alternating current generator converts the kinetic energy into electrical energy. An alternating current is the present preferred option when compared to direct current because transformers can be used to step-up and step-down the voltage in the power grid. However, this is slowly changing in some countries.

The turbines drive the alternators that produce three-phase electricity. Most generators use stationary electro-magnets to provide the magnetic fields. They have a rotating armature with hundreds of coils of copper wire wound around an iron core. With more coils a greater induced emf can be produced. These coils are arranged in sets. By having a number of electromagnets and three sets of armature coils for each magnetic pole of the electromagnets, separated by an angle of 120°, three e.m.fs can be produced during each revolution of the alternator. These three-phase generators are more energy efficient then a single-phase generator. Most power stations will have a number of generators with power ratings between 300 to 1000 MW.

Each alternator can produce voltages as high as 25kV. Step-up transformers increase the voltage to as high as 700 kV. This increased voltage results in a decreased current that reduces the heating losses in the power transmission. Additional transformers in the power grid further reduce energy losses and gradually lower the voltage to the required domestic or industrial level. Useful energy is lost due to eddy currents in the transformers in the form of thermal and sound energy. Thermal energy is also lost due to the current in the transmission cables.

Alternating current is the preferred transmission type. However, more and more high voltage direct current transmission (HVDC) is occurring. There are advantages in both systems. For example, dc currents travel through more of the cross-sectional area of a conducting cable whereas ac currents tend to travel through the outer portion of the cable – a phenomenon known as the "skin effect". Furthermore, three-phase ac requires multiples of 3 cables whereas dc only requires sets of 2. There are also problems synchronising generating stations to run at the same frequency. With increased globalisation and the selling of commodities such as electric power, dc does not have these problems.

ENERGY PRODUCTION

Exercise 8.1

1. The most efficient energy conversion occurs in

   A. tidal power stations
   B. diesel engines
   C. solar panels
   D. hydro-electric power stations

2. The original source of tidal power is the

   A. Moon
   B. Earth
   C. Sun
   D. water

3. Heat engines

   A. produce more work output than energy input
   B. take in thermal energy at a low temperature and exhaust it at high temperature
   C. convert heat into mechanical energy
   D. can be close to 100% efficient

4. All the following statements are correct EXCEPT

   A. generators convert mechanical energy into electrical energy
   B. nuclear reactors convert mass into energy
   C. chemical energy is a form of potential energy
   D. thermal energy and solar energy are the same

5. Two different objects that have different temperatures are in thermal contact with one another. It is the temperatures of the two objects that determines,

   A. the amount of internal energy in each object
   B. the process by which thermal energy is transferred
   C. the specific heat capacity of each object
   D. the direction of transfer of thermal energy between the objects

6. A generator takes in an amount E_k of kinetic energy. An amount W of useful electrical energy is produced. An amount Q of thermal energy is lost due to the moving parts of the generator. The law of conservation of energy and the efficiency of the generator are given by which of the following?

	Law of conservation of energy	Efficiency
A.	$E_k = W + Q$	W
B.	$E_k = W + Q$	W / E_k
C.	$E_k = W - Q$	W / Q
D.	$E_k = W - Q$	$W / (E_k - Q)$

7. The diagram below is a Sankey diagram for a typical oil-fired power station. It shows the useful electrical output after the energy has been transmitted to your home. Analyse the energy that is lost from the energy input to the final energy output. Determine the overall efficiency of the system.

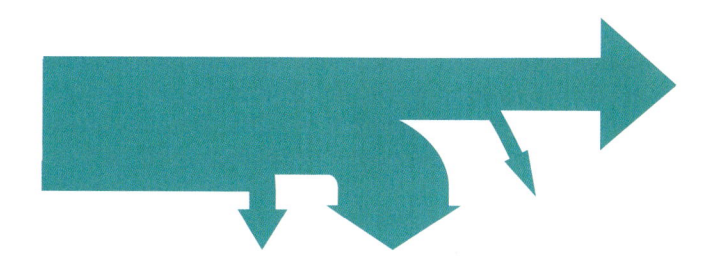

8. Copy and complete the following table to show the energy conversions for various devices. (*There could be more than one type of energy produced*).

Device	From	To
Cigarette lighter		
Human body		
Microphone		
Car engine		
Light bulb		
Light emitting diode		
Refrigerator		
Stereo speaker		
Thermocouple		
Atomic bomb		

CHAPTER 8

Specific energy and energy density of fuel sources

Energy density of a fuel is the amount of potential energy stored in a fuel per unit mass, or per unit volume depending on the fuel being discussed. In some cases it is obvious from context which quantity is most useful. For example, for coal, nuclear fuels and crude oil, energy per unit mass is the most important parameter. However, when discussing pressurized gases the energy per unit volume is more appropriate. When comparing, for example, the effectiveness of hydrogen fuel to petrol or gasoline both figures are appropriate. This is because hydrogen gas has a higher energy density per unit mass than does gasoline, but a much lower energy density per unit volume in most applications.

We are interested in the chemical potential energy storage within a fuel. Therefore, the energy density relates the mass of an energy store to its stored energy. The higher the energy density, the more energy may be stored or transported for the same amount of mass.

Energy Density Of A Fuel = Chemical Potential Energy ÷ Mass

It is measured in joules per gram ($J\,g^{-1}$). Bomb calorimetry is used to determine the value and this technique requires only small masses of a sample. However, the joule is a small quantity and therefore it is more common to use bigger units such as megajoules per kilogram $MJ\,kg^{-1}$.

Figure 805 gives some approximate energy density values for some fuels. When solving examination questions please consult the table that is provided in the *IB Data Booklet*.

Type Of Fuel	Gravimetric Energy Density MJ kg^{-1}	Volumetric Energy Density MJ dm^{-3}
Wood	15	
Coal	24	20
Crude Oil	42	27
LPG (Liquefied Petroleum Gas)	34.5	22
Compressed Natural Gas	55-56	10
Aviation Fuel	43	33
Ethanol	29.6	20
Plant Biomass	18	
Lead-acid Battery	0.1	
Nuclear Fission Of U-235	3 000 000 – 86 000 000	
Nuclear Fusion	645 000 000	

Figure 805   Energy density values for some fuels table

Figure 805 clearly shows that the **energy density** of uranium-235 would make it the best fuel per kilogram. However, nuclear power stations are very expensive to build and the cost for the production of electricity per kilowatt-hour is more expensive than the fossil fuels. There are also political, social and environmental factors to consider. The use of fossil fuels is the most cost effective way to produce energy.

Depending on the situation with regards to the use and storage of a fuel, the energy density will be an important consideration. For example, natural gas can be piped from the gas field to a site usually close to a shipping terminal where it can be placed in tanks as compressed natural gas and then be transported by ship to various countries in need of this fuel.

The energy density of coal is not the only consideration that has to be calculated. Its rank, chemical composition and heating ability are other important factors (*refer to Figure 802*). Coal with different rank advance will produce differing amounts of heat for a given mass. The grade of a sample of coal does not always indicate the chemical composition of that coal. However, with rank advance the purity of carbon content generally increases from peat → lignite (brown coal) → sub-bituminous coal → bituminous coal → anthracite. Chemical composition of the coal is defined in terms of the analysis of its moisture content, volatile matter (percentage of the coal that is lost as vapours when heated in the absence of air), ash, and fixed carbon. Table 806 lists some typical moisture values of some ranks of coal.

Rank Of Coal	Percentage Moisture Content
Peat	75 – 80
Lignite	50 – 70
Bituminous Coal	5 – 10
Anthracite	2 – 5

Figure 806   Different coals

If the coal is dried and analysed then the percentage by composition of carbon, hydrogen, oxygen and volatile matter changes. Figure 807 lists the approximate percentage composition by mass of different ranks of coal after they have been dried.

Component	Percentage By Mass			
	Peat	Lignite	Bituminous Coal	Anthracite
Carbon	50 – 60	60 – 75	80 – 90	90 – 95
Oxygen	35 – 40	20 – 30	10 – 15	2 – 3
Hydrogen	5 – 6	5 – 6	4 – 5	2 – 3
Volatile matter	60 – 65	45 – 55	20 – 40	5 – 7

Figure 807   Composition of coal

Because of the high composition of volatile matter in peat and lignite, they cannot be transported for long distances because of safety risks. Usually, a power station will be placed where the mine is located.

Other elemental analysis of nitrogen and sulfur within the coal will determine the choice when using various coals. The coal in some countries has a high sulfur content and this greatly affects air quality in some areas. On the other hand, Australian bituminous coal and anthracite is low in sulfur and it has become a top export fuel to Asia.

Example 1

A sample of lignite has a moisture content of 65%.

(i) Determine how much water is in a 10 tonne sample of this coal before crushing and drying?

(ii) Explain how the moisture content will reduce the amount of heat that can be obtained from combustion of the coal.

Solution

(i) There is 65% moisture content.

Therefore, the amount of water will be

$0.65 \times 10\,000$ kg = 6 500 kg

(ii) The more water present, the less coal there is to burn and some of the heat released has to be used to heat and evaporate the water.

Example 2

A sample of anthracite has a moisture content of 5% and when dried it has an energy density of 35 kJ g^{-1}. Assuming that during coal combustion the temperature of the water in the coal is raised from 20 °C to 100 °C and then vaporised at 100 °C, estimate the energy density of the coal as it is mined.

Solution

Energy required to convert the water to steam = $mc\Delta T + mL_V$

Since there is 5% moisture content, there is 5 g of water per 100g of coal. For 5 grams,

$Q = 5g \times 4.18\,Jg^{-1}K^{-1} \times (100\,°C - 20\,°C) + 5g \times (22.5 \times 10^2\,J\,g^{-1})$

= 12 922 J = 13 kJ

The energy in the other 95 grams of anthracite is
35 kJ g^{-1} × 95 g

= 3 325kJ

The total usable energy in 100 grams

= 3 325 – 13 kJ = 3 312 kJ

For one gram this would be 33.12 kJ

The energy density as it is mined will be 33.12 kJ per gram.

Primary energy sources

Figure 808 shows the approximate world consumption figures expressed as percentage of use.

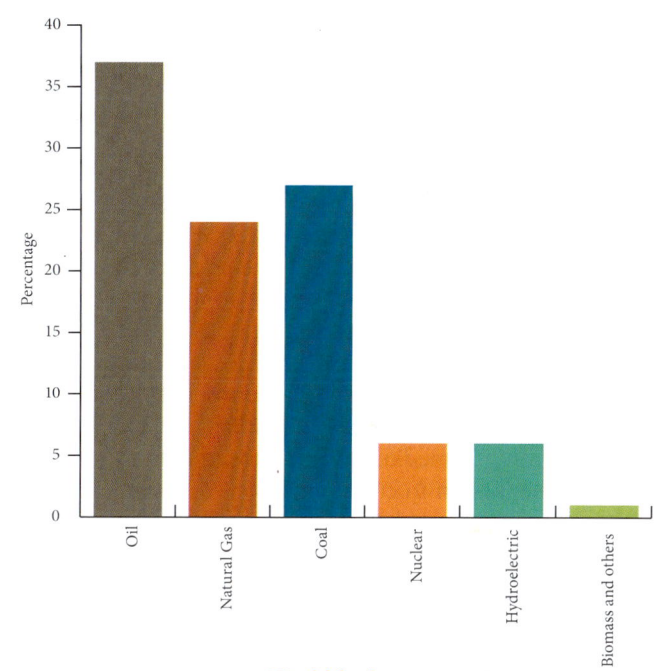

Figure 808   World fuel consumption

The 1% for biomass and others include biomass, solar, geothermal, wind power and other alternatives. The percentage use represents all uses of the fuels. Much of the crude oil and natural gas is used for the petrochemical industry for the manufacture of plastics, pharmaceuticals and many more synthetic materials. These 2 fuels are also used for transportation fuels. The next biggest use is for the production of electricity. Each country in the world has different resources and as such these percentage use will differ from country to country.

Most of the energy supply used today is derived from **fossil fuels** which are naturally occurring fuels that have been formed from the remains of plants and animals over millions of years. The common fossil fuels are peat, coal, crude oil, oil shale, oil tar and natural gas.

When these fuels are burnt in air (oxygen) they produce carbon dioxide (CO_2), water and varying amounts of energy per kilogram of fuel. There are also indirect means of producing thermal energy by the chemical and physical processing of fossil fuels such as coke and coal gas from coal, and petrol from crude oil.

Although the main products of the combustion of fossil fuels are carbon dioxide and water there are many other products that are released into the environment. Because of incomplete combustion, carbon (soot) and carbon monoxide (a poisonous gas in high concentrations) are also produced. Furthermore, pollution is caused by the emmision of sulfur dioxide, oxides of nitrogen, unburnt hydrocarbons and particulate matter.

Because of the abundance of carbon dioxide in the air from the carbon cycle, disposal of carbon dioxide into the air was believed to be harmless. However, it is a well known fact the the percentage of carbon dioxide in the air is increasing. Even if volcanic activity, forest fires and burning off to clear land has changed this composition, there is little doubt that man's use of fossil fuels has contributed to the increased concentration of carbon dioxide in the atmosphere. There has been an increased anxiety by many scientists, world leaders of governments and environmentalists as well as world citizens as to the long term effects of carbon dioxide increase. Many believe that this increase in concentration is increasing the temperature of the atmosphere and the oceans because of the enhanced greenhouse effect.

The primary source of world energy has directly or indirectly had its origin due to the radiant energy from the Sun. The Sun is a medium-size hydrogen/helium star that uses **nuclear fusion** to convert millions of tonnes of mass into energy each second. It has been in existence in excess of 6 billion years and it has enough hydrogen to last for at least another 8 billion years. It has been providing 90% of the thermal energy needed to heat the Earth to a temperature comfortable enough for a diversity of living things to exist.

For recorded history, the Sun has allowed plants and certain bacteria to convert solar energy into chemical energy stored in sugars and other carbon compounds. First-order consumers eat plants and other higher order consumers eat both plants and other animals to obtain the chemical energy necessary to survive. However, even today, up to 40% of the world's population does not get the minimum requirement of 8500 kJ of energy through food per day. Apart from the fact that the Sun produces 90% of the thermal energy needed to heat the Earth, it is also indirectly responsible for wind, ocean currents, wave action, water evaporation and precipitation, food, wood, biomass and the fossil fuels.

Some of these properties are now being used for alternative energy sources such as hydroelectricity, wind power, biogas, passive solar energy panels and photovoltaic cells.

Nuclear fission is the source of nuclear energy in reactors and geothermal plants. The process of nuclear fusion reactors is being developed. The Moon's gravitation is the cause of tides and there are a few tidal power stations in operation. Chemical energy is also used for supply of energy in batteries and fuel cells.

Non-renewable and renewable energy

A **non-renewable source** is one that is considered to be a temporary source that is depleted when it is used. (Nuclear energy could be considered to be a non-renewable source of energy also. However, modern breeder reactors can produce more fissionable material than they consume and it is estimated that reserves of nuclear fuels could last for several thousand years. We will not treat uranium as a non-renewable source and we never consider it to be a fossil fuel).

Coal is an organic material made up primarily of carbon, along with varying amounts of hydrogen, oxygen, nitrogen and sulfur. It is a sedimentary rock usually found as layers associated with sandstone and shale. It is the most abundant fossil fuel and has a high heat of combustion. It is formed by the partial decomposition of plant material. If you look at peat or brown coal under a microscope, bits of fossilised wood, leaves, and other plant material can be seen. When it was formed millions of years ago in the Carboniferous or younger geological periods such as the Permian period, there was abundant plant growth. The plant material was laid down in waters

- with a low flow rate
- free of excessive minerals
- where there was an absence of air
- that received little other sediment

Swamps would have been good environments for the beginning of coal formation because in stagnant water very little oxygen is present. In a swampy environment, anaerobic bacteria attack plant matter and partially decompose it. Carbon becomes concentrated in the remains. The bacteria gradually decrease, as they are killed by the poisonous acids produced in the decomposition process. At this stage, the plant matter is converted to peat.

Peat is a brownish material that looks like wood. Although it can be burnt as a fuel, it contains a lot of water, and is very smoky when burnt.

As the peat became buried beneath more plant matter, the pressure and temperature increased and the water was squeezed out of it. At a later stage in geological history, the swamp was covered rapidly with a sedimentary layer of rock. As the material became compacted the carbon content

ENERGY PRODUCTION

increased and the peat converted to **lignite**, then to **sub-bituminous** coal and finally **bituminous** coal. If the coal strata were then subjected to folding which increases the heat and pressure, a metamorphic rock called **anthracite** is formed. At each stage in the **rank advance,** the coal has a higher carbon content and a higher energy content per unit mass. Figure 809 demonstrates the formation of different coals.

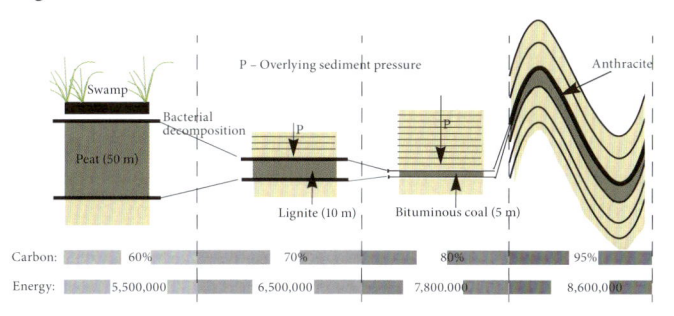

Figure 809   The formation of coal

Crude oil and **natural gas** are products of the decomposition of marine plants and animals that were rapidly buried in sedimentary basins where there was a lack of oxygen. The organic material was laid down in a body of stagnant water where the presence of the depositing organic matter created an acid environment. The organic matter was quickly buried beneath mud. The source rocks were buried under sufficient cover of overlaying strata so that the conditions were right for the conversion of the organic matter in the source rock into hydrocarbons and other organic compounds. When the hydrocarbons and other organic compounds were generated, they were dispersed as individual molecules in the source rock. With increased heat, pressure and Earth movement, these molecules migrated into other porous, permeable reservoir rocks. With further heat and pressure, these reservoir rocks became trapped between impervious (non-porous) cap rocks. The oil and gas in the reservoir rocks also became trapped with the less dense gas rising to the top and the more dense liquid and solid crude oil sinking to the bottom of the trapped reserve. Figure 810 shows the accumulation in a reservoir rock.

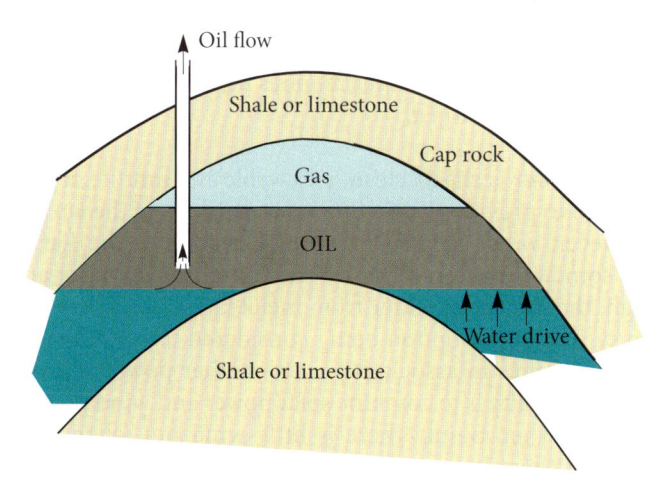

Figure 810   Accumulation of oil in reservoir rock

A borehole is drilled into the rock containing hydrocarbons using diamond core bits, and the reservoir is opened to the atmosphere and after the release of the natural gas the crude oil is pumped to the surface. If the pressure is not great enough or the liquid is too viscous to flow, the crude oil can be pumped to the surface.

The crude oil is separated into fractions in a fractionating column using fractional distillation. The crude oil is heated to about 400 °C to produce a hot liquid/vapour mixture. The lower boiling point components rise higher up in the fractionating column where they are condensed. The higher boiling point components are removed lower in the fractionating column.

Crude oil is found as a liquid or a vapour although some oil can exist as a solid. The solid/ liquid components are known as crude oil and the gaseous component is known as natural gas.

Although hydrocarbons have been found in rocks formed more than a billion years ago, it is thought that most reserves were formed less than 500 million years ago. Some formed as recently as 10 million years ago, and it is believed that some is being formed today.

Oil shale and **oil tar** deposits make up less than 2% of the world's fossil reserves. Oil shale is complex solid hydrocarbon material called kerogen. It is found in a fine-grained sedimentary rock called marl. Exploration and refining of this material began during the 1980s when oil prices escalated. It requires a high capital outlay to produce the fuel from the kerogen and at this stage it is not considered viable for economic reasons. Tar sands are deposits that contain a tar–like material called bitumen. It is found in the same geographic areas as oil is. It can be pyrolysed to produce crude oil. The extraction and refining of tar sands can be done at a price similar to crude oil. There are a number of tar oil plants in North America.

As fossil fuels are exhausted, we will come to rely on alternative renewable energy sources. A **renewable energy source** is one that is permanent or one that can be replenished as it is used. Renewable sources being developed for commercial use include nuclear fission, solar energy, biomass, wind energy, tidal energy, wave energy, hydro-electric energy and geothermal energy. Nuclear fusion is likely to be the energy source of the future but at this stage the reaction of the fusion elements can neither be contained nor controlled.

Figure 811 shows some of the processes whereby renewable energy is produced in industrialised societies.

253

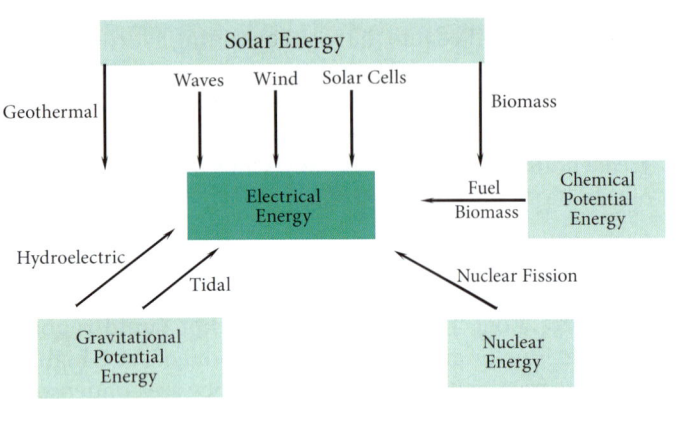

Figure 811 Renewable energy transformations

Advantages and disadvantages of fuels

The importance of fuels to meet the world's energy consumption cannot be overlooked. The fossil fuels are dirty fuels and they are the cause for environmental pollution and land degradation. There is little doubt that fossil fuels are partly to blame for the increased levels of carbon dioxide in the atmosphere that could be the cause of global warming. However, there appears to be another school of thought that argues that the increased carbon dioxide and melting of the ice caps may be due to natural cycles in the Earth's history. Ice core samples in Greenland have revealed that the Earth has been hotter in the past. Where one would expect low sea level countries like The Maldives to have rising sea levels, it has been found that the sea levels have actually decreased. However, the majority of scientists believe that combustion of fossil fuels is the major cause in global warming due to the greenhouse effect.

Nuclear energy power stations increased during 1950-1970. However, there was a lull in power production until the present century due to people being unsatisfied with the potential hazards of nuclear meltdowns as occurred in the USA (Three Mile Island in 1979) and the Ukraine (Chernobyl in 1986) and the hazards associated with radioactive wastes produced by nuclear fission. However, there has been a recent renaissance in nuclear power production as governments look for greener fuels to meet there electricity needs. France is a major producer of nuclear energy and sells electricity to many neighbouring nations.

Hydropower accounts for 19% of the world electricity supply, utilising one third of its economically exploitable potential. The advantages of hydroelectric power are numerous. It avoids emissions of greenhouse gases, sulfur dioxide and particulates. The efficiency is high and the running costs are low. There is no pollution and the energy resource is renewable. The main disadvantages are that large dams need to be built with population displacement and the flooding of plant and animal habitats, and farming land. There have been some dam failures resulting in the death of villagers downstream from the dam reservoir. Because of its advantages industrialised countries have already developed their hydroelectric potential.

Geothermal power is a cheap source of energy in past and present volcanic areas. There are large geothermal plants in Iceland, Italy, New Zealand, Japan, Hawaii, Mexico and Chile. In many places the underground temperature is greater than 150 °C at depths less than 3 kilometres and there is commonly circulation of underground water. The steam that is produced is piped under pressure to the surface to drive turbines and thus generate electricity. Geothermal power has great potential for many countries that have hot beds of rocks and an artesian water supply. The disadvantages of geothermal power include the release of polluting gases such as as sulfur dioxide and hydrogen sulfide into the atmosphere and groundwater pollution by chemicals including heavy metals.

The main advantage of **tidal power** is that there is no pollution and the energy is renewable. The main disadvantage is that there are few areas in the world that have the necessary tidal range. The construction costs are high. Because high tides occur approximately 50 minutes later each day, it is difficult to meet electricity demand during some high peak times. There are also long construction times, high capital intensity, and these factors are likely to rule out significant cost reductions in the near term.

Solar energy was neglected as an energy resource but today there is renewed interest because of its many advantages. It is a means of using a free, renewable energy resource. It is available to some extent everywhere in the world, unlike fossil fuels and nuclear power. It is exempt from rising energy prices. Few environmental problems are created. It can be used in a variety of energy transformations for heating, cooling, electricity, transportation, lighting and mechanical power. Its disadvantages are also evident. It produces a small energy output per surface area of the cell being used. For large-scale production, it requires thousands of mirrors or cells that take up a large area of land. It is intermittent with its output being upset by night and clouds. It is relatively expensive to set up and thus requires many years before the investment is returned.

Wind power is cheap, clean, renewable and infinite. It can be used to provide electricity to remote areas of the world. However, winds resulting from the heating of the Earth are somewhat unpredictable. The initial set-up costs are high, the structures suffer from metal fatigue and they are noisy. The better option being considered today is to use wind turbines in association with another power source. One possibility is to combine solar power and wind power. Usually, when Sunny there is little wind but when it is overcast there is more wind. Another possibility is to use the wind to pump water into high dams associated with hydroelectric power stations and then to run the water downhill during high electricity demand periods.

Wave and ocean current power has great potential. Recent favourable technological developments in Scotland, Australia, Denmark and the USA suggest that it could provide 10% of the current world electricity supply (if appropriately harnessed) – and the potential synergies with the offshore oil and gas industry. However, the cost for electricity using waves is too high at this stage. Ocean thermal energy conversion (OTEC) technology is also a high potential energy source but further R&D are still needed, in order to be able to build a representative pilot plant to demonstrate OTEC's advantages. The most promising wave power is the oscillating water column (OWC) wave energy converters that will be examined later.

Electricity as a secondary and versatile form of energy

A decision to develop a coal mine, an oil field or a gas field as a commercial venture requires that many factors have to be taken into account. Some of the considerations include

- the location of the reserve
- the availability of water
- construction costs
- the mining and treatment costs
- the cost of towns and services to house the workers and their families
- the cost of transportation systems
- the price obtained for the fossil fuel
- the energy density of a coal
- the type of light or heavy crude oil
- the lifetime of the reserve
- the environmental costs as imposed by government regulations.

Transportation and storage of fossil fuels can be undertaken using pipelines, railroads, trucks and ships. Natural gas is usually transported and stored in pipelines although exports of LNG are shipped in pressurised containers. Pipelines are a cost effective means of distributing natural gas and there are many agreements between countries who pipe their gas to other countries as is the case with the vast reserves of the Russian Federation. The disadvantages of piping gas include unsightly pipelines, possible leakage, explosions, governments holding other countries to ransom when they do not conform to political issues, and, possible terrorist activities.

Many oil refineries are located near the sea close to large cities so that the labour force and social infrastructures are available, and to ensure that the import or export of crude oil and it fractions can be transported by ships to their destinations throughout the world. The biggest disadvantage of shipping has proven to be the oil slicks that have threatened wildlife due to sinking and leaking ships. At the refineries, great care has to be taken in storing the crude oil fractions in tanks that have containment walls to protect leakage or explosion. Pipelines are common in many countries. However, they are unsightly and they are open to terrorist attack as has been the case in Iraq, Nigeria and Kuwait in recent times. Likewise, the transportation of petrol and diesel from the cities to the outlying provinces also has its associated hazards of leakages, transport accidents and explosions.

Power stations using coal are preferably located near coal mines to minimise transport costs. Furthermore, many steel mills are located near mines as much of the coal is required for coking (the removal of volatiles from the coal to produce coke). The coke is used in blast furnaces to aid in the reduction of iron ore to iron. In Australia, around 8% of the coal mined is used for coking and 25% is used as steaming coal in power stations. 45% is exported to Asia by ships. Again, it is not surprising that the major coal mines are located close to shipping docks. The coal is railed to the docks. Coal with high rank advance is relatively safe to transport and store.

There are three main factors which affect the efficiency of power stations:

- fuel type (lignite, black coal, gas, oil)
- the load factor (full load, part load)
- the employed technology (conventional, combined heat and power)

Power stations are becoming more efficient and many conventional ones have improved their efficiency by 5 to 10 percent over the last 30 years. This is mainly due to a better cooling water source allowing for a greater variation between the hot and cold reservoirs. A power station in Denmark obtains 45% efficiency, the world's highest for an operational, single cycle, large coal-fired plant.

Conventional coal-fired power stations can obtain efficiencies in the 33 – 39 percent range. Gas-powered power stations are in the 33 – 46 percentage efficiency range.

The **load factor** is an indication of electricity used by consumers and it is calculated by dividing the average load by the peak load over a certain period of time.

Residential homes tend to have low load factors because people only use electricity during certain hours of the day. Industrial consumers will have high load factors because they operate throughout the day and night.

The highest efficiencies are being obtained in combined cycle plants such as the cogeneration or CHP (combined heat and power) plants. In a combined cycle plant, surplus heat from a gas turbine is used to produce steam which in turn drives a steam turbine. Efficiencies of over 50% can be achieved. Furthermore, if the plants waste heat is utilized for the heating of houses or an industrial process, efficiencies of 80% are achievable.

With the advent of the Industrial Revolution in the 1750s, the **steam engine** and later the internal combustion energy became the principal sources of power. At their best steam engines were only 8% efficient. Large amounts of energy were needed to make and run machinery, to make roads and railway lines, to service cities as the rural population flocked to the cities for employment.

In the steam engine, water is heated in a boiler heat reservoir (high temperature) to produce steam. The steam passes into an open intake valve where it expands causing a piston to move outwards. As the piston returns to its original position, the intake valve is closed and the exhaust valve opens allowing the exhaust gas to be forced into a condenser heat reservoir (low temperature). The cycle is shown in Figure 812.

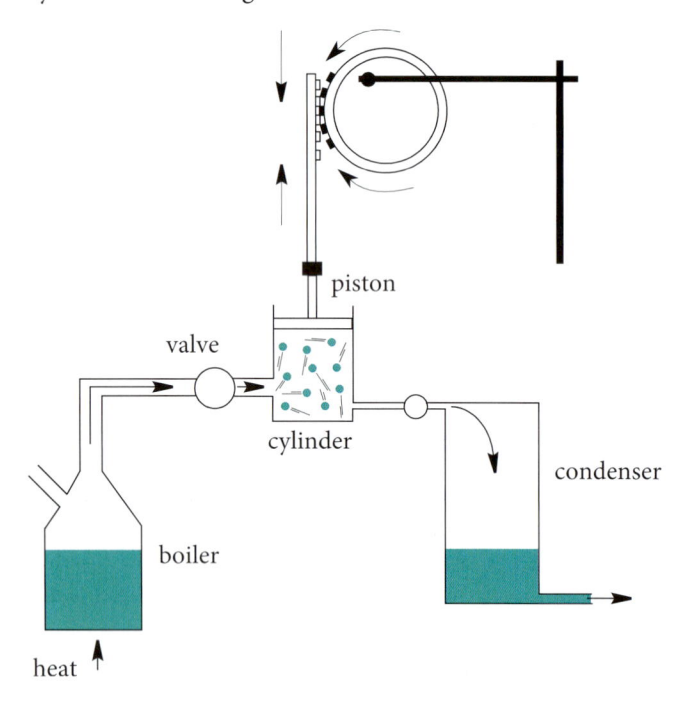

Figure 812   The steam engine

With the steam engine came the opportunity to transport agricultural products from the country to cities and industrial products to the country and other cities. However, the Industrial Revolution saw the production of large amounts of steel from iron ore and coke (a strong porous solid composed mainly of carbon).

Therefore, the advent of industrialisation saw a marked increase in the rate of energy usage that led to industries being established close to the source of the fuel in order to cut down on transportation costs. Coal is a versatile fuel that can be used directly as a solid fuel or modified to coke,

processed to liquid and gaseous fuels or manufactured into a wide range of chemicals. Every tonne of a 60% iron ore requires one tonne of coke. Before electricity was produced after the discovery of electromagnetic induction in 1832, town gas was produced for industry and homes for supplying heat energy for cooking and warming houses and for lighting. Those countries that were successful in modernising usually centred their manufacturing industries close to coalmines. This is especially true in the Ruhrgebiet of Germany and the coal mines of the USA, China, Australia, United Kingdom and Russia.

The steam engine (and steam turbine) are examples of external combustion engines. The fuel is burnt outside the engine and the thermal energy is transferred to a piston or a turbine chamber by means of steam.

Most electricity is produced today using steam turbines. The piston is replaced with a rotating turbine that contains blades. The rotating turbine converts mechanical energy to electrical energy via a generator. Heat from a fossil fuel (coal or natural gas) or nuclear fuel is added to water to form high-pressure steam. The steam performs work by expanding against the turbine blades. To be compressed at low pressure, the steam is condensed. This requires that heat is extracted from the water, and this is achieved by using cooling towers. Steam turbines are about 40% efficient.

The intensive use of electrical energy and fossil fuels for the use in turbines and heat engines is a fairly recent development in the history of our civilisation. These engines consume large amounts of energy to the extent that we consume up to one million kJ per person per day in the industrialised world. The industrial world consisting of one-third of the world's population consumes 80% of the world's energy.

Energy density of fossil fuels and power station demand

Each type of fuel used in fossil fuelled power stations has a different energy density. Although the energy density of lower ranked coals is not as high as bituminous coal and anthracite, many economic factors still make power stations in lignite areas feasible. The mines may be open cut mines rather than below the surface mines, they may be close to capital cities or they may be close to shipping terminals. Similarly, oil fields might be close to population areas. A big advantage of natural gas is that it does not have to be in situ because it can be piped over long distances thus not requiring transportation. Natural gas can be liquefied and shipped to countries that have the demand for this export commodity.

There are different demands on power stations throughout a day as there are peak and off-peak periods due to our lifestyle choices. While many industries never stop, humans have an appetite for power from around 18:00 hours until we go to sleep. The first hydro-electric power station at Niagara Falls in New York state and Canada designed by Nikola Tesla uses the off-peak electricity surplus to pump the water back into containment dams above the falls. During demand periods, the water is released from the dams to achieve more megawatts of energy per hour.

A schematic diagram of a typical coal-fired power station is shown in Figure 813.

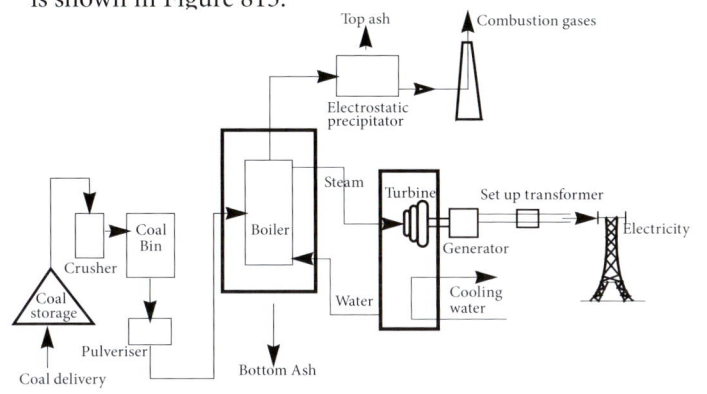

Figure 813   Coal-fired power station

Figure 814 lists typical energy density values of the major fuels used in fossil-fuelled power stations as well as petrol.

Fuel	Energy Density (Coals When Dried)		
	Solid Or Liquid		Gas
	MJ kg^{-1}	MJ dm^{-3}	MJ dm^{-3}
natural gas	55-56	23-24	0.038-0.039
propane	50.0	25.4	0.093
butane	49.5	28.7	0.124
gasoline	56.5	45-55	
peat	25		
lignite	25-30		
bituminous coal	30-35		
anthracite	35-38		

Figure 814   Typical energy densities for dried coals and other fuels (table)

Because of the higher energy density of natural gas, it is becoming the preferred choice of the new power stations that are coming into production. Natural gas is also cleaner and less polluting than coals and oils.

Figure 815 shows the quantities of fuel and air used hourly in a typical coal-fired power station and the hourly production of exhaust gases and ash.

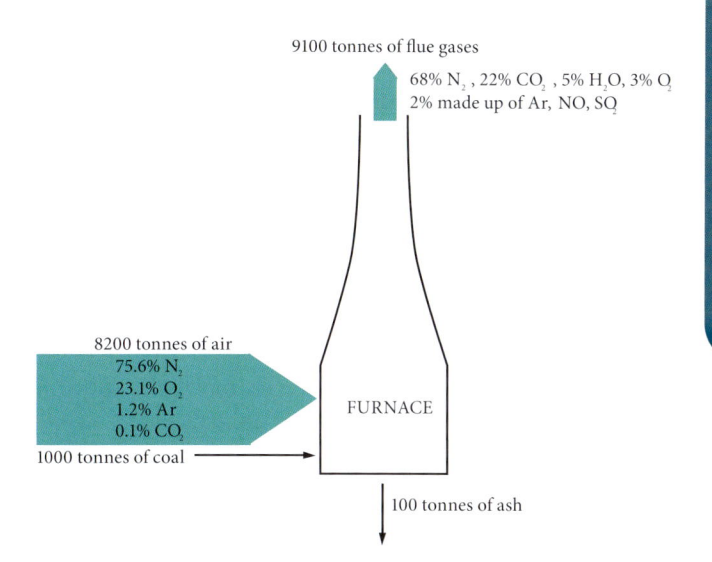

Figure 815   Hourly use of reactants and products of a coal power station

Pollutants are substances that have undesirable effects on living things and property. Air pollution occurs when these pollutants are introduced into the atmosphere.

Carbon dioxide and carbon monoxide are the major pollutants introduced into the air by fossil-fuelled power stations. Carbon monoxide is a poisonous gas that reduces the ability of the blood to transport oxygen from the lungs to the cells in the body. If the carbon monoxide is in a high enough concentration, the haemoglobin reacts with the carbon monoxide rather than the oxygen and poisoning or death can occur. Luckily, most of the carbon monoxide is quickly removed from the atmosphere by soil bacteria.

Carbon dioxide as already noted is increasing in the atmosphere and it is believed that infra-red radiation is being trapped in the atmosphere due to carbon dioxide increase. Through the enhanced greenhouse effect, the lower atmosphere temperature could increase by several degrees and this will contribute to global warming.

The sulfur dioxide results from the combustion of carbon compounds containing sulfur and the oxidation of metal sulfides in the coal. Attempts are made to minimise sulfur gases in a scrubber. Sulfur dioxide can combine with water in the atmosphere to form sulfurous acid (H_2SO_3), a mildly acidic solution that falls as acid rain. Many plants are sensitive to sulfur dioxide as it reduces the production of chlorophyll and their leaves turn yellow. At higher concentrations it can cause plants and trees to dry out, bleach and die. Humans who suffer from respiratory problems can have problems when the gas reacts with moisture above the larynx. The effects are increased further if particles are present as the sulfur dioxide adheres to them and they are carried into the bronchus and alveoli. Sulfur dioxide can also reduce the growth of nitrogen-fixing soil bacteria. Petroleum refining and the production of coke also contribute to atmospheric sulfur dioxide.

Oxides of nitrogen are formed either from the combustion of carbon compounds containing nitrogen or the reaction at high temperature between nitrogen and oxygen in the air. These oxides of nitrogen can also be an environmental hazard. The reactions of oxides of nitrogen and hydrocarbons can produce photochemical smog. In the presence of ultra-violet radiation nitrogen dioxide decomposes to an oxygen free radical that can react with oxygen molecules to form ozone which is harmful to plants and animals.

Power stations and factories produce particulate matter (small dust particles or ash) in the form of silica and metallic oxides, silicates and sulfates. By the use of electrostatics, much of this matter can be removed so that it does not enter the atmosphere, and the removed matter can be recycled to produce construction materials such as bricks. Figure 816 shows the basics of a typical electrostatic precipitator.

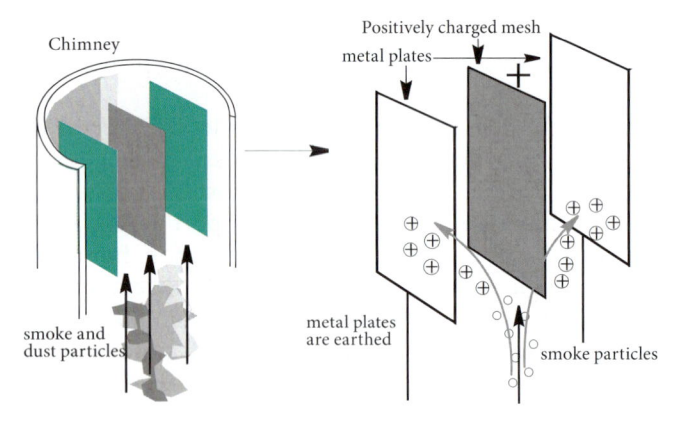

Figure 816 Schematic diagram of an electrostatic precipitator

A metal grid made of mesh is charged positively to about 50 000 V to cause the smoke surrounding it to be ionised to produce electrons and positive smoke ions. The positive ions are repelled from the mesh and attach themselves to some of the dust particles in the smoke. The charged dust particles are attracted to Earthed plates where they stick. A mechanical device hits the plates periodically and the ash falls into collecting bins.

Example

A coal-fired power station burns coal with 50% moisture content. The composition of the dried sample is found to contain on analysis 72% carbon, 5% hydrogen and 23% oxygen. If 500 tonnes is burnt hourly:

(a) Estimate the mass of water vapour emitted from the cooling towers each hour.

(b) Estimate the mass of water vapour produced in a week.

(c) Estimate the volume of condensed water vapour produced in a week.

Solution

(a) Assuming the hydrogen and oxygen is converted to steam, the total amount of steam = 50% + 28% of the remaining 50% = 64%

64% of 500 tonnes = 320 tonnes.

(b) $320 \times 24 \times 7 = 5.4 \times 10^4$ tonnes

(c) 1 tonne = 1000 kg $1 dm^3 = 1 kg$

5.4×10^4 tonnes $\times 1000 = 5.4 \times 10^7$ dm^3.

Exercise 8.2

1. Which one of the following is **not** considered to be a fossil fuel?

    A. wood
    B. uranium
    C. coal
    D. crude oil

2. The correct rank advance for rank advance of coal is:

    A. lignite, peat, bituminous coal, anthracite
    B. bituminous coal, peat, anthracite, lignite
    C. peat, lignite, bituminous coal, anthracite
    D. anthracite, lignite, bituminous coal, peat

3. The fuel below with the highest energy density value is:

    A. coal
    B. crude oil
    C. ethanol
    D. compressed natural gas

4. Why are energy density values of fuels usually expressed in J g^{-1} rather than kJ mol^{-1}?

5. Suggest a reason why coal is ground to a fine powder before combustion.

6. A sample of lignite has a moisture content of 65% and when dried it has an energy density of

258

28 kJ g^{-1}. Assuming that during coal combustion the temperature of the water in the coal is raised from 20 °C to 100 °C and then vaporised at 100 °C, estimate the energy density of the coal as it is mined.

7. A coal-fired power station burns coal with 30% moisture content. The composition of the dried sample is found to contain on analysis 70% carbon, 5% hydrogen and 25% oxygen. If 1000 tonnes is burnt hourly:

    (a) Estimate the mass of water vapour emitted from the cooling towers each hour?

    (b) Estimate the mass of water vapour produced in a week?

    (c) Estimate the volume of condensed water vapour produced in a week?

8. It has been suggested that crude oil should be used for other purposes rather than as a transportation fuel. Deduce the reasoning behind this statement.

9. Assume that a sample of coal has an empirical formula C_5H_4 and that a coal-fired power station burns a 1000 tonne of coal per hour.

    (a) Write an equation for the complete combustion of the coal.

    (b) Calculate the mass of oxygen required for this combustion each hour.

    (c) If 25 dm^3 of oxygen is required per mole, calculate the volume of oxygen that is required each hour for this combustion.

    (d) Air contains approximately 20% oxygen. What volume of air is required hourly.

10. A coal-fired power station has a power output of 500 MW and operates at an efficiency of 35%. The energy density of the coal being consumed during combustion is 31.5 MJ kg^{-1}.

    (a) Determine the rate at which heat is being produced by the burning coal.

    (b) Determine the rate at which coal is being burned.

    (c) The heat is discarded into the cooling towers of the power plant and is then stored in containment reservoirs. Determine the water flow rate needed to maintain the water temperature in the towers at 10 °C.

Energy transformations of a nuclear power station

The principle mechanisms involved in the production of electrical power can be demonstrated by looking at the energy conversions in a power station using an energy flow diagram as shown in Figure 817. The rectangles contain the different forms of energy, the circles show the conversion process, and the arrows show energy changes and energy outputs. The linked forms of energy can be said to form an energy chain.

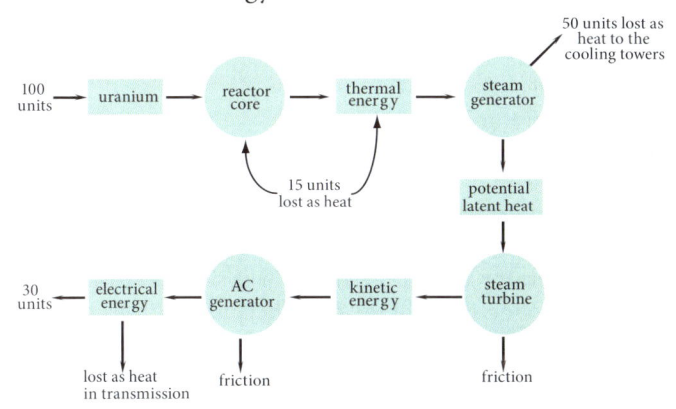

Figure 817   Principal mechanisms in a nuclear power station

The efficiency can be represented by the Sankey diagram as in Figure 818.

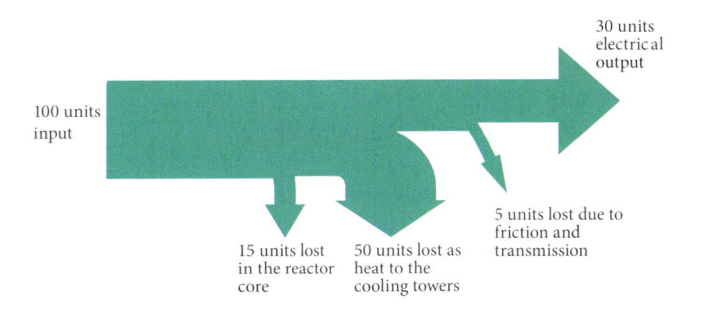

Figure 818   The efficiency of a power station

Heat energy is produced by the fission of uranium nuclei in the core of the nuclear reactor. Heavy water absorbs the heat energy in a heat exchanger under pressure, and it is turned into steam. The steam contains latent potential energy as it has been converted from a liquid to a gas. Steam under pressure is capable of doing mechanical work to supply the rotational kinetic energy to turn the steam turbines. The turbine is coupled to the generator that produces electrical energy.

Energy is lost to the surroundings at many stages. For example, if 100 units of energy are supplied from the primary energy source then only 30 units of useful energy is available. The majority of the useful energy is lost to water in the cooling towers as heat is evolved in the condensation component of the heat exchanger cycle. Other forms of energy loss are shown on the arrowed parts of the diagram.

Essential features of a thermal fission reactor

The essential features of a **thermal reactor** are:

- the fuel
- a moderator
- the control rods
- the coolant
- radiation shielding

The **moderator** is a material that will slow down the fast neutrons to the speed of the slow thermal neutrons needed for a self-sustained reaction without absorbing the neutrons when they collide with the moderator material. The moderator material is placed around the reactor core and in between each of the fuel assemblies. In order to be effective, the moderator must have a mass very close to the mass of a neutron so that the fast neutron can loose maximum energy in a single collision. Moderators include ordinary water, heavy water (D_2O), graphite, beryllium or liquid sodium. In the Chernobyl accident, the graphite core caught fire. Graphite fires are almost impossible to fully extinguish. The reactor core sank to the bottom of the reactor building and a theory called the 'China Syndrome' developed that the core could continue to burn out of control until it would eventually penetrate the Earth's surface.

If this had occurred, the core would have reached the water table and there would have been a massive radioactive cloud of steam many times the radioactivity of the original cloud. Fortunately, this did not happen.

The rate of nuclear fission in the reactor core can be controlled by inserting or removing the **control rods**. The control rods are constructed of materials that absorb neutrons. The rods are usually steel rods containing boron or cadmium that are said to have high neutron capture per cross-section. Most reactors have two sets of control rods: one set of regulating rods for routine control of the fission rate, the other as a safety measure in case they have to be lowered into the core during an emergency shut-down. The regulating rods can be added or removed, or partially or fully inserted into the core as needed. There can be a large number of control rods in or out of a reactor.

The **coolant** circulates through the reactor core and removes thermal energy transferring it to where it can do useful work by converting water into steam. The energy release in a single fission reaction is about 200 MeV or 3.2×10^{-11} J. Because the coolant is in direct contact with the reactor core, it can become radioactive and contaminated. Therefore, the coolant must exchange its heat with a secondary cooling circuit through a heat exchanger. The steam produced in the secondary loop drives a turbine to produce electricity. The low pressure steam that passes through the turbine is passed into cooling ponds or cooling towers where the excess heat is dissipated. Common coolants include air, helium gas, heavy water, liquid sodium or certain liquid organic compounds. In many reactors, the coolant is also the moderator.

The **radiation shielding** ensures the safety of personnel working inside and around the reactor from suffering the ill effects of radiation exposure. There are usually two shields: several metres of high-density concrete to protect the walls of the reactor core from radiation leakage and to help reflect neutrons back into the core and a biological shield to protect personnel made of several centimetres of high density concrete.

A typical schematic diagram of a thermal reactor is shown in Figure 819.

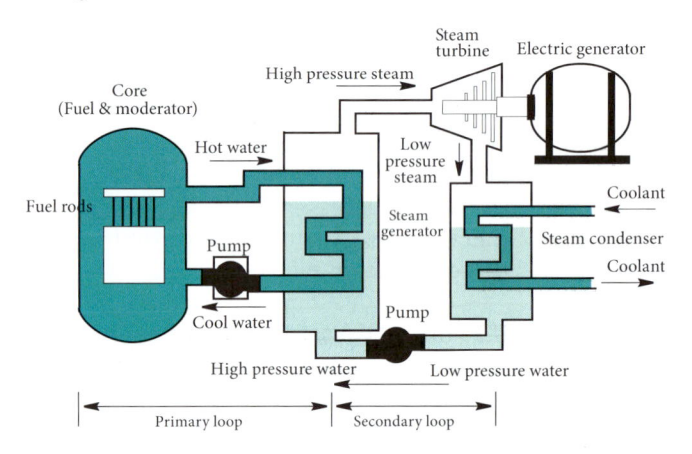

Figure 819   A schematic thermal reactor

The heat exchanger in nuclear reactors

Recall that we cannot convert all the random motion associated with the internal energy into useful work but we can at least extract some useful work from internal energy using a heat engine. To make a heat engine, we need a source of heat and a working fluid. Recall also that according to the Second Law of Thermodynamics that it is impossible to construct a heat engine operating in a cycle that extracts heat from a reservoir and delivers an equal amount of work.

A **heat exchanger** is a system basically acting as a heat engine driven by chemical reactions (the combustion of fossil fuels) or by nuclear reactions. The working fluid is water heated in a boiler that is converted to steam at high pressure. In a nuclear power station, the heat produced by the nuclear reactor is fed via the closed primary loop to a steam generator vessel.

The steam generated expands adiabatically against the blades of a turbine – a fan-like structure that spins when struck by the steam. The turbine is coupled to a generator that converts mechanical kinetic energy to electrical energy. The energy flow diagram of a heat exchanger is shown in Figure 820.

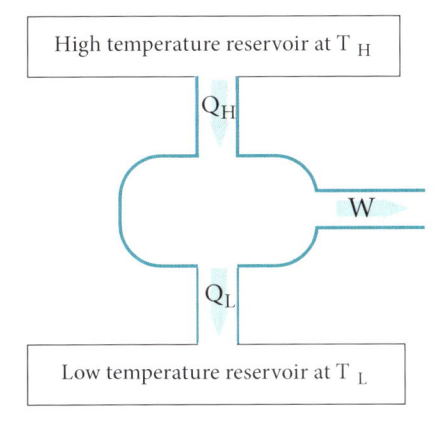

Figure 820   Energy flow diagram of a heat exchanger

When the steam leaves the turbine, it is in the gaseous state at a higher temperature than the water supplied to the boiler. If the steam had returned to the original liquid state all the energy acquired in the boiler would have been extracted as work and the Second Law of Thermodynamics would be violated.

To complete the cycle and to use the steam again, the steam is run through a condenser. The condenser is a coil of pipes in contact with a large volume of water, and carries the steam back to the boiler as cool water. The basic design features are shown in Figure 821.

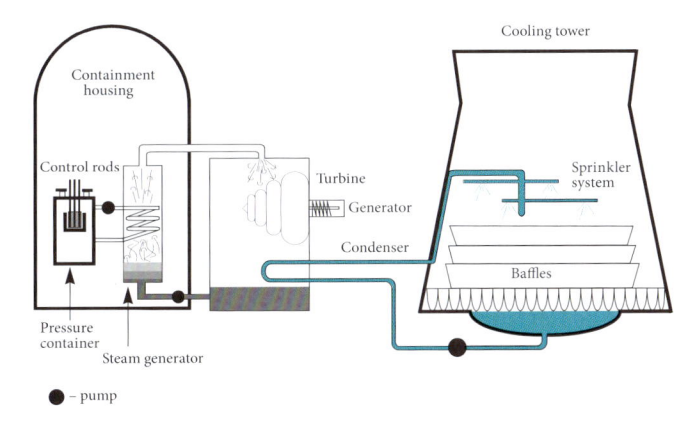

Figure 821   Design features of the heat exchanger and cooling tower

The temperature of the reactor is limited to a temperature of 570 K. Higher temperatures tend to damage the fuel rods. Typically, the water in the secondary loop is returned after condensation to the boiler at a temperature of 310 K. It can be shown that, the maximum possible efficiency of a nuclear power plant is:

$$\eta = 1 - \frac{Q_L}{Q_H} = 1 - \frac{310}{570} = 0.46 = 46\%$$

With further energy used to drive pumps and pollution control devices, the efficiency is usually reduced to 34%. Many of the original power plants used water in the reservoir surrounding the condenser from rivers and lakes to cool the steam back to liquid water. However, this caused thermal pollution and produced many ecological problems. Today, expensive cooling towers release the waste heat to the air. This form of thermal pollution has a decreased environmental effect.

Safety and risks of nuclear power

A significant shift to nuclear energy would reduce the need to use coal, natural gas and oil-based power plants. These fossil fuel power plants are more efficient than nuclear power plants but they too have their environmental impacts. As well as the thermal pollution produced in the heat exchange process, they contribute far more to the greenhouse effect and the contamination of ecosystems.

Nuclear power produces a significant **amount of energy** even though only 30% of the energy eventually "reaches" the electricity power grid with the other 70% being wasted in the production and transmission process. The reserves of uranium and thorium fuels are significant with some estimates of time of depletion as high as 2000 years. However, a significant shift to nuclear power could result in these being used up in a relatively short period of time, perhaps 100 years.

The main risks associated with the production of nuclear power are:

- Problems associated with the mining of uranium
- Problems with the disposal of nuclear waste
- The risk of a thermal meltdown
- The risk that nuclear power programmes may be used as a means to produce nuclear weapons

Uranium ore can be mined by open pit and underground mining, or solution mining where solutions are pumped underground to leach the uranium-bearing minerals from sand. Extraction of uranium from seawater has also been undertaken. Uranium mining is considerably more dangerous than other mining. The biggest risk is the exposure of miners and the environment to radon-222 gas and other highly radioactive daughter products as well as seepage water containing radioactive and toxic materials. In the 1950s, a significant number of American uranium miners developed small-cell lung cancer due to the radon that was shown to be the cancer causing agent.

The technology to build and operate fission reactors is significant. However, the main concern seems to be the effective disposal of the low-level (radioactive cooling water, laboratory equipment and protective clothing) intermediate-level (coolant) and high-level (fuel rods) waste. The products of fission called "ash" include isotopes of the elements strontium, caesium and krypton and these are highly radioactive with a half-life of 30 years or less. Perhaps the biggest concern is that plutonium-239, another highly radioactive product has a half-life of 24 600 years. This isotope is also a threat as it is used in nuclear warheads. Even though uranium-235 is only mildly radioactive, it becomes contaminated with the other highly reactive isotopes within the reactor. Presently, the disposal methods include storage in deep underground storage areas. If the present disposal methods fail, then the danger to the environment would be catastrophic. Radioactive waste would find its way into the food chain and underground water would be contaminated. A new method of disposal where the waste is ground into a powder and then made into a synthetic rock is having some success.

Provided the reactors are maintained and built to standard, no obvious pollutants escape into the atmosphere that would contribute to the "greenhouse effect". However, even with expensive cooling towers and cooling ponds, thermal pollution from the heat produced by the exchanger process could contribute to global warming.

Opposition to **nuclear fission** has grown extensively especially since the bad accident at Chernobyl in the Ukraine that upon explosion sent a cloud of radioactive dust and gases across Northern Europe. The engineers were carrying out some tests on the coolant and the control rods. The coolant was not recycling and when the alarm was raised, they turned the coolant back on. When the coolant interacted with the reactor there was a nuclear meltdown and its associated explosion. The thermal energy produced was so high that the graphite moderator caught on fire and it had to be encased in concrete. The immediate population were exposed to high doses of ionising radiation that destroyed body tissues and death occurred immediately or soon after exposure. The effects of lower doses did not show up for years after exposure but there was many cases of illness due to various changes in DNA molecules and chromosomes that caused cancer and other genetic effects leading to further deaths.

Farmers throughout Northern Europe had to stop selling grains, fruit, livestock and dairy products due to the chances of radioactive fallout entering the food chain. It took them years before the radiation levels had fallen to acceptable levels.

The disadvantage of possible nuclear power plant containment failure is always with us. Nuclear terrorism is always present. Australia, a big exporter of uranium is making plans to begin a nuclear energy program. Nuclear fission does provide large amounts of electrical power and the use of fission is likely to continue until solar energy and nuclear fusion become viable alternatives.

Problems unsolved with nuclear fusion

Recall the process of the nuclear fusion mechanism that you have already studied in Chapter 7. Controlled nuclear fusion could be the ideal energy source of the future because of its clean image. There are many technological problems that need to be overcome and some scientists believe that the costs involved in the technology would mean that the cost to produce electricity would be too high when compared to other energy sources. To date, more energy is required to produce the fusion than the energy produced by the nuclear reaction. However, if perfected, it would not require expensive mining and far less radioactive waste would be produced. It is said that one cubic kilometre of seawater would produce more energy than all the fossil fuels on the Earth.

The most probable way to produce electricity would be to fuse deuterium with tritium. Deuterium atoms can be extracted from seawater and tritium can be bred from lithium. When deuterium and tritium are fused, a neutron with very high speed is ejected. As it collides with other atoms most of its kinetic energy is converted to heat. In order to overcome the electrostatic forces of repulsion between the hydrogen isotopes they need to be heated to an extremely high temperature around 100 000 °C so that they can be ionised into electrons and positively charged ions in a **plasma** that expands in all directions.

In order for fusion to occur, the **plasma** has to be confined for 1 second with a density of about 500 trillion atoms per cubic centimetre. Because fusion is not a chain reaction, these temperature and density conditions have to be maintained for future fusions to occur. Because the plasma is electrically charged, a magnetic field surrounding the plasma (a "magnetic bottle") could lead to a confined and controlled plasma that can undergo nuclear fusion.

Example 1

Suppose that the average power consumption for a household is 500 W per day. Estimate the amount of uranium-235 that would have to undergo fission to supply the household with electrical energy for a year. Assume that for each fission, 200 MeV is released.

Solution

$200 \text{ MeV} = 200 \times 10^6 \text{ eV} \times 1.6 \times 10^{-19} \text{ C} = 3.2 \times 10^{-11} \text{ J}$.

$500 \text{ W} = 500 \text{ Js}^{-1}$.

The total number of seconds in a year

$= 60 \times 60 \times 24 \times 365.25 = 3.16 \times 10^7 \text{ s}$

Therefore, the total electrical energy per year

$= 3.16 \times 10^7 \text{ s} \times 500 \text{ Js}^{-1}$

$= 1.58 \times 10^{10} \text{ Jyr}^{-1}$.

1 fission produces 3.2×10^{-11} J. So for 1.58×10^{10} J there would be

1.58×10^{10} J / 3.2×10^{-11} J $= 4.9375 \times 10^{20}$ fissions.

Recall that $1u = 1.661 \times 10^{-27}$ kg

Mass of uranium-235

$= 235 \times 1.661 \times 10^{-27}$ kg $= 3.90335 \times 10^{-25}$ kg per fission

Mass of uranium-235 needed

$= 3.90335 \times 10^{-25}$ kg $\times 4.9375 \times 10^{20}$ fissions

$= 1.93 \times 10^{-4}$ kg or 0.193 g

Example 2

A fission reaction taking place in a nuclear power reactor is

$^{1}_{0}\text{n} + {}^{235}_{92}\text{U} \rightarrow {}^{144}_{56}\text{Ba} + {}^{89}_{36}\text{Kr} + 3{}^{1}_{0}\text{n}$.

Estimate the initial amount of uranium-235 needed to operate a 600 MW reactor for one year assuming 40% efficiency and that for each fission, 200 MeV is released.

Solution

10^{-11} J.

$600 \text{ MW} = 600 \times 10^6 \text{ Js}^{-1}$.

The total number of seconds in a year

$= 60 \times 60 \times 24 \times 365.25 = 3.16 \times 10^7 \text{ s}$

Per year the total electrical energy

$= 3.16 \times 10^7 \text{ s} \times 600 \times 10^6 \text{ Js}^{-1}$

$= 1.896 \times 10^{16} \text{ Jyr}^{-1}$.

Since 40% efficient, the total energy needed

$= \dfrac{1.896 \times 10^{16} \text{ Jyr}^{-1}}{0.4}$

$= 4.74 \times 10^{16} \text{ Jyr}^{-1}$.

1 fission produces 3.2×10^{-11} J. So for 4.74×10^{16} J there would be

4.74×10^{16} J / 3.2×10^{-11} J $= 1.48125 \times 10^{27}$ fissions.

Mass of uranium-235 $= 235 \times 1.661 \times 10^{-27}$ kg

$= 3.90335 \times 10^{-25}$ kg per fission

Mass of uranium-235 needed

$= 3.90335 \times 10^{-25}$ kg $\times 1.48125 \times 10^{27}$ fissions

$= \mathbf{578.2 \text{ kg}}$

Exercise 8.3

1. Identify the missing product in the reaction

   $^{239}_{94}\text{Pu} + {}^{1}_{0}\text{n} \rightarrow {}^{106}_{44}\text{Ru} + \ldots + 2{}^{1}_{0}\text{n}$

   A. $^{133}_{50}\text{Sn}$
   B. $^{134}_{50}\text{Sn}$
   C. $^{132}_{50}\text{Sn}$
   D. $^{131}_{50}\text{Sn}$

2. Fission is the process by which

   A. two light nuclei combine to form a heavier nucleus.
   B. a heavy nucleus splits to form two lighter nuclei.
   C. a heavy nucleus splits to form an alpha particle and another nucleus.
   D. a light nucleus splits to form an electron and another nucleus.

3. The term critical mass refers to

   A. the mass defect when a fissile nucleus decays.
   B. the mass of a fissile nucleus.
   C. the mass required for a self–sustaining fission reaction.
   D. the mass of uranium–235 required to fuel a nuclear reactor.

CHAPTER 8

4. The purpose of the control rods in a nuclear reactor is to:

   A. absorb excess neutrons
   B. slow down the neutrons
   C. provide a container for the fuel
   D. reduce the radioactivity of the fissile materials

5. Which of the following is a fission reaction?

   A. $^{235}_{92}U \rightarrow {}^{4}_{2}He + {}^{231}_{90}Th$

   B. $^{12}_{6}C + {}^{1}_{1}H \rightarrow {}^{13}_{7}N + \gamma$

   C. $^{4}_{2}He + {}^{14}_{7}N \rightarrow {}^{17}_{8}O + {}^{1}_{1}H$

   D. $^{235}_{92}U + {}^{1}_{0}n \rightarrow {}^{140}_{58}Ce + {}^{92}_{34}Se + 4{}^{1}_{0}n$

6. Why is a $^{238}_{92}U$ nucleus more likely to undergo alpha decay than fission as a means of attaining stability?

7. (a) Explain how fission reactions, once started, are considered to be self-sustaining.

   (b) How is the chain reaction in nuclear reactors controlled?

8. The thermal power from the reactor is 2400 MW and this is used to operate the steam generator and turbine. The mechanical power output of the generator and turbine is used to drive a generator. The generator is 60 % efficient and produces 600 MW of electrical power. This is represented by the energy flow diagram below.

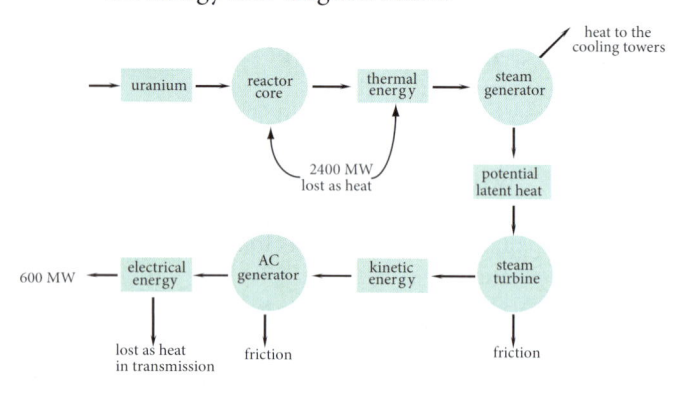

   (i) Calculate the power input to the generator.
   (ii) Calculate the power lost from the generator.
   (iii) Calculate the power lost by the heat engine.

9. (a) What are the strongest arguments in favour of pursuing nuclear fission as a source of energy?

   (b) What are the strongest arguments against using nuclear fission as source of energy?

10. Determine the number of fissions that will occur per second in a 500 MW nuclear reactor. Assume that 200 MeV is released per fission.

11. State three essential differences between chemical bond breaking and nuclear fission.

12. Estimate the initial amount of uranium-235 needed to operate a 500 MW reactor for one year assuming 35% efficiency and that for each fission, 200 MeV is released.

Solar power

Active solar heaters and photovoltaic cells

Solar or radiant energy can be converted directly to thermal energy or electrical energy. Solar heating can be achieved by both passive and active methods.

When the Sun's rays strike building materials with different specific heat capacities, the internal energy of the materials increases by the relevant conduction, convection or radiation method of heat transfer. If buildings are oriented correctly and materials are chosen that increase the insulation properties, then the solar energy can be captured passively, and energy costs can be reduced.

Active **solar heating** can be achieved by the use of solar collectors to convert solar energy into heat energy. Solar panels are mainly used to produce domestic hot water. The water can reach a temperature up to 70 °C. They can also used for space heating and heating swimming pools. Figure 822 is a schematic diagram of a typical solar panel.

Water is pumped through thin copper tubing that has been embedded in a blackened copper plate that is insulated on the bottom. This is covered with a glass plate, and the system is mounted in a metal frame. As the radiant energy enters the glass plate the infrared radiation is captured, and the cold water is warmed. The warmed water is pumped through a heat exchanger mounted in an insulated hot water system.

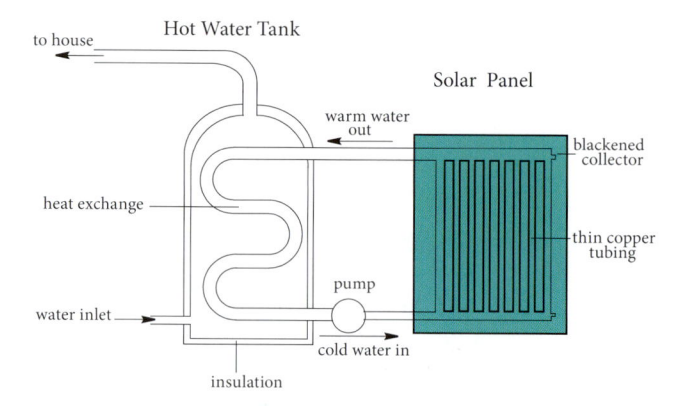

Figure 822 A solar collector

Parabolic dish collectors or solar furnaces are under construction in a number of countries. The Sun's rays are converged to a point by a parabolic mirror and temperatures greater than 3000°C can be produced. If a boiler is placed at the focus position, the steam that is generated can drive a turbine. A solar furnace that can generate 1 MW of thermal energy has been constructed in the French Pyrenees. It consists of a 45 m reflector made of 20 000 small mirrors moulded into a parabolic shape. 60 large computer controlled mirrors that follow the Sun reflect light onto the parabolic mirror.

Two common methods used for the production of electrical energy from solar energy are:

- photovoltaic solar cells
- thermoelectric devices

Photovoltaic devices use the photoelectric effect. Photons from radiant energy excite electrons in a doped semi-conducting material such as silicon or germanium, and the element becomes conducting allowing electrons to flow in an external circuit to produce electrical energy. The photons must have enough energy to cause electrons to move and this energy is available in the entire visible region of the electromagnetic spectrum.

Modern solar cells consist of thin circular wafers made of p-type and n-type silicon (4 valence electrons). Doping with Group 5 element (5 valence electrons) such as arsenic (As) produces an electron rich layer / n-type semiconductor. This electron is free to move about. Doping with Group 3 (3 valence electrons) element such as gallium (Ga) produces an electron deficient layer / p-type semiconductor. There are not enough electrons to form a covalent bond with a neighbouring atom. An electron from the n-type semiconductor can move into the hole. The electrons can move from hole to hole and produce a potential difference. The wafers are about 1 mm thick and they have a diameter of 5-8 cm and they are placed on top of each other to form a layer. Representations of p-type and p-type semiconductors are shown in Figure 823.

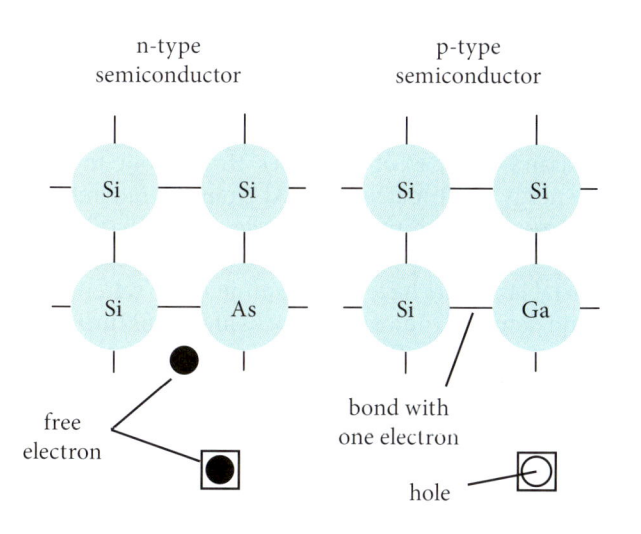

Figure 823 Types of semiconductors

Photovoltaic devices are a source of non-polluting, renewable energy that can be used in some ares of the world. Unfortunately, photovoltaic cells produce a very small voltage and provide very little current. They can be used to run electronic devices such as televisions and sound systems but cannot be used for high power rated appliances such washing machines, refrigerators and electric stoves. If connected in series the net voltage and current can be increased. The initial establishment costs are high and their efficiency at this stage is only 30%.

Thermoelectric converters appear to be a better option for the future. They not only use the visible region of the electromagnetic spectrum but also the infra-red region - the heating region of the em spectrum. Bars of doped silicon are again used to create an emf between the hot end and the cold end. By connecting p-type and n-type bars in series higher voltages can be obtained.

Its disadvantages are also evident. It produces a small energy output per surface area of the cell being used. For large-scale production, it requires thousands of mirrors or cells that take up a large area of land. It is intermittent with its output being upset by night and clouds. It is relatively expensive to set up and thus requires many years before the investment is returned.

Solar seasonal and regional variation

The main factors that account for solar seasonal and regional variation are:

- the solar constant
- the Earth's distance from the Sun
- the altitude of the Sun in the sky
- the length of night and day

The average radiant power radiated to an area placed perpendicular to the outer surface of the Earth's atmosphere while the Earth is at its mean distance from the Sun defines the irradiance or **solar constant** at a particular surface. This varies with Sunspot activity that has a cycle around 11 years.

Since the Earth-to-Sun distance varies over the course of a year from perihelion (nearest on Jan.3) to aphelion (furthest on Jul. 3) due to the elliptical orbit of the Earth, the solar constant varies about 6% from 1038 Wm^{-2} to 1398 Wm^{-2}. Furthermore, the energy radiated by the Sun has changed over its time of stellar evolution. As the solar constant applies perpendicular to the top of the atmosphere, and because the atmosphere reduces this flux considerably on a clear day, the value is reduced to about 1 kWm^{-2}. On an overcast day this value could be as low as a few watts per square metre.

Chapter 8

Solar radiation reaching the Earth will be different in regions at different latitudes because of the Sun's altitude in the sky. At the equator, solar radiation has to travel through a smaller depth of the atmosphere than at the poles. Each bundle of solar insolation (the energy received by the Earth as incoming short-wave radiation) has twice the area to heat up at 60° than at the equator. There is also less atmosphere near the equator and this means there will be less reflection and absorption of radiation.

Seasonal variations affect the amount of received radiation because the seasons will determine how spread out the rays become. Because the Earth is tilted on its axis by 23 ½°, at the poles there is no insolation for several months of the year.

In terms of the actual power that reaches an object on the Earth, many other factors have to be taken into account. The albedo (the fraction of incident light diffusely reflected from a surface) of the Earth is about 30%. On its way through the atmosphere, solar radiation is absorbed and scattered to a different degree depending on the altitude of the Sun at a particular place on Earth. The lower the Sun's altitude the greater is the zenith distance and thus the greater the degree of absorption and scattering.

Other factors that affect climate include changes in the Earth's orbit every 100 000 and 400 000 years, changes in the tilt of the Earth's axis, volcanic emissions, continental drift affecting the ocean currents and winds and human activity such as burning fossil fuels and deforestation.

Example 1

A solar panel with dimensions 2 m by 4 m is placed at an angle of 30° to the incoming solar radiation. On a clear day, 1000 Wm^{-2} reaches the Earth's surface. Determine how much energy can an ideal solar panel generate in a day.

Solution

Area of the solar panel = 8 m². Area in radiation terms = 8 cos 30°= 6.93 m²

1000 Wm^{-2}. = 1000 J s^{-1} m^{-2}

Energy produced / day

= (1000 J s^{-1} m^{-2})(6.93 m²)(24 h day^{-1})(60 min h^{-1})

(60 s min^{-1}) = 5.98 × 10^8 J

The energy produced per day = 6.0 × 10^8 J

Example 2

An active solar heater of volume 1.4 m³ is to provide the energy to heat water from 20 °C to 50 °C. The average power received from the Sun is 0.90 kWm^{-2}.

(a) Deduce that 1.8 × 10^8 J of energy is required to heat the volume of water in the tank from 20 °C to 50 °C.

(b) Estimate the minimum area of the solar panel needed to provide 1.8 × 10^8 J of energy in 2.0 hours.

Solution

(a) mass of water

= 1.4 × 10³ kg;

energy required

= 1.4 × 10³ kg × 4.18 × 10³ Jkg^{-1} °C × 30 °C

= 1.8 × 10^8 J.

(b) energy provided in 2 hours = 7 200 × 900 × A.

therefore A = (1.8 × 10^8 J) / (7200 s × 900 Js^{-1})

= 27.8 m².

Hydroelectric power

Types of hydroelectric schemes

Gravitational potential energy can be converted into electrical energy in **hydroelectric power** stations. There are three main schemes used to produce electricity:

- water storage in lakes
- tidal water storage
- pump storage

Hydroelectric power stations are widely used in mountainous areas of countries throughout the world. The energy is ultimately derived from the radiant energy of the Sun through the water cycle. Water that has fallen in the mountains is piped from rivers and stored in large artificial lakes that have been dammed. The water retained behind dams flows down penstocks and turns turbines that run the generators. The Three Gorges Dam on the Yangtze River is the largest hydroelectric dam in the world with a generating capacity of 18 200 MW. The dam is more than two kilometres wide and it has a height of 185 metres. Its reservoir stretches over 600 kilometres upstream.

Energy Production

Tides are produced by the interaction of the gravitational pull of the Moon and to some extent the Sun on the oceans. It is because of the Moon that so much energy is found in the oceans and the seas. It orbits the Earth on average in 27 days 7 hours 43 minutes and it rotates on its axis also in this time. The gravitational force of the Earth has slowed down the Moon's rotation about its axis until the rotational period exactly matches the revolution period about the Earth. As a result of this uncanny rotation there is a far side of the Moon that we do not see.

Tidal power stations can also be considered to be hydroelectric. They have been built in Russia and France, and other countries have stations in the developmental stage. The source of the energy can be assumed to be the kinetic energy of the Earth's rotation. Coastal estuaries that have a large vertical range in tides can be considered to be potential sites for tidal power stations. The station in France is situated on the estuary of the River Rance has a tidal range of 8.4 metres and generates 10 MW of electrical energy for each of the 24 turbines.

Stations consist of a dam to catch the high tide water and a series of sluices. The sluices are opened to let the high-tide water in. The water is released at low-tide, and the gravitational potential energy is used to drive turbines which produce electrical energy. During low peak electricity demand, the water is pumped into the holding dam by the generators that can act as a pump.

Pump storage systems can be used in off-peak electricity demand periods. The water is pumped from low reservoirs to higher reservoirs during this period. Even though the energy used in pumping the water uphill is more than the energy gained when the water flows downhill, this water can be used to produce more electricity during demand periods.

Energy transformations in hydroelectric schemes

The water trapped in reservoirs behind dams possesses gravitational potential energy. The water falls through a series of pipes where its potential energy is converted to rotational kinetic energy that drives a series of turbines. The rotating turbines drive generators that convert the kinetic energy to electrical energy by electromagnetic induction. A energy flow diagram is shown in Figure 824.

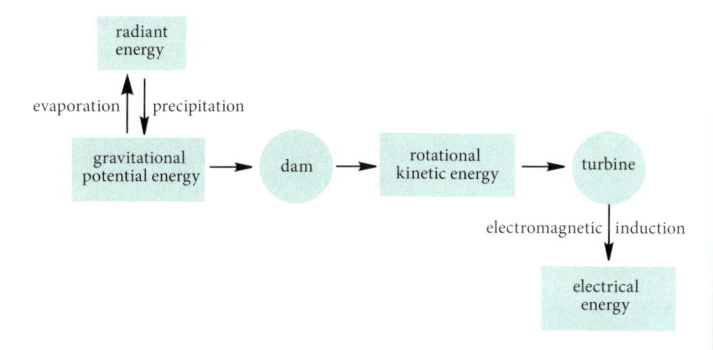

Figure 824 Energy flow for hydroelectricicty

The amount of energy available is directly proportional to the rate of flow of water and the height through which the water falls. Some dams rely on a small rate of flow, with a large fall and others have a large water flow through a smaller fall. With some hydro- electric stations, electricity is used to pump the water uphill to reservoirs during off-peak electricity use.

The power generated can be determined by knowing the rate of flow of a volume of freshwater or seawater. In order to determine the mass of water that is passing through the system, the density of the water will need to be considered.

Gravitational potential energy = $m\,g\,\Delta h$

Power = mass / volume × volume / time × g × Δh

This is dimensionally correct:

$J\,s^{-1} = kg\,m^{-3} \times m^3\,s^{-1} \times m\,s^{-2} \times m = kg\,m^2\,s^{-3}$

P = $\rho\,g\,\Delta h$ × volume per second

For tidal power, there is an area A of water at a height R. The mass of the water would be given by

$m = AR\rho$

The water level rises and falls between each tidal surge, so the centre of mass of the water would be at $R/2$. Therefore, the change in potential energy as the water runs out would be:

$\Delta E_p = (\rho AR) \times g \times R/2 = (\rho AgR^2)/2$

Example

A barrage is placed across the mouth of a river as shown in the diagram of a tidal power station. If the barrage height is 15 m and water flows through 5 turbines at a rate of 1.0×10^2 kg per second in each turbine, calculate the power that could be produced if the power plant is 70% efficient. Assume the density of seawater is 1030 kgm^{-3}.

267

Solution

Because the water level will change the average height of water = h/2.

Power = ρ g Δh × volume per second

= 1030 (kg m⁻³) × 100 (kg s⁻¹) × 9.8 (m s⁻²) × 7.5 (m) × 5 turbines

= 37.85 × 10⁶ J s⁻¹

If 60% efficient then the power produced
= 0.6 × 37.85 × 10⁶ J s⁻¹

Total power = 22.7 MW

Example

If water from a pumped storage dam fell through a pipe 150 m at a rate of 500 kg per second, calculate the power that could be produced if the power plant is 60% efficient. Assume the density of water is 1000 kgm⁻³.

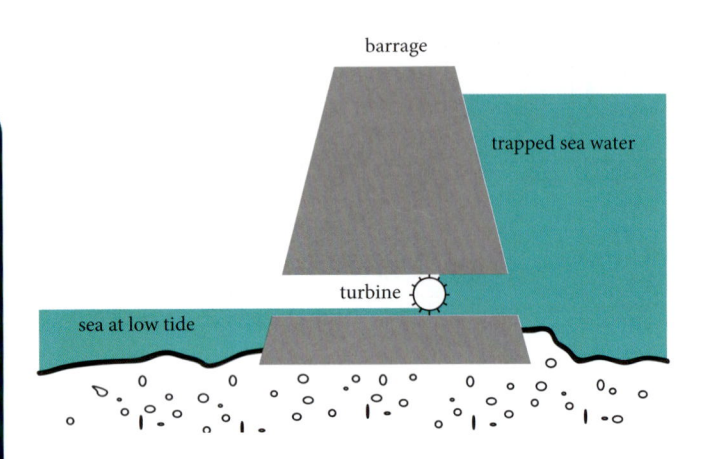

Solution

Power = ρ g Δh × volume per second

= 1000kg m⁻³ × 500 kg s⁻¹ × 9.8 ms⁻² × 150 m

= 735 × 106 J s⁻¹

If 60% efficient then the power produced
= 0.6 × 735 × 106 J s⁻¹

Total power = 441 MW.

Wind power

Basic features of wind generators

Winds are produced due to the uneven heating of the Earth's surface. The Sun's rays strike the equatorial regions at right angles but they approach the polar regions at an angle, large-scale convection currents are set up in the Earth's atmosphere. The inconsistency in the wind patterns is further compounded by the Earth's axial spin and the difference in local surface conditions (mountains, deserts, oceans, lakes, forests).

Wind power has been used in countries for centuries to run windmills for grinding grain and pumping water. The modern **wind turbines** used today in the world can have 60 metre blades that can drive a generator to produce 4 MW of electricity. The basic components of a wind power system are a tower to support the rotating blades (horizontal or vertical axis blades), a generator and a storage or grid system.

A horizontal axis wind turbine is shown in Figure 825 The blades can be rotated so that they can be steered into the wind. However, the generator usually has to be placed on the tower near the rotor shaft as a result of this.

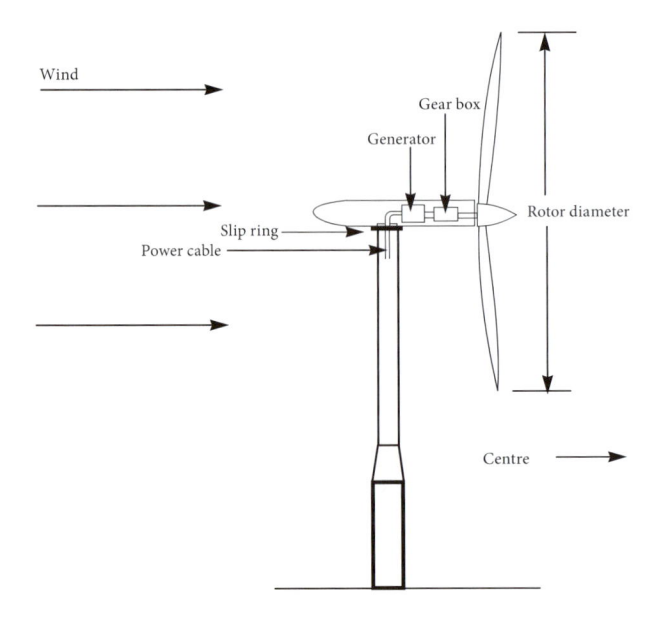

Figure 825 Horizontal axis wind turbine

The vertical axis wind turbine has the advantage that it does not have to be steered into the wind and as a result, the generator can be placed at the bottom of the system.

However, if the weight of the blades becomes too great, a lot of stress is put on the pivot to the generator.

Power produced by wind generators

It has been determined through experiment that the power output of a wind generator is:

- directly proportional to the blade area A.
- directly proportional to the cube of the wind speed v.

Consider a wind turbine that has a blade radius r as shown in Figure 826.

The area A swept out by the blades = πr^2

Wind speed v = distance d ÷ time t and $d = v\,t$

In one second the volume of air passing the turbine = $v\,A$

Therefore, the mass of air m passing the turbine in 1 second = $\rho\,v\,A$ where ρ is the air density.

The kinetic energy available each second

= $\tfrac{1}{2} m v^2 = \tfrac{1}{2} (\rho\,v\,A)\,v^2 = \tfrac{1}{2}\,\rho\,A\,v^3$

Power available = $\tfrac{1}{2}\,\rho\,A\,v^3$

Blade radius r

Air density

Wind speed v

d

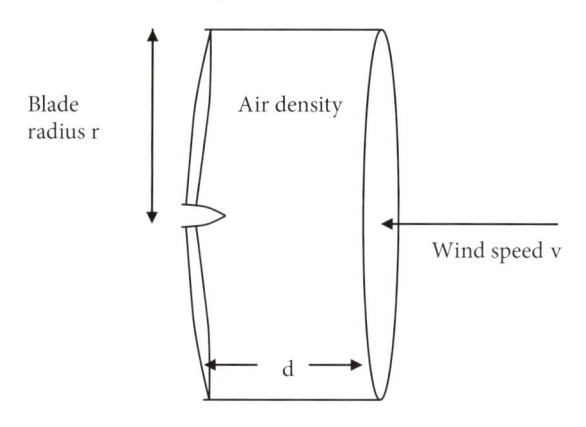

Figure 826 Power output of a wind generator

In reality, the power produced cannot strictly obey this equation as there are great fluctuations in wind speed and air density throughout the days and months of the year.

The electricity from wind power can be stored in batteries for later use but this storage method is very expensive. The most practical system is to set up a wind farm consisting of a large number of wind turbines interconnected to produce a power grid.

Wind power is cheap, clean, renewable and infinite. It can be used to provide electricity to remote areas of the world. However, winds resulting from the heating of the Earth are somewhat unpredictable. The initial set-up costs are high, the structures suffer from metal fatigue and they are noisy.

The better option being considered today is to use wind turbines in association with another power source. One possibility is to combine solar power and wind power.

Usually, when sunny there is little wind but when it is overcast there is more wind. Another possibility is to use the wind to pump water into high dams associated with hydroelectric power stations and then to run the water downhill during high electricity demand periods.

Example

A wind turbine has blades 20 m long and the speed of the wind is 25 ms^{-1} on a day when the air density is 1.3 kgm^{-3}. Calculate the power that could be produced if the turbine is 30% efficient.

Solution

Power = $\tfrac{1}{2}\,A\,\rho\,v^3$ and $A = \pi r^2$

= $0.3 \times 0.5 \times \pi \times 20^2\,m^2 \times 1.3\,kgm^{-3} \times 253\,m^3s^{-3}$

= 3.83×10^6 W

= 3.83 MW

Example

A wind generator is being used to power a solar heater pump. If the power of the solar heater pump is 0.5 kW, the average local wind speed is 8.0 ms^{-1} and the average density of air is 1.1 kgm^{-3}, deduce whether it would be possible to power the pump using the wind generator.

Solution

Power = $\tfrac{1}{2}\,A\,\rho\,v^3$ and $A = \pi r^2$

$500\,Js^{-1} = 0.5 \times \pi \times r^2\,(m^2) \times 1.1\,(kgm^{-3} \times 8.0^3\,m^3s^{-3})$

$r = \sqrt{\dfrac{2P}{\pi \rho v^3}} = \sqrt{\dfrac{1000}{\pi \times 1.1\,kgm^{-3} \times 512\,m^3s^{-3}}}$

= 0.75 m

This is a small radius so it could be feasible provided the wind speed was always present.

Chapter 8

Wave power

Oscillating water column ocean-wave energy converter

There are three types of wave energy collectors that are showing potential for harnessing wave energy. These can be grouped into:

- buoyant moored devices
- hinged contour devices
- oscillating water column (OWC) devices

The buoyant moored devices float above or below the water and are moored to the sea floor with cables. The **Salter Duck** is such an example as shown in Figure 827. As it bobs backwards and forwards matching the wave motion, it turns a generator.

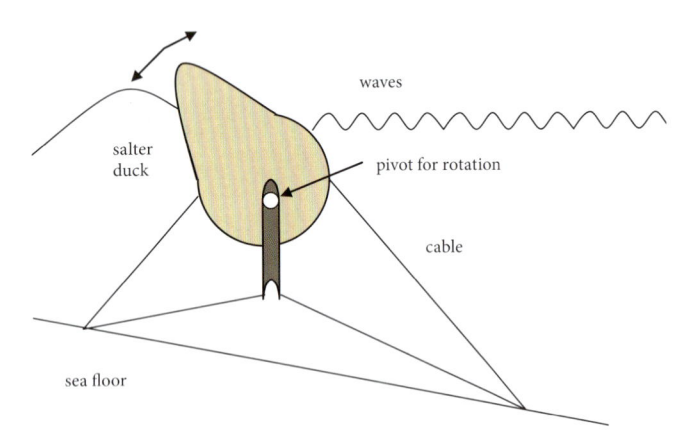

Figure 827 Salter duck

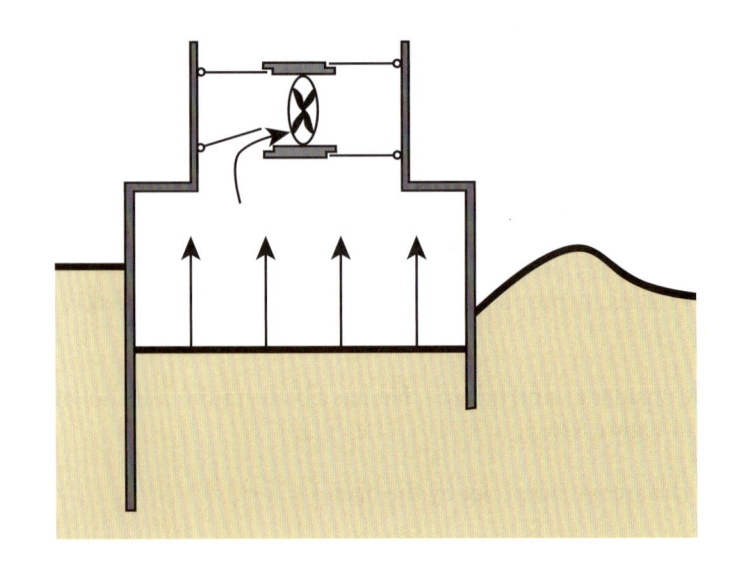

Figure 828 Offshore oscillating water column

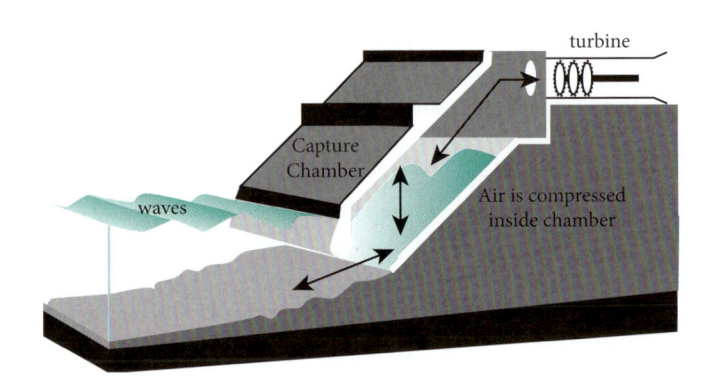

Figure 829 Onshore oscillating water column

Hinged contour devices consist of a series of floating 'mattresses' that are hinged together. As the sections move with the waves, the motion is resisted at the joints by hydraulic pumps that push high-pressure oil through hydraulic motors that produce electric power.

The most promising wave energy converters are the oscillating water column (OWC) devices because of their simple operation that uses conventional technology. These can be moored to the ocean floor or built into cliffs or ocean retainer walls as demonstrated in Figures 828 and 829 respectively. As the wave enters a capture chamber, the air inside the chamber is compressed and the high velocity air provides the kinetic energy needed to drive a turbine connected to a generator. Then as the captured water level drops, there is a rapid decompression of the air in the chamber which again turns the turbine that has been specially designed with a special valve system that turns in the same direction regardless of the direction of air flowing across the turbine blades.

The best places in the world for capturing wave power are the north and south temperate zones where the prevailing westerly winds are strongest in the winter. There are OWC wave power stations in England (average power of 7.5 kW) and Japan (average power of 6 kW) and there is a promising project on the Island of Islay off the west coast of Scotland being conducted by private enterprise in which the OWC feeds a pair of counter-rotating turbines each of which drives a 250 kW generator.

Wave size is determined by the speed of wind, the distance over which the wind excites the waves and by the depth and topography of the seafloor. Wave motion is highest at the surface and diminishes with depth.

The potential energy of a set of waves is directly proportional to wave amplitude squared and is also directly proportional to the wave velocity. The total energy of a wave is the sum of the potential energy and the kinetic energy created.

If one assumes that the wave is sinusoidal the potential energy in joules is given by the formula:

$$PE = g\frac{A^2}{2}\sin^2(kx - t)$$

where:

ρ = water density (kg/m^3)

w = wave width (m)
(assumed equal to the width of the chamber)

A = wave amplitude

k = the wave number $2\pi / \lambda$

λ = wavelength (m)

ω = wave frequency $2\pi / T$ (rad s^{-1})

T = wave period (s)

If the wave potential energy over one period is to be calculated this equation after differentiation becomes

$$PE = \tfrac{1}{4}\rho g A^2 \lambda$$

The total kinetic energy over one period is equal to the total potential energy:

$$KE = \tfrac{1}{4}\rho g A^2 \lambda$$

Type Of Power	Advantages	Disadvantages
Coal	Cheap to use High power output Can be used easily Large reserves still available	Lowest energy density of fossil fuels Non-renewable High CO_2 and SO_2 emissions Contribute to the enhanced greenhouse effect
Oil	Convenient in some oil producing countries Can be used in engines	Medium energy density Non-renewable High CO_2 and SO_2 emissions Contribute to the enhanced greenhouse effect
Natural gas	High energy density Cleaner and more efficient than other fossil fuels Can be used in engines	Medium CO_2 emissions Non-renewable Contribute to the enhanced greenhouse effect
Nuclear	High power output Reserves available	Expensive to build and run Radioactive materials have to be disposed of Possible nuclear accident
Passive solar	No fuel costs Renewable Non-polluting	Only works in daylight Not efficient when clouds present Power output is low
Photovoltaic solar	No fuel costs Renewable Non-polluting	Only works in daylight Not efficient when clouds present Power output is low Initial costs high Energy needs to be stored
Hydroelectric Tidal	No fuel costs Renewable Non-polluting	Need correct location Changes in the environment destroys ecosystems and can displace people Expensive to construct
Wind	No fuel costs Renewable Non-polluting	Need a windy location Power output is low Environmentally noisy High maintenance costs due to metal stress and strain
Wave	No fuel costs Renewable Non-polluting	High maintenance due to the power of waves High establishment costs.

Figure 830   Comparison of energy resources

Figure 830 summarises some of the advantages and disadvantages of the use of non-renewable and renewable energy sources.

Chapter 8

Example 1

If a wave is 3 m high and has a wavelength of 100 m and a frequency of 0.1 s⁻¹, estimate the power for each metre of the wave.

Solution

Power / λ = ½ ρ g A² f

Power

= 100 × 0.5 × 1020 kgm⁻³ × 10 ms⁻² × (1.5)² m² × 1m × 0.1s⁻¹

= 1.14 × 10⁶ kg m² s⁻³

= 114 kW per metre.

Example 2

If a wave is 3 m high and has a wavelength of 100 m and a period of 8 s, estimate the power over each metre of wavefront and calculate the wave speed.

Solution

Power = ½ ρ g A² λ /T

= 0.5 × 1020 kgm⁻³ × 10 ms⁻² × (1.5)² m² × 100m × 1m / 8s

= 143 × 10³ kg m² s⁻³

= 143 kW per metre.

Wave speed = wavelength / period = 100 m / 8 s = 12.5 ms⁻¹

Exercise 8.4

1. Doping a semiconductor to improve its conductivity means:

    A. adding elements with 3 valence electrons
    B. adding silicon or germanium to the semiconductor
    C. adding group 3 and group 5 elements
    D. adding elements with 5 valence electrons

2. Solar energy:

    A. is converted completely into electricity in a photovoltaic cell
    B. is not able to be stored
    C. is a renewable energy source
    D. is suitable only for heating water

3. All the following statements are correct EXCEPT

    A. generators convert mechanical energy into electrical energy
    B. nuclear reactors convert mass into energy
    C. chemical energy is a form of potential energy
    D. thermal energy and solar energy are the same

4. Photovoltaic cells can operate when the incident photons have

    A. frequencies above visible light
    B. infra-red frequencies
    C. microwave frequencies
    D. frequencies below visible light

5. In terms of energy transformations, distinguish between a solar panel and a solar cell.

6. A wind turbine farm is being designed for a town with a total required energy of 150 TJ per year. There is available space for 25 turbines and the average annual wind speed is 15 ms⁻¹.

    (a) Deduce that the average required output from one turbine is 0.19 MW.
    (b) Estimate the blade radius of the wind turbine that will give a power output of 0.19 MW. (Density of air = 1.3 kg m⁻³)

7. An active solar heater of volume 2.4 m³ is to provide the energy to heat water from 20 °C to 60 °C. The average power received from the Sun is 1000 Wm⁻².

    (a) Deduce that 4.0 × 10⁸ J of energy is required to heat the volume of water in the tank from 20 °C to 60 °C.
    (b) Estimate the minimum area of the solar panel needed to provide 4.0 × 10⁸ J of energy in 2.0 hours.

8. If a wave is 12 m high and has a wavelength of 30 m and a frequency of 0.1 s⁻¹, estimate the power for each metre of the wave.

9. If a wave is 12 m high and has a wavelength of 25 m and a period of 8 s, estimate the power over each metre of wavefront and calculate the wave speed.

10. In a hydro-electric power station, water falls through a 75 m pipe at the rate of 1500 kg s⁻¹. How many megawatts of electric power could be produced by the power plant if it is 80% efficient?

11. A photovoltaic cell can produce an average 40 Wm^{-2} of electrical energy if it is directly facing the Sun at the equator. If a house has an electrical consumption of 75 kW, what would be the required surface area of cells needed to provide the power requirements of the household?

12. The following table shows the power generated by a small wind turbine as a function of wind speed and radius of the blade. Plot graphs to show the linear relationships that exist between the power generated and these variables

Power / W	Blade radius / m	Power / W	Wind speed × 10^4 / km h^{-1}
300	0.5	20	12.6
400	0.7	80	15.9
500	0.8	200	20
620	0.9	370	25.2
805	1.0	580	29.2
1020	1.1	610	30.4
		1020	35

8.2 Thermal energy transfer

NATURE OF SCIENCE:

Simple and complex modelling: The kinetic theory of gases is a simple mathematical model that produces a good approximation of the behaviour of real gases. Scientists are also attempting to model the earth's climate, which is a far more complex system. Advances in data availability and the ability to include more processes in the models together with continued testing and scientific debate on the various models will improve the ability to predict climate change more accurately. (1.12)

© IBO 2014

Essential idea: For simplified modelling purposes the Earth can be treated as a black-body radiator and the atmosphere treated as a grey-body.

Understandings:
- Conduction, convection and thermal radiation
- Black-body radiation
- Albedo and emissivity
- The solar constant
- The greenhouse effect
- Energy balance in the Earth surface-atmosphere system

Conduction, convection and thermal radiation

At the macroscopic level, thermal energy (heat) can be transferred from one body to another by conduction, convection and radiation, or by combinations of these three.

Thermal **conduction** is the process by which a temperature difference causes the transfer of thermal energy from the hotter region of the body to the colder region by particle collision without there being any net movement of the substance itself.

Thermal **convection** is the process in which a temperature difference causes the mass movement of fluid particles from areas of high thermal energy to areas of low thermal energy (the colder region).

Thermal **radiation** is energy produced by a source because of its temperature that travels as electromagnetic waves. It does not need the presence of matter for its transfer.

Conduction can occur in solids, liquids and gases. In gases it occurs due to the collision between fast and slow moving particles where kinetic energy is transferred from the fast to the slow particle. The transfer of energy is very slow because the particles are far apart relative to solids and liquids. In liquids, a particle at higher temperature vibrates about its position with increased vibrational energy. Because the majority of the particles are coupled to other particles they also begin to vibrate more energetically. These in turn cause further particles to vibrate and thermal conduction occurs. This process is also slow because the particles have a large relative mass and the increase in vibrations is rather small. In solids, the transfer can occur in one of two ways. Most solids behave similarly to liquids.

However, solids are held in their fixed positions more rigidly than liquids and the transfer of vibrational energy is more effective. However, again their large masses do not allow for large energy transfer. If a substance in the solid or molten form has mobile electrons, as is the case for metals, these electrons gain energy due to the temperature rise and their speeds increase much more than those held in their fixed positions in the lattice. Metals are said to be good conductors of heat but most other solids are good insulators. Saucepans for cooking are usually made of copper or aluminium because these metals conduct heat quickly when placed on a stove. The handle is made from a good solid insulator to reduce the conduction of heat.

Figure 831 shows a potassium permanganate crystal placed in water inside a convection tube. Heat is applied for a short period of time and the direction of the purple trail is noted. Particles in a region of high thermal energy are further apart and hence their density is lower. In a region of low thermal energy the particles are closer together and the region is more dense. As a result, the less dense region rises as it is pushed out of the way by the more dense region and a convection current is produced.

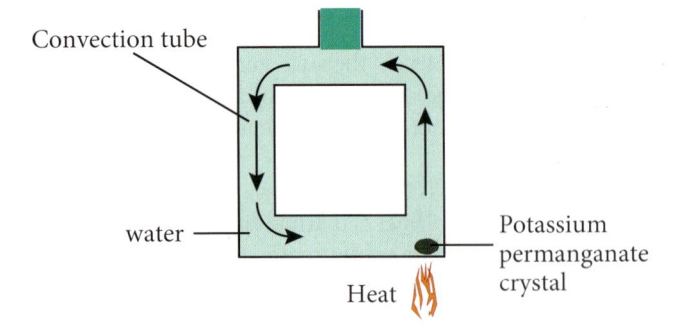

Figure 831   Convection current.

Another way in which a fluid can move is by forced convection. In this case, a pump or fan system maintains the movement of a fluid. The cooling system in nuclear reactors operates on this principle.

All thermal energy ultimately comes from the Sun in our solar system. It travels through 150 million km of mostly empty space. At the Earth's atmosphere the radiant energy is mainly reflected back into space. However, some is transmitted and absorbed causing a heating effect. Just as the Sun emits thermal radiation so does any source that produces heat such as a light bulb or an electric heater. Thermal radiation is mainly electromagnetic waves in the infrared region of the electromagnetic spectrum at temperatures below 1000 °C. Above this temperature, wavelengths of the visible and ultraviolet regions are also detected. Dull black bodies are better absorbers and radiators than transparent or shiny bodies.

Black body radiation

If a shiny metal tin and a dull black tin are placed in the Sun and their inside temperatures are recorded after a period of time, it is found that the air inside the dull black tin is at a higher temperature than the shiny one. Therefore, the black tin must absorb more radiation than the shiny one and as such the black tin must reflect and transmit less radiation than the shiny one. Furthermore, the dull black tin is a better radiator (emitter) than the shiny one and this is why the radiation grids on the back of a refrigerator are painted black.

When a piece of metal is heated, we can at some point detect the infrared radiation as heat being given off if we get close to the metal. At temperatures of the order of 1000 K, the metal start to glow red like the element in a toaster. At temperatures above 2000 K, the metal has a yellow-white glow like the filament in a light bulb. When an object is at even higher temperatures such as the temperature of a star, the energy radiated at any particular frequency or wavelength is related to its surface temperature according to the Planck equation $E = hf$.

The spectrum of light emitted from a glowing object is continuous because there is a range of visible colours. Therefore, it would absorb all the radiation falling on it and subsequently emit all the radiation leaving it.

Stars are not black bodies but their behaviour approximates a black body. Astronomers can use this fact to estimate the temperature of certain stars based upon their colour.

Black body radiation is the radiation emitted by a 'perfect' emitter. The radiation is sometimes called temperature radiation because the relative intensities of the emitted wavelengths are dependant only on the temperature of the black body.

An almost perfect black body can be made by painting the inside of an enclosed cylinder black and punching a small hole in the lid of the cylinder as shown in Figure 832. The hole looks black because the incident light ray is reflected many times off the walls of the cylinder. With the inside painted black the cylinder becomes as good a radiator and as bad a reflector as possible. In experiments performed by *Lummer* and *Pringsheim* in 1899, they used an enclosed porcelain sphere that was made black on the inside with soot. Porcelain can be heated to high temperatures which allows for a range of spectral colours, and infrared and ultraviolet spectra to be observed.

ENERGY PRODUCTION

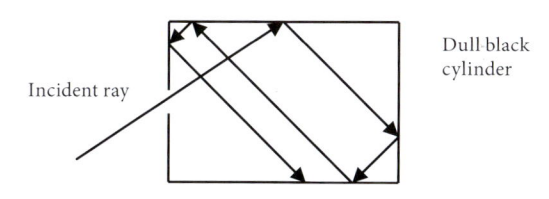

Figure 832 A black-body with multiple reflections

If the porcelain is heated to a given temperature black body radiation emerges from the hole. Depending on the temperature, the emerging radiation may appear red, yellow, blue or even white if the temperature is high enough. Emission occurs at every wavelength of light because it must be able to absorb every wavelength of all the incoming radiation. By using a suitable spectrometer, radiation intensity can be measured in the infrared, visible and ultra-violet regions of the electromagnetic spectrum.

The emission spectra of black bodies at different temperatures can be observed in Figure 833 which that shows the relative intensities of the different wavelengths.

It is evident that as the temperature rises:
- the intensity of every wavelength increases
- the total energy emitted increases because the power radiated increases
- the intensity of shorter wavelengths increases rapidly
- the curves are skewed to the left
- the curves move more into the red
- the peak wavelengths are all in the infrared region
- the peak wavelength moves to the left
- the curves get closer to the x-axis but never touch it.

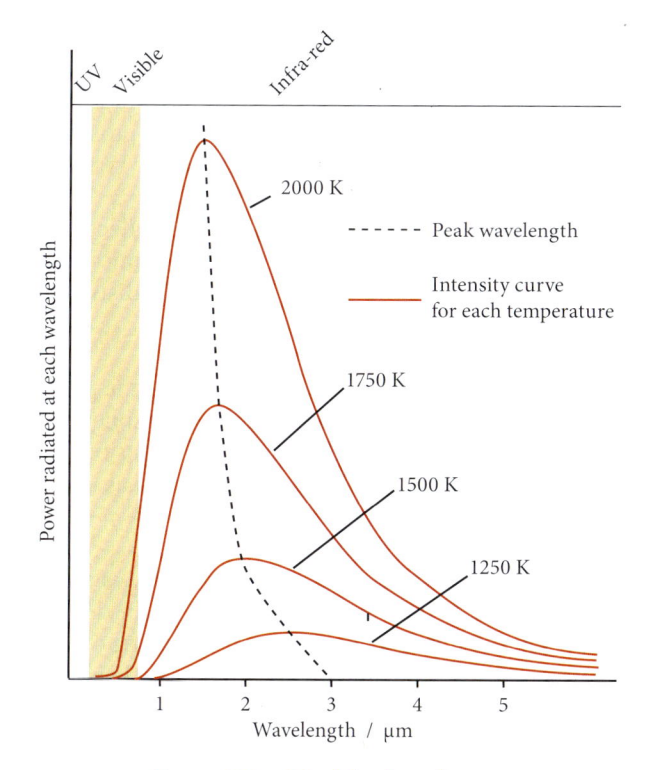

Figure 833 Black body radiation at different temperatures.

The temperature of the Sun is about 6000 K. Figure 834 shows emission spectra of black bodies at even higher temperatures. Note that sunlight has its peak at 500 nm and that all colours of the visible spectrum are present in the emission spectra thus accounting for the reason that the Sun appears white.

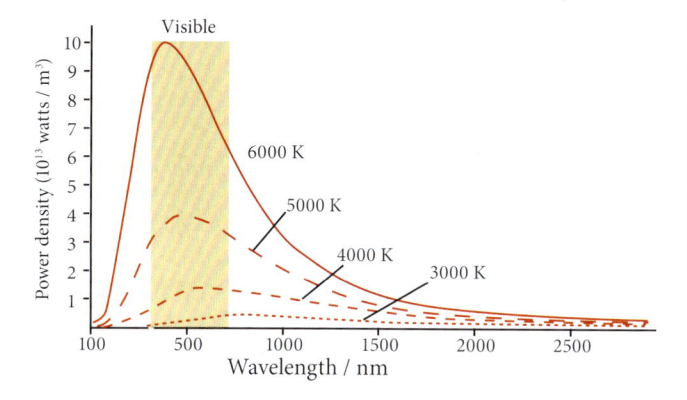

Figure 834 Intensity distribution in black body radiation at higher temperatures.

The total area under a spectral emission curve for a certain temperature T represents the total energy radiated per metre^2 per unit time E and for that assigned temperature it has been found to be directly proportional to the fourth power T^4. $E \propto T^4$

$$E = \sigma T^4$$

This relationship is called **Stefan's law**, and the constant is called the Stefan-Boltzmann constant and is equal to 5.67×10^{-8} W m^{-2} K^{-4}.

The power radiated by an area A of a black body radiator is represented by

$$P = A\sigma T^4$$

Looking at Figure 834, you can observe that the wavelength of the maximum intensity peak changes with temperature. In fact, the wavelength carrying the maximum intensity in a spectrum emitted by a hot body at a certain absolute temperature T is inversely proportional to that temperature.

$$\lambda_{max} \text{ (metres)} = W / T \text{ (Kelvin)}$$

This is known as Wien's displacement law where $W = 2.9 \times 10^{-3}$ m K.

Another form of this law is

$$\lambda_{max} T = 2.9 \times 10^{-3}$$

Example 1

The tungsten filament of a pyrometer (instrument for measuring high temperature thermal radiation) has a length of 0.50 m and an diameter of 5.0×10^{-5} m. The power rating

275

is 60 W. Estimate the steady temperature of the filament. Assume that the radiation from the filament is the same as a perfect black body radiator at that steady temperature.

Solution

Power radiated = power received = 60 W

$60 = P = A\sigma T^4$

The surface area of tungsten filament (cylinder) = $2\pi r h$

$= 2 \times \pi \times 5 \times 10^{-5}$ (m) $\times 0.5$ (m) $= 1.57 \times 10^{-4}$ (m²)

$P = 1.57 \times 10^{-4}$ (m²) $\times 5.67 \times 10^{-8}$ (W m⁻²) T^4 (K⁻⁴)

$T = [(60) \div (1.57 \times 10^{-4} \times 5.67 \times 10^{-8})]^{1/4}$

$= 1611\ K = 1600\ K$

Example 2

If one assumes that the Sun is a perfect black body with a surface temperature of 6000 K, calculate the energy per second radiated from its surface. (Assume the radius of the Sun = 7×10^8 m and $\sigma = 5.7 \times 10^{-8}$ W m⁻² K⁻⁴)

Solution

The surface area of a sphere = $4\pi r^2$

Energy per second = $P = A\sigma T^4 = 4\pi r^2 \sigma T^4$

$P = 4\pi \times (7 \times 10^8)^2$ (m²) $\times 5.7 \times 10^{-8}$ (W m⁻² K⁻⁴) $\times 6000^4$ (K⁴)

$P = 4.55 \times 10^{12}$ Wm⁻²

Example 3

The solar power received on the surface of the Earth at normal incidence is about 1400 Wm⁻². Deduce that the power output per square metre of the Sun's surface is about 7.5×10^7 Wm⁻². Comment on some assumptions that have been made in determining this answer. (Take the Sun's radius as 6.5×10^8 m and the radius of the Earth's orbit around the Sun as 1.5×10^{11}m).

Solution

Some assumptions are: the Sun and the Earth act as perfect black bodies, that the Earth's orbit around the Sun is circular rather than elliptical or all of the Sun's radiation falls on a sphere of this radius, that the Sun and the Earth are uniform spheres, and that the Earth's atmosphere absorbs no energy.

The surface area of a sphere = $4\pi r^2$

The total energy radiated by the Sun per second (power) = $A\sigma T^4 = 4\pi r_S^2 \sigma T^4$

This energy falls around a circular sphere equivalent to the Earth's orbit around the Sun equal to $4\pi r_E^2$.

Therefore, the power received per square metre on the Earth will be a fraction of that radiated by the Sun.

Power radiated by the Sun = $(4\pi r_E^2 \div 4\pi r_S^2) \times 1400$ Wm⁻²

$P = (1.5 \times 10^{11})^2 \div (6.5 \times 10^8)^2 \times 1400$ Wm⁻² $= 7.46 \times 10^7$ Wm⁻²

Albedo and emissivity

Albedo

The term albedo (α) (Latin for white) at a surface is the ratio between the incoming radiation and the amount reflected expressed as a coefficient or as a percentage.

Albedo = total scattered power / total incident power

Solar radiation is mainly radiated in the visible region of the electromagnetic spectrum (400 μm to 700 μm) and by incoming short-wave infrared radiation called insolation. This radiation can pass through the atmosphere to warm the land and sea by the so-called natural greenhouse effect.

Water vapour and clouds can absorb radiation in the 400 μm to 700 μm ranges, and carbon dioxide can absorb radiation in the 400 n m to 700 μm ranges. Between 700 nm to 1300 nm range, more than 70% of the radiation escapes into space. About 7% is radiated in the short-wave ultraviolet region around 500 μm. Outgoing longer wave infrared radiation in the night cools the Earth. The Earth receives approximately 1 kW m⁻² on a clear day at the rate of 1.7×10^7 W s⁻¹.

The incoming solar radiation is insolated, reflected and re-transmitted in various ways. Figure 835 demonstrates how 100 units of input solar radiation is distributed. It can be seen that:

- 30% is reflected back into space mostly by the polar icecaps and particulate matter in the atmosphere. This reflected radiant energy mainly consists of short wavelength electromagnetic radiation such as ultra-violet radiation.

- 51% is absorbed by the Earth during the day as thermal energy which is then radiated back into space during the night as radiant thermal energy consisting of long wavelength electromagnetic infra-red radiation.

- 23% of the 51% is used in the water cycle. The radiant energy is absorbed by water and evaporation occurs as enough energy is supplied to overcome the latent heat of vaporisation. The gaseous water vapour is carried by convection currents higher into the atmosphere, and clouds with high potential energy are formed.

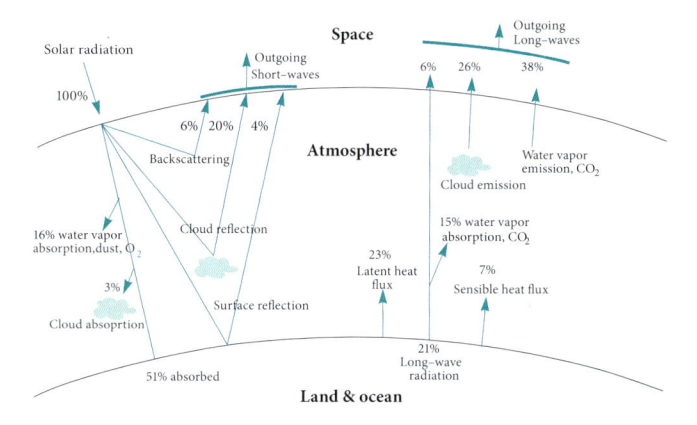

Figure 835 Solar radiation energy input and output

0.25% of the radiant energy is consumed in supplying the energy that drives the convection currents of the oceans and atmosphere, and only 0.025% is stored by photosynthesis in plants as chemical potential energy. This has been the main source of fossil fuels.

Solar or radiant energy can be converted indirectly to electrical energy by:

- biomass conversion
- wind power
- wave energy
- geothermal energy

The Earth's albedo varies daily and is dependent on season (cloud formations) and latitude. In thin clouds, the albedo varies from 30-40 per cent whereas in thicker cumulo-nimbus clouds the albedo can be 90 per cent. Albedos also vary over different land surfaces, water surfaces, urban areas, grasslands and forests and ice and snow surfaces. Some albedos are given in Figure 836.

The global annual mean albedo is 0.3 (30%) on Earth.

Surfaces	Albedo %
Oceans	10
Dark soils	10
Pine forests	15
Urban areas	15
Light coloured deserts	40
Deciduous forests	25
Fresh snow	85
Ice	90
Whole planet	31

Figure 836 Albedo percentages of different surfaces

Emissivity and emission rate from surfaces

Stefan's Law can also be written in the following way:

$P = eA\sigma T^4$

The factor e is called the emissivity of a material. **Emissivity** is the ratio of the amount of energy radiated from a material at a certain temperature and the energy that would come from a blackbody at the same temperature and as such would be a number between 0 and 1. Black surfaces will have a value close to 1 and shiny surfaces will have a value close to 0. Most materials are coloured and they reflect some wavelengths better than others. For example, a blue object will reflect blue and absorb the other colours of the visible spectrum and a black object will absorb nearly all spectral colours. Therefore effective emissivity is also affected by the surface emissivity and wavelength dependence.

Some approximate values are given in Figure 837.

Material	Emissivity	Material	Emissivity
mercury	0.05 – 0.15	snow	0.9
tungsten	0.1 – 0.6	ice	0.98
rusted iron	0.6 – 0.9	plate glass	0.85
water	0.6 – 0.7	coal	0.95
soil	0.4 – 0.95	black paint	0.92

Figure 837 Approximate emissivity values

Example

The Sun is at 50° to the horizontal on a clear day. Estimate how much radiation from the Sun is absorbed per hour by an animal that has a total area exposed to the Sun of 2.0 m².

(Assume $\sigma = 5.7 \times 10^{-8}$ W m^{-2} K^{-4} and the emissivity to be 0.8)

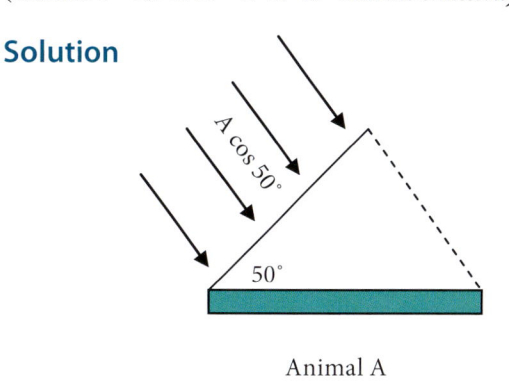

Solution

Take the average temperature to be 300 K.

Energy per second = $P = e A \sigma T^4 \cos 50$

$P = 0.8 \times 2.0 \, (m^2) \times 5.7 \times 10^{-8} \, (W \, m^{-2} \, K^{-4}) \times 300^4 \, (K^4) \times \cos 50$

277

CHAPTER 8

$P = 4.75 \times 10^2$ W

This is the energy absorbed per second. Multiply by 3600

$P = 1.7 \times 10^6$ W h^{-1}

The solar constant

Solar energy represents the greatest potential energy source available to man. It is produced by nuclear fusion reactions within the Sun.

The Earth, which is approximately 1.5×10^8 km from the Sun, receives the Sun's radiant energy over empty or nearly empty space. The temperature of the Sun's surface is about 6000 K. About 43% of its radiation is in the visible region of the electromagnetic spectrum with 49% in the infrared region and 8% in the ultra-violet region. The Earth intercepts only a small part of the Sun's total radiation (about 0.5 of a billionth).

The average radiant power radiated to an area placed perpendicular to the outer surface of the Earth's atmosphere while the Earth is at its mean distance from the Sun defines the flux density or solar constant at a particular surface.

It is defined as the amount of solar energy per second (power) that falls on an area of 1m² of the upper atmosphere perpendicular to the Sun's rays. Generally, this can be stated as:

Intensity = power / area

I = Power / A

The value of the solar constant is found to be equal to 1.35×10^3 Wm^{-2} or 1.35 kWm^{-2}. The total solar irradiance has been monitored with absolute radiometers since 1978 on board five satellites. The term "solar constant" is a misnomer because it is not constant. Since the Earth-to-Sun distance varies by about 6% over the course of a year from perihelion (nearest on January 3) to aphelion (furthest on July 3) due to the elliptical orbit of the Earth, the solar constant varies from 1038 Wm^{-2} to 1398 Wm^{-2}. Furthermore, the energy radiated by the Sun has changed over its time of stellar evolution.

As the solar constant applies perpendicular to the top of the atmosphere, and because the atmosphere reduces this flux considerably on a clear day, the value is reduced to about 1 kWm^{-2}. On an overcast day this value could be as low as a few watts per square metre.

In terms of the actual power that reaches an object on the Earth, many other factors have to be taken into account. The albedo (the fraction of incident light diffusely reflected from a surface) of the Earth is about 30%. On its way through the atmosphere, solar radiation is absorbed and scattered to a different degree depending on the altitude of the Sun at a particular place on Earth. The lower the Sun's altitude the greater is the zenith distance and thus the greater the degree of absorption and scattering.

The total solar radiation or irradiance reaching the top of the atmosphere is about 1.7×10^{17} W. Distributed over the whole globe, this amounts to about 170 Wm^{-2} averaged over a day and night.

Example 1

How much solar radiation does one square metre of the Earth's surface receive per day?

Solution

In one day, the solar radiation would be

$$170 \text{ W m}^{-2} \times 1 \text{ m}^2 \times 24 \text{ h} = 4080 \text{ W h}$$

For the land area of the USA, the solar radiation available over the total land surface is over 10^{17} kW h annually. This is about 600 times greater than the total energy consumption of the USA.

Example 2

Given that the mean Sun-Earth distance is 1.5×10^8 km and that the power received at the top of the Earth's atmosphere is given by the solar constant, determine the total power generated by the Sun.

Solution

Every square metre at an Earth–Sun mean distance receives 1.35 kW m^{-2}

The surface area of a sphere = $4 \pi r^2$

Total power received = $4 \pi (1.5 \times 10^{11} \text{ m})^2 (1.35 \times 10^3 \text{ J m}^{-2})$
= 3.8×10^{26} J.

The greenhouse effect

Describing the greenhouse effect

The Moon is able to provide information as to what it would be like if there was no atmosphere around the Earth. Being at approximately the same distance from the Sun as the Earth and without an atmosphere, the Moon's average temperature is -18 °C. But to say that the Moon and the Earth are at the right distance from the Sun is an oversimplified understanding of why Earth's average temperature is around 34 °C warmer than the Moon. The real reason is due to certain gases in our atmosphere causing the natural greenhouse effect. The Earth's atmosphere is around 78% nitrogen (N_2) and 21% oxygen (O_2). Being diatomic molecules with a triple bond and a double bond respectively between the two atoms, their vibration are restricted because they are tightly bound together and as a result they do not absorb heat and make no contribution to the natural greenhouse effect. However, there are 1% of natural greenhouse gases (water vapour, carbon dioxide, methane and nitrous oxide) in the atmosphere that make a major contribution to keeping the Earth at its current temperature.

So what is the natural greenhouse effect? Put simply, the 'natural' greenhouse effect is a phenomenon in which the natural greenhouse gases absorb the outgoing long wave radiation from the Earth and re-radiate it in all directions, including some of it back to the Earth. It is a process for maintaining an energy balancing process between the amount of long wave radiation leaving the Earth and the amount of energy coming in from the Sun.

Provided that the Sun's radiant energy remains constant and the percentage of greenhouse gases remains the same, then the established equilibrium will remain steady and the average temperature of the Earth will be maintained at 16 °C.

As mentioned earlier, about 43% of the radiation from the Sun is in the visible region of the electromagnetic spectrum with 49% in the infrared region and 8% in the ultra-violet region. The greenhouse gases of Earth's atmosphere absorb ultraviolet and infrared solar radiation before it penetrates to the surface. However, most of the radiation in the visible region passes through the atmosphere where it is absorbed by the land, waters and vegetation being transformed into heat and then re-radiated as long wave infrared radiation. The long wave infrared radiation is absorbed by greenhouse gases in the atmosphere and then re-radiated back to the biosphere.

The greenhouse gases and their sources

As just mentioned earlier, the main natural greenhouse gases are:

- carbon dioxide (CO_2)
- water vapour (H_2O)
- methane (CH_4)
- nitrous oxide (N_2O)

The concentration of these natural greenhouse gases has been affected by human activity in what is known as the enhanced greenhouse effect.

Carbon dioxide has always been the largest contributor to greenhouse gas concentration and in the 1700s it was thought to have a concentration of about 280 parts per million (ppm). The natural production is caused by respiration, organic decay of plants and animals, natural forest fires, dissolved carbon dioxide and volcanic activity.

Methane concentrations in the 1700s were believed to be around 0.7 parts per million. The main natural source is decaying vegetation. Even when we burp and pass wind we are removing methane and other gases from our body due to the fermentation of plants in our digestive system. Decaying vegetation is found in agriculture and in wetland peat bogs.

Water vapour is found in the atmosphere due to the water cycle in which water is evaporated from mainly the oceans and in the transpiration of plants (loosing water from their leaves).

Nitrous oxide exists in parts per billion and the concentration in the 1700s was thought to be about 250 parts per billion (ppb). Natural sources include forests, grasslands, oceans and soil cultivation.

The molecular mechanism for absorbing infrared radiation

Because ultraviolet radiation is more energetic than infrared radiation it tends to break bonds between atoms joined together. On the other hand, infrared radiation being less energetic tends to cause the atoms to vibrate in various ways. When the frequency of the infrared radiation is equal to the frequency of vibration then resonance occurs.

This means that the frequency of the radiation is equal to the natural frequency of vibration of two atoms bonded together. It just so happens that the natural frequency of vibration of the molecules of the greenhouse gases is in the infrared region.

If resonance occurs and the molecular dipole moment undergoes a change, then the greenhouse gas will absorb energy from the albedo infrared radiation coming from a surface. Only certain energies for the system are allowed and only photons with certain energies will excite molecular vibrations. Therefore vibrational motion is quantized and transitions can occur between different vibrational energy levels. The absorbed energy can then be re-radiated back into the biosphere.

In order to examine the vibrations of greenhouse gases and other molecules and compounds the analysis is carried out by infrared spectrophotometry. In a digital

IR spectrometer, a glowing filament produces infrared radiation in the form of heat and this is passed through an unknown sample held in a small transparent container. A detecting device then measures the amount of radiation at various wavelengths that is transmitted by the sample. This information is recorded as a spectrum showing the percentage transmission against the wavelength in micrometres (microns) (μm) or the frequency.

We have already learnt that energy is directly proportional to frequency or inversely proportional to wavelength (E = hf or E = hc / λ). If the wave number is the number of waves per centimetre (cm$_{-1}$) we have a variable that is directly proportional to energy. When the energy of the infrared radiation from the instrument matches the energy of vibration of a molecule in the sample, radiation is absorbed, and the frequency given in wave numbers (cm^{-1}) of the infrared radiation matches the frequency of the vibration. Each sample examined has its own individual spectrum and therefore a blueprint of the sample just like the DNA of an individual.

Let us take carbon dioxide as an example to examine the different modes of possible vibration. This molecule is a tri-atomic molecule that is linear in shape and it has two C=O bonds. The molecule can move (translate) in three-dimensional space and it can rotate in two directions. The most important factor if it is to absorb energy is that it must vibrate at the resonance frequency so as to change its molecular dipole moment. Imagine that the atoms are connected by a spring and when they absorb infrared radiation they can have different types of vibrations. Carbon dioxide being a linear molecule can have 3N – 5 vibrational modes (where N is the number of atoms). It has 3 atoms so it can have 3 × 3 – 5 = 4 vibrational modes and these vibrations are shown by arrows in Figure 838.

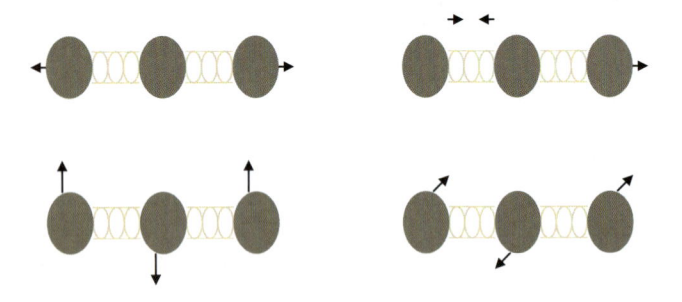

Figure 838 Vibrational modes of carbon dioxide

The top 2 vibrations represent the stretching of the C=O bonds, one in a symmetric mode with both C=O bonds lengthening and contracting in-phase. This symmetric stretch is infrared inactive because there is no change in the molecular dipole moment and so this vibration is not seen in infrared spectrum of CO_2. The top right diagram is in an asymmetric mode with one bond shortening while the other lengthens. The asymmetric stretch is infrared active due a change in the molecular dipole moment. Infrared radiation at 2349 cm^{-1} (4.26 μm) excites this particular vibration. The two bottom diagrams show vibrations of equal energy with one mode being in the plane of the paper and the other out of the plane of the paper. Infrared radiation at 667 cm^{-1} (15.00 μm) excites these vibrations. Figure 839 shows the IR spectrum for carbon dioxide at 4 kPa pressure with the 2 peaks clearly visible.

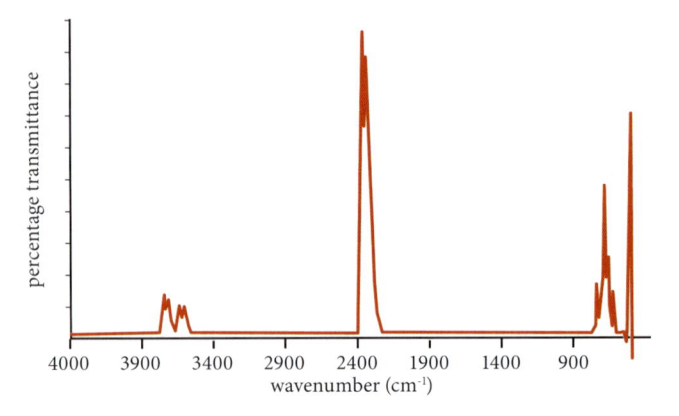

Figure 839 IR spectrum of carbon dioxide

Energy balance in the Earth surface-atmosphere system

Modeling is an ancient mathematical tool that allows us to simplify the real world to solve problems. We use models everyday of our lives without thinking about it. We use the internet to purchase products, to make airline and motel bookings and to find directions to our destinations. So we are solving problems using a system that has been set up using mathematical formulae in data spreadsheets.

There are 4 main types of climate models:

- energy balance models (EBM) use spreadsheets to study the incoming and outgoing global radiation balance and energy transfers using different latitudes from the equator to the poles
- one-dimensional radiative-convective models (RCM) simulate atmospheric environments by only cosisdering the radiation balance and heat transport by convection heat transfer
- two-dimensional statistical-dynamical models (STM) are a combination of the energy balance and radiative-convective models
- three-dimensional general circulation models (GCM) which try to simulate the global and the continental climate considerations.

General Circulation Models (GCM) are used to monitor the effects of the changing atmosphere as caused by the increased concentrations of the major greenhouse gases since pre-industrial times up to the present time and further into the future. GCMs use sophisticated computer models and complex mathematical equations to simulate and assess factors such as the winds, temperatures, pressures, cloud cover and precipitation to get a picture of what is the effect of increased greenhouse concentrations on the environment.

We will concentrate on the simplest atmospheric model - the **energy balance climate model.** The word "balance" infers that the system is in equilibrium with no energy being accumulated in the Earth's surface and atmosphere. This model attempts to account for the difference between the incoming radiation intensity and the outgoing radiation intensity, and the simplest energy balance model chooses temperature as the only variable to be considered. For the Earth's average temperature to remain steady, energy must leave a surface at the same rate as it arrived at the surface otherwise the Earth's temperature will rise. By taking a small region of latitude around the Earth called a **latitude belt** the temperature at the surface can be studied.

If we take the Earth and "slice it up" into grid regions by choosing latitude and longitude boundaries, we can study surface heat capacities and radiation differences in that geographical region. For example, at latitudes between 60 and 70 north and longitudes 140 and 165 west Alaska is located, and at latitudes between 0 and 10 south and longitudes 110 and 130 east Papua New Guinea is located. The two climate regions chosen will have considerable differences in the incoming and outgoing radiation intensity. In fact there is also considerable differences in the incoming and outgoing radiation intensity within each chosen region. Therefore, it may be more appropriate to choose a one-dimensional model such as only concentrating on a single latitude belt.

Recall that the albedo at a surface is the ratio between the incoming radiation and the amount reflected expressed as a coefficient or as a percentage. Also, recall that the Earth's albedo varies daily and is dependent on season (cloud formations) and latitude, different land surfaces, water surfaces, urban areas, grasslands and forests and ice and snow surfaces.

If all the radiation from the Sun reached the Earth then the incoming radiation would be equal to:

πR_E^2 (m²) × the solar constant (Wm⁻²)

to give the final units of J s⁻¹, where R_E is the mean average radius of the Earth.

Remember that the **solar constant α** at a particular surface was previously defined as the amount of solar energy per second that falls on an area of 1m² of the upper atmosphere perpendicular to the Sun's rays, and its value is equal to 1.35×10^3 Wm⁻².

Of course this is not the power that arrives on 1 m² of the Earth's surface because the planet's average incoming radiation (insolation) is reflected and scattered and absorbed. Of the 100 units of insolation, 28 units is reflected off the clouds and aerosols, 18 units are absorbed in the atmosphere and 4 units are reflected off land surfaces. In fact the fraction of power from the solar constant at the Earth's surface is 343 Wm⁻².

The temperature of the Earth can be determined by finding the Earth's average incoming insolation and subtracting the amount reflected back into space by the global albedo and adding in the the energy that is supplied to the surface by the greenhouse effect.

The incoming radiation only falls on an area equal to πR_E^2 because only one side of the Earth is facing the Sun to receive the incoming radiation. Now each 1 m² will have its own albedo and its own surface temperature and adjustments to the above value of the incoming radiation have to be made to account for these factors. It can be proposed in the balance climate model that the incoming radiation will therefore be equal to:

$\pi R_E^2 (1 - \alpha)$ (m²) × the solar constant (Wm⁻²)
where α is the albedo.

Assuming that the the Earth is radiating over its entire surface area then the **outgoing radiation** can be obtained using Stefan's Law and is determined to be equal to:

$4\pi R_E^2 \sigma T_E^4$

Therefore, assuming a balance in thermal equilibrium:

$\pi R_E^2 (1 - \alpha) \times$ solar constant $= 4\pi R_E^2 \sigma T_E^4$

$(1 - \alpha) \times$ solar constant $/ 4 = \sigma T_E^4$

Let us see if this equilibrium state can match some known quantities in the following example.

Example

If the long-wave radiation flux from the surface of the Earth has an average value of 240 Wm⁻² and the average temperature somewhere in the atmosphere is 255 K, determine the incoming and outgoing radiation of the Earth. Assume the global albedo is 0.3.

Solution

Incoming radiation $= (1 - \alpha) \times$ solar constant $/ 4$

$= (1 - 0.3) \times 1.35 \times 10^3$ Wm⁻² $\div 4 = 236.25$ Wm⁻²

Outgoing radiation $= \sigma T_E^4$

$= 5.7 \times 10^{-8}$ W m⁻² K⁻⁴ $\times (255)^4$ K⁴ $= 241$ Wm⁻²

These values are nearly equal to the average radiation flux value of 240 Wm^{-2}.

There are many energy balance climate models and they are only as good as the mathematics. If the mathematics is wrong or the wrong data is collected, then the model is flawed from the very beginning.

This model is highly oversimplified because the model is not global but rather restricted to a certain small latitudinal region and it doeas not account for the flow of energy from one zone to the next or the fact that each latitude zone will receive different amounts of incoming radiation depending on whether they are close to the equator or at the north and south extremes. Furthermore, it does not account for cloud cover. The model could be greatly improved by taking these factors into consideration or by using a more complex model such as three-dimensional general circulation model.

Exercise 8.5

1. All of the following are natural greenhouse gases **EXCEPT**:

    A. methane
    B. nitrogen
    C. water vapour
    D. nitrous oxide

The following information is about questions 2-5.

A perfectly black body sphere is at a steady temperature of 473 K and is enclosed in a container at absolute zero temperature. It radiates thermal energy at a rate of 300 Js-

2. If the temperature of the sphere is increased to 946 K it radiates heat at a rate of:

    A. 300 W
    B. 1200 W
    C. 3200 W
    D. 4800 W

3. If the radius of the sphere is doubled it radiates heat at a rate of:

    A. 300 W
    B. 1200 W
    C. 3200 W
    D. 4800 W

4. If the temperature of the enclosure is raised to 500 K it radiates heat at a rate of:

    A. 300 W
    B. 1200 W
    C. 3200 W
    D. 4800 W

5. If the enclosure is at 473 K the net rate of heat loss would be:

    A. 0 W
    B. 300 W
    C. 1200 W
    D. 100 000 W

6. The relative intensities of the emitted wavelengths of a perfect black body are dependant on the

    A. the surface area of the black body
    B. the temperature of the black body
    C. the radiation per square metre
    D. the radiation per second

7. By referring to Figures 859 and 860 827 and 828 answer the following questions.

    (a) What is the difference between a black body radiator and a non-black body radiator.

    (b) Explain why a body at 1500 K is "red hot" whereas a body at 3000 K is "white hot".

    (c) How can you use the information from the graphs to attempt to explain Stefan's law?

    (d) As the temperature increases, what changes take place to the energy distribution among the wavelengths radiated?

8. A very long thin-walled glass tube of diameter 2.0 cm carries oil at a temperature 40 °C above that of the surrounding air that is at a temperature of 27 °C. Estimate the energy lost per unit length.

9. (a) Estimate the mean surface temperature of the Earth if the Sun's rays are normally incident on the Earth. Assume the Earth is in radiative equilibrium with the Sun. The Sun's temperature is 6000 K and its radius is 6.5×10^8 m. The distance of the Earth from the Sun is 1.5×10^{11} m.

    (b) What assumptions have been made about the temperature obtained?

9. Wave phenomena

Contents

9.1 – Simple harmonic motion

9.2 – Single-slit diffraction

9.3 – Interference

Essential Ideas

The solution of the harmonic oscillator can be framed around the variation of kinetic and potential energy in the system.

Single-slit diffraction occurs when a wave is incident upon a slit of approximately the same size as the wavelength.

Interference patterns from multiple slits and thin films produce accurately repeatable patterns. © IBO 2014

… CHAPTER 9

9.1 Simple harmonic motion

NATURE OF SCIENCE:

Insights: The equation for simple harmonic motion (SHM) can be solved analytically and numerically. Physicists use such solutions to help them to visualize the behaviour of the oscillator. The use of the equations is very powerful as any oscillation can be described in terms of a combination of harmonic oscillators. Numerical modelling of oscillators is important in the design of electrical circuits. (1.11)

© IBO 2014

Essential idea: The solution of the harmonic oscillator can be framed around the variation of kinetic and potential energy in the system.

Understandings:
- The defining equation of SHM
- Energy changes

Recall from chapter 4 that if the **acceleration** a of a system is:

- **directly proportional to its displacement x from its equilibrium position**

and

- **directed towards the equilibrium position**

then the system will execute SHM.

This is the formal definition of SHM.

We expressed this definition mathematically as

$a = -\,\text{const} \times x$

The negative sign indicated that the acceleration was directed towards the equilibrium position. Mathematical analysis shows that the constant is in fact equal to ω^2 where ω is the angular frequency (defined above) of the system. Hence equation 4.5 becomes

$a = -\omega^2 x$

If a system is performing SHM, then to produce the acceleration, a force must be acting on the system in the direction of the acceleration. From our definition of SHM, the magnitude of the force F is given by

$F = -kx$

where k is a constant and the negative sign indicates that the force is directed towards the equilibrium position of the system. (Do not confuse this constant k with the spring constant. However, when dealing with the oscillations of a mass on the end of a spring, k will be the spring constant.)

To understand the solutions of the SHM equation, let us consider the oscillations of a mass suspended from a vertically supported spring. We shall consider the mass of the spring to be negligible and for the extension x to obey the rule $F = kx$. $F (= mg)$ is the force causing the extension. Figure 901(a) shows the spring and a suspended weight of mass m in equilibrium. In Figure 901(b), the weight has been pulled down a further extension x_0.

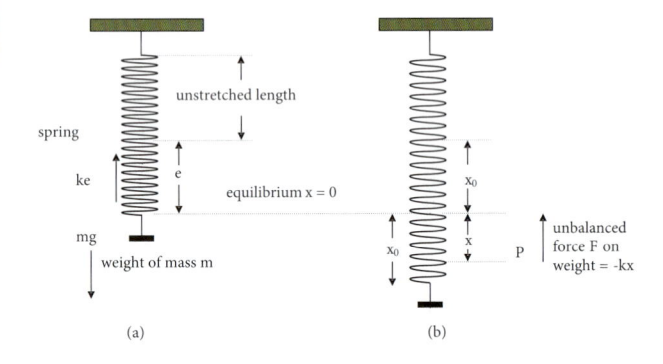

Figure 901   SHM of a mass suspended by a spring

In Figure 901 (a), the equilibrium extension of the spring is e and the net force on the weight is $mg - ke = 0$.

In Figure 408 (b), if the weight is held in position at $x = x_0$ and then released, when the weight moves to position P, a distance x from the equilibrium position $x = 0$, the net force on the weight is $mg - ke - kx$. Clearly, then the unbalanced force on the weight is $-kx$. When the weight reaches a point distance x above the equilibrium position, the compression force in the spring provides the unbalanced force towards the equilibrium position of the weight.

The acceleration of the weight is given by Newton's second law;

$F = -kx = ma$

i.e. $a = -\dfrac{k}{m} x$

This is of the form $a = -\omega^2 x$ where $\omega = \sqrt{\dfrac{k}{m}}$, that is, the weight will execute SHM with a frequency

$T = 2\pi \sqrt{\dfrac{m}{k}}$

The displacement of the weight x is given by

$x = x_0 \cos \omega t = x_0 \cos \sqrt{\dfrac{k}{m}}\, t$

This is the particular solution of the SHM equation for the oscillation of a weight on the end of a spring. This system is often referred to as *a harmonic oscillator*.

Alternatively, if the weight was at its equilibrium position when t = 0, then the trigonometric expression would be

$x = x_0 \sin \omega t$

The velocity v of the weight at any instant can be found by finding the **gradient** of the displacement-time graph at that instant. The displacement time graph is a cosine function and the gradient of a cosine function is a negative sine function. The gradient of

$x = x_0 \cos \omega t$ is in fact so

$v = -\omega x_0 \sin \omega t = -v_0 \sin \omega t$

where v_0 is the maximum and minimum velocity equal in magnitude to ωx_0.

Students familiar with calculus will recognise the velocity v as

$v = \dfrac{dx}{dt} = \dfrac{d}{dt}(x_0 \cos \omega t) = -\omega x_0 \sin \omega t$. Similarly,

$a = \dfrac{dv}{dt} = \dfrac{d}{dt}(-\omega x_0 \sin \omega t) = -\omega^2 x_0 \cos \omega t = -\omega^2 x$

which of course is just the defining equation of SHM.

However, we have to bear in mind that ωt varies between 0 and 2π where $\cos \omega t$ is negative for ωt for $\dfrac{\pi}{2} \to \dfrac{3\pi}{2}$ and $\sin \omega t$ is negative for ωt in the range π to 2π. This effectively means that when the displacement from equilibrium is positive, the velocity is negative and so directed towards equilibrium. When the displacement from equilibrium is negative, the velocity is positive and so directed away from equilibrium

The sketch graph in Figure 902 shows the variation with time t of the displacement x and the corresponding variation with time t of the velocity v. This clearly demonstrates the relation between the sign of the velocity and sign of the displacement.

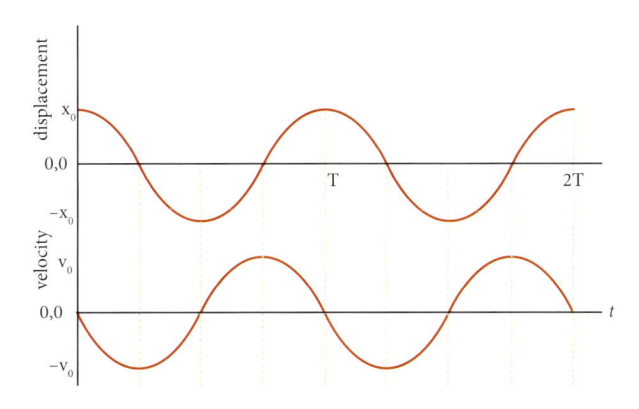

Figure 902   Displacement-time and velocity-time graphs

We can also see how the velocity v changes with displacement x.

We can express $\sin \omega t$ in terms of $\cos \omega t$ using the trigonometric relation

$\sin^2 \theta + \cos^2 \theta = 1$

From which it can be seen that

$\sin \theta = \sqrt{1 - \cos^2 \theta}$

Replacing θ with ωt we have

$v = -\omega x_0 \sqrt{1 - \cos^2 \omega t}$

Remembering that $x = x_0 \cos \omega t$ and putting x_0 inside the square root gives

$v = -\omega \sqrt{(x_0^2 - x^2)}$

Bearing in mind that v can be positive or negative, we must write

$v = \pm \omega \sqrt{(x_0^2 - x^2)}$

The velocity is zero when the displacement is a maximum and is a maximum when the displacement is zero.

The graph in Figure 903 shows the variation with x of the velocity v for a system oscillating with a period of 1 second and with an amplitude of 5 cm. The graph shows the variation over a time of any one period of oscillation.

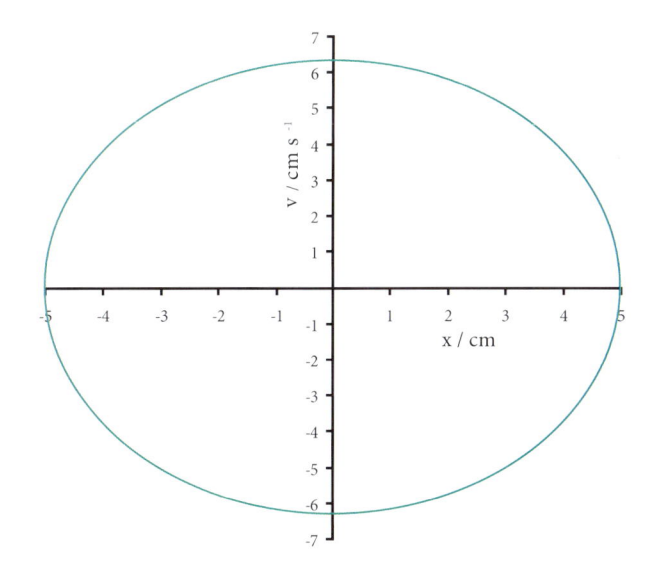

Figure 903   Velocity-displacement graph

CHAPTER 9

Boundary Conditions

The two solutions to the general SHM equation are $x = x_0 \cos \omega t$ and $x = x_0 \sin \omega t$. Which solution applies to a particular system depends, as mentioned above, on the boundary conditions for that system. For systems such as the harmonic oscillator and the simple pendulum, the boundary condition that gives the solution $x = x_0 \cos \omega t$ is that the displacement $x = x_0$ when $t = 0$. For some other systems it might turn out that $x = 0$ when $t = 0$. This will lead to the solution $x = x_0 \sin \omega t$. From a practical point of view, the two solutions are essentially the same; for example when timing the oscillations of a simple pendulum, you might decide to start the timing when the pendulum bob passes through the equilibrium position. In effect, the two solutions differ in phase by $\frac{\pi}{2}$.

The table in Figure 904 summarises the solutions we have for SHM.

$x = x_0 \cos \omega t$	$x = x_0 \sin \omega t$
$v = -v_0 \sin \omega t$	$v = v_0 \cos \omega t$
$v = -\omega x_0 \sin \omega t$	$v = -\omega x_0 \cos \omega t$
$v = \pm \omega \sqrt{x_0^2 - x^2}$	$v = \pm \omega \sqrt{x_0^2 - x^2}$

Figure 904   Common equations

We should mention that since the general solution to the SHM equation is $x = x_0 \sin \omega t + x_0 \cos \omega t$ there are in fact three solutions to the equation. This demonstrates a fundamental property of second order differential equations; that one of the solutions to the equation is the sum of all the other solutions. This is the mathematical basis of the so-called **principle of superposition**.

SHM is a very good example in which to apply the Newtonian method discussed in Chapter 2; i.e. if the forces that act on a system are known, then the future behaviour of the system can be predicted. Here we have a situation in which the force is given by $-kx$. From Newton's second law therefore $-kx = ma$, where m is the mass of the system and a is the acceleration of the system. However, the acceleration is not constant. For those of you who have a mathematical bent, the relation between the force and the acceleration is written as $-kx = m\frac{d^2x}{dt^2}$. This is what is called a "second order differential equation". The solution of the equation gives x as a function of t. The actual solution is of the SHM equation is

$x = P\cos\omega t + Q\sin\omega t$ where P and Q are constants and ω is the angular frequency of the system and is equal to $\sqrt{\frac{k}{m}}$.

Whether a particular solution involves the sine function or the cosine function, depends on the so-called 'boundary conditions'. If for example $x = x_0$ (the amplitude) when $t = 0$, then the solution is $x = x_0 \cos \omega t$.

The beauty of this mathematical approach is that, once the general equation has been solved, the solution for all systems executing SHM is known. All that has to be shown to know if a system will execute SHM, is that the acceleration of the system is given by Equation 4.5 or the force is given by equation 4.7. The physical quantities that ω will depend on is determined by the particular system. For example, for a weight of mass m oscillating at the end of a vertically supported spring whose spring constant is k, then $\omega = \sqrt{\frac{k}{m}}$ or, from equation 4.4 $T = 2\pi \sqrt{\frac{m}{k}}$. For a simple pendulum, $\omega = \sqrt{\frac{g}{l}}$ where l is the length of the pendulum and g is the acceleration of free fall such that $T = 2\pi \sqrt{\frac{l}{g}}$.

Examples

The graph in Figure 905 shows the variation with time t of the displacement x of a system executing SHM. (Some question parts were already done in the SL chapter 4).

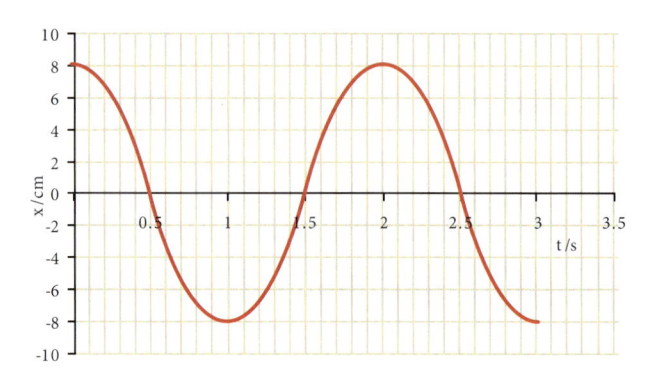

Figure 905   Displacement – time graph for SHM

Use the graph to determine the

(i) period of oscillation
(ii) amplitude of oscillation
(iii) maximum speed
(iv) speed at $t = 1.3$ s
(v) maximum acceleration

Solutions

(i) 2.0 s (time for one cycle)
(ii) 8.0 cm
(iii) Using $v_0 = \omega x_0$ gives $\frac{2\pi}{2} \times 8.0$

(remember that $\omega = \frac{2\pi}{T}$) $= 25$ cm s^{-1}

286

(iv) $v = -v_0 \sin\omega t = -25\sin(1.3\pi)$. To find the value of the sine function, we have to convert the 1.3π into degrees (remember ω and hence ωt, is measured in radians)

1 radian $= \dfrac{180}{\pi}$ deg therefore $1.3\pi = 1.3 \times 180 = 234°$

therefore $v_1 = -25\sin(234°) = +20$ cm s^{-1}.

Or we can solve using $\omega\sqrt{(x_0^2 - x^2)}$

from the graph at $t = 1.3$ s, $x = -4.8$ cm

therefore $v = \pi \times \sqrt{(8.0)^2 - (4.8)^2} = 20$ cm s^{-1}

(v) Using $a_{max} = \omega^2 x_0 = \pi^2 \times 8.0 = 79$ m s^{-2}

Energy changes

We must now look at the energy changes involved in SHM. To do so, we will again concentrate on the harmonic oscillator. The mass is stationary at $x = +x_0$ (maximum extension) and also at $x = -x_0$ (maximum compression). At these two positions the energy of the system is all potential energy and is in fact the elastic potential energy stored in the spring. This is the total energy of the system E_T and clearly

$E_T = \dfrac{1}{2}kx_0^2$ Equation 4.12

That is, that for any system performing SHM, the energy of the system is proportional to the square of the amplitude. This is an important result and one that we shall return to when we discuss wave motion.

At $x = 0$ the spring is at its equilibrium extension and the magnitude of the velocity v of the oscillating mass is a maximum v_0. The energy is all kinetic and again is equal to E_T. We can see that this is indeed the case as the expression for the maximum kinetic energy E_{max} in terms of v_0 is

$E_{max} = \dfrac{1}{2}mv_0^2$

Clearly E_T and E_{max} are equal such that

$E_T = \dfrac{1}{2}kx_0^2 = E_{max} = \dfrac{1}{2}mv_0^2$

From which

$v_0^2 = \dfrac{k}{m}x_0^2$ (as $v_0 \geq 0$)

Therefore

$v_0 = \sqrt{\dfrac{k}{m}}\, x_0 = \omega x_0$

which ties in with the velocity being equal to the gradient of the displacement-time graph (see 4.1.5). As the system oscillates there is a continual interchange between kinetic energy and potential energy such that the loss in kinetic energy equals the gain in potential energy and $E_T = E_K + E_P$.

Remembering that $\omega^2 = \dfrac{k}{m}$, we have that

$E_T = \dfrac{1}{2}m\omega^2 x_0^2$

Clearly, the potential energy E_P at any displacement x is given by

$E_P = \dfrac{1}{2}m\omega^2 x^2$

At any displacement x, the kinetic energy E_K is $E_K = \dfrac{1}{2}mv^2$ Hence remembering that

$v = -\omega\sqrt{(x_0^2 - x^2)}$, we have

$E_K = \dfrac{1}{2}m\omega^2(x_0^2 - x^2)$

Although we have derived these equations for a harmonic oscillator, they are valid for any system oscillating with SHM. The sketch graph in Figure 906 shows the variation with displacement x of E_K and E_P for one period.

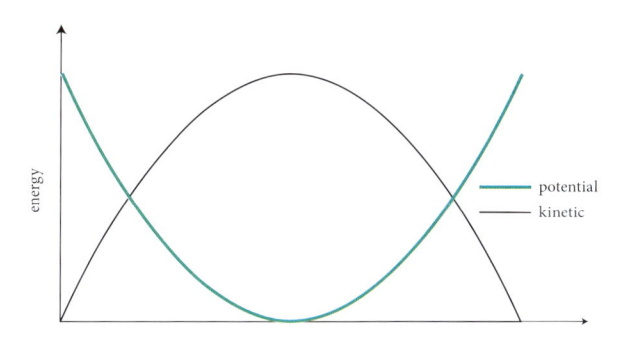

Figure 906   Energy and displacement

Example

The amplitude of oscillation of a mass suspended by a vertical spring is 8.0 cm. The spring constant of the spring is 74 N m^{-1}. Determine

(a) the total energy of the oscillator

(b) the potential and the kinetic energy of the oscillator at a displacement of 4.8 cm from equilibrium.

CHAPTER 9

Solution

(a) Spring constant $k = 74$ N m^{-1} and

$x_0 = 8.0 \times 10^{-2}$ m

$E_T = \frac{1}{2} kx_0^2$

$= \frac{1}{2} \times 74 \times 64 \times 10^{-4} = 0.24$ J

(b) At $x = 4.8$ cm

Therefore $E_P = \frac{1}{2} \times 74 \times (4.8 \times 10^{-2})^2$

$= 0.085 \times 10^{-4}$ J

$E_K = E_T - E_P = 0.24 - 0.085 = 0.16$ J

EXERCISE 9.1

1. Which graph shows the relationship between the acceleration a and the displacement x from the equilibrium position of an object undergoing simple harmonic motion?

   A. Acceleration vs Displacement
   B. Acceleration vs Displacement
   C. Acceleration vs Displacement
   D. Acceleration vs Displacement

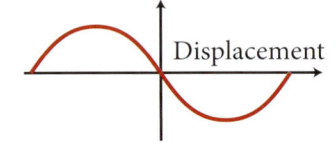

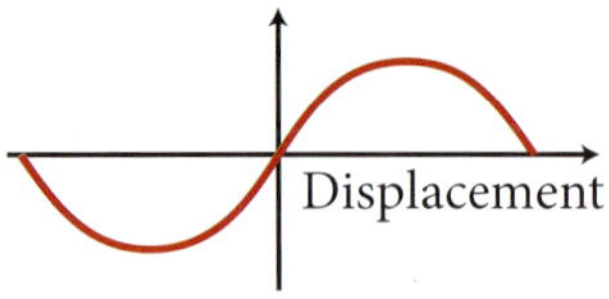

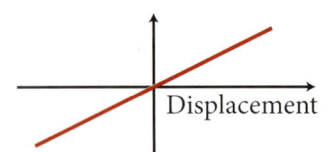

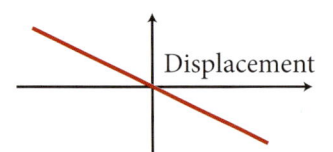

2. An object is undergoing simple harmonic motion about a fixed point P, and the magnitude of its displacement from P is x. Which one of the following statements is correct?

	Magnitude of the resultant force	Direction of the resultant force
A.	Proportional to x	Away from point P
B.	Inversely proportional to x	Away from point P
C.	Proportional to x	Towards point P
D.	Inversely proportional to x	Towards point P

3. The angular speed of the "minute" hand of an analogue watch is

   A. $\pi / 1800$ rad s^{-1}
   B. $\pi / 60$ rad s^{-1}
   C. $\pi / 30$ rad s^{-1}
   D. 120 rad s^{-1}

4. Which of the following is true of the magnitude of the acceleration of the object that is undergoing simple harmonic motion?

   A. It is greatest at the midpoint of the motion.
   B. It is greatest at the end points of the motion.
   C. It is uniform throughout the motion.
   D. It is greatest at the midpoints and the endpoints.

5. A particle oscillates with simple harmonic motion with a period T.

   At time $t = 0$, the particle has its maximum displacement. Which graph The variation with time t of the kinetic energy of the particle is shown in which diagram below?

   A. B. 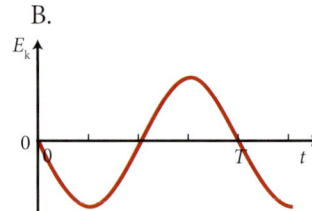 C. 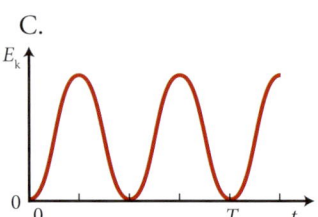 D.

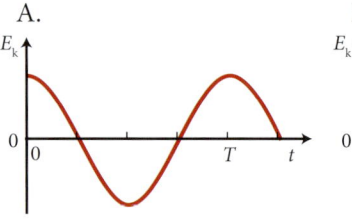

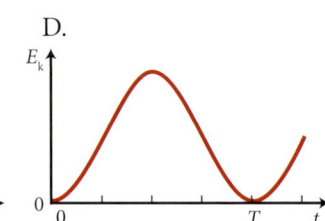

6. Figure 907 below shows how the displacement of the magnet of mass 0.3 kg hanging from a spring varies with time for two oscillations.

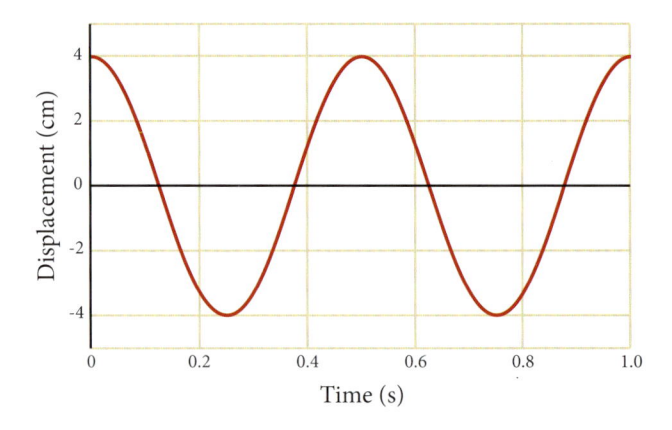

Figure 907

288

Using information from this graph calculate the

(a) value of the spring constant.
(b) maximum kinetic energy of the magnet.

7. A system is oscillating with SHM as described by the graph in the Figure 908 below.

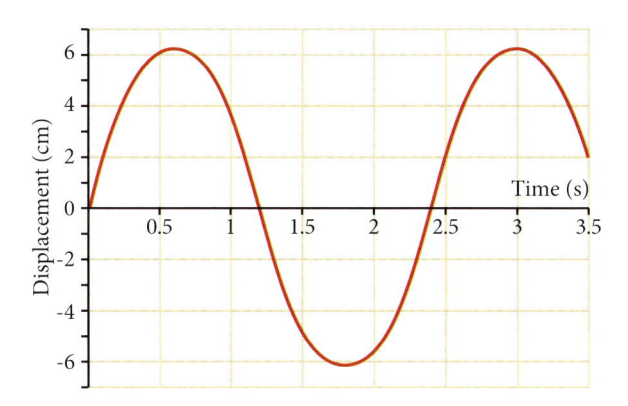

Figure 908

(a) Use the graph to determine the

(i) period of oscillation
(ii) amplitude of oscillation
(iii) maximum speed
(iv) speed at t = 1.3 s
(v) maximum acceleration

(b) State two values of t for when the magnitude of the velocity is a maximum and two values of t for when the magnitude of the acceleration is a maximum.

9. Figure 909 shows the relationship that exists between the acceleration and the displacement from the equilibrium position for a harmonic oscillator.

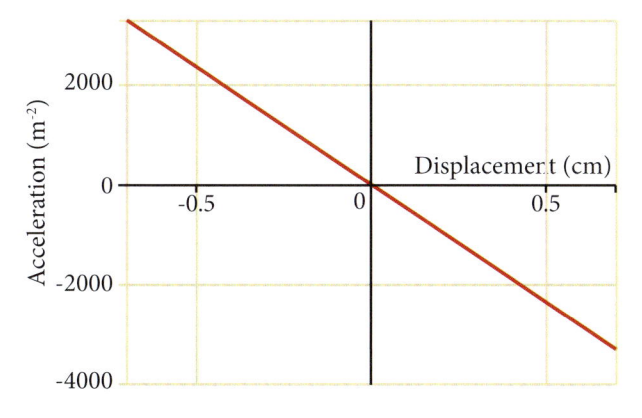

Figure 909

(a) State and explain two reasons why the graph opposite indicates that the object is executing simple harmonic motion.

(b) Determine the frequency of oscillation.

9.2 Single-slit diffraction

NATURE OF SCIENCE:

Development of theories: When light passes through an aperture the summation of all parts of the wave leads to an intensity pattern that is far removed from the geometrical shadow that simple theory predicts. (1.9)

© IBO 2014

Essential idea: Single-slit diffraction occurs when a wave is incident upon a slit of approximately the same size as the wavelength.

Understandings:
- The nature of single-slit diffraction

The nature of single-slit diffraction

Single slit diffraction intensity distribution

When plane wavefronts pass through a small aperture they spread out as discussed in chapter 4. This is an example of the phenomenon called diffraction. Light waves are no exception to this and ways for observing the diffraction of light have also been discussed previously.

However, when we look at the diffraction pattern produced by light we observe a fringe pattern, that is, on the screen there is a bright central maximum with "secondary" maxima either side of it. There are also regions where there is no illumination and these minima separate the maxima. If we were to actually plot how the intensity of illumination varies along the screen then we would obtain a graph similar to that as in Figure 910.

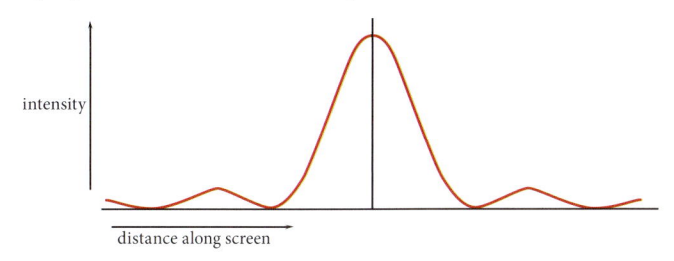

Figure 910 Intensity distribution for single-slit diffraction

We would get the same intensity distribution if we were to plot the intensity against the angle of diffraction θ. (See next section).

CHAPTER 9

This intensity pattern arises from the fact that each point on the slit acts, in accordance with Huygen's principle, as a source of secondary wavefronts. It is the interference between these secondary wavefronts that produces the typical diffraction pattern.

Obtaining an expression for the intensity distribution is mathematically a little tricky and it is not something that we are going to attempt here. However, we can deduce a useful relationship from a simple argument. In this argument we deal with a phenomenon called **Fraunhofer diffraction**, that is the light source and the screen are an infinite distance away form the slit. This can be achieved with the set up shown in Figure 911.

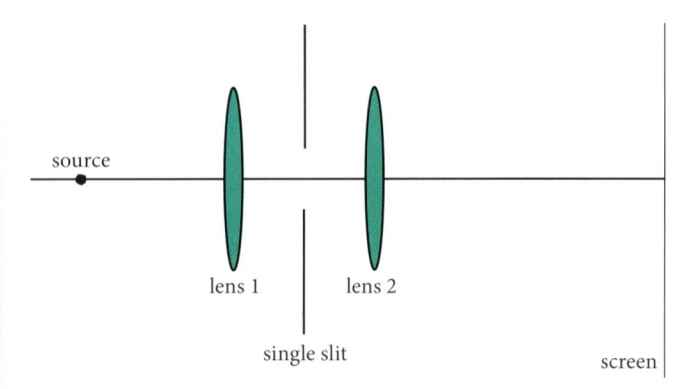

Figure 911 Apparatus for viewing Fraunhofer diffraction

The source is placed at the principal focus of lens 1 and the screen is placed at the principal focus of lens 2. Lens 1 ensures that parallel wavefronts fall on the single slit and lens 2 ensures that the parallel rays are brought to a focus on the screen. The same effect can be achieved using a laser and placing the screen some distance from the slit. If the light and screen are not an infinite distance from the slit then we are dealing with a phenomenon called **Fresnel diffraction** and such diffraction is very difficult to analyse mathematically. To obtain a good idea of how the single slit pattern comes about we consider the diagram Figure 912.

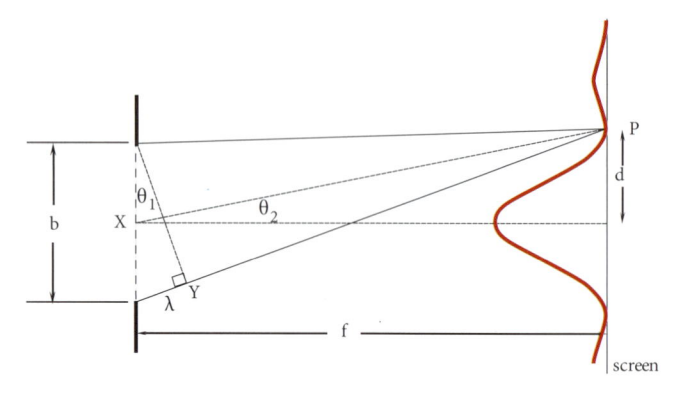

Figure 912 Single slit diffraction

In particular we consider the light from one edge of the slit to the point P where this point is just one wavelength further from the lower edge of the slit than it is from the upper edge. The secondary wavefront from the upper edge will travel a distance $\lambda/2$ further than a secondary wavefront from a point at the centre of the slit. Hence when these wavefronts arrive at P they will be out of phase and will interfere destructively. The wavefronts from the next point below the upper edge will similarly interfere destructively with the wavefront from the next point below the centre of the slit. In this way we can pair the sources across the whole width of the slit. If the screen is a long way from the slit then the angles θ_1 and θ_2 become nearly equal. (*If the screen is at infinity then they are equal and the two lines PX and XY are at right angles to each other*). From Figure 1116 we see therefore that there will be a minimum at P if

$$\lambda = b \sin \theta_1$$ where b is the width of the slit.

However, both angles are very small, equal to θ say, where θ is the angle of diffraction.

So it can be written $\theta = \dfrac{\lambda}{b}$

This actually gives us the half-angular width of the central maximum. We can calculate the actual width of the maximum along the screen if we know the focal length of the lens focussing the light onto the screen. If this is f then we have that

$$\theta = \dfrac{d}{f} \quad \text{Such that} \quad d = \dfrac{f\lambda}{b}$$

To obtain the position of the next maximum in the pattern we note that the path difference is $\dfrac{3}{2}\lambda$. We therefore divide the slit into three equal parts, two of which will produce wavefronts that will cancel and the other producing wavefronts that reinforce. The intensity of the second maximum is therefore much less than the intensity of the central maximum of the order of 0.05 or 1/20 of the original intensity. (Much less than one third in fact since the wavefronts that reinforce will have differing phases).

We can also see now how diffraction effects become more and more noticeable the narrower the slit becomes. If light of wavelength 430 nm was to pass through a slit of width say 10 cm and fall on a screen 3.0 m away, then the half angular width of the central maximum would be 0.13 µm.

$$d = \dfrac{f\lambda}{b} = \dfrac{3 \times 430 \times 10^{-9}}{0.1} = 0.13 \text{ µm}$$

There will be lots of maxima of nearly the same intensity and the maxima will be packed very closely together. (The first minimum occurs at a distance of 0.12 µm from the centre of the central maximum and the next occurs effectively at a distance of 0.24 µm). We effectively observe the geometric pattern. *Refer to Example.*

We also see now how for diffraction effects to be noticeable the wavelength must be of the order of the slit width. The width of the pattern increases in proportion to the wavelength and decreases inversely with the width of the slit. If the slit width is much greater than the wavelength then the width of the central maxima is very small.

In summary,

for **destructive interference** and **minima**

$d \sin\theta = n\lambda$ where n = 1, 2, 3…

for **constructive interference** and **maxima**

$d \sin\theta = (n + ½)\lambda$ where n = 1, 2, 3…

Note that waves from point sources along the slit arrive in phase at the centre of the fringe pattern and constructively interfere to produce the central maximum, called the **zero order maximum** (n = 0).

Example

Light from a laser is used to form a single slit diffraction pattern. The width of the slit is 0.10 mm and the screen is placed 3.0 m from the slit. The width of the central maximum is measured as 2.6 cm. Calculate the wavelength of the laser light?

Solution

Since the screen is a long way from the slit we can use the small angle approximation such that the f is equal to 3.0 m.

The half-width of the central maximum is 1.3 cm so we have

$$\lambda = \frac{(1.3 \times 10^{-2}) \times (1.0 \times 10^{-4})}{3.0}$$

To give λ = 430 nm.

This example demonstrates why the image of a point source formed by a thin converging lens will always have a finite width.

Exercise 9.2

1. A parallel beam of light of wavelength 500 nm is incident on a slit of width 0.25 mm. The light is brought to focus on a screen placed 1.50 m from the slit. Calculate the angular width and the linear width of the central diffraction maximum.

2. Light from a laser is used to form a single slit diffraction pattern on a screen. The width of the slit is 0.10 mm and the screen is 3.0 m from the slit. The width of the central diffraction maximum is 2.6 cm. Calculate the wavelength of the laser light.

3. Determine the angle that you would expect to find constructive interference (for n = 0, n = 1 and n = 2) from a single slit of width 2.0 μm and light of wavelength 600 nm.

9.3 Interference

NATURE OF SCIENCE:

Curiosity: Observed patterns of iridescence in animals, such as the shimmer of peacock feathers, led scientists to develop the theory of thin film interference. (1.5)

Serendipity: The first laboratory production of thin films was accidental. (1.5)

© **IBO 2014**

Essential idea: Interference patterns from multiple slits and thin films produce accurately repeatable patterns.

Understandings:
- Young's double-slit experiment
- Modulation of two-slit interference pattern by one-slit diffraction effect
- Multiple slit and diffraction grating interference patterns
- Thin film interference

Young's double slit experiment

Please refer to interference in chapter 4. However, we will reiterate the condition for the interference of waves from two sources to be observed. The two sources must be **coherent**, that is they must have the same phase or the phase difference between them must remain constant.

Also, to reinforce topics concerning the principle of superposition, and path difference and phase difference, included in Figure 913 is another example of two source interference.

We can obtain evidence for the wave nature of sound by showing that sound produces an interference pattern. Figure 913 shows the set up for demonstrating this.

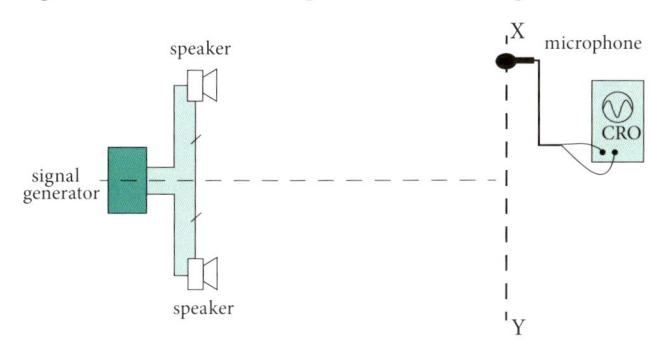

Figure 913   Interference using sound waves

The two speakers are connected to the same output of the signal generator and placed about 50 cm apart. The signal generator frequency is set to about 600 Hz and the microphone is moved along the line XY, that is about one metre from the speakers. As the microphone is moved along XY the trace on the cathode ray oscilloscope is seen to go through a series of maxima and minima corresponding to points of constructive and destructive interference of the sound waves.

An interesting investigation is to find how the separation of the points of maximum interference depends on the frequency of the source and also the distance apart of the speakers.

In the demonstration of the interference between two sound sources described above, if we were to move one of the speakers from side to side or backwards and forwards the sound emitted from the two speakers would no longer be in phase. No permanent points of constructive or destructive interference will now be located since the phase difference between the waves from the two sources is no longer constant, i.e. the sources are no longer coherent.

Light from an incandescent source is emitted with a completely random phase. Although the light from two separate sources will interfere, because of the randomly changing phase no permanent points of constructive or destructive interference will be observed. This is why a single slit is needed in the Young's double slit experiment. By acting as a point source, it essentially becomes a coherent light source. The light emitted from a laser is also very nearly coherent and this is why it is so easy to demonstrate optical interference and diffraction with a laser.

Young's double slit experiment is one of the great classic experiments of physics and did much to reinforce the wave theory of light. Thomas Young carried out the experiment in about 1830.

It is essentially the demonstration with the ripple tank and the sound experiment previously described, but using light. The essential features of the experiment are shown in Figure 914.

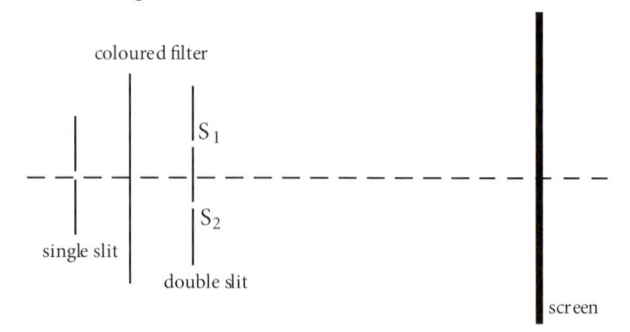

Figure 914   Young's double slit experiment

Young allowed sunlight to fall onto a narrow single slit. A few centimetres from the single slit he placed a double slit. The slits are very narrow and separated by a distance equal to about fourteen slit widths. A screen is placed about a metre from the double slits. Young observed a pattern of multicoloured "fringes" in the screen. When he placed a coloured filter between the single slit and double slit he obtained a pattern that consisted of bright coloured fringes separated by darkness.

The single slit essentially ensures that the light falling on the double slit is coherent. The two slits then act as the two speakers in the sound experiment or the two dippers in the ripple tank. The light waves from each slit interfere and produce the interference pattern on the screen. Without the filter a pattern is formed for each wavelength present in the sunlight. Hence the multicoloured fringe pattern.

You can demonstrate optical interference for yourself. Smoking a small piece of glass and then drawing two parallel lines on it can make a double slit. If you then look through the double slit at a single tungsten filament lamp you will see the fringe pattern. By placing filters between the lamps and the slits you will see the monochromatic fringe pattern.

You can also see the effects of optical interference by looking at net curtains. Each 'hole' in the net acts as a point source and the light from all these separate sources interferes and produces quite a complicated interference pattern.

A laser can also be used to demonstrate optical interference. Since the light from the laser is coherent it is very easy to demonstrate interference. Just point the laser at a screen and place a double slit in the path of the laser beam.

Let us now look at the Young's double slit experiment in more detail. The geometry of the situation is shown in Figure 915.

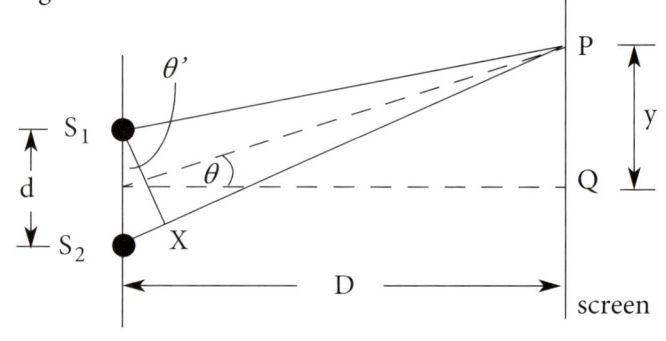

Figure 915   The geometry of Young's double slit experiment

S_1 and S_2 are the two narrow slits that we shall regard as two coherent, monochromatic point sources. The distance from the sources to the screen is D and the distance between the slits is d.

The waves from the two sources will be in phase at Q and there will be a bright fringe here. We wish to find the condition for there to be a bright fringe at P distance y from Q.

We note that that D ($\approx$ 1 metre) is very much greater than either y or d. ($\approx$ few millimetres). This means that both θ and θ' are very small angles and for intents and purposes equal.

From the diagram we have that

$$\theta = \frac{y}{D}$$

And

$$\theta' = \frac{S_2 X}{d}$$

(Remember, the angles are very small)

Since $\theta \approx \theta'$ then

$$\frac{y}{D} = \frac{S_2 X}{d}$$

But $S_2 X$ is the path difference between the waves as they travel to P. We have therefore that

$$\text{path difference} = \frac{yd}{D}$$

There will therefore be a bright fringe at P if

$$\frac{yd}{D} = n\lambda$$

Suppose that there is a bright fringe at $y = y_1$ corresponding to $n = n_1$ then

$$\frac{y_1 d}{D} = n_1 \lambda$$

If the next bright fringe occurs at $y = y_2$ this will correspond to $n = n_1 + 1$. Hence

$$\frac{y_2 d}{D} = (n_1 + 1)\lambda$$

This means that the spacing between the fringes $y_2 - y_1$ is given by

$$y_2 - y_1 = \frac{D}{d}\lambda$$

Young actually use this expression to measure the wavelength of the light he used and it is a method still used today.

We see for instance that if in a given set up we move the slits closer together then the spacing between the fringes will get greater. Effectively our interference pattern spreads out, that is there will be fewer fringes in a given distance. We can also increase the fringe spacing by increasing the distance between the slits and the screen. You will also note that for a given set up using light of different wavelengths, then "red" fringes will space further apart than "blue" fringes.

In this analysis we have assumed that the slits act as point sources and as such the fringes will be uniformly spaced and of equal intensity. A more thorough analysis should take into account the finite width of the slits.

Returning to Figure 915 we see that we can write the path difference $S_2 X$ as

$$S_2 X = d\tan\theta'$$

But since θ' is a small angle the sine and tangent will be nearly equal so that

$$S_2 X = d\sin\theta'$$

And since $\theta' \approx \theta$ then

$$S_2 X = d\sin\theta$$

The condition therefore for a bright fringe to be found at a point of the screen can therefore be written as

$$d\sin\theta = n\lambda$$

In summary,

for **constructive interference** and **minima**

d sinθ = nλ   where n = 1, 2, 3…

for destructive interference and maxima

d sinθ = (n + ½)λ   where n = 1, 2, 3…

Figure 916 shows the intensity distribution of the fringes on the screen when the separation of the slits is large compared to their width. The fringes are of equal intensity and of equal separation.

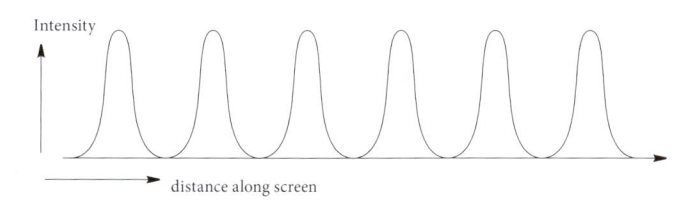

Figure 916   The intensity distribution of the fringes

It is worth noting that if the slits are close together, the intensity of the fringes is modulated by the intensity distribution of the diffraction pattern of one of the slits.

CHAPTER 9

Example

Light of wavelength 500 nm is incident on two small parallel slits separated by 1.0 mm. Determine the angle where the first maximum is formed?

If after passing through the slits the light is brought to a focus on a screen 1.5 m from the slits calculate the observed fringe spacing on the screen.

Solution

Using the small angle approximation we have

$$\theta = \frac{\lambda}{d} = \frac{5 \times 10^{-7}}{10^{-3}}$$

$$= 5 \times 10^{-4} \text{ rad}$$

The fringe spacing is given by

$$y = \frac{D\lambda}{d} = \frac{1.5 \times 5 \times 10^{-7}}{10^{-3}}$$

$$= 0.75 \text{ mm}$$

Modulation of two-slit interference pattern by one-slit diffraction effect

In practice the intensity pattern of the maxima as shown in Figure 917 is not constant, but fluctuates while decreasing symmetrically on either side of the central maximum as shown in Figure 917. The intensity pattern is a combination of both the single-slit **diffraction envelope** and the double slit pattern. That is the amplitude of the two-slit interference pattern is **modulated** (i.e. adjusted to) by a single slit diffraction envelope.

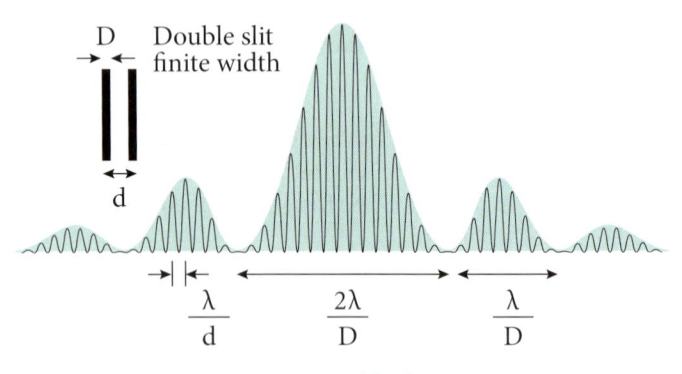

Figure 917 Double slit pattern

For a given slit separation d, wavelength of light, and fixed slit to screen distance, the variation in intensity depends on the width of the slit D. Although increasing the width of the slits increases the intensity of light in the fringes, it also makes them less sharp and they become blurred. As the slit width is widened, the fringes gradually disappear because numerous point sources along the widened slit give rise to their own dark and bright fringes that then overlap with each other.

Exercise 9.3

1. In Figure 913, the distance between the speakers is 0.50 m and the distance between the line of the speakers and the screen is 2.0 m. As the microphone is moved along the line XY, the distance between successive points of maximum sound intensity is 0.30 m. The frequency of the sound waves is 4.4×10^3 Hz. Calculate a value for the speed of sound.

2. Laser light of wavelength 610 nm falls on two slits 1.0×10^{-5} apart. Determine the separation of the first order maximum formed on a screen 2.0 m away.

3. Two parallel slits 0.12 mm apart are illuminated with red light with a wavelength of 600 nm. An interference pattern falls on a screen 1.5 m away.

   Calculate

   (a) the distance from the central maximum to the first bright fringe

   (b) the distance to the second dark line.

Multiple slit and diffraction grating interference patterns

If we examine the interference pattern produced when monochromatic light passes through a different number of slits we notice that as the number of slits increases the number of observed fringes decreases, the spacing between them increases and the individual fringes become much sharper. We can get some idea of how this comes about by looking at the way light behaves when a parallel beam passes through a large number of slits. The diagram for this is shown in Figure 918.

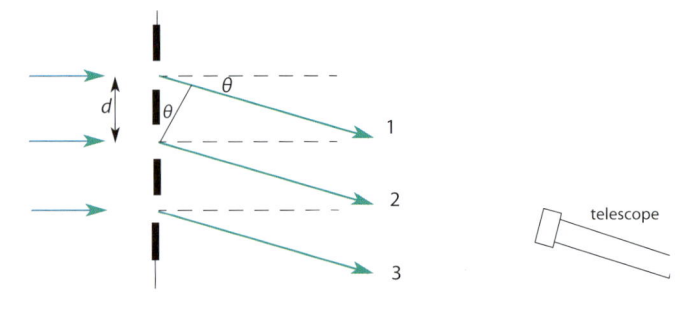

Figure 918 A parallel beam passing through several slits

The slits are very small so that they can be considered to act as point sources. They are also very close together such that d is small (10^{-6} m). Each slit becomes a source of circular wave fronts and the waves from each slit will interfere. Let us

294

consider the light that leaves the slit at an angle θ as shown. The path difference between wave 1 and wave 2 is $d\sin\theta$ and if this is equal to an integral number of wavelengths then the two waves will interfere constructively in this direction. Similarly wave 2 will interfere constructively with wave 3 at this angle, and wave 3 with 4 etc., across the whole grating. Hence if we look at the light through a telescope, that is bring it to a focus, then when the telescope makes an angle θ to the grating a bright fringe will be observed. The condition for observing a bright fringe is therefore

$$d\sin\theta = m\lambda$$

Suppose we use light of wavelength 500 nm and suppose that $d = 1.6 \times 10^{-6}$ m.

Obviously we will see a bright fringe in the straight on position $\theta = 0$ (the *zero order*).

The next position will be when $m = 1$ (the *first order*) and substitution in the above equation gives $\theta = 18°$.

The next position will be when $m = 2$ (the *second order*) and this give $\theta = 38°$.

For $m = 3$, $\sin\theta$ is greater than 1 so with this set up we only obtain 5 fringes, one zero order and two either side of the zero order.

The calculation shows that the separation of the orders is relatively large. At any angles other than 18° or 38° the light leaving the slits interferes destructively. We can see that the fringes will be sharp since if we move just a small angle away from 18° the light from the slits will interfere destructively.

An array of narrow slits is usually made by cutting narrow transparent lines very close together into the emulsion on a photographic plate (typically 200 lines per millimetre). Such an arrangement is called a **diffraction grating**.

The diffraction grating is of great use in examining the spectral characteristics of light sources.

All elements have their own characteristic spectrum. An element can be made to emit light either by heating it until it is incandescent or by causing an electric discharge through it when it is in a gaseous state. Your school probably has some discharge tubes and diffraction gratings. If it has, then look at the glowing discharge tube through a diffraction grating. If for example, the element that you are looking has three distinct wavelengths then each wavelength will be diffracted by a different amount in accordance with the equation $d\sin\theta = n\lambda$.

Also if laser light is shone through a grating on to a screen, you will see just how sharp and spaced out are the maxima. By measuring the line spacing and the distance of the screen from the laser, the wavelength of the laser can be measured.

If your school has a set of multiple slits say from a single slit to eight slits, then it is also a worthwhile exercise to examine how the diffraction pattern changes when laser light is shone through increasing numbers of slits.

Consider a grating of N parallel equidistant slits of width a separated by an opaque region of width D. We know that light diffracted from one slit will be superimposed and interfere with light diffracted from all the other slits. The **intensity profile** of the resultant wave will have a shape determined by

- the intensity of light incident on the slits
- the wavelength of this light
- the slit width
- the slit separation

Typical intensity profiles formed when a plane wave monochromatic light at normal incidence on one, two and six slits are shown in Figure 919. In this example the slit width is one quarter of the grating spacing.

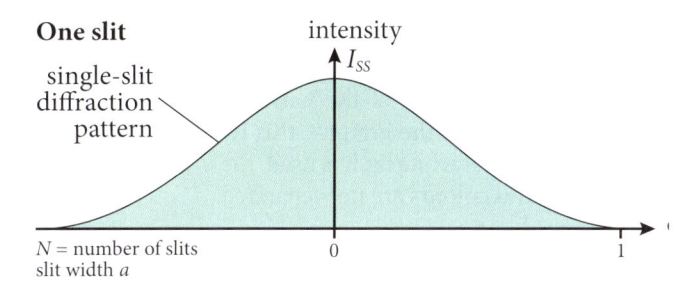

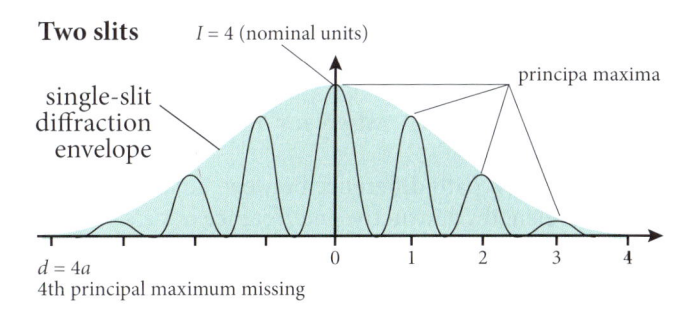

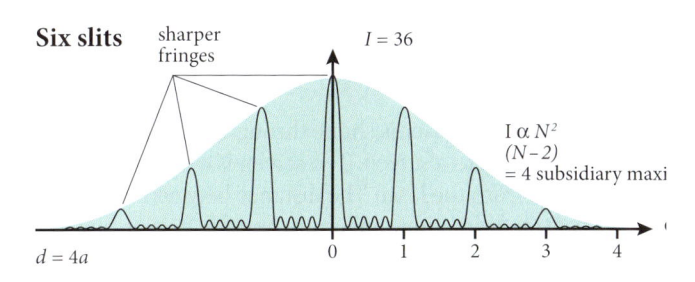

Figure 919   Intensity profiles

Chapter 9

1. For $N \geq 2$, the fringe pattern contains principal maxima that are modulated by the diffraction envelope.

2. For $N > 2$, the fringe pattern contains $(N - 2)$ subsidiary maxima.

3. Where the grating spacing d is four time the slit width a ($4 \times a = d$), then the fourth principal maximum is missing from the fringe pattern. Where the grating spacing d is three time the slit width a ($3 \times a = d$), then the third principal maximum is missing from the fringe pattern, and so on.

4. Increasing N increases the absolute intensities of the diffraction pattern ($I = N^2$)

If white light is shone through a grating then the central image will be white but for the other orders each will be spread out into a continuous spectrum composed of an infinite number of adjacent images of the slit formed by the wavelength of the different wavelengths present in the white light. At any given point in the continuous spectrum the light will be very nearly monochromatic because of the narrowness of the images of the slit formed by the grating. This is in contrast to the double slit where if white light is used, the images are broad and the spectral colours are not separated.

In summary, as is the case of interference at a double slit, there is a path difference between the rays from adjacent slits of d sinθ. Where there is a whole number of wavelengths, constructive interference occurs

For **constructive interference** and **minima**

d sinθ = mλ   where n = 1, 2, 3…

When there is a path difference between adjacent rays of half a wavelength, destructive interference occurs.

For and **maxima**

d sinθ = (m + ½)λ   where n = 1, 2, 3…

Exercise 9.4

1. Light from a laser is shone through a diffraction grating on to a screen. The screen is a distance of 2.0 m from the laser. The distance between the central diffraction maximum and the first principal maximum formed on the screen is 0.94 m. The diffraction grating has 680 lines per mm. Estimate the wavelength of the light emitted by the laser.

2. Monochromatic light from a laser is normally incident on a six-slit grating. The slit spacing is three times the slit width. An interference pattern is formed on a screen 2 m on the other side of the grating.

    (a) Determine and explain the order of the principal maxima missing from the fringe pattern.

    (b) How many subsidiary maxima are there between the principal maxima?

    (c) How many principal maxima are there

    (i) Between the two first order minima in the diffraction envelope?
    (ii) Between the first and second order minima in the diffraction envelope?

    (d) Draw the pattern formed by the bright fringes in (c).

    (e) Sketch the intensity profile of the fringe pattern.

Thin film interference

You might well be familiar with the coloured pattern of fringes that can be seen when light is reflected off the surface of water upon which a thin oil film has been spilt or from light reflected from bubbles. We can see how these patterns arise by looking at Figure 920.

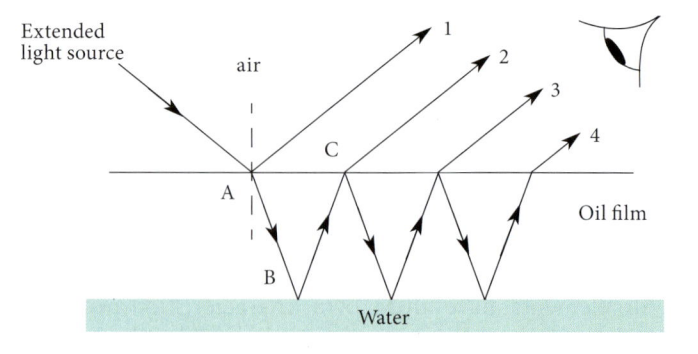

Figure 920   Reflection from an oil film

Consider light from an extended source incident on a thin film. We also consider a wave from one point of the source whose direction is represented by the ray shown. Some of this light will be reflected at A and some transmitted through the film where some will again be reflected at B (some will also be transmitted into the air). Some of the light reflected at B will then be transmitted at C and some reflected and so on. If we consider just rays 1 and 2 then these will be not be in phase when the are brought to a focus by the eye since they have travelled different distances. If the path difference between them is an integral number of half wavelengths then we might expect the two waves to be out of phase. However, ray 1 undergoes a phase change of π on reflection but ray 2 does not since it is reflected at a boundary between a more dense and less dense medium. (See chapter 4) Hence if the path difference is an integral number of half-wavelengths rays 1 and 2 will reinforce i.e. ray 1 and 2 are in phase. However, rays 3, 5, 7 etc. will be out of phase with rays 2, 4, 6 etc. but since ray 2 is more intense than ray 3 and ray 4 more intense than ray 5, these pairs will not cancel out so there will be a maximum of intensity.

If the path difference is such that wave 1 and 2 are out of phase, since wave 1 is more intense than wave 2, they will not completely annul. However, it can be shown that the intensities of waves 2, 3, 4, 5… add to equal the intensity of wave 1. Since waves 3, 4. 5… are in phase with wave 2 there will be complete cancellation.

The path difference will be determined by the angle at which ray 1 is incident and also on the thickness (and the actual refractive index as well) of the film. Since the source is an extended source, the light will reach the eye from many different angles and so a system of bright and dark fringes will be seen. You will only see fringes if the film is very thin (except if viewed at normal incidence) since increasing the thickness of the film will cause the reflected rays to get so far apart that they will not be collected by the pupil of the eye.

From the argument above, to find the conditions for constructive and destructive **interference** we need only find the path difference between ray 1 and ray 2. Figure 921 shows the geometry of the situation.

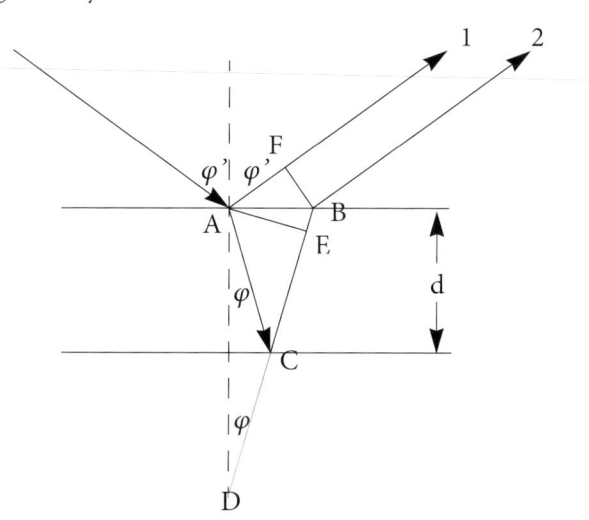

Figure 921 The geometry of interference

The film is of thickness d and refractive index n and the light has wavelength λ. If the line BF is perpendicular to ray 1 then the **optical path difference (opd)** between ray 1 and ray 2 when brought to a focus is

$opd = n(AC + CB) - AF$

We have to multiple by the refractive index for the path travelled by the light in the film since the light travels more slowly in the film. If the light travels say a distance x in a material of refractive index n then in the time that it takes to travel this distance, the light would travel a distance nx in air.

If the line CE is at right angles to ray 2 then we see that

$AF = n$BE

From the diagram AC = CD so we can write

$opd = n(CD + CB) - n$BE $= n$DE

Also from the diagram we see that, where ϕ is the angle of refraction

DE $= 2\cos\phi$

From which $\quad opd = 2nd\cos\phi$

Bearing in mind the change in phase of ray 1 on reflection we have therefore that the condition for constructive interference is

$$2nd\cos\phi = \left(m + \frac{1}{2}\right)\lambda, \; m = 1, 2, \ldots$$

And for destructive interference $\quad 2nd\cos\phi = m\lambda$

Each fringe corresponds to a particular *opd* for a particular value of the integer m and for any fringe the value of the angle ϕ is fixed. This means that it will be in the form of an arc of a circle with the centre of the circle at the point where the perpendicular drawn from the eye meets the surface of the film. Such fringes are called fringes of equal inclination. Since the eye has a small aperture these fringes, unless viewed at near to normal incidence ($\phi = 0$), will only be observed if the film is very thin. This is because as the thickness of the film increases the reflected rays will get further and further apart and so very few will enter the eye.

If white light is shone onto the film then we can see why we get multi-coloured fringes since a series of maxima and minima will be formed for each wavelength present in the white light. However, when viewed at normal incidence, it is possible that only light of one colour will under go constructive interference and the film will take on this colour.

CHAPTER 9

Thickness of oil films

The exercise to follow will help explain this use.

Non-reflecting films

A very important but simple application of thin film interference is in the production of non-reflecting surfaces.

A thin film of thickness d and refractive index n_1 is coated onto glass of refractive index n where $n_1 < n$. Light of wavelength λ that is reflected at normal incidence will undergo destructive interference if $2n_1 d = \frac{\lambda}{2}$, that is

$$d = \frac{\lambda}{4n_1}$$

(remember that there will now no phase change at the glass interface i.e. we have a rare to dense reflection)

The use of such films can greatly reduce the loss of light by reflection at the various surfaces of a system of lenses or prisms. Optical parts of high quality systems are usually all coated with non-reflecting films in order to reduce stray reflections. The films are usually made by evaporating calcium or magnesium fluoride onto the surfaces in vacuum, or by chemical treatment with acids that leave a thin layer of silica on the surface of the glass. The coated surfaces have a purplish hue by reflected light. This is because the condition for destructive interference from a particular film thickness can only be obtained for one wavelength. The wavelength chosen is one that has a value corresponding to light near the middle of the visible spectrum. This means that reflection of red and violet light is greater combining to give the purple colour. Because of the factor $\cos\phi$, at angles other than normal incidence, the path difference will change but not significantly until say about 30° (e.g. cos 25° = 0.90).

It should be borne in mind that no light is actually lost by a non-reflecting film; the decrease of reflected intensity is compensated by increase of transmitted intensity.

Non-reflecting films can be painted onto aircraft to suppress reflection of radar. The thickness of the film is determined by $nd = \frac{\lambda}{4}$ where λ is the wavelength of the radar waves and n the refractive index of the film at this wavelength.

Example

A plane-parallel glass plate of thickness 2.0 mm is illuminated with light from an extended source. The refractive index of the glass is 1.5 and the wavelength of the light is 600 nm. Calculate how many fringes are formed.

Solution

We assume that the fringes are formed by light incident at all angles from normal to grazing incidence.

At normal incidence we have $2nd = m\lambda$

From which,

$$m = \frac{2 \times 1.5 \times 2 \times 10^{-3}}{6 \times 10^{-7}} = 10,000$$

At grazing incidence the angle of refraction ϕ is in fact the critical angle.

Therefore, $\phi = \arcsin\left(\frac{1}{1.5}\right) = 42°$

i.e. $\cos\phi = 0.75$

At grazing incidence $2nd\cos\phi = m'\lambda$

From which (and using $\cos(\sin^{-1}(1/1.5)) = 0.75$),

$$m = \frac{2 \times 1.5 \times 2 \times 10^{-3} \times 0.75}{6 \times 10^{-7}} = 7500$$

The total number of fringes seen is equal to $m - m' = 2500$.

Exercise 9.5

When viewed from above, the colour of an oil film on water appears red in colour. Use the data below to estimate the minimum thickness of the oil film.

average wavelength of red light = 630 nm

refractive index of oil for red light = 1.5

refractive index of water for red light = 1.3

WAVE PHENOMENA

9.4 Resolution

NATURE OF SCIENCE:

Improved technology: The Rayleigh criterion is the limit of resolution. Continuing advancement in technology such as large diameter dishes or lenses or the use of smaller wavelength lasers pushes the limits of what we can resolve. (1.8)

© IBO 2014

Essential idea: Resolution places an absolute limit on the extent to which an optical or other system can separate images of objects.

Understandings:

- The size of a diffracting aperture
- The resolution of simple monochromatic two-source systems

The size of a diffracting aperture

Our discussion concerning diffraction so far has been for rectangular slits. When light from a point source enters a small circular aperture, it does not produce a bright dot as an image, but a circular disc known as Airy's disc surrounded by fainter concentric circular rings as shown in Figure 922.

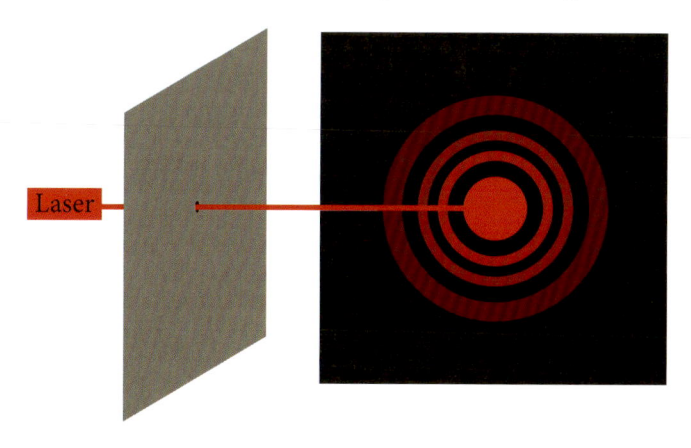

Figure 922   Diffraction at a circular aperture

Diffraction at an aperture is of great importance because the eye and many optical instruments have circular apertures.

What is the half-angular width of the central maximum of the diffraction formed by a circular aperture? This is not easy to calculate since it involves some advanced mathematics. The problem was first solved by the English Astronomer Royal, George Airy, in 1835 who showed that for circular apertures

$\theta = \dfrac{1.22\lambda}{b}$ where b is the diameter of the aperture.

Example 2

In the following diagram, parallel light from a distant point source (such as a star) is brought to focus on the screen S by a converging lens (*the lens is shown as a vertical arrow*).

The focal length (distance from lens to screen) is 25 cm and the diameter of the lens is 3.0 cm. The wavelength of the light from the star is 560 nm. Calculate the diameter of the diameter of the image on the screen.

Solution

The lens actually acts as a circular aperture of diameter 3.0 cm. The half angular width of central maximum of the diffraction pattern that it forms on the screen is

$\theta = \dfrac{1.22\lambda}{b} = \dfrac{1.22 \times 5.6 \times 10^{-7}}{3.0 \times 10^{-2}} = 2.3 \times 10^{-5}$ rad

The diameter of the central maxima is therefore

$25 \times 10^{-2} \times 2.3 \times 10^{-5}$

$= 5.7 \times 10^{-6}$ m.

Although this is small, it is still finite and is effectively the image of the star as the intensity of the secondary maxima are small compared to that of the central maximum.

The resolution of simple monochromatic two-source systems

The astronomers tell us that many of the stars that we observe with the naked eye are in fact binary stars. That is, what we see as a single star actually consists of two stars in orbit about a common centre. Furthermore, the astronomers tell us that if we use a "good" telescope then we will actually see the two stars. That is, we will resolve the single point source into its two component parts. So what is it that determines whether or not we see the two stars as a single point source i.e. what determines whether or not two sources can be resolved? It can't just be that the telescope magnifies the stars since if they are acting as point sources magnifying them is not going to make a great deal of difference.

In each of our eyes there is an aperture, the pupil, through which the light enters. This light is then focused by the eye lens onto the retina. But we have seen that when light passes through an aperture it is diffracted and so when we look at a point source, a diffraction pattern will be formed on the retina. If we look at two point sources then two diffraction patterns will be formed on the retina and these patterns will overlap. The width of our pupil and the wavelength of the light emitted by the sources will determine the amount by which they overlap. But the degree of overlap will also depend on the angular separation of the two point sources. We can see this from Figures 923.

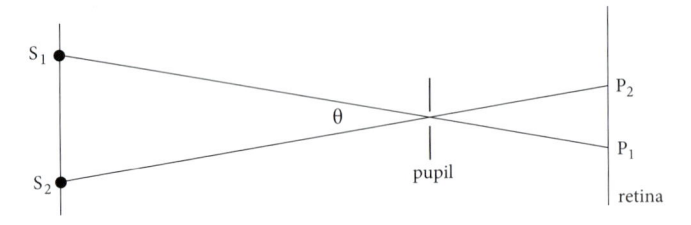

Figure 923

Light from the source S_1 enters the eye and is diffracted by the pupil such that the central maximum of the diffraction pattern is formed on the retina at P_1. Similarly, light from S_2 produces a maximum at P_2. If the two central maxima are well separated then there is a fair chance that we will see the two sources as separate sources. If they overlap then we will not be able to distinguish one source from another. From the diagram we see as the sources are moved closer to the eye, then the angle θ increases and so does the separation of the central maxima.

Figures 924, 925, 926 and 927 shows the different diffraction patterns and the intensity distribution, that might result on the retina as a result of light from two point sources

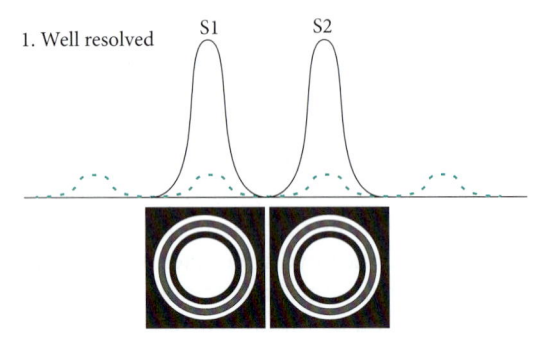

Figure 924   Very well resolved

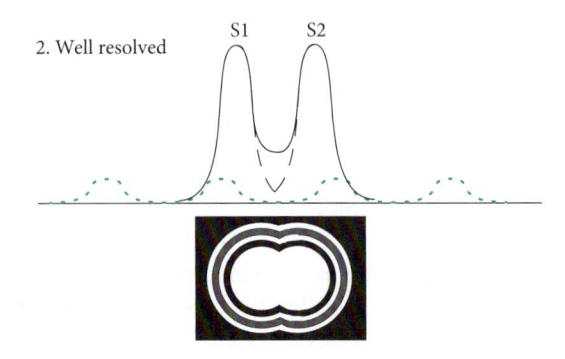

Figure 925   Well resolved

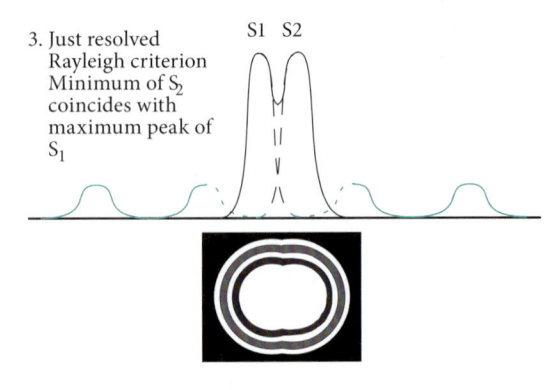

Figure 926   Just resolved

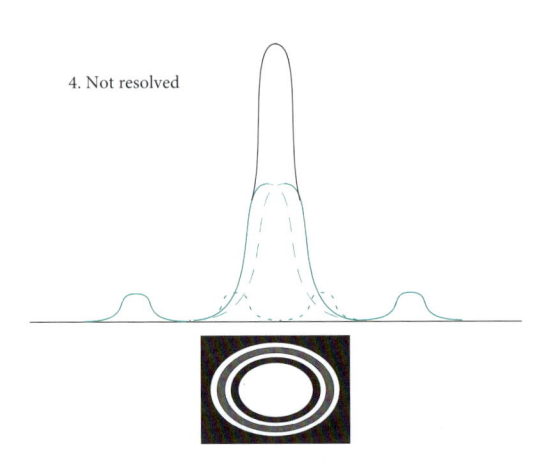

Figure 927   Not resolved

We have suggested that if the central maxima of the two diffraction patterns are reasonably separated then we should be able to resolve two point sources. In the late 19th century *Lord Rayleigh* suggested by how much they should be separated in order for the two sources to be just resolved. **If the central maximum of one diffraction pattern coincides with the first minima of the other diffraction pattern then the two sources will just be resolved.** This is known as the **Rayleigh Criterion**.

Figure 926 shows just this situation. The two sources are just resolved according to the Rayleigh criterion since the peak of the central maximum of one diffraction pattern coincides with the first minimum of the other diffraction pattern. This means that the angular separation of the peaks of the two central maxima formed by each source is just the half angular width of one central maximum i.e.

$$\theta = \frac{\lambda}{b}$$

where b is the width of the slit through which the light from the sources passes. However, we see from Figure 1117 that θ is the angle that the two sources subtend at the slit. Hence we conclude that two sources will be resolved by a slit if the angle that they subtend at the slit is greater than or equal to $\frac{\lambda}{b}$

So far we have been assuming that the eye is a rectangular slit whereas clearly it is a circular aperture and so we must use the formula

$$\theta = \frac{1.22\lambda}{b}$$

As mentioned above the angle θ is sometimes called the *resolving power* but should more accurately be called the **minimum angle of resolution** (θ_{min})

Clearly the smaller θ the greater will be the resolving power.

The diffraction grating is a useful for differentiating closely spaced lines in emission spectra. Like a prism spectrometer, it can disperse a spectrum into its components but it is better suited because it has higher resolution than the prism spectrometer.

If λ_1 and λ_2 are two nearly equal wavelengths that can barely be distinguished, the resolvance or resolving power of the grating is defined as:

$$R = \frac{\lambda}{(\lambda_2 - \lambda_1)} = \frac{\lambda}{\Delta\lambda}$$

Where $\lambda \approx \lambda_1 \approx \lambda_2$

Therefore, the diffraction grating with a high resolvance will be better suited in determining small differences in wavelength.

If the diffraction grating has N lines being illuminated, it can be shown that the resolving power of the mth order diffraction is given by

$R = mN$

Obviously, the resolvance becomes greater with the order number m and with a greater number of illuminated slits.

Example

A benchmark for the resolving power of a grating is the sodium doublet in the sodium emission spectrum. Two yellow lines in its spectrum have wavelengths of 589.00 nm and 589.59 nm.

(a) Determine the resolvance of a grating if the given wavelengths are to be distinguished from each other.

(b) How many lines in the grating must be illuminated in order to resolve these lines in the second order spectrum?

Solution

(a) $R = \dfrac{\lambda}{\Delta\lambda} = \dfrac{589.00 \text{ nm}}{(589.00 - 589.59) \text{ nm}}$

$= \dfrac{589 \text{ nm}}{0.59 \text{ nm}} = 1.0 \times 10^3$

(b) $N = \dfrac{R}{m} = \dfrac{1.0 \times 10^3}{2} = 5.0 \times 10^2$ lines.

It has been seen that diffraction effectively limits the resolving power of optical systems. This includes such systems as the eye, telescopes and microscopes. The resolving power of these systems is dealt with in the next section. This section looks at links between technology and resolving power when looking at the very distant and when looking at the very small.

Radio telescopes

The average diameter of the pupil of the human eye is about 2.5 mm. This means that the eye will just resolve two point sources emitting light of wavelength 500 nm if their angular separation at the eye is

$$\theta = 1.22 \times \frac{5.0 \times 10^{-7}}{2.5 \times 10^{-3}} = 2.4 \times 10^{-4} \text{ rad}$$

If the eye were to be able to detect radio waves of wavelength 0.15 m, then to have the same resolving power the pupil would have to have a diameter of about 600 m. Clearly this is nonsense, but it does illustrate a problem facing astronomers who wish to view very distant objects such as quasars and galaxies that emit radio waves. Conventional radio telescopes consist of a large dish, typically 25 m in diameter. Even with such a large diameter, the radio wavelength resolving power of the telescope is much less than the optical resolving power of the human eye. Let us look at an example.

Example

The Galaxy Cygnus A can be resolved optically as an elliptically shaped galaxy. However, it is also a strong emitter of radio waves of wavelength 0.15 m. The Galaxy is estimated to be 5.0×10^{24} m from Earth. Use of a radio telescope shows that the radio emission is from two sources separated by a distance of 3.0×10^{21} m. Estimate the diameter of the dish required to just resolve the sources.

Solution

The angle θ that the sources resolve at the telescope is given by

$$\theta = \frac{3.0 \times 10^{21}}{5.0 \times 10^{24}} = 6.0 \times 10^{-4} \text{ rad}$$

and $d = \dfrac{1.22\lambda}{\theta} = \dfrac{1.22 \times 0.15}{6.0 \times 10^{-4}} = 3000$ m $= 3.0$ km.

A radio telescope dish of this size would be impossible to make, let alone support. This shows that a single dish type radio telescope cannot be used to resolve the sources and yet they were resolved. To get round the problem, astronomers use two radio telescopes separated by a large distance. The telescopes view the same objects at the same time and the signals that each receive from the objects are simultaneously superimposed. The result of the superposition of the two signals is a two-slit interference pattern. The pattern has much narrower fringe spacing than that of the diffraction pattern produced by either telescope on its own, hence producing a much higher resolving power. When telescopes are used like this, they are called a **stellar interferometer.**

In Socorro in New Mexico there is a stellar interferometer that consists of 27 parabolic dishes each of diameter 25 m, arranged in a Y-shape that covers an area of 570 km². This is a so-called Very Large Array (VLA). Even higher resolution can be obtained by using an array of radio telescopes in observatories thousands of kilometres apart. A system that uses this so-called technique of 'very-long-baseline interferometry' (VLBI) is known as a 'very-long-baseline array' (VLBA). With VLBA, a radio wavelength resolving power can be achieved that is 100 times greater than the best optical telescopes. Even higher resolving power can be achieved by using a telescope that is in a satellite orbiting Earth. Japan's Institute of Space and Astronautical Science (ISAS) launched such a system in February 1997. The National Astronomical Observatory of Japan, the National Science Foundation's National Radio Astronomy Observatory (NRAO); the Canadian Space Agency; the Australia Telescope National Facility; the European VLBI Network and the Joint Institute back this project for Very Long Baseline Interferometry in Europe. This project is a very good example of how Internationalism can operate in Physics.

Electron microscope

Telescopes are used to look at very distant objects that are very large but, because of their distance from us, appear very small. Microscopes on the other hand, are used to look at objects that are close to us but are physically very small. As we have seen, just magnifying objects, that is making them appear larger, is not sufficient on its own to gain detail about the object; for detail, high resolution is needed.

Figure 928 is a schematic of how an **optical microscope** is used to view an object and Figure 929 is a schematic of a transmission **electron microscope** (TEM).

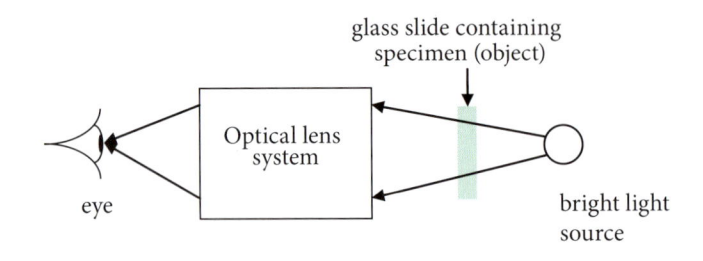

Figure 928   The principle of a light microscope

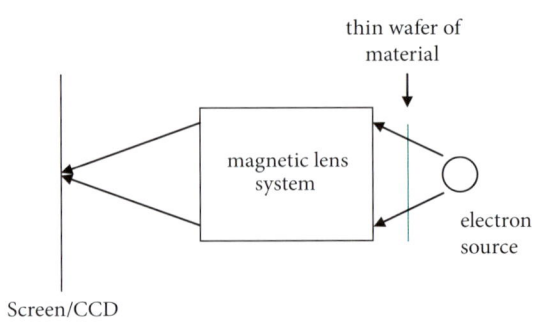

Figure 929   The principle of an electron microscope

In the optical microscope, the resolving power is determined by the particular lens system employed and the wavelength λ of the light used. For example, two points in the sample separated by a distance d will just be resolved if

$$d = \frac{\lambda}{2m}$$

where m is a property of the lens system know as the **numerical aperture**. In practice the largest value of m obtainable is about 1.6. Hence, if the microscope slide is illuminated with light of wavelength 480 nm, a good microscope will resolve two points separated by a distance $d \approx 1.5 \times 10^{-7}$ m ≈ 0.15 μm. Points closer together than this will not be resolved. However, this is good enough to distinguish some viruses such as the Ebola virus.

Clearly, the smaller λ the higher the resolving power and this is where the electron microscope comes to the fore. The electron microscope makes use of the wave nature of electrons (see 13.1.5). In the TEM, electrons pass through a wafer thin sample and are then focused by a magnetic field onto a fluorescent screen or CCD (charge coupled device see 14.2). Electrons used in an electron microscope have wavelengths typically of about 5×10^{-12} m. However, the numerical aperture of electron microscopes is considerably smaller than that of an optical microscope, typically about 0.02. Nonetheless, this means that a TEM can resolve two points that are about 0.25 nm apart. This resolving power is certainly high enough to make out the shape of large molecules.

Another type of electron microscope uses a technique by which electrons are scattered from the surface of the sample. The scattered electrons and then focused as in the TEM to form an image of the surface. These so-called scanning electron microscopes (SEM) have a lower resolving power than TEM's but give very good three dimensional images.

The eye

We saw in the last section that the resolving power of the human eye is about 2×10^{-4} rad. Suppose that you are looking at car headlights on a dark night and the car

WAVE PHENOMENA

is a distance D away. If the separation of the headlight is say 1.5 m then the headlights will subtend an angle $\theta = \frac{1.5}{D}$ at your eye. Assuming an average wavelength of 500 nm, your eye will resolve the headlights into two separate sources if this angle equals 2×10^{-4} rad. This gives $D = 7.5$ km. In other words if the car is approaching you on a straight road then you will be able to distinguish the two headlights as separate sources when the car is 7.5 km away from you. Actually because of the structure of the retina and optical defects the resolving power of the average eye is about 6×10^{-4} rad. This means that the car is more likely to be 2.5 km away before you resolve the headlights.

Astronomical telescope

Let us return to the example of the binary stars discussed at the beginning of this section on resolution. The stars Kruger A and B form a binary system. The average separation of the stars is 1.4×10^{12} m and their average distance from Earth is 1.2×10^{17} m. When viewed through a telescope on Earth, the system will therefore subtend an angle.

$$\theta = \frac{1.4 \times 10^{12}}{1.2 \times 10^{17}}$$

$= 1.2 \times 10^{-5}$ rad at the objective lens of the telescope. Assuming that the average wavelength of the light emitted by the stars is 500 nm, then if the telescope is to resolve the system into two separate images it must have a minimum diameter D where $1.2 \times 10^{-5} = \frac{1.22 \times 5.00 \times 10^{-7}}{D}$.

This gives $D = 0.050$ m, which is about 5 cm. So this particular system is easily resolved with a small astronomical telescope.

Exercise 9.6

1. It is suggested that using the ISAS, VLBA, it would be possible to "see" a grain of rice at a distance of 5000 km. Estimate the resolving power of the VLBA.

2. The distance from the eye lens to the retina is 20 mm. The light receptors in the central part of the retina are about 5×10^{-6} apart. Determine whether the spacing of the receptors will allow for the eye to resolve the headlights in the above discussion when they are 2.5 km from the eye.

3. The diameter of Pluto is 2.3×10^{6} m and its average distance from Earth is 6.0×10^{12} m. Estimate the minimum diameter of the objective of a telescope that will enable Pluto to be seen as a disc as opposed to a point source.

9.5 Doppler effect

NATURE OF SCIENCE:

Technology: Although originally based on physical observations of the pitch of fast moving sources of sound, the Doppler effect has an important role in many different areas such as evidence for the expansion of the universe and generating images used in weather reports and in medicine. (5.5)

© IBO 2014

Essential idea: The Doppler effect describes the phenomenon of wavelength/frequency shift when relative motion occurs.

Understandings:
- The Doppler effect for sound waves and light waves

The Doppler effect for sound waves and light waves

Consider two observers A and B at rest with respect to a sound source that emits a sound of constant frequency f. Clearly both observers will hear a sound of the same frequency. However, suppose that the source now moves at constant speed towards A. A will now hear a sound of frequency f_A that is greater than f and B will hear a sound of frequency f_B that is less than f. This phenomenon is known as the **Doppler Effect** or Doppler Principle after C. J. Doppler (1803-1853).

The same effect arises for an observer who is either moving towards or away from a stationary source.

Figure 930 shows the waves spreading out from a stationary source that emits a sound of constant frequency f. The observers A and B hear a sound of the same frequency.

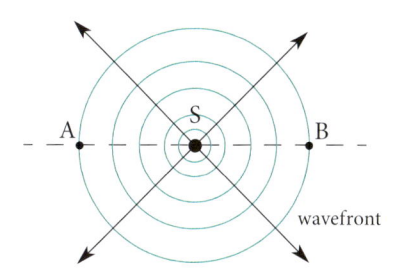

Figure 930 Sound waves from a stationary source

Suppose now that the source moves towards A with constant speed v. Figure 9310 (a) shows a snapshot of the new wave pattern.

303

CHAPTER 9

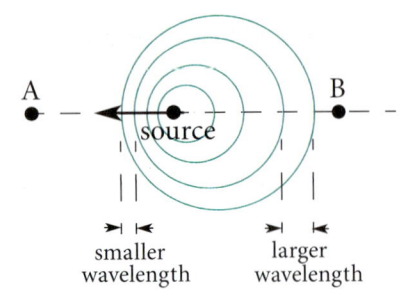

Figure 931 (a) Sound waves from a moving source

The wavefronts are now crowded together in the direction of travel of the source and stretched out in the opposite direction. This is why now the two observers will now hear notes of different frequencies. How much the waves bunch together and how much they stretch out will depend on the speed v. Essentially, $f_A = \frac{c}{\lambda_A}$ and $f_B = \frac{c}{\lambda_B}$ where $\lambda_A < \lambda_B$ and v is the speed of sound.

If the source is stationary and A is moving towards it, then the waves from the source incident on A will be bunched up. If A is moving away from the stationary source then the waves from the source incident on A will be stretched out.

Christian Doppler (1803–1853) actually applied the principle (incorrectly as it happens) to try and explain the colour of stars. However, the Doppler effect does apply to light as well as to sound. If a light source emits a light of frequency f then if it is moving away from an observer the observer will measure the light emitted as having a lower frequency than f. Since the sensation of colour vision is related to the frequency of light (blue light is of a higher frequency than red light), light emitted by objects moving way from an observer is often referred to as being red-shifted whereas if the object is moving toward the observer it is referred to as blue-shifted. This idea is used in Option E (Chapter 16).

We do not need to consider here the situations where either the source or the observer is accelerating. In a situation for example where an accelerating source is approaching a stationary observer, then the observer will hear a sound of ever increasing frequency. This sometimes leads to confusion in describing what is heard when a source approaches, passes and then recedes from a stationary observer. Suppose for example that you are standing on a station platform and a train sounding its whistle is approaching at constant speed. What will you hear as the train approaches and then passes through the station? As the train approaches you will hear a sound of *constant* pitch but increasing loudness. The pitch of the sound will be greater than if the train were stationary. As the train passes through the station you will hear the pitch change at the moment the train passes you, to a sound, again of constant pitch. The pitch of this sound will be lower than the sound of the approaching train and its intensity will decrease as the train recedes from you. What you do not hear is a sound of increasing pitch and then decreasing pitch.

The Doppler equations for sound

Although you will not be expected in an IB examination to derive the equations associated with aspects of the Doppler effect, you will be expected to apply them. For completeness therefore, the derivation of the equations associated with the Doppler effect as outlined above is given here.

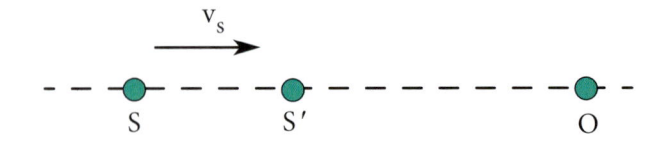

Figure 931 (b)

In Figure 931 (b) the observer O is at rest with respect to a source of sound S is moving with constan speed v_s directly towards O. The source is emitting a note of constant frequency f and the speed of the emitted sound is v.

S′ shows the position of the source Δt later. When the source is at rest, then in a time Δt the observer will receive $f\Delta t$ waves and these waves will occupy a distance $v\Delta t$. i.e

$$\lambda = \frac{v\Delta t}{f\Delta t} = \frac{v}{f}$$

(Because of the motion of the source this number of waves will now occupy a distance $(v\Delta t - v_s\Delta t)$. The 'new' wavelength is therefore

$$\lambda' = \frac{v\Delta t - v_s\Delta t}{f\Delta t} = \frac{v - v_s}{f}$$

If f' is the frequency heard by O then

$$f' = \frac{v}{\lambda'} \text{ or } \lambda' = \frac{v}{f'} = \frac{v - v_s}{f}$$

From which

$$f' = \frac{v}{v - v_s} f$$

Dividing through by v gives

$$f' = f \left(\frac{1}{1 - \frac{v_s}{v}} \right) \qquad \text{Equation 11.1}$$

If the source is moving away from the observer then we have

$$f' = \frac{v}{v - v_s} f = f \left(\frac{1}{1 + \frac{v_s}{v}} \right) \qquad \text{Equation 11.2}$$

Wave Phenomena

We now consider the case where the source is stationary and the **observer is moving towards the source** with speed v_0. In this situation the speed of the sound waves as measured by the observer will be $v_0 + v$. We therefore have that

$$v_0 + v = \frac{f'}{\lambda} = f' \times \frac{v}{\lambda}$$

From which

$$f' = \left(1 + \frac{v_0}{v}\right) f \qquad \text{Equation 11.3}$$

If the **observer is moving away from the source** then

$$f' = \left(1 - \frac{v_0}{v}\right) f$$

From equation (11.3), we have that

$$\Delta f = f' - f = \left(1 + \frac{v_0}{v}\right) f - f = \frac{v_0}{v} f \qquad \text{Equation 11.4}$$

The velocities that we refer to in the above equations are the velocities with respect to the medium in which the waves from the source travel. However, when we are dealing with a light source it is the relative velocity between the source and the observer that we must consider. The reason for this is that light is unique in the respect that the speed of the light waves does not depend on the speed of the source. All observers irrespective of their speed or the speed of the source will measure the same velocity for the speed of light. This is one of the cornerstones of the Special Theory of Relativity which is discussed in more detail in Option H (Chapter 18). When applying the Doppler effect to light we are mainly concerned with the motion of the source. We look here only at the situation where the speed of the source v is much smaller than the speed of light c in free space. ($v \ll c$). Under these circumstances, when the source is moving towards the observer, equation 11.1 becomes

$$f' - f = \Delta f = \frac{v}{c} f \qquad \text{Equation 11.5}$$

and when the source is moving away from the observer, equation 11.2 becomes $f' - f = \Delta f = -\frac{v}{c} f$

Provided that $v \ll c$, these same equations apply for a stationary source and moving observer

We look at the following example and exercise.

Example

A source emits a sound of frequency 440 Hz. It moves in a straight line towards a stationary observer with a speed of 30 m s⁻¹. The observer hears a sound of frequency 484 Hz. Calculate the speed of sound in air.

Solution

We use equation 11.1 and substitute $f' = 484$ Hz, $f = 440$ Hz and $v_s = 30$ m s⁻¹.

therefore $484 = 440 \left(\dfrac{1}{1 - \dfrac{30}{v}}\right)$ such that $1 - \dfrac{30}{v} = \dfrac{440}{484}$ to give $v = 330$ m s⁻¹.

Example

A particular radio signal from a galaxy is measured as having a frequency of 1.39×10^9 Hz. The same signal from a source in a laboratory has a frequency of 1.42×10^9 Hz.

Suggest why the galaxy is moving away from Earth and calculate its recession speed (i.e. the speed with which it is moving away from Earth).

Solution

The fact that the frequency from the moving source is less than that when it is stationary indicates that it is moving away from the stationary observer i.e. Earth.

Using $\Delta f = \dfrac{v}{c} f$ we have

$$v = \frac{c \Delta f}{f} = \frac{3 \times 10^8 \times (1.42 - 1.39) \times 10^9}{1.42 \times 10^9} = 6.34 \times 10^6 \text{ m s}^{-1}$$

It is usual when dealing with the Doppler effect of light to express speeds as a fraction of c. So in this instance we have $v = 0.021\, c$

Using the Doppler effect

We have seen in the above example and exercise how the Doppler effect may be used to measure the recession speed of distant galaxies. The effect is also used to measure speed in other situations. Here we will look at the general principle involved in using the Doppler effect to measure speed. Figure 932 shows a source (the transmitter) that emits either sound or em waves of constant frequency f. The waves

305

CHAPTER 9

from the source are incident on a reflector that is moving towards the transmitter with speed v. The reflected waves are detected by the receiver placed alongside the transmitter.

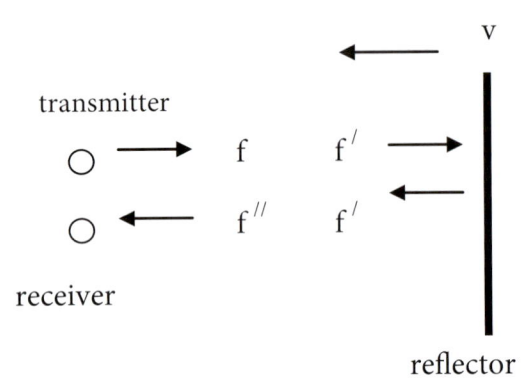

Figure 932 Using the Doppler effect to measure speed

We shall consider the situation where $v \ll c$ where c is the speed of the waves from the transmitter.

For the reflector receiving waves from the transmitter, it is effectively an observer moving towards a stationary source. From equation (11.4), it therefore receives waves that have been Doppler shifted by an amount

$$f' - f = \frac{v}{c} f \qquad \text{Equation 11.6}$$

For the receiver receiving waves from the reflector, it is effectively a stationary observer receiving waves from a moving source. From equation (11.5), it therefore receives waves that have been Doppler shifted by an amount

$$f'' - f' = \frac{v}{c} f' \qquad \text{Equation 11.7}$$

If we add equations (11.6) and (11.7) we get that the total Doppler shift at the receiver Δf is

$$f'' - f = \Delta f = f' \frac{v}{c} + f \frac{v}{c}$$

But $f' = \left(1 + \frac{v}{c}\right) f$ hence

$$\Delta f = f\left(1 + \frac{v}{c}\right)\frac{v}{c} + \frac{v}{c} f$$

But since $v \ll c$, we can ignore the term $\frac{v^2}{c^2}$ when we expand the bracket in the above equation.

Therefore we have

$$\Delta f = \frac{2v}{c} f \qquad \text{Equation 11.8}$$

If $v \approx c$ then we must use the full Doppler equations. However, for em radiation we will always only consider situations in which $v \ll c$.

Example

The speed of sound in blood is 1.500×10^3 m s^{-1}. Ultrasound of frequency 1.00 MHz is reflected from blood flowing in an artery. The frequency of the reflected waves received back at the transmitter is 1.05 MHz. Estimate the speed of the blood flow in the artery.

Solution

Using equation (11.8) we have

$$0.05 \times 10^6 = \frac{2v}{1.5 \times 10^3} \times 10^6$$

to give $v \approx 36$ m s^{-1}. (We have assumed that the ultrasound is incident at right angles to the blood flow.)

Exercise 9.7

1. Judy is standing on the platform of a station. A high speed train is approaching the station in a straight line at constant speed and is sounding its whistle. As the train passes by Judy, the frequency of the sound emitted by the whistle as heard by Judy, changes from 640 Hz to 430 Hz. Determine

   (a) the speed of the train

   (b) the frequency of the sound emitted by the whistle as heard by a person on the train. (Speed of sound = 330 m s^{-1})

2. A galaxy is moving away from Earth with a speed of 0.0500c. The wavelength of a particular spectral line in light emitted by atomic hydrogen in a laboratory is 6.56×10^{-7} m. Calculate the value of the wavelength of this line, measured in the laboratory, in light emitted from a source of atomic hydrogen in the galaxy.

10. Fields

Contents

10.1 – Describing fields

10.2 – Fields at work

Essential Ideas

Electric charges and masses each influence the space around them and that influence can be represented through the concept of fields.

Similar approaches can be taken in analysing electrical and gravitational potential problems. © IBO 2014

CHAPTER 10

10.1 Describing fields

NATURE OF SCIENCE:

Paradigm shift: The move from direct, observable actions being responsible for influence on an object to acceptance of a field's "action at a distance" required a paradigm shift in the world of science. (2.3)

© IBO 2014

Essential idea: Electric charges and masses each influence the space around them and that influence can be represented through the concept of fields.

Understandings:
- Gravitational fields
- Electrostatic fields
- Electric potential and gravitational potential
- Field lines
- Equipotential surfaces

Gravitational potential

We have seen that if we lift an object of mass m to a height h above the surface of the Earth then its gain in gravitational potential energy is mgh. However, this is by no means the full story. For a start we now know that g varies with h and also the expression really gives a difference in potential energy between the value that the object has at the Earth's surface and the value that it has at height h. So what we really need is a zero point. Can we find a point where the potential energy is zero and use this point from which to measure changes in potential energy?

The point that is chosen is in fact infinity. At infinity the gravitational field strength of any object will be zero. So let us see if we can deduce an expression for the gain in potential energy of an object when it is "lifted" from the surface of the Earth to infinity. This in effect means finding the work necessary to perform this task.

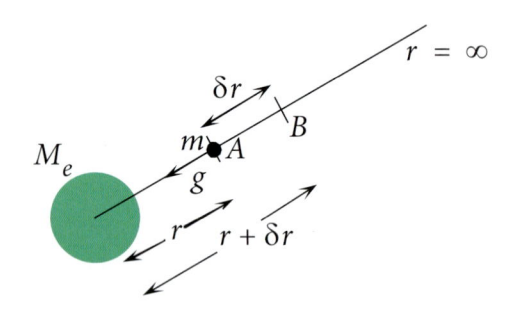

Figure 1001 Gravitational forces

In the diagram we consider the work necessary to move the particle of mass m a distance δr in the gravitational field of the Earth.

The force on the particle at A is $F = \dfrac{GM_e m}{r^2}$

If the particle is moved to B, then since δr is very small, we can assume that the field remains constant over the distance AB. The work δW done *against* the gravitational field of the Earth in moving the distance AB is

$$\delta W = -\dfrac{GM_e m}{r^2}\delta r$$

(remember that work done against a force is negative)

To find the total work done, W, in going from the surface of the Earth to infinity we have to add all these little bits of work. This is done mathematically by using integral calculus.

$$W = \int_R^\infty \left(-\dfrac{GM_e m}{r^2}\right) dr = -GM_e m \int_R^\infty \dfrac{1}{r^2} dr = -GM_e m \left[-\dfrac{1}{r}\right]_R^\infty$$

$$= -GM_e m \left[0 - \left(-\dfrac{1}{R}\right)\right]$$

$$= -\dfrac{GM_e m}{R}$$

Hence we have, where R is the radius of the Earth, that the work done by the gravitational field in moving an object of mass m from R (surface of the Earth) to infinity, is given by

$$W = -\dfrac{GM_e m}{R}$$

We can generalise the result by calculating the work necessary per unit mass to take a small mass from the surface of the Earth to infinity. This we call the **gravitational potential**, V, i.e.,

$$V = \dfrac{W}{m}$$

We would get exactly the same result if we calculated the work done by the field to bring the point mass from infinity to the surface of Earth. In this respect the formal definition of gravitational potential at a point in a gravitational field is therefore defined as **the work done per unit mass in bringing a point mass from infinity to that point.**

Clearly then, the gravitational potential at any point in the Earth's field distance r from the centre of the Earth (providing $r > R$) is

$$V = -\dfrac{GM_e}{r}$$

The potential is therefore a measure of the amount of work that has to be done to move particles between points in a gravitational field and its unit is the J kg^{-1}. We also note that the potential is negative so that the potential energy as we move away from the Earth's surface increases until it reaches the value of zero at infinity.

If the gravitational field is due to a point mass of mass m, then we have the same expression as above except that M_e is replaced by m and must also exclude the value of the potential at the point mass itself i.e. at $r = 0$.

We can express the gravitational potential due to the Earth (or due to any spherical mass) in terms of the gravitational field strength at its surface.

At the surface of the Earth we have

$$-g_0 R_e = -\frac{GM_e}{R_e}$$

So that,

$$g_0 R_e^2 = GM_e$$

Hence at a distance r from the centre of the Earth the gravitational potential V can be written as

$$V = -\frac{GM_e}{r} = -\frac{g_0 R_e^2}{r}$$

The potential at the surface of the Earth ($r = R_e$) is therefore $-g_0 R_e$

It is interesting to see how the expression for the gravitational potential ties in with the expression mgh. The potential at the surface of the Earth is $-g_0 R_e$ (see the example above) and at a height h will be $-g_0(R_e + h)$ if we assume that g_0 does not change over the distance h. The difference in potential between the surface and the height h is therefore $g_0 h$. So the work needed to raise an object of mass m to a height h is mgh, i.e., $m \times$ difference in gravitational potential

This we have referred to as the gain in gravitational potential energy (see 2.3.5).

However, this expression can be extended to any two points in any gravitational field such that if an object of mass m moves between two points whose potentials are V_1 and V_2 respectively, then the change in gravitational potential energy of the object is $m(V_1 - V_2)$.

Gravitational potential gradient

Let us consider now a region in space where the gravitational field is constant. In Figure 912 the two points A and B are separated by the distance Δx.

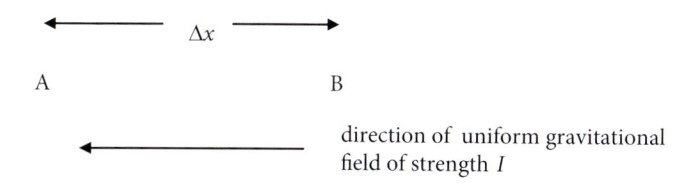

Figure 1002   The gravitational potential gradient

The gravitational field is of strength I and is in the direction shown. The gravitational potential at A is V and at B is $V + \Delta V$.

The work done is taking a point mass m from A to B is $F\Delta x = mI\Delta x$.

However, by definition this work is also equal to $-m\Delta V$.

Therefore $mI\Delta x = -m\Delta V$

or $I = -\dfrac{\Delta V}{\Delta x}$

Effectively this says that the magnitude of the gravitational field strength is equal to the negative gradient of the potential. If I is constant then V is a linear function of x and I is equal to the negative gradient of the straight line graph formed by plotting V against x. If I is not constant (as usually the case), then the magnitude of I at any point in the field can be found by find the gradient of the V-x graph at that point. An example of such a calculation can be found in Section 9.2.9.

For those of you who do HL maths the relationship between field and potential is seen to follow from the expression for the potential of a point mass viz:

$$V = -G\frac{m}{r}$$

$$-\frac{dV}{dr} = +G\frac{m}{r^2} = I$$

Potential due to one or more point masses

Gravitational potential is a scalar quantity so calculating the potential due to more than one point mass is a matter of simple addition. So for example, the potential V due to the Moon and Earth and a distance x from the centre of Earth, on a straight line between them, is given by the expression

$$V = -G\left(\frac{M_E}{x} + \frac{M_M}{r-x}\right)$$

where M_E = mass of Earth, M_M = mass of Moon and r = distance between centre of Earth and Moon.

CHAPTER 10

Equipotentials and field lines

If the gravitational potential has the same value at all points on a surface, the surface is said to be an **equipotential surface**. So for example, if we imagine a spherical shell about Earth whose centre coincides with the centre of Earth, this shell will be an equipotential surface. Clearly, if we represent the gravitational field strength by field lines, since the lines "radiate" out from the centre of Earth, then these lines will be at right angles to the surface. If the field lines were not normal to the equipotential surface then there would be a component of the field parallel to the surface. This would mean that points on the surface would be at different potentials and so it would no longer be an equipotential surface. This of course holds true for any equipotential surface.

Figure 1003 shows the field lines and equipotentials for two point masses m.

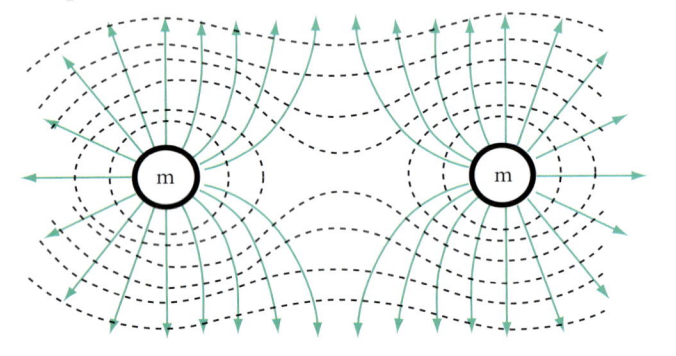

Figure 1003   Equipotentials for two point masses

It is worth noting that we would get exactly the same pattern if we were to replace the point masses with two equal point charges. (See 9.3.5)

Escape speed

The potential at the surface of Earth is $-G\frac{M}{R}$ which means that the energy required to take a particle of mass m from the surface to infinity is equal to $-G\frac{Mm}{R}$

But what does it actually mean to take something to infinity? When the particle is on the surface of the Earth we can think of it as sitting at the bottom of a "potential well" as in figure 1004.

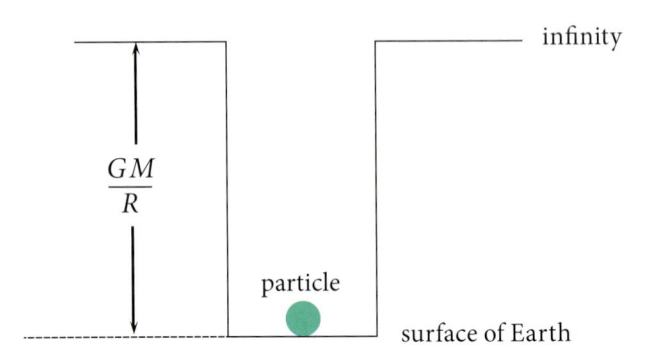

Figure 1004   A potential well

The "depth" of the well is $\frac{GM}{R}$ and if the particle gains an amount of kinetic energy equal to $\frac{GMm}{R}$ where m is its mass then it will have just enough energy to "lift" it out of the well.

In reality it doesn't actually go to infinity it just means that the particle is effectively free of the gravitational attraction of the Earth. We say that it has "escaped" the Earth's gravitational pull. We meet this idea in connection with molecular forces. Two molecules in a solid will sit at their equilibrium position, the separation where the repulsive force is equal to the attractive force. If we supply just enough energy to increase the separation of the molecules such that they are an infinite distance apart then intermolecular forces no longer affect the molecules and the solid will have become a liquid. There is no increase in the kinetic energy of the molecules and so the solid melts at constant temperature.

We can calculate the escape speed of an object very easily by equating the kinetic energy to the potential energy such that

$$\frac{1}{2}mv_{escape}^2 = \frac{GM_e m}{R_e}$$

$$\Rightarrow v_{escape} = \sqrt{\frac{2GM_e}{R_e}} = \sqrt{2g_0 R_e}$$

Substituting for g_0 and R_e gives a value for v_{escape} of about 11 km s^{-1} from the surface of the Earth.

You will note that the escape speed does not depend on the mass of the object since both kinetic energy and potential energy are proportional to the mass.

In theory, if you want to get a rocket to the moon it can be done without reaching the escape speed. However, this would necessitate an enormous amount of fuel and it is likely that the rocket plus fuel would be so heavy that it would never get off the ground. It is much more practical to accelerate the rocket to the escape speed from Earth orbit and then, in theory, just launch it to the Moon.

Example

Use the following data to determine the potential at the surface of Mars and the magnitude of the acceleration of free fall

mass of Mars   = 6.4 × 10^{23} kg

radius of Mars   = 3.4 × 10^6 m

Determine also the gravitational field strength at a distance of 6.8 × 10^6 m above the surface of Mars.

Solution

$$V = -G\frac{M}{R} = -6.7 \times 10^{-11} \times \frac{6.4 \times 10^{23}}{3.4 \times 10^6} = -1.3 \times 10^7 \text{ N kg}^{-1}$$

But $V = -g_0 R$

Therefore $g_0 = -\dfrac{V}{R} = \dfrac{1.3 \times 10^7}{3.4 \times 10^6} = 3.8 \text{ m s}^{-2}$

To determine the field strength g_h at 6.8×10^6 m above the surface, we use the fact that $g_0 = G\dfrac{M}{R^2}$ such that $GM = g_0 R^2$

Therefore $g_h = \dfrac{GM}{R_h^2} = \dfrac{g_0 R^2}{R_h^2} = \dfrac{3.8 \times (3.4)^2}{(10.2)^2} = 0.42 \text{ m s}^{-2}$

(the distance from the centre is $3.4 \times 10^6 + 6.8 \times 10^6 = 10.2 \times 10^6$ m)

Exercise 10.1

1. The graph below shows how the gravitational potential outside of the Earth varies with distance from the centre.

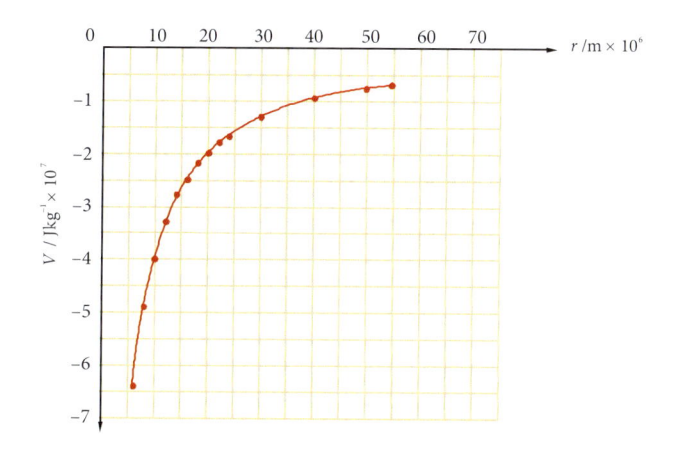

    (a) Use the graph to determine the gain in gravitational potential energy of a satellite of mass 200 kg as it moves from the surface of the Earth to a height of 3.0×10^7 m above the Earth's surface.

    (b) Calculate the energy required to take it to infinity?

    (c) Determine the slope of the graph at the surface of the Earth, m? Comment on your answer.

Electric potential energy

The concept of electric potential energy was developed with gravitational potential energy in mind. Just as an object near the surface of the Earth has potential energy because of its gravitational interaction with the Earth, so too there is electrical potential energy associated with interacting charges.

Let us first look at a case of two positive point charges each of 1µC that are initially bound together by a thread in a vacuum in space with a distance between them of 10 cm as shown in Figure 1005. When the thread is cut, the point charges, initially at rest would move in opposite directions, moving with velocities v_1 and v_2 along the direction of the electrostatic force of repulsion.

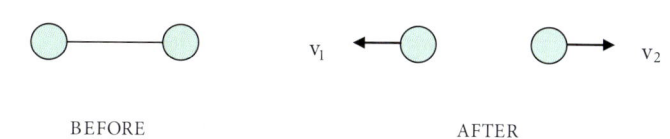

BEFORE   **AFTER**

Figure 1005   Interaction of two positive particles

The **electric potential energy** between two point charges can be found by simply adding up the energy associated with each pair of point charges. For a pair of interacting charges, the electric potential energy is given by:

$$\Delta U = \Delta E_p + \Delta E_k = \Delta W = \Delta F_r = \frac{kqQ}{r^2} \times r = \frac{kqQ}{r}$$

Because no external force is acting on the system, the energy and momentum must be conserved. Initially, $E_k = 0$ and $E_p = kqQ/r = 9 \times 10^9 \times 1 \times 10^{-12} / 0.1 \text{ m} = 0.09$ J. When they are a great distance from each other, E_p will be negligible. The final energy will be equal to ½ mv_1^2 + ½ mv_2^2 = 0.09 J. Momentum is also conserved and the velocities would be the same magnitude but in opposite directions.

Electric potential energy is more often defined in terms of a point charge moving in an electric field as:

'the electric potential energy between any two points in an electric field is defined as <u>negative of the work done</u> by an electric field in moving a <u>point electric charge between two locations</u> in the electric field.'

$$\Delta U = \Delta Ep = -\Delta W = -Fd = qEx$$

where x is the distance moved along (or opposite to) the direction of the electric field.

Electric potential energy is measured in joule (J). Just as work is a scalar quantity, so too electrical potential energy is a **scalar quantity**. The negative of the work done by an electric field in moving a unit electric charge between two points is independent of the path taken. In physics, we say the electric field is a 'conservative' field.

CHAPTER 10

Suppose an external force such as your hand moves a small positive point test charge in the direction of a uniform electric field. As it is moving it must be gaining kinetic energy. If this occurs, then the electric potential energy of the unit charge is changing.

In Figure 1006 a point charge $+q$ is moved between points A and B through a distance x in a uniform electric field.

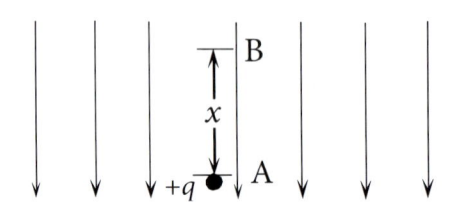

Figure 1006 Movement of a positive point charge in a uniform field

In order to move a positive point charge from point A to point B, an external force must be applied to the charge equal to qE ($F = qE$).

Since the force is applied through a distance x, then negative work has to be done to move the charge because energy is gained, meaning there is an increase **electric potential energy** between the two points. Remember that the work done is equivalent to the energy gained or lost in moving the charge through the electric field. The concept of electric potential energy is only meaningful as the electric field which generates the force in question is conservative.

$$W = F \times x = Eq \times x$$

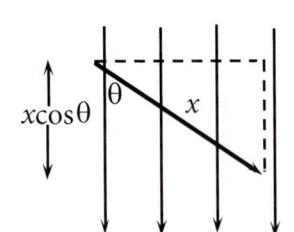

Figure 1007 Charge moved at an angle to the field

If a charge moves at an angle θ to an electric field, the component of the displacement parallel to the electric field is used as shown in Figure 918

$$W = Fx = Eq \times x \cos\theta$$

The electric potential energy is stored in the electric field, and the electric field will return the energy to the point charge when required so as not to violate the Law of conservation of energy.

Electric potential

The electric potential at a point in an electric field is defined as being the work done per unit charge in bringing a small positive point charge from infinity to that point.

$$\Delta V = V_\infty - V_f = -\frac{W}{q}$$

If we designate the potential energy to be zero at infinity then it follows that electric potential must also be zero at infinity and the electric potential at any point in an electric field will be:

$$\Delta V = -\frac{W}{q}$$

Now suppose we apply an external force to a small positive test charge as it is moved towards an isolated positive charge. The external force must do work on the positive test charge to move it towards the isolated positive charge and the work must be positive while the work done by the electric field must therefore be negative. So the electric potential at that point must be positive according to the above equation. If a negative isolated charge is used, the electric potential at a point on the positive test charge would be negative. Positive point charges of their own accord, move from a place of high electric potential to a place of low electric potential. Negative point charges move the other way, from low potential to high potential. In moving from point A to point B in the diagram, the positive charge $+q$ is moving from a low electric potential to a high electric potential.

In the definition given, the term "work per unit charge" has significance. If the test charge is $+1.6 \times 10^{-19}$ C where the charge has a potential energy of 3.2×10^{-17} J, then the potential would be 3.2×10^{-17} J / $+1.6 \times 10^{-19}$ C = 200 JC^{-1}. Now if the charge was doubled, the potential would become 6.4×10^{-17} J. However, the potential per unit charge would be the same.

Electric potential is a scalar quantity and it has units JC^{-1} or volts where 1 volt equals one joule per coloumb. The volt allows us to adopt a unit for the electric field in terms of the volt.

Previously, the unit for the electric field was NC^{-1}.

$W = qV$ and $F = qE$, so $\dfrac{W}{V} = \dfrac{F}{E}$

$$E = \frac{FV}{W} = \frac{FV}{Fm} \text{ V m}^{-1}.$$

That is, the units of the electric field, E, can also be expressed as V m^{-1}.

The work done per unit charge in moving a point charge between two points in an electric field is again independent of the path taken.

312

Electric potential due to a point charge

Let us take a point r metres from a charged object. The potential at this point can be calculated using the following:

$$W = -Fr = -qV \text{ and } F = -\frac{q_1 q_2}{4\pi\varepsilon_0 r^2}$$

Therefore,

$$W = -\frac{q_1 q_2}{4\pi\varepsilon_0 r^2} \times r = -\frac{q_1 q_2}{4\pi\varepsilon_0 r} = -q_1 \times \frac{q_2}{4\pi\varepsilon_0 r} = -q_1 V$$

That is

$$V = \frac{q}{4\pi\varepsilon_0 r}$$

Or, simply

$$V = \frac{kq}{r}$$

Example

Determine how much work is done by the electric field of point charge 15.0 μC when a charge of 2.00 μC is moved from infinity to a point 0.400 m from the point charge. (Assume no acceleration of the charges).

Solution

The work done by the electric field is $W = -qV$
$= -1/4\pi\varepsilon_0 \times q \times (Q/r_\infty - Q/r_{0.400})$

$W = (-2.00 \times 10^{-6} \text{ C} \times 9.00 \times 10^9 \text{ NmC}^{-2} \times 15.0 \times 10^{-6} \text{ C}) \div 0.400 \text{ m} = -0.675 \text{ J}$

An external force would have to do +0.675 J of work.

Electric field strength and electric potential gradient

Let us look back at Figure 1006. Suppose again that the charge $+q$ is moved a small distance by a force F from A to B so that the force can be considered constant. The work done is given by:

$$\Delta W = F \times \Delta x$$

The force F and the electric field E are oppositely directed, and we know that:

$F = -qE$ and $\Delta W = q \Delta V$

Therefore, the work done can be given as:

$q \Delta V = -q \, E \, \Delta x$

Therefore

$$E = -\frac{\Delta V}{\Delta x}$$

The rate of change of potential ΔV at a point with respect to distance Δx in the direction in which the change is maximum is called the **potential gradient**. We say that the electric field = - the potential gradient and the units are Vm^{-1}. From the equation we can see that in a graph of electric potential versus distance, the gradient of the straight line equals the electric field strength.

In reality, if a charged particle enters a uniform electric field, it will be accelerated uniformly by the field and its kinetic energy will increase. This is why we had to assume no acceleration in the last worked example.

$$\Delta E_k = \frac{1}{2}mv^2 = q \cdot E \cdot x = q \cdot \frac{V}{x} \cdot x = q \cdot V$$

Example

Determine how far apart two parallel plates must be situated so that a potential difference of 1.50×10^2 V produces an electric field strength of 1.00×10^3 NC^{-1}.

Solution

Using $E = -\frac{\Delta V}{x} \Leftrightarrow x = \frac{\Delta V}{E} = \frac{1.5 \times 10^2 \text{ V}}{1.00 \times 10^3 \text{ N C}^{-1}}$

$= 1.50 \times 10^{-1}$

The plates are $\mathbf{1.50 \times 10^{-1}}$ **m apart.**

The electric field and the electric potential at a point due to an evenly distributed charge $+q$ on a sphere can be represented graphically as in Figure 1008.

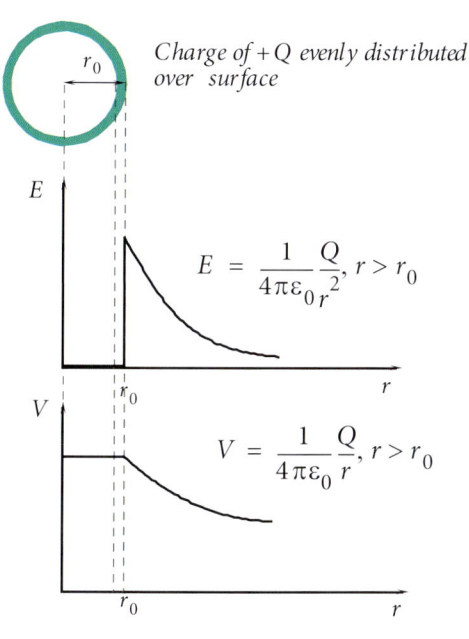

Figure 1008   Electric field and potential due to a charged sphere

CHAPTER 10

When the sphere becomes charged, we know that the charge distributes itself evenly over the surface. Therefore every part of the material of the conductor is at the same potential. As the electric potential at a point is defined as being numerically equal to the work done in bringing a unit positive charge from infinity to that point, it has a constant value in every part of the material of the conductor.

Since the potential is the same at all points on the conducting surface, then $\Delta V / \Delta x$ is zero. But $E = - \Delta V / \Delta x$. Therefore, the electric field inside the conductor is zero. There is no electric field inside the conductor.

Some further observations of the graphs in Figure 1009 are:

- Outside the sphere, the graphs obey the relationships given as $E \alpha 1 / r^2$ and $V \alpha 1 / r$
- At the surface, $r = r_0$. Therefore, the electric field and potential have the minimum value for r at this point and this infers a maximum field and potential.
- Inside the sphere, the electric field is zero.
- Inside the sphere, no work is done to move a charge from a point inside to the surface. Therefore, there is no potential **difference** and the potential is the same as it is when $r = r_0$.

Similar graphs can be drawn for the electric field intensity and the electric potential as a function of distance from conducting parallel plates and surfaces, and these are given in Figure 1009.

Figure 1009 Electric field and electric potential at a distance from a charged surface

Potential due to one or more point charges

The potential due to one point charge can be determined by using the equation formula

$V = kq / r.$

Example 1

Determine the electric potential at a point 2.0×10^{-1} m from the centre of an isolated conducting sphere with a point charge of 4.0 pC in air.

Solution

Using the formula

$V = kq / r$, we have

$$V = \frac{(9.0 \times 10^9) \times (4.0 \times 10^{-12})}{(2.0 \times 10^{-1})} = 0.18 \text{ V}$$

the potential at the point is 1.80×10^{-1} V.

The potential due to a number of point charges can be determined by adding up the potentials due to individual point charges because the electric potential at any point outside a conducting sphere will be the same as if all the charge was concentrated at its centre.

Example 2

Three point charges of are placed at the vertices of a right-angled triangle as shown in the diagram below. Determine the absolute potential at the + 2.0 µC charge, due to the two other charges.

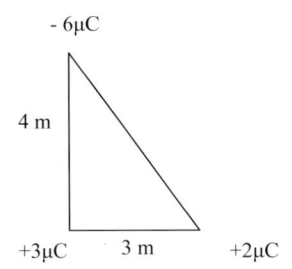

Solution

The electric potential of the +2 µC charge due to the – 6 µC charge is:

$V = (9 \times 10^9 \ Nm^2C^{-2} \times -6 \times 10^{-6} \ C) \div (\sqrt{3^2 + 4^2}) \ m =$
$- 1.08 \times 10^4$ V

The electric potential of the +2 μC charge due to the +3 μC charge is:

$V = (9 \times 10^9\ Nm^2C^{-2} \times 3 \times 10^{-6}\ C) \div 3m = 9 \times 10^3\ V$

The net absolute potential is the sum of the 2 potentials
$-1.08 \times 10^4\ V + 9 \times 10^3\ V =$
$-1.8 \times 10^3\ V$

The absolute potential at the point is **-1.8 × 10³ V**.

Equipotential surfaces

Regions in space where the electric potential of a charge distribution has a constant value are called **equipotentials**. The places where the potential is constant in three dimensions are called **equipotential surfaces,** and where they are constant in two dimensions they are called **equipotential lines**.

They are in some ways analogous to the contour lines on topographic maps. In this case, the gravitational potential energy is constant as a mass moves around the contour lines because the mass remains at the same elevation above the Earth's surface. The gravitational field strength acts in a direction perpendicular to a contour line.

Similarly, because the electric potential on an equipotential line has the same value, an electric force can do no work when a test charge moves on an equipotential. Therefore, the electric field cannot have a component along an equipotential, and thus it must be everywhere perpendicular to the equipotential surface or equipotential line. This fact makes it easy to plot equipotentials if the lines of force or lines of electric flux of an electric field are known.

For example, there are a series of equipotential lines between two parallel plate conductors that are perpendicular to the electric field. There will be a series of concentric circles (each circle further apart than the previous one) that map out the equipotentials around an isolated positive sphere. The lines of force and some equipotential lines for an isolated positive sphere are shown in Figure 1010.

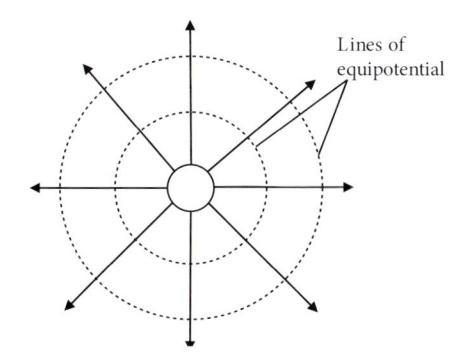

Figure 1010 Equipotentials around an isolated positive sphere

In summary, we can conclude that

- **No work is done to move a charge along an equipotential.**
- **Equipotentials are always perpendicular to the electric lines of force.**

Figure 1011 and 1012 show some equipotential lines for two oppositely charged and identically positive spheres separated by a distance.

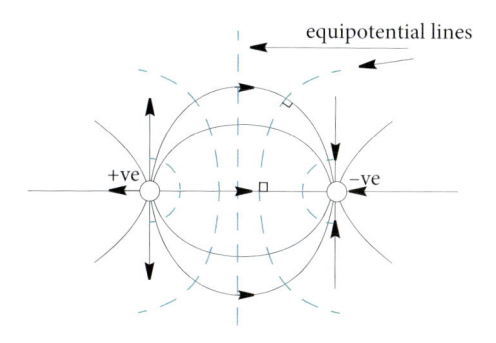

Figure 1011 Equipotential lines between two opposite charges

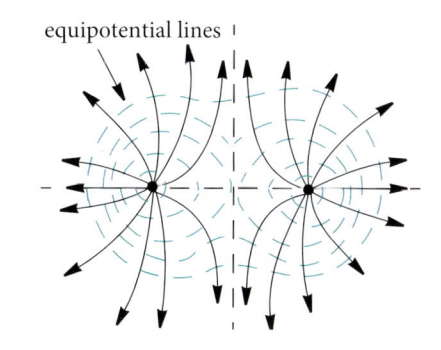

Figure 1012 Equipotential lines between two charges which are the same

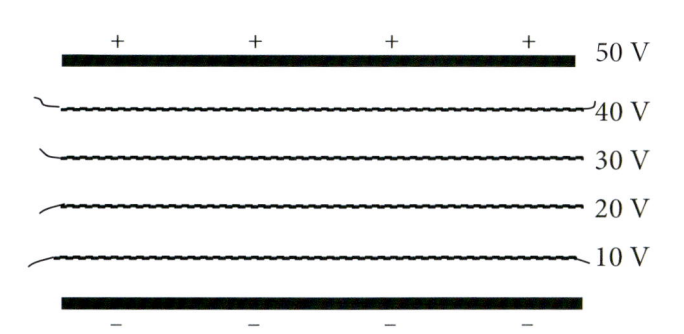

Figure 1013 Equipotential lines between charged parallel plates

Figure 1013 shows the equipotential lines between charged parallel plates. Throughout this chapter the similarities and differences between gravitational fields and electric fields have been discussed. The relationships that exists between gravitational and electric quantities and the effects of point masses and charges is summarised in Table 1014

315

CHAPTER 10

	Gravitational quantity	Electrical quantity
Quantities	$V = \dfrac{W}{m}$	$V = \dfrac{W}{q}$
	$g = \dfrac{F}{m}$	$E = \dfrac{F}{q}$
	$g = -\dfrac{\Delta V}{\Delta x}$	$E = -\dfrac{\Delta V}{\Delta x}$
Point masses and charges	$V = -G\dfrac{m}{r}$	$V = \dfrac{1}{4\pi\varepsilon_0}\dfrac{q}{r}$
	$g = -G\dfrac{m}{r^2}$	$E = \dfrac{1}{4\pi\varepsilon_0}\dfrac{q}{r^2}$
	$F = G\dfrac{m_1 m_2}{r^2}$	$F = \dfrac{1}{4\pi\varepsilon_0}\dfrac{q_1 q_2}{r^2}$

Figure 1014   Formulas

Example

Deduce the electric potential on the surface of a gold nucleus that has a radius of 6.2 fm.

Solution

Using the formula

$V = kq / r$, and knowing the atomic number of gold is 79. We will assume the nucleus is spherical and it behaves as if it were a point charge at its centre (relative to outside points).

$V = 9.0 \times 10^9$ $Nm^2C^{-2} \times 79 \times 1.6 \times 10^{-19}$ C $\div$ 6.2×10^{-15} m
$= 1.8 \times 10^7$ V

The potential at the point is **18 MV**.

Example

Deduce the ionisation energy in electron-volts of the electron in the hydrogen atom if the electron is in its ground state and it is in a circular orbit at a distance of 5.3×10^{-11} m from the proton.

Solution

This problem is an energy, coulombic, circular motion question based on Bohr's model of the atom (not the accepted quantum mechanics model). The ionisation energy is the energy required to remove the electron from the ground state to infinity. The electron travels in a circular orbit and therefore has a centripetal acceleration. The ionisation energy will counteract the coulombic force and the movement of the electron will be in the opposite direction to the centripetal force.

Total energy = E_k electron + E_p due to the proton-electron interaction

$\Sigma F = kqQ / r^2 = mv^2 / r$ and as such $mv^2 = = kqQ / r$.

Therefore, E_k electron = ½ kqQ / r.

E_p due to the proton-electron interaction = $- kqQ / r$.

Total energy = ½ $kqQ / r + - kqQ / r = -½ kqQ / r$

$= - 9.0 \times 10^9$ $Nm^2C^{-2} \times (1.6 \times 10^{-19}$ $C)^2 \div 5.3 \times 10^{-11}$ m $=$ -2.17×10^{-18} J

$= -2.17 \times 10^{-18}$ J $\div 1.6 \times 10^{-19} = -13.6$ eV.

The ionisation energy is **13.6 eV**.

Exercise 10.2

1. A point charge P is placed midway between two identical negative charges. Which one of the following is correct with regards to electric field and electric potential at point P?

	Electric field	Electric potential
A	non-zero	zero
B	zero	non-zero
C	non-zero	non-zero
D	zero	zero

2. Two positive charged spheres are tied together in a vacuum somewhere in space where there are no external forces. A has a mass of 25 g and a charge of 2.0 μC and B has a mass of 15 g and a charge of 3.0 μC. The distance between them is 4.0 cm. They are then released as shown in the diagram.

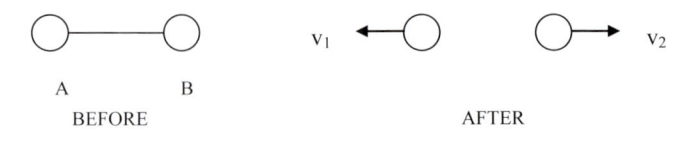

   BEFORE   AFTER

   (a) Determine their initial electric potential energy in the before situation.
   (b) Determine the speed of sphere B after release.

3. The diagram below represents two equipotential lines in separated by a distance of 5 cm in a uniform electric field.

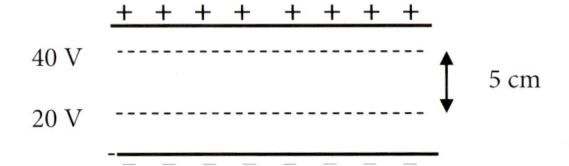

   40 V
   20 V
   5 cm

316

Determine the strength of the electric field.

4. This question is about the electric field due to a charged sphere and the motion of electrons in that field. The diagram below shows an isolated, metal sphere in a vacuum that carries a negative electric charge of 6.0 µC.

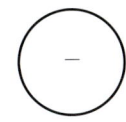

(a) On the diagram draw the conventional way to represent the electric field pattern due to the charged sphere and lines to represent three equipotential surfaces in the region outside the sphere.

(b) Explain how the lines representing the equipotential surfaces that you have sketched indicate that the strength of the electric field is decreasing with distance from the centre of the sphere.

(c) The electric field strength at the surface of the sphere and at points outside the sphere can be determined by assuming that the sphere acts as a point charge of magnitude 6.0 µC at its centre. The radius of the sphere is 2.5×10^{-2} m. Deduce that the magnitude of the field strength at the surface of the sphere is 8.6×10^7 Vm^{-1}.

An electron is initially at rest on the surface of the sphere.

(d) (i) Describe the path followed by the electron as it leaves the surface of the sphere.

(ii) Calculate the initial acceleration of the electron.

5. Determine the amount of work that is done in moving a charge of 10.0 nC through a potential difference of 1.50×10^2 V.

6. Three identical 2.0 µC conducting spheres are placed at the corners of an equilateral triangle of sides 25 cm. The triangle has one apex C pointing up the page and 2 base angles A and B. Determine the absolute potential at B.

7. Determine how far apart two parallel plates must be situated so that a potential difference of 2.50×10^2 V produces an electric field strength of 2.00×10^3 NC^{-1}.

8. The gap between two parallel plates is 1.0×10^{-3} m, and there is a potential difference of 1.0×10^4 V between the plates. Calculate

i. the work done by an electron in moving from one plate to the other
ii. the speed with which the electron reaches the second plate if released from rest.
iii. the electric field intensity between the plates.

9. An electron gun in a picture tube is accelerated by a potential 2.5×10^3 V. Determine the kinetic energy gained by the electron in electron-volts.

10. Determine the electric potential 2.0×10^{-2} m from a charge of -1.0×10^{-5} C.

11. Determine the electric potential at a point midway between a charge of –20 pC and another of + 5 pC on the line joining their centres if the charges are 10 cm apart.

12. During a thunderstorm the electric potential difference between a cloud and the ground is 1.0×10^9 V. Determine the magnitude of the change in electric potential energy of an electron that moves between these points in electron-volts.

13. A charge of 1.5 µC is placed in a uniform electric field of two oppositely charged parallel plates with a magnitude of 1.4×10^3 NC^{-1}.

(a) Determine the work that must be done against the field to move the point charge a distance of 5.5 cm.

(b) Calculate the potential difference between the final and initial positions of the charge.

(c) Determine the potential difference between the plates if their separation distance is 15 cm.

14. During a flash of lightning, the potential difference between a cloud and the ground was 1.2×10^9 V and the amount of transferred charge was 32 C.

(a) Determine the change in energy of the transferred charge.

(b) If the energy released was all used to accelerate a 1 tonne car, deduce its final speed.

(c) If the energy released could be used to melt ice at 0 °C, deduce the amount of ice that could be melted.

15. Suppose that when an electron moved from A to B in the diagram along an electric field line that the electric field does 3.6×10^{-19} J of work on it.

    Determine the differences in electric potential:

    (a) $V_B - V_A$
    (b) $V_C - V_A$
    (c) $V_C - V_B$

16. Determine the potential at point P that is located at the centre of the square as shown in the diagram below.

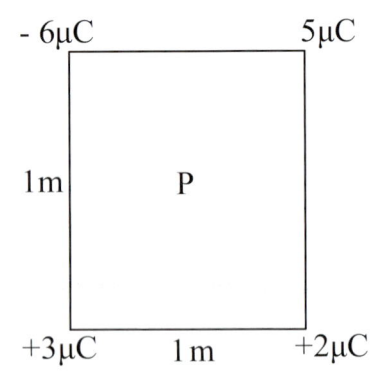

```
- 6μC              5μC
  ┌─────────────┐
  │             │
1m│      P      │
  │             │
  └─────────────┘
+3μC    1 m    +2μC
```

10.2 Fields at work

NATURE OF SCIENCE:

Communication of scientific explanations: The ability to apply field theory to the unobservable (charges) and the massively scaled (motion of satellites) required scientists to develop new ways to investigate, analyse and report findings to a general public used to scientific discoveries based on tangible and discernible evidence. (5.1)

© IBO 2014

Essential idea: Similar approaches can be taken in analysing electrical and gravitational potential problems.

Understandings:
- Potential and potential energy
- Potential gradient
- Potential difference
- Escape speed
- Orbital motion, orbital speed and orbital energy
- Forces and inverse-square law behaviour

Orbital motion, orbital speed and orbital energy

Although orbital motion may be circular, elliptical or parabolic, this sub-topic only deals with circular orbits. This sub-topic is not fundamentally new physics, but an application that synthesizes ideas from gravitation, circular motion, dynamics and energy.

The Moon orbits the Earth and in this sense it is often referred to as a satellite of the Earth. Before 1957 it was the only Earth satellite. However, in 1957 the Russians launched the first man made satellite, Sputnik 1. Since this date many more satellites have been launched and there are now literally thousands of them orbiting the Earth. Some are used to monitor the weather, some used to enable people to find accurately their position on the surface of the Earth, many are used in communications, and no doubt some are used to spy on other countries. Figure 1015 shows how, in principle, a satellite can be put into orbit.

The person (whose size is greatly exaggerated with respect to Earth) standing on the surface on the Earth throws some stones. The greater the speed with which a stone is thrown the further it will land from her. The paths followed by the thrown stones are parabolas. By a stretch of the imagination we can visualise a situation in which a stone is thrown with such a speed that, because of the curvature of the Earth, it will not land on the surface of the Earth but go into "orbit". (Path 4 on Figure 1015).

318

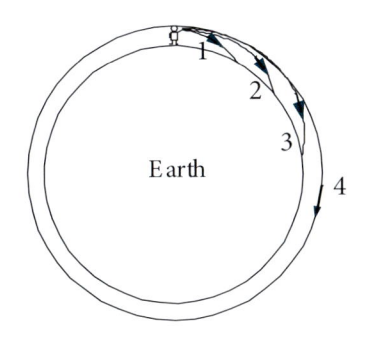

Figure 1015   Throwing a stone into orbit

The force that causes the stones to follow a parabolic path and to fall to Earth is gravity and similarly the force that keeps the stone in orbit is gravity. For circular motion to occur we have seen that a force must act at right angles to the velocity of an object, that is there must be a centripetal force. Hence in the situation we describe here the centripetal force for circular orbital motion about the Earth is provided by gravitational attraction of the Earth.

We can calculate the speed with which a stone must be thrown in order to put it into orbit just above the surface of the Earth.

If the stone has mass m and speed v then we have from Newton's 2nd law

$$\frac{mv^2}{R_E} = G\frac{M_E m}{R_E^2}$$

where R_E is the radius of the Earth and M_E is the mass of the Earth.

Bearing in mind that $g_0 = G\frac{M_E}{R_E^2}$, then

$$v = \sqrt{gR_E} = \sqrt{10 \times 6.4 \times 10^6} = 8 \times 10^3$$

That is, the stone must be thrown at 8×10^3 m s^{-1}.

Clearly we are not going to get a satellite into orbit so close to the surface of the Earth. Moving at this speed the friction due to air resistance would melt the satellite before it had travelled a couple of kilometres. In reality therefore a satellite is put into orbit about the Earth by sending it, attached to a rocket, beyond the Earth's atmosphere and then giving it a component of velocity perpendicular to a radial vector from the Earth. See Figure 1016.

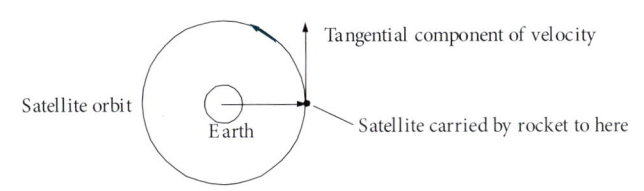

Figure 1016   Getting a satellite into orbit

Kepler's third law

(This work of Kepler and Newton's synthesis of the work is an excellent example of the scientific method and makes for a good TOK discussion)

In 1627 *Johannes Kepler* (1571-1630) published his laws of planetary motion. The laws are empirical in nature and were deduced from the observations of the Danish astronomer *Tycho de Brahe* (1546-1601). The third law gives a relationship between the radius of orbit R of a planet and its period T of revolution about the Sun. The law is expressed mathematically as

$$\frac{T^2}{R^3} = \text{constant}$$

We shall now use Newton's Law of Gravitation to show how it is that the planets move in accordance with Kepler's third law.

In essence Newton was able to use his law of gravity to predict the motion of the planets since all he had to do was factor the F given by this law into his second law, $F = ma$, to find their accelerations and hence their future positions.

In Figure 1017 the Earth is shown orbiting the Sun and the distance between their centres is R.

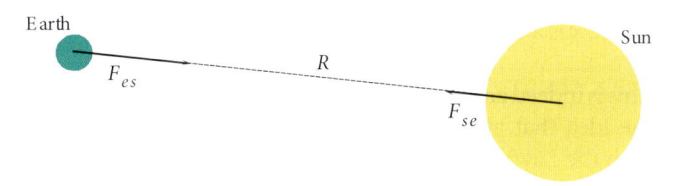

Figure 1017   Planets move according to Kepler's third law

F_{es} is the force that the Earth exerts on the Sun and F_{se} is the force that the Sun exerts on the Earth. The forces are equal and opposite and the Sun and the Earth will actually orbit about a common centre. However since the Sun is so very much more massive than the Earth this common centre will be close to the centre of the Sun and so we can regard the Earth as orbiting about the centre of the Sun. The other thing that we shall assume is that we can ignore the forces that the other planets exert on the Earth. (This would not be a wise thing to do if you were planning to send a space ship to the Moon for example.). We shall also assume that we have followed Newton's example and indeed proved that a sphere will act as a point mass situated at the centre of the sphere.

Kepler had postulated that the orbits of the planets are elliptical but since the eccentricity of the Earth's orbit is small we shall assume a circular orbit.

319

The acceleration of the Earth towards the Sun is $a = R\omega^2$

where $\omega = \dfrac{2\pi}{T}$

Hence,
$$a = R \times \left(\dfrac{2\pi}{T}\right)^2 = \dfrac{4\pi^2 R}{T^2}$$

But the acceleration is given by Newton's Second Law, $F = ma$, where F is now given by the Law of Gravitation. So in this situation

$$F = ma = \dfrac{GM_s M_e}{R^2},$$ but, we also have that

$a = \dfrac{4\pi^2 R}{T^2}$ and $m = M_e$ so that

$$\dfrac{GM_s M_e}{R^2} = M_e \times \dfrac{4\pi^2 R}{T^2} \Rightarrow \dfrac{GM_s}{4\pi^2} = \dfrac{R^3}{T^2}$$

But the quantity $\dfrac{GM_s}{4\pi^2}$

is a constant that has the same value for each of the planets so we have for all the planets, not just Earth, that

$$\dfrac{R^3}{T^2} = k$$

where k is a constant. Which is of course Kepler's third law.

This is indeed an amazing breakthrough. It is difficult to refute the idea that all particles attract each other in accordance with the Law of Gravitation when the law is able to account for the observed motion of the planets about the Sun.

The gravitational effects of the planets upon each other should produce perturbations in their orbits. Such is the predictive power of the Universal Gravitational Law that it enabled physicists to compute these perturbations. The telescope had been invented in 1608 and by the middle of the 18th Century had reached a degree of perfection in design that enabled astronomers to actually measure the orbital perturbations of the planets. Their measurements were always in agreement with the predictions made by Newton's law. However, in 1781 a new planet, Uranus was discovered and the orbit of this planet did not fit with the orbit predicted by Universal Gravitation. Such was the physicist's faith in the Newtonian method that they suspected that the discrepancy was due to the presence of a yet undetected planet. Using the Law of Gravitation the French astronomer J.Leverrier and the English astronomer. J. C. Adams were able to calculate just how massive this new planet must be and also where it should be. In 1846 the planet Neptune was discovered just where they had predicted. In a similar way, discrepancies in the orbit of Neptune led to the prediction and subsequent discovery in 1930 of the planet Pluto. Newton's Law of Gravitation had passed the ultimate test of any theory; it is not only able to explain existing data but also to make predictions.

Satellite energy

When a satellite is in orbit about a planet it will have both kinetic energy and gravitational potential energy. Suppose we consider a satellite of mass m that is in an orbit of radius r about a planet of mass M.

The gravitational potential due to the planet at distance r from its centre is

$$-\dfrac{GM_e}{r}.$$

The gravitational potential energy of the satellite V_{sat}

is therefore $-\dfrac{GM_e m}{r}$.

That is, $V_{sat} = -\dfrac{GM_e m}{r}$.

The gravitational field strength at the surface of the planet is given by

$$g_0 = \dfrac{GM_e}{R_e^2}$$

Hence we can write

$$V_{sat} = -\dfrac{g_0 R_e^2 m}{r}$$

The kinetic energy of the satellite K_{sat} is equal to $\tfrac{1}{2}mv^2$, where v is its orbital speed.

By equating the gravitational force acting on the satellite to its centripetal acceleration we have

$$\dfrac{GM_e m}{r^2} = \dfrac{mv^2}{r} \Leftrightarrow mv^2 = \dfrac{GM_e m}{r}.$$

$$\tfrac{1}{2}mv^2 = \dfrac{1}{2} \times \dfrac{GM_e m}{r}$$

$$= \dfrac{g_0 R_e^2 m}{2r}$$

Which is actually quite interesting since it shows that, irrespective of the orbital radius the KE is numerically equal to half the PE, Also the **total** energy E_{tot} of the satellite is always negative since

$$E_{tot} = K_{sat} + V_{sat} = \dfrac{1}{2} \times \dfrac{GM_e m}{r} + \left(-\dfrac{GM_e m}{r}\right) = -\dfrac{1}{2}\dfrac{GM_e m}{r}$$

The energies of an orbiting satellite as a function of radial distance from the centre of a planet are shown plotted in Figure 1018.

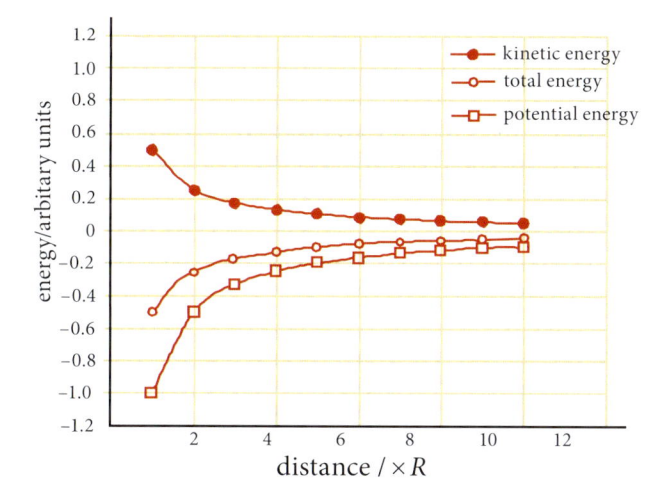

Figure 1018 Energy of an orbiting satellite as a function of distance from the centre of a planet

Weightlessness

Suppose that you are in an elevator (lift) that is descending at constant speed and you let go of a book that you are holding in your hand. The book will fall to the floor with acceleration equal to the acceleration due to gravity. If the cable that supports the elevator were to snap (a situation that I trust will never happen to any of you) and you now let go the book that you are holding in your other hand, this book will not fall to the floor - it will stay exactly in line with your hand. This is because the book is now falling with the same acceleration as the elevator and as such the book cannot "catch" up with the floor of the elevator. Furthermore, if you happened to be standing on a set of bathroom scales, the scales would now read zero - you would be apparently weightless. It is this idea of free fall that explains the apparent weightlessness of astronauts in an orbiting satellite. These astronauts are in free fall in the sense that they are accelerating towards the centre of the Earth.

It is actually possible to define the weight of a body in several different ways. We can define it for example as the gravitational force exerted on the body by a specified object such as the Earth. This we have seen that we do in lots of situations where we define the weight as being equal to *mg*. If we use this definition, then an object in free fall cannot by definition be weightless since it is still in a gravitational field. However, if we define the weight of an object in terms of a "weighing" process such as the reading on a set of bathroom scales, which in effect measures the contact force between the object and the scales, then clearly objects in free fall are weightless. One now has to ask the question whether or not it is possible. For example, to measure the gravitational force acting on an astronaut in orbit about the Earth.

We can also define weight in terms of the net gravitational force acting on a body due to several different objects. For example for an object out in space, its weight could be defined in terms of the resultant of the forces exerted on it by the Sun, the Moon, the Earth and all the other planets in the Solar System. If this resultant is zero at a particular point then the body is weightless at this point.

In view of the various definitions of weight that are available to us it is important that when we use the word "weight" we are aware of the context in which it is being used.

Example

Calculate the height above the surface of the Earth at which a geo-stationary satellite orbits.

Solution

A geo-stationary satellite is one that orbits the Earth in such a way that it is stationary with respect to a point on the surface of the Earth. This means that its orbital period must be the same as the time for the Earth to spin once on its axis i.e. 24 hours.

From Kepler's third law we have $\dfrac{GM_s}{4\pi^2} = \dfrac{R^3}{T^2}$.

That is,

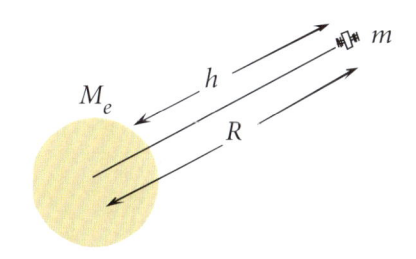

using the fact that the force of attraction between the satellite and the Earth is given by

$$F = \dfrac{GM_e m}{R^2}$$

and that F = ma

where m is the mass of the satellite and $a = \dfrac{4\pi^2 R}{T^2}$

we have,

$$\dfrac{GM_e m}{R^2} = m \times \dfrac{4\pi^2 R}{T^2} \Rightarrow \dfrac{GM_e}{4\pi^2} = \dfrac{R^3}{T^2}$$

CHAPTER 10

Now, the mass of the Earth is 6.0×10^{24} kg and the period, T, measured in seconds is given by T = 86,400 s.

So substitution gives R = 42×10^6 m

The radius of the Earth is 6.4×10^6 m so that the orbital height, h, is about 3.6×10^7 m.

Example

Calculate the minimum energy required to put a satellite of mass 500 kg into an orbit that is as a height equal to the Earth's radius above the surface of the Earth.

Solution

We have seen that when dealing with gravitational fields and potential it is useful to remember that

$$g_0 = \frac{GM}{R_e^2} \text{ or, } g_0 R_e^2 = GM$$

The gravitational potential at the surface of the Earth is

$$-g_0 R_e = -\frac{GM}{R_e}.$$

The gravitational potential at a distance R from the centre of the Earth is $-\frac{GM}{R}$

$$= -\frac{g_0 R_e^2}{R}$$

The difference in potential between the surface of the Earth and a point distance R from the centre is therefore

$$\Delta V = g_0 R_e \left(1 - \frac{R_e}{R}\right)$$

If $R = 2R_e$ then $\Delta V = \frac{g_0 R_e}{2}$

This means that the work required to "lift" the satellite into orbit is $g_0 R m$ where m is the mass of the satellite. This is equal to

$10 \times 3.2 \times 10^6 \times 500 = 16000$ MJ.

However, the satellite must also have kinetic energy in order to orbit Earth. This will be equal to

$$\frac{g_0 m R_e^2}{2R} = \frac{g_0 m R_e^2}{4} = 8000 \text{ MJ}$$

The minimum energy required is therefore

24000 MJ.

Exercise 10.3

1. The speed needed to put a satellite in orbit does not depend on

   A. the radius of the orbit.
   B. the shape of the orbit.
   C. the value of g at the orbit.
   D. the mass of the satellite.

2. Estimate the speed of an Earth satellite whose orbit is 400 km above the Earth's surface. Also determine the period of the orbit.

3. Calculate the speed of a 200 kg satellite, orbiting the Earth at a height of 7.0×10^6 m.

   Assume that g = 8.2 m s^{-2} for this orbit.

4. The radii of two satellites, X and Y, orbiting the Earth are 2r and 8r where r is the radius of the Earth. Calculate the ratio of the periods of revolution of X to Y.

5. A satellite of mass m kg is sent from Earth's surface into an orbit of radius 5R, where R is the radius of the Earth. Write down an expression for

   (a) the potential energy of the satellite in orbit.

   (b) the kinetic energy of the satellite in orbit.

   (c) the minimum work required to send the satellite from rest at the Earth's surface into its orbit.

6. A satellite in an orbit of 10r, falls back to Earth (radius r) after a malfunction. Determine the speed with which it will hit the Earth's surface?

7. The radius of the moon is ¼ that of the Earth Assuming Earth and the Moon to have the same density, compare the accelerations of free fall at the surface of Earth to that at the surface of the Moon.

8. Use the following data to determine the gravitational field strength at the surface of the Moon and hence determine the escape speed from the surface of the Moon.

   Mass of the Moon = 7.3×10^{22} kg,

   Radius of the Moon = 1.7×10^6 m

11. Electromagnetic induction

Contents

11.1 – Electromagnetic induction

11.2 – Power generation and transmission

11.3 – Capacitance

Essential Ideas

The majority of electricity generated throughout the world is generated by machines that were designed to operate using the principles of electromagnetic induction.

Generation and transmission of alternating current (ac) electricity has transformed the world.

Capacitors can be used to store electrical energy for later use. © IBO 2014

CHAPTER 11

11.1 Electromagnetic induction

Essential idea: The majority of electricity generated throughout the world is generated by machines that were designed to operate using the principles of electromagnetic induction.

Understandings:
- Electromotive force (emf)
- Magnetic flux and magnetic flux linkage
- Faraday's law of induction
- Lenz's law

NATURE OF SCIENCE:

Experimentation: In 1831 Michael Faraday, using primitive equipment, observed a minute pulse of current in one coil of wire only when the current in a second coil of wire was switched on or off but nothing while a constant current was established. Faraday's observation of these small transient currents led him to perform experiments that led to his law of electromagnetic induction. (1.8)

© IBO 2014

TOK INTRODUCTION

In the chapter on electrostatics, and in this chapter on electromagnetism, mention is made of *Michael Faraday*. The laws of electricity and magnetism owe more perhaps to the experimental work of Faraday than any other person. There were great theoreticians like Ampère and Maxwell but Faraday was a real experimenter. He invented the first dynamo, electric motor and transformer. It was Faraday who originated the use of electric fields lines that he called "lines of force" even before the concept of the electric field was clearly understood. He along with Joseph Henry discovered electromagnetic induction, and this concept will be expanded on in this chapter. Electromagnetic induction has revolutionised the way we live. This phenomenon has had a huge impact on society and it has become the basis for the generation of electric power that we so often take for granted in our everyday life.

Michael Faraday (1791-1867), the son of a blacksmith, was born in Newington, Surrey. He had little formal education as a child and at the age of 14, he took up an apprenticeship as a bookbinder. While rebinding a copy of the Encyclopaedia Britannica, he happened to read an article on electricity, and to his own admission, this article gave him a lifelong fascination with science.

He started to attend lectures given by Sir *Humphry Davy*, a famous electrochemist and publicist. Faraday became interested in electrolysis and he prepared a set of lecture notes that greatly impressed Davy. By good fortune or misfortune, when Davy was temporarily blinded in a laboratory accident at the Royal Institution in 1812, he needed a laboratory assistant and he requested that Faraday be given the position. During this time, Faraday discovered and described the organic compound benzene as well as some other chloro-carbon compounds. He did research on steel, optical glass and the liquefaction of gases. Faraday's work was impressive, and he eventually became director of the *Royal Institution*.

Faraday had a great talent for explaining his ideas to both children and adults. He gave many "wizz-bang" lectures to the young, and his book addressed at their level called "*The Chemical History of the Candle*" is still in print. He introduced the *Friday Evening Discourses* and the Christmas lectures for children at the Royal Institution, and these lectures still continue to this day. In 1865, he retired from the Royal Institution after 50 years service.

In the 1830s, Faraday became interested in electrochemistry and he was the first to use the term "**electrolysis**" in 1832. Furthermore, he introduced the use of the terms 'electrolyte', 'cell', 'electrodes' and 'electrochemical reaction' so commonly used in the subject of electrochemistry. He subjected electrolysis to the first quantitative experimentation and in 1834 was able to establish that the amount of chemical compound decomposed at the electrodes was proportional to the amount of electricity used – Faraday's First and Second Law of Electrolysis. He devised the terminology, "ions", for the part of the compound discharged at the electrodes.

Once *Oersted* had discovered that a current flowing in a conductor produced a magnetic field in 1819, scientists were convinced that a moving magnetic field should be able to produce a current in the conductor. It took eleven years before, in 1831, the American, *Joseph Henry* (1797-1878), and the Englishman, *Michael Faraday* (1791-1867), while working independently, explained the cause and effect of an induced current/emf being produced by a changing magnetic field. Henry is credited with the discovery but *Faraday* was the first to publish, introducing the concept of line of magnetic flux in his explanations.

In his notebooks in the early 1830s, *Faraday* described how he placed wires near magnets looking for current in the wire but without success. However, as he moved

Electromagnetic Induction

the apparatus he noticed a brief pulse of current but the current immediately fell back to zero. Perhaps the missing ingredient was motion.

The solution came in 1831 when he set up an apparatus similar to that in Figure 1101 that he called an "induction ring". The apparatus may look familiar to us with its battery, coils and galvanometer (a meter to detect current). We also know that a soft iron core increases the strength of a magnetic field.

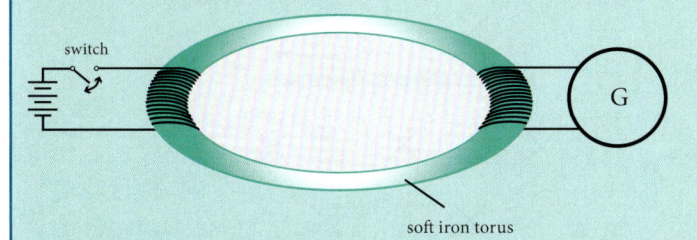

Figure 1101 Faraday's induction ring

To his initial disappointment, when he closed the switch to allow steady current to flow, only a slight twitch was observed in the galvanometer before the needle fell back to zero. This twitch could have been due to mechanical vibration. However, using his intuition, he noticed that when he slowly opened and closed the switch, a current was produced in one direction, then fell to zero, then a current was produced in the opposite direction. He called the current produced by a changing magnetic flux an **induced current**, and he called the general phenomenon **electromagnetic induction**. He assumed that the magnetic flux must be changing, but how? Was the iron ring really necessary to produce induction or did it merely strengthen an effect? Was it necessary to have two coils or could an induced current be produced by simply moving a magnet in and out of a single coil of wire?

Remembering back to his earlier experiment where he considered that motion could be a factor, he quickly set up experiments using a coil and a magnet, and proved that the iron ring was not essential, and that motion inside one coil could produce an induced current. Furthermore, he found that a rotating copper disc inserted between the poles of a magnet could be used (instead of a coil) to produce an induced current. It didn't matter whether the magnet was in motion or the coil (or disc) was in motion, an induced current was produced provided there was a change in magnetic flux. Faraday presented his findings to the *Royal Society* in November 1831 and January 1832 in his '*Experimental researches into electricity*' in which he gave his "*Law which governs the evolution of electricity by magneto-electric induction*" – a change in magnetic flux through any surface bounded by closed lines causes an ε.m.f around the lines.

Within no time, this brilliant experimental scientist invented the dynamo, the generator and the transformer. We will expand on the principles of electro-magnetic induction in the remainder of this chapter.

Electromotive force (emf)

Faraday used an apparatus shown simply in Figure 1102. If a conductor is held stationary in a magnetic field of a U-shaped magnet that is connected in series with a very sensitive, zero-centred galvanometer, no reading is observed. However, if the conductor is moved across the magnetic field, then a deflection occurs in the needle of the galvanometer in one direction. After a very short period of time, the needle returns to zero on the scale.

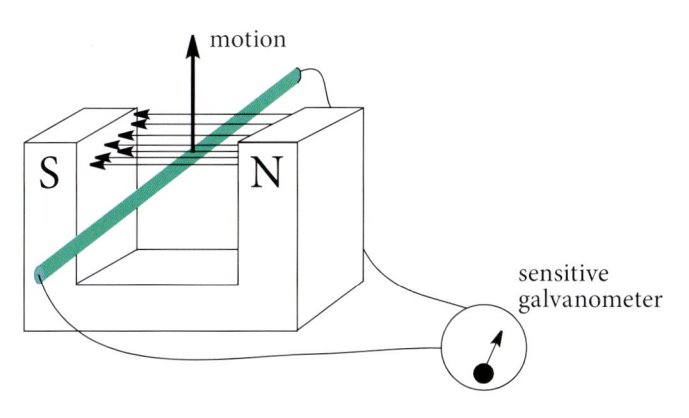

Figure 1102 Producing an induced current

The current produced is called an **induced current**.

As work is done in moving the current from one end of the conductor to the other, an electrical potential difference exists, and an **induced emf** is produced.

If the conductor is then moved in the opposite direction, the needle of the galvanometer deflects in the opposite direction before then falling to zero again. If the conductor is moved in the same direction as the magnetic field then no deflection occurs.

The direction of the induced current can be obtained by using the **left-hand palm rule** (refer to the palm rules discussed for the motor effect in Chapter 5). Using the magnetic field and direction of movement of the wire, if the palm of your left hand points in the direction of motion of the conductor, and your fingers point in the direction of the magnetic field, then the thumb gives the direction of the induced current. In this case, the current is in an anti-clockwise direction.

Alternatively, you can continue to use the **right-hand palm rule** BUT your palm points in the opposite direction

325

to the applied force. Fleming's right-hand rule can also be used. The right-hand palm rule for the direction of an induced current and Fleming's right-hand rule are shown in Figures 1103 (a) and (b) respectively.

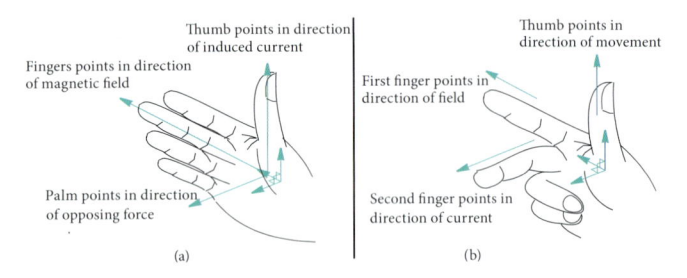

Figure 1103 (a) & (b) Palm rules for electromagnetic induction

The simple apparatus in Figure 1102 can detect the induced current, but the readings on the galvanometer are small (a zero-centred micro-ammeter is better). *Faraday* improved the apparatus by moving different magnetic flux densities into and out of different sized solenoids at different speeds. He found that the strength of the induced emf was dependent on

1. The speed of the movement
2. The strength of the magnetic flux density
3. The number of turns on the coil
4. The area of the coil

Faraday realised that the magnitude of the induced e.m.f was not proportional to the rate of change of the magnetic field **B** but rather proportional to the rate of change of magnetic flux Φ for a straight conductor or flux linkage NΦ. *This will be discussed further in the next section.*

Example

Determine the direction of the induced current for each situation given below.

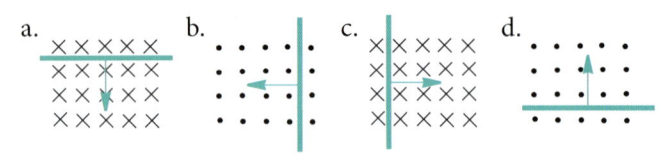

a. b. c. d.

Solution

Using a hand rule, the direction of the induced current for each situation is indicated by the arrows as shown in the diagram below:

a. b. c. d.

Derivation of an induced emf in a conductor

Consider a conductor of length l that moves with velocity v perpendicular to a magnetic flux density or induction **B** as shown in Figure 1104.

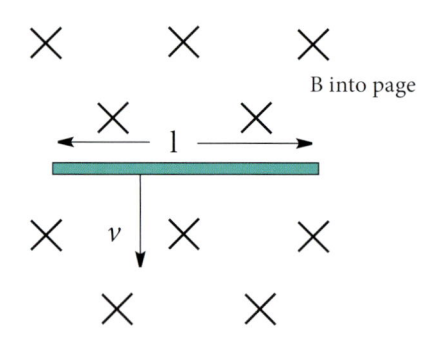

Figure 1104 Cause of an induced emf.

When the wire conductor moves in the magnetic field, the free electrons experience a force because they are caused to move with velocity v as the conductor moves in the field.

$$F = e \times v \times B$$

This force causes the electrons to drift from one end of the conductor to the other, and one end builds-up an excess of electrons and the other a deficiency of electrons. This means that there is a potential difference or emf between the ends. Eventually, the emf becomes large enough to balance the magnetic force and thus stop electrons from moving.

$$ev \times B = e \times E \Leftrightarrow E = B \times v$$

If the potential difference (emf) between the ends of the conductor is ε then

$$\varepsilon = E \times l$$

By substitution, we have,

$$\varepsilon = B \times l \times v$$

If the conducting wire was a tightly wound coil of N turns of wire the equation becomes:

$$\varepsilon = NB\, l\, v$$

Magnetic flux and magnetic flux linkage

Consider a small planar coil of a conductor for simplicity as shown in Figure 1105 (it could be any small shape). Now imagine it is cut by magnetic lines of flux. It would be reasonable to deduce that the number of lines per unit cross-sectional area is equal to the magnitude of the magnetic flux density B × the cross-sectional area A. This product is the **magnetic flux** Φ.

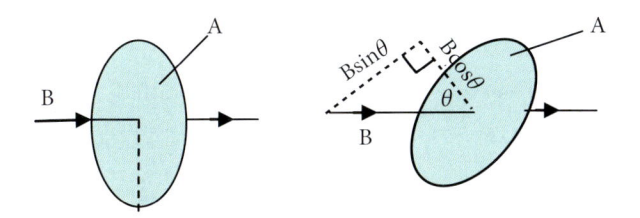

Figure 1105 (a) and (b)   Flux through a small, plane surface

The **magnetic flux** Φ through a small plane surface is the product of the flux density normal to the surface and the area of the surface.

$$\Phi = BA$$

The **unit of magnetic flux is the weber Wb**.

Rearranging this equation it can be seen that:

$B = \Phi / A$ which helps us understand why B can be called the flux density. So the unit for flux density can be the tesla T, or the weber per square metre Wbm^{-2}. So, 1T = 1 Wbm^{-2}.

If the normal shown by the dotted line in Figure 1105 (b) to the area makes an angle θ with B, the magnetic flux is given by:

$$\Phi = BA\cos\theta$$

where A is the area of the region and θ is the angle of movement between the magnetic field and a line drawn perpendicular to the area swept out. *(Be careful that you choose the correct vector component and angle because questions on past IB examinations give the correct answers of BA sin θ or BA cos θ depending on components supplied in the diagrams).*

If Φ is the flux density through a cross-sectional area of a conductor with N coils, the total flux density will be given by:

$$\Phi = NBA\cos\theta$$

This is called the **flux linkage**.

So it should now be obvious that we can increase the magnetic flux by:

- Increasing the conductor area
- Increasing the magnetic flux density B
- Keeping the flux density normal to the surface of the conductor

Faraday's law of induction

Now examine Figure 1106 that shows the shaded area of the magnetic flux density swept out in one second by a conductor of length l moving from the top to the bottom of the figure through a distance d.

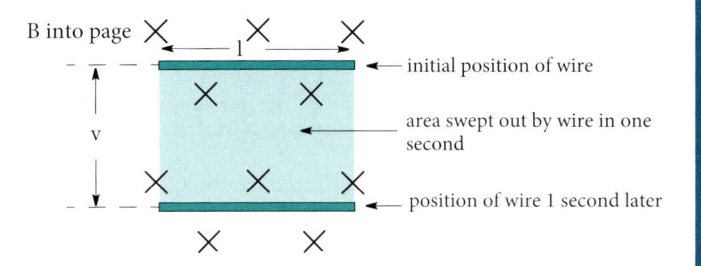

Figure 1106   Rate of area swept out.

We have already derived that $\varepsilon = Blv$

The area swept out in a given time is given by $(l \times d) / t$. But $v = d / t$. So that the area swept out = lv / t.

That is,

where A is the area in m^2.

For a single conductor in the magnetic flux density, it can be seen that:

$$\varepsilon \propto \frac{\Delta\Phi}{\Delta t} \Leftrightarrow \varepsilon = -\frac{\Delta\Phi}{\Delta t}$$

where the constant equals −1. The negative sign will be explained in the next section.

If there are N number of coils, then:

$$\varepsilon = -N \times \frac{\Delta\Phi}{\Delta t}$$

Faraday's Law can therefore be stated as:

The magnitude of the induced emf in a circuit is directly proportional to rate of change of magnetic flux or flux-linkage.

Lenz's Law

Figure 1107 shows some relative movements of a bar magnet at various positions relative to a coil with many turns.

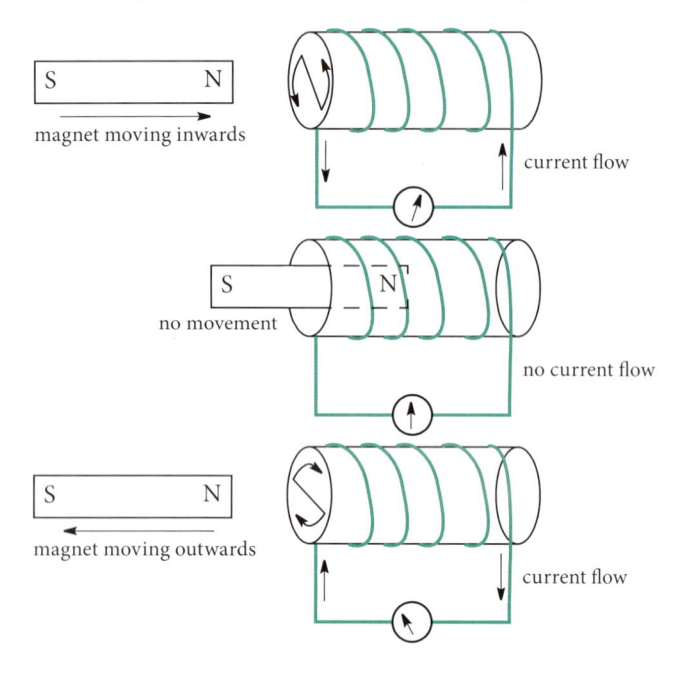

Figure 1107 Lenz's Law applied to a solenoid.

When the North Pole is moved toward the core of the solenoid, an induced current flows in the external circuit as indicated by a zero-centred galvanometer or a micro-ammeter. The pointer moves to the right meaning that the conventional induced current is flowing anti-clockwise at the end of the solenoid nearest the magnet. This end is acting like a north pole. When the magnet is stationary the meter reads zero. This suggests that the induced current is dependent on the speed of the movement. When the bar magnet is removed from the solenoid, the induced current flows in the opposite direction, and a south pole is created in the end that was previously a north pole.

In 1834, a Russian physicist *Heinrich Lenz* (1804-1865) applied the Law of Conservation of Energy to determine the direction of the induced emf for all types of conductors. It is known as the Second Law of Electromagnetic Induction and it can be stated as:

The direction of the induced emf is such that the current it causes to flow opposes the change producing it.

In the above case, the current induced in the coil creates a north pole to oppose the incoming north pole of the magnet. Similarly, when the magnet is withdrawn its north pole creates a south pole in the solenoid to oppose the change.

It can be reasoned that the *Law of Conservation of Energy* must apply. If the solenoid in Figure 1108 had an induced south pole when the north pole of the magnet was moved towards it, the magnet would accelerate as it would experience a force of attraction. More induced current would be produced creating more acceleration. The kinetic energy would increase indefinitely – energy would be created. As this is impossible, it makes sense that the induced current must oppose the change producing it.

Lenz's Law can be applied to straight conductors as well as solenoids. Figure 1108 shows the magnetic lines of force for a bar magnet and a current-carrying wire directed into the page before and during interaction. Suppose the conductor is carrying an induced current initially.

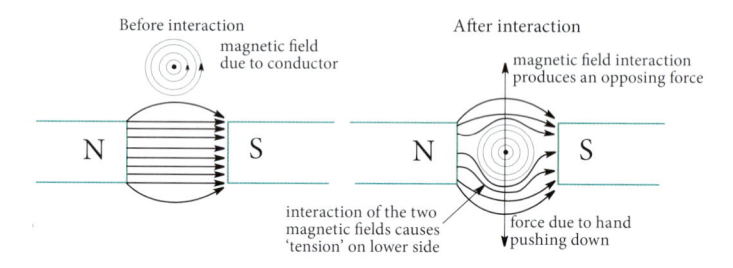

Figure 1108 Lenz's Law in a straight conductor.

The straight conductor is then pushed downwards say with your hand. Your energy source induces the current but the combined magnetic fields tend to push the conductor upwards (a force is applied in the direction from the region of most flux density to the region of least flux density). Therefore, the induced current will be in such a direction that tries to stop the conductor through the field.

If we now combine Faraday's and Lenz's Laws of electromagnetic induction into the equation, we can now understand the significance of the negative sign.

$$\varepsilon = -N \times \frac{\Delta \Phi}{\Delta t}$$

Example 1

A metal conductor 2.5 m long moves at right angles to a magnetic field of 4.0×10^{-3} T with a velocity of 35 m s^{-1}. Calculate the emf of the conductor.

Solution

Using the formula, $\varepsilon = \mathbf{B}\, l\, \mathbf{v}$, we have,

$\varepsilon = (4.0 \times 10^{-3} \text{ T}) \times (2.5 \text{ m}) \times (35 \text{ ms}^{-1})$

$= 0.35$ V

*The potential difference between the ends of the conductor is **0.35 V**.*

ELECTROMAGNETIC INDUCTION

Example 2

A square solenoid with 120 turns and sides of 5.0 cm is placed in air with each turn perpendicular to a uniform magnetic flux density of 0.60 T. Calculate the induced emf if the field decreases to zero in 3.0 s.

Solution

This time we need to use the formula, $\Phi = A\mathbf{B}$, so that

$$\Phi = (0.05 \text{ m})^2 \times (0.60 \text{ T})$$
$$= 1.5 \times 10^{-3} \text{ Wb}$$

Next, we make use of the formula,

$$\varepsilon = -N \times \frac{\Delta\Phi}{\Delta t} = -120 \times \frac{(0 - 1.5 \times 10^{-3} \text{ Wb})}{3.0 \text{ s}}$$
$$= 0.060 \text{ V}$$

The induced emf is **6.0×10^{-2} V**.

Example 3

(a) A coil with 20 turns has an area of 2.0×10^{-1} m². It is placed in a uniform magnetic field of flux density 1.0×10^{-1} T so that the flux links the turns normally. Calculate the average induced emf in the coil if it is removed from the field in 0.75 s.

(b) The same coil is turned from its normal position through an angle of 30° in 0.3 s in the field. Calculate the average induced emf.

Solution

(a) This time we need to use the formula, $\Phi = N\mathbf{B}A$, so that

$\Phi = 20 \text{ turns} \times 2.0 \times 10^{-1} \text{ m}^2 \times 1.0 \times 10^{-1} \text{ T} = 4.0 \times 10^{-1} \text{ Wb}$

Next, we make use of the formula,

$$emf = -\frac{\Delta\Phi}{\Delta t} = \frac{4.0 \times 10^{-1} \text{ Wb}}{0.75 \text{ s}}$$
$$= 0.533 \text{ V}$$

The induced emf is **0.53 V**.

(b) The flux change through the coil $= N\mathbf{B}A - N\mathbf{B}A\cos\theta$

$= 4.0 \times 10^{-1} \text{ Wb} - 4.0 \times 10^{-1} \text{ Wb} \cos 30 = 0.054 \text{ Wb}$

Average induced emf $= \frac{0.54 \text{ Wb}}{0.3 \text{ s}} = 0.179 \text{ V}$

The induced emf is **0.18 V**.

Exercise 11.1

1. Consider a coil of length l, cross-sectional area A, number of turns n, in which a current I is flowing. The magnetic flux density of the coil depends on

   A. I, l, n but not A
   B. I, n, A but not l
   C. I, A, l but not n
   D. A, l, n but not I

2. The magnetic flux Φ through a coil having 200 turns varies with time t as shown below.

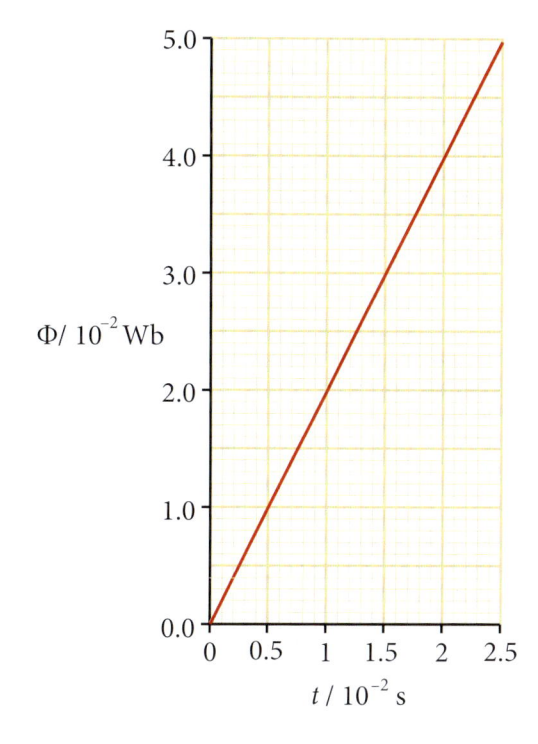

   The magnitude of the emf induced in the coil is:

   A. 0.5 V
   B. 2 V
   C. 100 V
   D. 400 V

3. A metal ring falls over a bar magnet as shown

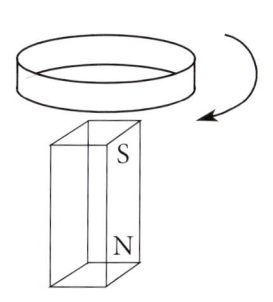

329

CHAPTER 11

The induced current is directed

- A. always opposite to the direction of the arrow.
- B. always in the same direction as the arrow.
- C. first in the opposite direction to the arrow, then as shown by the arrow.
- D. first as shown by the arrow, then in the opposite direction to the arrow.

4. The magnitude of an induced emf produced by the relative motion between a solenoid and a magnetic field is dependent upon:

- A. the strength of the magnetic flux density
- B. the number of turns on the coil
- C. the area of the coil
- D. all of the above

5. Which of the following is a suitable unit to measure magnetic flux density?

- A. $A\, m\, N^{-1}$
- B. $Kg\, A^{-1} s^{-2}$
- C. $A\, N^{-1} m^{-1}$
- D. $T\, m^{-1}$

6. Faraday's law of electromagnetic induction states that the induced emf is

- A. equal to the change in magnetic flux
- B. equal to the change in magnetic flux linkage
- C. proportional to the change in magnetic flux linkage
- D. proportional to the rate of change of magnetic flux linkage

7. A uniform magnetic field of strength B completely links a coil of area A. The field makes an angle θ to the plane of the coil.

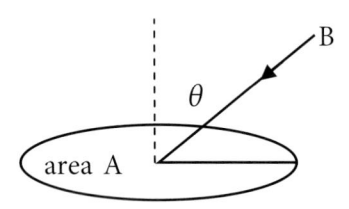

The magnetic flux linking the coil is

- A. $BA\cos\theta$
- B. BA
- C. $BA\sin\theta$
- D. $BA\tan\theta$

8. What factors determine the magnitude of an induced emf?

9. Refer to Figure 1204. Use Lenz's Law to explain what would happen if the solenoid was moved rather than the magnet.

10. What effect would the following have on the magnitude of the induced emf in a conductor moving perpendicular to a magnetic field?

    (a) Doubling the velocity of movement of the conductor.
    (b) Halving the magnetic flux density and velocity.
    (c) Changing the conductor from copper to iron.

11. Explain in detail the difference between magnetic flux density and magnetic flux.

12. The magnetic flux through a coil of wire containing 5 loops changes from -25 Wb to $+15$ Wb in 0.12 s. What is the induced emf in the coil?

13. The wing of a Jumbo jet is 9.8 m long. It is flying at 840 km h^{-1}. If it is flying in a region where the earth's magnetic field has a vertical component of 7.2×10^{-4} T, what potential difference could be produced across the wing?

14. Find the total flux through an area of 0.04 m^2 perpendicular to a uniform magnetic flux density of 1.25 T.

15. If the total flux threading an area of 25 cm^2 is 1.74×10^{-2} Wb, what would be the magnetic flux density?

16. A coil of area 5 cm^2 is in a uniform magnetic field of flux density 0.2 T. Determine the magnetic flux in the coil when:

    (a) The coil is normal to the magnetic field
    (b) The coil is parallel to the magnetic field
    (c) The normal to the coil and the field have an angle of 60°

17. A metal conductor 2.5 m long moves at right angles to a magnetic field of 4.0×10^{-3} T with a velocity of 35 m s^{-1}. Calculate the emf of the conductor.

18. A square solenoid with 120 turns and sides of 5.0 cm is placed in air with each turn perpendicular to a uniform magnetic flux density of 0.60 T. Calculate the induced emf if the field decreases to zero in 3.0 s.

19. A coil with 1500 turns and a mean area of 45 cm^2 is placed in air with each turn perpendicular to a uniform magnetic field of 0.65 T. Calculate the induced emf if the field decreases to zero in 5.0 s.

20. The radius of the copper ring is 0.15 m and its resistance is 2.0×10^{-2} Ω. A magnetic field strength is increasing at rate of 1.8×10^{-3} T s^{-1}. Calculate the value of the induced current in the copper ring.

ELECTROMAGNETIC INDUCTION

11.2 Power generation and transmission

NATURE OF SCIENCE:

Bias: In the late 19th century Edison was a proponent of direct current electrical energy transmission while Westinghouse and Tesla favoured alternating current transmission. The so called "battle of currents" had a significant impact on today's society. (3.5)

© IBO 2014

Essential idea: Generation and transmission of alternating current (ac) electricity has transformed the world.

Understandings:
- Alternating current (ac) generators
- Average power and root mean square (rms) values of current and voltage
- Transformers
- Diode bridges
- Half-wave and full-wave rectification

Alternating current (ac) generators

The induced emf in a coil rotated within a uniform magnetic field is sinusoidal if the rotation is at constant speed.

The most important practical application of the Laws of Electromagnetic Induction was the development of the electric generator or dynamo.

The frequency of the generator cycle used in power stations can be investigated with the use of a cathode ray oscilloscope (CRO). A CRO can be used to measure the voltage output of an ac source. If a low safe ac voltage (9V) from a power pack is connected to the CRO then its source is the power stations of your community supplier. If the time-base is adjusted to obtain a sine curve trace on the screen, the mains supply frequency can be determined.

Example

In Figure 1109, if the potentiometer is set on 2 V/division and the time base is set a 5 ms/cm, what is the voltage and frequency of the ac generator?

Solution

The amplitude of the wave is 3 divisions and each division is 2 V. Therefore, the emf would be **6 V**.

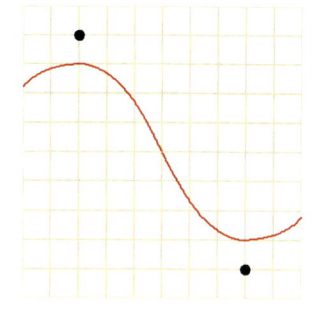

Figure 1109   CRO trace

Between the two dots there are 6 divisions. Therefore, the wavelength is equivalent to 12 divisions. Now there are 5 milliseconds/ cm. So the period of the wave is 60 ms, that is, T = 60 ms.

Therefore, the frequency of the source is given by,

$$f = \frac{1}{T} = \frac{1}{60 \times 10^{-3}} = 16.67 \text{ Hz}$$

The frequency of the source is **17 Hz**.

A **generator** is essentially a device for producing electrical energy from mechanical energy. (*Remember that an electric motor did the opposite energy conversion*). Generators use mechanical rotational energy to provide the force to turn a coil of wire, called an armature, in a magnetic field (the magnet can also be turned while the coil remains stationary). As the armature cuts the magnetic flux density, emfs are induced in the coil. As the sides of the coil reverse direction every half turn, the emf alternates in polarity. If there is a complete circuit, alternating current ac is produced.

The induced currents are conducted in and out by way of the slip-rings and the carbon brushes.

Turbines driven by steam or water are the commonest devices used for the generation of electricity in modern day society.

Figure 1110 shows a simple generator in which a coil is rotating clockwise. The circuit is completed with a lamp acting as the load.

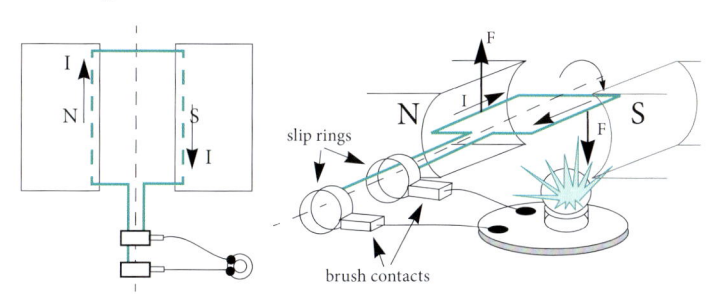

Figure 1110   AC generator

331

Chapter 11

To determine the direction of the induced current produced as the coil rotates, we must apply Lenz's Law. As the left hand side of the coil (nearest the north-pole of the magnet) moves upward, a downward magnetic force must be exerted to oppose the rotation. By applying the right-hand palm rule for electromagnetic induction, you can determine the direction of the induced current on that side of the coil. The direction of the current in the right-side of the coil can also be determined.

The magnitude of the emf and current varies with time as shown in Figure 1111. Consider a coil ABCD rotating clockwise initially in the horizontal position. From the graph of current versus time, you can see that the current reaches a maximum when the coil is horizontal and a minimum when the coil is vertical. If more lines of magnetic flux are being cut, then the induced current will be greater. This occurs to the greatest extent when the coil is moving at right angles to the magnetic field. When the coil moves parallel to the field, no current flows.

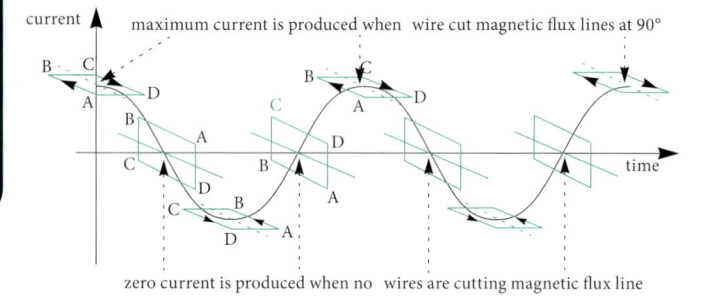

Figure 1111   The changes of current with time

Each complete cycle of the sinusoidal graph corresponds to one complete revolution of the generator.

The emf of a rotating coil can be calculated at a given time. If a coil of N turns has an area A, and its normal makes an angle θ with the magnetic field **B**, then the flux-linkage Φ is given by:

$$\Phi = N \times A \times B \cos\theta$$

The emf varies sinusoidally (sin and cos graphs have the same shape) with time and can be calculated using

$$\varepsilon = -\frac{\Delta\Phi}{\Delta t} = -N \times A \times B \times \frac{\Delta\cos\theta}{\Delta t}$$

Using calculus and differentiating $\cos\theta$, this relationship becomes

$$\varepsilon = N \times A \times B \times \sin\theta \times \frac{\Delta\theta}{\Delta t}$$

Remember from your knowledge of rotational motion that

$\Delta\theta \div \Delta t$ = the angular velocity in rad s^{-1} = $2\pi f$

Also

$$\theta = \omega t = 2\pi f t$$

so that

$$\varepsilon = \omega \times N \times A \times B \times \sin(\omega t)$$

So that,

$$\varepsilon = 2\pi f \times N \times A \times B \times \sin(2\pi f t)$$

When the plane of the coil is parallel to the magnetic field, $\sin\omega t$ will have its maximum value as $\omega t = 90°$, so $\sin\omega t = 1$. This maximum value for the emf ε_0 is called the peak voltage, and is given by:

$$\varepsilon_0 = \omega NBA$$

Therefore:

$$\varepsilon = \varepsilon_0 \sin\omega t$$

The frequency of rotation in North America is 60 Hz but the main frequency used by many other countries is 50 Hz.

Note that if the speed of the coil is doubled then the frequency and the magnitude of the emf will both increase as shown in Figures 1112 and 1113 respectively.

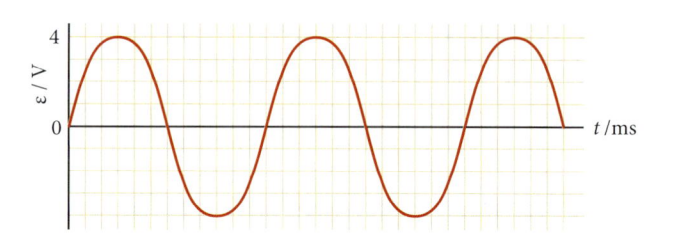

Figure 1112   Normal frequency

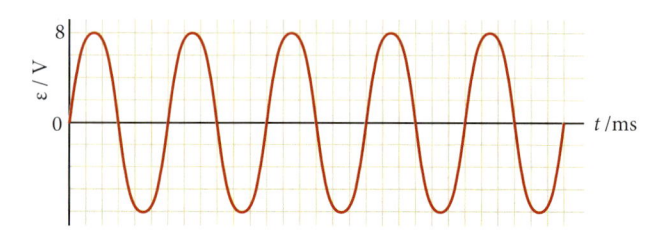

Figure 1113   Doubled frequency

Example 1

Calculate the peak voltage of a simple generator if the square armature has sides of 5.40 cm and it contains 120 loops. It rotates in a magnetic field of 0.80 T at the rate of 110 revolutions per second.

Electromagnetic Induction

Solution

Making use of the formula above, that is,

$$\varepsilon = \omega \, N \, A \, B \, \sin(\omega t)$$
$$= 2\pi f \, N \, A \, B \, \sin(2\pi f t)$$

We can see that the maximum emf will occur when $\sin \omega t = 1$, so that, $\varepsilon_{max} = \omega NAB$

But, $\omega = 2\pi f$, so that

$\varepsilon_0 = (2\pi) \times (110.0 \text{ Hz}) \times (120 \text{ turns}) \times (5.4 \times 10^{-4} \text{ m}^2) \times (0.80 \text{ T})$

$= 35.8 \text{ V}$

That is, the output voltage is **36 V**.

Example 2

Suppose a coil with 1200 turns has an area of 2.0×10^{-2} m^2 and is rotating at 50 revolutions per second in a magnetic field of magnitude 0.50 T. Draw graphs to show how the magnetic flux, the emf and the current change as a function of time. (Assume the current flows in a circuit with a resistance of 25 Ω).

Solution

The magnetic flux in the coil changes over time as shown in Figure 1114.

$\Phi = NBA = 1200 \text{ turns} \times 0.5 \text{ T} \times 2 \times 10^{-2} \text{ m}^2 = 12 \text{ Wb}$

50 revolutions per second would have 1 revolution in 0.02 seconds = 20 ms.

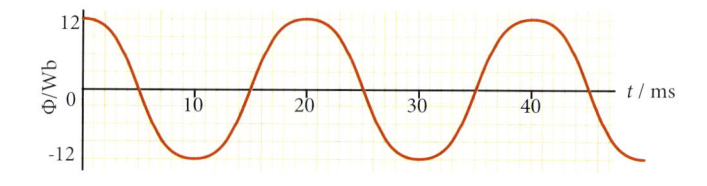

Figure 1114   Changing flux linkage over time

We can see that the maximum emf will occur when $\sin \omega t = 1$, so that,

$\varepsilon_{max} = \omega NAB$

But, $\omega = 2\pi f$, so that

$\varepsilon_0 = (2\pi) \times (50 \text{ Hz}) \times (1200 \text{ turns}) \times (2 \times 10^{-2} \text{ m}^2) \times (0.50 \text{ T})$

$= 75.4 \text{ V} = 75 \text{ V}$

The appropriate graph is shown in Figure 1115.

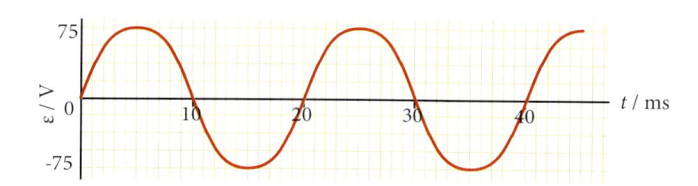

Figure 1115   Induced emf over time

The current flows in a circuit with a resistance of 25 Ω.

$I = \dfrac{\varepsilon}{R} = \dfrac{\varepsilon_0 \sin \omega t}{R} = I_0 \sin \omega t$

where $I_0 = \dfrac{75.4 \text{ V}}{25 \text{ Ω}} = 3.0 \text{ A}$

The appropriate graph is shown in Figure 1116.

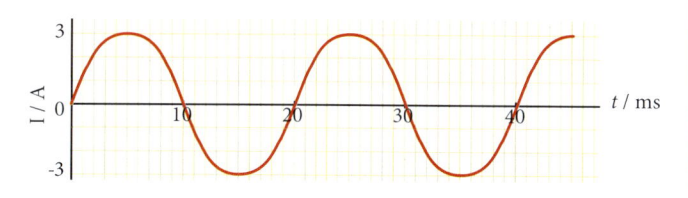

Figure 1116   Induced current over time

Average power and root mean square (rms) values of current and voltage

An alternating current varies sinusoidally and can be represented by the equation:

$I = I_0 \sin(\omega t)$

where I_0 is the maximum current called the **peak current** as shown in Figure 1117 for a 50 Hz mains supply.

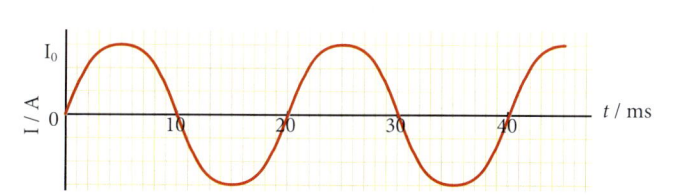

Figure 1117   Peak current and current over time

In commercial practice, alternating currents are expressed in terms of their **root-mean-square (r.m.s.) value.**

Consider 2 identical resistors each of resistance R, one carrying d.c. and the other a.c. in an external circuit. Suppose they are both dissipating the same power as thermal energy.

333

CHAPTER 11

The r.m.s. value of the alternating current that produces the power is equal to the d.c. value of the direct current.

For the maximum value in a.c., the power dissipated is given by

$$P = V_0 \sin(\omega t) \times I_0 \sin(\omega t) = V_0 I_0 \sin^2(\omega t)$$

This means that the power supplied to the resistor in time by an alternating current is equal to the average value of $I^2 R$ multiplied by time.

$$P = I^2_{ave} \times R = I_0^2 R \sin^2 \omega t$$

Because the current is squared the, the value for the power dissipated is always positive as shown in Figure 1118.

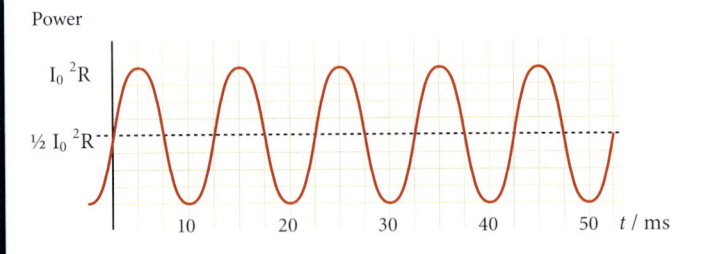

Figure 1118 Power delivered to a resistor in an alternating current circuit.

The value of $\sin^2 \omega t$ will therefore vary between 0 and 1.

Therefore its average value $= \dfrac{(0+1)}{2} = ½$.

Therefore the average power that dissipates in the resistor equals:

$$P_{ave} = ½ \, I_0^2 \, R = \dfrac{I_0}{\sqrt{2}} \times \dfrac{I_0}{\sqrt{2}} \times R$$

or

$$P_{ave} = ½ \, v_0^2 / R = \dfrac{\dfrac{V_0}{\sqrt{2}} \times \dfrac{V_0}{\sqrt{2}}}{R}.$$

So the current dissipated in a resistor in an a.c. circuit that varies between I_0 and $-I_0$ would be equal to a current $\dfrac{I_0}{\sqrt{2}}$ dissipated in a d.c circuit. This d.c current is known as **r.m.s. equivalent current** to the alternating current.

It can be shown that

$$I_{r.m.s.} = \dfrac{I_0}{\sqrt{2}} \text{ and } V_{r.m.s.} = \dfrac{V_0}{\sqrt{2}}$$

Provided a circuit with alternating current only contains resistance components, it can be treated like a direct current circuit.

Example 1

In the USA, the r.m.s value of the "standard line voltage" is 110 V and in some parts of Europe, it is 230 V. Calculate the peak voltage for each region.

Solution

$V_0 = \sqrt{2} \, V_{rms}$

In the USA = $1.414 \times 110 \text{ V} = 170 \text{ V}$

In Europe = $1.414 \times 230 \text{ V} = 325 \text{ V}.$

Example 2

The domestic standard line voltage in Australia is 240 V. Calculate the current and resistance in a 1200 W electric jug and compare these values with the same electric jug used in the USA.

Solution

$$I_{rms} = \dfrac{P_{ave}}{V_{rms}} \text{ and } R = \dfrac{V_{rms}}{I_{rms}}$$

In Australia:

$$I = \dfrac{1200 \text{ W}}{240 \text{ V}} = 5.0 \text{ A and } R = \dfrac{240 \text{ V}}{5.0 \text{ A}} = 48 \, \Omega$$

In the USA:

$$I = \dfrac{1200 \text{ W}}{110 \text{ V}} = 11 \text{ A and } R = \dfrac{110 \text{ V}}{10.9 \text{ A}} = 10 \, \Omega$$

The current would have a greater heating effect in the USA than Australia but the element in the electric jug would need to be made of a conductor with a a greater cross-sectional area.

Transformers

A useful device that makes use of electromagnetic induction is the **ac transformer** as it can be used for increasing or decreasing ac voltages and currents.

It consists of two coils of wire known as the primary and secondary coils. Each coil has a laminated (thin sheets fastened together) soft iron core to reduce eddy currents (currents that reduce the efficiency of transformers). The coils are then enclosed with top and bottom soft iron bars that increase the strength of the magnetic field. Figure 1119 shows a typical circuit for a simple transformer together with the recommended circuit symbol.

ELECTROMAGNETIC INDUCTION

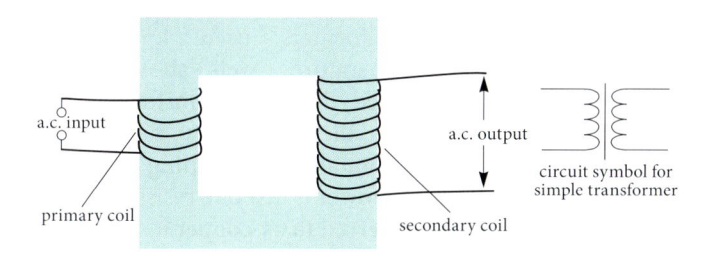

Figure 1119   A simple transformer

When an ac voltage is applied to the primary coil, an ac voltage of the same frequency is induced in the secondary coil. This frequency in most countries is 50 Hz.

When a current flows in the primary coil, a magnetic field is produced around the coil. It grows quickly and cuts the secondary coil to induce a current and thus to induce a magnetic field also. When the current falls in the primary coil due to the alternating current, the magnetic field collapses in the primary coil and cuts the secondary coil producing an induced current in the opposite direction.

The size of the voltage input/output depends on the number of turns on each coil. It is found that

$$\frac{V_p}{V_s} = \frac{N_p}{N_s} = \frac{I_s}{I_p}$$

Where N = the number of turns on a designated coil and I is the current in each coil.

It can be seen that if **Ns is greater than Np** then the transformer is a **step-up transformer**. If the reverse occurs and **Ns is less than Np** it will be a **step-down transformer**.

If a transformer was 100% efficient, the power produced in the secondary coil should equal to the power input of the primary coil. In practice the efficiency is closer to 98% because of eddy currents and heating in the core.

$$V_p I_p = V_s I_s \Leftrightarrow \frac{V_p}{V_s} = \frac{I_s}{I_p}$$

This means that if the voltage is stepped-up by a certain ratio, the current in the secondary coil is stepped-down by the same ratio.

Example

The figure below shows a step-down transformer that is used to light a filament globe of resistance 4.0 Ω under operating conditions.

Calculate

(a) the reading on the voltmeter with S open

(b) the current in the secondary coil with an effective resistance of 0.2 Ω with S closed

(c) the power dissipated in the lamp

(d) the power taken from the supply if the primary current is 150 mA

(e) the efficiency of the transformer.

Solution

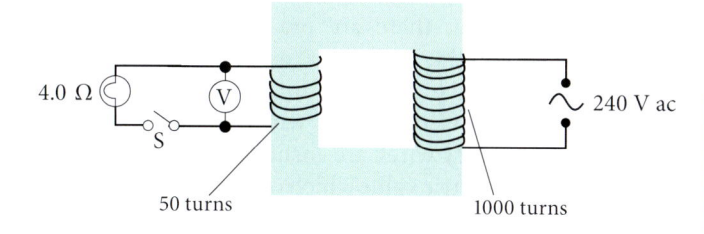

(a) Using the formula, $V_p / V_s = N_p / N_s$, with V_p = 240 V, N_p = 1000 turns and N_s = 50 turns we have,

$$\frac{240}{V_s} = \frac{1000}{50} \Rightarrow V_s = 240 \times \frac{50}{1000} = 12$$

That is,

$$V_s = 12 \text{ V}$$

(b) Total resistance = 0.2 Ω + 4 Ω = 4.2 Ω.

From the formula, $I = V / R$, we have

$$I = \frac{12 \text{ V}}{4.2 \text{ Ω}} = 2.86 \text{ A}$$

(c) From the formula, $P = VI = (IR) \times I = I^2 R$, we have that

$$P = 2.86^2 \times 4.0 = 32.7 \text{ W}$$

(d) Using, $P = VI$, we have,

$$P = 240 \times 150 \times 10^{-3} = 36 \text{ W}$$

(e) Efficiency

$$= \frac{32.7 + 1.6}{36} \times 100\% = 95\%$$

335

Chapter 11

There are a number of reasons for power losses in transmission lines such as:

- Heating effect of a current
- Resistance of the metal used
- Dielectric losses
- Self-inductance

The main heat loss is due to the heating effect of a current. By keeping the current as low as possible, the heating effect can be reduced. The resistance in a wire due to the flow of electrons over long distances also has a heating effect. If the thickness of the copper wire used in the core of the transmission line is increased, then the resistance can be decreased. However, there are practical considerations such as weight and the mechanical and tensional strength that have to be taken into account. The copper wire is usually braided (lots of copper wires wound together) and these individual wires are insulated. The insulation material has a dielectric value which can cause some power loss. Finally, the changing electric and magnetic fields of the electrons can encircle other electrons and retard their movement on the outer surface of the wire through **self-inductance**. This is known as the 'skin effect'. The size of the power loss depends on the magnitude of the transmission voltage, and power losses of the order of magnitude of 10^5 watts per kilometre are common.

Power losses in real transformers are due to factors such as:

- Eddy currents
- Resistance of the wire used for the windings
- Hysteresis
- Flux leakage
- Physical vibration and noise of the core and windings
- Electromagnetic radiation
- Dielectric loss in materials used to insulate the core and windings.

As already mentioned, any conductor that moves in a magnetic field has emf induced in it, and as such current, called eddy currents, will also be induced in the conductor. This current has a heating effect in the soft iron core of the transformer which causes a power loss termed an iron loss. There is also a magnetic effect in that the created magnetic fields will oppose the flux change that produces them according to *Lenz's Law*. This means that eddy currents will move in the opposite direction to the induced current causing a braking effect. Eddy currents are considerably reduced by alloying the iron with 3% silicon that increases the resistivity of the core. To reduce the heating effect due to eddy currents, the soft-iron core is made of sheets of iron called laminations that are insulated from each other by an oxide layer on each lamination. This insulation prevents currents from moving from one lamination to the next.

Copper wire is used as the windings on the soft-iron core because of its low resistivity and good electrical conductivity. Real transformers used for power transmission reach temperatures well above room temperature and are cooled down by transformer oil. This oil circulates through the transformer and serves not only as a cooling fluid but also as a cleaning and anti-corrosive agent. However, power is lost due resistance and temperature commonly referred to as 'copper loss'.

Hysteresis is derived from the Greek word that means "lagging behind" and it becomes an important factor in the changes in flux density as a magnetic field changes in ferromagnetic materials. Transformer coils are subject to many changes in flux density. As the magnetic field strength increases in the positive direction, the flux density increases. If the field strength is reduced to zero, the iron remains strongly magnetised due to the retained flux density. When the magnetic field is reversed the flux density is reduced to zero. So in one cycle the magnetisation lags behind the magnetising field and we have another *iron loss* that produces heat. Hysteresis is reduced again by using silicon iron cores.

The capacity for the primary coil to carry current is limited by the insulation and air gaps between the turnings of the copper wire and this leads to flux leakage. This can be up to 50% of the total space in some cases. Because the power is being delivered to the transformer at 50Hz, you can often hear them making a humming noise. Minimal energy is lost in the physical vibration and noise of the core and windings.

Modern transformers are up to 99% efficient.

Long-distance AC alternating current transmission is affected by a transmission line's reactive power that is actually 90 degrees out of phase with the flow of real current to a load at the other end of the line. For short transmission lines, the effect is not as significant. Direct current transmission does not have reactive power once the voltage has been raised to the normal level. The power losses are considerably less than alternating current.

For economic reasons, there is no ideal value of voltage for electrical transmission. Electric power is generated at approximately 11 000 V and then it is stepped-up to the highest possible voltage for transmission. Alternating current transmission of up to 765 kV are quite common. For voltages higher than this, direct current transmission at up to 880 kV is used. Ac can be converted to dc using rectifiers and this is what is done in electric train and tram operations. Dc can be converted to ac using inverters. There are a number of dc transmission lines such as the underground cross-channel link between the UK and France. The New Zealand high-voltage direct current scheme has around 610 km of overhead and submarine transmission lines.

There are 3 conductors on a transmission line to maximize the amount of power that can be generated. Each high voltage circuit has three phases. The generators at the power station supplying the power system have their coils connected through terminals at 120° to each other.

When each generator at the power station rotates through a full rotation, the voltages and the currents rise and fall in each terminal in a synchronized manner.

Once the voltage has been stepped-up, it is transmitted into a national supergrid system from a range of power stations. As it nears a city or town it is stepped-down into a smaller grid. As it approaches heavy industry, it is stepped down to around 33 – 132 kV in the UK, and when it arrives at light industry it is stepped-down to 11-33 kV. Finally, cities and farms use a range of values down to 240V from a range of power stations.

When the current flows in the cables, some energy is lost to the surroundings as heat. Even good conductors such as copper still have a substantial resistance because of the significant length of wire needed for the distribution of power via the transmission cables. To minimise energy losses the current must be kept low.

Example

An average of 120 kW of power is delivered to a suburb from a power plant 10 km away. The transmission lines have a total resistance of 0.40 Ω. Calculate the power loss if the transmission voltage is

(a) 240 V
(b) 24 000 V

Solution

(a) From the formula, $P = VI$, we have that

$$I = \frac{P}{V} = \frac{1.2 \times 10^5 \text{ W}}{240 \text{ V}} = 500 \text{ A}$$

So that

$$P = I^2 \times R = (500 \text{ A})^2 \times (0.40 \text{ Ω}) = 100 \text{ kW}$$

(b) Again, using $P = VI$, we have,

$$I = \frac{P}{V} = \frac{1.2 \times 10^5 \text{ W}}{2.4 \times 10^4 \text{ V}} = 5.0 \text{ A}$$

So that,

$$P = I^2 \times R = (5.0 \text{ A})^2 \times (0.40 \text{ Ω}) = 10 \text{ W}$$

Exercise 11.2

1. Which of the following could correctly describe a step-up transformer?

	power supply	Core	primary coil	secondary coil
A.	dc	Steel	10 turns	100 turns
B.	dc	Iron	100 turns	10 turns
C.	ac	Steel	10 turns	100 turns
D.	ac	Iron	100 turns	10 turns

2. An alternating current with a root-mean-square value of 2 A is compared with the direct current I flowing through a given resistor. If both currents generate heat at the same rate, the value of I would be

A. 4 A
B. $2\sqrt{2}$ A
C. 2 A
D. $\sqrt{2}$ A

3. An ideal transformer has a primary coil of 5000 turns and a secondary coil of 250 turns. The primary voltage produced is 240 V. If a 24 W lamp connected to the secondary coil operates at this power rating, the current in the primary coil is:

A. 0.05 A
B. 0.1 A
C. 12 A
D. 20 A

4. If an alternating e.m.f has a peak value of 12 V then the root-mean-square value of this alternating e.m.f would be:

A. 0 V
B. $\sqrt{6}$ V
C. 3.5 V
D. $6\sqrt{2}$ V

5. The figure below below shows the variation with time t of the emf ε generated in a coil rotating in a uniform magnetic field.

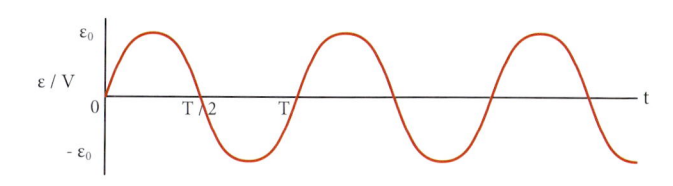

CHAPTER 11

What is the root-mean-square value ε_{rms} of the emf and also the frequency f of rotation of the coil?

	ε_{rms}	f
A.	ε_0	$\dfrac{2}{T}$
B.	ε_0	$\dfrac{1}{T}$
C.	$\varepsilon_0/\sqrt{2}$	$\dfrac{2}{T}$
D.	$\varepsilon_0/\sqrt{2}$	$\dfrac{1}{T}$

6. A load resistor is connected in series with an alternating current supply of negligible internal resistance. If the peak value of the supply voltage is V_0 and the peak value of the current in the resistor is I_0, the average power dissipation in the resistor would be:

    A. $\dfrac{V_0 I_0}{2}$

    B. $\dfrac{V_0^2 I_0}{\sqrt{2}}$

    C. $2 V_0 I_0$

    D. $V_0 I_0$

7. Explain why a soft-iron core is used in the construction of a transformer.

8. Explain why a transformer will not work with direct current.

9. If there are 1200 turns in the primary coil of an ac transformer with a primary voltage of 240 V, calculate the secondary voltage if the secondary coil has

    (a) 300 turns
    (b) 900 turns
    (c) 1800 turns

10. The armature of a 30 Hz a.c. generator contains 120 loops. The area of each loop is 2.0×10^{-2} m^2. It produces a peak output voltage of 120 V when it rotates in a magnetic field. Calculate the strength of the magnetic field.

11. Calculate the r.m.s value of the following currents:

    2A, 4A, 6A, 3A, -5A, 1A, 6A, 8A, -9A and 10A.

12. Calculate the peak current in a 2.4×10^3 Ω resistor connected to a 230 V domestic a.c source.

13. This question deals with the production and transmission of electric power, electricity costs and efficiency, and fuse systems.

    (a) Is it feasible to transmit power from a power station over long distances using direct current rather than alternating current? Justify your answer.

    (b) An aluminum transmission cable has a resistance of 5.0 Ω when 10 kW of power is transmitted in the cable. Justify why it is better to transmit the power at 100 000V rather than 1000 V by comparing the power that would be wasted in the transmission at both of these voltages.

    (c) Many step-up and step-down transformers are used in the electricity transmission from the power station to the home. In order to increase the efficiency of the transformers, eddy currents have to minimised. Describe how this achieved in the transformer design.

    (d) If the fuse controlling the maximum power for lighting in your house is rated at 8 A, calculate the maximum number of 60 W light bulbs that can be operated in parallel with a 110 V power supply so as not to blow the fuse?

    (e) A stainless steel calorimeter with a mass of 720 g was used to heat 2.5 kg of water. If the current / voltage in a heating element supplying the power was 30.2 A / 110V, and it took 2.5 minutes to heat the water from 25 °C to 98 °C, determine the specific heat capacity of the steel. (Assume no heat loss to the surroundings. The specific heat capacity of water is 4.18×10^3 J kg^{-1}K^{-1}).

    (f) How much does it cost to run the following appliances at the same time for one hour if electricity costs 10.5 cent per kilowatt-hour: a 6 kW oven, two 300 W colour televisions and five 100 W light globes?

14. Name four factors that affect the magnitude of an induced emf in a generator.

15. If there are 4 laminations in a transformer core, what fraction of the flux is in each lamination and what fraction of the power is dissipated in each?

Diode bridges

A semiconductor is a material with conductivity somewhere between a conductor and an insulator. They are usually made from the Group 4 elements silicon or germanium doped with impurity atoms as already discussed in chapter 8. They are made of p-type and n-type silicon or germanium (4 valence electrons). Doping with Group 5 element (5 valence electrons) such as arsenic (As) produces an electron rich layer / n-type semiconductor. This electron is free to move about. Doping with Group 3 (3 valence electrons) element such as gallium (Ga) produces an electron deficient layer / p-type semiconductor. There are not enough electrons to form a covalent bond with a neighbouring atom. An electron from the n-type semiconductor can move into the hole. The electrons can move from hole to hole and produce a potential difference. The wafers are about 1 mm thick and they are placed on top of each other to form a layer. Representations of p-type and p-type semiconductors are shown in Figure 1120.

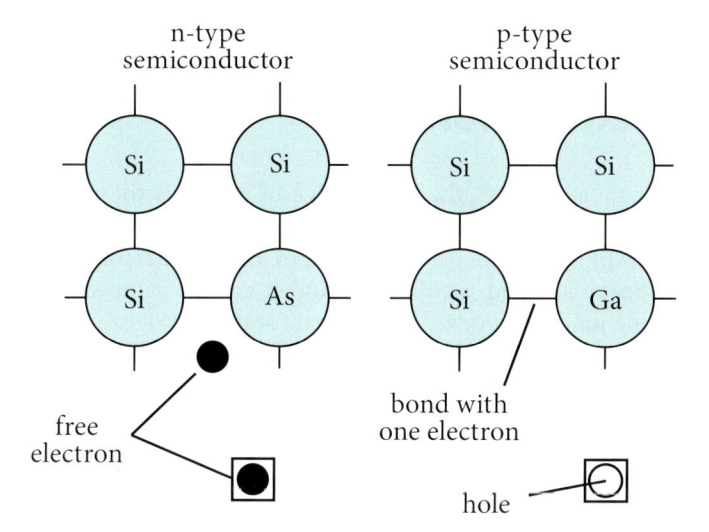

Figure 1120 Types of semiconductors

Semiconductors whether pure or doped, p-type or n-type, allow current to flow equally in both directions. However, when a piece of p-type semiconductor is placed next to a region of n-type, it has unidirectional properties. That is, current flows easily in one direction but none or almost none flows in the opposite direction. This type of semiconductor device is referred to as a P-N junction diode. Therefore, a **diode is a semiconductor device** that will allow current to go in one direction only. They can be used to rectify alternating current into direct current.

When a p-type and n-type semiconductors are in contact, some of the electrons in the n-type semiconductor diffuse into the p-type semiconductor eliminating some of the holes near the boundary between them. Similarly, holes from the p-type region diffuse into the n-type region and recombine with electrons near the boundary. Therefore, there is a positive charge build up at the n-type junction and cause a negative charge in the area to which they move near the p-type junction as shown in Figure 1121.

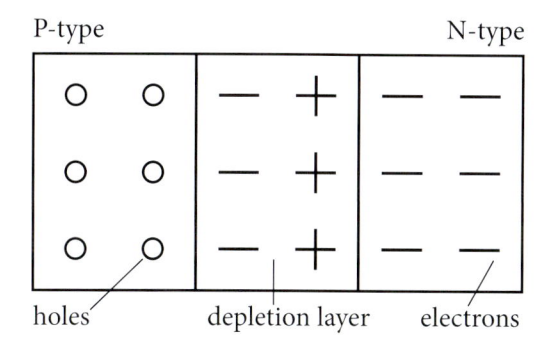

Figure 1121 p-n junction with a depletion region

Eventually, the build up of charge stops further movement of electrons resulting in the formation of a region around the junction between the semiconductors called the "depletion layer, and this results in an internal electric field opposing further diffusion.

Suppose an emf is connected to a p-n junction as shown in Figure 1122. The depletion layer will become wider because the potential difference applied across the p-n junction creates an electric field that adds to that in the depletion layer increasing the resistance of the p-n junction.

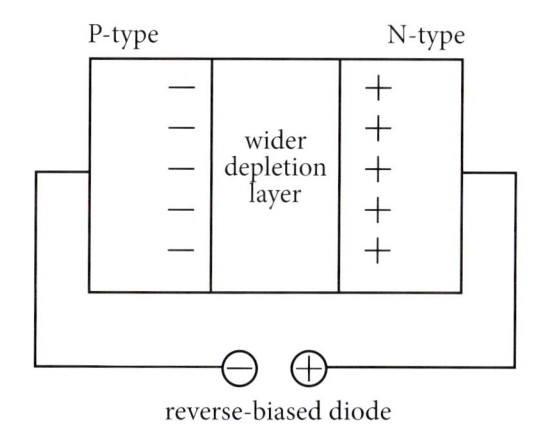

Figure 1122 Forward-biased p-n junction

In this case, the diode is said to be reverse-biased, and only small currents of the order of micro-amps can flow through the diode. This very small current is called the leakage current.

If the terminals of the source of emf are now reversed as shown in Figure 1123, the depletion layer becomes narrower because the potential difference applied across the p-n junction creates an electric field opposite to that in the depletion layer and the resistance of the p-n junction decreases.

Chapter 11

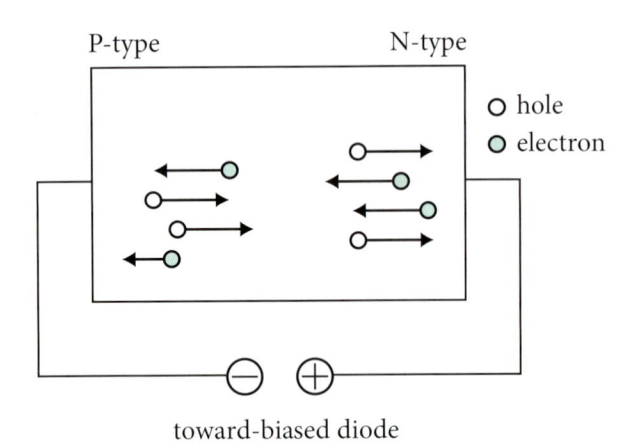

Figure 1123 *Reverse-biased p-n junction*

In this case, the diode is said to be forward-biased, and the depletion layer is removed thus allowing a current to flow through the diode.

If an ac power supply is connected to a circuit containing a junction diode and a load resistor as shown in Figure 1124 (a), the trace of the current wave over time that will be seen on the cathode ray oscilloscope (CRO) is shown in Figure 1124(b). When the diode is forward-biased, the diode conducts as shown by XY, and when it is reverse-biased only a small leakage current is produced as shown by YZ. So, only half of the available input is available and the other half is wasted.

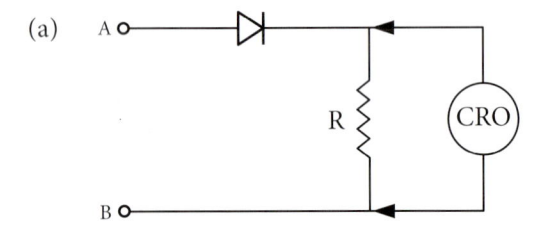

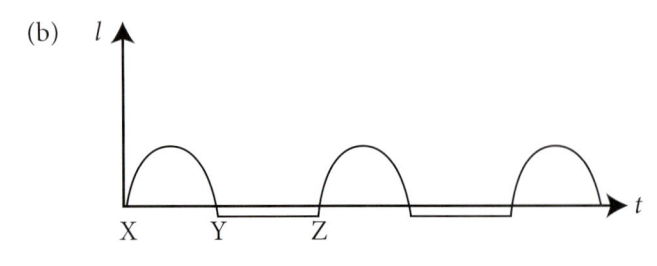

Figure 1124 *(a) and (b)*

In order to have most of the available power as input, 2 or 4 diodes are used in the circuit. This is known as a diode bridge and a typical example is shown in Figure 1125.

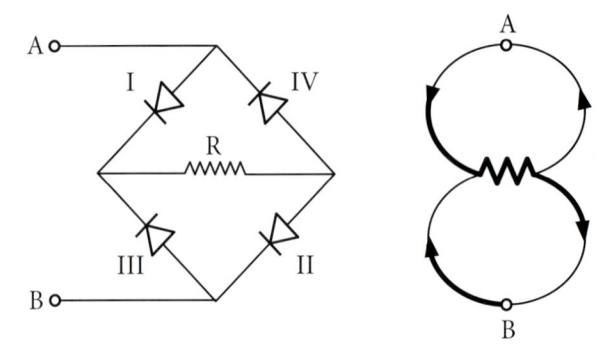

Figure 1125 *Bridge rectifier*

When A is positive and B is negative, diode I and II and current will flow from left to right through the resistor R. When A is negative and B is positive, diodes III and IV the current again flows from left to right though R. Hence, the magnitude of the current changes but its direction of flow is the same for each half cycle of the alternating current.

Half-wave and full-wave rectification

For most devices, the best source of power is the mains supply that supplies an alternating current. The alternating current needs to be converted to direct current. This process is called rectification and it can be achieved by using junction diodes. In the circuit shown in Figure 1126, the diode is forward-biased on the positive pulse of the input signal, and so conducts. The current flows through the load resistor and produces a voltage the same as the positive input pulse.

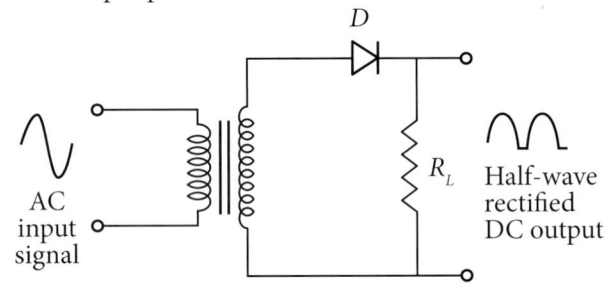

Figure 1126 *Half-wave rectifier*

On the negative pulse, the diode is reverse-biased and no current flows through the load resistor. The output voltage is zero. The voltage appearing across the load resistor is only half the input signal. It is said to be half-wave rectified.

Full-wave rectification can be achieved by using two diodes in a circuit called a centre tap rectifier, or by using 4 diodes in a circuit called a bridge rectifier as shown in Figure 1127.

ELECTROMAGNETIC INDUCTION

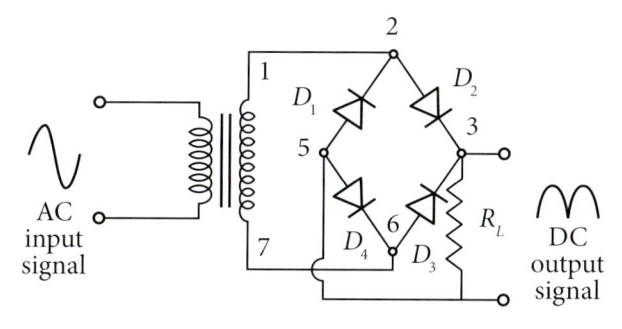

Figure 1127 Bridge rectifier

The current produced in the half-wave and full-wave rectifiers discussed are unidirectional but the amplitude of the current varies a lot with time and hence it is often unsuitable for a DC supply. A more constant current can be obtained by connecting a capacitor in parallel with the output as shown in Figures 1128(a) for a half-wave rectifier and Figure 1128(b) for a full-wave rectifier.

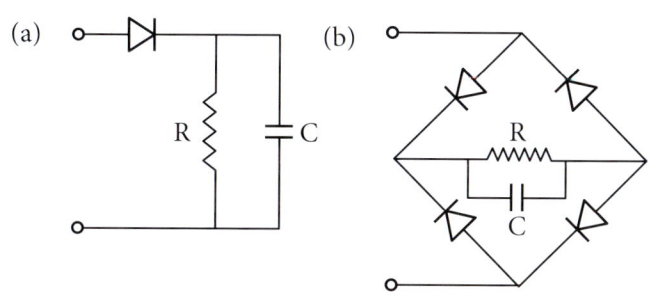

Figures 1128 (a) and (b) showing smoothing capacitors

When the voltage across the load resistor builds up, charge is stored in the capacitor, and as the voltage across the load resistor decreases, the capacitor discharges. Therefore, the current in the load resistor does not drop off as fast. Figures 1129(a) and (b) shows the effect of the smoothing capacitor on the rectified current.

(a) Half-wave rectification   (b) Full-wave rectification
— with smoothing capacitor
--- without smoothing capacitor

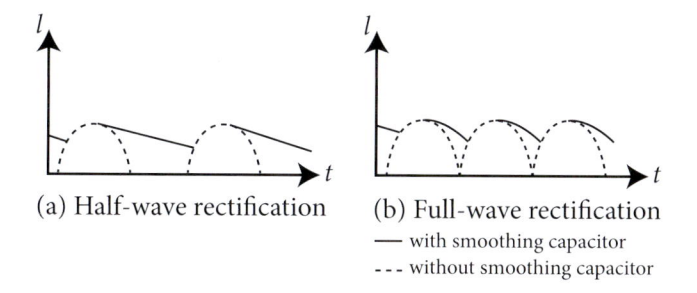

Figures 1129 (a) and (b) Capacitor smoothing

Please note that one of the compulsory laboratory exercises is to investigate a diode bridge rectification circuit experimentally.

11.3 Capacitance

NATURE OF SCIENCE:

Relationships: Examples of exponential growth and decay pervade the whole of science. It is a clear example of the way that scientists use mathematics to model reality. This topic can be used to create links between physics topics but also to uses in chemistry, biology, medicine and economics. (3.1)

© IBO 2014

Essential idea: Capacitors can be used to store electrical energy for later use.

Understandings:
- Capacitance
- Dielectric materials
- Capacitors in series and parallel
- Resistor-capacitor (RC) series circuits
- Time constant

Capacitance

A capacitor (condenser) is a device for storing electrical energy in the form of an electric field. It allows alternating current to flow but it blocks direct current.

In Chapter 10 we learnt that the potential V of a sphere of radius r is which has a charge Q is given by the expression

$$V = \frac{Q}{4\pi\varepsilon_0 r}$$

From this we see that the ratio $\frac{Q}{V}$ is equal to $4\pi\varepsilon_0 r$. This in effect means that the quantity of charge that can be stored per unit of potential of the sphere depends only on its geometry. This is in fact true for any system that can store electric charge and we call this property of the system **capacitance**.

A capacitor consists essentially of two parallel conducting plates separated by an insulator material called a **dielectric**. When a potential difference is applied across the plates of the capacitor electrical work is being done on the capacitor. This work

(W = q V) that establishes an electric field between the plates is stored as energy in the field. The two plates can be charged so that one becomes positive and the other becomes negative. The charging continues until the resulting potential difference across the capacitor equals that due to the electrical source such as a dry cell.

Chapter 11

The charge Q on a capacitor is proportional to the potential difference V across it.

$q \propto V$

$q = CV$

where C, the constant of proportionality, is the capacitance.

$C = q / V$

The SI unit of capacitance C V^{-1} is called the farad (F). A device that is manufactured to have a specific capacitance is called a capacitor.

A farad is an enormous unit and typical values of capacitance of commercial capacitors range from pF to several μF. To show just how large a unit the farad is, the capacitance of Earth is only about

10^{-3} F ($4\pi \times 8.85 \times 10^{-12} \times 6.4 \times 10^6$).

Capacitors are made from different materials such as metal foil, ceramic, plastic film and electrolytic capacitors. In order that they take up less space, the plates are coiled.

Dielectric materials

The capacitance of a parallel plate capacitor is proportional to the area A of the plates and inversely proportional to their distance d apart.

$C \propto \dfrac{A}{d}$

If air is the dielectric between the plates, the expression becomes:

$C = \varepsilon_0 \dfrac{A}{d}$

where ε_0 is the permittivity of free space

$= 8.85 \times 10^{-12}$ C^2 N^{-1} m^{-2}.

This expression can also be determined to be:

$C = K \varepsilon_0 \dfrac{A}{d}$

where K is a constant determined by the the nature of the dielectric and is called the *dielectric constant*.

Capacitors are named after the material between their plates. The dielectric constant for various materials is given in Figure 1130.

Material	Dielectric constant
Air	1
Polystyrene	2.56
Paper	3.7
Quartz	4.3
Oil	4
Pyrex glass	5
Mica	7
Strontium titanate	300
Barium titanate	1143

Figure 1130 *Dielectric constants for some materials*

The capacity and maximum voltage of a capacitor is usually printed on it or indicated by a colour code.

Sometimes a conducting medium called an electrolyte is used. Electrolytic capacitors have a very thin metal oxide dielectric soaked in a conducting paste and they have a high capacitance between 1 μF and 104 μF.

Example

A parallel plate capacitor has plates of area 250 cm2 with a distance between the plates of 2.0 mm. The dielectric between the plates is air. The capacitor is charged to a potential difference of 3.0 V.

(a) Determine the magnitude of the electric field between the plates.

(b) Find the capacitance.

(c) Determine the charge on the positive plate.

Solution

(a) $E = \dfrac{\Delta V}{d} = \dfrac{3.0}{2.0 \times 10^{-3}} = 1.5 \times 10^3$ V m^{-1}

(b) $C = \varepsilon_0 \dfrac{A}{d} = 8.85 \times 10^{-12}$ C^2 N^{-1} m^{-2} $\left(\dfrac{2.5 \times 10^{-2} \, m^2}{2.0 \times 10^{-3} \, m}\right)$

$= 1.1 \times 10^{-10} = 11 nF$

(c) $q = C.\Delta V = 1.1 \times 10^{-10} \times 3.0 = 3.32 \times 10^{-10}$ C

Capacitors in series and parallel

Capacitors in parallel

Figure 1131 shows three capacitors connected in parallel to a battery.

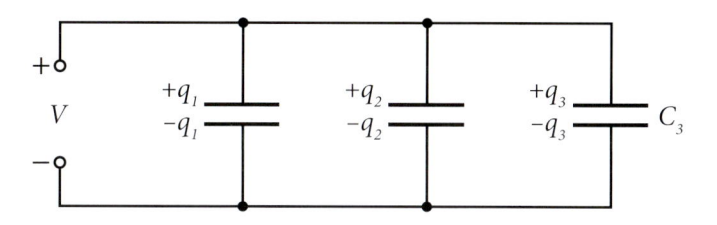

Figure 1131   Capacitors in parallel

In parallel here means that each capacitor is directly wired together to one plate and directly wired together to the other plate such that the potential difference across each capacitor is the same. That is, when a potential difference is applied across a number of capacitors connected in parallel, the total charge stored on the capacitors is the sum of the charges stored on each capacitor.

If we have three capacitors

$q_1 = C_1 V \qquad q_2 = C_2 V \qquad$ and $\qquad q_3 = C_3 V$

The total charge on the parallel combination would be

$q = q_1 + q_2 + q_3 = (C_1 + C_2 + C_3) V$

The equivalent capacitance C_{eq} of a parallel combination equals the sum of the individual capacitances.

$C_{eq} = = \dfrac{q}{V} = C_1 + C_2 + C_3$

Capacitors in series

Figure 1132 shows three capacitors connected in series to a battery.

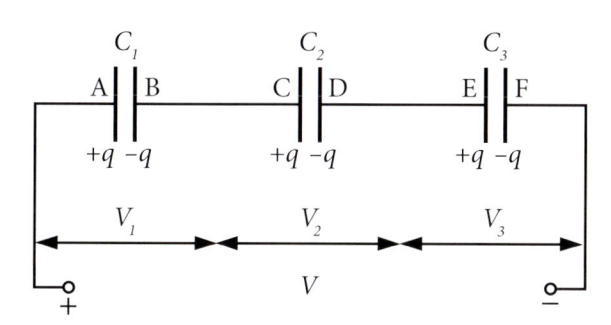

Figure 1132   Capacitors in series

They have capacitances of C_1, C_2 and C_3. Suppose a potential difference V across the capacitors causes the motion of charge from plate F to plate A so that a charge of $+q$ appears on A and $-q$ appears on F. This charge $-q$ induces a charge of $+q$ on plate E. Then plate $+q$ on plate E induces a charge of $-q$ on D and so on.

So capacitors in series all have the same charge and the potential differences across each capacitor are given by:

$$V_1 = \dfrac{q}{C_1} \qquad V_2 = \dfrac{q}{C_2} \qquad V_3 = \dfrac{q}{C_1}$$

The total potential difference across the combination is

$$V = V_1 + V_2 + V_3$$

Therefore

$$V = \dfrac{q}{C_1} + \dfrac{q}{C_2} + \dfrac{q}{C_3} = q\left(\dfrac{1}{C_1} + \dfrac{1}{C_2} + \dfrac{1}{C_3}\right)$$

Now if the 3 capacitors were replaced by a single equivalent capacitor, it would have charge q when the potential difference across it was V. Therefore,

$$V = \dfrac{q}{C_{eq}}$$

So

$$\dfrac{q}{C_{eq}} = q\left(\dfrac{1}{C_1} + \dfrac{1}{C_2} + \dfrac{1}{C_3}\right)$$

$$\dfrac{1}{C_{eq}} = \dfrac{1}{C_1} + \dfrac{1}{C_2} + \dfrac{1}{C_3}$$

Example

Use Figure 1133, parts (a) and (b) to answer the following questions based on the following information. $C_1 = 6.0\ \mu F$, $C_2 = 4.0\ \mu F$ and $C_3 = 8.0\ \mu F$. If the capacitors are charged with a 12 volt battery

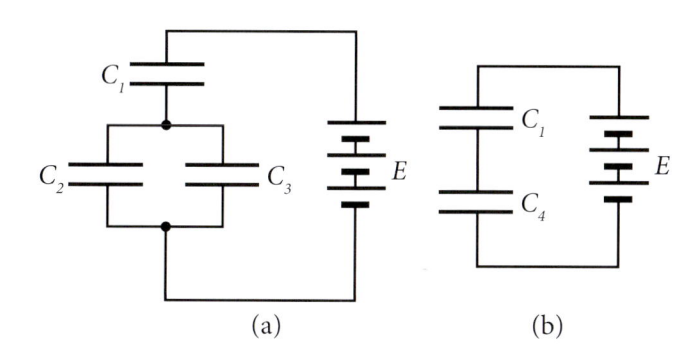

(a)   (b)

Figure 1133   (a) and (b)

Chapter 11

(a) determine the equivalent capacitance of the combination

(b) find the potential difference across each capacitor

(c) determine the charge on each capacitor

Solution

(a) C_2 and C_3 are in parallel and can be replaced with a single capacitor

$C_{23} = C_2 + C_3 = 4.0\ \mu F + 8.0\ \mu F = 12.0\ \mu F$

C_{23} is in series with C_1. Therefore

$$\frac{1}{C_{eq}} = \frac{1}{C_1} + \frac{1}{C_{23}} = \frac{1}{6.0}\mu F + \frac{1}{12.0}\mu F$$

$$= \frac{3}{12.0}\mu F$$

Therefore, $C_{eq} = \frac{12.0\ \mu F}{3} = 4.0\ \mu F$

(b) The circuit can now be redrawn as in Figure 1133(b) and C_1 and C_4 are in series. The charge that flows from the battery is

$q = CV = (4.0 \times 10^{-6}\ F) \times (12\ V) = 4.8 \times 10^{-5}\ C$.

C_1 and C_4 both carry this charge. Therefore, the potential difference across C_1 is

$V_1 = \frac{q}{C_1} = \frac{4.8 \times 10^{-5}\ C}{6.0 \times 10^{-6}\ F} = 8.0\ V$.

The potential difference across the combination C4 is

$V_1 = \frac{q}{C_4} = \frac{4.8 \times 10^{-5}\ C}{12.0 \times 10^{-6}\ F} = 4.0\ V$

Since the actual capacitors C_2 and C_3 are in parallel, this represents the potential difference across each of them.

$V_2 = V_3 = 4.0\ V$.

(c) The charge on C1 has already been determined in part (b) = 4.8×10^{-5} C.

The charges on C2 and C3 are

$q_2 = C_2 V_2 = (4.0 \times 10^{-6}\ F) \times (4.0) = 1.6 \times 10^{-5}\ C$

$q_3 = C_3 V_3 = (8.0 \times 10^{-6}\ F) \times (4.0) = 3.2 \times 10^{-5}\ C$

Energy of a charged capacitor

Suppose we had a 10 000 μF electrolytic capacitor that had been charged from a 3V supply and then discharged through a filament lamp, and its brightness was noted. Now suppose the capacitor was charged with a 6V supply and discharged through four filament lamps in parallel. It can be observed that the four lamps can be lit with the same brightness and for the same period of time as the one lamp in the original circuit.

Therefore, doubling the potential difference to which a capacitor is charged quadruples the energy stored in the capacitor. That is, the energy stored is proportional to V 2.

We can also determine the energy stored graphically as in Figure 1134 that shows charge as a function of potential difference.

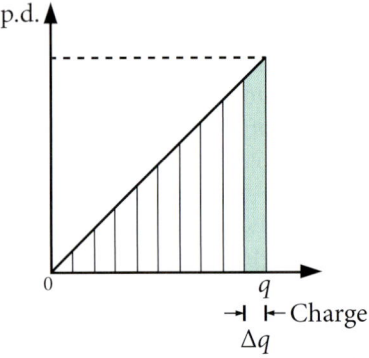

Figure 1134 Charge as a function of potential difference

We have already seen that the charge on a capacitor is directly proportional to the potential difference across it where q = C V where C is a constant. So we can draw a straight line that passes through the origin. Suppose at the top of the graph a capacitor of capacitance when the potential difference is V it has a charge q. Now when the capacitor starts to discharge, a small charge Δ q passes from the negative to the positive plate. Provided Δ q is small, the resulting energy loss is V.

Therefore, from the graph the energy lost will be the area of the hatched portion. So, if the capacitor is completely discharge so the potential difference and charge are zero, we can deduce that

Total energy loss = area under the graph

E = ½ q V

But

q = C V

Therefore,

E = ½ C V²

Example

A 6-volt battery is connected to a 20μF capacitor. Determine the energy that can be stored in the capacitor.

344

Solution

Energy = ½ C V² = ½ (20 × 10⁻⁶ F) (6V)² = 3.6 × 10⁻⁴ J.

Resistor-capacitor (RC) series circuits

There are many electronic devices that involve capacitors charging and discharging through resistors. The process is very similar to growth (bacteria growth) and decay (radioactive decay) curves. Here we will study the growth and decay of a capacitor over time. This process can be described both graphically and mathematically.

A basic RC series circuit is shown in Figure 1135(a) and is compared with a basic resistor circuit in Figure 1135(b).

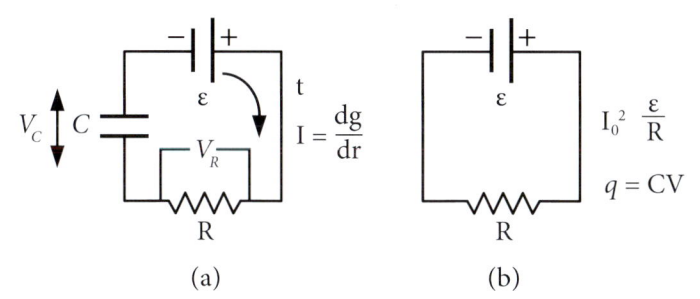

Figure 1135 (a) and (b) Comparison of a RC circuit with a resistor circuit.

In a basic series circuit (b), the charge is immediately available to be used by the resistor and it has a constant value. That is

$$I = \frac{q}{t} = \frac{\varepsilon}{R}$$

Charging a capacitor

However, in a RC series circuit (a), the charge builds up over time. Let us compare some values when t = 0, when t = a very long time and t between these two instances Δt. Figure 1136 summarises some circuit values.

When t = 0	When t approaches infinity	Δt during charging
$I_0 = \varepsilon / R$	I = 0	I = Δq /Δt
$V_R = \varepsilon$	$V_R = 0$	
$V_C = 0$	$V_C = \varepsilon$	
$q_C = 0$	q_C = Qfinal	
	q_C = CVC	

Figure 1136 Component values over a period of time.

When the capacitor begins to be charged, the battery begins to charge the plates of the capacitor and charge passes through the resistor, and the circuit carries a changing current. The charging process continues until the capacitor is charged to its maximum value and q = Cε where ε is the maximum voltage across the capacitor. If we assume that the capacitor is uncharged at t = 0, once the capacitor is fully charged, the current in the circuit will be zero. Figure 1137 shows the charging process as time progresses.

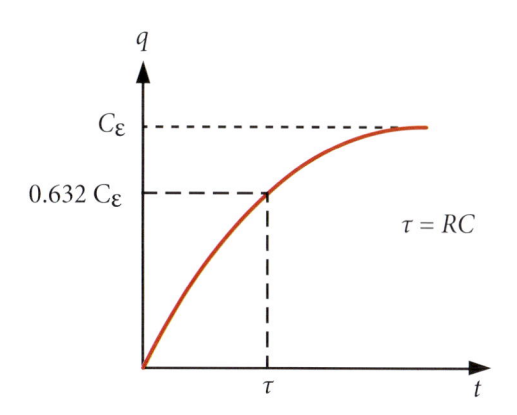

Figure 1137 Charge on a capacitor versus time.

By using calculus, it can be shown that the charge on the capacitor varies with time according to the equation

$$q = Q_{final} (1 - e^{-t/RC})$$ where e = 2.718

However, the study guide states "problems involving the charging of a capacitor will only be treated graphically".

Time constant

The product of resistance R and capacitance C is called the time constant τ.

τ = RC

The time constant tells us how fast the circuits charge and discharge (the rate of charging and discharging of a capacitor).

It can be shown that the larger the values of R and C, the slower will be the charging and discharging processes.

Discharging a capacitor

Consider the circuit as shown in Figure 1138, and let's say the capacitor has C = 500 μF and R = 100 kΩ, and the capacitor is charged to about 10 volts by touching the wire to the capacitor circuit. When the capacitor is fully charged, the wire is then removed. We can obtain a decay curve by taking readings on the microammeter (0 – 100 μA) at 10-second intervals during the discharging process.

CHAPTER 11

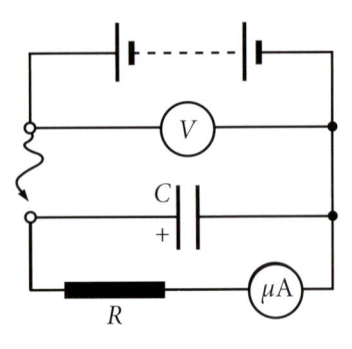

Figure 1138 Discharging a capacitor circuit.

It can be seen from the decay curve in Figure 1139 that it decreases by the same fraction in successive time intervals.

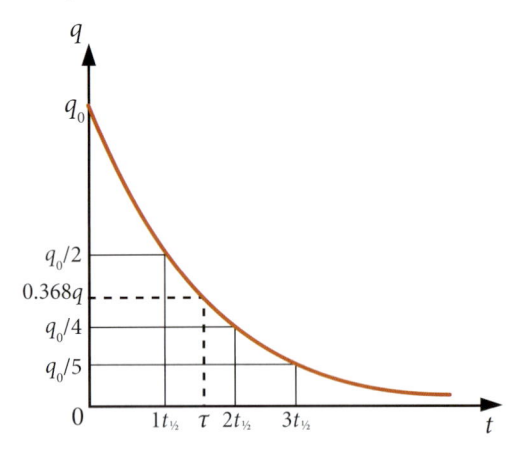

Figure 1139 Decay curve for the discharging of a capacitor.

You can see that it falls from Q_0 to $\frac{Q_0}{2}$ in time $t_{1/2}$, and from $\frac{Q_0}{2}$ to $\frac{Q_0}{4}$ in the next time interval $2t_{1/2}$, and so on.

The time $t_{1/2}$ is known as the half-life of the decay process. That is $t_{1/2}$ is the time for the charge to fall to by a half.

It can be shown using calculus that

$$q = q_0\, e^{\frac{-t}{RC}} = q_0\, e^{-\frac{t}{\tau}}$$

Example

For the values given for R and C given in the circuit scenario, determine the half-life of the decay.

Solution

$\tau = RC = (100 \times 10^3\ \Omega) \times (500 \times 10^{-6}\ F) = 50$ s

The time t for the half-life $t_{1/2}$ would be equal to

$$\frac{q_0}{2} = q_0\, e^{\frac{-t_{1/2}}{50}}$$

$$t_{1/2} = e^{\frac{-t_{1/2}}{50}}$$

$t_{1/2} = CR \log_e 2 = 50 \log_e 2 = 34.657 \approx$ **35 s.**

The decay curve of Q against t will have the same shape as the decay curve of I against t because q = C V and V = I R. Therefore, combining these two equations will give q = RC I. But R and C are constants, so q is proportional to I.

Therefore

$$I = I_0\, e^{\frac{-t}{RC}} = I_0\, e^{\frac{-t}{\tau}}$$

Similarly, it can be shown that

$$V = V_0\, e^{\frac{-t}{RC}} = V_0\, e^{\frac{-t}{\tau}}$$

Exercise 11.3

1. Refer to Figure 1133 to answer the following. If $C_1 = 2.0$ μF, $C_2 = 0.5$ μF and $C_3 = 0.5$ μF. If the capacitors are charged with E = 6 V

   (a) determine the equivalent capacitance of the combination
   (b) find the potential difference across each capacitor
   (c) determine the charge on the 12.0 capacitor.

2. Four capacitors with values of 3.0 μF, 6.0 μF, 12.0 μF and 24.0 μF are connected in series to an 18 V battery.

   (a) determine the equivalent capacitance of the combination
   (b) find the potential difference across 12.0 μF capacitor
   (c) determine the charge on the 12.0 μF capacitor.

3. A parallel plate capacitor has a plate area of 2.50×10^{-3} m² and a distance x between the plates of 2.00 mm. Find the maximum charge if the dielectric is polystyrene.

4. An uncharged capacitor with C = 5 μF and a resistor R = 8 x 10⁵ Ω are connected in series to a 12 V battery. Determine

   (a) the time constant of the circuit
   (b) the maximum charge on the capacitor
   (c) the charge on the capacitor after 6 s.

5. A 10 μF is charged from a 30 V power supply. It is then connected across an uncharged 50 μF capacitor. Determine

   (a) the final potential difference across the combination
   (b) the initial and final enegy of the combination.

346

12. Atomic, nuclear and particle physics

Contents

12.1 – The interaction of matter with radiation

12.2 – Nuclear physics

Essential Ideas

The microscopic quantum world offers a range of phenomena whose interpretation and explanation require new ideas and concepts not found in the classical world.

The idea of discreteness that we met in the atomic world continues to exist in the nuclear world as well. © IBO 2014

CHAPTER 12

12.1 The interaction of matter with radiation Photons

Essential idea: The microscopic quantum world offers a range of phenomena whose interpretation and explanation require new ideas and concepts not found in the classical world.

Understandings:
- Photons
- The photoelectric effect
- Matter waves
- Pair production and pair annihilation
- Quantization of angular momentum in the Bohr model for hydrogen
- The wave function
- The uncertainty principle for energy and time and position and momentum
- Tunnelling, potential barrier and factors affecting tunnelling probability

NATURE OF SCIENCE:

Observations: Much of the work towards a quantum theory of atoms was guided by the need to explain the observed patterns in atomic spectra. The first quantum model of matter is the Bohr model for hydrogen. (1.8)

Paradigm shift: The acceptance of the wave-particle duality paradox for light and particles required scientists in many fields to view research from new perspectives. (2.3)

© IBO 2014

TOK WHAT IS QUANTUM MECHANICS?

This topic raises fundamental philosophical problems relating to the nature of observation and measurement. The concept of a 'paradigm shift' can be discussed here and in particular how this paradigm shift led to the downfall of Newtonian Determinism.

The advent of quantum mechanics meant that the determinism of classical physics was a thing of the past. In classical physics, it was believed that if the initial state of a system is known precisely, then the future behaviour of the system could be predicted for all time. However, according to quantum mechanics, because of the inability to define the initial data with absolute precision, such predictions can no longer be made. In classical physics it was thought that the only thing that limited knowing the initial state of the system with sufficient precision was determined basically by precision of the measuring tool. The Uncertainty principle put paid to this idea – uncertainty is an inherent part of Nature.

There have been many attempts to understand what quantum mechanics is really all about. On a pragmatic level, many physicists accept that it works and get on with their job. Others worry about the many paradoxes to which it leads.

One of the true mysteries (apart from the ever famous Schrödinger cat) is the double slit experiment and its interpretation. Fire electrons at a double slit and just like light waves, an observable interference pattern can be obtained. However, if you observe through which slit each electron passes, the interference pattern disappears and the electrons behave like particles. If you can interpret this then you can "understand" what quantum mechanics is all about. However, remember what Richard Feynman had to say on this topic.

Finally, if quantum mechanics is the "correct" physics, why do we in this course spend so much time learning classical physics? An example might suffice to answer this question.

If you apply Newtonian mechanics to the motion of a projectile, you get the "right" answer, if you apply it to the motion of electrons, you get the "wrong" answer. However, if you apply quantum mechanics to each of these motions, in each case you will get the right answer. The only problem is, that using quantum mechanics to solve a projectile problem is like using the proverbial sledgehammer to crack a walnut. We must move on.

ATOMIC, NUCLEAR AND PARTICLE PHYSICS

Photons

Physics prior to the 1900s is generally termed *classical physics* and physics after 1900 is referred to as *quantum physics*.

In 1900, Max Planck (1858-1947) was studying the heat and light radiated by hot solids. He developed a new theory called the quantum theory to explain the relationship between the energy radiated by a hot body, and the frequency of light emitted by that same body. He could not get agreement between theory and experiment without making a fundamental change to the laws of physics. He proposed that a hot object did not emit energy continuously.

Planck suggested that atoms, when radiating, change their energy in separate quantised amounts, which he called quanta. He developed an equation relating energy to frequency

$$E = hf$$

where E is measured in joules and represents the energy released by a hot body, f, is the frequency of light emitted by the body, measured in hertz, and h is a constant which is now accepted to be 6.63×10^{-34} J s. The constant came to be known as Planck's constant. Planck's equation and the quantum theory he introduced earned Max Planck the Nobel Prize in Physics in 1918.

Einstein extended the Planck theory in 1905 by deriving an equation that explained fully the laws of photoelectric emission. He proposed that not only light and other forms of electromagnetic radiation were emitted in whole number of quanta but that they were also absorbed as quanta, called **photons**. He also proposed that the energy of light was not spread out evenly over the wavefront but was concentrated in tiny particles or photons. The photon could give up *all* its energy to *one* electron, but it cannot give part of it – the all or nothing principle.

The photoelectric effect

In 1888, *Heinrich Hertz* carried out an experiment to verify *Maxwell's* electromagnetic theory of radiation. Whilst performing the experiment *Hertz* noted that a spark was more easily produced if the electrodes of the spark gap were illuminated with ultra-violet light. *Hertz* paid this fact little heed, and it was left to one of his pupils, *Hallswach*, to investigate the effect more thoroughly. Hallswach noticed that metal surfaces became charged when illuminated with ultra-violet light and that the surface was always positively charged. He concluded therefore that the ultra-violet light caused negative charge to be ejected from the surface in some manner. In 1899 Lenard showed that the negative charge involved in the **photo-electric effect** consisted of particles identical in every respect to those isolated by J. J. Thomson two years previously, namely, electrons.

Figure 1201 shows schematically the sort of arrangement that might be used to investigate the photo-electric effect in more detail.

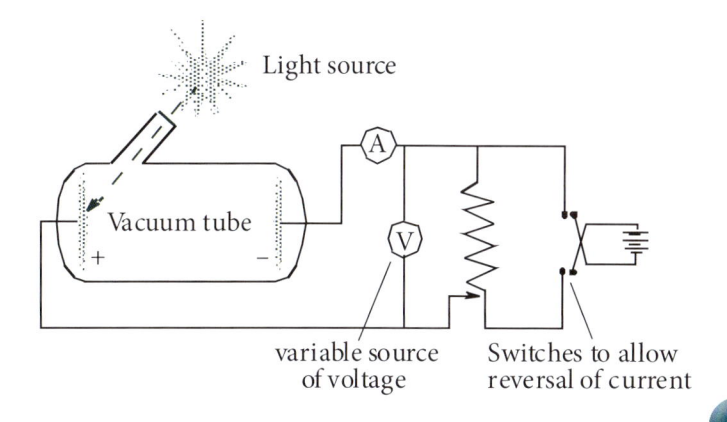

Figure 1201   Apparatus for detecting the photoelectric effect

The tube B is highly evacuated, and a potential difference of about 10 V is applied between anode and cathode. The cathode consists of a small zinc plate, and a quartz window is arranged in the side of the tube such that the cathode may be illuminated with ultra-violet light. The current measured by the micro-ammeter gives a direct measure of the number of electrons emitted at the cathode.

When the tube is dark no electrons are emitted at the cathode and therefore no current is recorded. When ultra-violet light is allowed to fall on the cathode electrons are ejected and traverse the tube to the anode, under the influence of the anode-cathode potential. A small current is recorded by the micro-ammeter.

In Figure 1202, the photoelectrons travel through the vacuum tube towards the anode, where they enter a circuit. The current created by the photoelectrons is measured using a microammeter. The circuit also contains a variable potential source, which can be used to make the anode negative. This potential is also known as a retarding potential, as it is used to impede the passage of the electrons, or in other words, to diminish and eventually stop the photocurrent.

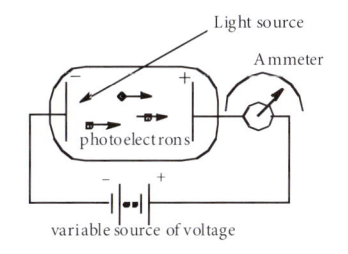

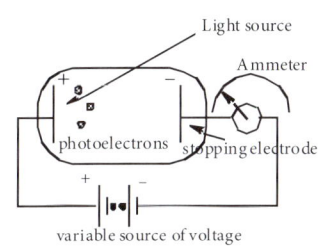

By reversing the direction of current flow and varying the potential, the photoelectrons can be 'held back'.

Figure 1202   Apparatus for photoelectric effect

349

CHAPTER 12

The following are observations typical of this type of experiment:

1. For every metal, there is a certain frequency of light, below which no electrons are emitted, no matter how intense the light as shown in Figure 1203. This frequency, fo, became known as the threshold frequency. Each metal has its own characteristic threshold frequency.

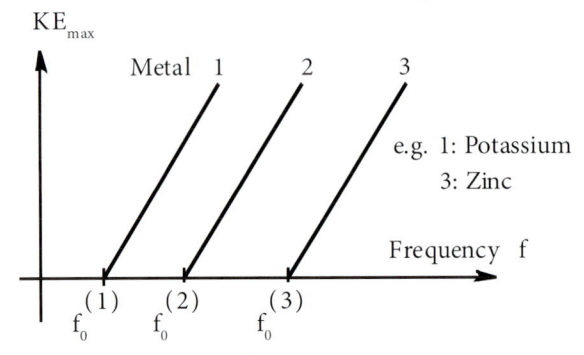

Graph: KE_{max} vs Frequency f, showing three lines for Metal 1, 2, 3 with thresholds $f_0^{(1)}, f_0^{(2)}, f_0^{(3)}$. e.g. 1: Potassium, 3: Zinc

Figure 1203

2. Above the threshold frequency, photoelectrons are emitted. The greater the intensity of the light, the greater the photoelectron current. This suggests that increasing the incident light intensity increases the number of electrons emitted, since current is directly proportional to the amount of charge. See Figure 1204.

Intensity 1 (In_1) > Intensity 2 (In_2)

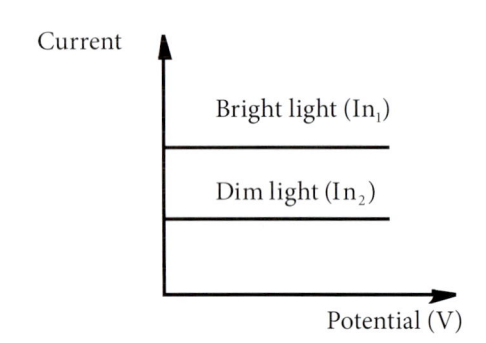

Graph: Current vs Potential (V), showing Bright light (In_1) and Dim light (In_2)

Figure 1204

3. Increasing the retarding potential applied to the anode results in a decrease in the photocurrent, no matter how intense the light. This suggests the photoelectrons are all emitted with the same kinetic energy regardless of the intensity of the incident light. The potential at which the photocurrent no longer flows is called the stopping or cut-off potential. See Figure 1205.

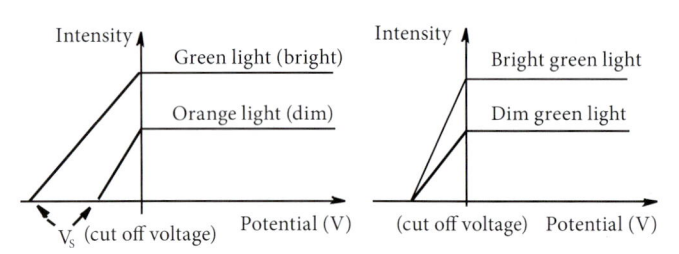

Two graphs of Intensity vs Potential (V): Left shows Green light (bright) and Orange light (dim) with V_s (cut off voltage). Right shows Bright green light and Dim green light with (cut off voltage).

Figure 1205

4. Different frequencies of light directed at the same metal surface have different stopping potentials. That is, the greater the frequency of the incident light, the greater the stopping potential. See Figure 1206.

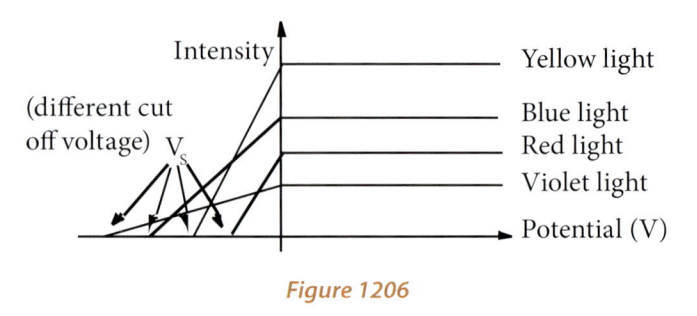

Graph: Intensity vs Potential (V), showing Yellow light, Blue light, Red light, Violet light with different cut off voltages V_s

Figure 1206

The above observations were in direct conflict with the prevailing wave theory of light that suggested that light existed as continuous waves. To begin with, if light does exist as continuous waves, then electrons should absorb energy continuously, allowing them to be emitted as soon as they absorb sufficient energy. This means that there should be no threshold frequency, since any light should eventually allow electrons to be ejected. Also, there should be no limit to the energy of the electrons, so that at high intensities, the stopping potential should also be higher. It wasn't until 1905 that Albert Einstein developed a plausible explanation of the photoelectric effect, using Planck's quantum theory.

The existence of a threshold frequency and spontaneous emission even for light of a very low intensity cannot be explained in terms of a wave theory of light. In 1905 Einstein proposed a daring solution to the problem. Planck had shown that radiation is emitted in pulses, each pulse having an energy hf where h is a constant known as the Planck constant and f is the frequency of the radiation. Why, argued Einstein, should these pulses spread out as waves? Perhaps each pulse of radiation maintains a separate identity throughout the time of propagation of the radiation. Instead of light consisting of a train of waves we should think of it as consisting of a hail of discrete energy bundles. On this basis the significance of light frequency is not so much the frequency of a pulsating electromagnetic field, but a measure of the energy of each 'bundle' or 'particle of light'. The name given to these tiny bundles of energy is photon or quantum of radiation.

Einstein's interpretation of a threshold frequency is that a photon below this frequency has insufficient energy to remove an electron from the metal. The minimum energy required to remove an electron from the surface of a metal is called the **work function** of the metal. The electrons in the metal surface will have varying kinetic energy and so at a frequency above the threshold frequency the ejected electrons will also have widely varying energies. However, according to Einstein's theory there will be a definite upper limit to the energy that a **photoelectron** can have. Suppose we have an electron in the metal surface that needs just φ units of energy to be ejected, where **φ is the work function** of the metal. A photon of energy hf strikes this electron and so the electron absorbs hf units of energy. If hf ≥ φ the electron will be ejected from the metal, and if energy is to be conserved it will gain an amount of kinetic energy E_K given by

$$E_K = hf - \phi$$

Since ϕ is the minimum amount of energy required to eject an electron from the surface, it follows that the above electron will have the maximum possible kinetic energy. We can therefore write that

Or
$$E_{K_{max}} = hf - \phi$$
$$E_{K_{max}} = hf - hf_0$$

Where f_0 is the **threshold frequency**. Either of the above equations is referred to as the Einstein Photoelectric Equation.

It is worth noting that Einstein received the Nobel prize for Physics in 1921 for "his contributions to mathematical physics and especially for his discovery of the law of the photoelectric effect".

In 1916 *Robert Millikan* verified the Einstein photoelectric equation using apparatus similar to that shown in Figure 1202 (a). Millikan reversed the potential difference between the anode and cathode such that the anode was now negatively charged. Electrons emitted by light shone onto the cathode now face a 'potential barrier' and will only reach the anode if they have a certain amount of energy. The situation is analogous to a car freewheeling along the flat and meeting a hill. The car will reach the top of the hill only if its kinetic energy is greater than or equal to its potential energy at the top of the hill. For the electron, if the potential difference between cathode and anode is V_s ('stopping potential') then it will reach the anode only if its kinetic energy is equal to or greater than $V_s e$ where e is the electron charge.

In this situation, the *Einstein equation* becomes

$$V_s e = hf - hf_0$$

Millikan recorded values of the stopping potential for different frequencies of the light incident on the cathode. For Einstein's theory of the photoelectric effect to be correct, a plot of stopping potential against frequency should produce a straight line whose gradient equals $\frac{h}{e}$. A value for the *Planck constant* had been previously determined using measurements from the spectra associated with hot objects. The results of Millikan's experiment yielded the same value and the photoelectric effect is regarded as the method by which the value of the *Planck constant* is measured. The modern accepted value is $6.62660693 \times 10^{-34}$ J s. The intercept on the frequency axis is the threshold frequency and intercept of the V_s axis is numerically equal to the work function measured in electron-volt.

Figure 1207 shows the typical results of Millikan's experiment which shows the variation with frequency f of the maximum kinetic energy E_k.

It is left to you as an exercise, using data from this graph, to determine the *Planck constant*, and the threshold frequency and work function of the metal used for the cathode.

(Ans 6.6×10^{-34} J s, 4.5 eV)

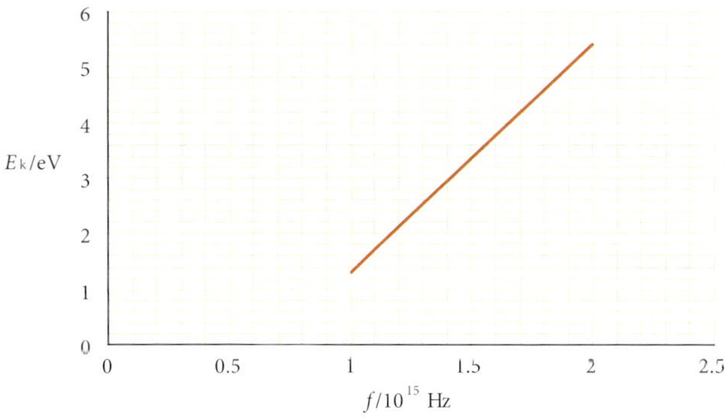

Figure 1207 *The results of Millikan's experiment*

In summary, Figure 1208 shows the observations associated with the photoelectric effect and why the Classical theory, that is the wave theory, of electromagnetic radiation is unable to explain the observations i.e. makes predictions inconsistent with the observations.

Observation	Classical theory predictions
Emission of electrons is instantaneous no matter what the intensity of the incident radiation.	Energy should be absorbed by the electron continuously until it has sufficient energy to break free from the metal surface. The less the intensity of the incident radiation, the less energy incident of the surface per unit time, so the longer it takes the electron to be ejected.
The existence of a threshold frequency.	The intensity of the radiation is independent of frequency. Emission of electrons should occur for all frequencies.

Figure 1208 *Observations associated with the photoelectric effect*

The fact that the photoelectric effect gives convincing evidence for the particle nature of light, raises the question as to whether light consists of waves or particles. If particulate in nature, how do we explain such phenomena as interference and diffraction?. This is an interesting area of discussion for TOK and it is worth bearing in mind that

351

CHAPTER 12

Newton wrote in his introduction to his book *Optics* 'It seems to me that the nature of light be particulate'.

Example 1

Calculate the energy of a photon in light of wavelength 120 nm.

Solution

$$f = \frac{c}{\lambda} = \frac{3.0 \times 10^8}{1.2 \times 10^{-7}} = 2.5 \times 10^{15} \text{ Hz}$$

$$E = hf = 6.6 \times 10^{-34} \times 2.5 \times 10^{15} = 1.7 \times 10^{-18} \text{ J}$$

Example 2

The photoelectric work function of potassium is 2.0 eV. Calculate the threshold frequency of potassium.

Solution

$$f_0 = \frac{\phi}{h} = \frac{2.0 \times 1.6 \times 10^{-19}}{6.6 \times 10^{-34}} = 4.8 \times 10^{14} \text{ Hz}$$

Example 3

The work function for metallic sodium is 2.28 eV. Calculate the maximum kinetic energy and speed of an emitted electron when sodium is illuminated by violet light of frequency 7.00 x10^{14} Hz?

Solution

The work function, ϕ, is given in eV and must first be converted to joules, as follows:

$$2.28 \text{ eV} \times 1.60 \times 10^{-19} \frac{J}{eV} = 3.65 \times 10^{-19} \text{ J}$$

The kinetic energy is given by

$$E_K = \tfrac{1}{2} mv^2_{max} = hf - \phi = (6.63 \times 10^{-34} \text{ Js}) \times$$

$$(7.00 \times 10^{14} \text{ Hz}) - 3.65 \times 10^{-19} \text{ J}$$

$$= 9.91 \times 10^{-20} \text{ J}$$

Using the formula, $E_K = \tfrac{1}{2} mv^2_{max}$

we have $v_{max} = \sqrt{\frac{2EK}{m}}$.

So that $v_{max} = \sqrt{\frac{2 \times 9.91 \times 10^{-20} \text{ J}}{9.11 \times 10^{-31}}}$ kg = $4.66 \times 10^5 \text{ ms}^{-1}$

EXERCISE 12.1

1. In order for a photoelectron to be emitted from a metal surface, the light illuminating that surface must

   A. exceed a certain minimum wavelength.
   B. exceed a certain minimum frequency.
   C. not exceed a certain maximum frequency.
   D. be equal to a certain given frequency.

2. The photoelectric effect provides evidence of

   A. the wave nature of light.
   B. the wave nature of electrons.
   C. the particle nature of light.
   D. the particle nature of electrons.

3. The work function of molybdenum is 4.2 eV. To what frequency of light does this correspond?

   A. 6.33×10^{33} Hz
   B. 9.87×10^{-16} Hz
   C. 2.78×10^{-33} Hz
   D. 1.01×10^{15} Hz

4. The maximum kinetic energy of the photoelectrons emitted from a metal surface depends on

   A. the frequency of the incident light.
   B. the intensity of the incident light.
   C. the speed of the incident light.
   D. the binding energy of the metallic atoms.

5. If the work function for copper is 4.5 eV, calculate the maximum kinetic energy of electrons emitted when ultraviolet light of frequency 2.0×10^{15} Hz illuminates a copper surface.

6. The work function for metallic calcium is 5.11×10^{-19} J. Calculate

   a. the minimum frequency of light required to emit photoelectrons from calcium.
   b. the maximum kinetic energy of photoelectrons emitted by incident ultraviolet light of frequency 1.50×10^{15} Hz.
   c. the maximum velocity of the photoelectrons emitted by ultraviolet light of frequency 1.50×10^{15} Hz.

7. How does the cut–off frequency in the photoelectric effect support the photon theory of light?

8. Use data from example 2 to calculate the maximum kinetic energy in electron volts of electrons emitted from the surface of potassium when illuminated with light of wavelength 120 nm.

10. In an experiment to measure the Planck constant, light of different frequencies f was shone on to the surface of silver and the stopping potential Vs for the emitted electrons was measured.

The results are shown below. Uncertainties in the data are not shown.

Vs / V	f / 10¹⁴ Hz
0.33	6.0
0.79	7.1
1.2	8.0
1.49	8.8
1.82	9.7

Plot a graph to show the variation of Vs with f. Draw a line of best fit for the data points.

Use the graph to determine

(i) a value of the Planck constant
(ii) the work function of silver in electron-volt.

Matter waves

Given that the photoelectric effect demonstrated the particle nature of light, the next logical question seemed to be whether matter, which is made up of particles, might have a wave nature. This question was first addressed by the French physicist, Prince Louis-Victor de Broglie (1892-1987), in his PhD thesis in 1923. He suggested that since Nature should be symmetrical, that just as waves could exhibit particle properties, then what are considered to be particles should exhibit wave properties.

We have seen that as a consequence of Special relativity the total mass-energy of a particle is given by

$E = mc^2$

We can use this expression to find the momentum of a photon by combining it with the *Planck equation* $E = hf$ such that

$E = hf = \dfrac{hc}{\lambda} = mc^2$

from which

$mc = \dfrac{h}{\lambda}$

But mc is the momentum p of the photon, so that

$p = \dfrac{h}{\lambda}$

Based on this result, the *de Broglie hypothesis* is that any particle will have an associated wavelength given by $p = \dfrac{h}{\lambda}$

The waves to which the wavelength relates are called **matter waves**.

For a person of 70 kg running with a speed of 5 m s⁻¹, the wavelength λ associated with the person is given by

$\lambda = \dfrac{h}{p} = \dfrac{6.6 \times 10^{-34}}{70 \times 5} \approx 2 \times 10^{-36}$ m

This wavelength is minute to say the least. However, consider an electron moving with speed of 10⁷ m s⁻¹, then its associated wavelength is

$\lambda = \dfrac{h}{p} = \dfrac{6.6 \times 10^{-34}}{9.1 \times 10^{-31} \times 10^7} \approx 7 \times 10^{-11}$ m

Although small this is measurable.

In 1927 *Clinton Davisson* and *Lester Germer* who both worked at the Bell Laboratory in New Jersey, USA, were studying the scattering of electrons by a nickel crystal. A schematic diagram of their apparatus is shown in Figure 1209.

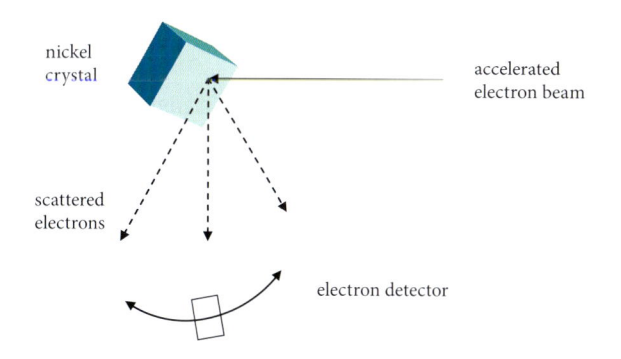

Figure 1209 *The scattering of electrons by a nickel crystal*

Their vacuum system broke down and the crystal oxidized. To remove the oxidization, *Davisson* and *Germer* heated the crystal to a high temperature. On continuing the experiment they found that the intensity of the scattered electrons went through a series of maxima and minima- the electrons were being diffracted. The heating of the nickel crystal had changed it into a single crystal and the electrons were now behaving just as scattered X-rays do. Effectively, that lattice ions of the crystal act as a diffraction grating whose slit width is equal to the spacing of the lattice ions.

Davisson and *Germer* were able to calculate the de Broglie wavelength λ of the electrons from the potential difference V through which they had been accelerated.

Using the relationship between kinetic energy and momentum, we have

$E_k = Ve = \dfrac{p^2}{2m}$

Therefore

$p = \sqrt{2mVe}$

353

CHAPTER 12

Using the de Broglie hypothesis $p = \frac{h}{\lambda}$, we have

$$\lambda = \frac{h}{\sqrt{2mVe}}$$

They knew the spacing of the lattice ions from X-ray measurements and so were able to calculate the predicted diffraction angles for a wavelength equal to the de Broglie wavelength of the electrons. The predicted angles were in close agreement with the measured angles and the de Broglie hypothesis was verified – particles behave as waves.

To put it another way, light rays passing through a slit of width d form maxima at angles given by the formula:

n λ = d sin θ

Since Davisson and Germer knew the value of d and the positions of the diffraction maxima for the electron beam, they were able to calculate the wavelength of the electron using the formula

λ = d sin θ / n

To calculate the predicted de Broglie wavelength, they started with the kinetic energy of the electrons as determined by the formula ½ m v² = eV where V is the potential difference through which the electrons were accelerated.

The velocity of the electrons is then given by $v = \sqrt{\frac{2eV}{m}}$ and their momentum is given by

$p = mv = \sqrt{2eVm}$ since $E_K = \frac{p^2}{2m}$

The equation for the de Broglie wavelength is therefore

$$\lambda = \frac{h}{\sqrt{2mVe}}$$

Davisson and Germer discovered that the wavelength calculated using the light diffraction equation was the same as that calculated using de Broglie's equation. This suggests that electrons are themselves diffracted in the same way, as one would expect their De Broglie waves to be diffracted. In other words, electrons have wave properties!

The electron orbiting in a ground–state hydrogen atom will have its circumference measuring exactly one de Broglie wavelength

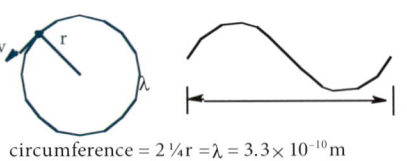

circumference = 2¼r = λ = 3.3× 10⁻¹⁰ m

That is, the electron wave is a standing wave on the circumference of the classical orbit.
For example, a standing circular wave for 4 wavelengths. That is, C = 2¼r₄ = 4λ.

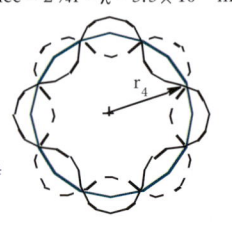

Figure 1210 *Electron orbits as standing waves.*

G. P. Thomson, who did similar experiments bombarding gold foil with an electron beam, confirmed Davisson and Germer's results in 1927. He was able to produce a similar diffraction pattern to the one observed for light travelling through thin slices of material, thus providing further evidence of the wave nature of electrons. It is somewhat interesting that Thomson was the only son of J. J. Thomson, who discovered the electron by studying its particle nature.

De Broglie's hypothesis had far reaching implications for the very nature of matter. He suggested that electron orbits are actually standing waves as shown in Figure 1210, which can only be sustained if the circumference of the orbit contains a whole number of wavelengths. If the circumference of a Bohr orbit of radius rn is 2π rn, the circumference can also be written in terms of the de Broglie condition as

$$2\pi r_n = n \lambda \text{ where } n = 1, 2, 3, ...$$

Substituting $\lambda = \frac{h}{\sqrt{2mVe}}$ for λ yields the equation

$2\pi r_n = \frac{nh}{mv}$

or

$mv\, r_n = \frac{nh}{2\pi}$

This is Bohr's quantum condition as we will see shortly, suggesting that it is the wave nature of an electron that determines its Bohr orbit!

It is now known that electrons are not the only particles with a wave nature. All particles, either charged or uncharged, exhibit wave like characteristics.

For his discovery of the wave nature of electrons, Louis de Broglie was awarded the Nobel Prize in Physics in 1929. Davisson shared the 1937 Nobel Prize in Physics with Thomson, for their joint discovery of the diffraction of electrons by crystals.

Of course we now have a real dilemma; waves behave like particles and particles behave like waves. How can this be? This so-called wave-particle duality paradox was not resolved until the advent of Quantum Mechanics in 1926-27. There are plenty of physicists today who argue that it has still not really been resolved. To paraphrase what the late Richard Feynman once said, 'If someone tells you that they understand Quantum Mechanics, they are fooling themselves'.

ATOMIC, NUCLEAR AND PARTICLE PHYSICS

Example

Calculate the de Broglie wavelength of an electron after acceleration through a potential difference of 75 V.

Solution

Use $\lambda = \dfrac{h}{\sqrt{2mVe}}$

$\lambda = \dfrac{6.6 \times 10^{-34}}{\sqrt{2 \times 9.1 \times 10^{-31} \times 75 \times 1.6 \times 10^{-19}}} = 0.14$ nm.

Example

Baseball pitches have been clocked in excess of 90 mph (about 145 km h^{-1}). Calculate the wavelength of a 0.150 kg baseball travelling at 145 km h^{-1}.

Solution

The speed of the baseball in m/s is $145 \times \dfrac{1000}{3600} = 40.3$ ms^{-1}

The wavelength of this baseball is $\lambda = \dfrac{h}{mv} = 1.10 \times 10^{-34}$ m.

This wavelength is so small in comparison with the dimension of the baseball, no wave behaviour is to be expected, and the wave is virtually undetectable.

EXERCISE 12.2

1. An electron in a television tube might have a speed of 6.50×10^6 m s^{-1}.

   The de Broglie wavelength associated with the electron is

   A. 8.93×10^{-9} m
   B. 4.73×10^{3} m
   C. 1.12×10^{-10} m
   D. 9.29×10^{-17} m

2. Davisson and Germer's experiment provided evidence of

   A. the wave nature of light.
   B. the wave nature of electrons.
   C. the particle nature of light.
   D. the particle nature of electrons.

3. As an electron travels faster, its de Broglie wavelength

   A. increases.
   B. decreases.
   C. remains the same.
   D. many of the above, depending on how quickly the electron is accelerated.

4. Given the mass of the moon as 7.34×10^{22} kg, its orbital radius as 3.8×10^8 m and its orbital period as 2.36×10^6 s, calculate the wavelength associated with the moon's motion about the earth.

5. Electrons incident on an aluminum crystal at an angle of 15° to the planes form a first order diffraction pattern. If the crystal planes are 2.33×10^{-8} m apart, calculate the wavelength of the electrons.

6. An electron and a proton travel at the same speed. Which has the longer wavelength and why?

7. Electrons are accelerated through a potential difference of 54 V. If they are directed at a beryllium crystal with planes of spacing $d = 7.32 \times 10^{-9}$ m, calculate

   a. their de Broglie wavelength.
   b. their first order angle of diffraction.

8. Calculate the de Broglie wavelength Repeat the example above but for a proton after acceleration through a potential difference of 75 V.

9. Determine the ratio of the de Broglie wavelength of an electron to that of a proton accelerated through the same magnitude of potential difference.

CHAPTER 12

Pair production and pair annihilation

All particles have antiparticles which are identical to the particle in mass and half-integral spin but are opposite in charge to their corresponding particle. In 1928, Paul Dirac, while trying to combine special relativity and electromagnetic theories mathematically, came to the conclusion that particles with the same mass but opposite charge might exist somewhere in the Universe. In 1932, Carl Anderson found that cosmic radiation travelling through a bubble chamber during pair production split up into an electron path and a path with the same mass as an electron but in an opposite direction. Thus the positron had been discovered. In Figure 1211, incoming gamma radiation (from the bottom) produces two circular tracks of a positron (to the left) and an electron (to the right), and a secondary electron-positron track.

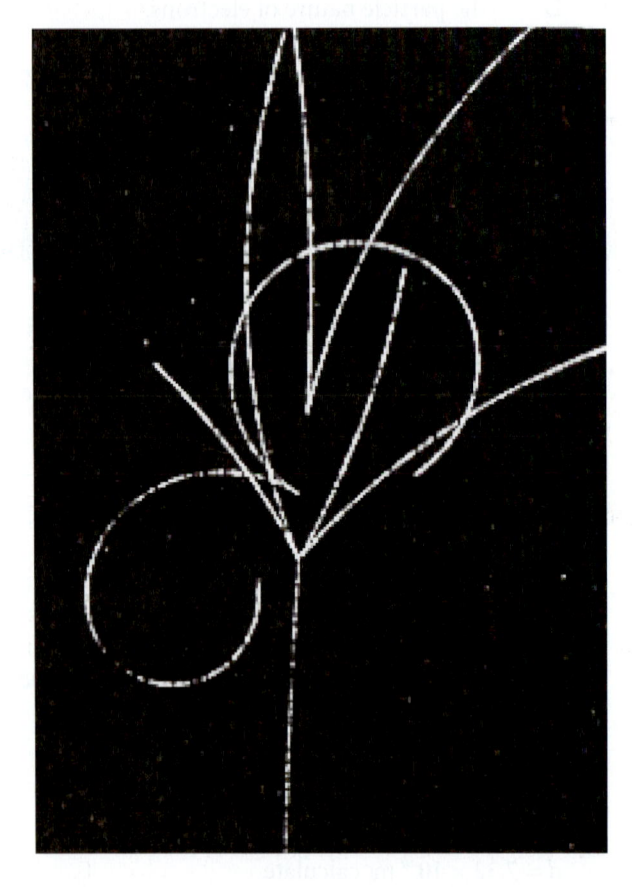

Figure 1211 Electron-positron pair creation

Further evidence came with the discovery of the antineutrino during beta-minus decay. It was not until 1955 that the antiproton was discovered. In particle accelerators, beams of electrons and positrons or protons and antiprotons are accelerated in opposite directions and are then collided with each other in detectors. These collisions produce enormous amounts of energy and the products of the collisions are studied. Today, most antiparticles have been identified in this way. The question still remains as to where has all the antimatter gone that must have been created during the Big Bang.

Although antiparticles have the same mass as their particle pair, they have opposite charge, lepton number, baryon number and strangeness. Some electrically neutral bosons and mesons are their own antiparticle.

Example

Figure 1212 is a sketch of the path of the pair production of an a proton and an antiproton. There is a magnetic field pointing out of the page.

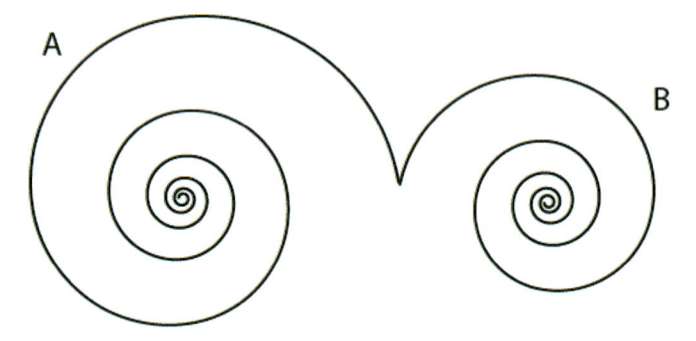

Figure 1212

(a) Explain which of the tracks is due to the antiproton, A or B.

(b) Deduce whether the particles have the same energy.

(c) Calculate the minimum energy required for the pair production in GeV.

Solution

(a) Use a hand rule to show A is an antiproton.

(b) No, because of their different radii. Therefore, they have different speeds and thus different kinetic energy.

(c) $E_{min} = 2m_0c^2 = 2 \times 1.673 \times 10^{-27}$ kg $\times (3 \times 10^8$ ms$^{-1})^2$

$= 3.01 \times 10^{-10}$ J

$= 1.9$ GeV.

When matter (such as an electron) collides with its corresponding antimatter (such as a positron), both particles are annihilated, and 2 gamma rays with the same energy but with a direction at 180° to each other are produced. This is called **pair annihilation**. The direction of the gamma rays produced is in accordance with the law of conservation of momentum and the electron-positron annihilation gives energy equal to $E = mc^2$. This is depicted in Figure 1213.

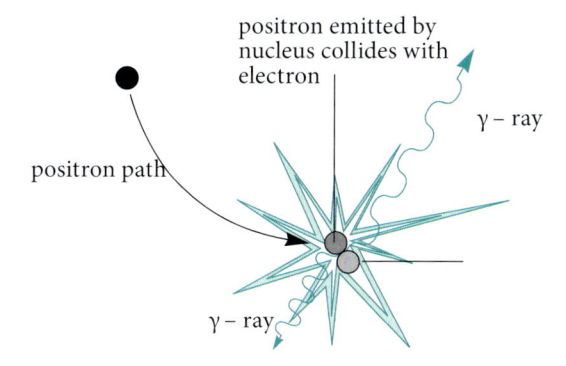

Figure 1213   Electron-positron annihilation

When particles and antiparticles collide, they can annihilate one another, releasing energy in the form of gamma rays. The gamma rays result from conversion of matter to energy, according to Einstein's famous equation, $E = mc^2$. For example, an electron and positron annihilating one another at rest release an amount of energy equal to:

$E = mc^2 = (2 \times 9.11 \times 10^{-31} \text{ kg}) \times (3 \times 10^8 \text{ ms}^{-1})^2$

$= 1.64 \times 10^{-13}$ J $= 1.03$ MeV.

The energy released is higher if the particles annihilate in a collision where one or both contributes kinetic energy to the process.

Particle–antiparticle pairs can also be produced when a gamma ray with sufficient energy passes close by a nucleus. The process is the reverse of annihilation and is called **pair production**. The law of conservation of mass-energy requires particles and antiparticles to be produced in pairs.

Figure 1214 gives the Feynman diagrams for pair annihilation and pair production.

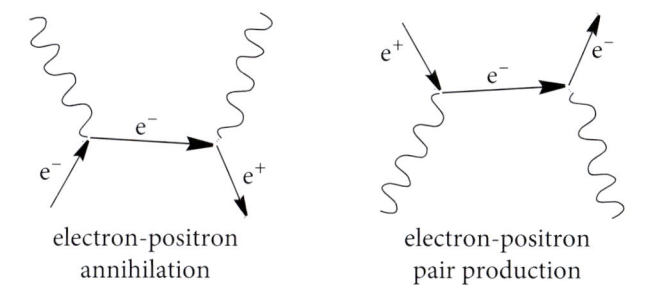

Figure 1214   Feynman diagrams for electron-positron annihilation and production

Note that the backward arrow is an antiparticle, in this case a positron. Also remember the space-time concept of the Feynman diagram with time progressing upwards.

Even though the arrow of the antiparticle is downward, the antiparticle is still progressing upwards in time.

Providing sufficient energy is available, particles other than photons can be produced.

Quantization of angular momentum in the Bohr model for hydrogen

In 1913 Neils Bohr (1885-1962) was working in the Manchester laboratory of Ernest Rutherford. Accordingly, he was fascinated with the planetary model of the atom, and he looked for an explanation as to why the electron in the hydrogen was not drawn into the nucleus by the electrostatic forces of attraction to the protons in the nucleus. He eventually put forward a number of postulates:

1. The electron travels in circular orbits around the positively charged nucleus, but only certain orbits are allowed. The electron will not radiate energy as long as it is in one of these orbits.

2. The allowed orbits have radii rn. For the hydrogen atom, the formula for the radii is
$r_n = (0.53 \times 10^{-10} \text{ m}) n^2$ where $n = 1, 2, 3…$

3. The allowable orbits have angular momentum (mvr) given by

$L = mvr_n = \dfrac{nh}{2\pi}$

Where m = the mass of the electron, v is the velocity, rn is the radius of the orbit, n = 1, 2, 3… and h is Planck's constant.
This is a particularly interesting because it is clear that the allowable energy levels have angular momentum is quantised.

4. The allowable orbits have energy given by

$E = \dfrac{k}{n^2}$

For the hydrogen atom, the constant k = -2.18 × 10⁻¹⁸ J or -13.6 eV, and *n* is any integer = 1, 2, 3… All energy values are negative because the energy of an electron at rest outside an atom is taken to be zero, and when an electron "falls" into the atom, energy is lost as electromagnetic radiation. The electron is passing from a higher energy level to a lower energy level, and the lower the energy level the larger the negative energy value. The amount of energy required to remove an electron from an energy level to infinity is called the ionisation energy.

357

The maximum ionisation energy is the energy required to remove an electron in its ground state to infinity that is given by

$$E = \frac{13.6 \text{ eV}}{n^2}$$

5. If an electron falls from one orbit, or energy level, to another, it loses energy in the form of a photon of light. The energy of the photon equals the difference between the energies of the two orbits. The frequency of the radiation can be calculated using E = hf. These calculated frequencies correspond to the lines in the emission spectra of the atom.

The wave function

We can understand the origin of the existence of discrete energy levels within the atom if we consider the wave nature of the electron. To simplify matters, we shall consider the electron to be confined by a one dimensional box rather than a three dimensional "box" whose ends follow a r^{-1} shape.

In classical wave theory, a wave that is confined is a standing wave. If our electron box is of length L then the allowed wavelengths λ_n are give by (see Topic 4.5)

$$\lambda_n = \frac{2L}{n} \quad \text{where } n = 1, 2, 3 \dots$$

However from the de Broglie hypothesis we have that

$$p_n = \frac{h}{\lambda_n} = \frac{nh}{2L}$$

Using $E_k = \frac{p^2}{2m}$ we have therefore that the energy of the electron E_n with wavelength λ_n

$$E_n = \frac{n^2 h^2}{8 m_e L^2}$$

Hence we see that the energy of the electron is quantized.

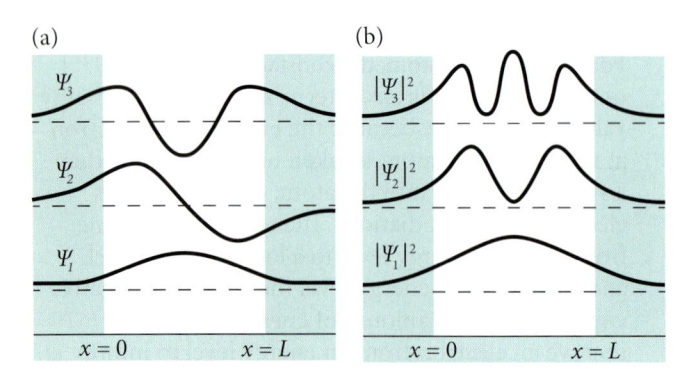

(a) (b)

Figure 1215   Wave functions

A **potential well** is the region surrounding a local minimum of potential V. In normal classical physics, energy captured in a potential well is unable to convert to other types of energy because it is captured in the local minimum of the potential well. However, in quantum physics, potential energy may escape a potential well without added energy due to the **probability characteristics** P of quantum particles; in these cases a particle may be imagined to tunnel through the walls of a potential well as we will shortly examine.

In 1926 *Erwin Schrödinger* (1887-1961) proposed a model of the hydrogen atom based on the wave nature of the electron and hence the de Broglie hypothesis. This was actually the birth of Quantum Mechanics. Quantum mechanics and General Relativity are now regarded as the two principal theories of physics. Schrödinger's major contribution to quantum mechanics was his mathematical description of the dual nature of matter.

The mathematics of Schrödinger's so-called **wave mechanics** is somewhat complicated so at this level, the best that can be done is to outline his theory. Essentially, he proposed that the electron in the hydrogen atom is described by a wave function. This wave function Ψ is described by an equation known as the Schrödinger wave equation, the solution of which give the values that the wave function can have.

P (r) = | Ψ² | ΔV

Where P is probability, r is a radial place as in the Bohr model, | Ψ² | is the probability density and ΔV is the change in potential.

He reasoned that the absolute value of the probability density | Ψ |² for a given particle at a given time or place is proportional to the probability of finding that particle at that place and time. By comparing the probability densities in Figure 1215(a) to the possible states of a particle in a finite potential well in Figure 1215(b), a larger value of Ψ2 is indicative of a high probability of finding a particle at a given place and time.

If the equation is set up for the electron in the hydrogen atom, it is found that the equation will only have solutions for which the energy E of the electron is given by E = (n + ½) hf. Hence the concept of quantization of energy is built into the equation. Of course we do need to know what the wave function is actually describing. In fact the square of the amplitude of the wave function measures the probability of locating the electron in a specified region of space.

When the equation is used to plot a probability region for the position of the electron at a given time, the region takes a spherical shape that is more dense near the nucleus and less dense farther from the nucleus. This probability region came to be known as an electron cloud or orbital.

The solution of the equation predicts exactly the line spectra of the hydrogen atom. If the relativistic motion of the electron is taken into account, the solution even predicts the fine structure of some of the spectral lines. (For example, the red line on closer examination is found to consist of seven lines close together.)

The Schrödinger equation is not an easy equation to solve and to get exact solutions for atoms other than hydrogen or singly ionised helium is well-nigh impossible. Nonetheless, Schrödinger's theory changed completely the direction of physics and opened whole new vistas and posed a whole load of new philosophical problems. Eventually, the equation facilitated the description of a series of electron orbitals, allowing every electron surrounding a given atom to be assigned a space within which it was likely to be found at any given time.

The uncertainty principle for energy, time, position and momentum

The Heisenberg Uncertainty Principle

In 1927 *Werner Heisenberg* proposed a principle that went along way to understanding the interpretation of the *Schrödinger* wave function. Suppose the uncertainty in our knowledge of the position of a particle is Δx and the uncertainty in the momentum is Δp, then the **Uncertainty Principle** states that the product $\Delta x \Delta p$ is at least the order of h, the Planck constant. A more rigorous analysis shows that

$$\Delta x \Delta p \geq \frac{h}{4\pi}$$

To understand how this links in with the de Broglie hypothesis and wave functions, consider a situation in which the momentum of a particle is known precisely. In this situation, the wavelength is given by $\lambda = \frac{h}{p}$ and is completely defined. But for a wave to have a single wavelength it must be infinite in time and space. For example if you switch on a sine-wave signal generator and observe the waveform produced on an oscilloscope, you will indeed see a single frequency/wavelength looking wave. However, when you switch off the generator, the wave amplitude decays to zero and in this decay there will be lots of other wavelengths present. So if you want a pure sine wave, don't ever switch the generator off and, conversely, never switch it on. For our particle what this means is that if its momentum is defined precisely, then its associated probability wave is infinite in extent and we have no idea where it is. Effectively, the more precisely we know the momentum of a particle, the less precisely we know its position and vice versa.

From the argument above, we see that in the real world waves are always made up of a range of wavelengths and form what is called a wave group. If this group is of length Δx, then classical wave theory predicts that the wavelength spread $\Delta \lambda$ in the group is given by

$$\Delta\left(\frac{1}{\lambda}\right)\Delta x \approx 1$$

If we consider the wave group to be associated with a particle i.e. a measure of the particle momentum, then

$$\Delta\left(\frac{1}{\lambda}\right) = \frac{\Delta p}{h}$$

that is

$$\frac{\Delta p}{h}\Delta x \approx 1$$

or

$$\Delta p \Delta x \approx h$$

We have in fact used the classical idea of a wave but we have interpreted the term $\Delta\left(\frac{1}{\lambda}\right)$ as a measure of the uncertainty in the momentum of a particle and this leads to the idea of the Uncertainty Principle. You will not be expected to produce this argument in an IB examination and is presented here only as a matter of interest.

The Principle also applies to energy and time. If ΔE is the uncertainty in a particle's energy and Δt is the uncertainty in the time for which the particle is observed then

$$\Delta E \Delta t = \frac{h}{4\pi}$$

This is the reason why spectral lines have finite width. For a spectral line to have a single wavelength, there must be no uncertainty in the difference of energy between the associated energy levels. This would imply that the electron must make the transition between the levels in zero time.

Tunnelling, potential barrier and factors affecting tunnelling probability

It is possible to show that a particle such as an electron can exist outside the potential well or potential barrier. It forms a sinusoidal wave and thus the wave function Ψ is not zero. This contradicts classical physics. Furthermore, outside the potential well, it is found that the potential energy is greater than the total energy, and this violates the conservation of energy. If you refer back to Figure 1215(b), you can see that the electron can spend some time outside the classical potential barrier but not by much as $|\Psi|^2$ drops off exponentially with distance from either wall.

If one reasons from the uncertainty principle in the form

$\Delta E \Delta t \geq h$

one can accept this nonconservation of energy because the energy is uncertain by an amount ΔE for a very short time Δt.

It can also be shown that a particle such as an electron can penetrate a barrier into a region not allowed in classical physics. There are many applications of this tunneling phenomenon especially when the barrier wall is thin. Without the tunneling effect most computers would not work because they use semiconductor tunnel diodes. Quantum tunelling is also important in the scanning tunneling electron microscope, nuclear fusion reactors and fusion in stars.

Refer to the following Figure 1216(a). Consider a particle with mass m and energy E travelling along the x axis to the right in free space. Along this axis the potential energy U is zero and $E = EK$. Now suppose the particle encounters a thin potential barrier of thickness L whose height U_0 in energy units is greater than E.

In classical physics, one would expect that upon encountering the barrier, the particle would not penetrate the barrier and it would be reflected in the opposite direction from which it came. This still holds true for macroscopic objects.

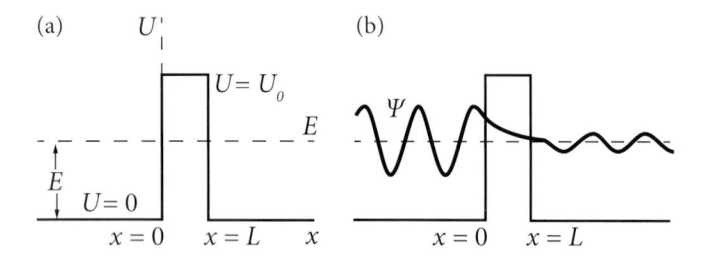

Figure 1216   Tunelling through a barrier.

However, quantum mechanics predicts the probability for finding the particle on the other side of the barrier. Referring to Figure 1216(b) and the wave function, the approaching wave has a sinusoidal wave function. As predicted by the Schrödinger Equation there is an exponential decay occurring within the barrier. However, it does not drop to zero by the time it reaches $x = L$. After L you can again see a sinusoidal wave that has been greatly reduced in amplitude.

This process is called tunneling through the barrier. The particle cannot be observed within the barrier but it can be detected after it has passed through the barrier.

12.2 Nuclear physics

NATURE OF SCIENCE:

Theoretical advances and inspiration: Progress in atomic, nuclear and particle physics often came from theoretical advances and strokes of inspiration.

Advances in instrumentation: New ways of detecting subatomic particles due to advances in electronic technology were also crucial.

Modern computing power: Finally, the analysis of the data gathered in modern particle detectors in particle accelerator experiments would be impossible without modern computing power. (1.8)

© IBO 2014

Essential idea: The idea of discreteness that we met in the atomic world continues to exist in the nuclear world as well.

Understandings:
- Rutherford scattering and nuclear radius
- Nuclear energy levels
- The neutrino
- The law of radioactive decay and the decay constant

Rutherford scattering and nuclear radius

In earlier chapters an outline was given as to how the Geiger-Marsden experiment, in which α-particles were scattered by gold atoms, provided evidence for the nuclear model of the atom. The experiment also enabled an estimate of the nuclear diameter to be made.

Figure 1217 shows an α-particle that is on a collision course with a gold nucleus and its subsequent path. Since the gold nucleus is much more massive than the α-particle we can ignore any recoil of the gold nucleus.

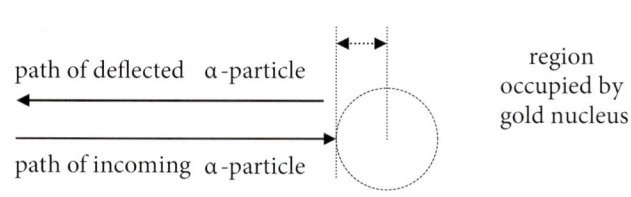

Figure 1217   An α-particle colliding with a gold nucleus

Atomic, Nuclear and Particle Physics

The kinetic energy of the α-particle when it is a long way from the nucleus is E_k. As it approaches the nucleus, due to the Coulomb force, its kinetic energy is converted into electrostatic potential energy. At the distance of closest approach all the kinetic energy will have become potential energy and the α-particle will be momentarily at rest. Hence we have that

$$E_k = \frac{Ze \times 2e}{4\pi\varepsilon_0 d} = \frac{Ze^2}{2\pi\varepsilon_0 d}$$

where Z is the proton number of gold such that the charge of the nucleus is Ze. The charge of the α-particle is $2e$.

For an α-particle with kinetic energy 4.0 MeV we have that

$$d = \frac{Ze^2}{2\pi\varepsilon_0 E_k} = \frac{79 \times 1.6 \times 10^{-19}}{2 \times 3.14 \times 8.85 \times 10^{-12} \times 4.0 \times 10^6}$$

$$= 5.6 \times 10^{-14} \quad 5.7 \times 10^{-14} \text{ m} = 0.57 \text{ fm}$$

(1×10^{-15} m = 1 femtometre = 1 fm = 1 fermi)

The distance of closest approach will of course depend on the initial kinetic energy of the α-particle. However, as the energy is increased a point is reached where Coulomb scattering no longer take place. The above calculation is therefore only an estimate. It is has been demonstrated at separations of the order of 10^{-15} m, the Coulomb force is overtaken by the strong nuclear force.

Nuclear forces are attractive and short ranged, and the size of the force between nucleons is dependent on their nucleon separation. Figure 1218 demonstrates how the nuclear force **F** varies with distance d between nucleons.

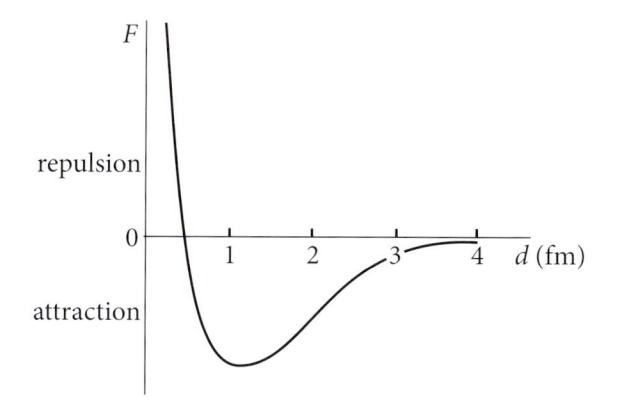

Figure 1218 Nuclear force varies with distance between nucleons.

When the nucleons are very close together they repel each other. At a distance of 1 fermi apart, that is 1×10^{-15} m, the nucleons experience the strong nuclear force. At a distance of 4 fm apart, the nuclear forces are negligible, and electrostatic forces of repulsion between the nucleons predominate and make the nucleus unstable.

Experiments carried out in the 1920s concerning the scattering of alpha particles by various nuclei enabled physicists to calculate the closest approach of alpha particles to the nucleus.

Electrons can be used in collisions to indirectly identify protons inside the nucleus, and we now know that at high energies the de Broglie wavelength of an electron is small enough to resolve particles inside the proton. Modern direct measurements are based on the scattering of electrons by nuclei.

If electrons are accelerated to very high energies, their wave-like properties become important according to the de Broglie wavelength. The electron waves can be diffracted around the nucleus when their wavelengths are similar to nuclear sizes, and their electron diffraction patterns can be studied.

$$\lambda = \frac{h}{p} \text{ and } E = \frac{hc}{\lambda}$$

Therefore

$$\lambda = \frac{h}{p} = \frac{hc}{E}$$

For electrons with energy of 300 MeV, the de Broglie wavelength would be

$$\lambda = \frac{6.64 \times 10^{-34} \text{ Js} \times 3.0 \times 10 \text{m s}^{-1}}{300 \times 10^6 \text{ eV} \times 1.6 \times 10^{-19} \text{ C}}$$

$= 4.15 \times 10^{-15}$ m.

From diffraction theory

$$\sin \theta = \frac{1.22 \lambda}{D}$$

where D is the nuclear diameter of the target atom.

Let us say that the first minimum of the diffraction pattern is found when θ = 750. Then

$$\sin 75 = \frac{1.22 \times 4.15 \times 10^{-15} \text{ m}}{D}$$

$D = 5.24 \times 10^{-15}$ m

It was shown experimentally that the maximum radius R of a nucleus is given by:

R = R$_0$ A^{⅓}

where R$_0$ is a constant called the Fermi radius and is equal to 1.20×10^{-15} m, and A is the nucleon number (mass number).

Chapter 12

Example

Determine the radius of a uranium-238 nucleus.

Solution

$R = R_0 A^{1/3}$

$R = (1.2 \text{ fm}) \times (238)^{1/3} = 7.437 \text{ fm}$.

Various types of scattering experiments suggest that nuclei are roughly spherical and appear to have essentially the same density. "Students should be aware that nuclear densities are approximately the same for all nuclei and that the only macroscopic objects with the same density as nuclei are neutron stars".

Nuclear energy levels

The α-particles emitted in the radioactive decay of a particular nuclide do not necessarily have the same energy. For example, the energies of the α-particles emitted in the decay of nuclei of the isotope thorium-C have six distinct energies, 6.086 MeV being the greatest value and 5.481 MeV being the smallest value. To understand this we introduce the idea of nuclear energy levels. For example, if a nucleus of thorium emits an α-particle with energy 6.086 MeV, the resultant daughter nucleus will be in its ground state. However, if the emitted α-particle has energy 5.481 MeV, the daughter will be in an excited energy state and will reach its ground state by emitting a gamma photon of energy 0.605 MeV. Remember that the energy of photons emitted by electron transitions in the hydrogen atom which are only of the order of several eV.

The existence of nuclear energy levels receives complete experimental verification from the fact that γ-rays from radioactive decay have discrete energies consistent with the energies of the α-particles emitted by the parent nucleus. Not all radioactive transformations give rise to γ-emission and in this case the emitted α-particles all have the same energies.

The neutrino

It was earlier stated that β- decay results from the decay of a neutron into a proton and that β+ decay results from the decay of a proton in a nucleus into a neutron. For example

$_0^1 n \rightarrow {}_1^1 p + {}_{-1}^0 e + \bar{\nu}$

It is found that the **energy spectrum of the β-particles is continuous** whereas that of **any γ-rays involved is discrete**. This was one of the reasons that the existence of the neutrino was postulated otherwise there is a problem with the conservation of energy. α-decay clearly indicates the existence of nuclear energy levels so something in β-decay has to account for any energy difference between the maximum β-particle energy and the sum of the γ-ray plus intermediate β-particle energies. We can illustrate how the neutrino accounts for this discrepancy by referring to Figure 1219 that show the energy levels of a fictitious daughter nucleus and possible decay routes of the parent nucleus undergoing β+ decay.

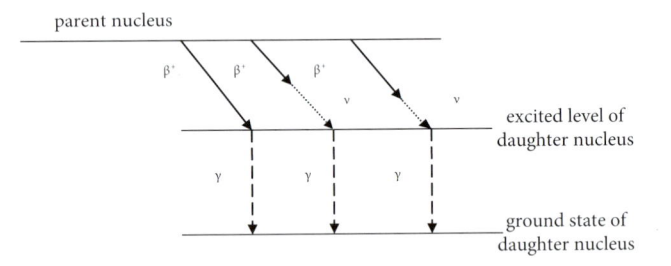

Figure 1219 *Neutrinos and the conservation of energy*

The figure shows how the neutrino accounts for the continuous β spectrum without sacrificing the conservation of energy. An equivalent diagram can of course be drawn for β- decay with the neutrino being replaced by an anti-neutrino.

The law of radioactive decay and the decay constant

We have seen in chapter 7 that radioactive decay is a random process. However, we are able to say that the activity of a sample element at a particular instant is proportional to the number of atoms of the element in the sample at that instant. If this number is N we can write that

$$\frac{\Delta N}{\Delta t} = -\lambda N$$

where λ is the constant of proportionality called the **decay constant** and is defined as 'the probability of decay of a nucleus per unit time'.

The above equation should be written as a differential equation i.e.

$$\frac{dN}{dt} = -\lambda N$$

This equation is solved by separating the variables and integrating viz,

$$\frac{dN}{N} = -\lambda dt$$

such that

$\ln N = -\lambda t + \text{constant}$

It is then said that at time $t = 0$ the number of atoms is N_0 such that

$\ln N_0 = \text{constant}$

therefore

$\ln N - \ln N_0 = -\lambda t$

or

$N = N_0 e^{-\lambda t}$

This is the radioactive decay law and verifies mathematically the exponential nature of radioactive decay that we introduced in chapter 7.

The radioactive decay law enables us to determine a relation between the **half-life** of a radioactive element and the decay constant.

If a sample of a radioactive element initially contains N_0 atoms, after an interval of one half-life the sample will contain $\frac{N_0}{2}$ atoms. If the half-life of the element is $T_{1/2}$ from the decay law we can write that

$$\frac{N_0}{2} = N_0 e^{-\lambda T_{1/2}}$$

or

$e^{-\lambda T_{1/2}} = \frac{1}{2}$

that is

$T_{1/2} = \frac{\ln 2}{\lambda}$

The method used to measure the half-life of an element depends on whether the half-life is relatively long or relatively short. If the activity of a sample stays constant over a few hours it is safe to conclude that it has a relatively long half-life. On the other hand if its activity drops rapidly to zero it is clear it has a very short half-life.

Elements with long half-lives

Essentially the method is to measure the activity of a known mass of a sample of the element. The activity can be measured by using a Geiger counter, and the decay equation in its differential form is used to find the decay constant. An example will help understand the method.

A sample of the isotope uranium-234 has a mass of 2.0 µg. Its activity is measured as 3.0×10^3 Bq. The number of atoms in the sample is

$\frac{2.0 \times 10^{-6}}{N_A} = \frac{2.0 \times 10^{-6}}{6.0 \times 10^{23}} \quad 3.3 \times 10^{-16}$

Using $\frac{\Delta N}{\Delta t} = -\lambda N$ we have

$\lambda = \frac{\frac{\Delta N}{\Delta t}}{N} = \frac{3.0 \times 10^3}{3.3 \times 10^{16}} = 9.0 \times 10^{-14} \text{ s}^{-1}$

Using the relation between half-life and decay constant we have

$T_{1/2} = \frac{\ln 2}{\lambda} = \frac{0.69}{9.0 \times 10^{-14}} = 7.6 \times 10^{12} \text{ s} \approx 2.4 \times 10^5 \text{ years}$

Elements with short half-lives

For elements that have half-lives of the order of hours, the activity can be measured by measuring the number of decays over a short period of time (minutes) at different time intervals. A graph of activity against time is plotted and the half-life read straight from the graph. Better is to plot the logarithm of activity against time to yield a straight line graph whose gradient is equal to the negative value of the decay constant.

For elements with half-lives of the order of seconds, the ionisation properties of the radiations can be used. If the sample is placed in a tube across which an electric field is applied, the radiation from the source will ionise the air in the tube and thereby give rise to an ionisation current. With a suitable arrangement, the decay of the ionisation current can be displayed on an oscilloscope.

The above is just an outline of the methods available for measuring half-lives and is sufficient for the HL course. Clearly in some cases the actual measurement can be very tricky. For example, many radioactive isotopes decay into isotopes that themselves are radioactive and these in turn decay into other radioactive isotopes. So, although one may start with a sample that contains only one radioactive isotope, some time later the sample could contain several radioactive isotopes.

CHAPTER 12

EXERCISE 12.3

1. The activity of a certain radioactive isotope decreases from 1000 Bq to 800 Bq in 1 hour. The half-life of the isotope is

    A. 2.5 hours
    B. 1.6 hours
    C. 6.2 hours
    D. 3.1 hours

2. The half-life of carbon-14 is 5730 years. A fossilized tree is found to have 21.0% of its original carbon–14 activity. How long ago did the tree die?

    A. 12 900 years
    B. 11 460 years
    C. 17 190 years
    D. 8 595 years

3. Strontium-90 has a half-life of 29.1 years. If a newborn baby ingests milk containing strontium–90, what percentage of the original strontium will remain in her bones if she dies at age 75?

    A. none
    B. 7.5%
    C. 17%
    D. 25%

4. Pair production results when

    A. particles and antiparticles annihilate one another.
    B. particles of sufficient energy pass close to a nucleus.
    C. gamma rays of sufficient energy pass close to a nucleus.
    D. gamma rays and particles annihilate one another.

5. The neutrino belongs to the same family as the

    A. neutron.
    B. proton.
    C. electron.
    D. baryon.

6. Sodium-24 (proton number = 11) undergoes β decay with a half–life of about 15 hours.

    a. What is the decay constant for sodium–24?
    b. If the initial activity of a sample is 3.6 x 103 Bq, what will its activity be after 6 hours?

7. Calculate the energy released when a proton and antiproton annihilate one another

    a. at rest.
    b. in a collision where each has kinetic energy of 25 MeV (assume no energy is lost).

8. If a γ-ray is to produce a neutron-antineutron pair, what is the minimum energy, in MeV, that it must have?

9. The isotope radium-223 has a half-life of 11.2 days. Determine

    (i) the decay constant for radium-223
    (ii) the fraction of a given sample that will have decayed after 3 days.

10. The isotope technetium-99 has a half-life of 6.02 hours. A freshly prepared sample of the isotope has an activity of 640 Bq. Calculate the activity of the sample after 8.00 hours.

11. A radioactive isotope has a half-life of 18 days. Calculate the time it takes for of the atoms in a sample of the isotope to decay.

12. A nucleus of potassium-40 decays to a stable nucleus of argon-40. The half-life of potassium-40 is 1.3 × 109 yr.

    In a certain lump of rock, the amount of potassium-40 is 2.1 µg and the amount of trapped argon-40 is 1.7 µg. Estimate the age of the rocks.

13. Option A: Relativity

Contents

A.1 – The beginnings of relativity
A.2 – Lorentz transformations
A.3 – Spacetime diagrams
A.4 – Relativistic mechanics
A.5 – General relativity

Essential Ideas

Einstein's study of electromagnetism revealed inconsistencies between the theory of Maxwell and Newton's mechanics. He recognized that both theories could not be reconciled and so choosing to trust Maxwell's theory of electromagnetism he was forced to change long-cherished ideas about space and time in mechanics.

Observers in relative uniform motion disagree on the numerical values of space and time coordinates for events, but agree with the numerical value of the speed of light in a vacuum. The Lorentz transformation equations relate the values in one reference frame to those in another. These equations replace the Galilean transformation equations that fail for speeds close to that of light.

Spacetime diagrams are a very clear and illustrative way to show graphically how different observers in relative motion to each other have measurements that differ from each other.

The relativity of space and time requires new definitions for energy and momentum in order to preserve the conserved nature of these laws.

General relativity is applied to bring together fundamental concepts of mass, space and time in order to describe the fate of the universe. © IBO 2014

CHAPTER 13 (OPTION A)

A.1 The beginnings of relativity

> **NATURE OF SCIENCE:**
>
> Paradigm shift: The fundamental fact that the speed of light is constant for all inertial observers has far-reaching consequences about our understanding of space and time. Ideas about space and time that went unchallenged for more than two thousand years were shown to be false. The extension of the principle of relativity to accelerated frames of reference leads to the revolutionary idea of general relativity that the mass and energy that spacetime contains determines the geometry of spacetime. (2.3)
>
> © IBO 2014

Essential idea: Einstein's study of electromagnetism revealed inconsistencies between the theory of Maxwell and Newton's mechanics. He recognized that both theories could not be reconciled and so choosing to trust Maxwell's theory of electromagnetism he was forced to change long-cherished ideas about space and time in mechanics.

Understandings:
- Reference frames
- Galilean relativity and Newton's postulates concerning time and space
- Maxwell and the constancy of the speed of light
- Forces on a charge or current

> **TOK INTRODUCTION**
>
> This is another opportunity, as with Quantum Mechanics to discuss the concept of a 'paradigm shift' and its relation to the development of scientific understanding. In particular, the Einstein Theory of Relativity gives rise to the completely different view of space and time than that put forward by Newton.

Reference frames

All measurement must be made relative to some **frame of reference**. Usually this frame of reference in physics experiments will be your school laboratory and in mathematics, the conventional Cartesian reference frame consists of three mutually perpendicular axes x, y and z.

To understand why the concept of a frame of reference is so important, suppose that you were asked to carry out an experiment to measure the acceleration due to gravity at the Earth's surface by timing the period of oscillation of a simple pendulum whilst on a fair-ground merry-go-ground. You would certainly expect to get some very unusual results. Your overall perspective of the world would in fact be very different from the world observed by somebody not on the ride. You might well expect the laws of physics to be different. Yet we all live on a merry-go-round. The Earth spins on its axis as it orbits the sun. Fortunately the Earth spins relatively slowly compared to the merry-go-round so most of the time we can ignore the effects. However, because of the Earth's rotation, the acceleration due to gravity has a different value at the poles than that at the equator. Also you certainly cannot ignore the rotation of the Earth when making astronomical observations or measurements.

Newton was well aware of the complications produced by making measurements relative to the Earth. Furthermore, he felt that, for the laws of physics to be precisely valid, then all observations must be made in a reference system that it is at rest or in a reference system that is moving with uniform speed. Based on the work of Galileo he stated a theory of "relativity" as follows: "The motions of bodies included in a given space are the same among themselves, whether the space is at rest or moves uniformly forward in a straight line". The question of course arises, at rest with respect to what or at uniform speed with respect to what?

In this option we shall see how the search for a reference system that truly is at rest led to a radical re-think about the nature of space and time culminating in the two relativity theories of Einstein. We shall in fact see that the Laws of Physics are always true even if you live on a fair ground merry-go-round.

A frame of reference which is moving with uniform velocity or which is at rest is known as an **inertial reference frame**. When your maths teacher draws the conventional x and y axes on the board he or she is in fact expecting you to regard this as an inertial reference frame and ignore any effects of the Earth's motion. Following in your maths teacher's footsteps we will for the time being assume in this section on Special Relativity that we are dealing with truly inertial reference frames. (A space ship far away from any gravitational effects drifting along with constant velocity is a pretty good approximation of an inertial reference frame).

Galilean relativity and Newton's postulates concerning time and space

It is generally accepted that the study of motion that involves the concepts of *space* and *time* began with Galileo because he proposed that motion must be *relative* – it involves the displacement of objects relative to some reference system. He was a supporter of the Copernican heliocentric model that stated that the earth revolved around the Sun that lay at the centre of our solar system. Most scientists in Galileo's time could not accept that the earth was *rotating* on its axis and *revolving* around the Sun because they argued that if this was true, a stone dropped from a high tower would "lag behind" the spinning earth, and this did not happen.

Galileo explained this paradox by arguing that the stone and the earth shared the same rotating motion. Projectile motion describes that the horizontal and vertical motion are independent of each other, and Galileo proposed that the stone would fall straight down relative to the tower.

Galileo performed a number of experiments including dropping objects from the mast of a moving ship, and these experiments confirmed in his mind that the motion of the ship did not affect the way the object fell but depended on the **frame of reference** that was being examined. He summarised his ideas in what has come to be called the principle of Galilean relativity:

- The laws of mechanics are the same for a body at rest and a body moving with constant velocity.

The idea of a frame of reference came important in Newtonian relativity that states:

- It is impossible to do any mechanical experiment wholly within an inertial frame of reference that can tell you whether the frame is at rest or moving with constant velocity.

In Figure 1301 Paul regards his reference system to be at rest and Mary's reference system to be moving away from him at a constant speed v.

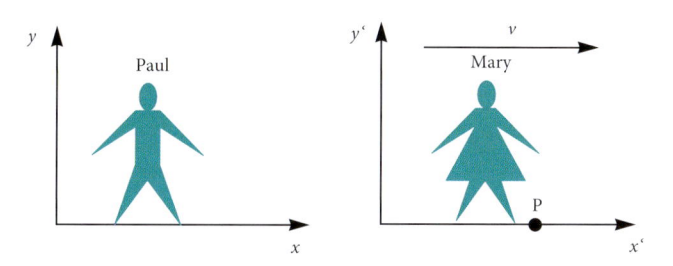

Figure 1301   Reference systems

Mary measures the point P to be at the point x'. Paul on the other hand will measure the point to be at the point x as measured in his reference system where $x = x' + vt$ and t is the time that has elapsed from the moment when the two reference systems were together. We have therefore that, $x' = x - vt$. (*Note that in all that follows we will only consider motion in the x–direction*)

Suppose now that Mary observes an object in her reference system to be moving with speed u' in the x direction, then clearly Paul will observe the object to be moving with speed $u = u' + v$, hence we have $u' = u - v$. It is not difficult to show that if Mary were to measure the acceleration of an object as a' then Paul would measure the acceleration of the object as $a = a'$. In this respect they would both interpret Newton's Second Law (in it's basic form, $F = ma$) identically. This means that there is no mechanics experiment that Mary or Paul could carry out to determine whether they were at rest or whether they were moving at constant speed in a straight line.

Exercise 13.1

1. Figure 1302 below shows three buoys A, B and C in a river. The river flows in the direction shown and the speed of the current is 2.0 m s^{-1}.

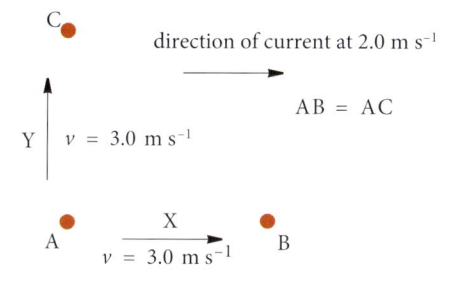

   Figure 1302

   The distance AB is equal to the distance AC. A swimmer X swims from A to B and back with a steady speed of 3.0 m s^{-1} relative to the water. At the same time X leaves A, a swimmer Y sets off to C and back with the same steady speed relative to the water.

   (a) Determine, for an observer on the bank the speed of

   i.   X as she swims towards B.
   ii. X as she swims back to A.
   iii. Y as he swims towards C.
   iv. Y as he swims back to A.

   (b) Calculate the ratio of the times for the two journeys.

CHAPTER 13 (OPTION A)

Maxwell and the constancy of the speed of light

In 1864 *Clerk Maxwell* published his electromagnetic theory of light in which he united the separate studies of Optics and Electromagnetism. His theory was one of the great unification points in physics comparable with the *Galileo/Newton* idea that the laws of celestial and terrestrial mechanics were the same. The source of the electric and magnetic fields is electric charge. If an electric charge is stationary with respect to you then you will measure an electrostatic field. If the charge is moving with constant velocity relative to you then you will measure a magnetic field as well. However, if the charge is accelerated then the fields are no longer static and according to *Maxwell* they radiate out from the charge, moving through space with the speed of light.

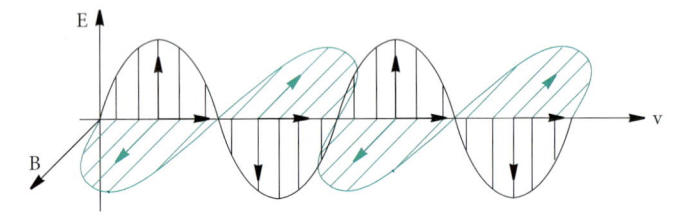

Figure 1303   The nature of EM waves

In its simplest form, the speed of light is given by

$$c = \frac{1}{\sqrt{\varepsilon_0 \mu_0}}$$

where ε_0 is the permittivity of free space and is equal to 8.85×10^{-12} C^2 N^{-1} m^{-2}, and μ_0 is the permeability of free space and is equal to 4.25×10^{-7} T m A^{-1}.

Suppose that the charge in question is a small charged sphere suspended by a string. If it oscillates with a frequency of 1000 Hz then it is a source of long-wave radio waves, at a frequency of 10^9 Hz it becomes a source of television signals. If it oscillates with a frequency of 10^{12} Hz it is a source of infrared radiation, at about 10^{15} Hz it would look yellow and at 10^{18} Hz it would be emitting x-rays. Of course all this is a bit absurd but it does illustrate the fact that the source of all radiation in the electromagnetic spectrum is the accelerated motion of electric charge.

Forces on a charge or current

Maxwell's theory also showed that the speed with which electromagnetic waves travel depends only on the electric and magnetic constants of the medium through which they travel. In a vacuum this means that the speed depends only on ε_0, the permittivity of free space and μ_0, the permeability of free space (see Chapters 7 & 9)

This is fact means that the speed of light (or any other electromagnetic wave) is independent of the speed of the observer. This has far reaching consequences.

Let us return to our two observers Paul and Mary above. Suppose that Mary were to measure the speed of a light pulse as 3×10^8 m s^{-1} in her reference system and that the relative velocity between Paul and Mary is 2×10^8 m s^{-1}. This means that Paul would measure the speed of the pulse to be 5×10^8 m s^{-1}. This, according to Maxwell, is not possible. Also it would seem that by measuring the speed of light pulses Mary and Paul have a means of determining whether they are at rest or not. This seems odd.

We have also seen that the laws of mechanics are the same for all inertial observers so we would expect the laws of electromagnetism to be the same for all inertial observers. If, however, we apply the rules for a Galilean transformation (something that is beyond the scope of this book) we find that the laws of electromagnetism are different for different observers. Clearly something is wrong.

Furthermore, if light is an electromagnetic wave, what is the nature of the medium that supports its motion? To answer this question, physicists of the time proposed that an invisible substance called the **ether** permeates all of space. It was also proposed that this medium is at rest absolutely and is therefore the reference system with respect to which the laws of physics would hold exactly for all inertial observers. Putting it another way, an inertial observer should be able to determine his of her speed with respect to the ether. Meanwhile we shall see how *Einstein's* Special Theory of Relativity resolves the problem we seem to have with Maxwell's theory and the speed of light.

To *Einstein* it seemed that not only the **Laws of Mechanics** should be the same for all observers but all the Laws of Physics including Maxwell's laws. This was an idea first proposed by *Poincaré*. However, *Einstein* realised that this idea must alter completely the then accepted notions of time and space. Einstein was able to see the discrepancy between Maxwell' theory and Newtonian mechanics because in Maxwell's theory the speed of light was independent of its source whereas Newtonian mechanics inferred the speed of light was dependent on the speed of the source of light. In the Special Theory (or restricted theory as it is sometimes called) he confined himself to inertial frames of reference, that is non-accelerating reference frames. We have mentioned inertial reference frames earlier. Another way of describing an inertial frame is to note that it is a frame in which Newton's first law hold true. Clearly, if a reference frame is accelerating, then objects in the reference frame will not remain at rest or continue to move with uniform motion in a straight line.

A.2 Lorentz transformations

> **NATURE OF SCIENCE:**
>
> Pure science: Einstein based his theory of relativity on two postulates and deduced the rest by mathematical analysis. The first postulate integrates all of the laws of physics including the laws of electromagnetism, not only Newton's laws of mechanics. (1.2)
>
> © IBO 2014

Essential idea: Observers in relative uniform motion disagree on the numerical values of space and time coordinates for events, but agree with the numerical value of the speed of light in a vacuum. The Lorentz transformation equations relate the values in one reference frame to those in another. These equations replace the Galilean transformation equations that fail for speeds close to that of light.

Understandings:
- The two postulates of special relativity
- Clock synchronization
- The Lorentz transformations
- Velocity addition
- Invariant quantities (spacetime interval, proper time, proper length and rest mass)
- Time dilation
- Length contraction
- The muon decay experiment

The two postulates of special relativity

There are two postulates of the **Special Theory**:

1. the laws of physics are the same for all inertial observers

2. all inertial observers will measure the same value for the free space velocity of light irrespective of their velocity relative to the source.

Another way of looking at the second postulate is to recognise that it means that speed of light is independent of the speed of the source.

We have seen that the laws of mechanics are the same for all inertial observers. However, the **laws of electromagnetism** do not appear to be so. As we have already seen, if they were, then we would have a means of finding an absolute reference system. The laws of electromagnetism had been verified by careful experiment and were certainly not in error. *Einstein* realised that in fact what was in error was the not the laws of electromagnetism but the nature of the Galilean transformation itself. A Galilean transformation does not take into account the second postulate and it also assumes that time is an absolute quantity. It assumes in fact that all inertial observers will measure the same value for given time intervals. But if, as we shall see, the second postulate of relativity is correct then this cannot be the case.

To see why we can no longer regard time as being absolute, let us see how two events that occur at different points in space and which are simultaneous for one observer, cannot be simultaneous for another observer who observes the events from a different frame of reference. In Figure 1304 observer Y is in a train which is moving with constant speed v as measured by the observer X who is standing by the side of the railway tracks.

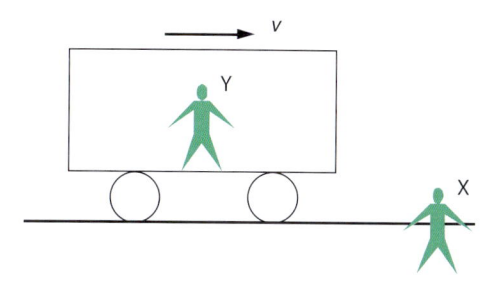

Figure 1304   Situation A

Consider observer, Y, who is at the mid-point of the train. Just as the train reaches a point where she is opposite X, lightning strikes both ends of the train. X sees these two events to take place simultaneously. Refer to Figure 1305.

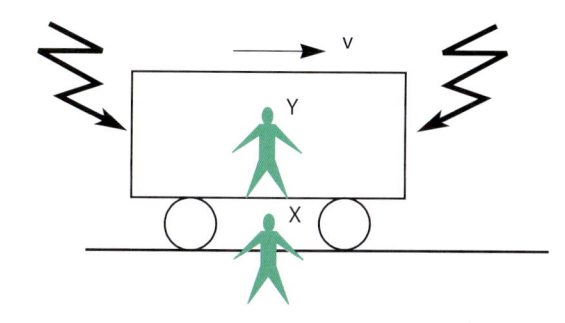

Figure 1305   Situation B

But this will not be the case for Y. Since the speed of light is independent of the speed of the source, by the time the light from each of the strokes reaches Y the train will have moved forward, Figure 1306.

Chapter 13 (Option A)

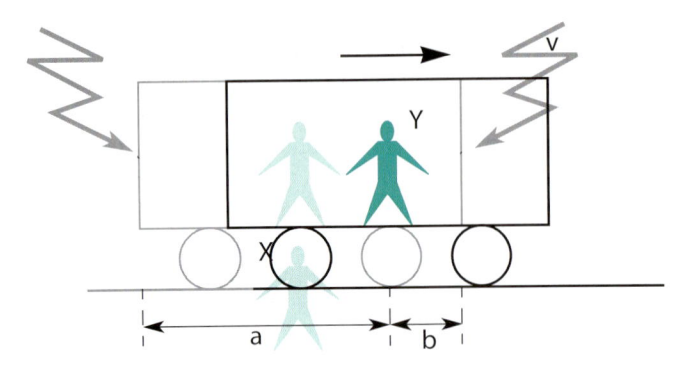

Figure 1306   Situation C

So in effect the light from the strike at the front of the train will reach Y before the light from the strike at the rear of the train. That is, Y has moved forward and, in doing so, has moved closer to where the lightning first hit, so that the light travels a shorter distance in getting to Y. Whereas the light from the back of the cart needs to travel further ($a > b$) and so takes more time to get to Y.

Y will not see the two events as occurring simultaneously.

We might ask 'which observer is correct? Are the two events simultaneous?' In fact both observers are correct. What is simultaneous for one observer is not simultaneous for the other; there is no preferred reference frame. The interpretation of any sequence of events will depend on an individual's frame of reference.

Einstein proposed that the three dimensions of space and the one dimension of time describe a four dimensional space-time continuum and that different observers will describe the same event with different space time co-ordinates. We shall see later on in the chapter how this idea is developed further.

The Lorentz transformations

Shortly after Maxwell's equations of electromagnetism where published it was found, unlike Newton's Laws, that they did not keep their same form under a Galilean transformation. *Lorentz* found that, for them to keep the same form, the transformations in the following table have to be applied.

Galilean	Lorentz
$x' = x - vt$	$x' = \gamma (x - vt)$
$\Delta t' = \Delta t$	$\Delta t' = \gamma \Delta t$

Figure 1307   Galilean–Lorentz transformation equations

Where c is the free space velocity of light, and

$$\gamma = \frac{1}{\sqrt{1 - \frac{v^2}{c^2}}}$$

Δt and $\Delta t'$ refer to a time interval in the respective reference systems.

The Lorentz transformation equations are embedded in the Maxwell equations – the equations that express the behaviour of electric and magnetic fields. There is a certain amount of irony here. Newton was well aware of the concept of relativity and as we have seen, took steps to address the issue in terms of the Galilean transformations. Maxwell on the other hand put his equations together without addressing the relativity issue. When Lorentz addressed the issue, this led to a complete re-assessment of how we think of time and space. The major contribution that Einstein made was to realise that the Lorentz transformation equations can also be derived from the second postulate of Special Relativity. This is not difficult to do but we shall not do so here.

According to the Special Theory all the laws of physics must transform according to the Lorentz transformation equations. The constancy of the speed of light is contained within the laws of electromagnetism but not within Newton's laws under a Galilean transformation. Hence Newton's laws must transform according to the Lorentz equations.

Clock synchronization

To understand the Lorentz equations and to see the effect that they have on our conventional understanding of space and time, let us first look at how our understanding of time is affected.

Let us return to the observer in the moving train. The observer has set up an experiment in which she times how long it takes for a light pulse to bounce back and forth between two mirrors separated by a vertical distance d as shown in Figure 1308. This set up is effectively a **light clock**.

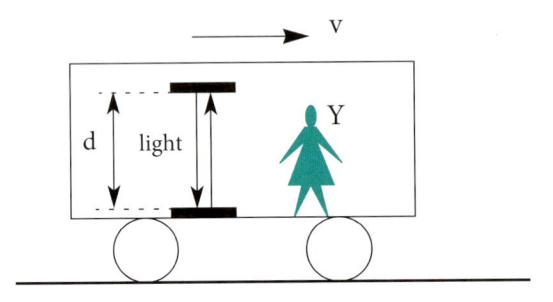

Figure 1308   A simple light clock

RELATIVITY

As measured by Y, the pulse leaves the top mirror at time t'_1 and reaches the bottom mirror at time t'_2. The time interval $t'_2 - t'_1$ is given by:

$$t'_2 - t'_1 = \frac{d}{c}, \text{ where } c \text{ is the speed of light.}$$

This time measured by Y is known as the **proper time**. In general, the term proper time refers to the time interval between two events as measured in the reference system in which the two events occur at the same place.

Invariant quantities (spacetime interval, proper time, proper length and rest mass)

- A space-time system is a coordinate system cinsisting of three dimensions of space and one dimension of time. Inorder to represent the motion of an object we use a space-time diagram. For any given event we will therefore have a space-time interval.

- Proper time is the time interval between two events as measured by an observer that sees the events take place at the same point in space.

- Proper length is the length of an object as measured by an observer at rest with repect to the object.

- Rest mass is the mass of an object as measured by an observer at rest with respect to the object.

Time dilation

How will the observer X who is by the side of the tracks see the light clock?

Suppose that at the time that the light pulse leaves the top mirror, X and Y are directly opposite each other. X measures this time as t_1 and the time that the pulse reaches the bottom mirror as t_2

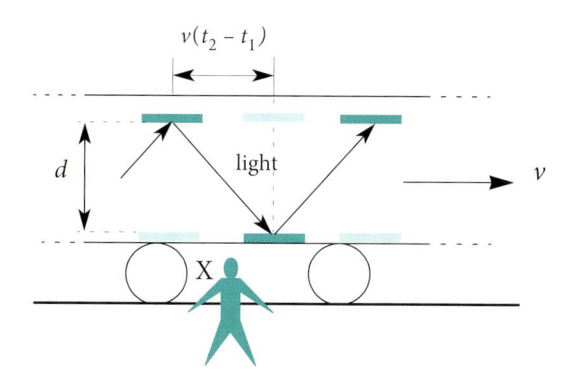

Figure 1309

By the time that the light pulse reaches the bottom mirror, the train will have moved forward a distance $v(t_2 - t_1)$ as measured by X and he will see the pulse follow the path as shown in Figures 1909 and 1910.

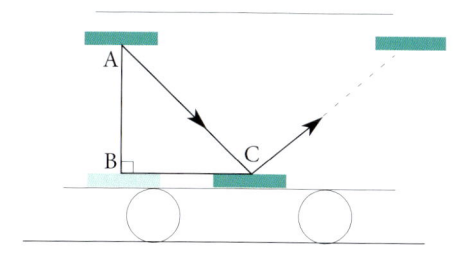

Figure 1310

If we let $t_2 - t_1 = \Delta t$ and $t'_2 - t'_1 = \Delta t'$, then, the distance that the train travels is $BC = v\Delta t$.

The distance that X measures for the path of the pulse is $AC = c\Delta t'$.

The distance that Y measures for the path of the pulse is $AB = c(t'_2 - t'_1)$.

Applying Pythagoras's theorem to the triangle, we have

$$AC^2 = AB^2 + BC^2 \Rightarrow [c\Delta t]^2 = [c\Delta t']^2 + [v\Delta t]^2$$

$$\Leftrightarrow c^2(\Delta t)^2 = c^2(\Delta t')^2 + v^2(\Delta t)^2$$

From which we have $(c^2 - v^2)\Delta t^2 = c^2(\Delta t')^2$

So that, $(\Delta t)^2 = \left(\frac{c^2}{c^2 - v^2}\right)(\Delta t')^2$

Rearranging this equation, we have

$$(\Delta t)^2 = \frac{1}{\left(1 - \frac{v^2}{c^2}\right)}(\Delta t')^2$$

Therefore (taking the positive square root) we have that

$$\Delta t = \frac{1}{\sqrt{1 - \frac{v^2}{c^2}}} \Delta t'$$

Or, letting $\gamma = \frac{1}{\sqrt{1 - \frac{v^2}{c^2}}}$, we have $\Delta t = \gamma \Delta t'$.

The equation $\Delta t = \gamma \Delta t'$ is stated in the IB data booklet as $\Delta t = \gamma \Delta t_0$ where t_0 is the proper time.

What this effectively means is that to observer X, the light pulse will take a longer time to traverse the distance between the mirrors compared to the time measured by observer Y. This phenomenon is known as **time dilation**.

371

Chapter 13 (Option A)

As stated above, the "bouncing" light pulse can effectively be regarded as a clock. We reach the conclusion therefore that any type of clock in an inertial reference systems which is moving relative to an observer in another inertial reference system will appear to run slower as measured by this observer. This effectively means that time is slowed for the moving observer. It has been said, tongue in cheek, that if you want to live longer, then keep running. If there were a clock on the train then to X it would not tell the same time as a clock on the ground nor would the clock on the ground tell the same time to Y as the clock in the train. Time can no longer be regarded as absolute. We must also bear in mind that the situation is symmetric and to observer Y, a clock in observer X's system will appear to run slower than a clock in her own system.

You might also like to ponder the following: is the speed of light the same for all inertial observers because time is not absolute or is time not absolute because the speed of light is the same for all inertial observers? This is the sort of question that keeps the philosophers happy for ages. From the physicist's point of view, what is of importance is that the Special Theory has been verified experimentally and that all the predictions that it makes have also been verified experimentally. The Special Theory of Relativity is part of the accepted framework of Physics.

In terms of predicted time dilations we are actually looking at some very small differences when considering everyday situations. For our two observers X and Y above, if for example the train is moving at 30 m s^{-1} relative to observer X then a time interval of 1 second as recorded by Y will be recorded as 1.00001 seconds by X. For this time dilation to be noticeable then clearly we must have relative velocities between observers which are close to the speed of light. Figure 1311 shows the variation with relative speed between inertial observers of the γ function.

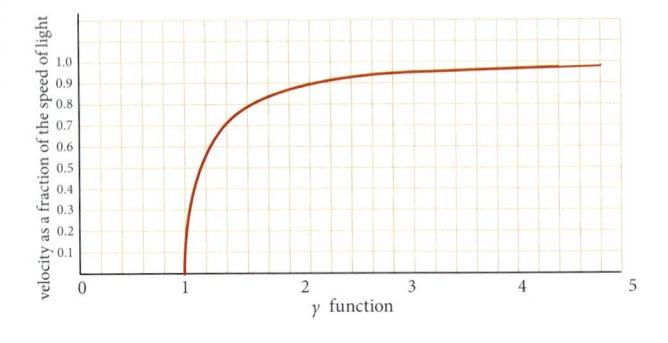

Figure 1311

From this graph you can see that, for two observers with a relative speed of about $0.98c$, then a time interval of one second as measured by one observer will be measured as a time interval of nearly 5 seconds by the other observer.

Example

An observer sets up an experiment to measure the time of oscillation of a mass suspended from a vertical spring. He measures the time period as 2.0 s. To another observer this time period is measured as 2.66 s. Calculate the relative velocity between the two observers.

Solution

Since the measured event occurs at the same place in which the time period is measured as 2.00 s, then

$2.66 = \gamma \times 2.00$

such that $\gamma = 1.33 = \dfrac{1}{\sqrt{1 - \dfrac{v^2}{c^2}}}$ or

$1.33^2 = \dfrac{1}{1 - \dfrac{v^2}{c^2}} \Rightarrow \left(1 - \dfrac{v^2}{c^2}\right) = \dfrac{1}{1.33^2} (= 0.57)$

Therefore $\dfrac{v^2}{c^2} = 0.43$ so that $v = 0.66c$.

This is a very high speed to say the least and in one time period the observers will have moved about 4×10^9 m apart. Which makes the whole thing strange. However we will look at a more realistic example later in the chapter.

Length contraction

We now consider how the Special Theory affects our concept of space.

In Figure 1312 Mary is moving at a constant velocity v in the x direction relative to Paul who regards himself as being at rest. AB is a rod, which is at rest with respect to Mary.

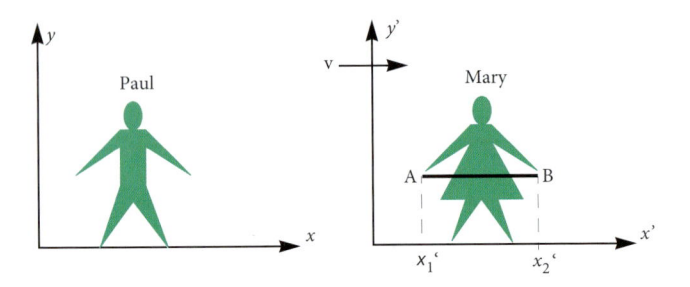

Figure 1312

In terms of Galilean relativity Mary and Paul would both obtain the same value for the length AB of the rod. But this is not the case if we apply the Lorentz transformations.

We define the **proper length** of the rod as the length of the rod measured by the observer at rest with respect to the rod.

Suppose that end A of the rod is at x'_1 and B is at x'_2 as measured by Mary. The rod is at rest in Mary's system therefore the proper length of the rod is $(x'_2 - x'_1) = L'$ (say).

If we apply the Lorentz transformation then x'_1 and x'_2 are given by $x'_1 = \gamma(x_1 - vt_1)$ and $x'_2 = \gamma(x_2 - vt_2)$ where x_2 and x_2 are the respective ends of the rod as measured by Paul.

The length of the rod as measured by Paul can therefore have many different values depending on the choice of t_1 and t_2, the times when the rod is measured. To make any sense, the length of the moving rod (from Paul's point of view) is defined as the length when the ends are measured simultaneously i.e. $t_1 = t_2$.

Hence the proper length of the rod

$$L' = (x'_2 - x'_1) = \gamma(x_2 - x_1).$$

Hence the length L of the rods measured by Paul will be given by $L' = \gamma L$

This equation is written as $L = \dfrac{L_0}{\gamma}$ in the data booklet where L_0 is the proper length. (You will not be expected to derive this equation in an examination).

Since γ is always greater than unity, L will always be less than the proper length. To Paul the rod will appear contracted in the direction of motion.

Example 1

Mary is travelling in a space ship, which is not accelerating. To her the length of the space ship is 100 m. To Paul who is travelling in another space ship, which is also not accelerating, Mary's space ship has a length of 98 m. Calculate the relative velocity of Paul and Mary.

Solution

The proper length is 100 m therefore $\gamma = \dfrac{100}{98} = 1.02$

so that $\gamma = 1.02 = \sqrt{\dfrac{1}{1 - \dfrac{v^2}{c^2}}}$.

Therefore $\dfrac{1}{1 - \dfrac{v^2}{c^2}} = \dfrac{1}{(1.02)^2} = 0.96$

giving $v^2 = 0.04c^2$,

so that $v = 0.2c$.

Example 2

To Paul, his space ship measures 150m. Determine the length that it will appear to Mary?

Solution

The γ factor is the same for both Paul and Mary but at this time 150 m is the proper length so Mary will see the ship contracted by a factor $1/\gamma$ i.e. 147 m.

Since both observers are inertial observers both will observe the length contraction and both observers will be correct.

Velocity addition

In Figure 1313 (a) Mary is in a train which is moving with speed v relative to Paul who is standing at the side of the tracks.

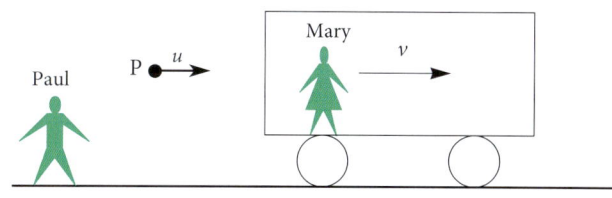

Speed of object relative to Mary $u' = u - v$

Speed of object relative to Mary $u' = u + v$

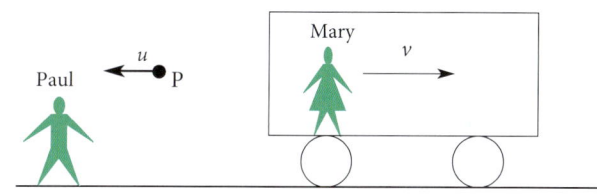

Figure 1313 (a) and (b)

The object P is moving horizontally with a speed u relative to Paul. According to a Galilean transformation, Mary would measure the speed of the object as $u' = u - v$. If the object is moving in the other direction, Figure 1313 (b), then the speed would be, as measured by Mary, given by $u' = u + v$.

If P were a light beam, then Mary would measure the speed of the beam to be $c + v$.

But if we are to believe the Special Theory then this can not be so since all inertial observers must measure the same value for the speed of light. The Galilean transformations for velocity cannot therefore be correct. It is not difficult to show that, if we apply the Lorentz transformations for displacement and time, then the velocity transformation equations become

Chapter 13 (Option A)

$$u'_x = \frac{u_x - v}{1 - \frac{u_x v}{c^2}} \quad \text{for } u_x \text{ in the positive } x \text{ direction}$$

$$u'_x = \frac{u_x + v}{1 + \frac{u_x v}{c^2}} \quad \text{for } u_x \text{ in the negative } x \text{ direction}$$

Exercise 13.2

1. Use the definition of the γ function to confirm its value as shown by the previous graph when the velocity is 0.5c, 0.8c and 0.95c (Figure 1909).

2. Mary measures the time between two events as being separated by an interval of 2.0 ms. Paul measures the interval to be 2.5 ms. Calculate the relative velocity of Mary and Paul.

3. A spaceship is travelling away from the Earth with a speed of 0.6c as measured by an observer on the Earth. The rocket sends a light pulse back to Earth every 10 minutes as measured by a clock on the spaceship.

    (a) Calculate the distance that the rocket travels between light pulses as measured by
      i. the observer on Earth.
      ii. somebody on the spaceship?

    (b) If the Earth observer measures the length of the spaceship as 60 m, determine the proper length of the spaceship.

4. Calculate the relative velocity between two inertial observers which produces a 50% reduction in the proper length.

5. Use the Lorentz velocity transformation equations to show that any two inertial observers will always measure the same value for the velocity of light.

The muon decay experiment

Direct experimental confirmation for time dilation and length contraction as predicted by the Special Theory is found in the decay of muons. Muons are sub-atomic particles, which can be created in high-energy particle accelerators. The muons are unstable and decay with a half-life of 3.1×10^{-6} s as measured in a reference frame in which they are at rest. (Or at least moving with relatively low speeds). However, muons are also created in the upper atmosphere of the Earth from cosmic ray bombardment and these muons can, and do, have very high velocities. If we consider muons that are formed at a height of say 10 km and with a velocity of 0.98c, then they will take 3.4×10^{-5} s to reach the ground. This is about 10 half lives and in this time the majority of the muons would have decayed. However, large numbers of muons are in fact detected at the Earth's surface. Because they are moving at such high speeds, time dilation becomes important and at 0.98c the γ factor is 5. To the Earth observer, the half-life is not 3.1×10^{-6} s but γ times this i.e. 15.5×10^{-6} s. The time to reach the ground is now only about 2.2 half-lives and so plenty of muons will be detected. To the muons the distance to the Earth is not 10 km but $\frac{10}{\gamma}$ kilometres = 2 km and from their point of view, the time to travel this distance (6.8×10^{-6} s) is just 2.2 half lives. In experiments carried out at the surface of the Earth, the half-lives of muons moving with different velocities have been measured and the values have been found to agree closely with the values predicted by the Special Theory.

The Michelson-Morley experiment

We mentioned early in this chapter that it was proposed that electromagnetic waves were propagated through a substance called the ether and that this ether permeated all space. It was further proposed that the ether was absolutely at rest and this was the reference frame against which all measurements should be made.

In 1887 two American physicists, Michelson and Morley, devised and carried out an experiment to measure the absolute velocity of the Earth with respect to the ether. The essential principle of the experiment was to find the difference in time that it took light to travel along two paths, one in the direction of travel of the Earth through the ether and one at right angles to the direction of travel of the Earth through the ether. Figure 1314 shows the basic set up of the experiment.

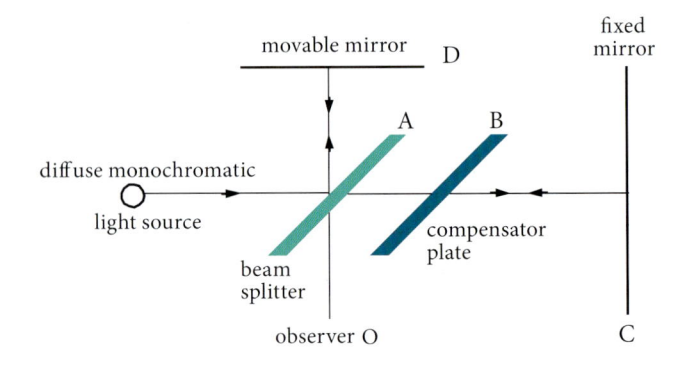

Figure 1314   The Michelson – Morley experiment

Light from a diffuse monochromatic source is incident on the half silvered mirror A. This acts as a beam splitter such that some of the light goes on to the moveable mirror and some on to the fixed mirror C. On reflection from these mirrors, light from both mirrors arrives at the observer O. B is a compensator plate equal in thickness to the beam splitter.

If the different path length of the two rays on reaching the observer is a multiple number of wavelengths, then a bright spot of light will be observed at O and if it is an odd number of half wavelengths, a dark spot will be observed at O. However, since a diffuse source is used a great many paths of slightly differing lengths will occur and the overall effect will be to observe a series of light and dark fringes at O. An interference pattern in fact.

If the mirror D is moved backwards or forwards the interference pattern will be shifted and the amount that it is shifted will depend on the amount that the mirror D is moved.

The actual apparatus used by *Michelson* and *Morley* was very large. The effective length of the two arms was about 10 m. One arm was aligned to be parallel to the direction of motion of the Earth in its orbit such that the other arm was at right angles. If the ether through which the light travels is at rest relative to the motion of the Earth, then the light will take slightly different times to traverse the two paths. Hence there will be an observed shift in the interference pattern. The situation is analogous with two swimmers in a river both setting off from the same point. One swims parallel to the direction of the current, and one swims at right angles to the direction of the current. They each swim the same distance from the starting point and then return. A little thought will show that the trip for the swimmer who sets off at right angles is going to take longer than the trip for the swimmer who swims parallel to the current.

The result of the Michelson-Morley experiment was spectacular in as much as no shift in the interference pattern was observed. Many attempts were made to explain this non-result before Einstein recognised that there was no ether and therefore no absolute reference frame and that all inertial observers will measure the same value for the speed of light.

The decay of a neutral pion (see Topic J.1) into two photons suggests that the speed of light is independent of the speed of the source. However more direct evidence to support this comes from astronomical observations of so-called **gamma ray bursters** (GRBs). Bursts of gamma rays have been observed to come from seemingly random parts of the Universe and are thought to be due to the collapse of a rapidly rotating neutron star see (E.5.6) that is expanding and contracting. Gamma radiation reaches us from the expanding near side and the receding far side of the object. The near and far sides are expected to be moving apart at speeds close to that of the speed of light. If the speed of light were dependent on the speed of the source, then gamma pulses reaching Earth from the two sides of the GRB would result in a separation of the pulses. To within 1 part in 10^{20}, no separation has been observed.

A.3 Space-time diagrams

NATURE OF SCIENCE:

Visualization of models: The visualization of the description of events in terms of spacetime diagrams is an enormous advance in understanding the concept of spacetime. (1.10)

© IBO 2014

Essential idea: Spacetime diagrams are a very clear and illustrative way to show graphically how different observers in relative motion to each other have measurements that differ from each other.

Understandings:
- Space-time diagrams
- Worldlines
- The twin paradox

Space-time diagrams

In the Special Theory, space and time are intimately linked. A very useful way of envisaging this is to represent the motion of particles in a **space–time diagram**. Space can be represented by the **Cartesian co-ordinates, x, y and z** and **time t is represented by an axis at right angles to the other three axes.** This is obviously impossible to draw or even to picture mentally so here we will only concern ourselves with motion in the x-direction. However, mathematically, there is nothing intrinsically wrong in having four lines mutually at right angles describing a space–time continuum.

In Figure 1315, space is represented by the conventional x-axis and time t by an axis perpendicular to the x-axis. The y-axis is quite often labelled ct rather than t.

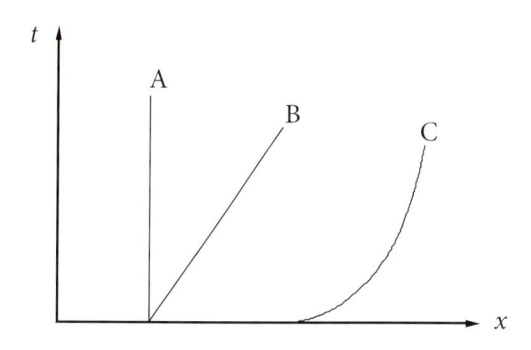

Figure 1315   Time – space graph

Chapter 13 (Option A)

The line A represents a stationary particle and the line B represents a particle that starts from the point where A is at rest and moves with constant velocity away from this point. The line C represents a particle that has certain velocity at some different point in space and is slowing down as it moves away from this point.

Suppose a person travels from the earth into space at 0.5c for 1 year and then turns around and travels back to earth at the 0.5c. A schematic space-time sketch would look like Figure 1316.

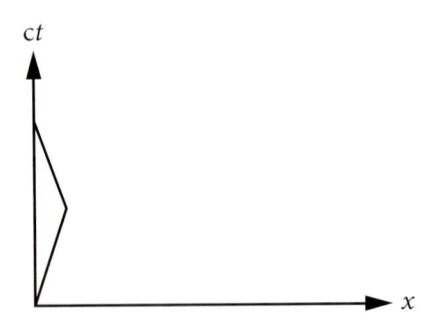

Figure 1316   Space-time for a spaceship

We need to be careful with the angle between the x-axis and a straight diagonal line as nothing can travel faster than the speed of light. Therefore, the angle cannot be greater than 45° to the vertical because when

$$\theta = \tan^{-1}\left(\frac{v}{c}\right)$$

if $\theta = 45°$ then $\tan \theta = 1$ and $v = 1c$

if $\theta = 50°$ then $\tan 50 = 1.19$ and $v = 1.19c$

Figure 1317 shows the space-time graph with diagonal lines at 45° to the vertical. The point at the origin represents "now" of an event. The top "cone" represents the future in space-time and the bottom "cone" represents the past in space-time. The left and right areas cannot be time events as the event would have to occur at speeds greater than the speed of light.

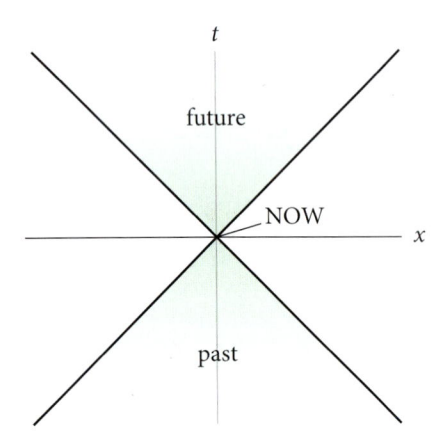

Figure 1317   General space-time diagram

World lines

Now suppose I went to a supermarket that was 3 km south and 4 km east of my starting point. And it took 10 minutes to get there. A straight line on a space-time graph would not be a true representation of the motion as it is simply a measure of the space-time distance between the two events. The actual graph would be a complicated series of curves because you would be accelerating and decelerating and changing direction in your journey. This complicated wavy graph of the journey in relativistic terms is called the **world line**. An example is shown in Figure 1318 that depicts the world line in a space-time of three dimensions that is sometimes called a Minkowski three-space.

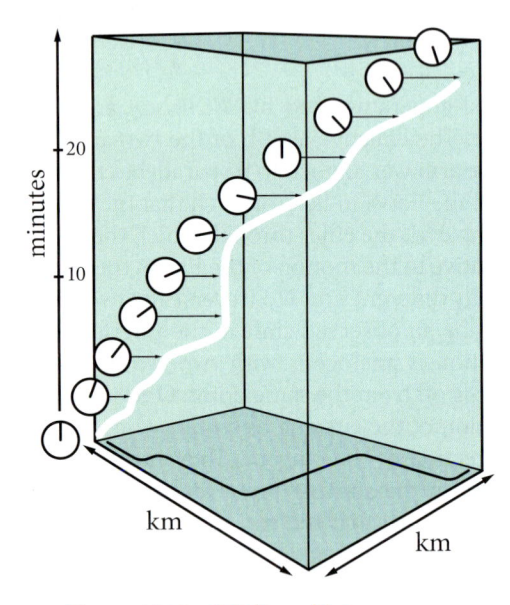

Figure 1318   Minkowski three-space

The axes labelled x and ct in Figure 1319 are the coordinate system of an object. If the object is considered to be at rest with respect to the space-time coordinates, it is still travelling through the dimension of time. Its world line would be a straight line parallel with the time axis.

Each parallel line to this axis would correspond also to an object at rest but at another position.

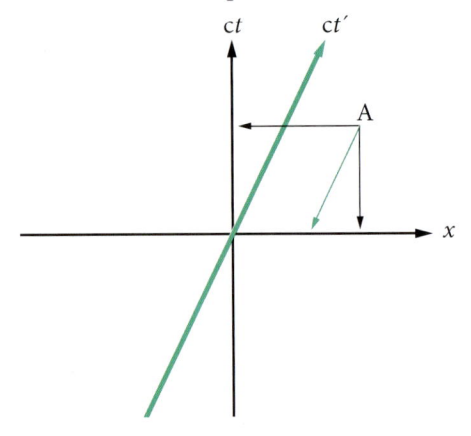

Figure 1319

Relativity

The diagonal line, however, describes an object moving with constant speed v to the right, such as a moving observer. Its world line would be a straight line that is not parallel with the time axis.

The line labelled ct' could represent the time axis for the second observer. Together with the path axis (labeled x, that is identical for both observers), it represents his coordinate system. Both observers agree on the location of the origin of their coordinate systems. The axes for the moving observer are not perpendicular to each other and the scale on his time axis is stretched. To determine the coordinates of a certain event, two lines parallel to the two axes must be constructed passing through the event as shown, and their intersections with the axes read off.

Now let us look closely at the next two equations:

$$\Delta x' = \Delta x - \frac{v\Delta t}{\sqrt{1 - \frac{v^2}{c^2}}}$$

and

$$\Delta t' = \Delta t - \frac{\frac{v\Delta x}{c^2}}{\sqrt{1 - \frac{v^2}{c^2}}}$$

So what does $(c\Delta t')^2 - (\Delta x')^2$ equal when these equations are combined?

Substituting for $\Delta x'$ and $\Delta t'$ we obtain:

$$c^2\left(\Delta t - \frac{v\Delta x}{c^2}\right)^2 \gamma^2 - (\Delta x - v\Delta t)^2 \gamma^2 \text{ where } \gamma = \frac{1}{\sqrt{1 - \frac{v^2}{c^2}}}$$

Rearranging we can obtain

$$\gamma^2 \left[\left(c^2\Delta t^2 - 2\Delta t\,\Delta x\,v + \frac{\Delta x^2 v^2}{c^4}\right) - (\Delta x^2 - 2\Delta x\,v\,\Delta t + v^2\Delta t^2)\right]$$

If we multiply the first part equation by c^2, we obtain

$$\gamma^2 \left[\left(c^2\Delta t^2 - 2\Delta t\,\Delta x\,v + \frac{\Delta x^2 v^2}{c^2}\right) - (\Delta x^2 - 2\Delta x\,v\,\Delta t + v^2\Delta t^2)\right]$$

Now $2\Delta t\,\Delta x\,v$ cancels out and we can gather terms together to give

$$\gamma^2 \left[\Delta t^2 (c^2 - v^2) + \Delta x^2\left(\frac{v^2}{c^2} - 1\right)\right]$$

Take out the front c^2 in the first term and we have

$$\gamma^2 \left[c^2 \Delta t^2 \left(1 - \frac{v^2}{c^2}\right) - \Delta x^2\left(1 - \frac{v^2}{c^2}\right)\right]$$

This leaves

$c^2 \Delta t^2 - \Delta x^2$

Therefore

$(c\Delta t')^2 - (\Delta x')^2 = (c\Delta t)^2 - \Delta x^2$

Both statements agree with each other. Therefore the value $(c\Delta t)^2 - \Delta x^2$ must equal a constant. The constant is given the symbol S^2. Therefore, it does not matter what speed something moves in time, the only changing variables are the values for the distance and the time.

On a space-time diagram, the worldline can be drawn as a hyperbole as shown in Figure 1320.

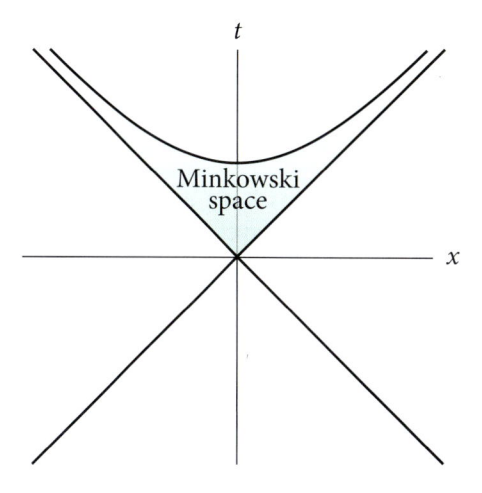

Figure 1320   Minkowski space

Therefore, the worldliness will be a series of hyperbolas that allow for the cone shape to be formed as depicted in Figure 1321.

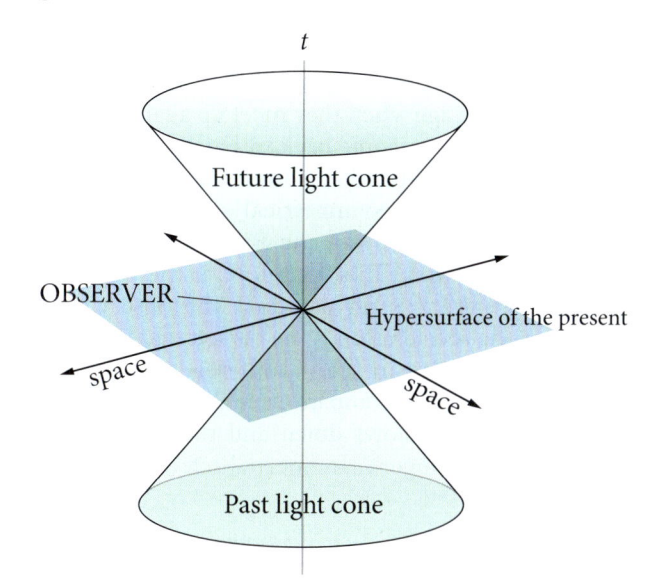

Figure 1321   Worldline outline to form a cone shape

377

Chapter 13 (Option A)

The twin paradox

It might well have occurred to some of you during the discussion of time dilation that if there is no preferred reference system why doesn't the observer on the train perceive that clocks on the other side of the tracks run slower than his own clocks? In fact, we could just as well have described the experiment the other way round so that the observer on the tracks was the one carrying out the experiment with the bouncing light pulse. Proper time would then be the time measure in his reference system. The Special Theory tells us that both observers are correct. This introduces the famous so called "twin paradox".

Paul and Mary are twins born sometime in the future when inter-stellar star travel is commonplace. When they are 30 years old, Mary sets off in a space ship to travel to Alpha Centauri and back, a total distance of 8.6 light years. Her space ship travels at an average speed of 0.98c. The round trip for her will take about 9 years. However, to Paul all Mary's clocks will appear to run slower, her heart will beat slower, in fact everything will be slowed down from Paul's point of view. If you refer to the previous time dilation graph you will see that 1 second for Mary will appear as about 5 seconds for Paul. So when Mary returns to Earth she will be about 39 years old and Paul will be about 77 years old. But if all motion is relative, why can't we regard Paul as being the person in motion and Mary at rest? As we shall see, it has been verified experimentally that time does go slower in moving reference frames. So this is the paradox. By symmetry the only possible result is that Paul and Mary must be the same age when they meet up again on Earth. However, for Mary to come back to Earth from the star, she has to turn round and this involves an acceleration so this is no longer a symmetrical situation. She also has to accelerate away from the Earth and slow down to land back on Earth. This, therefore, is the way round the paradox - there is no paradox. The observer that experiences the acceleration is the person that is moving. If both observers are in inertial reference frames then they can never meet to compare their ages or their clocks unless one of them slows down and the observer that slows down is no longer in an inertial reference system.

The Hafele-Keating experiment

In order to test Einstein's theory of relativity with actual clocks rather than "light clocks", in October 1971, Hafele and Keating took four cesium atomic beam clocks aboard commercial airliners. The airliners flew around the world twice, once eastward and once westward. The times read by the clocks were then compared with clocks at U.S. Naval Observatory. According to Einstein's theory the clock should have lost 40 ns during the eastward trip and should have gained 275 ns during the westward trip as compared with the atomic time scale of the U.S. Naval Observatory. The "flying" clocks actually lost 59 ns during the eastward trip and gained 273 ns during the westward trip. When the errors and the corresponding standard deviations were taken into account, the agreement between the theory and experiment was excellent, providing unambiguous evidence for the resolution of the twin paradox.

The event can also be solved using a space-time diagram as shown in Figure 1322. Suppose the twin on the spaceship is travelling at 0.6c and 10 years passes for the twin on earth.

Then

$$t = t' \sqrt{1 - \frac{9}{25}}$$

Then $t = t' \frac{4}{5}$

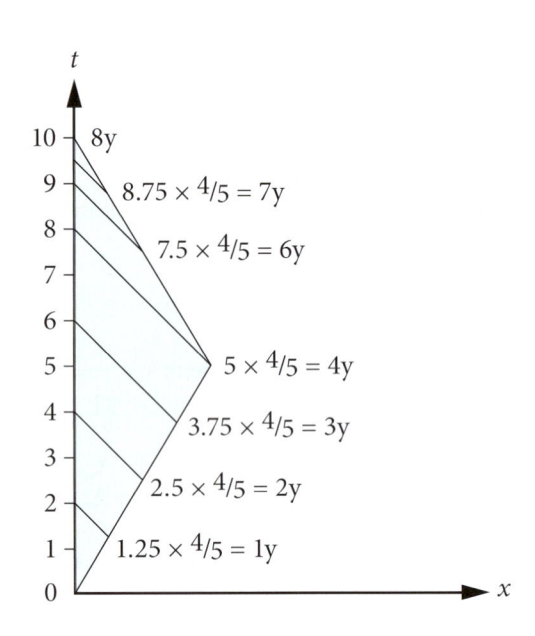

Figure 1322 Twin paradox space-time diagram

Therefore, if 5 seconds passed for the twin on the earth, then 4 seconds will have passed for the twin on the spaceship. Furthermore, if 5 years passed for the twin on the earth, then 4 years will have passed for the twin on the spaceship.

After 2 years for the twin year on earth, the equivalent time value for the twin in the spaceship is 1.25. So, 1.25 multiplied by $\frac{4}{5}$ = 1 year. Keep doing this along the space-time graph as shown.

A.4 Relativistic mechanics

> **NATURE OF SCIENCE:**
>
> Paradigm shift: Einstein realized that the law of conservation of momentum could not be maintained as a law of physics. He therefore deduced that in order for momentum to be conserved under all conditions, the definition of momentum had to change and along with it the definitions of other mechanics quantities such as kinetic energy and total energy of a particle. This was a major paradigm shift. (2.3)
>
> © IBO 2014

Essential idea: The relativity of space and time requires new definitions for energy and momentum in order to preserve the conserved nature of these laws.

Understandings:
- Total energy and rest energy
- Relativistic momentum
- Particle acceleration
- Electric charge as an invariant quantity
- Photons
- MeV c^{-2} as the unit of mass and MeV c^{-1} as the unit of momentum

Total energy and rest energy

If a body of mass m is subjected to a force F then, according to Newton's Second Law, it's acceleration a can be computed from the equation $F = ma$.

A prediction of the Special Theory, (in order to ensure that the conservation of momentum holds for all inertial observers,) is that **as the speed of body increases then, to an observer at rest relative to the body, the mass of the body will increase.**

If m_0 is the mass of the body when it is at rest with respect to the observer, the so-called **rest mass**, then its mass m when moving at a speed v relative to the observer is given by $m = \gamma m_0$

$$\left[\text{remember that } \gamma = \frac{1}{\sqrt{1 - \frac{v^2}{c^2}}}\right].$$

We have seen that, as a consequence of the Special Theory, the mass of a moving object increases. A net force is needed to accelerate any object and this force does work on the object. In Newtonian mechanics the conservation of energy leads to the idea that the work done by the net force is equal to the increase in kinetic energy of the object such that

Work done = force × displacement = gain in kinetic energy

i.e., $F \times \Delta s = \Delta(KE) = \frac{1}{2}m(\Delta v^2)$

This equation will not apply in Special Relativity. In fact, if we are still to believe in the conservation of energy we must look for a different relationship between the work done and the energy transferred. It might be tempting to substitute m as γm_0 but this does not in fact lead to the correct physical interpretation.

In thinking along these lines *Einstein* was led to the idea of mass and energy being interchangeable such that the gain in mass of an accelerated body could be equated to a gain in energy. This led him to the celebrated equation

$E = mc^2$

In this equation, E is the total energy of the object and m is its relativistic mass. If the object is at rest then it has a **rest mass energy** given by

$E_0 = m_0 c^2$

If we combine these two equations in terms of the work done when a force accelerates an object from rest then, at a certain speed v, the object will have a total mass–energy mc^2. Its mass–energy will have changed from rest by an amount, E_k, where

$E_k = E - m_0 c^2 = mc^2 - m_0 c^2$

It is this change in energy which is the gain in the kinetic energy of the object and is equal to the work done on the object. It is important to understand the significance of these equations, namely that energy and mass are entirely equivalent.

Lets us now look at an example using the idea of mass-energy equivalence.

Example

A coal-fired power station has a power output of 100 MW. Calculate the mass of coal that is converted into energy in one year (3.15×10^7 s).

Solution

The energy produced in one year =

$100 \times 10^6 \times 3.15 \times 10^7 = 3.15 \times 10^{15}$ J

379

this represents a mass change of

$$\Delta m = \frac{energy}{c^2} = \frac{3.15 \times 10^{15}}{9 \times 10^{16}} = 3.5 \times 10^{-2} \text{ kg}.$$

This actually means that if the coal is weighed before it is burnt and all the ashes and fumes could be weighed after burning there would be a mass deficiency of about 35 g. Although, we must not forget that some oxygen would have been added.

In view of the equivalence of mass energy we can no longer talk about the conservation of mass and the conservation of energy as two separate laws of physics but instead we have just one law, the **conservation of mass-energy**. However, for many chemical reactions we can use the separate laws since the mass deficiency involved is usually very small. For example, when one gram mole of carbon combines with two gram moles of oxygen, the mass deficiency is only about 10^{-9} g which is far too small to detect. However, the mass energy equation plays a very important role in nuclear reactions.

Newton's Second Law does not put any upper limit on the velocity that a body can attain. For example we could envisage a force of 1000 N acting on an object of mass 1 g for a period of 1000 s. After this period of time the speed of the object would be 10^9 m s^{-1} that is much greater than the speed of light. However, the Lorentz transformation equations show us that if the relative velocity between two observers is c then the time dilation is infinite and the length contraction is zero. Putting it another way, time stays still for objects travelling at the speed of light and also the object has no length in the direction of travel. Furthermore, the equation for the relativistic increase of mass shows that the mass of the object would be infinite. So not only does the Special Theory put an upper limit on the velocity that objects can attain (the speed of light) it predicts that this velocity is unattainable. The only thing that can travel at the speed of light would seem to be light itself.

Relativistic momentum

In Classical Physics, we saw in Topic 2.3 that there is a useful relationship between the momentum p and the kinetic energy E_k of a particle, namely

$$E_k = \frac{p^2}{2m}$$

where m is the mass of the particle.

In Special Relativity, we can find an equally useful relationship between momentum and energy, but in this instance the energy E is the *total energy* of the particle.

We have that

$$m = \frac{m_0}{\sqrt{1 - \frac{v^2}{c^2}}}$$

such that if we square both sides and rearrange we have

$$m^2 c^2 = m_0^2 c^2 + m^2 v^2$$

If we now multiply through by c^2 we have

$$(mc^2)^2 = m_0^2 c^4 + (mv)^2 c^2$$

But, mv is the momentum p of the particle and mc^2 is equal to E the total energy. Hence

$$E^2 = m_0^2 c^4 + p^2 c^2$$

This is the relativistic equivalent to the classical equation

$$E_k = \frac{p^2}{2m}$$ and is useful in many different situations.

Electric charge as an invariant quantity

Example

Find the momentum and speed of an electron after it has been accelerated through a potential difference of 1.8×10^5 V.

Solution

The total energy E is given by

$$E = Ve + m_0 c^2.$$

If we bear in mind that

$m_0 c^2$ is equal to 0.51 MeV, and $Ve = 0.18$ MeV, then in this instance we have that

$$E = (0.18 + 0.51) \times 10^6 \text{ eV} \quad E^2 = 0.4761 \times 10^{12} \text{ eV}^2$$

Also, from the relativistic relation between total energy and momentum, we have

$$p^2 = \frac{E^2 - m_0^2 c^4}{c^2} \quad (1)$$

And $m_0^2 c^4 = (0.51 \times 10^6 \text{ eV})^2 = 0.2601 \times 10^{12} \text{ eV}^2$,

hence,

$$p^2 = \frac{0.216 \times 10^{12} e^2 V^2}{c^2} = 0.46 \text{ MeV c}^{-1}.$$

Or, $p = 2.5 \times 10^{-22}$ N s

Since particle physicists are often dealing with energies measured in electron volts, they often express momentum in the units MeV c^{-1} (energy/speed).

To find the speed, we need to find the mass of the electron after acceleration.

We have that $E_{Total} = mc^2 = Ve + m_o c^2$

Therefore

$m = (Ve + m_o c^2) c^{-2} = (0.18 + 0.51)$ MeV c^{-2} = 0.68 MeV c^{-2}

And $v = p/m = 0.46$ MeV c^{-1}/0.68 MeV c^{-2} = 0.68 c.

This demonstrates how much easier are relativistic dynamic calculations when we deal in the units MeV for energy, MeV c^{-2} for mass and MeV c^{-1} for momentum.

Photons

A gamma photon that is travelling close to a lead atom, materialises into an electron-positron pair

e.g. $\gamma \rightarrow e^- + e^+$

The initial energy of the photon is 3.20 MeV. Neglecting the recoil of the lead atom, calculate the speed and mass of the electron and positron.

Solution

For one particle $E_{tot} = 1.60 = E_K + m_o c^2 = E_K + 0.511$

$E_K = 1.09$ MeV = $(1 - \gamma) m_o c^2$

To give $\gamma = 3.13$

Which gives $v = 0.948c$

$m = \gamma m_o c^2 = 1.6$ MeV c^{-2}

Particle acceleration

Let us look at another example of mass-energy equivalence that brings in the idea of work done and changes in kinetic energy.

Example

An electron is accelerated through a potential difference of 2.0 V. Calculate, after acceleration, the velocity of the electron applying

(a) classical Newtonian mechanics
(b) relativistic mechanics.

Solution

(a) We use the equation $v = \sqrt{\frac{2eV}{m}}$ (Ve = gain in kinetic energy) to give

$$v = \sqrt{\frac{4 \times 10^6 \times 1.6 \times 10^{-19}}{9.1 \times 10^{-31}}} \approx 8.4 \times 10^8 \text{ m s}^{-1}.$$

A value which is clearly greater than the speed of light.

(b) Using relativistic mechanics we have that the energy supplied i.e. the work done = V e and this equals the gain in KE of the electron such that

$$Ve = mc^2 - m_0 c^2 = \gamma m_0 c^2 - m_0 c^2$$

So that $\gamma m_0 c^2 = m_0 c^2 + Ve \Rightarrow \gamma = \frac{m_0 c^2 + Ve}{m_0 c^2}$

To give $\gamma = 1 + \frac{Ve}{m_0 c^2}$ — Equation (1)

(We can actually express $m_0 c^2$ in units of eV =

$$\frac{9.1 \times 10^{-31} \times 9 \times 10^{16}}{1.6 \times 10^{-19}} = 0.51 \text{ MeV}.$$

(We can therefore express the mass of the electron as 0.5 MeV c^{-2}.)

So substituting 0.5 MeV into equation (1) we have

$$\gamma = 1 + \frac{2 \text{ MeV}}{0.51 \text{ MeV}} = 4.9$$

The total energy of the particle after acceleration is $\gamma m_o c^2$

= 4.9 × 0.51 = 2.5 MeV

Using the equation for the gamma factor, a value of 5 gives the velocity of the electron as 0.98c.

MeV c⁻² as the unit of mass and MeV c⁻¹ as the unit of momentum

In the equation $E = mc^2$, if m is measured in kilograms and c as m s⁻¹, then the unit of E is clearly joules. However, the theory of relativity only becomes significant for speeds close to that of c and this usually means we are dealing with the acceleration or movement of atomic or sub-atomic particles. For a example, an electron accelerated from rest through a potential difference of 10^6 volt will attain an energy of 1M eV as measured in electron-volts. It is much more convenient to express the energy of particles in eV (or multiples thereof) such as MeV. Similarly, it is much more convenient to express their mass in units of MeV c⁻². So for example the rest mass of a proton is 938 MeV c⁻²

which equals $\dfrac{938 \times 10^6 \times 1.6 \times 10^{-19}}{9 \times 10^{16}}$ kg = 1.67×10^{-27} kg.

Lets us now look at an example using the idea of mass-energy equivalence.

A.5 General relativity

NATURE OF SCIENCE:

Creative and critical thinking: Einstein's great achievement, the general theory of relativity, is based on intuition, creative thinking and imagination, namely to connect the geometry of spacetime (through its curvature) to the mass and energy content of the spacetime. For years it was thought that nothing could escape a black hole and this is true but only for classical black holes. When quantum theory is taken into account a black hole radiates like a black body. This unexpected result revealed other equally unexpected connections between black holes and thermodynamics. (1.4)

© IBO 2014

Essential idea: General relativity is applied to bring together fundamental concepts of mass, space and time in order to describe the fate of the universe.

Understandings:
- The equivalence principle
- The bending of light
- Gravitational redshift and the Pound-Rebka-Snider experiment
- Schwarzschild black holes
- Event horizons
- Time dilation near a black hole
- Applications of general relativity to the universe as a whole

Gravitational mass and inertial mass

The Special Theory of Relativity is special in the sense that it apples only to inertial reference systems. A logical question to ask would be "is non-uniform motion also relative?" For example, when a Jumbo Jet takes off does it make any difference whether we consider the Earth to be at rest with respect to the jet or whether we consider the jet to be at rest and the Earth accelerating away from the jet. If accelerated motion is relative then it should be possible to choose the jet as the fixed reference system. But how then do we account for the inertial forces that act on the passengers during take-off? (As the jet accelerates the passengers are "pushed" back into their seats). After the publication of the Special Theory, physicists were happy to believe that uniform motion was indeed a special case and that non-uniform motion was not relative. Einstein could not accept this viewpoint and in 1916 he published the General Theory of Relativity in which he postulated that all motion is relative.

Central to the General Theory is the so-called Mach's Principle. Towards the end of the last century Ernst Mach suggested that inertial force and gravitational force are equivalent. There was evidence for this assumption, which has been around since the time that Galileo reportedly dropped objects of different masses from the Leaning Tower of Pisa and concluded that the acceleration of free fall is the same for all objects. This is something that is taught very early in all High School Physics courses. However, the reason why this should be so is rarely mentioned because High School physics generally makes no attempt to distinguish between 'inertial' and 'gravitational' mass.

As we have seen, the concept of mass arises in two very different ways in Physics. You met it for the first time in connection with the property of inertia - all objects are "reluctant" to change their state of motion. This "reluctance" is measured by a property of the object called its **inertial mass**. The concept of inertial mass is quantified in Newton's Second law, $F = ma$.

But the concept of mass also arises in connection with Newton's gravitational law in which the force between two point masses m_1 and m_2 separated by a distance r is given by

$$F = \frac{Gm_1m_2}{r^2}$$

In this respect, mass can be thought of as the property of an object, which gives rise to the gravitational force of attraction between all objects and is therefore called **gravitational mass**.

Since gravitational and inertial mass measure entirely different properties there is no reason why we should consider them to be identical quantities. However, consider an object close to the surface of the Earth which has a gravitational mass m_g and an inertial mass m_I. If the gravitational mass of the Earth is M_g then the magnitude of the gravitational force exerted on the object is given by

$$F = \frac{GM_g m_g}{R^2}$$

Where R is the radius of the Earth.

The object will accelerate according to Newton's Second law such that

$$F = \frac{GM_g m_g}{R^2} = m_I a$$

Letting $\frac{GM_g}{R^2} = k$ (= constant) so that $a = k \times \frac{m_g}{m_I}$

All the experimental evidence points to the fact that the value of a (the acceleration of free fall at the surface of the Earth) is the same for all objects.

Hence we conclude that $m_g = m_I$. i.e. gravitational and inertial mass are equivalent.

The equivalence principle

The General Theory of Relativity gives the interpretation of the equivalence between gravitational and inertial mass.

In Figure 1323, a person is in a lift (elevator), far from any mass, which is accelerating upwards with acceleration g.

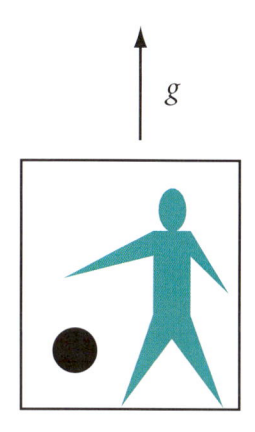

Figure 1323

If he releases an object as shown he will observe that it "falls" to the ground with acceleration g. An outside observer would say that the ball stays where it is but the lift floor is accelerating upwards towards the ball with acceleration g.

Another interesting situation arises here in which we can consider a lift in free fall close to the surface of the Earth. When the person in the lift releases the object in this situation then the object will stay where it is. The object is in fact "weightless". This is the reason that astronauts in orbit around the Earth are "weightless" – they are in free fall and although they are in a gravitational field, because of their acceleration, they will feel no gravitational force.

Figure 1324

In Figure 1324, the lift is at the surface of the Earth and again the person drops the ball and observes it to accelerate downwards with an acceleration g.

An outside observer would say that this is because of the gravitational attraction of the Earth.

The two results (Figures 1323 and 1324) are identical and according to *Einstein* there is no physical experiment that

383

an observer can carry out to determine whether the force acting on the object arises from inertial effects due to the acceleration of the observer's frame of reference or whether it arises because of the gravitational effects of a nearby mass.

This is the so-called **Einstein Principle of Equivalence** and it can be stated

- **there is no way in which gravitational effects can be distinguished from inertial effects**.

In this respect *Einstein* concluded that all motion is relative. If we consider the situation in Figure 1325 we can choose the lift to be the fixed reference system and it is the rest of the Universe that must be considered to be accelerating. It is this acceleration of the Universe, in Einstein's interpretation, that generates what *Newton* called a gravitational field. According to *Einstein* there is no absolute choice of a reference system, only relative motion can be considered.

The bending of light

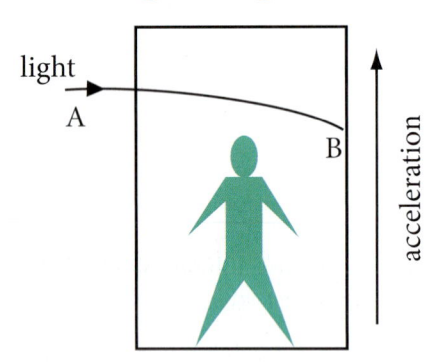

Figure 1325   The space ship is accelerating

The General Theory predicts that light will be bent by gravity. In Figure 1326 a person is in an accelerating space ship far away from any mass. A ray of light enters through a window at A. Because of the acceleration of the ship the light will strike the opposite wall at point B that is below A. To the person in the ship the path of the light ray will therefore appear to be bent.

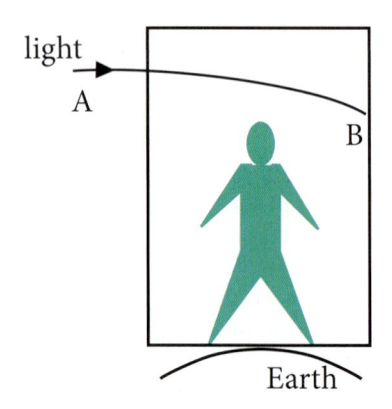

Figure 1326   The space ship is stationary

In Figure 1326 a space ship is at rest on the surface of the Earth. If the Einstein principle of equivalence is correct then this situation cannot be distinguished from the situation described in Figure 1323. The path of a light beam entering a window of this space ship will therefore also appear to be bent. The prediction is therefore that light that passes close to large masses will have its path altered.

Gravitational redshift and the Pound-Rebka-Snider experiment

Figure 1327 (a) shows a space ship which is accelerating with acceleration a. On the floor of the spaceship is a light source, 1, which emits light of a well-defined frequency f. The observer O is at the "top" of the spaceship and another light source 2, identical to light source 1, is placed next to this observer. At time $t = 0$ the spaceship starts to accelerate and at the same instant the two light sources emit light. Bearing in mind that the speed of light is invariant, the light from source 1 will appear to O to be emitted from a source that is moving away from him at speed v, where $v = at$, t being the time it takes the light from source 1 to reach him. Compared with the light from source 2 the light from source 1 will appear to be Doppler shifted i.e. it will be of a lower frequency than the light emitted from source 2.

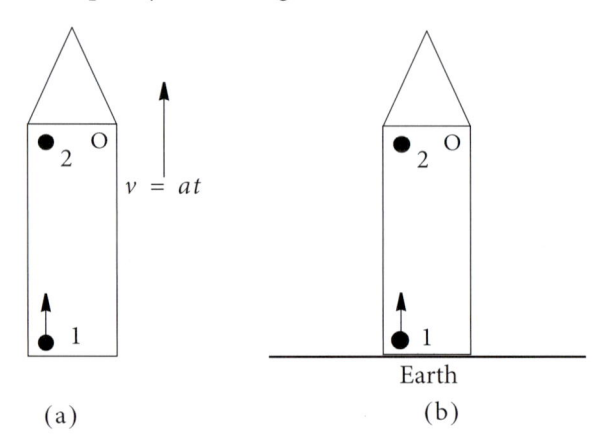

Figure 1327 (a) and (b)

If we consider the space ship to be at rest on the surface of the Earth as shown in Figure 1327 (b), then because of the principle of equivalence, the same effects will be observed as in diagram (a). This means that the observed frequency of light emitted from a source depends upon the position of the source in a gravitational field. For example, light emitted from the surface of a star will be red-shifted as seen by an observer on Earth. Light emitted from atoms in the star's corona will not be as red-shifted as much.

Since frequency is essentially a measure of time this means a consequence of General Relativity is that, to an observer on the top floor of a building, clocks on the ground floor will appear to run more slowly. The conclusion is that time slows in the presence of a gravitational field. As mentioned above, a remarkable consequence of this is that at the event horizon, of a black hole, to an outside observer, time stops.

You will not be expected to prove the red-shift equation in an examination.

However, for completeness sake, an elementary proof is given here based on the Doppler effect.

In this proof we return to the accelerating space ship in which there are two light sources, one on the floor of the ship and the other on the "ceiling". (Figure 1328)

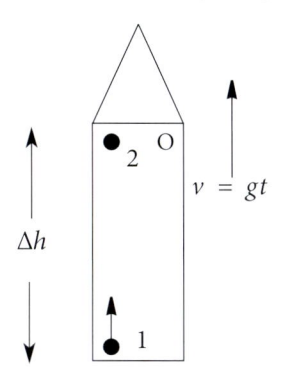

Figure 1328

Let the acceleration $a = g$ and let the distance between the two sources be Δh. The time t for the light from source 1 to reach the observer is therefore

$$t = \frac{\Delta h}{c}$$

In this time the speed gained by the O is

$$v = gt = \frac{g\Delta h}{c}$$

Because of the effective Doppler shift of the light from source 2 (remember, to the observer O this source will effectively be moving away from him at speed $v = gt$ when the light from the source reaches him), the frequency f' measured by O will be given by the Doppler shift equation, i.e.,

$$f' = f \times \left(1 - \frac{v}{c}\right)$$

If we now substitute for v then

$$f' = f \times \left(1 - \frac{\frac{g\Delta h}{c}}{c}\right) = f\left(1 - \frac{g\Delta h}{c^2}\right)$$

$$f' - f = \Delta f = \frac{fg\Delta h}{c^2}$$

so that

$$\frac{\Delta f}{f} = \frac{g\Delta h}{c^2}$$

This equation can also be derived on the basis of Einstein's principle of equivalence by considering the loss in energy of a photon from source 1 as it moves to source 2 in the gravitational field of the Earth. The two separate proofs of the gravitational redshift equation again show that there is no difference between inertia and gravity.

Example 1

Calculate the redshift that is observed between radiation emitted from the surface of a neutron star and radiation emitted from the centre of the star if the mass of the star is 10^{31} kg and its radius is 10 km.

Solution

We need to calculate g for the star which we do from

$$g = \frac{GM}{r^2}$$

and then substitute in the equation $\frac{\Delta f}{f} = \frac{g\Delta h}{c^2}$

with Δh = 10 km. You can do this to show that

$$\frac{\Delta f}{f} = 0.74.$$

Which is indeed an enormous redshift.

Note that we have assumed that g is constant.

Example 2

A satellite communication signal has a frequency of 100 MHz at the surface of the Earth. What frequency will be measured by an astronaut in a satellite which is in orbit 200 km above the surface of the Earth?

Solution

Again, assuming that g is constant we can use the red shift equation in the form

$$\frac{\Delta f}{f} = \frac{g\Delta h}{c^2}$$

and substitute to get $\frac{g\Delta h}{c^2} = \frac{9.8 \times 200 \times 10^3}{(3 \times 10^8)^2}.$

So that,

$$\frac{\Delta f}{f} = 2.2 \times 10^{-11} \therefore \frac{\Delta f}{100 \times 10^6} = 2.2 \times 10^{-11}$$

$$\Leftrightarrow \Delta f = 0.002 \ Hz$$

Chapter 13 (Option A)

Schwarzschild black holes

In 1939 *Oppenheimer* and *Snyder* pointed out that the General Theory predicts the existence of **black holes**. During the life time of some stars, there is a period when they undergo collapse. As they collapse, their density increases and therefore their surface gravity increases. Radiation leaving the surface will not only be redshifted by this gravitational field but, as the surrounding space-time becomes more and more warped, as the gravitational field increases, the path of the radiation will become more and more curved. If the surface gravity increases sufficiently there will come a point when the path of the radiation is so curved that none of the radiation will leave the surface of the star. The star has effectively become a black hole.

From a classical point of view we can think of a black hole in terms of escape velocity. A star becomes a black hole when the escape velocity at the surface becomes equal to the speed of light. The radius at which a star would become a black hole is known as the Schwarzchild Radius after the person who first derived the expression for it value.

General Relativity enables a value for the radius of a particular star for this to happen to be derived, but the derivation is beyond the scope of HL physics.

The Schwarzchild radius R_{sch} is given by

$$R_{sch} = \frac{2GM}{c^2}$$

(Coincidentally, Newtonian mechanics gives the same value.)

Event horizons

The surface of a black hole as defined by the Schwarzchild radius is called the **event horizon** since inside the surface all information is lost.

It is left as an exercise for you to show that if our Sun were to shrink until its radius was 3000 m then it would become a black hole.

Of course, if no radiation can leave a black hole and all radiation falling on it will also be "trapped", we have to ask how can such things be detected, should they exist. One possibility is to observe a black hole as a companion to a binary star system. Another way is to observe the effect that a black hole has on high frequency gamma radiation as it passes close to a black hole. It is sufficient to say at this point that astronomers do not doubt the existence of black holes.

This also relates to gravitational red-shift that will be described shortly.

If a person is outside the gravitational field of a black hole and observes an event that takes place a distance r from a black hole, then the time for the event will be dilated according to the equation

$$\Delta t = \frac{\Delta t_0}{\sqrt{1 - \frac{R_s}{r}}}$$

where R_S is the Schwarzschild radius of the black hole.

This effectively means that, if the person where to observe a clock approaching a black hole, the motion of the hands of the clock would appear to get slower and slower the nearer the clock gets to the event horizon of the black hole. At the event horizon, they would stop moving and time would stand still.

Example

A person a distance of 3 R_S from the event horizon of a black hole measures an event to last 4.0 s. Calculate how long the event would appear to last for a person outside the field of the black hole.

Solution

Substituting in to the above formula gives $\Delta t = 4.9$ s.

Applications of general relativity to the universe as a whole

Introduction

For any physical theory to be accepted it must not only explain known phenomena but also make predictions that can be verified experimentally.

The Special Theory introduced a completely new way of thinking about time and it was able to account for the Lorentz transformations encountered in Maxwell's theory. It also made several predictions all of which have been born out by experiment. We therefore accept that the Special Theory tells us the 'correct' way in which to think about time.

The General Theory introduces a completely new way of thinking about space, time and gravity and if the theory is to be accepted then it too must account for known phenomena and make predictions that can be verified by experiment. In this section we look at some of the evidence that supports the predictions made by General Relativity.

Gravitational attraction

The General Theory of Relativity essentially does away with the concepts of gravitational mass and gravitational force. How then do we account for the gravitational force of attraction between objects? We have seen that, in Special Relativity, space and time are intimately linked and an event is specified by four co–ordinates of space-time. Einstein proposed that space-time is curved by the presence of mass. An analogy is to think of a stretched elastic membrane onto which is placed a heavy object. In the vicinity of the object, the membrane will no longer be flat but will be curved, see Figure 1329. The curvature will be greatest close to the object and the general curvature will also increase as the mass of the object increases.

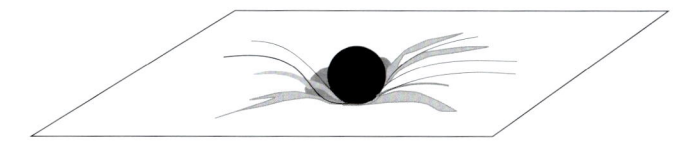

Figure 1329   Spacetime and gravity

We can explain gravitational attraction in terms of this "warping" of space.

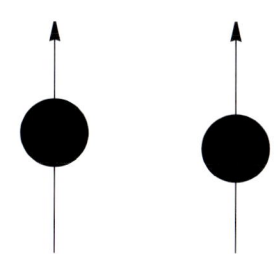

Figure 1330

Consider two objects moving in the direction shown in Figure 1330. Each object curves its local space and will therefore move towards the other object. As the objects get closer, the local space-time will become more curved and they will behave as though they were experiencing an ever-increasing force of attraction. If we choose the appropriate geometry to describe the curvature of space, then the objects will move just as if there were an inverse square force between them. You can think of the analogy of two ships at the equator sailing due North. Because of the curvature of the Earth's surface they will get closer and closer together even though they are following a "straight line path". However, do not take this analogy too literally since, in Einstein's theory, it is the objects themselves that curve the space.

Einstein proposed that all objects would take the shortest possible distance between two events in space-time. Such a distance is known as a **geodesic**. The geodesic for a plane surface is a straight line and for a sphere, a great circle. In this sense, the planets are actually following geodesics in the particular geometry of the space-time produced by the mass of the sun.

Gravitational lensing

Einstein suggested a method by which the effect of gravity on the path of light could be detected. The position of a star can be measured very accurately relative to the position of other stars. Einstein suggested measuring the position of a particular star, say in June, and then again six months later when the Earth is on the other side of the Sun relative to the star.

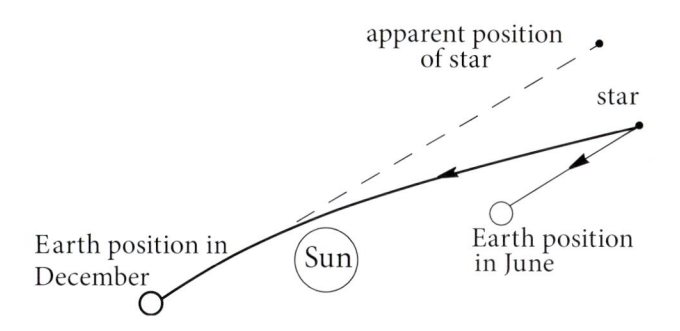

Figure 1331 Gravitational lensing

The path of the light from the star reaching the Earth according to Einstein should now be bent as it passes close to the Sun. This will cause an apparent shift in the position of the star. See Figure 1331.

Einstein predicted that the path of starlight should be deflected by 1.75 seconds of arc as it passes by the Sun. To observe starlight that passes close to the Sun, then the stars must be observed during the day and the only way that this can be done is during a total eclipse of the Sun. The General Theory was published in 1917 and by good fortune a total eclipse of the Sun was predicted for 29th March 1919 near the Gulf of Guinea and Northern Brazil. Expeditions were mounted to both destinations and scientists were able to collect enough photographs of suitable stars to test Einstein's prediction. The location of the stars indicated that the path of the light had been deviated by 1.64 seconds of arc, a result that compared very favourably with the Einstein prediction. However, recently doubt has been cast on Eddington's interpretation of the data.

More recently observations have been made of quasar images indicating that galaxies or even clusters of galaxies on its passage to Earth have deflected the light from the quasars. Even more accurate work has been carried out on radio signals transmitted from Earth and reflected from the planets. Solar space-time curvature will effect the delay times for the echo signals. The experimental results are in excellent agreement with the General Theory. This bending of light by large gravitational masses is often referred to as "gravitational lensing" in analogy with the bending of light by optical lenses.

387

Chapter 13 (Option A)

Redshift experiment

In 1960 *Pound* and *Rebka* performed an experiment that verified gravitational redshift to a high degree of accuracy. Instead of light sources they used gamma radiation emitted from two cobalt–57 sources mounted vertically 22 m above each other. The wavelength of the used gamma radiation emitted by the lower of the two sources was "red-shifted" with respect to the other source by an amount predicted by the General Theory.

Recently even more accurate experiments have been done with atomic clocks. In 1976 an atomic clock was placed in a rocket and sent to an altitude of 10,000 km. The frequency at which this clock ran was then compared with the frequency of a clock on the Earth. The observed differences in frequency confirmed the gravitational redshift predicted by the General Theory to an accuracy of 0.02 %.

Conclusion

The General Theory of Relativity is now accepted as being the correct interpretation of gravity and as such the correct model for our view of space and time. General Relativity and Quantum Theory form the two great theories upon which the whole of Physics rests. To date, all attempts to unify then into one complete theory have been unsuccessful.

Summary

Special Theory

Postulates

I. The Laws of Physics are the same for all inertial observers irrespective of their relative velocity.

II. All inertial observers will measure the same value for the free space velocity of light irrespective of their relative velocity.

The Lorentz Transformation equations

Given that

$$\gamma = \left(1 - \frac{v^2}{c^2}\right)^{-\frac{1}{2}} = \frac{1}{\sqrt{1 - \frac{v^2}{c^2}}}$$

then

Time dilation

$$\Delta t = \gamma \Delta t_0$$

Length contraction

$$L = \frac{L_0}{\gamma}$$

Mass increase

$$m = \gamma m_0$$

Rest mass energy

$$E_0 = m_0 c^2$$

Total energy

$$E = mc^2$$

Velocity transformation

$$u'_x = \frac{u_x - v}{1 - \frac{u_x v}{c^2}}$$

Relativistic momentum

$$p = \gamma m_0 u$$

Total energy

$$E = \gamma m_0 c^2 = E_k + m_0 c^2$$

Relativity

Energy-momentum equation

$$E^2 = p^2c^2 + m_0^2c^4$$

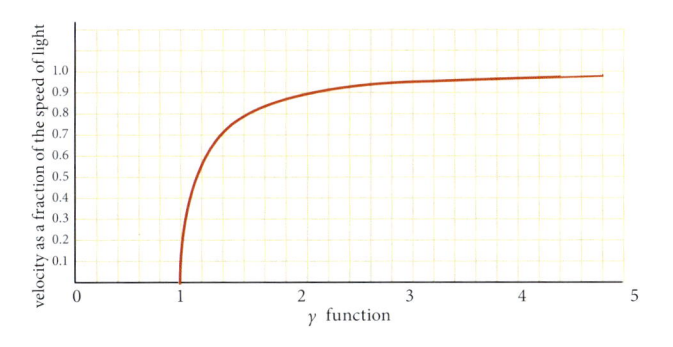

Figure 1332

The graph shows how the γ function varies with velocity and should be referred to in problems in order to verify if a particular value of the γ function has been calculated correctly.

General Relativity

Postulates

I. Mach's principle - Inertial and gravitational forces are indistinguishable.

II. Four dimensional space-time is curved as a result of the presence of mass.

III. Objects take the shortest path between two points in space-time.

The Einstein Principle of Equivalence

There is no way in which gravitational effects can be distinguished from inertial effects.

The Gravitational redshift equation

$$\frac{\Delta f}{f} = \frac{g \Delta h}{c^2}$$

Gravitational Time Dilation Equation

$$\Delta t = \frac{\Delta t_0}{\sqrt{1 - \frac{R_s}{r}}}$$

Experimental evidence

Special Theory

The Special Theory is well supported by experimental evidence.

The invariance of the velocity of light in respect of source and observer's relative motion.

Measured mass increase of accelerated electrons.

The arrival of muons at the surface of the Earth and the measurement of their respective half-lives gives evidence of time dilation.

Pair production gives evidence of the conservation of mass-energy.

Nuclear binding energy and nuclear processes all verify the conservation of mass-energy.

General Theory

Except for some minor perturbations the orbits of the planets are ellipses and the major axis of the elliptical orbit is fixed. However it was observed that the major axis of the orbit of Mercury shifts its plane by some 5.75 seconds of arc per century.

By considering the gravitational effects of all the other planets on the orbit of Mercury, Newton's theory accounted for all but 43 seconds of arc of the precession of the major axis. However, the General Theory accounts for the entire precession.

The apparent displacement of the measured position of stars gives evidence of the bending of the path of a ray of light by a gravitational field.

The Pound-Rebka experiment gives evidence of gravitational redshift.

The existence of black holes gives further evidence of the warping of space by the presence of matter.

CHAPTER 13 (OPTION A)

Exercise 13.3

In the following exercises, should a particular value of γ be required, then refer to the graph, Figure 1331.

1. State the two postulates of the Special Theory of Relativity and explain, with the use of appropriate diagrams, how two events that are simultaneous to one observer need not necessarily be simultaneous to another observer in a different reference frame.

2. Show that 1 atomic mass unit is equivalent to about 930 MeV.

3. An electron is moving at a constant velocity of $0.90c$ with respect to a laboratory observer X.

    (a) Determine the mass of the electron as measured by X?

    (b) Another observer Y is moving at a constant velocity $0.50c$ with respect to X in a direction opposite to that of the electron in X's reference frame. Determine the mass of the electron has measured by Y

4. Use the time dilation (page 477) graph to find the γ function when $v = 0.5c$ and $0.8c$ and hence calculate the relativistic mass *increase* of an electron when travelling at these speeds.

5. A proton is accelerated from rest through a potential difference of 8.00×10^8 V. Calculate as measured in the laboratory frame of reference after acceleration the

    i. proton mass.
    ii. velocity of the proton.
    iii. momentum of the proton (HL only).
    iv. total energy of the proton (HL only).

6. Estimate how far you would have to push a ball of mass 2.0 kg with a force of 50 N until its mass was 4.0 kg.

7. If another beam of protons is accelerated at the same time through the same potential difference as in question 5, but in the opposite direction, calculate, after acceleration, the relative velocity of a proton in one beam with respect to a proton in the other beam.

8. Calculate the de Broglie wavelength of electrons that have been accelerated through a potential difference of 1.8×10^5 V.

9. Explain what is meant by Einstein's principle of equivalence.

10. Describe how Einstein's description of the gravitational attraction between two particles differs from that offered by Newton.

11. Summarise the evidence that supports the General Theory of Relativity.

12. γ-rays are emitted from a source placed in a ground floor laboratory. They are measured to have a wavelength of 0.05 nm. If the source is moved to a laboratory on the top floor of the building, they are measured to have a frequency shift of 3.3×10^4 Hz. Estimate the height of the building.

13. Use a spreadsheet to plot a graph that shows the variation with distance r from a black hole of the time dilation Δt of an event.

14. Option B: Engineering physics

Contents

B.1 – Rigid bodies and rotational dynamics
B.2 – Thermodynamics
B.3 – Fluids and fluid dynamics
B.4 – Forced vibrations and resonance

Essential Ideas

The basic laws of mechanics have an extension when equivalent principles are applied to rotation. Actual objects have dimensions and they require the expansion of the point particle model to consider the possibility of different points on an object having different states of motion and/or different velocities.

The first law of thermodynamics relates the change in internal energy of a system to the energy transferred and the work done. The entropy of the universe tends to a maximum.

Fluids cannot be modelled as point particles. Their distinguishable response to compression from solids creates a set of characteristics which require an in-depth study.

In the real world, damping occurs in oscillators and has implications that need to be considered. © IBO 2014

Chapter 14 (Option B)

B.1 Rigid bodies and rotational dynamics

> **NATURE OF SCIENCE:**
>
> Modelling: The use of models has different purposes and has allowed scientists to identify, simplify and analyse a problem within a given context to tackle it successfully. The extension of the point particle model to actually consider the dimensions of an object led to many groundbreaking developments in engineering. (1.2)
>
> © IBO 2014

Essential idea: The basic laws of mechanics have an extension when equivalent principles are applied to rotation. Actual objects have dimensions and they require the expansion of the point particle model to consider the possibility of different points on an object having different states of motion and/or different velocities.

Understandings:
- Torque
- Moment of inertia
- Rotational and translational equilibrium
- Angular acceleration
- Equations of rotational motion for uniform angular acceleration
- Newton's second law applied to angular motion
- Conservation of angular momentum

Rotational and translational equilibrium

A very common type of motion that an object can undergo involves a combination of translational and rotational motion. Translational motion involves the movement of an object form one place to another such as motion along a straight line or along a curve. In translational motion, the mass is a measure of an object's inertia – an object's ability to resist changes in its translational motion. **Rotational motion** involves the motion of an object about an axis. In rotational motion, the analogous quantity to mass is called the ***moment of inertia***. This is a measure of an object's rotational inertia – an object's tendency to resist changes in its rotational motion.

Forces can change the velocity, direction and shape of an object. In order to simplify the study of rotation, we will restrict our study to rigid bodies. A **rigid body** is one whose separate parts retain fixed distances from each other when subjected to external forces.

For a body to be in equilibrium, two conditions must be met:

- The vector sum of the forces must be zero thereby ensuring zero translational motion.
  $\Sigma F = 0$

- The algebraic sum of the moments of the forces about any point must be zero thereby ensuring zero rotational motion.
  $\Sigma M = 0$

For a system of coplanar forces or couples to maintain a body in equilibrium, the algebraic sum of the clockwise moments about any point equals the algebraic sum of the anticlockwise moments about that same point.

Σ clockwise moments = Σ anticlockwise moments

The centre of gravity of a body is the point where the entire weight of the body is considered to be concentrated. Therefore, the weight of a uniform body would act through the geometrical centre of the body. For example, the weight of a uniform metre rule acts through the 50 cm mark.

Torque

The moment of a force F about any axis is a measure of the effectiveness of the force in producing rotation about that axis. Another word equivalent to a moment is the word torque τ. We will use the word torque in the rest of this unit.

Torque is measured as the product of the applied force and the perpendicular distance r between the line of action of the force and the axis of rotation.

When the turning force is perpendicular to the line joining the point where the force is acting to a pivot point as shown in Figure 1401, the magnitude of the torque is given by:

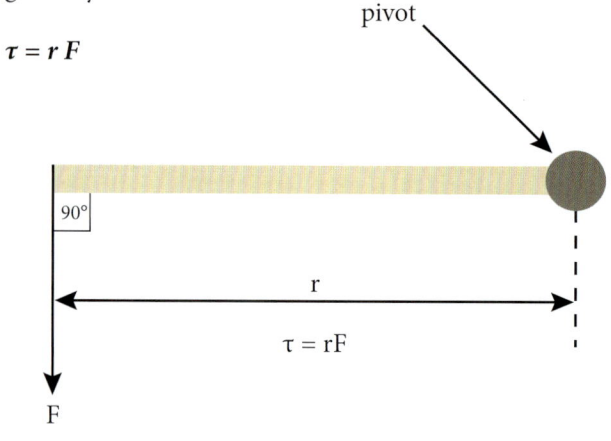

$\tau = r F$

Figure 1401   A uniform rod being turned about a pivot.

Engineering Physics

It is a vector quantity (in this case the multiplication of 2 vector quantities) with units Nm. Experience shows that the torque of the force is greater with greater magnitude of the force and the greater the distance of its point of application from the pivot. For example, a door handle is usually placed far away from the door hinges.

When the turning force is not perpendicular to the line joining the point where the force is acting to the pivot as shown in Figure 1402, the magnitude of the force is given by:

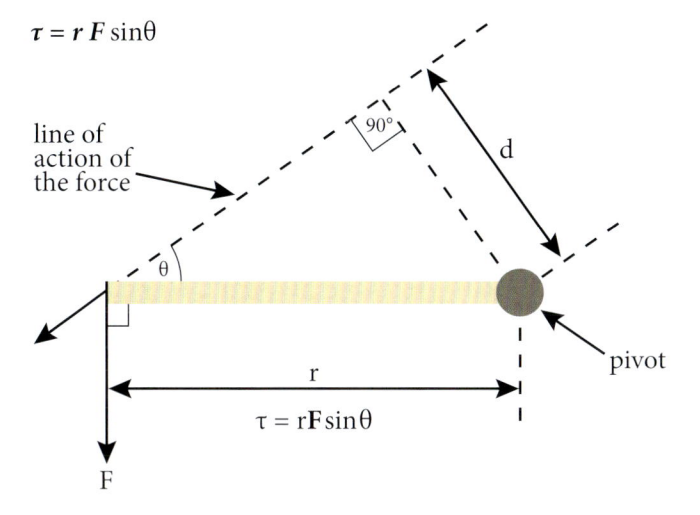

$\tau = rF\sin\theta$

Figure 1402 A uniform rod being turned by a force that is not perpendicular to the line joining the point where the force is acting to the pivot.

It should be noted that $r\sin\theta$ to the distance d in Figure 1402.

Example

A light horizontal beam of length 14 m, supported at its mid-point is held in equilibrium by two vertical downward forces, one of 16 N acting 3.7 m from one end as shown in Figure 1403. Determine where the other force of 12 N is acting.

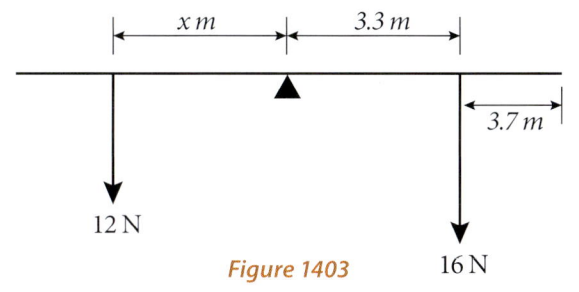

Figure 1403

Solution

Let us firstly find the torques about the fulcrum.
Σ *clockwise torque* = Σ *anticlockwise torque*

$12 \times x = 16 \times 3.3$
$x = 16 \times 3.3 / 12$
$x = 4.4$ m

The 12 N force must act 4.4 m from the mid-point.

An alternative method to solve this problem could be to consider the torque about the point through which the 12 N force acts as shown in Figure 1404.

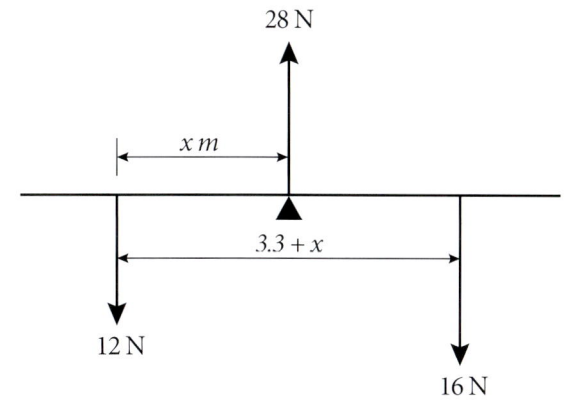

Figure 1404

Σ *clockwise torque* = Σ *anticlockwise torque*
$28 \times x = 16 \times (3.3 + x)$
$28x = 52.8 + 16x$
$12x = 52.8$
$x = 4.4$ m

Again, the 12 N force must act 4.4 m from the mid-point.

Example

A truck of weight 2.0×10^5 N is driving across a steel bridge of weight 5.0×10^6 N. All the distances are shown in Figure 1405. Determine the two supporting contact forces C_1 and C_2 provided by the pylons.

weight of truck $W_2 = 200$ kN
weight of bridge $W_1 = 5000$ kN

clockwise moment M_1
anticlockwise moment M_2
anticlockwise moment M_3

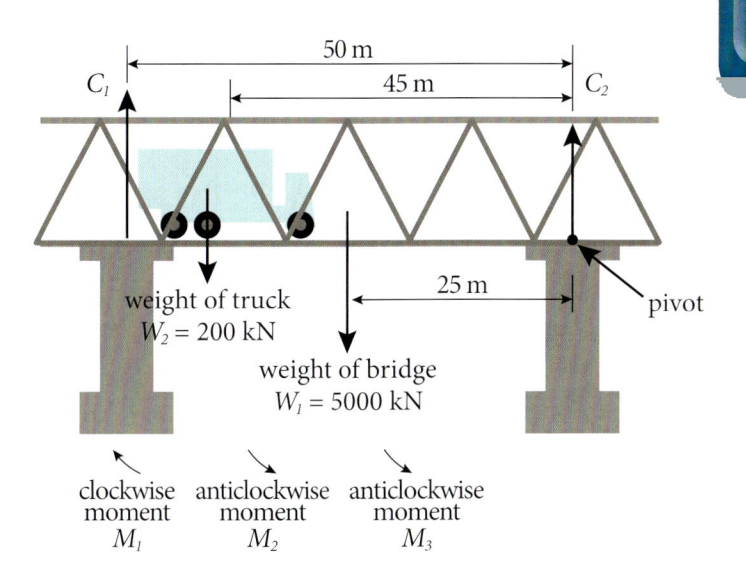

Figure 1405

393

CHAPTER 14 (OPTION B)

Solution

$C_1 + C_2 = W_1 + W_2$

$\qquad = 5200 \text{ kN}$

Let us consider the torque about the point C_2
Σ clockwise torque = Σ anticlockwise torque

$C_1 \times 50 = (W_2 \times 45) + (W_1 \times 25)$

$50C_1 = (200 \text{ kN} \times 45) + (5000 \text{ kN} \times 25)$

$50C_1 = 9000 \text{ kNm} + 125000 \text{ kNm}$

$C_1 \quad = 2680 \text{ kN}$

$C_1 + C_2 = 5200 \text{ kN}$

$2680 \text{ kN} + C_2 = 5200 \text{ kN}$

$C_2 = 2520 \text{ kN}.$

Example

A uniform ladder weighing 4.00×10^2 N rests against a **smooth** wall at a point 6.00 m above the ground as shown in Figure 1406. The other end of the ladder is on the ground 4.00 m from the wall. The ground exerts a force on this end of the ladder. Determine the magnitude and direction of this reaction force.

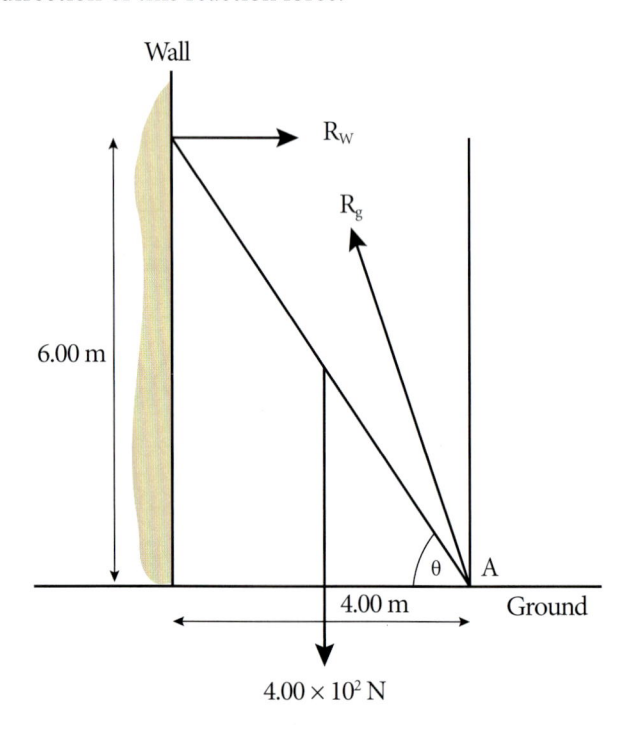

Figure 1406

Solution

There are two important points to consider in this problem.

1. The ladder is at rest as the frictional force between the ground and the bottom of the ladder prevents it sliding. Because of the horizontal and vertical forces, this reaction force is not vertical.

2. The reaction force of the wall on the top of the ladder is perpendicular to the wall because the smooth wall provides no frictional forces. ("smooth" means that the wall can only exert a normal force directed perpendicular to the wall and cannot exert a frictional force parallel to it).

To determine the reaction of the wall on the ladder, let us take torques about A.

Anticlockwise torque due to weight - Clockwise torque due to the reaction of the wall = 0

$4.00 \times 10^2 \times 2.00 - R_W \times 6.00 = 0$

$8.00 \times 10^2 = 6.00 \, R_W$

$R_W = 1.33 \times 10^2 \text{ N}$

The reaction force R_g is the resultant vector (equilibrant) of the other reaction force and the weight of the ladder as shown in Figure 1407.

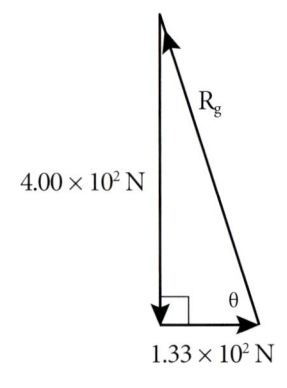

Figure 1407

$R_g^2 = (4.00 \times 10^2)^2 + (1.33 \times 10^2)^2$

$R_g = 4.22 \times 10^2 \text{ N}$

$\sin\theta = \dfrac{4.00 \times 10^2}{4.22 \times 10^2}$

$\theta = 71.6^0$

The reaction force has a magnitude of 4.22×10^2 at 18.4^0 to the vertical.

ENGINEERING PHYSICS

Exercise 14.1

1. The 4.0 kg mass as shown in Figure 1408 is in equilibrium. Determine the magnitude of the tension forces T_1 and T_2.

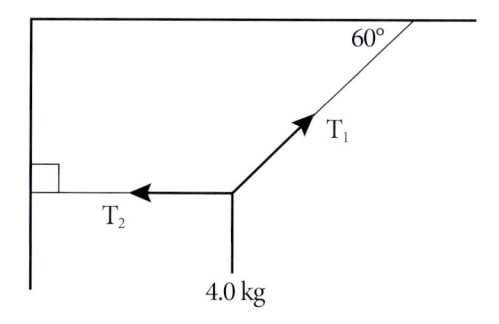

Figure 1408

2. A light horizontal beam is 16 m long. It is supported at its mid-point and it is held in equilibrium by two vertical downwards forces of 14 N acting 2.6 m from one end, and a 27 N force at another location. Determine where the 27 N force is acting.

3. A crowbar of length 1.40 m is used to lift a cement slab of mass 36.0 kg. If the fulcrum is 20.0 cm from the load, determine the effort required to lift the slab.

4. A uniform metre rule is pivoted at its mid-point. A 1.2 N force and a 2.4 N force are acting downwards at the 30.0 cm and the 90.0 cm mark respectively. Where must an upward force of 3.6 N be applied to balance the rule?

5. A uniform scaffold plank 4.80 m long weighs 118 N and rests on trestles at each end of the plank. A bricklayer weighing 686 N stands 1.20 m from one end of the plank. Determine the forces exerted by each trestle on the plank.

6. A uniform plank is 4.00 m long and weighs 54.0 kg. It is supported at its ends by two ladders. Two painters with equal masses of 90.0 kg stand on the plank. One painter is standing 1.00 m from one end and the other painter stands 1.50 m from the other end. Determine the forces exerted by the ladders.

7. An object of mass 125 kg is being lifted out of a cargo ship by two light ropes, one making an angle of 45.0° with the vertical and the other making an angle of 30.0° to the vertical. Determine the tension in each rope.

8. A uniform ladder of length 6.5 m and weight 2.0×10^2 N rests against a smooth wall and reaches 6.0 m above the ground. Calculate:

   (a) the reaction of the wall
   (b) the reaction at the ground assuming that it is rough enough to prevent slipping.

9. A diving board of negligible mass is tightly bolted down at the left end and hangs over the diving pool and is supported by a fulcrum 1.40 m from the bolted end. A diver with weight of 530 N is poised at the right end of the board. Find the forces that the bolt and the fulcrum exert on the board.

10. An electric light fitting is suspended over a table by two light wires that make angles of 60.0° and 30.0° to the vertical respectively. Find the tension in each wire.

Moment of inertia

Consider a series of particles $m_1, m_2, \ldots m_i$ situated distances $r_1, r_2, \ldots r_i$ respectively from an axis as shown in Figure 1409.

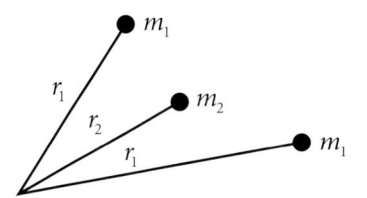

Figure 1409   Mass distribution

By definition:

Moment of inertia $(I) = m_1 r_1^2 + m_2 r_2^2 + \ldots m_i r_i^2$

$I = \Sigma m\, r^2$

The units of I are kg m².

For example, a "governor" used on some engines to stop drivers from exceeding speed limits can schematically be drawn as in Figure 1410. It consists of two 5kg masses rotating about a vertical axis.

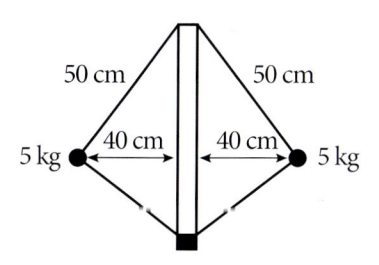

Figure 1410   A governor.

The moment of inertia would be:

$I = \Sigma\, m\, r^2$
$I = 5 \times 0.4^2 + 5 \times 0.4^2 = 1.6$ kg m².

395

Chapter 14 (Option B)

The moment of inertia of a body depends on:

1. the mass of the body
2. the distribution of the mass about the axis
3. the positioning of the axis.

Some moments of inertia are given in Figure 1411. For examination purposes, the equation for the moment of inertia of a specific shape will be provided when necessary.

Object	Axis of rotation	I
Thin rod length $2a$ mass M	Through centre of mass	$I_{rod} = 1/3\, Ma^2$
Solid sphere radius a mass M	Through centre of mass	$I_{sphere} = 2/5\, Ma^2$
Ring radius a mass M	Through centre at right angles to plane of ring	$I_{ring} = Ma^2$
Disc radius a mass M	Through centre and perpendicular to the plane of disc	$I_{disc} = ½\, Ma^2$

Figure 1411 Moment of inertia of some bodies.

Angular acceleration

In most cases the angular velocity of a rotating body is not constant. For example, if you roll a cylinder down an incline plane, the angular velocity will change.

By analogy with linear acceleration uniformly accelerated rotational motion (angular acceleration) α is defined as:

Angular acceleration = time rate of change of angular velocity.

$$\alpha = \frac{\omega}{t}$$

Therefore, instantaneous acceleration would be:

$$\alpha = \frac{\Delta\omega}{\Delta t}$$

Angular acceleration is a vector quantity whose direction is the same as the direction of angular velocity. The units of α are rad s^{-2}.

Now from the earlier definition of the radian, we know that

$$\Delta s = r\, \Delta\theta$$

If we divide both sides by Δt (since the linear and angular changes occur at the same time), then:

$$\frac{\Delta s}{\Delta t} = \frac{r\Delta\theta}{\Delta t}$$

That is

$$v = r\omega$$

If the object is accelerating, then

$$\Delta v = r\, \Delta\omega$$

If we divide both sides by Δt we have:

$$\frac{\Delta v}{\Delta t} = r \frac{\Delta\omega}{\Delta t}$$

Therefore,

$$a_T = r\alpha$$

Equations of rotational motion for uniform angular acceleration

For uniformly accelerated motion, it follows from the definition of angular acceleration that

$$\alpha = \frac{\Delta\omega}{\Delta t} = \frac{\omega - \omega_0}{t}$$

That is

$$\omega = \omega_0 + \alpha t \qquad 1$$

Angular displacement = average angular velocity x time

$$\theta = \left(\frac{\omega + \omega_0}{2}\right) t$$

$$\theta = \left(\frac{\omega_0 + \alpha t + \omega_0}{2}\right) t$$

$$\theta = \omega_0 t + ½\, \alpha t^2 \qquad 2$$

If we square both sides in equation 1 and substitute equation 2, we get:

$$\omega^2 = (\omega_0 + \alpha t)^2$$

$$= \omega_0^2 + 2\alpha\omega_0 t + \alpha^2 t^2$$

$$\omega^2 = \omega_0^2 + 2\alpha(\omega_0 t + ½\, \alpha t^2)$$

$$\omega^2 = \omega_0^2 + 2\alpha\theta \qquad 3$$

These equations are analogous to those for translational motion as shown in Figure 1412.

Translational motion	Rotational motion
$v = u + at$	$\omega = \omega_0 + \alpha t$
$s = ut + \frac{1}{2} at^2$	$\theta = \omega_0 t + \frac{1}{2} \alpha t^2$
$v^2 = u^2 + 2as$	$\omega^2 + 2\alpha\theta$

Figure 1412 Comparison of translational and rotational equations.

Figure 1413 shows some graphs that compare translational and rotational quantities with respect to time.

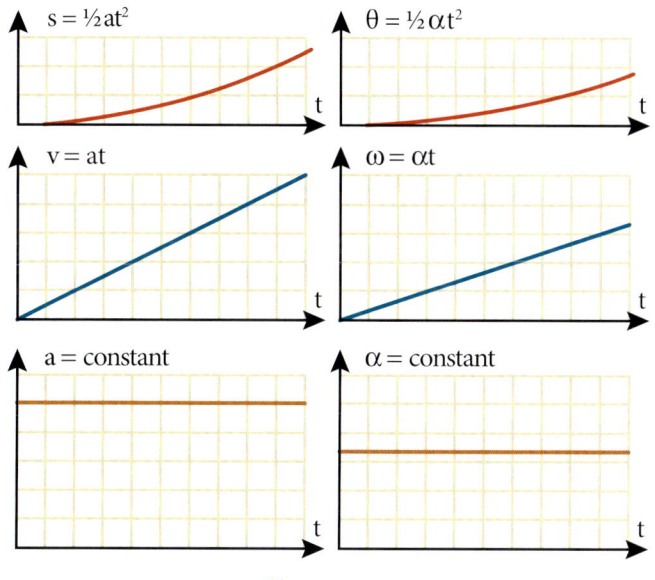

Figure 1413

Newton's second law applied to angular motion

Newton's second law for translational motion is given by:

$F = ma$

By analogy, the angular counterparts of force, mass and acceleration are torque, moment of inertia and angular acceleration. Therefore:

Torque = moment of inertia x angular acceleration

$\tau = I\alpha$

The direction of the torque is the same as the angular acceleration.

Example

A uniform disc of radius 30.0 cm and mass of 1.00 kg is pivoted about its axis. Determine what torque would generate an angular velocity of 25.0 rad s^{-1} in 5.00 seconds.

Solution

$\alpha = \omega - \dfrac{\omega_0}{t} = \dfrac{25.0}{5.00} = 5.00$ rad s^{-2}

The moment of inertia for a disc is given by:

$I_{disc} = \frac{1}{2} Ma^2 = \frac{1}{2} \times 1.00 \times 0.300^2 = 0.045$ kg m^2

$\tau = I\alpha = 0.045 \times 5.00 = 0.225$ Nm

Example

A flywheel with a radius of 20 cm and a mass of 5.0 kg is pivoted on an axle as shown in Figure 1414.

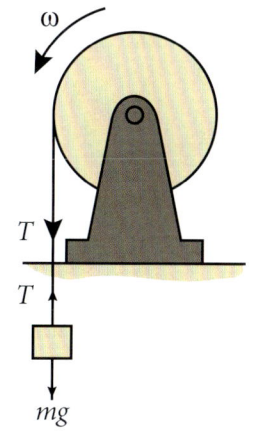

Figure 1414

A mass of 10.0 kg is attached to one end of a light inextensible cord and the other end is wrapped around the rim of the flywheel. Determine the velocity of the flywheel if the weight is allowed to fall a distance of 50.0 cm.

Solution

The forces acting on the system are:
- the weight of the flywheel and the normal reaction of the support (neither of which contribute to the motion).
- The tension in the string and the weight.

Let us take downwards as the positive direction. Therefore, from Newton's second law we obtain:

$mg - T = ma$ (1)

For the wheel:
$\tau = I\alpha$ (2)

397

Chapter 14 (Option B)

$TR = I\alpha$ or $T = I\alpha / R$

But $a = R\alpha$ and $I = \frac{1}{2} MR^2$ so

$T = \frac{1}{2} Ma$ (3)

Substituting (3) into (1)

$mg - \frac{1}{2} Ma = ma$

$mg = (m + \frac{1}{2} M)a$

$a = \dfrac{mg}{m + \frac{1}{2} M}$

Assuming the weight falls from rest:

$v^2 = u^2 + 2as$

$v = \sqrt{\dfrac{2mgh}{m + \frac{1}{2}M}}$

$= \sqrt{\dfrac{2 \times 10 \times 9.8 \times 0.5}{10 + 2.5}}$

$= 2.8 \text{ ms}^{-1}$

Conservation of angular momentum

Rotational kinetic energy

The kinetic energy of a particle with mass m and speed is given by:

$E_k = \frac{1}{2} m v^2$

For a rotating body, consider a single particle of mass m_i and speed v_i as shown in Figure 1415 that is situated at a distance r_i from the axis of rotation.

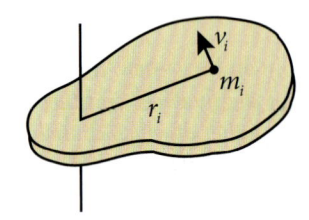

Figure 1415   Rotational kinetic energy

$E_i = \frac{1}{2} m_i v_i^2$

But $v = r\omega$

$E_i = \frac{1}{2} m_i r_i^2 \omega_i^2$

The total kinetic energy could then be given as:

$E_{total} = \Sigma \frac{1}{2} m_i r_i^2 \omega^2 = \frac{1}{2} (\Sigma m_i r_i^2) \omega^2$

But $I = \Sigma m r^2$ = moment of inertia

Therefore:

$E_k = \frac{1}{2} I\omega^2$

Angular momentum

Consider a small element of mass m_i and linear velocity v_i as shown in Figure 1416 that is situated at a distance r_i from the axis of rotation.

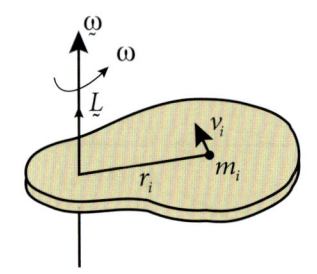

Figure 1416   Angular momentum

The angular momentum is equal to the moment of the linear momentum.

$L_i = m_i v_i r_i = m_i r_i^2 \omega$

$L = \Sigma\, m_i r_i^2 \omega = \omega \Sigma\, m_i r_i^2$

$L = I\omega$

Angular momentum is a vector quantity because I is a scalar quantity and it is multiplied by the vector ω. The units of angular momentum are kg m² s⁻¹. Its direction is the same as for the angular velocity.

Torque and angular momentum

$\tau = I\alpha = \dfrac{I\Delta\omega}{\Delta t} = \dfrac{\Delta(I\omega)}{\Delta t}$

Since I is constant:

$\tau = \dfrac{\Delta(I\omega)}{\Delta t}$

That is

$\tau = \dfrac{\Delta L}{\Delta t}$

Therefore, torque equals the time rate of change of angular momentum.

ENGINEERING PHYSICS

Conservation of angular momentum

If $\tau = 0$, then $\frac{\Delta L}{\Delta t}$ must also equal zero.

Therefore, **L** is a constant.

If no external *torques* are applied to a system then angular momentum is conserved.

This is statement of the Law of conservation of momentum.

$L = I\omega$ = constant

$I_1 \omega_1 = I_2 \omega_2$

Example

An artificial satellite is placed in an elliptical orbit about the earth. The satellite's distance from the centre of the earth varies from 8.37×10^6 m to 25.1×10^6 m. At its closest approach, the satellite has a speed of 8450 m s^{-1}. Calculate the speed of the satellite when it is furthest from the centre of the earth.

Solution

The only significant force on the satellite is the gravitational force of the earth. This act through the centre of mass of the earth, so there is no resultant torque on the satellite about the earth. Therefore, the angular momentum of the satellite remains constant and is equal in both positions.

$I_1 \omega_1 = I_2 \omega_2$

$m r_1^2 \left(\frac{v_1}{r_1}\right) = m r_2^2 \left(\frac{v_2}{r_2}\right)$

$v_1 = \frac{r_2 v_2}{r_1}$

$v_1 = \frac{8.37 \times 10^6 \times 8450}{25.1 \times 10^6}$

$= 2820$ m s^{-1}

Note that the answer is independent of the mass of the satellite.

When an ice skater is spinning with her arms and a leg outstretched, and then she pulls her arms and leg inwards, her motion changes dramatically. Bringing the arms and leg closer to the body lowers her moment of inertia and so her angular velocity increases. Her inward movement involves internal, not external torques and, therefore does not change her angular momentum.

Energy conservation

Just as energy conservation makes many problems easy to solve for translational motion, so is the case for rotational motion.

In general

Loss in gravitational potential energy = increase in translational energy + increase in rotational kinetic energy.

$mgh = \frac{1}{2} mv^2 + \frac{1}{2} I\omega^2$

Example

A solid sphere with a mass of 2.0 kg rolls without slipping from a height of 25 cm down a ramp inclined at an angle of 30° to the horizontal as shown in Figure 1417. Determine the speed of the centre of mass at the bottom of the ramp.

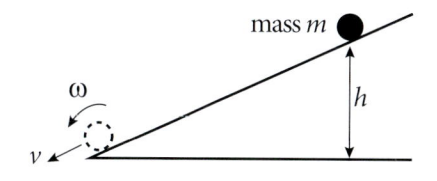

Figure 1417   Sphere rolling down an incline.

Solution

$mgh = \frac{1}{2} mv^2 + \frac{1}{2} I\omega^2$

Moment of inertia for a sphere $= \left[\frac{2}{5}\right] mr^2$

$mgh = \frac{1}{2} mv^2 + \frac{1}{2} \left(\frac{2}{5} mr^2\right) \left(\frac{v}{r}\right)^2$

$gh = \frac{1}{2} v^2 + \frac{1}{5} v^2$

$v^2 = \frac{10}{7} gh$

$v = 7.0$ m s^{-1}

Figure 1418 compares the quantities of translational and rotational motion.

399

Chapter 14 (Option B)

Quantity	Translation	Rotation	Relation
Displacement	s	θ	$s = r\theta$
Velocity	$v = \Delta s / \Delta t$	$\omega = \Delta\theta / \Delta t$	$v = r\omega$
Acceleration	$a_T = \Delta v / \Delta t$	$\alpha = \Delta\omega / \Delta t$	$a_T = r\alpha$
Force/torque	F	τ	$\tau = Fp$
Equations of motion	$v = u + at$	$\omega = \omega_0 + \alpha t$	
	$s = ut + \tfrac{1}{2}at^2$	$\theta = \omega_0 t + \tfrac{1}{2}\alpha t^2$	
	$v^2 = u^2 + 2as$	$\omega^2 + 2\alpha\theta$	
Inertia	m	I	$I = \Sigma\, m_i r_i^2$
Newton's 2nd Law	$F = ma$	$\tau = I\alpha$	
Impulse	$F\Delta t = \Delta p$	$\tau \Delta t = \Delta L$	
Kinetic energy	$E_k = \tfrac{1}{2}mv^2$	$E_k = \tfrac{1}{2}I\omega^2$	
Momentum	$p = mv$	$L = I\omega$	

Figure 1418

Exercise 14.2

1. A steel ball bearing has a mass of 33 g and a radius of 0.01 m. It rolls 25 cm from rest down a ramp inclined at 10^0 to the horizontal. Determine the time elapsed to move the 25 cm.

2. A wooden disc has a radius of 30 cm and a mass of 1.0 kg, and it is spinning at 50 revolutions per minute. Determine its

    (a) rotation frequency
    (b) angular velocity
    (c) rotational kinetic energy
    (d) angular momentum

3. Refer back to Figure 1414 concerning the falling weight being used to cause the rotation of the flywheel. Using the conservation of energy, determine its velocity.

4. A bicycle wheel of radius of 0.32 m and moment of inertia of 0.035 kg m² is spun with a frequency of 10 Hz horizontally on its axle. It comes to rest after 3.0 minutes. Determine

    (a) the angular deceleration of the wheel
    (b) the average frictional torque acting on the wheel.

5. A circular saw blade is being turned by an electric motor has a moment of inertia of 1.41×10^{-3} kg m². The motor brings it to its rated angular speed of 80.0 rev s^{-1} in 240.0 revolutions. Determine the net torque that must be applied by the motor to the blade.

6. Determine the angular speed of the "second hand" on an analogue watch.

7. A drum rotates about its axis at an angular velocity of 12.6 rad s^{-1}. If the drum then slows down at a constant rate of 4.20 rad s^{-2}, determine:

    (a) the time it takes to come to rest
    (b) the total angle through which it rotates in coming to rest.

8. Determine the rotational inertia of a wheel that has a kinetic energy of 2.44×10^4 J when rotating at 602 revolutions per minute.

ENGINEERING PHYSICS

B.2 Thermodynamics

NATURE OF SCIENCE:
Variety of perspectives: With three alternative and equivalent statements of the second law of thermodynamics, this area of physics demonstrates the collaboration and testing involved in confirming abstract notions such as this. (4.1)

© IBO 2014

Essential idea: The first law of thermodynamics relates the change in internal energy of a system to the energy transferred and the work done. The entropy of the universe tends to a maximum.

Understandings:
- The first law of thermodynamics
- The second law of thermodynamics
- Entropy
- Cyclic processes and pV diagrams
- Isovolumetric, isobaric, isothermal and adiabatic processes
- Carnot cycle
- Thermal efficiency

The first law of thermodynamics

As already mentioned in Chapter 3, an **ideal gas** is a theoretical gas that obeys the equation of state of an ideal gas exactly. They obey the equation $PV = nRT$ when there are no forces between molecules at all pressures, volumes and temperatures.

Remember from **Avogadro's hypothesis** that one mole of any gas contains the Avogadro number of particles N_A equal to 6.02×10^{23} particles. It also occupies 22.4 dm³ at 0 °C and 101.3 kPa pressure (STP).

The internal energy of an ideal gas would be entirely kinetic energy as there would be no intermolecular forces between the gaseous atoms. As temperature is related to the average kinetic energy of the atoms, the kinetic energy of the atoms would depend only on the temperature of the ideal gas.

From the combined gas laws, we determined that:

$$\frac{PV}{T} = k \text{ or } PV = kT$$

If the value of the **universal gas constant** is compared for different masses of different gases, it can be demonstrated that the constant depends not on the size of the atoms but rather on the number of particles present (the number of moles). Thus for n moles of any ideal gas:

$$\frac{PV}{nT} = R$$

or $PV = nRT$

This is called the 'equation of state' of an ideal gas, where R is the universal gas constant and is equal to 8.31 J mol⁻¹ K⁻¹. The equation of state of an ideal gas is determined from the gas laws and Avogadro's law. The ideal gas equation can also be stated in terms of the Boltzmann constant as

$$PV = nkT$$

where k = the Boltzmann constant = R / N_A and its value is 1.38×10^{-23} J K⁻¹.

The internal energy for an ideal gas U can be written as

$U = K_{AV} = n N_A \times$ the average kinetic energy = $n N_A \frac{3}{2} k T$.

But $k = \dfrac{R}{NA}$

Therefore:

$$U = \frac{3}{2} n R T$$

It can also be shown that for a monatomic gas:

$$p V^{\frac{5}{3}} = \text{a constant.}$$

Work done in volume change

Consider a mass of gas with pressure p enclosed in a cylinder by a piston of cross-sectional area A as in Figure 1419.

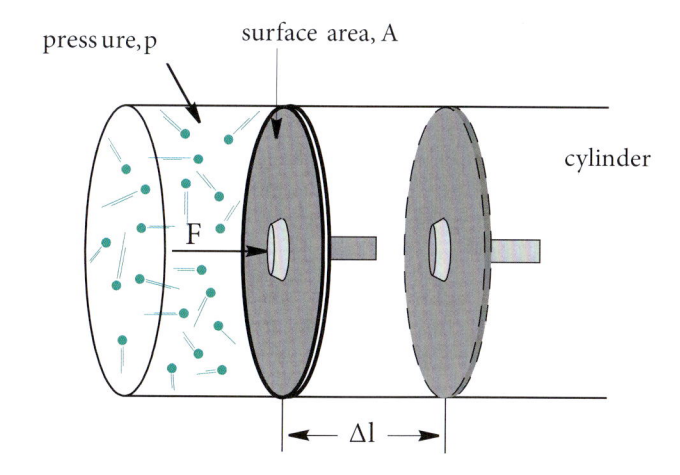

Figure 1419 *Expansion of a gas at constant pressure*

The pressure, p, on the piston = force per unit area

So that,

$$p = \frac{F}{A}$$

401

Therefore, the force on the piston, F, is given by

$$F = pA$$

Suppose the piston is moved a distance l when the gas expands. Normally, if the gas expands, the volume increases and the pressure decreases, as was determined from Boyle's Law for ideal gases in the previous section. However, if the distance l is a small Δl, then the pressure can be considered constant. If the pressure is constant then the force F will be constant. The work done by the gas is:

$\Delta W = F \Delta l = pA \Delta l$ since pressure p = Force F / Area A

$\quad\quad = p\Delta V$ since volume $\Delta V = A l$

That is, (work done / J) = (pressure / Nm^{-2}) × (volume change / m^3)

So that,
$$\Delta W = p \cdot \Delta V = p(V_2 - V_1)$$

The sign of the work done by the gas depends on whether volume change is positive or negative. When a gas expands, as is the case for Figure 1419, then **work is done by** the gas, and the volume increases. As V is positive, then W **is positive**.

This equation is also valid if the gas is compressed. In the compression, **work is done on** the gas and the volume is decreased. Therefore, ΔV is negative which means that W will be **negative**. From the first law of thermodynamics this means that positive work is done on the gas.

Thermodynamics is the name given to the study of processes in which thermal energy is transferred as heat and as work. It had its foundations with engineers in the 19th century who wanted to know what were the limitations of the Laws of Physics with regard to the operation of steam engines and other machines that generate mechanical energy. Thermodynamics treats thermal energy from the macroscopic point of view in that it deals with the thermodynamic variables of pressure, volume, temperature and change in internal energy in determining the state of a system.

Heat can be transferred between a system and its environment because of a temperature difference. Another way of transferring energy between a system and its environment is to do work **on** the system or allow work to be done **by** the system on the surroundings.

In order to distinguish between thermal energy (heat) and work in thermodynamic processes

- If a system and its surroundings are at different temperatures and the system undergoes a process, the energy transferred by non-mchanical means is referred to as thermal energy (heat).

- **Work** is defined as the process in which thermal energy is transferred by means that are independent of a temperature difference.

In thermodynamics the word system is used often. A **system** is any object or set of objects that is being investigated. The surroundings will then be everything in the Universe apart from the system. For example, when a volume of gas in a cylinder is compressed with a piston, then the system is the cylinder-gas-piston apparatus and the surroundings is everything else in the Universe. A closed system is one in which no mass enters or leaves the system. It is an **isolated system** if no energy of any kind enters or leaves the system. Most systems are open systems because of the natural dynamic processes that occur in the Universe.

In Chapter 3, **internal energy** U was defined as the sum total of the potential energy and kinetic energy of the particles making up the system. From a microscopic viewpoint, the internal energy of an ideal gas is due to the kinetic energy of the thermal motion of its molecules. There are no intermolecular forces and thus there cannot be any increase in potential energy. Therefore a change in the temperature of the gas will change the internal energy of the gas.

From the macroscopic point of view of thermodynamics, one would expect that the internal energy of the system would be changed if:

- work is done on the system
- work is done by the system
- thermal energy is added to the system
- thermal energy is removed from the system

Internal energy is a property of the system that depends on the "state" of the system. In thermodynamics, a **change of state of an ideal gas occurs if some macroscopic property of the system has changed** eg. phase, temperature, pressure, volume, mass, internal energy.

Heat and work can change the state of the system but they are not properties of the system. They are not characteristic of the state itself but rather they are involved in the thermodynamic process that can change the system from one state to another.

The absolute value for internal energy is not known. This does not cause a problem as one is mainly concerned with changes in internal energy, denoted by ΔU, in thermodynamic processes.

The **first law of thermodynamics** is a statement of the **Law of Conservation of Energy** in which the equivalence of work and thermal energy transfer is taken into account. It can be stated as:

the heat added to a closed system equals the change in the internal energy of the system plus the work done by the system.

That is,
$$Q = \Delta U + W = \Delta U + p\Delta V$$

or,
$$\Delta U = Q - W$$

where '+Q' is the thermal energy **added** to the system and '+W' is the work done **by** the system.

If thermal energy leaves the system, then Q is negative. If work is done on the system, then W is negative.

For an isolated system, then $W = Q = 0$ and $\Delta U = 0$.

Isovolumetric, isobaric, isothermal and adiabatic processes

Isobaric processes

A graph of pressure as a function of volume change when the pressure is kept constant is shown in Figure 1420. Such a process is said to be **isobaric**. Note that the work done by the gas is equal to the area under the curve.

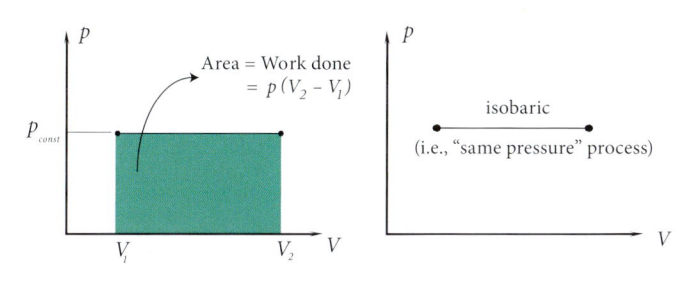

Figure 1420 Work done by a gas expanding at constant pressure

An isobaric transformation requires a volume change at constant pressure, and for this to occur, the temperature needs to change to keep the pressure constant.

p = constant, or V / T = constant.

For an isobaric expansion, work is done by the system so ΔW is positive. Thermal energy is added to cause the expansion so ΔQ is positive. This means that ΔU must be positive. For an isobaric compression, all terms would be negative.

Isochoric (isovolumetric) processes

A graph of pressure as a function of volume change when the volume is kept constant is shown in Figure 1421. Such a process is said to be **isochoric**. When the volume is kept fixed, the line of the transformation is said to be an isochore.

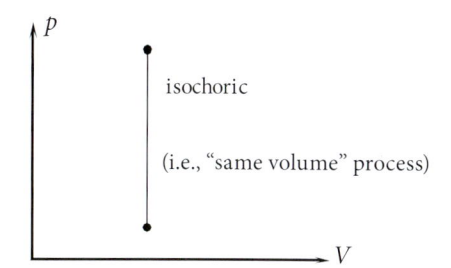

Figure 1421 A Constant volume transformation

Note that the work done by the gas is equal to zero as $\Delta V = 0$. There is zero area under the curve on a p–V diagram. However, the temperature and pressure can both change and so such a transformation will be accompanied by a thermal energy change.

V = constant, or p / T = constant.

For an isochoric process, no work is done by the system so ΔW is zero. Thermal energy leaves the system so ΔQ is negative. This means that ΔU must be negative. For an isobaric process, ΔW is zero, and ΔQ and ΔU are positive.

Isothermal processes

A thermodynamic process in which the pressure and the volume are varied while the temperature is kept constant is called an **isothermal** process. In other words, when an ideal gas expands or is compressed at constant temperature, then the gas is said to undergo an isothermal expansion or compression.

Figure 1422 shows three isotherms for an ideal gas at different temperatures where

$T_1 < T_2 < T_3$.

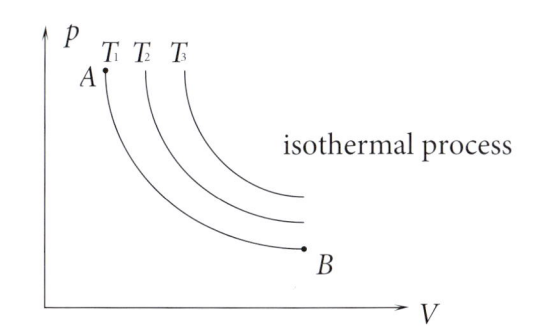

Figure 1422 Isotherms for an ideal gas

The curve of an isothermal process represents a Boyle's Law relation

T = constant, or pV = constant = nRT

The moles of gas n, the molar gas constant R, and the absolute temperature T are constant.

403

Chapter 14 (Option B)

For an isothermal expansion, temperature is constant so ΔU is zero. Work is done by the system so ΔW is positive. This means that ΔU must be positive.

In order to keep the temperature constant during an isothermal process

- the gas is assumed to be held in a thin container with a high thermal conductivity that is in contact with a heat reservoir – an ideal body of large mass whose temperature remains constant when heat is exchanged with it. eg. a constant–temperature water bath.

- the expansion or compression should be done slowly so that no eddies are produced to create hot spots that would disrupt the energy equilibrium of the gas.

Consider an ideal gas enclosed in a thin conducting vessel that is in contact with a heat reservoir, and is fitted with a light, frictionless, movable piston. If an amount of heat Q is added to the system which is at point A of Figure 1008, then the system will move to another point on the graph, B. The heat taken in will cause the gas to expand isothermally and will be equivalent to the mechanical work done by the gas. Because the temperature is constant, there is no change in internal energy of the gas.

That is,

$$\Delta T = 0 \text{ and } \Delta U = 0 \Rightarrow Q = W$$

If the gas expands isothermally from A to B and then returns from B to A following exactly the same path during compression, then the isothermal change is said to be reversible. The conditions described above would follow this criterion.

Adiabatic processes

An **adiabatic** expansion or contraction is one in which no heat Q is allowed to flow into or out of the system. For the entire adiabatic process, Q = 0.

To ensure that no heat enters or leaves the system during an adiabatic process it is important to

- make sure that the system is extremely well insulated.
- carry out the process rapidly so that the heat does not have time to leave the system.

The compression stroke of an automobile engine is essentially an adiabatic compression of the air-fuel mixture. The compression occurs too rapidly for appreciable heat transfer to take place.

In an adiabatic compression the work done on the gas will lead to an increase in the internal energy resulting in an increase in temperature.

$$\Delta U = Q - W \text{ but } Q = 0 \Rightarrow \Delta U = -W$$

In an adiabatic expansion the work done by the gas will lead to a decrease in the internal energy resulting in a decrease in temperature.

Figure 1423 shows the relationship that exists between an adiabatic and three isothermals. Note that the adiabatic curve is steeper than the isotherm AB because the adiabatic process has to occur rapidly so that the heat does not have time to leave the system. The gas expands isothermally from point A to point B, and then it is compressed adiabatically from B to C. The temperature increases as a result of the adiabatic process from T_1 to T_3. If the gas is then compressed at constant pressure from the point C to A, the net amount of work done on the gas will equal the area enclosed by ABC.

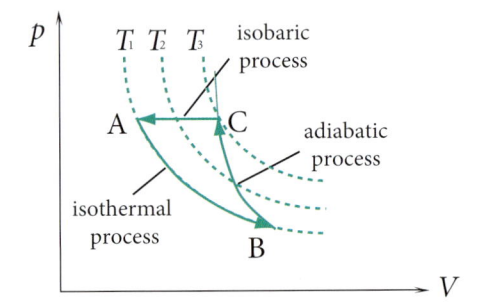

Figure 1423   An isothermal and adiabatic process

For an adiabatic compression, no heat enters or leaves the system so ΔQ is zero. Work is done on the system so ΔW is negative. This means that ΔU must be negative. For an adiabatic expansion, ΔQ is zero, and ΔW and ΔU are positive.

In Figure 1424 the area ABDE = work done by the gas during isothermal expansion.

The area ACDE = work done by the gas during an adiabatic expansion.

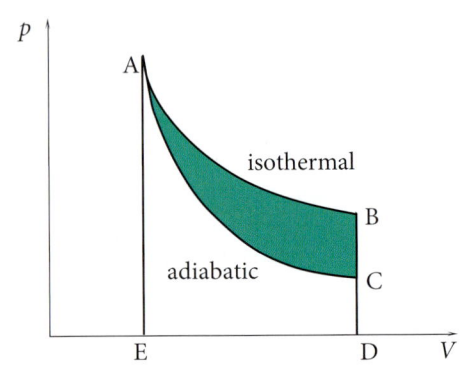

Figure 1424   Work done during expansion of a gas

The work done by a gas and the work done on a gas can be seen using the following graphical representation of a pressure–volume diagram (Figure 1425).

ENGINEERING PHYSICS

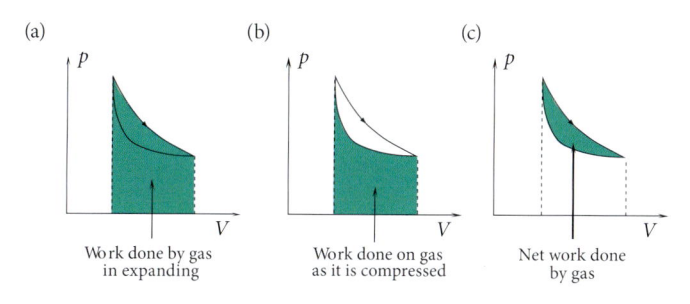

(a) Work done by gas in expanding
(b) Work done on gas as it is compressed
(c) Net work done by gas

Figure 1425 *Pressure volume diagrams*

Cyclic processes and *pV* diagrams

A **thermodynamic engine** is a device that transforms thermal energy to mechanical energy (work) or mechanical energy to thermal energy such as in refrigeration and air-conditioning systems. Cars, steam trains, jets and rockets have engines that transform fuel energy (chemical energy) into the kinetic energy of their motion. In all presently manufactured engines, the conversion is accompanied by the emission of exhaust gases that waste some of the thermal energy. Consequently, these engines are not very efficient as only part of the thermal energy is converted to mechanical energy. An engine has two crucial features:

- It must work in cycles to be useful.
- The cyclic engine must have more than one heat reservoir.

A **thermodynamic cycle** is a process in which the system is returned to the same state from which it started. That is, the initial and final states are the same in the cyclic process. A cycle for a simple engine was shown in Figure 1423. The net work done in the cycle is equal to the area enclosed by the cycle.

Suppose a piston was placed on a heat reservoir, such as the hot plate of a stove. Thermal energy is supplied by the thermal reservoir, and the gas inside the piston does work as it expands. But this is not an engine as it only operates in one direction. The gas cannot expand indefinitely, because as the volume of the piston increases, the pressure decreases (Boyle's Law). Some point will be reached when the expanding gas will not be able to move the piston. For this simple engine to function, the piston must eventually be compressed to restore the system to its original position ready to do work.

For a cycle to do net work, thermal contact with the original heat reservoir must be broken, and temperatures other than that of the original heat reservoir must play a part in the process. In the above example, if the piston is returned to its original position while in contact with the hot plate, then all the work that the gas did in the expansion will have to be used in the compression. On a $p-V$ diagram, one would draw an isotherm for the expansion and an isotherm for the compression lying on top of the expansion isotherm but in the opposite direction. Therefore, the area enclosed by the cycle would be zero. However, if the gas is compressed at a lower temperature the internal pressure of the system will be lower than during the expansion. Less work will be needed for the compression than was produced in the expansion, and there will be net work available for transformation to mechanical energy.

Heat engines and heat pumps

A **heat engine** is any device that converts thermal energy into work. Examples include petrol and diesel engines, fossil-fuelled (coal, oil and natural gas) and nuclear power plants that use heat exchangers to drive the blades in turbines, and jet aircraft engines.

Although we cannot convert all the random motion associated with the internal energy into useful work, we can at least extract some useful work from internal energy using a heat engine. To make a heat engine, we need a source of heat such as coal, petrol (gasoline), diesel fuel, aviation fuel or liquefied petroleum gas (LPG) and a **working fluid**. A working fluid is a substance that undergoes a thermodynamic change of state and in the process does work on the surroundings. Common working fluids are water that is heated and converted to steam as used in power stations and petrol-air mixtures as used in car engines.

Most heat engines contain either pistons with intake and exhaust valves that work in thermodynamic cycles as in a car engine, or rotating turbines used in heat exchanger systems in power stations.

The purpose of a heat engine is to convert as much of the heat input Q_H from a high temperature reservoir into work. Figure 1426 shows an energy flow diagram of a typical heat engine.

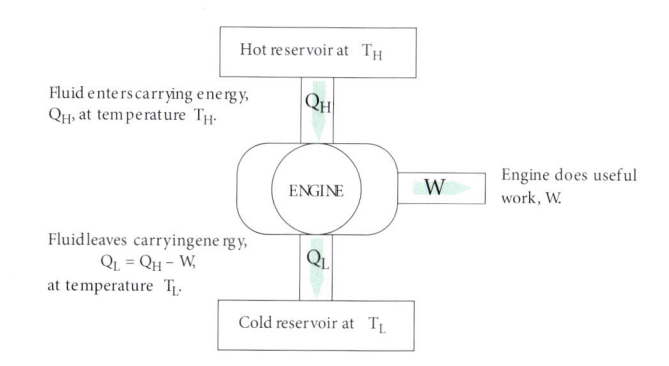

Hot reservoir at T_H
Fluid enters carrying energy, Q_H, at temperature T_H.
Q_H
ENGINE
W — Engine does useful work, W.
Fluid leaves carrying energy, $Q_L = Q_H - W$, at temperature T_L.
Q_L
Cold reservoir at T_L

Figure 1426 *Energy flow diagram of a heat engine*

The heat input Q_H is represented as coming from the high temperature reservoir T_H which is maintained at a constant temperature. Thermal energy Q_L is taken from the hot reservoir. This thermal energy is used to do work in the heat engine. Then thermal energy can be given to the low temperature reservoir T_L without increasing its temperature.

405

Chapter 14 (Option B)

If a "perfect" engine completed a cycle, the change in internal energy ΔU would be zero because all the heat would be converted to work. However, there is no perfect heat engine and the flow diagram in Figure 1426 is more the reality. At this stage, we will assume that the change in internal energy is zero. From the First Law of Thermodynamics

$$\Delta U = 0 = \Delta Q - W, \text{ so that } W = \Delta Q$$

That is,

$$Q_H - Q_L = W$$

Thus for a cycle, the heat added to the system equals the work done by the system plus the heat that flows out at lower temperature.

An ideal gas can be used as a heat engine as in the simple cycle in Figure 1427.

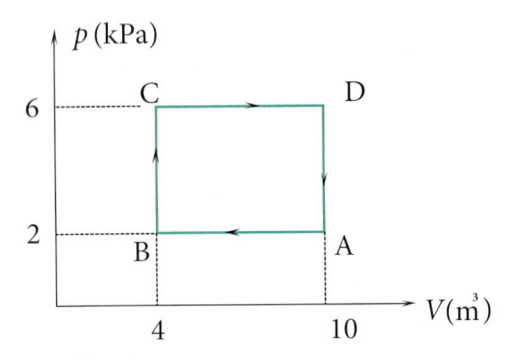

Figure 1427 *Behaviour of an ideal gas*

From A to B, the gas is compressed (volume decreases) while the pressure is kept constant – an isobaric compression. The amount of work done by the gas is given by the area under the 2 kPa isobar.

Using the fact that $W = p\Delta V$, we have that

$W = 2 \text{ kPa} \times (4 - 10) \text{ m}^3$

$= -1.2 \times 10^4 \text{ J}$

From B to C, the volume is kept constant as the pressure increases – an isochoric increase in pressure. This can be achieved by heating the gas. Since $\Delta V = 0$, then no work is done by the gas, $W = 0$.

From C to D, the gas expands (volume increases) while the pressure is kept constant – an isobaric expansion. The amount of work done by the gas is given by the area under the 6 kPa isobar.

Now, we have that $W = p\Delta V$, so that

$W = 6 \text{ kPa} \times (10 - 4) \text{ m}^3$

$= 3.6 \times 10^4 \text{ J}$

From D to A, the gas is cooled to keep the volume constant as the pressure is decreased – an isochoric decrease in pressure. Again $\Delta V = 0$ and no work is done by the gas, $W = 0$.

That is, the net work done by the gas is therefore

$3.6 \times 10^4 \text{ J} - 1.2 \times 10^4 \text{ J} = \mathbf{2.4 \times 10^4 \text{ J}}$.

The internal combustion engine

Figure 1428 shows a series of schematic diagrams for the cycle of an **internal combustion engine** as used in most automobiles.

- With the exhaust valve closed, a mixture of petrol vapour and air from the carburettor is sucked into the combustion chamber through the inlet valve as the piston moves down during the intake stroke.

- Both valves are closed and the piston moves up to squeeze the mixture of petrol vapour and air to about 1/8 th its original volume during the compression stroke.

- With both valves closed, the mixture is ignited by a spark from the spark plug.

- The mixture burns rapidly and the hot gases then expand against the piston in the power stroke.

- The exhaust valve is opened as the piston moves upwards during the exhaust stroke, and the cycle begins again.

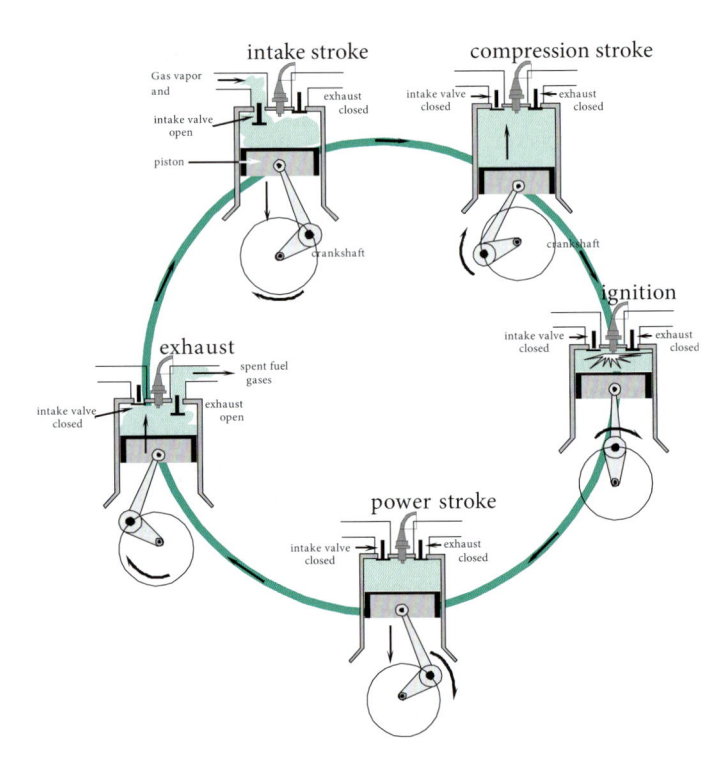

Figure 1428 *Four-stroke internal combustion engine*

Motor cars usually have four or six pistons but five and eight cylinders are also common. The pistons are connected by a crankshaft to a flywheel which keeps the engine turning over during the power stroke. Automobiles are about 25% efficient.

Any device that can pump heat from a low-temperature reservoir to a high-temperature reservoir is called a **heat pump**. Examples of heat pumps include the refrigerator and reverse cycle air-conditioning devices used for space heating and cooling. In the summer component of Figure 1429, the evaporator heat exchanger on the inside extracts heat from the surroundings. In the winter component, the evaporator heat exchanger is outside the room, and it exhausts heat to the inside air. In both cases, thermal energy is pumped from a low-temperature reservoir to a high- temperature reservoir.

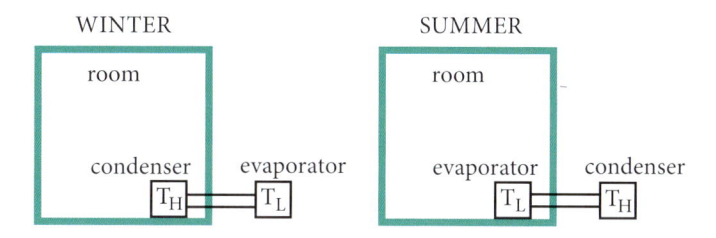

Figure 1429   A reverse cycle heat pump

Figure 1430 shows the energy flow that occurs in a heat pump cycle. By doing work on the system, heat Q_L is added from the low temperature T_L reservoir, being the inside of the refrigerator. A greater amount of heat Q_H is exhausted to the high temperature T_H reservoir.

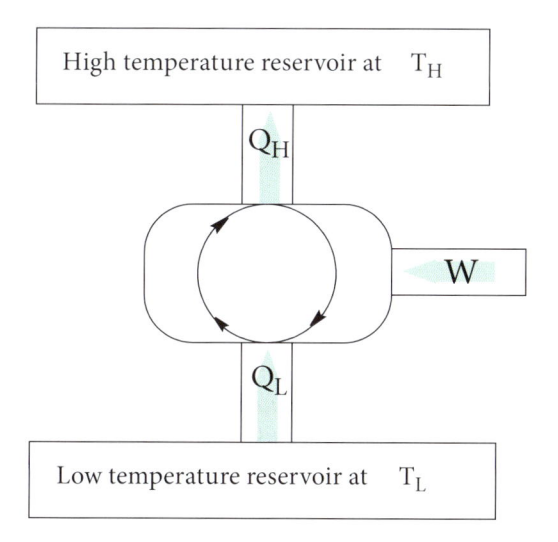

Figure 1430   Energy flow diagram for a heat pump

An ideal gas can be used as a heat pump as in the simple cycle in Figure 1431.

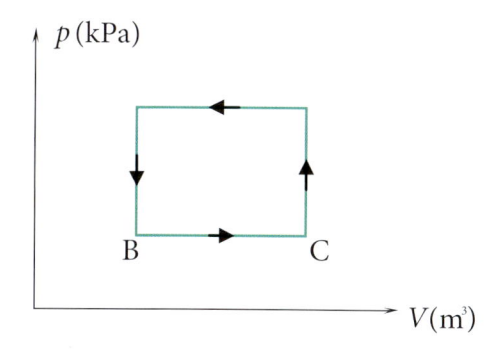

Figure 1431   An ideal gas as a heat pump

Because the cycle is traced in an anticlockwise direction, the net work done on the surroundings is negative.

The refrigerator – a heat pump

A **refrigerator** is a device operating in a cycle that is designed to extract heat from its interior to achieve or maintain a lower temperature inside. The heat is exhausted to the surroundings normally at a higher temperature. A typical refrigerator is represented in Figure 1432.

A motor driving a compressor pump provides the means by which a net amount of work can be done on the system for a cycle. Even though refrigerator cabinets are well insulated, heat from the surroundings leaks back inside. The compressor motor can be heard to switch on and off as it pumps this heat out again.

A volatile liquid called HFC (tetrafluoroethane) is circulated in a closed system of pipes by the compressor pump, and, by the process of evaporative cooling, the vaporised HFC is used to remove heat. Evaporative cooling was discussed in Chapter 3.

The compressor maintains a high-pressure difference across a throttling valve. Evaporation of the HFC occurs in several loops called the evaporator pipes that are usually inside the coldest part of the fridge. As the liquid evaporates on the low-pressure, low-temperature side, heat is added to the system. In order to turn from a liquid to a gas, the HFC requires thermal energy equal to the latent heat of vaporisation. This energy is obtained from the contents of the fridge.

Chapter 14 (Option B)

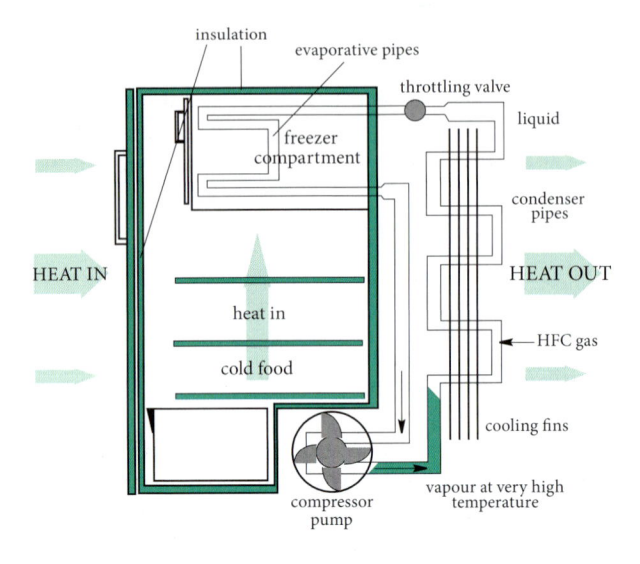

Figure 1432 The typical small refrigerator

On the high-pressure, high-temperature side of the throttling valve, thermal energy is removed from the system. The vaporised HFC in the compressor pipes is compressed by the compressor pump, and gives up its latent heat of vaporisation to the air surrounding the compressor pipes. The heat fins act as a heat sink to radiate the thermal energy to the surroundings at a faster rate. The fins are painted black and they have a relatively large surface area for their size.

Example 1

If 22 J of work is done on a system and 3.4×10^2 J of heat is added, determine the change internal energy of the system.

Solution

Using the formula, $Q = \Delta U + W$,
we have that 340 J = ΔU + (-22) J

340 J = ΔU + (−22) J

so that ΔU = 340 J + 22 J

= 362 J

That is, the change in internal energy of the system is **3.6×10^2 J.**

Example 2

6.0 dm³ of an ideal gas is at a pressure of 202.6 kPa. It is heated so that it expands at constant pressure until its volume is 12 dm³. Determine the work done by the gas.

Solution

Using the formula $W = p \Delta V$, we have that

$$W = 202.6 \text{ kPa} \times (12 - 6.0) \text{ dm}^3$$

$$= 202.6 \times 10^3 \text{ Pa} \times (12 - 6.0) \times 10^{-3} \text{ m}^3$$

$$= 1.216 \times 10^3 \text{ J}$$

That is, the work done by the gas in the expansion is **1.2×10^3 J.**

Example 3

A thermal system containing a gas is taken around a cycle of a heat engine as shown in the Figure below.

(a) Starting at point A, describe the cycle.

(b) Label the diagram fully showing the maximum and minimum temperature reservoirs.

(c) Estimate the amount of work done in each cycle.

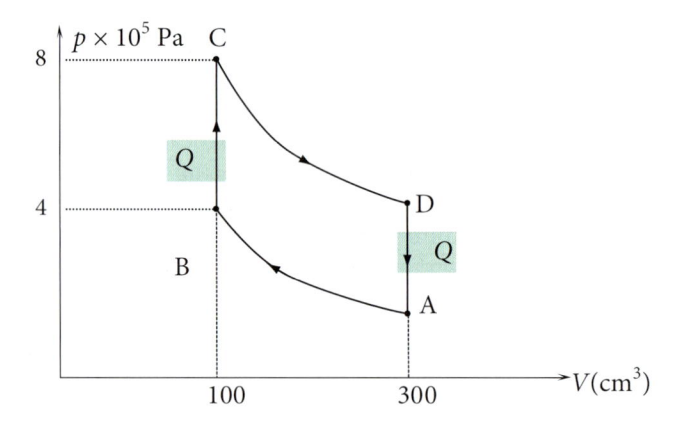

Solution

(a) The fuel-air mixture enters the piston at point A. The compression AB is carried out rapidly with no heat exchange making it an adiabatic compression. The ignition and combustion of the gases introduces a heat input Q_H that raises the temperature at constant volume from B to C. The power stroke is an adiabatic expansion from C to D. Thermal energy Q_L leaves the system during the exhaust stroke, and cooling occurs at constant volume from D to A.

(b) The Figure Below shows the changes that occur for each process in the cycle.

408

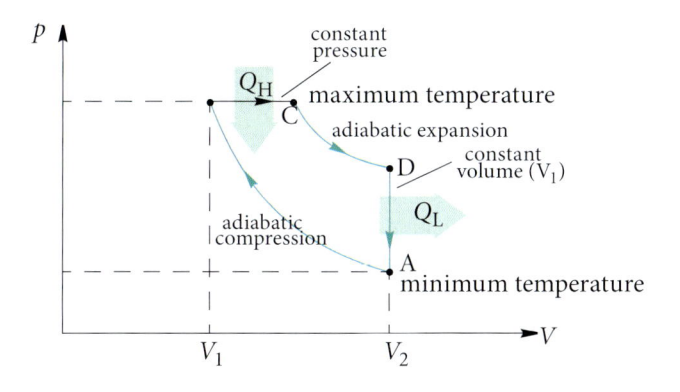

(c) The net work is represented by the enclosed area ABCD. If we assume that the area is approximately a rectangle with sides of 4×10^5 Pa and 200 cm³, we have:

4×10^5 Nm⁻² $\times 200 \times 10^{-6}$ m³ = 80 J.

Example 4

For the compression stroke of an experimental diesel engine, the air is rapidly decreased in volume by a factor of 15, the compression ratio. The work done on the air-fuel mixture for this compression is measured to be 550 J.

(a) What type of thermodynamic process is likely to have occurred?

(b) What is the change in internal energy of the air-fuel mixture?

(c) Is the temperature likely to increase or decrease?

Solution

(a) Because the compression occurs rapidly appreciable heat transfer does not take place, and the process can be considered to be adiabatic, $Q = 0$.

(b) $\Delta U = Q - W = 0 - (-550)$ J

Therefore, the change in internal energy is **550J**.

(c) The temperature rise will be very large resulting in the spontaneous ignition of the air-fuel mixture.

Exercise 14.3

1. An ideal gas was slowly compressed at constant temperature to one quarter of its original volume. In the process, 1.5×10^3 J of heat was given off.

   The change in internal energy of the gas was

   A. 1.5×10^3 J
   B. 0 J
   C. -1.5×10^3 J
   D. 6.0×10^3 J

2. When an ideal gas in a cylinder is compressed at constant temperature by a piston, the pressure of the gas increases. Which of the following statement(s) best explain the reason for the pressure increase?

   I. the mean speed of the molecules increases
   II. the molecules collide with each other more frequently
   III. the rate of collision with the sides of the cylinder increases.

   A. II only
   B. III only
   C. I and II only
   D. II and III only

3. An ideal gas in a thermally insulated cylinder is compressed rapidly. The change in state would be:

   A. isochoric
   B. isothermal
   C. adiabatic
   D. isobaric

4. The Figure below shows the variation of pressure p with volume V during one complete cycle of a simple heat engine.

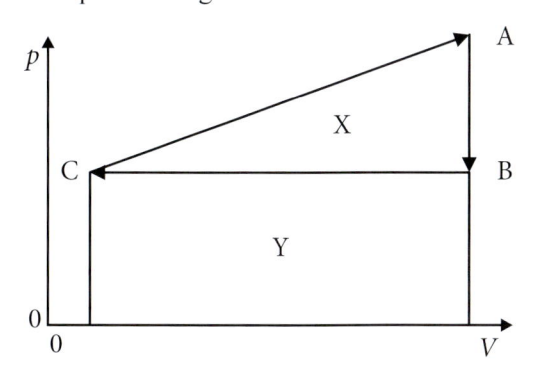

   The total work done is represented by the area:

   A. X + Y   C. X
   B. X − Y   D. Y

Chapter 14 (Option B)

5. The Figure below shows the variation of the pressure p with volume V of a gas during one cycle of the Otto engine.

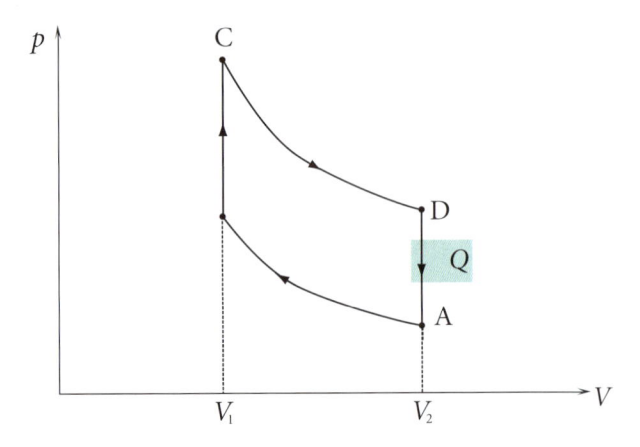

   During which process does the gas do external work?

   A. AB
   B. CD
   C. BC and CD
   D. AB and CD

6. A system absorbs 100 J of thermal energy and in the process does 40 J of work. The change in internal energy is:

   A. 60 J
   B. 40 J
   C. 100 J
   D. 140 J

7. Work is done when the volume of an ideal gas increases. During which of the following state processes would the work done be the greatest?

   A. isochoric
   B. isothermal
   C. isobaric
   D. adiabatic

8. How much heat energy must be added at atmospheric pressure to 0.50 kg of ice at 0 °C to convert it to steam at 100 °C?

9. If 1.68×10^5 J of heat is added to a gas that expands and does 8.1×10^5 J of work, what is the change in internal energy of the gas?

10. 6.0 m³ of an ideal gas is cooled at constant normal atmospheric pressure until its volume is 1/6 th its original volume. It is then allowed to expand isothermally back to its original volume. Draw the thermodynamic process on a p–V diagram.

11. A system consists of 3.0 kg of water at 75 °C. Stirring the system with a paddlewheel does 2.5×10^4 J of work on it while 6.3×10^4 J of heat is removed. Calculate the change in internal energy of the system, and the final temperature of the system.

12. A gas is allowed to expand adiabatically to four times its original volume. In doing so the gas does 1750 J of work.

    (a) How much heat flowed into the gas?
    (b) Will the temperature rise or fall?
    (c) What is the change in internal energy of the gas?

13. For each of the processes listed in the following table, supply the symbol +, −, or 0 for each missing entry.

Process	Q	W	ΔU
Isobaric compression of an ideal gas			+
Isothermal compression of an ideal gas			
Adiabatic expansion			
Isochoric pressure drop			
Free expansion of a gas			

14. Helium gas at 312 K is contained in a cylinder fitted with a movable piston. The gas is initially at 2 atmospheres pressure and occupies a volume of 48.8 L. The gas expands isothermally until the volume is 106 L. Then the gas is compressed isobarically at that final pressure back to the original volume of 48.8L. It then isochorically returns back to its original pressure. Assuming that the helium gas behaves like an ideal gas

    (a) Calculate the number of moles of helium gas in the system.
    (b) Determine the pressure after the isothermal expansion.
    (c) Draw a diagram of the thermodynamic cycle.
    (d) Assuming that the isotherm is a diagonal line rather than a curve, estimate the work done during the isothermal expansion.
    (e) Determine the work done during the isobaric compression.
    (f) Determine the work done during the isochoric part of the cycle.

(g) Calculate the net work done by the gas.

(h) Calculate the final temperature of the helium.

15. (a) Distinguish between an *isothermal* process and an *adiabatic* process as applied to an ideal gas.

(b) A fixed mass of an ideal gas is held in a cylinder by a moveable piston and thermal energy is supplied to the gas causing it to expand at a constant pressure of 1.5×10^2 kPa as shown in the Figure below.

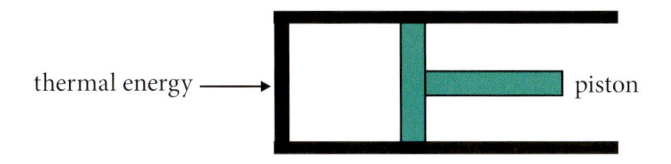

The initial volume of the gas in the container is 0.040 m^3 and after expansion the volume is 0.10 m^3. The total energy supplied to the gas during the process is 7.0 kJ.

(i) State whether this process is isothermal, adiabatic or neither of these processes.
(ii) Determine the work done by the gas.
(iii) Calculate the change in internal energy of the gas.

16. This question is about a diesel engine cycle as shown in the Figure below. Mark on the diagram each of the state changes that occur at AB, BC, CD and DA. Identify the maximum and minimum temperature reservoirs and label Q_H and Q_L.

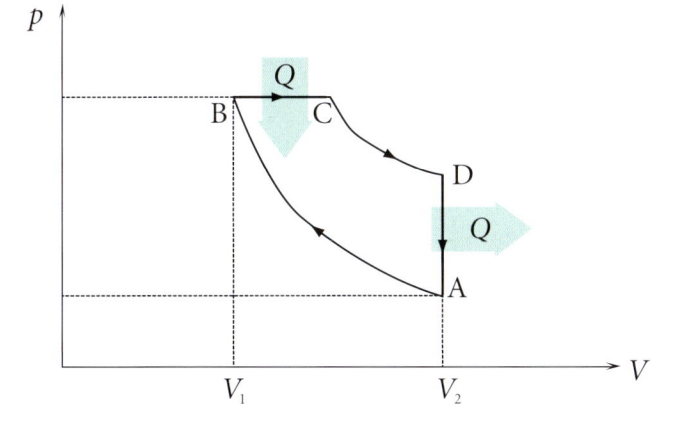

The second law of thermodynamics

We are always told to conserve energy. But according to the First Law of Thermodynamics, in a closed system, energy is conserved, and the total amount of energy in the Universe does not change no matter what we do. Although the First Law of Thermodynamics is correct, it does not tell the whole story.

How often have you seen a movie played in reverse sequence. Views of water flowing uphill, demolished buildings rising from the rubble, people walking backwards. In none of the natural Laws of Physics studied so far have we encountered time reversal. If all of these Laws are obeyed, why then does the time-reversed sequence seem improbable? To explain this reversal paradox, scientists in the latter half of the nineteenth century came to formulate a new principle called the Second Law of Thermodynamics. This Law allows us to determine which processes will occur in nature, and which will not.

There are many different but equivalent ways of stating the Second Law of Thermodynamics. Much of the language used for the definitions had its origins with the physicists who formulated the Law, and their desire to improve the efficiency of steam engines. These statements of the Second Law of Thermodynamics will be developed within this section.

The second law of thermodynamics implies that thermal energy cannot spontaneously transfer from a region of low temperature to a region of high temperature.

The Kelvin – Planck statement of the second law of thermodynamics

All attempts to construct a heat engine that is 100% efficient have failed. The Kelvin – Planck statement of the Second Law of Thermodynamics is a qualitative statement of the impossibility of certain types of processes.

- It is impossible for an engine working in a cycle to transform a given amount of heat from a reservoir completely into work.

  or

- Not all the thermal energy in a thermal system is available to do work.

It is possible to convert heat into work in a non-cyclic process. An ideal gas undergoing an isothermal expansion does just that. But after the expansion, the gas is not in its original state. In order to bring the gas back to its original state, an amount of work will have to be done on the gas and some thermal energy will be exhausted.

411

CHAPTER 14 (OPTION B)

The **Kelvin–Planck statement** formulates that if energy is to be extracted from a reservoir to do work, a colder reservoir must be available in which to exhaust a part of the energy.

The Clausius statement of the second law of thermodynamics

Just as there is no cyclic device that can convert a given amount of heat completely into work, it follows that the reverse statement is also not possible. The Clausius statement of the Second Law of Thermodynamics can be stated as:

- It is impossible to make a cyclic engine whose only effect is to transfer thermal energy from a colder body to a hotter body.

There is no perfect refrigerator and no perpetual motion machine.

Entropy

Recall that in thermodynamics, a system in an equilibrium state is characterised by its state variables (p, V, T, U, n …). The change in a state variable for a complete cycle is zero. In contrast, the net thermal energy and net work factors for a cycle are not equal to zero.

In the latter half of the nineteenth century, Rudolf Clausius proposed a general statement of the Second Law in terms of a quantity called **entropy**. Entropy is a thermodynamic function of the state of the system and can be interpreted as the amount of order or disorder of a system. As with internal energy, it is the change in entropy that is important and not its absolute value.

Change in entropy

The change in entropy ΔS of a system when an amount of thermal energy Q is added to a system by a reversible process at constant **absolute temperature** T is given by

$$\Delta S = \frac{Q}{T}$$

The units of the change in entropy are $J\,K^{-1}$.

Example

A heat engine removes 100 J each cycle from a heat reservoir at 400 K and exhausts 85 J of thermal energy to a reservoir at 300 K. Compute the change in entropy for each reservoir.

Solution

Since the hot reservoir loses *heat*, we have that

$$\Delta S = -\frac{Q}{T} = \frac{100\ J}{400\ K} = 0.25\ J\ K^{-1}$$

For the cold reservoir we have

$$\Delta S = -\frac{Q}{T} = \frac{85\ J}{300\ K} = 0.283\ J\ K^{-1}$$

The change in entropy of the hot reservoir is **–0.25 J K⁻¹** and the change in entropy of the cold reservoir is **0.28 J K⁻¹**.

The change in entropy of the cold reservoir is greater than the decrease for the hot reservoir. The total change in entropy of the whole system equals 0.033 J K⁻¹.

That is,

$$\Delta S = \Delta S_H + \Delta S_L = -0.25 + 0.283 = 0.033\ J\ K^{-1}$$

So that the net change in entropy is positive.

In this example and all other cases, it has been found that the total entropy increases. This infers that total entropy increases in all natural systems. The entropy of a given system can increase or decrease but the change in entropy of the system ΔS_S plus the change in entropy of the environment ΔS_{env} must be greater than or equal to zero.

i.e.,

$$\Delta S = \Delta S_S + \Delta S_{env} \geq 0$$

In terms of entropy, the Second Law of Thermodynamics can be stated as

- The total entropy of any system plus that of its environment increases as a result of all natural processes.

or

- The entropy of the Universe increases.

or

- Natural processes tend to move toward a state of greater disorder.

Although the local entropy may decrease, any process will increase the total entropy of the system and surroundings, that is, the universe. Take for example a chicken growing inside an egg. The entropy of the egg and its contents decreases because the inside of the egg is becoming more ordered. However, the entropy of surroundings increases by a greater factor because the process gives off thermal energy. So the total energy of the Universe is increasing.

In the beginning of Section B.2, irreversible processes were discussed. A block of ice can slide down an incline plane if the frictional force is overcome, but the ice cannot spontaneously move up the incline of its own accord. The conversion of mechanical energy to thermal energy by friction as it slides is irreversible. If the thermal energy could be converted completely to mechanical energy, the Kelvin-Planck statement of the second Law would be violated. In terms of entropy, the system tends to greater disorder, and the entropy increases.

In another case, the conduction of thermal energy from a hot body to a cold body is irreversible. Flow of thermal energy completely from a cold body to a hot body violates the Clausius statement of the Second Law. In terms of entropy, a hot body causes greater disorder of the cold body and the entropy increases. If thermal energy was given by a cold body to a hot body there would be greater order in the hot body, and the entropy would decrease. The Second Law does not allow this.

Irreversibility can also occur if there is turbulence or an explosion causing a non-equilibrium state of the gaseous system. The degree of disorder increases and the entropy increases.

Entropy indicates the direction in which processes occur. Hence entropy is often called the 'arrow of time'.

Ludwig Boltzmann first applied a statistical approach to the definition of entropy. Boltzmann (1844 –1906), an Austrian physicist, was also concerned with the "heat death" of the Universe and irreversibilty. He concluded that the tendency toward dissipation of heat is not an absolute Law of Physics but rather a Statistical Law.

Consider 10^{22} air molecules in a container. At any one instant, there would be a large number of possibilities for the position and velocity of each molecule – its microstate and the molecules would be disordered. Even if there is some momentary order in a group of molecules due to chance, the order would become less after collision with other molecules. Boltzmann argued that probability is directly related to disorder and hence to entropy. In terms of the Second Law of Thermodynamics, probability does not forbid a decrease in entropy but rather its probability of occurring is extremely low.

If a coin is flipped 100 times, it is not impossible for the one hundred coins to land heads up, but it is highly improbable. The probability of rolling 100 sixes from 100 dice is even smaller.

A small sample of a gas contains billions of molecules and the molecules have many possible microstates. It is impossible to know the position and velocity of each molecule at a given point in time. The probability of these microstates suddenly coming together into some improbable arrangement is infinitesimal. In reality, the macrostate is the only measurable part of the system.

The Second Law in terms of probability does not infer that a decrease in entropy is not allowed but it suggests that the probability of this occurring is low.

A final consequence of the Second Law is the heat degradation of the Universe. It can be reasoned that in any natural process, some energy becomes unavailable to do useful work. An outcome of this suggests that the Universe will eventually reach a state of maximum disorder. An equilibrium temperature will be reached and no work will be able to be done. All change of state will cease as all the energy in the Universe becomes degraded to thermal energy. This point in time is often referred to as the 'heat death' of the Universe.

Carnot cycle

Before the First Law of Thermodynamics was even established, Nicolas Léonard Sadi Carnot (1796-1832), a young engineer, was able to establish the theoretical maximum efficiency that was possible for an engine working between two heat reservoirs. In 1824, he formulated that:

No engine working between two heat reservoirs can be more efficient than a reversible engine between those reservoirs.

Carnot argued that if thermal energy does flow from a cold body to a hot body then work must be done. Therefore, no engine can be more efficient than an ideal reversible one and that all such engines have the same efficiency. This means that if all engines have the same efficiency then only a simple engine was needed to calculate the efficiency of any engine.

Consider an ideal perfectly insulated, frictionless engine that can work backwards as well as forwards. The p–V diagram would have the form of that shown in Figure 1433.

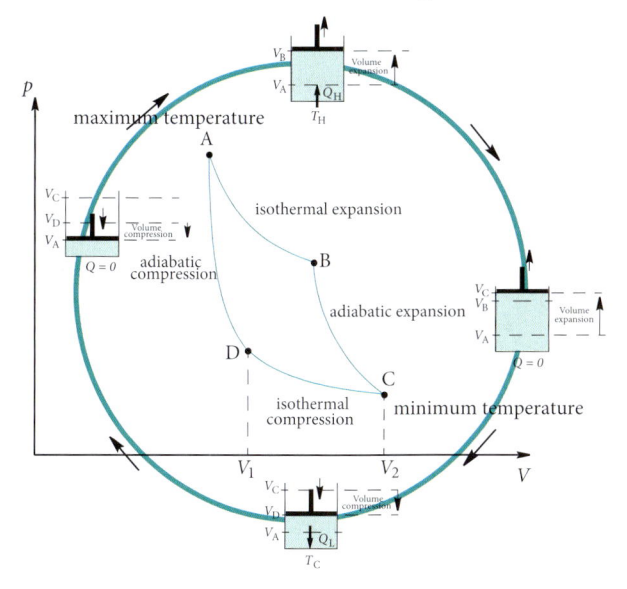

Figure 1433 The Carnot engine

# Chapter 14 (Option B)

The nett work is the area enclosed by ABCDA. In the case given, the Carnot engine is working in a clockwise cycle ABCDA. Thermal energy is absorbed by the system at the high temperature reservoir T_H and is expelled at the low temperature reservoir T_L. Work is done by the system as it expands along the top isotherm from A to B, and along the adiabat from B to C. Work is then done on the system to compress it along the bottom isotherm from C to D and along the left adiabat from D to A.

The efficiency of the Carnot cycle depends only on the absolute temperatures of the high and low temperature reservoirs. The greater the temperature difference, the greater the efficiency will be.

As a result of the Carnot efficiency, many scientists list a **Third Law of Thermodynamics** which states:

- **It is impossible to reach the absolute zero of temperature, 0 K.**

The efficiency of the Carnot cycle would be 100% if the low temperature reservoir was at absolute zero. Therefore absolute zero is unattainable.

Generally speaking, the efficiency of a energy conversion is the ratio of the useful energy output to the total energy input expressed as a percentage. It can also be stated as the ratio of the useful work done to the energy input.

$$\eta = \frac{\text{useful work done}}{\text{energy input}}$$

The efficiency of a Carnot engine like any heat engine is given by:

$$\eta_{carnot} = 1 - \frac{T_L}{T_H} = 1 - \frac{Q_L}{Q_H}$$

Exercise 14.4

1. The efficiency of a heat engine is the ratio of

   A. the thermal energy input to the thermal energy output
   B. the thermal energy output to the thermal energy input
   C. the work output to the thermal energy input
   D. the work output to the thermal energy output

2. A heat engine is most efficient when it works between objects that have a

   A. large volume
   B. large temperature difference
   C. large surface area
   D. small temperature difference

3. The four-stroke engine is often said to consist of the 'suck, squeeze, bang and blow' strokes. Describe what these terms relate to.

4. Explain the difference between internal and external combustion engines, and give an example of each.

5. A car engine operates with an efficiency of 34% and it does 8.00×10^3 J of work each cycle.

   Calculate

   (a) the amount of thermal energy absorbed per cycle at the high-temperature reservoir.
   (b) the amount of exhaust thermal energy supplied to the surroundings during each cycle.

6. On a hot day, a person closed all the doors and windows of the kitchen and decided to leave the door of the refrigerator open to cool the kitchen down. What will happen to the temperature of the room over a period of several hours. Give a full qualitative answer.

7. Modern coal-fired power plants operate at a temperature of 520 °C while nuclear reactors operate at a temperature of 320 °C. If the waste heat of the two plants is delivered to a cooling reservoir at 21 °C, calculate the Carnot efficiency of each type of plant. (*optional question*)

8. It takes 7.80×10^5 J of thermal energy to melt a given sample of a solid. In the process, the entropy of the system increases by 1740 J K^{-1}. Find the melting point of the solid in °C.

9. If 2.00 kg of pure water at 100 °C is poured into 2.00 kg of water at 0 °C in a perfectly insulated calorimeter, what is the net change in entropy. (Assume there is 4.00 kg of water at a final temperature of 50 °C).

10. Use the concepts of entropy and the arrow of time to explain the biological growth of an organism.

11. You are given six coins which you shake and then throw onto a table. Construct a table showing the number of microstates for each macrostate.

12. Describe the concept of energy degradation in terms of entropy.

13. Using an example, explain the meaning of the term reversal paradox.

14. What is meant by the 'heat death' of the universe?

B.3 Fluids and fluid dynamics

> **NATURE OF SCIENCE:**
> Human Understandings: Understanding and modelling fluid flow has been important in many technological developments such as designs of turbines, aerodynamics of cars and aircraft and measurement of blood flow. (1.1)
>
> © IBO 2014

Essential idea: Fluids cannot be modelled as point particles. Their distinguishable response to compression from solids creates a set of characteristics which require an in-depth study.

Understandings
- Density and pressure
- Buoyancy and Archimedes' principle
- Pascal's principle
- Hydrostatic equilibrium
- The ideal fluid
- Streamlines
- The continuity equation
- The Bernoulli equation and the Bernoulli effect
- Stokes' law and viscosity
- Laminar and turbulent flow and the Reynolds number

Density and pressure

A **fluid** is a substance that can flow. Therefore, liquids and gases are fluids. The study of fluids at rest is called hydrostatics and the study of moving fluids is called hydrodynamics.

The **density** of a substance ρ is defined as its mass per unit volume and is measured in kg m^{-3} or g cm^{-3}.

$$\rho = \frac{m}{V}$$

Some density values are shown in Figure 1434.

Substance		Density	
Liquid	Gas	kg m^{-3}	g cm^{-3}
Water		1000	1
	Air	1.3	0.0013
Petrol		800	0.80
Ethanol		810	0.81

Figure 1434

Density values are often compared to the density of water. This ratio is called the *specific gravity* (SG) or *relative density* (RD), where:

RD = density of liquid / density of water

Example

Determine the density and specific gravity of a piece of lead that has a mass of 285.0 kg and a volume of 0.0250 m^3.

Solution

$$\rho = \frac{285.0 \text{ kg}}{0.0250 \text{ m}^3} = 1.14 \times 10^4 \text{ kg m}^{-3}$$

$$SG = \frac{1.140 \times 10^4 \text{ kg m}^{-3}}{1.00 \times 10^4 \text{ kg m}^{-3}} = 11.4$$

Pressure is defined as the force per unit area:

$$P = \frac{F}{A}$$

The SI derived unit of pressure is the pascal (Pa) where 1 Pa = 1 N m^{-2}.

In liquids:
- Pressure is transmitted throughout the liquid.
- Pressure acts in all directions.
- Pressure increases with depth below the air-fluid interface.
- Pressure decreases with height.
- Pressure depends on the density of the liquid.

Consider a point at a depth d below the surface of a liquid as shown in Figure 1435.

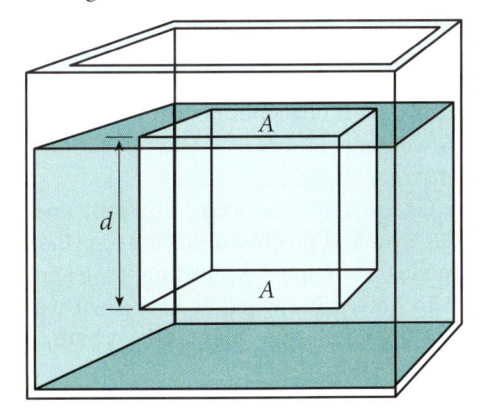

Figure 1435 Pressure at a depth d.

The pressure at this point is due to the weight force of the column of water above that point. Therefore, the force due to the weight of liquid acting on an area A would be:

$$F = mg = (\rho V)g = \rho g A d$$

where Ad is the volume of the column of liquid.

$$P = \frac{F}{A} = \frac{\rho g A d}{A} = \rho g d$$

So, the area A does not affect the pressure at a given depth, and the pressure is directly proportional to the of the density and the depth within the liquid.

The pressure experienced by divers and mountaineers is generally called hydrostatic pressure.

The pressure in a liquid of density ρ at a depth d is given by:

$$P = P_0 + \rho g d$$

Where P_0 is the atmospheric pressure at the top of the liquid surface.

The atmospheric pressure at sea level is
1.01×10^5 Pa = 1atm = 760 mmHg.

Buoyancy and Archimedes' principle

When an object is wholly or partially immersed in a fluid at rest, the fluid will exert a pressure on every part of the object that it is in contact with. When you weigh the object when immersed in a fluid, it weighs less than when it is weighed outside the liquid. Since pressure increases with depth, it follows that the pressure will be greater on the deeper parts of the object. Therefore, there is a resultant upwards force (upthrust) called the **buoyancy force** on the object.

Take a ping-pong ball and push it down into a fluid. When you release the force on the ball it bobs up and down. When it stabilizes, the partially submerged ping-pong ball has a buoyancy force on it due to the resultant of all the forces exerted by the fluid pressure over the submerged surface of the ball. It floats because the buoyancy is at least equal to its weight.

It was the Greek mathematician, physicist and engineer Archimedes who first proposed this idea in what is known as Archimedes Principle. According to legend, a king asked him to determine whether his crown was made of pure gold or a gold alloy. While taking a bath, he arrived at a "eureka" solution.

- **Any object completely or partially submerged in a fluid is buoyed up by a force with magnitude equal to the weight of the fluid displaced by the object.**

That is:

Buoyant force = mass of fluid displaced x gravity

= density of the fluid × volume of the fluid × gravity

$$B = \rho_f V_f g$$

Example

A person purchased a "gold" statue for a bargain price in an antique fair. She weighed it in air and it was 7.84 N. She weighed it in water and found it to be 6.86 N. Determine whether the statue is genuine or a fake. (Take the density of gold to be 19.3×10^3 kg m^{-3}).

Solution

Using Newton's second law:

$T_{AIR} - mg = 0$ *when weighed in air*
    1

$T_{WATER} - mg + B = 0$ *when weighed in water*
    2

Substitute for mg in equation 1 into equation 2 and making B the subject:

$B = T_{AIR} - T_{WATER} = 7.84 - 6.86 = 0.980$ N

$B = \rho_{WATER} V_{WATER} g = 0.980$ N

Now the buoyant force is equal to the weight of the displaced water. Find the water volume then the mass of the statue:

$$V_{WATER} = \frac{0.980}{9.8 \times 1000} = 1.00 \times 10^{-4} \text{ m}^3$$

$$m = \frac{7.84}{9.8} = 0.800 \text{ kg}$$

Determine the density

$$\rho = \frac{0.800}{1.00 \times 10^{-4} \text{ m}^3} = 8.00 \times 10^3 \text{ kg m}^{-3}$$

*Therefore, **the statue is a fake**.*

Pascal's principle

The pressure in a fluid can be used to work machinery hydraulic systems such as a hydraulic car hoist or hydraulic disc braking systems.

These systems work on the fact that external force acting on a fluid is transmitted throughout the fluid. This is based on the ideas of the French the philosopher and scientist *Blaise Pascal* (1623-1662) in what is known as Pascal's principle.

Pascal's principle states that:

- **if an external force is applied to a confined fluid, the pressure at every point within the fluid increase by that amount.**

ENGINEERING PHYSICS

In the hydraulic disc brake system shown in Figure 1436, a small input force on a small "master" cylinder is used to exert a large output force on the pair of larger pistons near the disc.

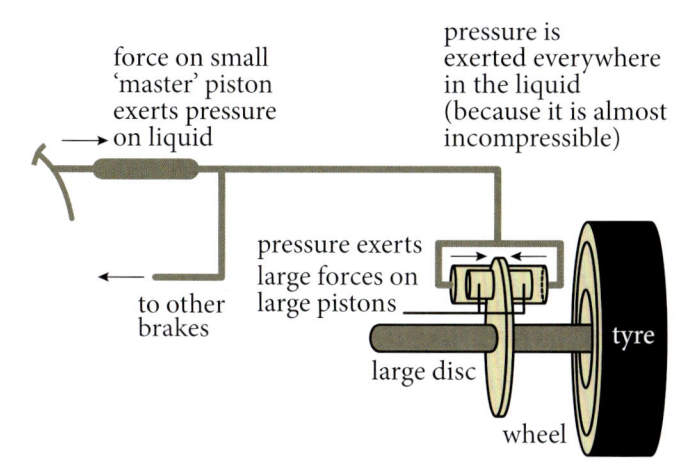

Figure 1436   Hydraulic disc brakes

The driver applies a force on the small piston to exert a pressure on the fluid. The pressure is transmitted equally throughout the fluid in the braking system. The pressure makes the pistons squeeze the disc to slow the car down. The pistons are ideally at the same height.

$P_{OUT} = P_{IN}$

$F_{OUT} / A_{OUT} = F_{IN} / A_{IN}$

The quantity F_{OUT} / F_{IN} is called the *mechanical advantage* of the braking system, and it is equal to the ratio of the areas – the force has been magnified.

If the pistons at the discs are twice the area of the master piston, they will exert twice the force that the driver applies with her foot.

Example

Figure 1437 shows the principle of the hydraulic car hoist. A downward force of 50 N is applied to the small cylinder.

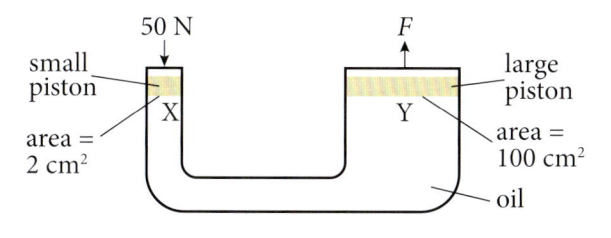

Figure 1437   Hydraulic lift

(a)   Calculate the pressure in the oil at X.
(b)   What is the pressure exerted by the oil at Y?
(c)   Calculate the upward force F acting on the large piston.

Solution

(a) $P = \dfrac{F}{A} = \dfrac{50}{2} = 25$ N cm^{-2}

(b)   The pressure is transmitted equally throughout the fluid = 25 N cm^{-2}

(c) $F = 50 \times 50 = 2500$ N.

Hydrostatic equilibrium

For a fluid to be in hydrostatic equilibrium

- all portions of the fluid need to be at rest with respect to an observer.
- All points at the same depth must be at the same pressure.

For these conditions to be met, a volume of fluid, either a gas or a liquid, is in hydrostatic equilibrium when an upward force exerted by the pressure of the fluid balances the downward force exerted by gravity.

The piling up of fluid (liquid and gas) above the earth's surface causes atmospheric pressure. The atmosphere is pulled downwards by the gravitational force, and towards the surface, the air is compressed by the weight of all the air above. Therefore, the air's density increases from the top of the atmosphere to the Earth's surface.

Because of this density difference, air pressure decreases with altitude so that the upward pressure from below is greater than the downward pressure from above. This net upward force balances the downward force of gravity, thus keeping the earth's atmosphere at a fairly constant height.

When a volume of fluid is not in hydrostatic equilibrium it must contract if the gravitational force exceeds the pressure, or expand if the internal pressure is greater. This is the case for many thermodynamic and stellar processes.

This concept can be expressed as the hydrostatic equilibrium equation:

$\dfrac{\Delta P}{\Delta x} = -g\rho$ where x is the thickness of a layer of air.

The ideal fluid

An ideal fluid is an imaginary fluid that lacks viscosity and thermal conductivity.

As a result, its flow is:

- Steady – it moves at the same speed.
- Incompressible – liquids are incompressible. Compression changes the speed.

417

Chapter 14 (Option B)

- Non-viscous – there is no drag so the flow maintains constant speed.
- Linear – it does not form vortices or rotate but respond to the push of the molecules behind them.

There is no internal friction in an ideal fluid—that is, there are no tangential stresses between two neighboring layers. The flow of an ideal fluid makes it possible to find theoretical solutions to a number of problems of the motion of liquids and gases in channels of various shapes, in the outflow of jets, and in flow around bodies.

Streamlines

We will now turn from our study of fluids at rest to the more complex subject of fluids in motion – hydrodynamics. The mechanics is complicated and many aspects of fluids in motion are not fully understood. With the onset of turbulence, the smooth flow is disrupted and vortices and eddies arise. In some circumstances turbulence is good as in the mixing of air and aviation fuel in a jet engine. However, in many circumstances, turbulence is disastrous as could be the case in air flight. We will restrict most of our study to steady flow of fluids.

If the flow is smooth such that neighbouring elements of fluid slide by each other smoothly, the flow is said to be **streamline** or **laminar**. Each particle of the fluid follows a smooth path and the paths do not cross over each other.

The path followed by an elemental portion of a moving fluid is called a **flow line**. In general, the velocity of the element changes in both magnitude and direction along the flow line. The flow is said to be **steady** if successive elements of fluid follow the same flow line. In steady flow, the velocity of the fluid *at any given point* within the fluid remains constant even though the speed of each individual particle may vary along its path.

A line drawn such that its tangent at any point is in the direction of the fluid velocity at that point is called a **streamline** as shown in Figure 1438.

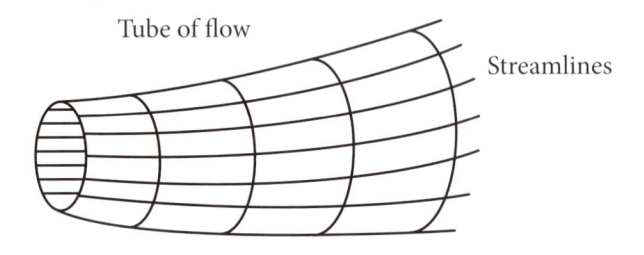

Figure 1438   Streamlines and tube of flow

In steady flow, the streamlines and flow lines coincide. From the definition of streamlines, it follows that an element of fluid cannot cross from one streamline to another.

Suppose that the flow of the fluid is such that the velocities of the fluid elements over any cross-section perpendicular to the flow are constant. The streamlines then form a **tube of flow** that may be considered a *bundle* of streamlines as shown in Figure 1438. Where the speed is the greatest, streamlines are close together.

The continuity equation

Let us now examine the steady streamline flow of a tube of flow and determine how the speed of the fluid varies with the size of the tube.

If we consider any fixed, closed surface in a moving fluid, then fluid flows into the volume enclosed by the surface at some points and flows out at other points.

The statement of continuity is that the *net* rate of flow of mass *inward* across any closed surface is equal to the rate of increase of mass within the surface. The rate of flow of fluid into a system equals the rate of flow out of the system.

For an incompressible fluid in steady flow with a tube of flow with cross-sectional areas A_1 and A_2 and speeds v_1 and v_2 in these sections as shown in Figure 1439, the equation of continuity states:

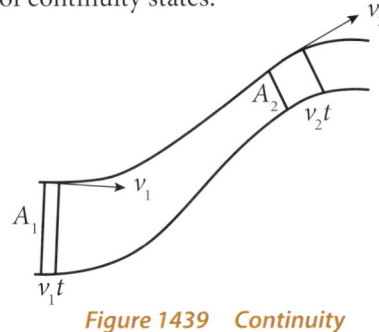

Figure 1439   Continuity

$A_1 v_1 = A_2 v_2$

$A v$ = **constant**

From the continuity equation it can be seen that as the **cross-sectional area decreases, the speed of flow increases**.

The Bernoulli equation and the Bernoulli effect

As a consequence of the continuity equation, the speed of flow of an incompressible fluid in a tube of varying cross-section will very – that is, it *accelerates* or decelerates. This acceleration implies a resultant **force** on the fluid. This means the pressure along the tube must vary even though the elevation of the tube may not change.

ENGINEERING PHYSICS

In 1738 Daniel Bernoulli derived an equation relating the speed and density of flow with the pressure change.

Consider a liquid of density ρ and a tube of flow of pressure P_1 and velocity v_1 over the cross-sectional area A_1, and a pressure P_2 and velocity v_2 over the cross-sectional area A_2 as in Figure 1440.

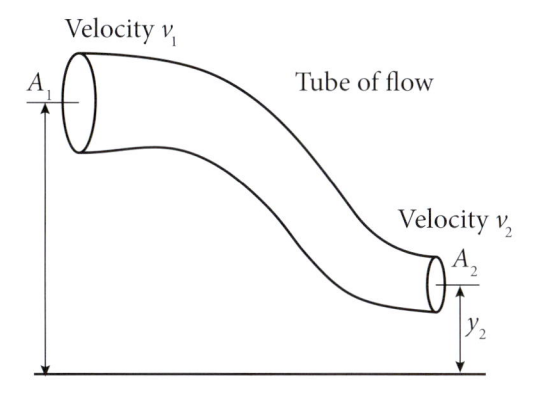

Figure 1440 Bernoulli's principle.

Let the average height of A_1 above some horizontal level be y_1 and let y_2 be the *mean* height of A_2 above this level.

If the area A_1 is greater than the area A_2 then the equation of continuity shows that v_1 is less than v_2 – that is, the liquid accelerated between A_1 and A_2. This acceleration results from:

- A reduction in pressure between A_1 and A_2.
- A reduction in the vertical height of the liquid.

Therefore, the liquid gains *kinetic energy* at the expense of the *pressure energy* and the *gravitational potential energy*.

Bernoulli was able to show that:

$$P_1 + \rho g y_1 + \tfrac{1}{2} \rho v_1^2 = P_2 + \rho g y_2 + \tfrac{1}{2} \rho v_2^2$$

That is:

$$P + \rho g y + \tfrac{1}{2} \rho v^2 = \text{constant}$$

Dynamic lift

A body moving through an ideal fluid experience no net force provided the flow is *symmetrical*.

However, if the flow is such that the speed above the body is increased and the speed below is decreased, then applying the Bernoulli effect, the pressure at the upper surface is *decreased* and that at the lower surface is *increased*. Therefore, a resultant *upward* force must act on the body, giving rise to *dynamic lift* as shown in Figure 1441.

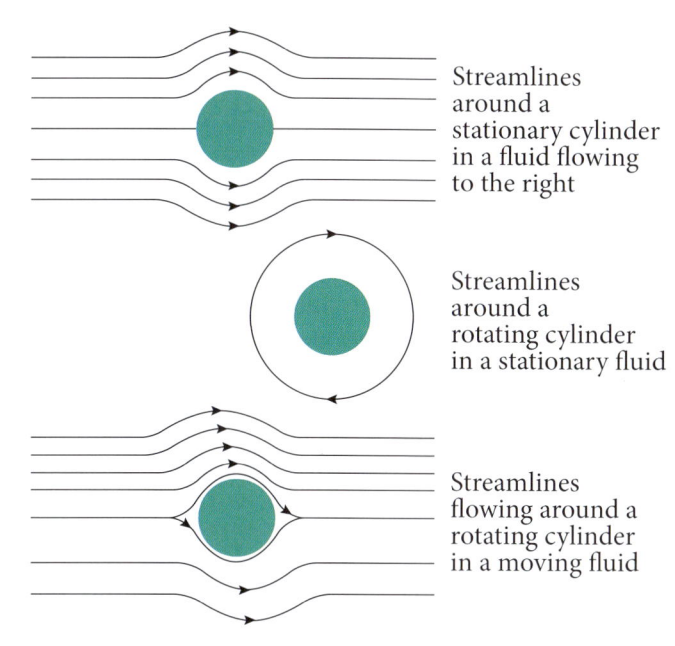

Figure 1441 Dynamic lift.

The lift is in addition to the static lift produced by buoyancy effects.

Dynamic lift is used to give lift to aircraft – the shape and the angle of the wing creating speed differences of the air and hence the pressure difference as shown in Figure 1442.

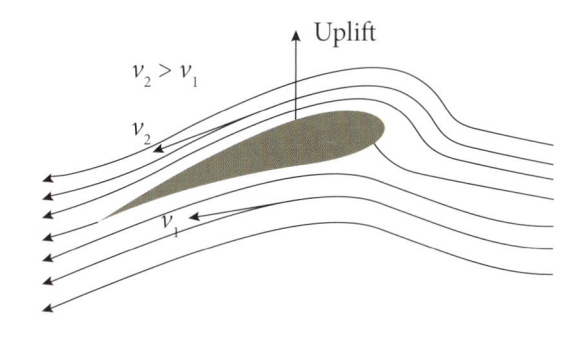

Figure 1442 Dynamic lift on an aircraft wing.

The Venturi tube

A schematic diagram of a Venturi tube is shown in Figure 1443. It consists of a constriction or throat inserted in a pipeline and having properly designed tapers at the inlet and outlet to avoid the creation of turbulence.

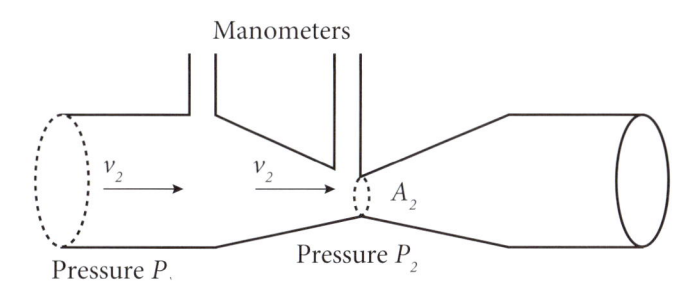

Figure 1443 Venturi tube.

419

Chapter 14 (Option B)

Because the elevation is constant for the Venturi, Bernoulli's equation becomes:

$P_1 + \frac{1}{2}\rho v_1^2 = P_2 + \frac{1}{2}\rho v_2^2$

The equation of continuity shows that speed v_2 is greater than speed v_1 because A_2 is greater than A_1. Hence, pressure P_2 in the throat is less than P_1. The pressures P_1 and P_2 can be measured by means of the open-ended manometers as shown in the schematic diagram. By knowing the pressure values and the cross-sectional areas, the speed and flow rate can be calculated.

The Pitot tube

The Pitot tube is another device used for measuring the speed of liquids. One type is schematically shown in Figure 1444.

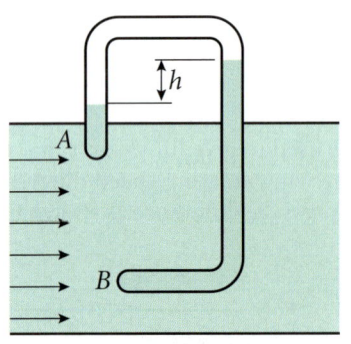

Figure 1444   Pitot tube.

Opening A is perpendicular to the flow and opening B faces the flow. Therefore, the liquid levels in the two arms are separated by a height h. Along the streamline in line with tube B, the velocity of the liquid falls rapidly to zero. The manometer reading gives the value of $P + \frac{1}{2}\rho v^2$.

Opening A registers the static pressure in the liquid. Therefore:

$\rho g h = \frac{1}{2}\rho v^2$

Hence,

$v = \sqrt{2gh}$

A Pitot tube is frequently attached to an airplane for measuring the speed of the air relative to the plane – an air speed meter.

Stokes' law and viscosity

Viscosity is the internal friction of a fluid. It is because of viscosity that a force must be exerted to cause one layer of a fluid to slide past another.

All fluids exhibit viscosity. However, liquids are much more viscous than gases.

When a fluid flows past an object, no matter how smooth it may appear, on the molecular level it is quite rough and the fluid is dragged to rest on the surface. The viscosity of the fluid therefore determines the velocity of the fluid at various regions within the liquid.

Stoke's Law allows us to determine the velocity and the coefficient of viscosity η. The law states that a sphere of radius r moving with a constant speed v through a fluid with a coefficient of viscosity η experience a force F given by:

$F = 6\pi \eta r v$

If the sphere is allowed to fall through a viscous fluid, a **terminal** velocity (v_T) is reached, when the weight of the sphere is balanced by the viscous force **and** the buoyant force.

Weight of sphere = mass × gravity

= volume × density x gravity

$= \frac{4}{3}\pi r^3 \times \rho \times g$

Buoyant force $= \frac{4}{3}\pi r^3 \times \rho' \times g$

Therefore:

$\frac{4}{3}\pi r^3 \rho' g + 6\pi \eta r v_T = \frac{4}{3}\pi r^3 \rho g$

$\eta = \frac{2r^2}{9v_T} \times (\rho' - \rho)$

All the quantities on the right hand side are known and hence η can be calculated.

Laminar and turbulent flow and the Reynolds number

As mentioned earlier, if the flow is smooth such that neighbouring elements of fluid slide by each other smoothly, the flow is said to be **laminar flow**. Each particle of the fluid follows a smooth path and the path do not cross over each other.

420

It is much easier for water to flow than honey because honey is a more viscous than water. It is very difficult for layers of a viscous material to slide past each other. In some sense, viscosity of a liquid is a measure of its resistance to shearing. When an ideal fluid flows through a pipe, the fluid layers slide past each other and offer little resistance.

A liquid with low viscosity can go from smooth laminar flow to turbulent flow at low velocities, and the streamlines behave in an unpredictable manner. However, a liquid of higher viscosity will remain laminar at higher flow velocities.

At sufficiently high velocities, fluid flow can abruptly change from laminar flow to turbulent flow. The onset of turbulence is not fully understood but it can cause sudden bad effects such as plane suddenly stalling because its angle of ascent is too steep.

Experimentally, the onset of turbulence in a tube is determined by a dimensionless factor called the Reynold's number R that is given by:

$$R \text{ or } R_e = \frac{2 v r \rho}{\eta} = \frac{v d \rho}{\eta}$$

Where v is the average speed of the fluid along the direction of flow, ρ is the density of the fluid, d is the diameter of the tube and η is the viscosity of the fluid.

If the Reynold's number is below about 2000, the flow of the fluid through a tube is laminar. Turbulence occurs for a number above 3000, and numbers between 2000 and 3000 measure unstable flow.

Example

The average speed of blood in the aorta with a radius of 1.0 cm during the resting part of the heart's cycle is about 30 cm s^{-1}. The density of blood is 1.05×10^3 kg m^{-3} and its coefficient of viscosity is 4.0×10^{-3} N s m^{-2}. Determine whether the flow of blood is laminar or turbulent.

Solution

$R = 2 \times 0.01$ m $\times 1.05 \times 10^3$ kg m^{-3} / 4.0×10^{-3} N s m^{-2} = 1600

Therefore, the flow is probably laminar but it is unstable and it could be turbulent.

Exercise 14.4

1. A cube of aluminium has sides of 2.00 cm and a density of 2.70×10^3 kg m^{-3}. Calculate the mass.

2. The density of copper is 8.90×10^3 kg m^{-3}. Determine the radius of a sphere of copper that has a mass of 0.150 kg.

3. An airplane is taking off on a runway. The pair of airplane wings has a total area of 50.0 m^2. The pressure on the upper surface is 1.20×10^2 kPa and the pressure on the lower surface is 2.10×10^2 kPa. Determine the lift to the airplane wings.

4. In a hydraulic brake a force of 500 N is applied to an area of 5cm^2.

    (a) What is the pressure transmitted throughout the liquid?
    (b) If the other piston has an area of 20 cm^2, what is the force exerted on it?

5. A hydraulic press uses a piston of 5.00×10^{-3} m in diameter. If a 40.0 N force is applied to this piston to lift a weight of 2.00×10^3 N on a second piston, what must the diameter of the second piston be?

6. A raft has a surface area of 5.70 m^2 and a volume of 0.60 m^3. It is made of wood having a density of 6.00×10^2 kg m^{-3}. When the raft is placed in fresh water, to what depth will the bottom of the raft submerge in the water?

7. A Hershey airship contains 5.40×10^3 m^3 of helium whose density is 0.179 kg m^{-3}. Determine the weight of the load that it can carry in equilibrium at an altitude where the density of air is 1.20 kg m^{-3}.

8. A garden hose has an opening of cross-sectional area 2.85×10^{-4} m^2, and it fills a bucket of 8.00×10^{-3} m^3 in 30.0 s.. Determine the speed of the water that leaves the opening.

9. A large water pipe with a cross-sectional area of 1.00 m^2 descends 5.00 m and narrows to 0.500 m^2. If the pressure at the top is atmospheric pressure, determine the speed of the water leaving the pipe.

Chapter 14 (Option B)

B.4 Forced vibrations and resonance

NATURE OF SCIENCE:

Risk assessment: The ideas of resonance and forced oscillation have application in many areas of engineering ranging from electrical oscillation to the safe design of civil structures. In large-scale civil structures, modelling all possible effects is essential before construction. (4.8)

© IBO 2014

Essential idea: In the real world, damping occurs in oscillators and has implications that need to be considered.

Understandings:
- Natural frequency of vibration
- Q factor and damping
- Periodic stimulus and the driving frequency
- Resonance

Natural frequency of vibration

Consider a small child sitting on a swing. If you give the swing a single push, the swing will oscillate. With no further pushes, that is energy input, the oscillations of the swing will die out and the swing will eventually come to rest. This is an example of damped harmonic motion. The frequency of oscillation of the swing under these conditions is called the **natural frequency** of oscillation (vibration). So far in this topic, all the systems we have looked at have been systems oscillating at their natural frequency.

Suppose now when each time the swing returns to you, you give it another push. See Figure 1445.

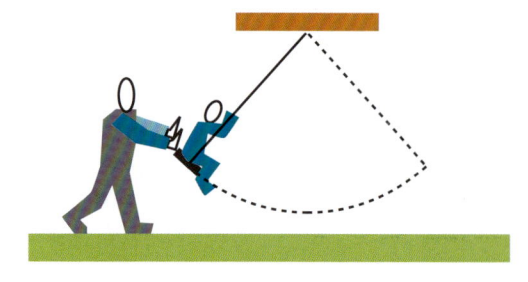

Figure 1445 A forced oscillation

The amplitude of the swing will get larger and larger and if you are not careful your little brother or sister, or who ever the small child might be, will end up looping the loop.

The frequency with which you push the swing is exactly equal to the natural frequency of oscillation of the swing and importantly, is also in phase with the oscillations of the spring. Since you are actually forcing the swing to oscillate, the swing is said to be undergoing **forced oscillations.** In this situation the frequency of the so-called *driver* (in this case, you) is equal to the natural frequency of oscillation of the system that is being driven (in this case, the swing). If you just push the swing occasionally when it returns to you, then the swing is being forced at a different frequency to its natural frequency. In general, the variation of the amplitude of the oscillations of a driven system with time will depend on the

- frequency of the driving force
- frequency of natural oscillations
- amplitude of the driving force
- phase difference between driving force frequency and natural frequency
- amount of damping on the system

(There are many very good computer simulations available that enable you to explore the relation between forced and natural oscillations in detail.)

The driving force and system are in phase if, when the amplitude of system is a maximum, it receives maximum energy input from the driver. Clearly this is when the amplitude of the driver is a maximum.

What is of particular interest is when the forced frequency is close to and when it equals the natural frequency. This we look at in the next two sections.

Q factor and damping

In this section, we look at oscillations of real systems. In Chapter 9, we described an arrangement by which the oscillations of a pendulum could be transcribed onto paper. Refer to Figure 1446.

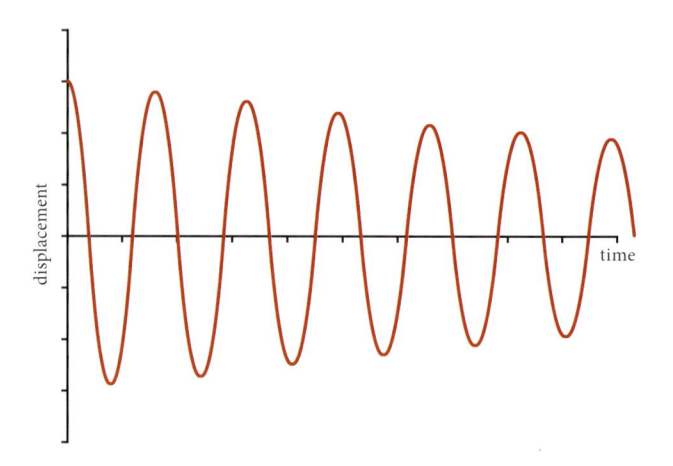

Figure 1446 Damping

Engineering Physics

The amplitude of the oscillations gradually decreases with time, whereas for SHM, the amplitude stays at the same value forever. Clearly, the pendulum is losing energy as it oscillates. The reason for this is that dissipative forces are acting that oppose the motion of the pendulum. As mentioned earlier, these forces arise from air resistance and though friction at the support. Oscillations, for which the amplitude decreases with time, are called **damped oscillations.**

All oscillating systems are subject to damping as it is impossible to completely remove friction. Because of this, oscillating systems are often classified by the degree of damping. The oscillations shown in Figure 1446 are said to be **lightly damped**. The decay in amplitude is relatively slow and the pendulum will make quite a few oscillations before finally coming to rest. Whereas the amplitude of the oscillations shown in Fig 1447 decay very rapidly and the system quickly comes to rest. Such oscillations are said to be **heavily damped**.

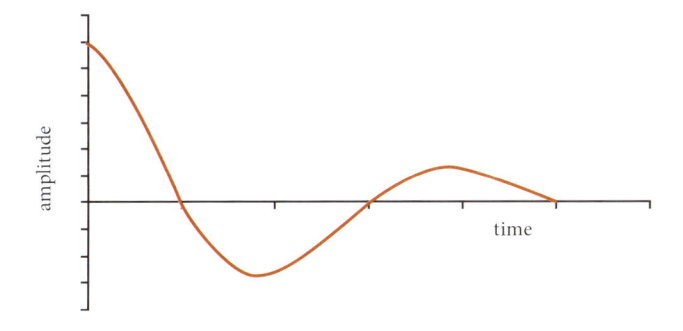

Figure 1447   Heavily damped oscillations

Consider a harmonic oscillator in which the mass is pulled down and when released, and the mass comes to rest at its equilibrium position without oscillating. The friction forces acting are such that they prevent oscillations. However, suppose a very small reduction in the friction forces would result in heavily damped oscillation of the oscillator, then the oscillator is said to be **critically damped.**

The graph in Figure 1448 shows this special case of damping known as **critical damping.**

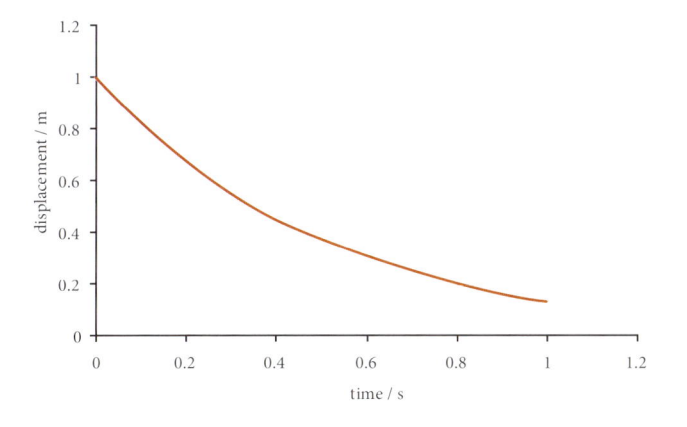

Figure 1448   Critical damping

A useful way of classifying oscillating systems, is by a quantity known as the quality factor or Q-factor. In physics and engineering the quality factor or Q factor is a dimensionless parameter that describes how underdamped an oscillator or resonator is. It is used to describe simple harmonic oscillators such as the pendulum or a mass oscillating on a spring but also many types of resonators such as electronic RLC circuits, acoustic instruments and the muffler on an automobile.

The Q-factor does have a formal definition but it is approximately equal in value to the number of oscillations that occur before all the energy of the oscillator is dissipated.

$$Q = 2\pi \times \frac{\text{energy stored}}{\text{energy dissipated per cycle}}$$

or

$$Q = 2\pi \times \frac{\text{resonant frequency} \times \text{energy stored}}{\text{power loss}}$$

Higher Q indicates a lower rate of energy loss relative to the stored energy of the oscillator; the oscillations die out more slowly. For example, a simple pendulum has a Q-factor of about 1000.

When $Q < \frac{1}{2}$, the system is overdamped. When $Q > \frac{1}{2}$, the system is underdamped. When $Q = \frac{1}{2}$, the system is critically damped.

The oscillations (vibrations) made by certain oscillatory systems can produce undesirable and sometimes, dangerous effects. Critical damping plays an important role in these situations. For example, when a ball strikes the strings of a tennis racquet, it sets the racquet vibrating and these vibrations will cause the player to lose some control over his or her shot. For this reason, some players fix a "damper" to the springs. If placed on the strings in the correct position, this has the effect of producing critically damped oscillations and as a result the struck tennis racquet moves smoothly back to equilibrium. The same effect can be achieved by making sure that the ball strikes the strings at a point known as the 'sweet spot' of which there are two, one of which is know as the 'centre of percussion (COP)'. Cricket and baseball bats likewise have two sweet spots.

Another example is one that involves vibrations that may be set up in buildings when there is an earthquake. For this reason, in regions prone to earthquakes, the foundations of some buildings are fitted with damping mechanisms. These mechanisms insure any oscillations set up in the building are critically damped.

Chapter 14 (Option B)

Exercise 14.5

Identify which of the following oscillatory systems are likely to be lightly damped and which are likely to be heavily damped.

1. Atoms in a solid
2. Car suspension
3. Guitar string
4. Harmonic oscillator under water
5. Quartz crystal
6. A cantilever that is not firmly clamped
7. Oil in a U-tube
8. Water in a U-tube

Periodic stimulus and the driving frequency

We now look at how the amplitude of an oscillating system varies with the frequency of the driving force.

The graph in Figure 1449 shows the variation with frequency f of the driving force of the amplitude A of three different systems to which the force is applied.

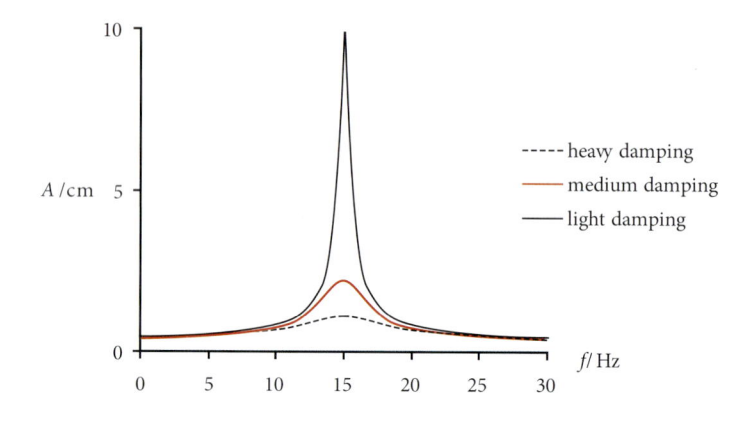

Figure 1449 Forced frequency

Each system has the same frequency of natural oscillation, $f_0 = 15$ Hz. The thing that is different about the systems is that they each have a different degree of damping: heavy (low Q), medium (medium Q), light (large Q).

For the heavily damped system we see that the amplitude stays very small but starts to increase as the frequency approaches f_0 and reaches a maximum at $f = f_0$; it then starts to fall away again with increasing frequency.

For the medium damped system, we see that as f approaches f_0, the amplitude again starts to increase but at a greater rate than for the heavily damped system. The amplitude is again a maximum at $f = f_0$ and is greater than that of the maximum of the heavily damped system.

For the lightly damped system, again the amplitude starts to increase as f approaches f_0, but at a very much greater rate than for the other two systems; the maximum value is also considerably larger and much more well-defined i.e. it is much easier to see that the maximum value is in fact at $f = f_0$.

If there were such a thing as a system that performs SHM, then if this system were driven at a frequency equal to its natural frequency, its amplitude would be infinite. Figure 1450 shows how the amplitude A for a driven system with very little damping and whose natural frequency of oscillation $f_0 = 15$ Hz, varies with the frequency f of the driving force.

ENGINEERING PHYSICS

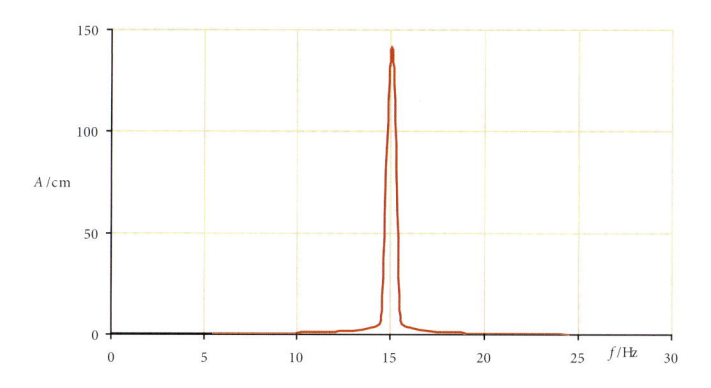

Figure 1450 Amplitude-frequency graph for a lightly damped oscillator

We see that the maximum amplitude is now very large and also very sharply defined. Also, either side of f_0, the amplitude drops off very rapidly.

Resonance

(Note: As well as the availability of a large number of computer simulations that demonstrate resonance, there are also many laboratory demonstrations and experiments that can be done to demonstrate it).

We have seen that when an oscillatory system is driven at a frequency equal to its natural frequency, the amplitude of oscillation is a maximum. This phenomenon is known as **resonance**. The frequency at which resonance occurs is often referred to as the **resonant frequency**.

In the introduction to this Topic, we referred to resonance phenomena without actually mentioning the term resonance. For example we can now understand why oscillations in machinery can be destructive If a piece of machinery has a natural frequency of oscillation and moving parts in the machine act as a driver of forced oscillations and have a frequency of oscillation equal to the natural frequency of oscillation of the machine, then the amplitude of vibration set up in the machine could be sufficient to cause damage.

Similarly, if a car is driven along a bumpy road, it is possible that the frequency, with which the car crosses the bumps, will just equal the natural frequency of oscillation of the chassis of the car. If this is the case then the result can be very uncomfortable.

We now also see why it is important that systems such as machines, car suspensions, suspension bridges and tall buildings are critically or heavily damped.

Resonance can also be very useful. For example, the current in a particular type of electrical circuit oscillates. However, the oscillating current quickly dies out because of resistance in the circuit. Such circuits have a resonant frequency and if driven by an alternating current supply, the amplitude of the current may become very large, particularly if the resistance of the circuit is small. Circuits such as this are referred to as *resonant circuits*. Television and radios have resonant circuits that can be tuned to oscillate electrically at different frequencies. In this way they can respond to the different frequencies of electromagnetic waves that are sent by the transmitting station as these waves now act as the driving frequency.

Earlier we mentioned the use of quartz crystal as timing devices. If a crystal is set oscillating at its natural frequency, electric charge constantly builds up and dies away on it surface in time with the vibration of the crystal (This is known as the *piezoelectric effect*.). This makes it easy to maintain the oscillations using an alternating voltage supply as the driving frequency. The vibrations of the crystal are then used to maintain the frequency of oscillation in a resonant circuit. It is the oscillations in the resonant circuit that control the hands of an analogue watch or the display of a digital watch.

Chapter 14 (Option B)

15. Option C: Imaging

Contents

C.1 – Introduction to imaging

C.2 – Imaging instrumentation

C.3 – Fibre Optics

C.4 – Medical imaging

Essential Ideas

The progress of a wave can be modelled via the ray or the wavefront. The change in wave speed when moving between media changes the shape of the wave.

Optical microscopes and telescopes utilize similar physical properties of lenses and mirrors. Analysis of the universe is performed both optically and by using radio telescopes to investigate different regions of the electromagnetic spectrum.

Total internal reflection allows light or infra-red radiation to travel along a transparent fibre. However the performance of a fibre can be degraded by dispersion and attenuation effects.

The body can be imaged using radiation generated from both outside and inside. Imaging has enabled medical practitioners to improve diagnosis with fewer invasive procedures. © IBO 2014

CHAPTER 15 (OPTION C)

C.1 Introduction to imaging

NATURE OF SCIENCE:

Deductive logic: The use of virtual images is essential for our analysis of lenses and mirrors. (1.6)

© IBO 2014

Essential idea: The progress of a wave can be modelled via the ray or the wavefront. The change in wave speed when moving between media changes the shape of the wave.

Understandings:
- Thin lenses
- Converging and diverging lenses
- Converging and diverging mirrors
- Ray diagrams
- Real and virtual images
- Linear and angular magnification
- Spherical and chromatic aberrations

Thin lenses

A **lens** is a transparent object with at least one curved surface but more commonly two curved faces. The amount of refraction is determined by the refractive index. Most lenses are made of glass but perspex (lucite) and quartz lenses are common. They are used to correct defects of vision using spectacles and in optical instruments such as cameras, microscopes and refracting telescopes. The curved surfaces of lenses may be spherical, parabolic or cylindrical but we will restrict our discussion to thin lenses with spherical surfaces.

In optics, a **thin lens** is a lens with a thickness (distance along the principal axis between the two surfaces of the lens) that is negligible compared to the radii of curvature of the lens surfaces.

Converging and diverging lenses

Lenses are either convex (converging) or concave (diverging) as shown by the ray diagrams that locate the focus in Figure 1501.

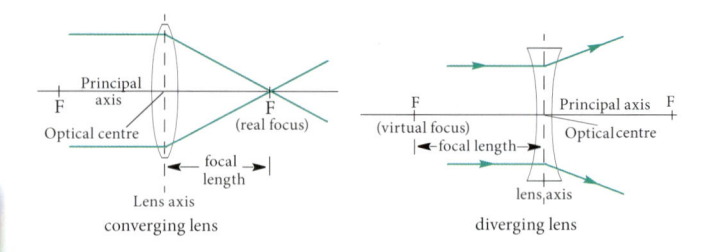

Figure 1501 Converging and diverging lenses

Lenses come in various shapes as shown in Figure 1502.

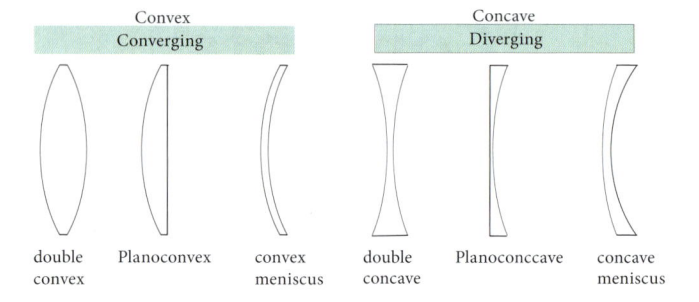

Figure 1502 Some lenses

It is important to understand the meaning of the terms used with lenses when describing the geometrical optics, and constructing ray diagrams.

Centre of curvature 'C'

the centre of the sphere of which the lens is made.

Radius of curvature 'R'

the radius of the sphere from which the lens is made.

Pole 'P'

central point of the refracting surface.

Principal axis

line that passes through the centre of curvature and the centre of the refracting surface.

Principal focus 'F'

point through which rays parallel and close to the principal axis pass after refraction if the lens is convex, or appear to come from if the lens is concave.

Focal length 'f'

the distance between the principal focus and the centre of the refracting surface. From the definition it can be seen that

$$f = \frac{1}{2}R$$

IMAGING

Aperture

the length of the refracting surface on which the incident rays can be refracted.

Principal focal plane

the plane that passes through the principal focus and is perpendicular to the principal axis.

Power of a convex lens and dioptre

The power of a **convex lens** P is the reciprocal of the focal length. It is a measure of the strength of a lens as used by optometrists and opthalmologists.

$P = \dfrac{1}{f}$

The unit for the lens power is the **dioptre** D with the unit m^{-1}.

So if a lens has a focal length of 40 cm, the power

$= 1/0.40 \text{ m} = 2.5 \text{ D}$.

The power of a converging lens is positive and the power of a diverging lens is negative as we will find out later in this section.

Converging and diverging mirrors

A spherical mirror is one in which the reflecting surface is a portion of a sphere. A parabolic mirror is one in which the reflecting surface is a portion of a parabola.

The two types of spherical mirrors are shown in Figure 1503.

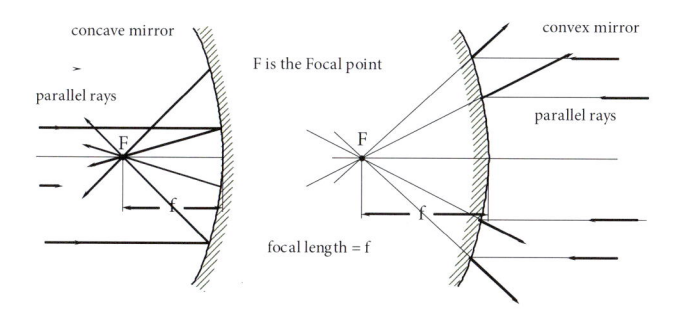

Figure 1503   Types of spherical mirrors.

Mirrors that curve in, like a cave, are called a concave mirrors. They are called converging mirrors because parallel rays of light close to its centre are reflected through the principal focus F.

Mirrors that curve outwards are called convex mirrors. They are called diverging mirrors because parallel rays of light close to its centre are reflected so that they appear to come from its principal focus F.

It is easy to see the image formed by a simple curved mirror by looking at yourself in both sides of a shiny metal spoon. The image formed is blurred because of spherical aberration.

It is important to understand the meaning of the terms used with spherical mirrors when describing geometrical optics and constructing ray diagrams. Figure 1504 illustrates the following definitions.

- centre of curvature C

- the centre of the sphere of which the mirror is made.

- radius of curvature R

- the radius of the sphere from which the mirror is made.

- pole P

- central point of the reflecting surface.

- principal axis

- that line that passes through the centre of curvature and the centre of the reflecting surface.

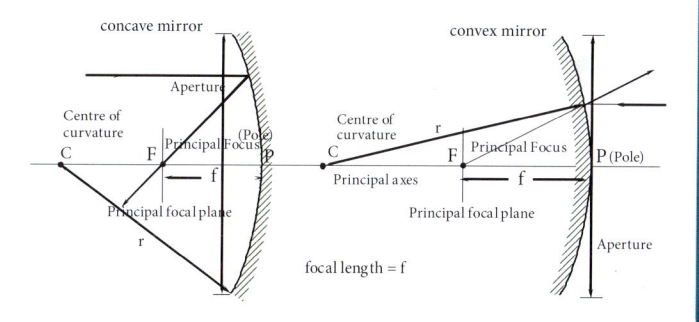

Figure 1504   Terms used in ray optics.

- principal focus F

- that point through which rays parallel and close to the principal axis pass

- after reflection if the mirror is concave, or appear to come from if the mirror is convex.

- focal length f

- the distance between the principal focus and the centre of the reflecting surface. From definition it can be seen that

- aperture

429

Chapter 15 (Option C)

- the length of the reflecting surface on which the incident rays can be reflected.

- principal focal plane

- the plane that passes through the principal focus and is perpendicular to the principal axis.

- paraxial rays

- rays close to the principal axis that are reflected through, or appear to come from, the image point. Non-paraxial rays converge or diverge to different points near the image point causing blurring of the image, an effect already mentioned as spherical aberration.

Ray diagrams

It is possible to use graphical techniques to locate images formed by mirrors. At least two rays are drawn of the three construction rules as shown in Figure 1505.

1. An incident ray that is parallel to the principal axis will always reflect through the principal focus in the case of a concave mirror and away from the principal focus in the case of a convex mirror.

2. An incident ray directed through the principal focus is always reflected parallel to the principal axis.

3. An incident ray directed through the centre of curvature will always be reflected back along its original path.

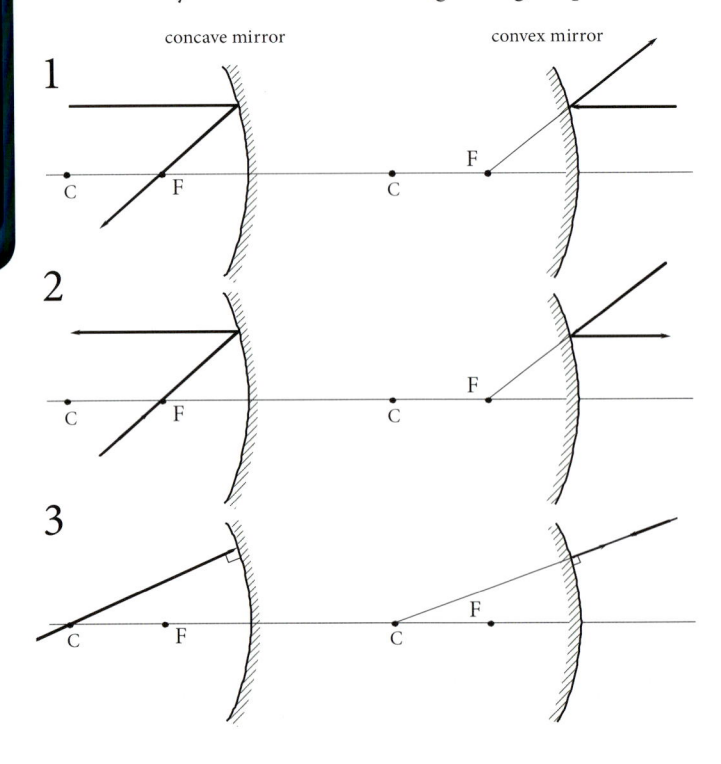

Figure 1505 Curved mirror construction rules.

There are three important properties of images. They can be:

1. Real or virtual

2. Upright (erect) or inverted

3. Diminished, the same size or enlarged (magnified).

Figure 1506 (on opposite page) shows the ray diagrams for an object placed at different positions along the principal axis, and describes the nature of the image.

Note that a convex mirror gives only one type of image, so only one ray diagram need be drawn.

A convex mirror gives an erect, diminished virtual image. A plane mirror gives an erect virtual image but the image is the same size. As a result, convex mirrors with diminished images create a wider field of view making them useful as driving mirrors on cars, or as traffic mirrors on the road that increase the field of vision on secluded lanes and roads.

Exercise 15.1

1. When you run toward a vertical plane mirror at 5 m s^{-1}, you

    A. approach your image at 5 m s^{-1}
    B. recede from your image at 5 m s^{-1}
    C. approach your image at 10 m s^{-1}
    D. stay a constant distance from your image

2. Why can't you see an image of yourself in this page?

3. An object 4.0 cm high is placed 24.0 cm from a concave mirror of focal length 8.0 cm. On graph paper, draw a ray diagram to find the position and nature of the image.

4. An object 4.0 cm tall is placed 12.0 cm from a concave mirror whose radius of curvature is 30.0 cm. What are the position, nature, magnification and size of the image?

5. An object is placed 4.0 m in front of a concave mirror and a virtual image is formed 6.0 m behind the mirror. What is the focal length of the mirror?

6. A dentist's mirror can magnify 5.0 times when it is placed 1.5 cm from a tooth. What mirror type and focal length would be suitable for this purpose?

IMAGING

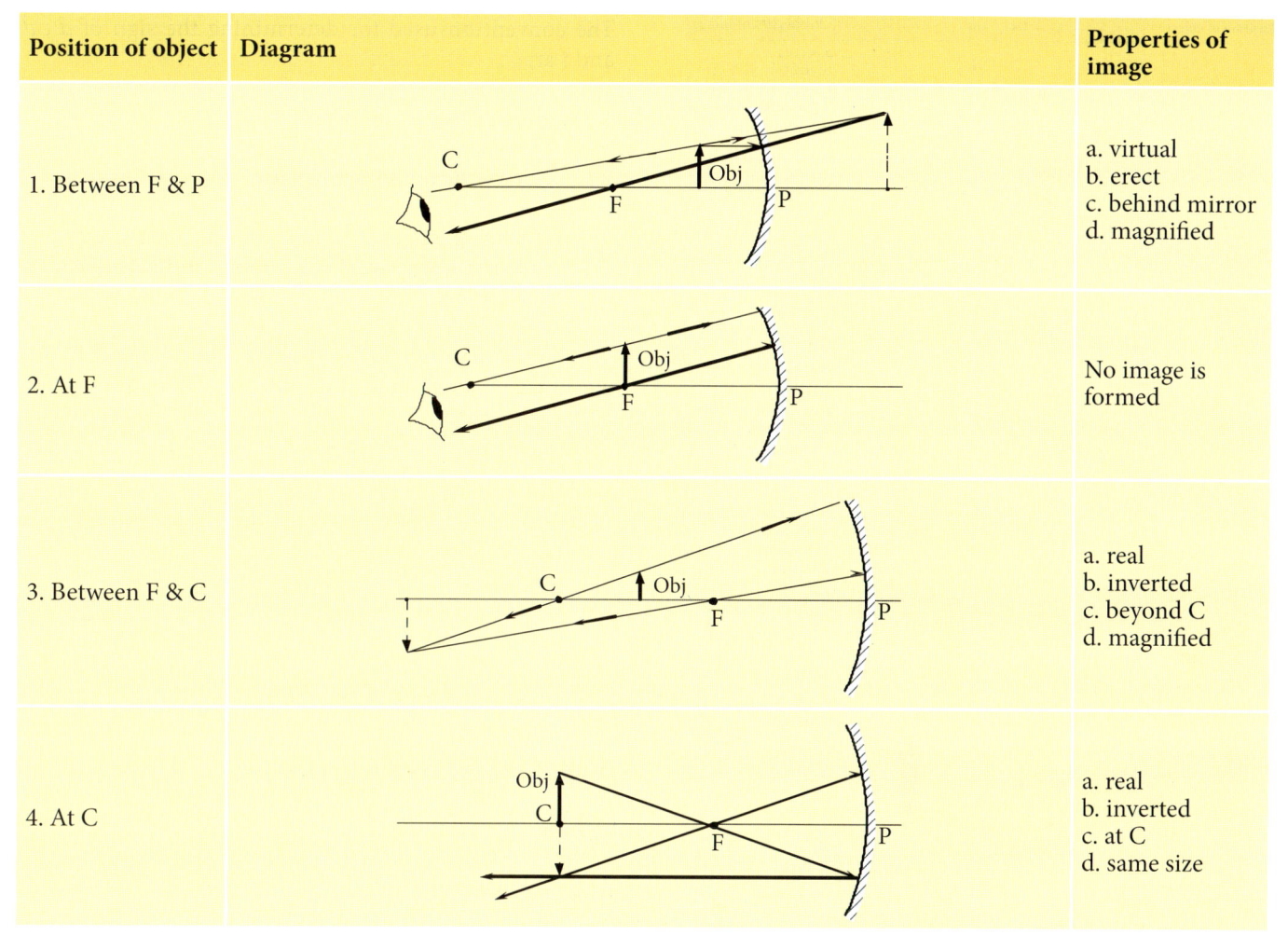

Position of object	Diagram	Properties of image
1. Between F & P		a. virtual b. erect c. behind mirror d. magnified
2. At F		No image is formed
3. Between F & C		a. real b. inverted c. beyond C d. magnified
4. At C		a. real b. inverted c. at C d. same size

Figure 1506 *Images formed by mirrors.*

Ray diagrams to find images of lenses

It is possible to use graphical techniques to locate images formed by lenses. At least two rays are drawn for the three construction rules as shown for a converging and a diverging lens in Figure 1507.

1. A ray passes through the optical centre without any deviation

2. A ray parallel to the principal axis refracts so that it passes through the principal focus if convex, or appear to diverge from the principal focus if concave

3. A ray passing through, or appearing to come, from the principal focus refracts so that it travels parallel to the principal axis.

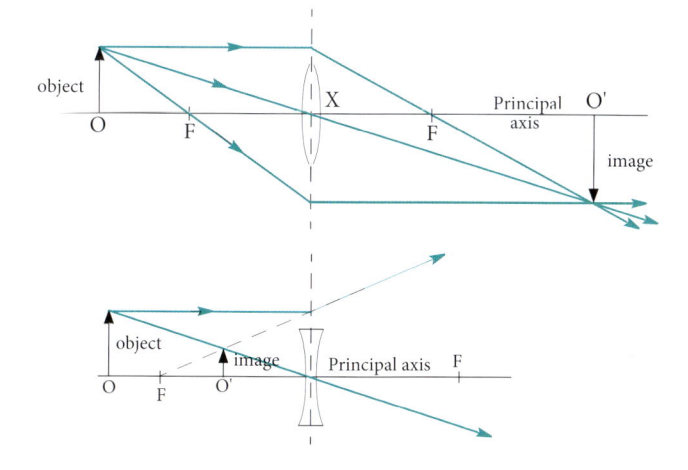

Figure 1507 *Spherical lens construction rules*

Figure 1508 shows the ray diagrams for an object placed at different positions along the principal axis, and gives examples of the uses of the lens for each situation.

431

Chapter 15 (Option C)

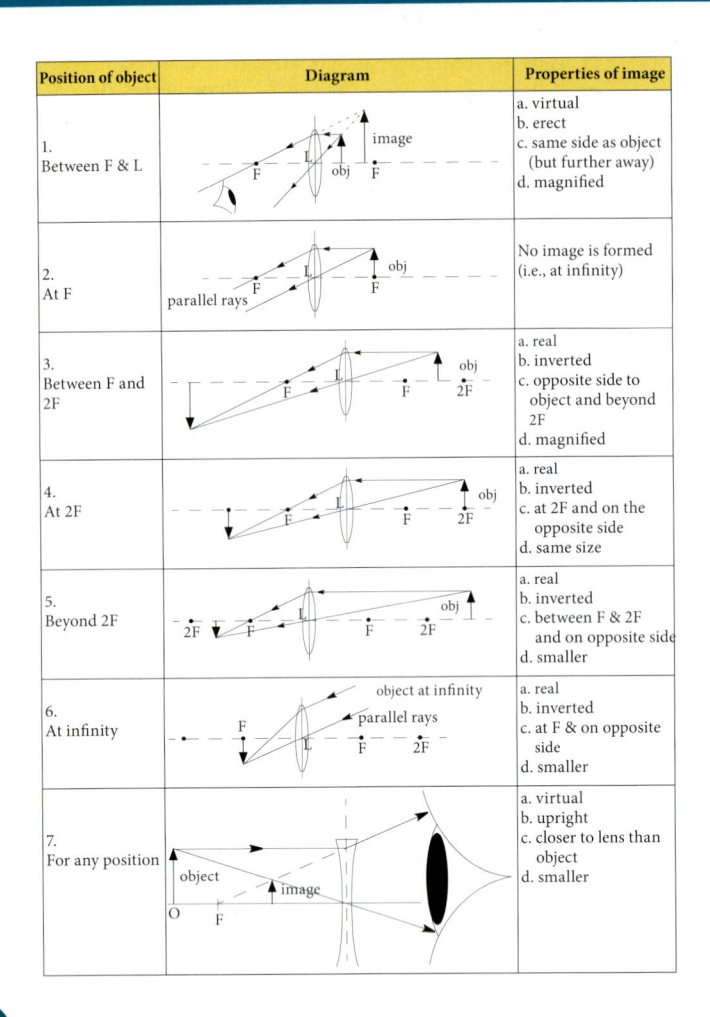

Position of object	Diagram	Properties of image
1. Between F & L		a. virtual b. erect c. same side as object (but further away) d. magnified
2. At F		No image is formed (i.e., at infinity)
3. Between F and 2F		a. real b. inverted c. opposite side to object and beyond 2F d. magnified
4. At 2F		a. real b. inverted c. at 2F and on the opposite side d. same size
5. Beyond 2F		a. real b. inverted c. between F & 2F and on opposite side d. smaller
6. At infinity		a. real b. inverted c. at F & on opposite side d. smaller
7. For any position		a. virtual b. upright c. closer to lens than object d. smaller

Figure 1508 Images formed by lenses

Note that a concave lens gives only one type of image, so only one ray diagram need be drawn. A concave lens gives an erect, diminished virtual image.

Real and virtual images

A **virtual image** is an image that appears to come from a single point when rays are extrapolated to that point as shown by the dashed lines in the first and last case of Figure 1509. A **real image** is an image that can be seen on a screen that has been put at the point where the rays intersect at a single point.

The thin lens equation

In terms of the focal length f, the lens equation is stated as
$$\frac{1}{d_o} + \frac{1}{d_i} = \frac{1}{f}$$

where d_o is the object distance and d_i is the image distance.

The convention used for determining the sign of d_o, d_i and f are:

1. real distances are positive

2. virtual distances are negative

3. the focal length and the radius of curvature of a lens are positive if converging, and negative if diverging.

The lens equation can be derived using geometry and algebra. Consider Figure 1510, the following ray diagram:

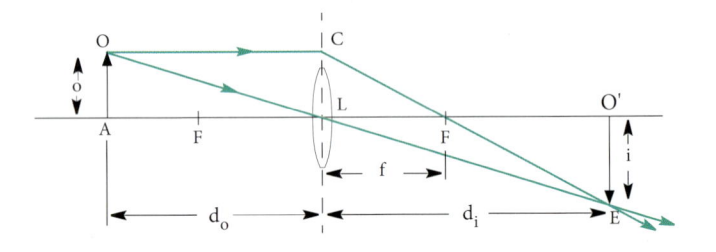

Figure 1510 A ray diagram

Δ AOL is similar to Δ EIL. Therefore,

$$\frac{AO}{IE} = \frac{AL}{IL} \Rightarrow \frac{o}{i} = \frac{d_o}{d_i}$$

This is equation **1**.

Δ CLF is similar to Δ FIE. Therefore,

$$\frac{CL}{IE} = \frac{LF}{FI}$$

Because

$$CL = EB \text{ then } \frac{CL}{IE} = \frac{AO}{IE}$$

$$\therefore \frac{AO}{IE} = \frac{LF}{FI}$$

but $LF = f$ and $FI = LI - LF = d_i - f$

$$\therefore \frac{AO}{IE} = \frac{f}{d_i - f}$$

This is equation **2**.

From equations (**1**) and (**2**), we get,

$$\frac{d_o}{d_i} = \frac{f}{d_i - f} \text{ and } d_i f = d_o(d_i - f)$$

Divide both sides by $d_o d_i \times f$:

$$\frac{d_i f}{d_o d_i f} = \frac{d_o(d_i - f)}{d_o d_i f} \Rightarrow \frac{1}{d_o} = \frac{1}{f} - \frac{1}{d_i}$$

So that,

$$\frac{1}{d_o} + \frac{1}{d_i} = \frac{1}{f}$$

This equation is quite often written in textbooks as:

$1/f = 1/u + 1/v$

where v is the image distance and u is object distance.

This equation can also be used for mirrors in what is called the mirror equation. The conventions used for mirrors are:

- Object distances are always positive

- The image distance is positive for a real image and negative for a virtual image

- The focal length is positive for a concave mirror and negative for a convex mirror.

Many optical devices contain more than one thin lens. The final image produced by the system of lenses is determined by firstly finding the image distance of the first lens and using this value along with the distances between lenses to find the object distance for the second lens.

Many optical devices contain more than one thin lens. The final image produced by the system of lenses is determined by firstly finding the image distance of the first lens and using this value along with the distances between lenses to find the object distance for the second lens.

Linear and angular magnification

The linear or lateral **magnification** m of a lens is given by the ratio of the height of an image to the height of its object or the ratio of the image distance to the object distance

$$m = -\frac{d_i}{d_o} = \frac{h_i}{h_o}$$

where d_i and d_o are the image and object distances respectively and h_i and h_o the image and object heights respectively. Linear magnification has no units.

A negative magnification when both d_o and d_i are positive indicates that the image is inverted.

Example

A small object is 15.0 cm from a concave lens with a focal length of 10.0 cm. Locate the image and determine its magnification.

Solution

Using the formula $\frac{1}{d_o} + \frac{1}{d_i} = \frac{1}{f}$, we have that

$$\frac{1}{15.0} + \frac{1}{d_i} = \frac{1}{10.0} \Leftrightarrow d_i = -6.0 \text{ cm}$$

So that the image is a virtual image located 6.0 cm in front of the lens.

The magnitude is given by

$$m = -\frac{d_i}{d_o} = -\frac{(-6.0 \text{ cm})}{10.0 \text{ cm}} = +0.40$$

The virtual image is **erect** and has a magnification of **0.40**.

Example

An object 1.2 cm high is placed 6.0 cm from a double convex lens with a focal length of 12.0 cm. Locate the image and determine its magnification.

Solution

From the formula, $\frac{1}{d_o} + \frac{1}{d_i} = \frac{1}{f}$, we have,

$$\frac{1}{6.0} + \frac{1}{d_i} = \frac{1}{12.0} \Leftrightarrow d_i = -12.0 \text{ cm}$$

The image is a virtual image located 12.0 cm in front of the lens.

Then, for the magnitude, we use the formula,

$$m = -\frac{d_i}{d_o} = -\frac{(-12.0 \text{ cm})}{6.0 \text{ cm}} = +2$$

The virtual image is **erect** and has a magnification of **2.0**.

Example

Two converging lenses each of focal length 10 cm are 15 cm apart. Find the final image of an object that is 15 cm from one of the lenses.

Solution

This problem can be solved either graphically or algebraically. In any case, a diagram as in the figure below can help to roughly see the position of the final image.

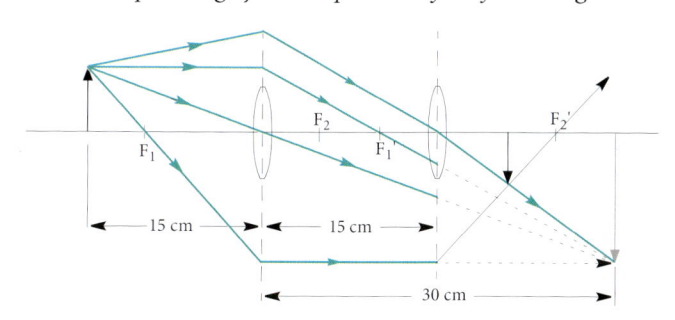

Again, we use the expression, $\frac{1}{d_o} + \frac{1}{d_i} = \frac{1}{f}$ so that,

$\frac{1}{15} + \frac{1}{d_i} = \frac{1}{10} \Leftrightarrow d_i = 30$ cm

The image is not formed because the light rays strike the second lens before reaching the image position. However, the unformed image acts as a virtual object for the second lens. Since the unformed image is on the transmission side of the second lens, it is a virtual object.

So, we now have,

$\frac{1}{-15} + \frac{1}{d_i} = \frac{1}{10} \Leftrightarrow d_i = 6$ cm

The image is **6 cm on the transmission side of the second lens.**

Exercise 15.2

1. An object 1.0 cm high is placed 5.0 cm from a double convex lens with a focal length of 12.0 cm. Locate the image and describe its nature and its magnification.

2. Using geometric construction rules, draw the following ray diagrams to show the position and properties of the image formed when:

    (a) An object is placed outside the principal focus using a convex lens
    (b) An object is placed outside the principal focus using a concave lens

3. An small object is 16.0 cm from a concave lens with a focal length of 12.0 cm. Locate the image and determine its magnification.

4. An object 1.2 cm high is placed 6.0 cm from a double convex lens with a focal length of 12.0 cm. Locate the image, and determine its magnification.

Far point and near point for the unaided eye

The main purpose for the manufacture of magnifying optical instruments that operate in the visible region of the electromagnetic spectrum is to increase the angular size of the object or its angular magnification. We rely on an eventual sharp image being formed on the retina of the eye.

Linear magnification was previously defined as the ratio of the height of an image to the height of its object, or the ratio of the image distance to the object distance. In some circumstances this means of describing magnification can be misleading because it is assumed in such cases that the object is bigger than its image. However, sometimes the image size is bigger than the object size but appears smaller because it is further away (it makes a smaller angle at the eye than does the object). In this case, linear magnification does not always give the measure of the ratio of the apparent size of the image. Therefore, we need a more useful term for these occasions and the concept of **angular magnification** is used.

The size of any image formed on the retina of the eye depends on the angle subtended by the object at the eye. The closer the object is to the eye the greater will be the angle and thus the angular magnification. However, if an object is too close to the eye then there is difficulty focusing the image.

The range over which an eye can sharply focus an image is determined by what are known as the near point and far point of the eye. The **near point** is the position of the closest object that can be brought into focus by the unaided eye. The near point varies from person to person but it has been given an arbitrary value of 25 cm. The **far point** is the position of the furthest object that can be brought into focus by the unaided eye. The far point of a normal eye is at infinity.

The ability of the eye to focus over this range is called **accommodation**, and the ciliary muscles pulling or relaxing in order to change the focal length of the flexible eye lens control this. The eye has most accommodation for prolonged viewing when viewing at the far point.

The apparent size of an object can be increased by using a converging lens to allow the object to be brought closer to the eye, thus increasing the size of the image on the retina. This is the basis behind the simple magnifier.

Angular magnification

Consider the two diagrams of Figure 1510. In case (a), a small object y is at the near point X_{np}. The size of the image on the retina depends on the angular magnification θ_0.

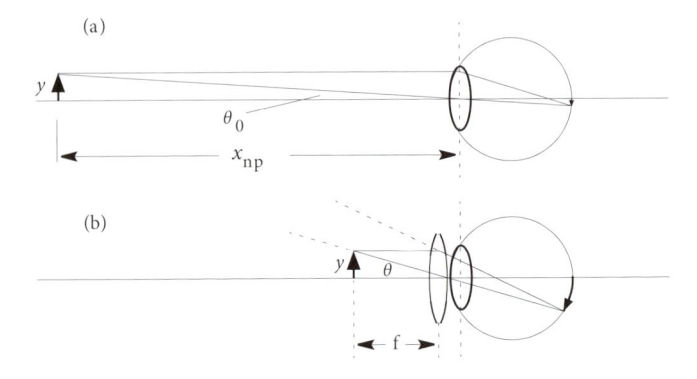

Figure 1510   The simple magnifier

In case (b), a converging lens of focal length f is placed in front of the eye. The lens allows the eye to be closer than the near point to the object. Therefore, the focal length of the double convex lens can be **less** than the near point. This results in a greater angular magnification θ at the retina.

When the object is at the focal point of the lens, parallel rays emerge from the lens and enter the eye as if they came from an object at infinity. This is the most comfortable state for prolonged viewing.

The ratio θ / θ_0 is called the angular magnification M or **magnifying power** of the lens.

$$M = \frac{\theta}{\theta_0}$$

where θ = the angle subtended at the eye by the image and θ_0 = the angle subtended at the unaided eye by the object when it is at the near point.

The difference between the **angular magnification** and the linear magnification is that the angular magnification is the ratio of the apparent sizes of the object and the image whereas the linear magnification is the real size of the object and the image.

Expression for angular magnification

From Figure 1511 (b), for small angles $\tan \theta \approx \theta$

But $\tan \theta = y / L$ Therefore $\theta = y / L$

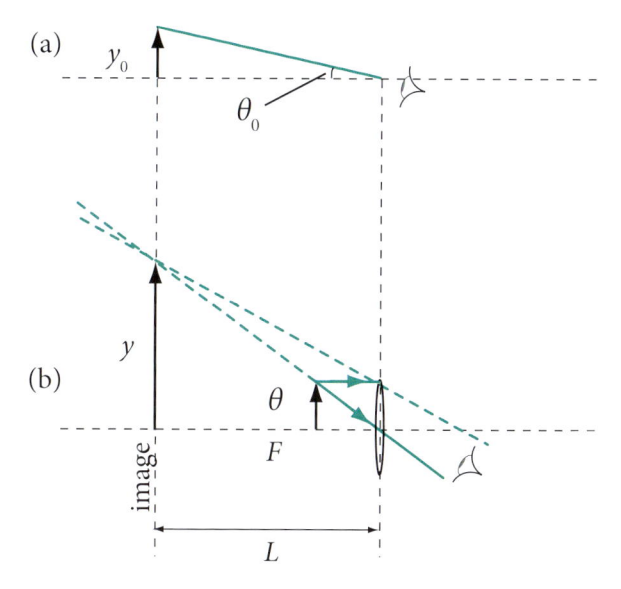

Figure 1511

From Figure 1822 (a), $\tan \theta_0 = \frac{y_0}{L}$ Therefore $\theta_0 = \tan^{-1}\left(\frac{y_0}{L}\right)$

But $M = \frac{\theta}{\theta_0} = \left(\frac{y}{L}\right) \div \left(\frac{y_0}{L}\right) = \frac{y}{y_0}$

So, in the case of a magnifying glass:

M (angular magnification) = m (linear magnification)

This means that $M = \frac{d_i}{d_o}$

If we multiply $\frac{1}{d_o} - \frac{1}{d_i} = \frac{1}{f}$ by d_i

$\frac{1}{d_i}$ is negative because the image is virtual.

we get:

$$\frac{d_i}{d_o} = \frac{d_i}{f} + 1$$

therefore,

$$M = \frac{d_i}{f}$$

So if the focal length is 3.0 cm and the image distance is at the near point of the eye, then the image distance will be negative. So:

$$M = -\frac{25}{3} + 1 = -8.33 + 1 = -7.3$$

Therefore, the angular magnification or the magnifying power of the lens is 7.3.

So it should be clear that a lens of a small focal length is required and this is why a simple magnifying glass has smooth, curved faces. However, there is an upper limit to the angular magnification since images become distorted if the radius of curvature becomes too great.

Simple **magnifiers** are used as the eyepiece in compound microscopes and both reflecting and refracting telescopes to view the image formed by another lens system.

Example 1

(a) Draw a ray diagram for an object that is inside the focal length of a double convex lens.

(b) For this lens, the focal length is 8.0 cm and the distance of the object from the lens forms an image at the near point of the eye. If the distance of the lens from the eye is 5.0 cm, calculate the distance of the object from the lens.

Solution

(a) The ray diagram is shown in the following Figure.

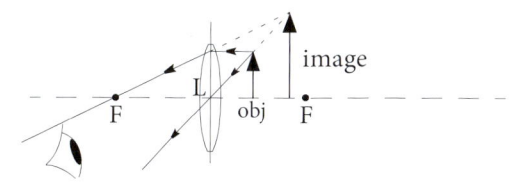

(b) Remember the near point is the position of the closest object that can be brought into focus by the unaided eye, and that virtual distances are negative.

435

Therefore,

$d_i = -(25 - 5) = -20$ cm.

We use the expression, $\frac{1}{d_o} = -\frac{1}{d_i} + \frac{1}{f}$ where

$\frac{1}{d_o} = -\frac{1}{-20} + \frac{1}{8} = 0.175$

$d_o = 5.7$ cm

Example 2

A double convex lens has a focal length of 8.0 cm. An object is placed 5.0 cm from the lens.

(a) Draw a ray diagram to show the position of the image and describe the nature of the image.

(b) Determine the magnifying power of the lens if the image is at the near point of the eye

Solution

(a) The following figure shows the correct construction.

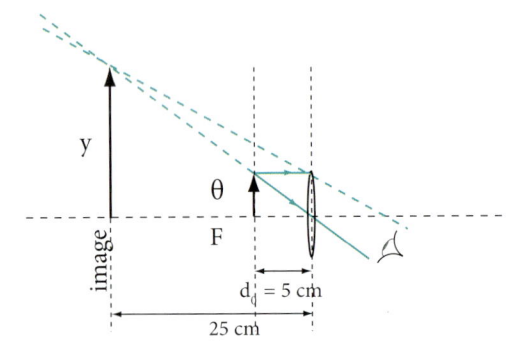

The image virtual, erect and magnified.

(b) $M = -\frac{25}{8.0} - 1 = -3.125 - 1 = -4.125$

The magnifying power of the lens is 4.1

An **aberration** is an image defect of which blurring and distortion are the most common image defects. Aberrations can occur with the use of both lenses and mirrors.

Spherical aberration is most noticeable in mirrors with large apertures. The curve or envelope shown in Figure 1512 is called a caustic curve. Rays close to the principal axis (called paraxial rays) are all reflected close to the principal focus. Those rays that are not paraxial tend to blur this image causing spherical aberration.

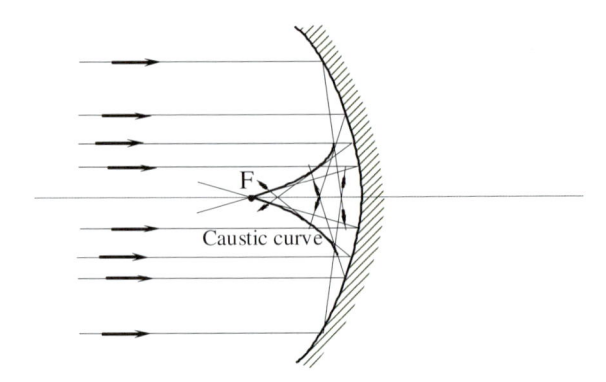

Figure 1512 Caustic curve due to spherical aberration

If the aperture of the mirror is less than 10°, spherical aberration is greatly reduced but not eliminated.

To reduce spherical aberration, parabolic mirrors that have the ability to focus parallel rays are used in car headlights and reflecting telescopes.

In practice it is found that a single converging lens with a large aperture is unable to produce a perfectly sharp image because of two inherent limitations:

1. spherical aberration

2. chromatic aberration

As with spherical mirrors, the nonparaxial rays do not allow for a sharp image.

Spherical aberration occurs because the rays that refract at the outer edges of a lens will have a different focal length to those rays that refract near the principal focus. To put it another way, spherical aberration occurs because the rays incident near the edges of a converging lens are refracted more than the paraxial rays as shown in Figure 1513.

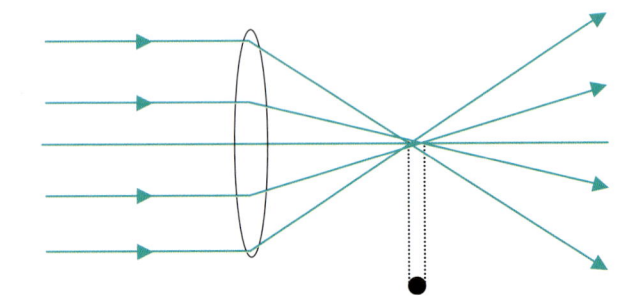

Figure 1513 The circle of least confusion

This produces an area of illumination rather than a point image even when monochromatic light is used called the **circle of least confusion**. Spherical aberration causing curving of the image at its edges. So if the object was a series of square grids as shown in Figure 1514 then the image would be distorted at the edges.

436

object image

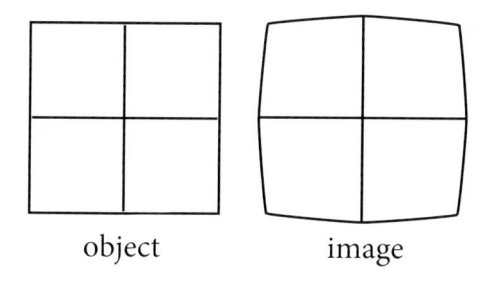

Figure 1514   Spherical aberration

To reduce spherical aberration, the lens can be ground to be slightly non-spherical to adjust for the circle of least confusion or by using different combinations of lenses put together. Alternatively, a stop (an opaque disc with a hole in it) is inserted before the lens so that the aperture size can be adjusted to allow only paraxial rays to enter. However, this reduces the light intensity and introduces diffraction of light.

Recall from your studies that the property refractive index is a function of wavelength

$$_1n_2 = \frac{\sin i}{\sin r} = \frac{\lambda_1}{\lambda_2}$$

Because visible light is a mixture of wavelengths, the refractive index of the lens is different for each wavelength or colour of white light. Consequently, different wavelengths are refracted by different amounts as they are transmitted in the medium of the lens. For example, blue light is refracted more than red light as shown in Figure 1515.

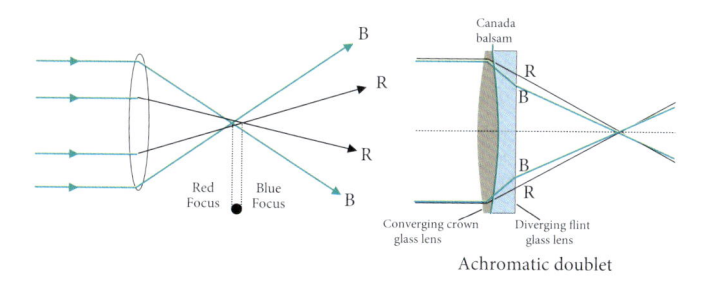

Figure 1515   Refraction of red and blue light

Each colour must therefore have a different focal length and it further follows that focal length is a function of wavelength.

Chromatic aberration produces coloured edges around an image. It can be minimised by using an **achromatic doublet** that is made from a converging crown glass lens and a diverging flint glass lens that are adhered together by canada balsam as drawn in the right of Figure 1515. Since the chromatic aberration of converging and diverging lenses is opposite, a combination of these two lenses will minimise this effect.

C.2 Imaging instrumentation

NATURE OF SCIENCE:

Improved instrumentation: The optical telescope has been in use for over 500 years. It has enabled humankind to observe and hypothesize about the universe. More recently, radio telescopes have been developed to investigate the electromagnetic radiation beyond the visible region. Telescopes (both visual and radio) are now placed away from the Earth's surface to avoid the image degradation caused by the atmosphere, while corrective optics are used to enhance images collected at the Earth's surface. Many satellites have been launched with sensors capable of recording vast amounts of data in the infrared, ultraviolet, X-ray and other electromagnetic spectrum ranges. (1.8)

© IBO 2014

Essential idea: Optical microscopes and telescopes utilize similar physical properties of lenses and mirrors. Analysis of the universe is performed both optically and by using radio telescopes to investigate different regions of the electromagnetic spectrum.

Understandings:
- Optical compound microscopes
- Simple optical astronomical refracting telescopes
- Simple optical astronomical reflecting telescopes
- Single-dish radio telescopes
- Radio interferometry telescopes
- Satellite-borne telescopes

Optical compound microscopes

The schematic diagram for a compound microscope is shown in Figure 1516. It is used to see very small objects at close distance. In its simplest form, it consists of two converging lenses.

Chapter 15 (Option C)

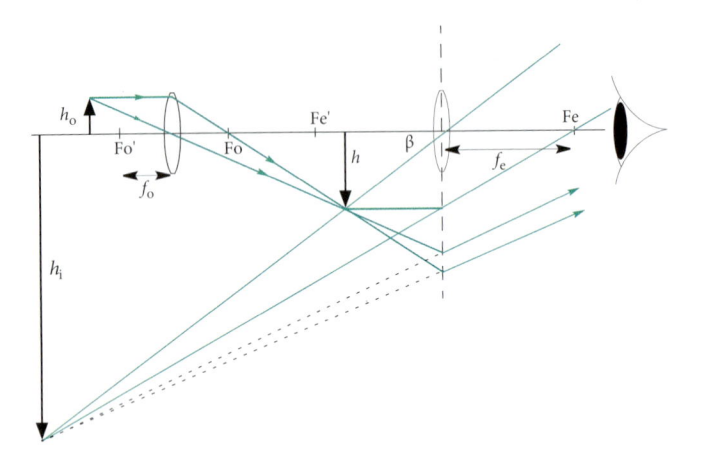

Figure 1516 *Schematic diagram of a compound microscope*

2. The astronomical refracting telescope

3. The terrestrial refracting telescope.

A refracting astronomical telescope shown in Figure 1517 uses the properties of a two converging lens combination - the objective lens, and the eyepiece that acts as a simple magnifier.

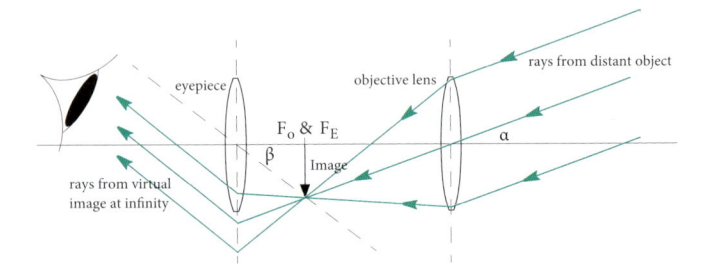

Figure 1517 *A reflecting astronomical telescope*

The objective lens has a short focal length. The object is placed just outside the focal length of the objective. The image produced by the objective is real, magnified and inverted. (This image is slightly coloured due to chromatic aberration as explained in the last section). The image acts as a real object for the eyepiece.

The eyepiece is placed close to the eye and has a longer focal length. It acts as a simple magnifier so that the final image is an inverted, magnified and virtual image that is positioned at the near point.

If the eyepiece is placed so that the image of the objective falls at the first principal focus of the eyepiece, then the image can be viewed at infinity. However, to gain a greater angular magnification, the eyepiece is placed a little inside the first principal focus of the eyepiece so that the final image is at the near point of the viewer. This virtual image is fairly free of colour.

The overall magnification is given by the product of the angular magnification of the eyepiece and the linear magnification of the objective.

$$M = \frac{h_i}{h} \times \frac{h}{h_o}$$

The manufacturer normally prints these values on the microscope. For example, if a microscope has a 20× eyepiece and the objective being used is 40×, then the magnification is 800 times.

Simple optical astronomical refracting telescopes

Telescopes are used to view objects that are often large and that are far away. There are three basic types of telescopes

1. The reflecting telescope

The objective lens has a long focal length, and a large diameter so that large quantities of light from a distant object can enter the telescope. The object distance, being very far away, is much larger than the focal length of the lens, and this produces an image that is very small (diminished), real and inverted at the focal length of the objective F_o. This real image is placed just inside the focal length of the eyepiece F_E.

The eyepiece has a short focal length. It is placed in position to produce an inverted, virtual image at infinity.

As a rough estimate (allowing for accommodation), the diagram shows that the objective and the eyepiece need to be separated by a distance equivalent to the sum of their focal lengths, $F_o + F_E$

The angular magnification is given by:

$$M = \frac{\beta}{\alpha} = -\frac{F_O}{F_E}$$

The above expression for M is only true for normal adjustment – that is when the separation of the objective and the eyepiece is $F_O + F_E$. A telescope is in normal adjustment when the final image is formed at infinity. A microscope is in normal adjustment when the final image is at the near point.

The negative sign for the focal length ratio indicates that the image is inverted. The equation also indicates that the angular magnification is optimum when an objective of large focal length and an eyepiece of small focal length are used. When specifying a refracting telescope, the diameter of the objective lens is frequently quoted. The bigger the diameter, the more light is collected thus allowing for greater resolution of what is being viewed.

For distant objects such as the Moon, it does not cause major problems when viewing the inverted image. The terrestrial refracting telescope as shown in Figure 1518 incorporates a third converging lens called the inverting lens so that the image is upright.

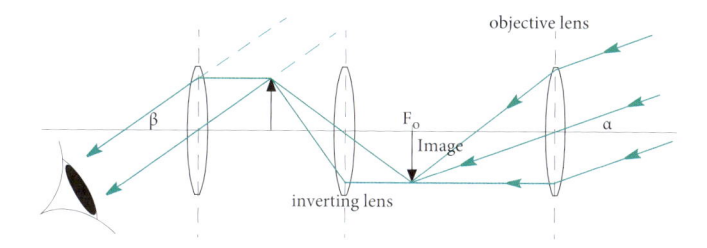

Figure 1518   A refracting terrestrial telescope

An image that forms the object for the inverter is at twice the focal length of the inverter. Therefore, the image is real and upright and very diminished, and is formed at twice the focal length (on the transmission side) of the inverter. This image is just inside the focal length of the eyepiece as before and is viewed as an enlarged, virtual and upright image.

Overall, it should be understood that the magnifying powers of telescopes are not as crucial as their light-gathering power. They increase the light-gathering power of the eye. The object being viewed such as a star forms a brighter image. Therefore, when the diameter of the objective lens is doubled, the telescope collects four times more light from a given star. However, there is a limit to the size of the objective lens before aberrations are produced, and it is difficult to support large, heavy lenses by their edges before they sag under their own weight.

Example

A compound microscope consists of an objective lens with a focal length of 1.50 cm and an eyepiece with a focal length of 10.0 cm. The lenses are separated by a distance of 15.0 cm as shown in the Figure 1519 (not to scale). An object of height 0.3 cm is placed at a distance of 2.00 cm from the objective lens.

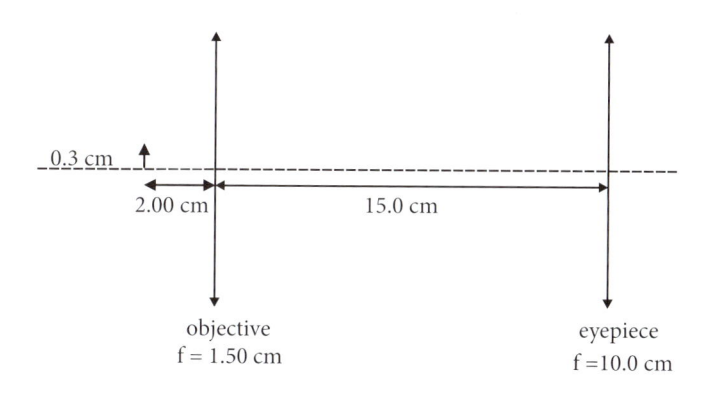

Figure 1519

(a) Determine the image distance and the magnification of the objective lens.

(b) Determine the object distance of the eyepiece.

(c) Calculate the image distance of the eyepiece

(d) Calculate the magnification of the microscope.

Solution

(a) $\dfrac{1}{d_i} = \dfrac{1}{f} - \dfrac{1}{d_o} = \dfrac{1}{1.50} - \dfrac{1}{2.00} = 0.167\ cm^{-1}$

$d_i = 6.00\ cm$

Magnification $= -\dfrac{6\ cm}{2\ cm} = -3$

(the negative sign shows us that the image is inverted).

(b) d_o of the eyepiece $= 15.0 - 6.00 = 9.0\ cm$.

(c) $\dfrac{1}{d_i} = \dfrac{1}{f} - \dfrac{1}{d_o} = \dfrac{1}{10.0} - \dfrac{1}{9.00} = -0.011\ cm^{-1}$

$d_i = -90.00\ cm$

(the negative sign indicates the image is virtual).

(d) Magnification = magnification of the objective lens × magnification of the eyepiece

$= -3 \times \left(-\dfrac{90}{9}\right) = 30\ times$

Example

The following Figure shows 3 rays of light coming from a distant star and passing through the objective lens of a telescope. The focal length of the objective lens and the eyepiece are f_O and f_E.

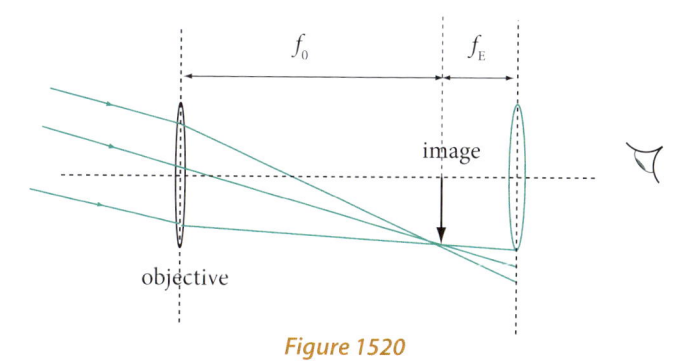

Figure 1520

(a) Complete the ray diagram to show the formation of the final image.

(b) Label the principal focus of the eyepiece lens f_E and the image of the star formed by the objective lens. State where the final image is formed by the telescope.

(c) The telescope has a magnification of 65.0 and the lenses are 70.0 cm apart. Determine the focal length of the two lenses.

Solution

(a) and (b) The Figure above has now been completed. The image formed is at infinity.

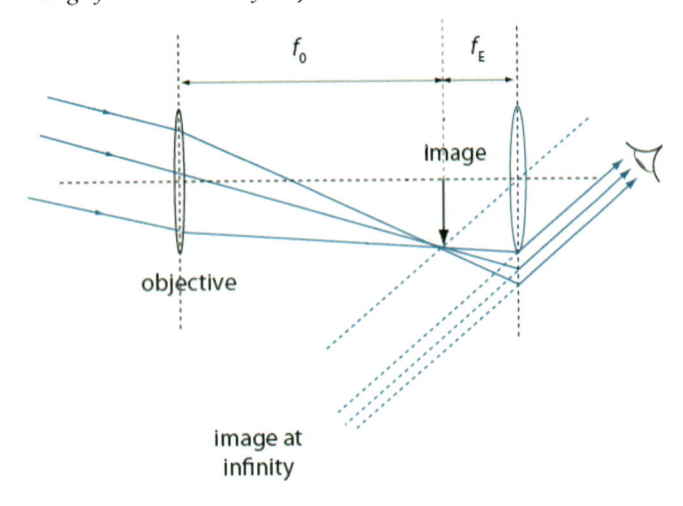

(c) $M = -f_O/f_E = -65 f_O = 65 f_E$ (1) Also, $f_O + f_E = 70.0$ cm. (2) Substituting (1) into (2) we have:

So, $65 f_E + f_E = 70.0$ $66 f_E = 70$ $f_E = 1.06$ cm and $f_O = 68.9$ cm.

Activity

Background:

The purpose of a refracting telescope is to make distant objects appear closer and therefore larger. This is achieved by placing two converging lenses with different focal lengths in certain positions on an optics bench or similar device as shown in Figure 1521.

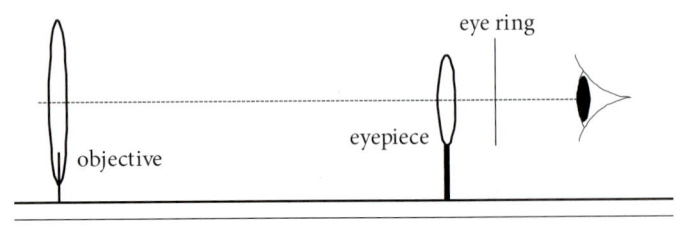

Figure 1521

The first lens, the objective, has a long focal length and forms a real image in its focal plane when a distant object is viewed through it alone. When a second lens, the eyepiece is placed at a distance so that the real image of the objective acts as a real object in the focal plane of the eyepiece, then the eyepiece magnifies the real image and produces a virtual image at infinity. The normal position of the separation of the lenses is equal to the sum of their focal lengths.

The brightness and area of the image can vary if the eye is too near or too far from the eyepiece. When the eye is in the correct position the full image can be seen and no halo is observed. If the eye is too near or too far, then part of the image is obscured and a halo is formed around part of the image. The method used for locating the correct eye position is called 'locating the eye ring'. You will need to practise this part of the investigation. The position of the eye ring in theory is the position of the image of the objective lens itself, produced by the eyepiece.

It can be shown that the angular magnification, M, for a set of appropriate lenses is given by:

$$M = -f_O/f_e$$

where f_o is the focal length of the objective and f_e is the focal length of the eyepiece.

Aim:

To construct a refracting telescope, and measure and compare its magnification.

Equipment:

Two converging lenses with focal lengths of 10cm and 50cm, optics bench, metre rule, lamp and power unit, greaseproof paper, screen, a card with a hole in it (the eye ring).

Method:

Part A:

1. Determine the focal length of each lens by viewing a distant object and focusing its image on a screen.

2. Determine the angular magnification of the telescope using the equation given in the Background section of this investigation.

Part B:

3. Place the lamp and power unit on the far side of the laboratory and turn the lamp on.

4. Mount the lens with the longer focal length on one end of the optics bench.

5. Point the lens in the direction of the lamp and use a piece of greaseproof paper to locate the image of the lamp.

6. Record the position of the image using the metre rule.

7. Attach the second lens to the other end of the optics bench.

8. Point the optics bench in the direction of the lamp and position the second lens so that it acts as a magnifying glass for the image on the greaseproof paper that you located and recorded for the first lens.

9. Remove the greaseproof paper and view the image formed of the lamp by the telescope.

10. Record the distance to the image from each lens.

Part C:

11. Illuminate with a lamp a piece of greaseproof paper placed close to the objective.

12. Place the card with a hole behind the eyepiece so that you observe a circle of light.

13. Move the card and your eye until the circle has a sharp outline.

14. Measure the diameter of this image and the diameter of the objective lens.

15. Record these values and find a ratio between the two quantities that is numerically equal to the angular magnification.

Data analysis:

1. Draw a fully labelled ray diagram of the refracting telescope that you constructed.

2. Calculate the angular magnification of the telescope by a variety of methods.

Simple optical astronomical reflecting telescopes

The largest optical telescopes in the world are reflecting telescopes. The Mount Palomar Observatory in California has a reflecting telescope containing a concave paraboloidal mirror 5m in diameter. It is made of low expansion glass that took 6 years to grind, and the reflecting surface is coated with aluminium. Photographs of nebulae up to a distance of 1010 light-years can be obtained.

A schematic diagram of a Newtonian reflecting astronomical telescope, as seen in Figure 1522, uses the properties of a parabolic mirror to view celestial objects. The mirror collects the light from a source, and converges it to a single point on a small plane mirror inclined at 45°. An eyepiece lens is placed between the mirror and the eye.

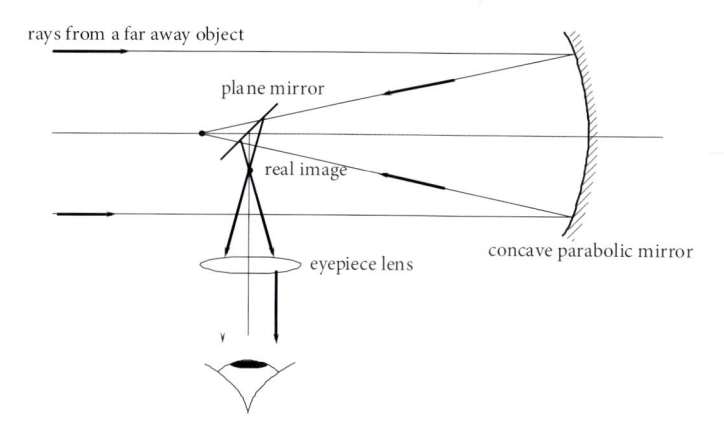

Figure 1522

In normal adjustment, the magnifying power is given by

f_O / f_E.

Another reflecting telescope is the Cassegrain reflector as seen in Figure 1523. It uses a combination of a primary concave mirror and a secondary convex mirror, often used in optical telescopes and radio antennae.

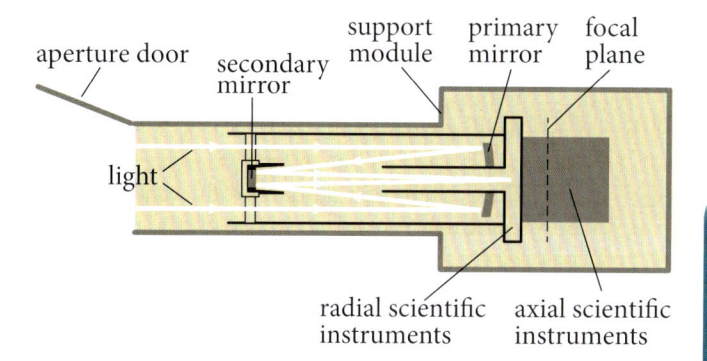

Figure 1523

In a symmetrical Cassegrain, both mirrors are aligned about the optical axis. The primary mirror usually contains a hole in the centre that permits the light to reach an eyepiece, or a camera, or a light detector. In many radio telescopes, the final focus may be in front of the primary. In an asymmetrical Cassegrain, the mirror(s) may be tilted saving the need for a hole in the primary mirror.

For general astronomical work, refracting telescopes are more easily handled than large reflecting telescopes. Reflecting telescopes are preferred when high resolving power is required.

Reflecting telescopes also have some major advantages over refracting telescopes, as there is no chromatic aberration since no refraction occurs at the objective. If a parabolic mirror is used, there is also minimal spherical aberration. Furthermore, a mirror can have a much larger diameter because it can be supported on its back side, and, on one surface of glass needs to be ground.

CHAPTER 15 (OPTION C)

Exercise 15.38

1. A simple optical device is placed inside a box as shown in the diagram. An object O, placed as shown, produces a real magnified image I. Refer to the following Figure.

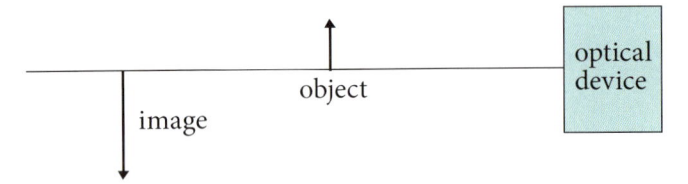

   The optical device is

   A. a convex lens
   B. a concave lens
   C. a convex mirror
   D. a concave mirror

2. Which of the following is an incorrect statement?

   A. the magnifying power of an astronomical telescope can be increased by substituting an eyepiece of greater focal length
   B. with a simple astronomical telescope things look upside down
   C. the eye lens of a human produces a real, diminished inverted image on the retina
   D. refracting telescopes produce less chromatic aberration than reflecting telescopes.

3. An object 4.0 cm high is placed 15.0 cm from a convex lens of focal length 5.0 cm. On graph paper draw a ray diagram to determine the position and nature of the image.

4. An object 4.0 cm high is placed 15.0 cm from a concave lens of focal length 5.0 cm. On graph paper draw a ray diagram to find the position and nature of the image.

5. Place the following into the convex lens or concave lens category:

   Magnifying glass, eye lens, camera lens, the objective lens of a microscope, a lens to correct short-sightedness, spotlight lens.

6. Determine the position, nature and magnification of the image of an object placed 15 cm from a convex lens of focal length 10 cm.

7. A convex lens with a focal length of 8 cm is to be used to form a virtual image that is four times the size of the object. Where must the lens be placed?

8. A slide projector is place 5.0 m from a screen. Determine the focal length of a lens that would be used to produce an image that is five times as large as the object?

9. A convex lens with a focal length of 4.0 cm is placed 20.0 cm from a concave lens with a focal length of 5.0 cm. Find the position of the image when the object is placed 12.0 cm in front of the convex lens.

10. A double concave lens has a refractive index of 1.5 and radii of curvature of 10.0 cm and 15 cm. Determine its focal length.

11. A refracting telescope has an objective lens with a diameter of 102 cm and a focal length of 19.5 m. If the focal length of the eyepiece is 10.0 cm, calculate the angular magnification of the telescope.

12. Determine the angle of minimum deviation for a prism with an apex angle of 60° and a refractive index of 1.6.

13. Can a rectangular prism be used to disperse white light? Explain your answer.

14. Describe the meaning of spherical and chromatic aberration, and a method to reduce the effect of each.

15. Refer to the following table. Choose the correct combination of lenses that are used in either the compound microscope or the refracting astronomical telescope.

	optical instrument	objective lens	Eyepiece
A	compound microscope	long focal length	Long focal length
B	compound microscope	long focal length	Short focal length
C	refracting astronomical telescope	short focal length	Long focal length
D	refracting astronomical telescope	long focal length	Short focal length

Single-dish radio telescopes

Radio astronomy was developed around 80 years ago. Thomas Edison suggested that that celestial objects might be a source of radio waves. In 1933, Karl Jansky set up an antenna he was testing. He noticed that there was variation in the radio noise, and that it was most intense in the galactic centre – the place where there is the greatest concentration of stars in the sky. In 1937, an radio amateur, Grote Reber, noted sources of radio waves had increasing flux for lower frequencies – synchrotron emission. Research carried out by Reber and many scientists during the Second World War led to an increase in interest in radio astronomy.

W e say there are two "windows" concerning the percentage of radiation that can get through the atmosphere and reach sea level:

- The optical window (about 70%)
- The radio window (about 90%)

Of the two, optical astronomy is the most affected by atmospheric effects and pollution, and this is why most telescope observatories are situated at high altitudes and remote areas free of pollution and climate variations favour plenty of clear nights.

Radio waves have longer wavelengths and are not affected by water droplets or particulate matter that scatter visible light. Therefore, the location of radio telescopes is not as critical as it is in the optical window.

The radio window known as the radio regime covers wavelengths from 1 mm to

10 m but the best atmospheric transmission is over 1-20 cm. This is the range over which the atmosphere is almost transparent. The longer radio waves are reflected back into space by the ionosphere and, water molecules and oxygen in the atmosphere absorb the shorter wavelengths.

A typical model of a basic radio telescope is shown in Figure 1524.

The incoming incident parallel radio waves are reflected by the parabolic reflector, and focus the radiation onto a small radio receiver at the focal point. The signal is pre-amplified at the masthead to boost the faint received signal. It is then amplified and tuned to particular wavelengths in the receiver building. The data is then recorded in the computer building.

The parabolic shape of the metal prevents spherical aberration. The dish is usually tens of metres in diameter, and this large diameter increases the signal strength and the resolving power. The power is proportional to the area of the collecting dish and thus doubling the diameter quadruples the power received.

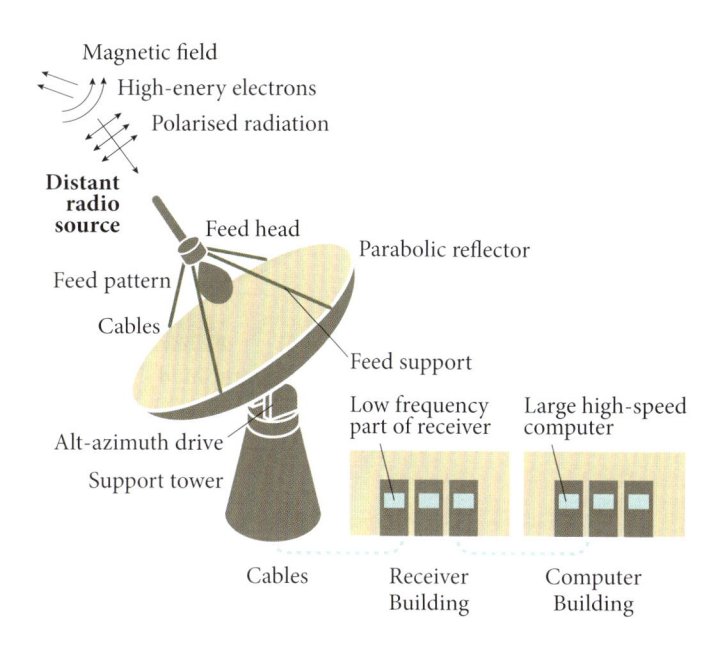

Figure 1524

However, the larger the dish, the greater the weight of the dish and some have been known to collapse under the strain. The weight can put strains on the metal and distorts the structure and thus the image. To overcome this problem, the solid metal reflector can be replaced with a fine-wire mesh. Of course they are easier to steer if there is less weight.

The largest steerable single dish telescope is the 100m x 110m Green Bank telescope (GBT) in West Virginia, USA. At 305m, the Arecibo telescope in Puerto Rico is the world's largest dish but it is not steerable.

The unit of intensity for radio measurements is the Jansky. 1 Jy = 10^{-26} W m^{-2} Hz^{-1}.

Radio interferometry telescopes

An astronomical interferometer is an array of telescopes, mirrors or more commonly radio dishes acting together to increase spatial resolution by means of interferometry. If the signals from two or more radio telescopes are combined they interfere, and the pattern of interference can be used to construct radio images that show finer detail than the individual telescope. This type of arrangement is called an interferometer.

The simplest type feeds the signals from each dish a certain distance apart (a baseline) to a single receiver as shown in Figure 1525.

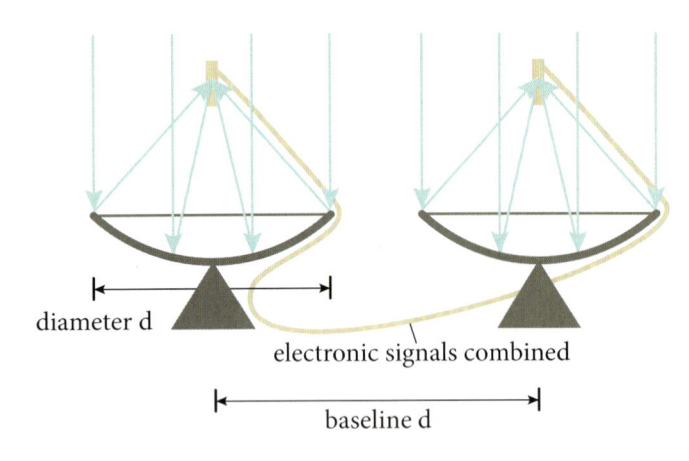

Figure 1525 Radio interferometry.

The path difference between two parallel rays from a distant source will vary with the rotation of the earth. This variation appears as an interference pattern consisting of a series of maxima and minima patterns in the combined signal. Constructive interference occurs when the path difference equals $d \sin \theta = n\lambda$ as shown in Figure 1526. The signal moves from one maximum to the next when $\theta = \lambda / d$.

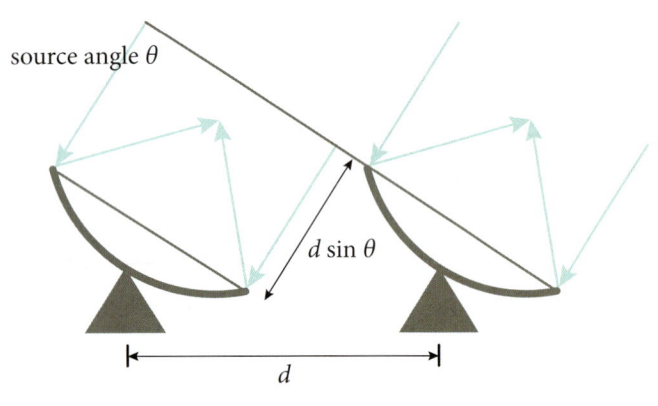

Figure 1526 Constructive interference

In interferometry, the resolution is not determined by the size of individual collectors such as mirrors or radio dishes but rather by the maximum separation of the collecting elements. The Australia Telescope Compact Array (ATCA) is a radio interferometer that has six-22m dishes. Five of the dishes are on a 3km long wide gauge railway track allowing them to be placed at different stations and separations. The sixth is a further 3 km distance, providing a maximum baseline of 6 km.

Even this array has less resolution than a single, large diameter radio telescope. However, if we have a larger array called a VLA then the resolution can be increased. The Very Large Array near Socorro, New Mexico, one of the world's premier astronomical radio observatories, consists of 27 radio antennas in a Y-shape. Each antenna is 25 meters in diameter. The data from the antennae is combined electronically to give the resolution of an antenna 36km across, with the sensitivity of a dish 130 meters in diameter.

Increasing the baseline length improves the resolution an interferometer can achieve. The sensitivity, however, is still a function of the total collecting area. The best resolution on earth-based arrays can be achieved by configuring radio telescopes in different continental regions to form a VLBI (very long baseline interferometer). Further resolution can be achieved by linking an earth array with a space observatory such as the 8m Japanese VSOP satellite (VLBI Space Observatory Programme). Using VLBI, radio astronomers can achieve spatial resolutions in a fraction of an arscecond range, higher than that currently obtainable by optical telescopes.

Example

What resolution can Australia Telescope Compact Array (ATCA) achieve for the 21cm hydrogen line?

Solution

1 radian is equal to 206264.806247 arcsecond. Therefore:

$$\theta = 2.1 \times \frac{10^5 \lambda}{d}$$

The array has a maximum baseline of 6 km. Therefore

$$\theta = 2.1 \times 10^5 \times \frac{0.21}{6000}$$

$$\theta = 7.35 \text{ arcseconds}.$$

Optical interferometers are also being developed for the largest telescopes currently operating. The twin 10 m Keck telescopes in Hawaii are being linked to form an interferometer with an effective diameter of 100 m. Even more powerful is the Very Large Telescope Interferometer (VLTI) European answer to Keck. It will link the four VLT 8.3 m telescopes with four moveable 1.8 m auxiliary telescopes being built in Chile. Once fully operational, the VLTI will provide both a high sensitivity as well as milli-arcsecond angular resolution provided by baselines of up to 200 m.

Satellite-borne telescopes

Satellite-borne telescopes provide the best environment for optical telescopes. However, the early space astronomy missions concentrated on the observations on almost all wavelengths of the electromagnetic spectrum.

Space telescopes have many advantages over ground-based telescopes. Although their aperture may be smaller, space telescope resolution is better because it does not have the problems connected with the terrestrial environment. Some of the disadvantages that land-based telescopes have are:

- **Atmospheric turbulence and refraction**

  Variations in the chemical and physical properties of the atmosphere cause variation in the refractive index of light. These variations cause random and systematic deviations of light rays referred to as atmospheric turbulence and atmospheric refraction respectively. These cause a smearing of the image.

- **Atmospheric extinction**

  Incoming photons from space are scattered and absorbed by molecules and dust in the atmosphere and this leaves only two small "windows" in the optical and radio regions, the others areas of the spectrum being absorbed. Even in the optical window, absorption and scattering dim and redden celestial bodies.

- **Mechanical flexure**

  The effect of the earth's gravitational field causes bending of the truss structure of the telescope causing the image to become distorted.

- **Thermal stability**

  Temperature variations in ground-based telescopes cause expansion and compression of the instrument parts resulting in image distortion.

- **Sky background**

  Even in the best earth environment, the sky is not completely dark.

The most famous of all telescopes is the Cassegrain-type reflector called the Hubble space telescope (HST). It orbits the earth at an altitude around 600 km. It has an aperture (primary mirror) of 2.4 m and it collects light from a region of the sky equivalent to the angle subtended by a full moon as observed from the earth. This light is then directed to the 0.3 m secondary mirror where it is focused and reflected back through a 0.6 m hole in the primary mirror. It is then projected onto an area about the size of a dinner plate in the focal plane 1.5 m behind the primary mirror. The radiation is then analysed by five instruments. To achieve sharp images it has to be pointed at a celestial objects to within 2×10^{-6} degrees over a 24 hour period.

It is left to you to research some other space-borne satellites such as IRAS, ROSAT and COBE.

Example

The HST has an aperture of diameter 2.4 m.

(a) Determine the limiting angle of resolution at a wavelength of 600 nm.

(b) What is the smallest crater it could resolve on the moon if the moon is at a distance of 3.84×10^8 m from the telescope.

Solution

(a) $\theta = \dfrac{1.22\lambda}{d}$.

$\theta = 1.22 \times \dfrac{6 \times 10^{-9}}{2.4} = 3.05 \times 10^{-7}$ rad

(b) $s = r\theta = 3.84 \times 10^8 \, m \times 3.05 \times 10^{-7} \, rad = 117 \, m.$

Chapter 15 (Option C)

C.3 Fibre Optics

NATURE OF SCIENCE:

Applied science: Advances in communication links using fibre optics have led to a global network of optical fibres that has transformed global communications by voice, video and data. (1.2)

© IBO 2014

Essential idea: Total internal reflection allows light or infra-red radiation to travel along a transparent fibre. However the performance of a fibre can be degraded by dispersion and attenuation effects.

Understandings:
- Structure of optic fibres
- Step-index fibres and graded-index fibres
- Total internal reflection and critical angle
- Waveguide and material dispersion in optic fibres
- Attenuation and the decibel (dB) scale

Structure of optic fibres

If light enters the end of a solid glass rod at an angle greater than the critical angle then total internal reflection occurs. The light emerges from the other end with very little loss of intensity. This is the principle of fibre optics.

A bundle of glass fibres made of fused silica of high purity each with a cross-sectional area of a human hair (50 μm) form a light pipe. Slight changes to the refractive index are made by the addition low concentrations of doping materials (titanium, boron or germanium) to give a refractive index in the range 1.44 to 1.46 depending on the wavelength Because light can travel along the light pipe even when it is bent, this makes optical fibres ideal for carrying light around bends, and into other difficult locations as shown in Figure 1527.

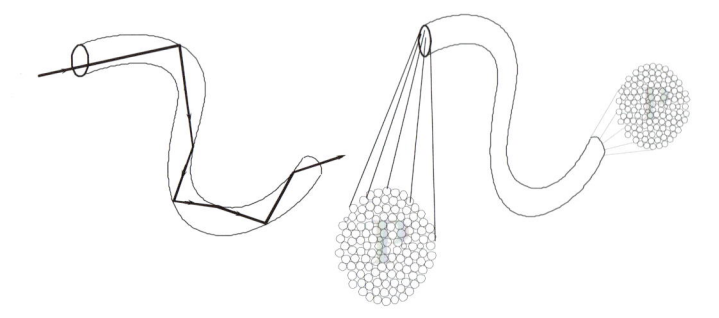

Figure 1527   A bundle of optical fibres.

A bundle of glass fibres allow light from an object to be transported to form an image of the object at the other end. If the image is not to be scrambled, then the fibres must be in a fixed position relative to one another along the light pipe. For example, the letter "**P**" in the figure for the image would be scrambled if this were not the case.

The optical fibre consists cylindrical inner **core** that carries the light, and an outer concentric shell called the **cladding**. The cladding is also made of silica that has a relatively low refractive index. Light enters one end of the core, strikes the core/cladding interface at an angle of incidence greater than the critical angle, is reflected back into the core, and travels down the fibre with a zigzag path. Little light is lost due to absorption of the core, so the light can travel many kilometres. Fibres are bundled together to form cables. A laser beam travelling through a single fibre can carry tens of thousands of telephone calls and several television programs simultaneously.

The use of fibre optics in telecommunications, medicine and technology is a fairly recent phenomenon. In medicine, the endoscope is used to view the image of any of a patient's large cavities such as the stomach and the heart. In telecommunications, fibre optic cables are replacing expensive copper cables. A laser light signal is allowed to vary rapidly to represent information such as conversation, fax, e-mail and television pictures.

Step-index fibres and graded-index fibres

Fibres with **constant refractive indices** of the core and the cladding are called **step-index fibres**.

In a waveguide, light propagates in the form of modes. One of the difficulties using a multimode fibre is that the modes tend to travel at slightly different speed along the fibre. This results in a variety of travel times so that the light pulses are broadened as they travel through the fibre in what is known as modal dispersion.

An ingenious way to overcome this modal dispersion is by grading the refractive index of the core from a maximum value at the centre to a minimum value at the core/cladding interface. This type of fibre is called a graded-index fibre. In this type of fibre, the speed increases with distance from the core axis (this is because the refractive index decreases). Although rays of greater inclination to the fibre axis have to travel further, they travel faster so that the travel time different signals are equalised.

Total internal reflection and critical angle

If we consider, for example, light waves incident on the surface of a glass block in air, then some of the light will be absorbed at the surface, some reflected and some transmitted. However, let us just concentrate on the transmitted light. In Figure 1528(a), the direction of the incident and transmitted waves are represented by the rays labelled I and T. Figure 1528(b) shows the waves travelling from glass to air.

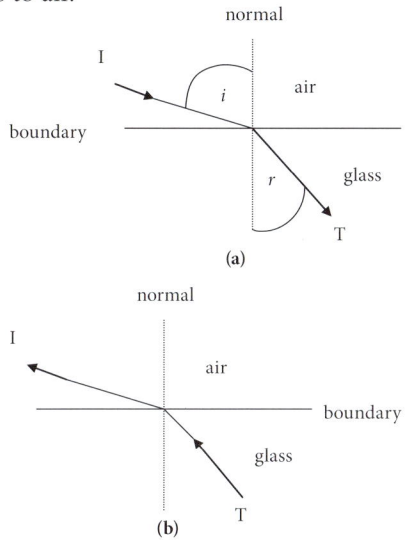

Figure 1528   Refraction: (a) air to glass transmission (b) glass to air transmission

In Figure 1528(b) the light clearly just follows the same path as in Figure 1528(a) but in the opposite direction. However, whereas any angle of incidence between 0 and 90° is possible when the light is travelling from air to glass, this is not the case when travelling from glass to air. When angle $i = 90°$ in (a), this will correspond to the maximum possible angle of refraction when the light is travelling from glass to air as is shown in Figure 1529.

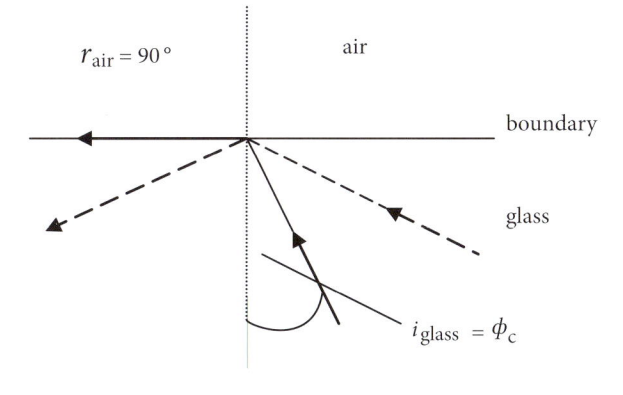

Figure 1529   Total internal reflection

For light travelling from air to glass, we have from Snell's law that the refractive index of the glass n is given by

$$n = \frac{\sin i}{\sin r}$$

Therefore travelling from glass to air, we have

$$\frac{1}{n} = \frac{\sin i_{glass}}{\sin r_{air}}$$

For the maximum value of $r_{air} = 90°$ therefore

$$\frac{1}{n} = \frac{\sin i_{glass}}{1} = \sin \phi_c$$

where ϕ_c is the angle of incidence in the glass that corresponds to an angle of refraction in air of 90°.

The dotted rays in Figure 1529 show what happens to light that is incident on the glass-air boundary at angles greater than ϕ_c; the light is **totally internally reflected**. The reflection is total since no light is transmitted into the air. The angle of incidence at which total reflection just occurs is the angle ϕ_c and for this reason ϕ_c is called the **critical angle**.

Example

The critical angle for a certain type of glass is 40.5°. Determine the refractive index of the glass.

Solution

$$\frac{1}{n} = \sin \phi_c, \text{ therefore: } n = \frac{1}{\sin 40.5°} = 1.54$$

The phenomenon of total internal reflection along with the principles of modulation and analogue to digital conversion discussed above, form the basis for data transmission using optic fibres. In optic fibres, the carrier wave is light. There are essentially three types of optic fibres but we will only concern ourselves with the so-called **step-index** fibre. Figure 1530 shows the essential structure of a single step fibre and of light transmission along the fibre by total internal reflection.

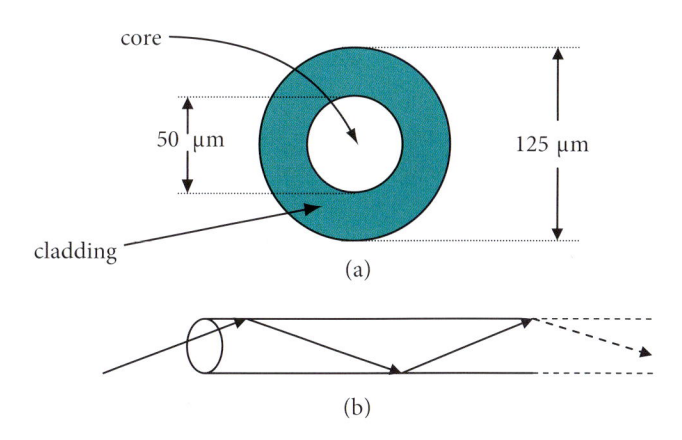

Figure 1530   Optical fibre (a) structure (b) transmission of light by TIR

447

Chapter 15 (Option C)

Figure 1529 (a), shows that there is a two-layer structure of the fibre, the core and the cladding. The diameter of the core is constant, at approximately 50 to 60 μm and a typical refractive index index would be 1.440. The surface of the core is kept as smooth as possible. The outer layer, the cladding, is bonded at all points to the surface of the core and a typical refractive index would be 1.411. The cladding ensures that the refractive index of the outside of the core is always less than the inside. A layer of plastic that protects and strengthens the fibre usually surrounds the cladding

Waveguide and material dispersion in optic fibres

A **waveguide** is some sort of structure that guides wave down the fibre. The waveguide used in fibre optics is **dielectric** material with high permittivity and thus a high refractive index, surrounded by a material with low permittivity. The structure guides the light by total internal reflection.

Although all electromagnetic waves have the same speed in a vacuum, the speed of the wave in a medium depends on the wavelength of the wave, a phenomenon known as **dispersion**. Another way to regard dispersion is to recognise that the refractive index of a medium depends on wavelength. The implication of this for optical fibres is that light of different wavelengths will travel different distances along a fibre. This means that the pulses in the fibre will spread out as they travel along the fibre and the information carried by the waves will be distorted. Figure 1531 shows how a square wave pulse might be distorted by this so-called **material dispersion**.

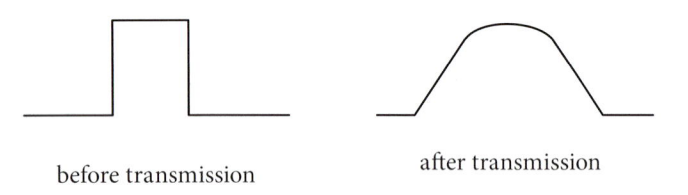

Figure 1531   Material dispersion

Using laser light as the source of the carrier wave greatly reduces material dispersion since the bandwidth of the light emitted by a laser is only about 5 nm. Even using light emitted by light emitting diodes (LED) reduces material dispersion since the light emitted still as a bandwidth of only about 20 nm.

Modal (multipath) dispersion

The light travelling down a fibre from the source to the detector can be incident at a variety of different angles to the core-cladding interface. This means that different waves can travel different paths (or **modes** as they are called) in the fibre. Therefore pulses associated with different waves will arrive at the detector at different times. This is called **modal dispersion**.

Attenuation

The intensity of the carrier wave in a fibre will decrease with the distance travelled along the fibre. This phenomenon is called attenuation and is due to energy being carried by the wave being lost. The energy loss is due to a variety of reasons such as scattering and absorption within the core. The **attenuation** is often measured in decibel per kilometre. The decibel scale is a logarithmic scale (see Topic I.1.6 to see how it applies to hearing loss) and in optic fibres the attenuated power is defined below.

loss of power (attenuation) in decibels

$$= 10 \log \frac{\text{initial power}}{\text{output power}} = 10 \log \frac{I_1}{I_2}$$

Example

The power loss between source and detector in a particular optic fibre of length 1.5 km is 50%. Calculate the power loss in dB km^{-1}.

Solution

$power\ loss\ in\ dB = 10 \log_{10}\left(\frac{1}{2}\right) = -3.0\ dB$

$power\ loss\ (attenuation) = 2.0\ dB\ km^{-1}$.

A monomode fibre is one in which there is only one transmission axis thereby eliminating modal dispersion. In monomode fibres, the core is about 5 μm in diameter which is the same order of magnitude as the wavelength of the carrier light. The graph in Figure 1532 shows the variation with wavelength of the power loss in a monomode silica optic fibre.

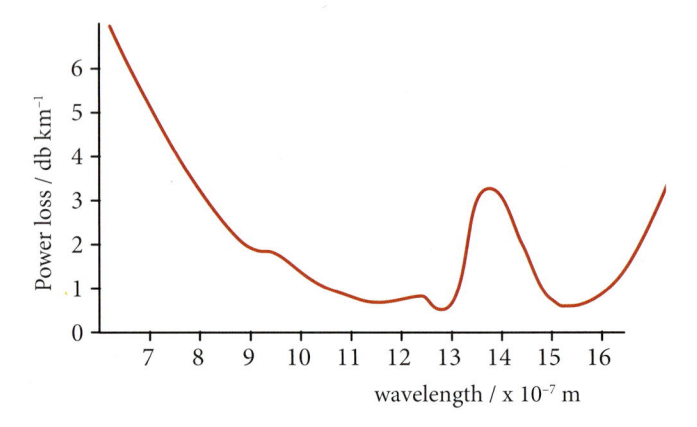

Figure 1532   Attenuation loss in a monomode optic fibre

448

We see that attenuation decreases with increasing wavelength and that there are two distinct minima at about 1300 and 1500 nm. Monomode fibres using these frequencies of carrier wave make them the best choice for long-distance communication such as in telecommunication.

One great advantage of using optic fibres for transmitting data is that they are not very susceptible to noise. Any noise that does affect the signals arises from stray light entering the fibre at the transmitting and receiving ends of the fibre. At the receiving end, photodiodes are used to convert the light pulses into electrical pulses and photodiodes are subject to random noise. The noise to signal power ratio for an optic fibre is typically in the range 10^{-17} to 10^{-18}.

Reshapers (regenerators)

Monomode fibres effectively eliminate modal dispersion and although lasers reduce material dispersion the latter is still present. Suppose for example, data is transmitted at the rate of 1 Gb s^{-1}, then, to ensure that the pulses remain distinct from each other, they need to be separated by at least 0.5 ns. Over long distances the pulses can become quite spread out such that even with laser light, after about 50 km, individual pulses will be starting to merge together. So every 40-60 km, the pulses are detected and then reshaped. The reshaped pulses are then encoded onto a new laser beam for continued transmission. This is the function of the reshaper.

Amplifier

Even if pulses have been re-shaped, the carrier wave and signal still undergo attenuation.

In 1987 David Payne and his co-workers at the University of Southampton in England developed the first practical optical **amplifier** suitable for optic fibre communication systems.

These amplifiers enable the attenuated carrier and pulses to be amplified at various points along the fibre.

The combination of **reshapers** and amplifiers means that data can be transmitted along optic fibres for vast distances. The longest optic fibre links are those that span the World's oceans, including several across the Atlantic from North America to Europe and several from the United States to Japan. The transatlantic cable laid in 1988 contains eight fibres and each fibre has a bit rate of 560 Mb s^{-1} enabling the cable to carry 40,000 telephone calls at one time. The next generation of cables which came into service in 1992, are able to carry double this number of calls.

Exercise 15.4

1. The cladding of an optic fibre has a refractive index of 1.46 and the core a refractive index of 1.48. The fibre is 1.80×10^2 m long.

   (a) Calculate the critical angle between the core and cladding.

   (b) Show that the difference between the transmission time for an axial signal and a signal that is incident to the cladding at an angle that is just greater than the critical angle is about 9 ns. (*HINT: consider the geometry of the situation.*)

2. The input power to an optic fibre is 10 mW and the signal noise is 1.0×10^{-20} W. The attenuation loss in the fibre is 2.5 dB km^{-1}.

   (a) Calculate the ratio of input power to signal noise in decibels.

   (b) The input signal needs to be amplified when its power is attenuated to 1.0×10^{-18} W. Determine the maximum separation of the amplifiers in the cable.

3. Determine the critical angle for light travelling from air to water of refractive index 1.33.

4. Calculate the length that will result in a power loss of 80% for of an optical fibre with an attenuation of 2.0 dB km^{-1}.

CHAPTER 15 (OPTION C)

C.4 Medical imaging

> **NATURE OF SCIENCE:**
>
> Risk analysis: The doctor's role is to minimize patient risk in medical diagnosis and procedures based on an assessment of the overall benefit to the patient. Arguments involving probability are used in considering the attenuation of radiation transmitted through the body. (4.8)
>
> © IBO 2014

Essential idea: The body can be imaged using radiation generated from both outside and inside. Imaging has enabled medical practitioners to improve diagnosis with fewer invasive procedures.

Understandings:
- Detection and recording of X-ray images in medical contexts
- Generation and detection of ultrasound in medical contexts
- Medical imaging techniques (magnetic resonance imaging) involving nuclear magnetic resonance (NMR)

Detection and recording of X-ray images in medical contexts

Production of X-rays

In 1895 whilst experimenting with a discharge tube William Rontgen discovered X-rays. He had completely covered a discharge tube with black cardboard and was working in a darkened room. About a metre from the tube he noticed a weak light shimmering on a bench nearby. The source of this strange light was a hexa-cyanoplatinate coated screen and when the discharge tube was turned off the fluorescence from this screen disappeared. During the next seven weeks, Rontgen conducted a series of very extensive experiments to determine the nature of the rays emanating from the discharge tube and which gave rise to the fluorescence at the screen. He concluded that the rays originated from the point where the electrons in the discharge tube (at this time the electron had not been discovered and in his published paper Rontgen referred to cathode rays) struck the side of the tube or the anode. Furthermore, the rays travelled in straight lines from their point of production and were capable of great penetrating power, quite a thick sheet of aluminium being necessary to stop them entirely; they could pass through a 1000 page book without any noticeable decrease in intensity. Perhaps more striking was their ability to 'photograph' the bone structure of the hand and other parts of the body.

In 1912, seventeen years after their discovery, Von Laue demonstrated that X-rays are very short wavelength electromagnetic radiation ($\approx 10^{-10}$ to 10^{-11} m). So in fact X-rays are high-energy photons.

Figure 1533 is a schematic representation of a modern X-ray(Coolidge) tube.

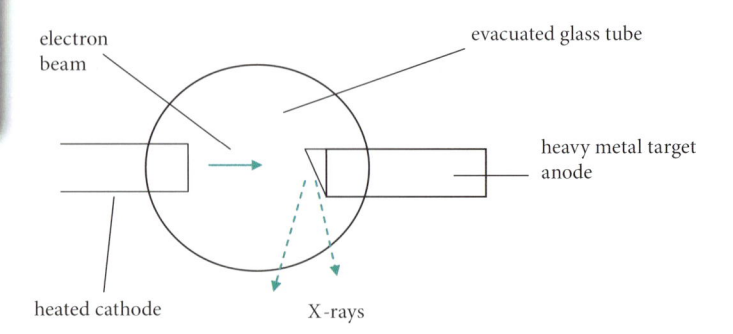

Figure 1533 *The principle of a modern X-ray tube*

Electrons are produced by the heated cathode. The potential difference between cathode and anode may range from about 10 kV to 50 kV. The anode, which is often oil cooled because of the large amount of thermal energy produced is faced with a heavy metal such as tungsten or molybdenum.

As a result of the electrons striking the metal anode, X-rays are ejected from the metal. The production of X-rays is sometimes referred to as the **inverse photoelectric effect**.

A typical X-ray spectrum and understanding its origin

A great deal of work has been done in measuring the wavelengths of X-rays produced when electron beams of various energy strike targets made from different elements. The results obtained using a molybdenum target are shown in Figure 1534 The electrons have been accelerated through 25 kV and also through 15 kV. There are several features of these curves which are immediately apparent. For both the spectra produced by electrons of 25 keV and 15 keV energies there is a minimum wavelength λ_{min} produced. The 25 keV curve also shows two distinct peaks called the K_α and the K_β lines.

Let us see if we can understand these curves.

IMAGING

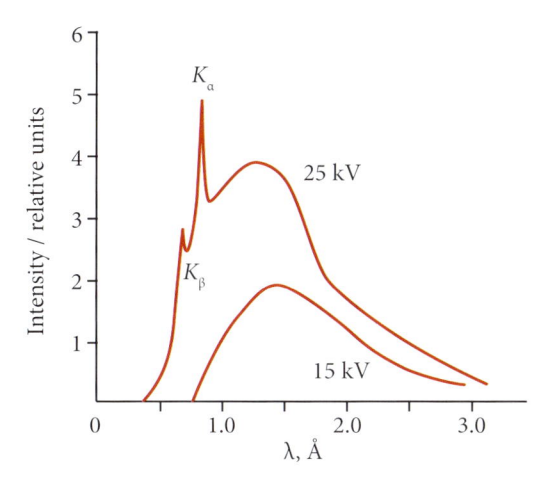

Figure 1534   A typical X-ray spectrum

When electrons strike a target most of them usually lose their energy gradually by making 'glancing' collisions with the atoms of the target material. The effect of this is to increase the average kinetic energy of the atoms and so increase the temperature of the target. It is for this reason the target in a high energy X-ray tube must be cooled, usually by a flow of oil. If it were not cooled the temperature rise could be sufficient to melt the target. About 99 per cent of the energy of the electron beam goes into heating the target. However, each glancing collision will result in the electron emitting radiation due to its acceleration. It is these glancing collisions that result in the continuous part of the spectrum. The continuous spectrum is sometimes referred to as 'bremsstrahlung', the German for 'braking radiation', a very descriptive name indeed. A few electrons will lose all their energy in one collision and this rapid acceleration of the electron results in an energetic pulse of electromagnetic radiation, i.e. a high energy photon. If an electron is accelerated from rest through a potential difference V then the maximum energy of the photon that is produced when it is brought to rest is Ve such that

$$Ve = hf = \frac{hc}{\lambda_{min}}$$

where λ_{min} is the minimum wavelength of the photon produced, hence

$$\lambda_{min} = \frac{hc}{Ve}$$

There is, however, another very important mechanism for the production of X-rays. If an electron has a sufficiently high energy it can ionize an atom of the target not by removing one of the electrons in an outer shell but by removing an electron from one of the inner electron energy levels. The ground state energy level is often referred to as the K-shell ($n = 1$) and the next energy level, the M-shell ($n = 2$) For example, suppose an incident electron removes a K-shell electron, the vacancy in this shell can now be filled by an electron of the L-shell, or the M-shell or other shells, making a transition to the K-shell. A transition from the L-shell to the K-shell gives rise to the K_α peak shown in Figure 1533.

An electron that makes a transition from the M-shell to a vacancy created in the K-shell gives rise to the K_β line. If an incident electron ionizes a target atom by ejecting an L-shell electron then electron transitions from the M-shell and N-shell to fill the vacancy give rise to X-ray lines called the L_α and L_β respectively. It is apparent that the wavelength of the lines in an X-ray spectrum will be characteristic of a particular element. In fact in 1913 *Moseley*, who measured the X-ray spectra produced by many different elements, showed that the frequency of a given line (say the K_β line) was related to the proton number Z. This was of vital importance since at this time Z was just regarded as a number that referred to the position of the element in the periodic table. *Moseley* was in fact able to show that the position of certain elements should be reversed. He was also able to fill in several gaps in the table by predicting the existence of new elements, such as Technetium and Promethium. Unfortunately, *Moseley* was killed in the ill-fated Dardanelles expedition of World War I in August 1915.

Introduction

There are a number of X-ray tubes used for medical purposes. The tubes can be classified as either diagnostic (medical imaging) or therapeutic (radiation therapy). We will concentrate in this section on diagnostic details because we are interested in medical imaging at the moment. Most diagnostic X-ray machines use a rotating anode X-ray tube as shown in Figure 1535.

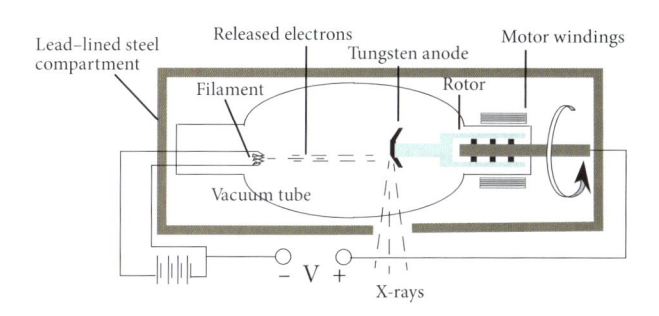

Figure 1535   Rotating-anode X-ray tube.

Electrons with a very high potential difference (typically around 15 000V in hospital machines) are accelerated between the cathode and the anode. A focusing cup usually made of molybdenum contains a tungsten filament. Electrons are released from the tungsten filament by thermionic emission. Because the large number of electrons released experience forces of repulsion, the electron stream tends to spread out. To prevent this spreading out so that a small area of the anode target material is bombarded, the cathode-focusing cup produces electrical forces that cause the electron stream to converge onto the anode target at a focal spot. Most of the energy of the electrons is converted to heat in this collision due to their sudden deceleration with less than 1% being converted to X-radiation. Because large amounts of heat are produced at the focal spot of the electron beam, the tungsten disc anode is made to rotate using an induction motor so that the heat loading on any particular point on the disc is reduced.

X-Ray attenuation coefficient and half-value thickness

The attenuation of an X-ray beam is the reduction in its intensity due to its passage through matter. When a beam of X-rays passes through a material such as the soft tissue of the body or bone, some of the X-rays will be absorbed.

There are four attenuation mechanisms where energy can be lost due to absorption in matter:

1. simple coherent scattering
2. the photoelectric effect
3. Compton scattering
4. pair production.

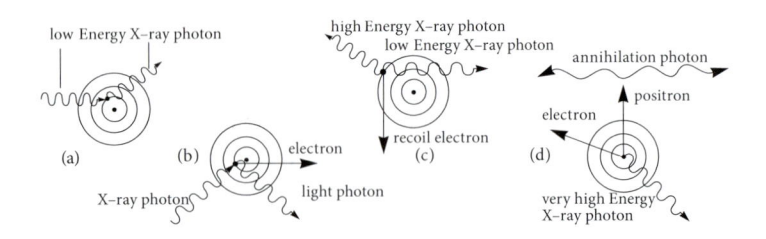

Figure 1536 The mechanisms of attenuation

Simple coherent scattering see Figure 1536(a) occurs when the energy of the incoming X-ray photon is smaller than the energy required to remove inner-shell electrons from an atom. When the incident X-ray photon interacts with an atom, it is scattered in a new direction without a loss of energy. It is the dominant mechanism in soft tissue in the 1-30 keV range.

In the photoelectric effect mechanism Figure 1536(b) the incoming X-ray photon has an energy greater than the energy required to remove inner-shell electrons, and photoelectrons and positive ions are produced. As other electrons in the atom fill the vacant spots of the ejected photoelectrons, characteristic lower-energy photon emission occurs. It is the dominant mechanism in soft tissue in the 1-100 keV range. The optimum photon energy for diagnostic radiography is around 30 000 kV where the photoelectric effect predominates because this gives the maximum contrast between body tissues and bones.

Compton scattering Figure 1536(c) occurs when the X-ray photon ejects outer-shell recoil electrons and the X-ray photon moves off in a different direction with a slightly lower energy. It is the dominant mechanism in soft tissue in the 0.5-5 MeV range. High energy X-ray photons can produce electron-positron pairs. It is the dominant mechanism in soft tissue above 5 MeV. The Compton scatter is used in therapeutic radiology where higher energies are preferred. When matter (such as an electron) collides with its corresponding antimatter (such as a positron), both particles are annihilated, and 2 gamma rays with the same energy but with a direction at 180 degrees to each other are produced. The direction of the gamma rays produced is in accordance with the law of conservation of momentum and the electron- positron annihilation gives energy equal to $E = mc^2$ (0.51 MeV each). This is depicted in Figure 1536(d).

The attenuation (reduction in intensity) of X-rays occurs in two ways:

1. the intensity of the X-ray beam may decrease with distance from the source (tungsten target) as they diverge or spread out in spherical wavefronts.

2. the intensity of the X-ray beam decreases as the X-ray photons are scattered or absorbed by a material.

The radiation emitted by an X-ray tube is heterogeneous because it is made up of photons with a range of energies. The filters already mentioned filter out the low energy photons to improve the X-ray quality. These photons would only be absorbed by the skin or surface tissue of a person being X-rayed, and one aim is to minimise any excess radiation dose because of the invasive characteristics of X-radiation.

A beam of homogeneous, monoenergetic X-rays contains photons of only one energy and thus only one wavelength. When a beam of monoenergetic X-rays of intensity I_0 passes through a medium with a thickness x, the attenuation or fractional reduction in intensity I is given by:

$$I = I_0 e^{-\mu x}$$

where μ = the constant of proportionality called the **linear attenuation coefficient**. Its value depends on the X-ray energy concerned and the nature of the absorbing material. It has units m^{-1}.

The intensity of the monoenergetic beam decreases exponentially with absorber thickness. The value of the attenuation coefficient increases as the X-ray energy decreases and higher absorption results. Figure 1537 shows a small thickness of lead absorbing X-rays.

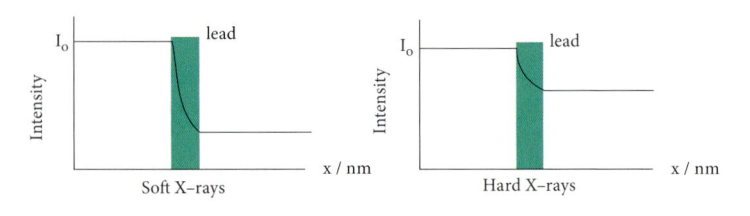

Figure 1537 Attenuation in a slab of lead.

The most penetrating radiation with short wavelengths (~ 0.01 nm) are termed hard X-rays. Very little absorption occurs when they pass through the lead slab. Long wavelength (~1 nm) X-rays are easily absorbed by the lead slab and these are called soft X-rays. They are less penetrating and more absorbing than hard X-rays.

The quality (penetrating power) of a monoenergetic beam of X-rays can be described in terms of the half-value thickness (HVT) in a given material.

The **half-value thickness** is the thickness of a material that reduces the intensity of a monoenergetic X-ray beam to half its original value.

Figure 1538 shows the exponential decay of attenuation and the corresponding half-value thickness of an absorbing material.

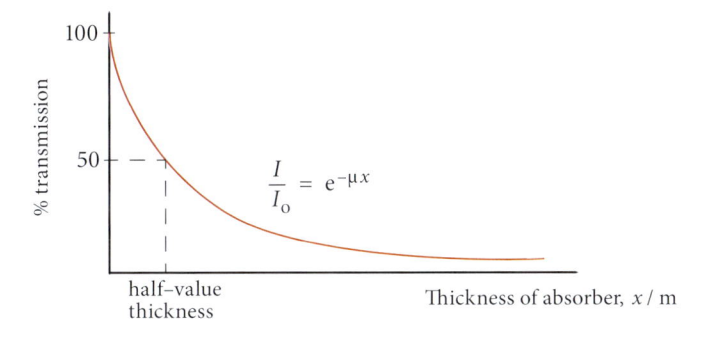

Figure 1538 The exponential decay of attenuation

Since the **half-value thickness** is the thickness of a material that reduces the intensity of a monoenergetic X-ray beam to half its original value, then in this instance:

If $x = x_{1/2}$ then $I = \frac{1}{2}I_0$.

Using $I = I_0 e^{-\mu x}$, we have:

$\frac{1}{2}I_0 = I_0 e^{-\mu x_{1/2}} \Leftrightarrow \frac{1}{2} = e^{-\mu x_{1/2}} \Leftrightarrow \ln(0.5) = -\mu x_{1/2}$

$\Leftrightarrow -\ln(0.5) = \mu x_{1/2}$

$\Leftrightarrow 0.6931 = \mu x_{1/2}$

That is,

$$x_{1/2} = \frac{0.6931}{\mu}$$

We can also determine a value for the linear attenuation coefficient for a monoenergetic beam by plotting a graph of ln I against thickness:

$I = I_0 e^{-\mu x}$. Take $\log_e$ of both sides.

$\ln I = \ln I_0 + \ln_e e^{-\mu x}$

$\ln I = -\mu x + \ln I_0$

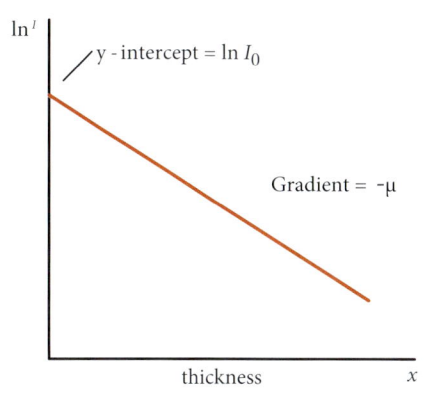

Figure 1539 The linear attenuation coefficient for a monoenergetic beam

From Figure 1539 we can see that the gradient of the straight line is equal to $-\mu$ and the y-intercept is equal to $\ln I_0$.

Example

The half-value thickness of a 30 keV X-ray photon in aluminium is 2.4 mm. If the initial intensity of the X-ray beam is 4.0×10^2 kW m^{-2}.

(a) What is the intensity after passing through 9.6 mm of aluminium?

(b) Calculate the linear attenuation coefficient of the aluminium.

(c) What is the intensity of the beam after passing through 1.5 mm of aluminium?

Solution

(a) *Intensity after passing through 2.4 mm would be half the initial intensity.*

Intensity after passing through 4.8 mm would be a quarter of the initial intensity.

Intensity after passing through 7.2 mm would be an eighth of the initial intensity.

Intensity after passing through 9.6 mm would be one-sixteenth of the initial intensity

New intensity $= \frac{1}{16}(4.0 \times 10^2 \text{ kWm}^{-2}) = 25 \text{ kWm}^{-2}$

Chapter 15 (Option C)

(b) $x_{\frac{1}{2}} = \frac{0.6931}{\mu} \Leftrightarrow \mu = \frac{0.6931}{x_{\frac{1}{2}}} = \frac{0.6931}{2.4} = 0.29$

Therefore, the linear attenuation coefficient is 0.29 mm^{-1} or 2.9×10^2 m^{-1}. Be careful of the units here because if the value is 0.29 per mm then it is 290 per m.

(c) $I = I_0 e^{-\mu x} = 4.0 \times 10^2 \times e^{-(290 \times 0.0015)} = 2.59 \times 10^2$

The intensity is 2.59×10^2 kW m^{-2}.

Example

(a) What is meant by the term attenuation when referring to X-rays?

(b) Name two mechanisms responsible for the attenuation of X-rays by matter.

(c) Name two ways in which the attenuation of X-rays occur.

(d) Define the term attenuation coefficient.

(e) State the two factors upon which the value of the attenuation constant depends.

(f) The transmission of X-rays by matter can depend upon the thickness of the material in the path of the X-rays.

(i) Sketch a graph of percentage transmission versus the thickness of the absorbing material.

(ii) By using your graph, explain the meaning of the term half-value thickness.

Solution

(a) The attenuation of an X-ray beam is the reduction in its intensity due to its passage through matter. (When a beam of X-rays passes through a material such as the soft tissue of the body or bone, some of the X-rays will be absorbed).

(b) There are four attenuation mechanisms where energy can be lost due to absorption in matter:

1. simple coherent scattering
2. the photoelectric effect
3. Compton scattering
4. pair production

(c) The attenuation (reduction in intensity) of X-rays occurs in two ways:

1. the intensity of the X-ray beam may decrease with distance from the source (tungsten target) as they diverge or spread out in spherical wavefronts.
2. the intensity of the X-ray beam decreases as the X-ray photons are scattered or absorbed by a material.

(d) A beam of homogenous, monoenergetic X-rays contains photons of only of one energy and thus only one wavelength. When a beam of monoenergetic X-rays of intensity I_0 passes through a medium with a thickness x, the attenuation or fractional reduction in intensity I is given by :

$$I = I_0 e^{-\mu x}$$

Where μ = the constant of proportionality called the linear attenuation coefficient. It has units m^{-1}.

(e) Its value depends on the energy of the X-ray photons and the nature of the absorbing material.

(f) (i) See the graph below which shows the exponential attenuation by an absorbing material.

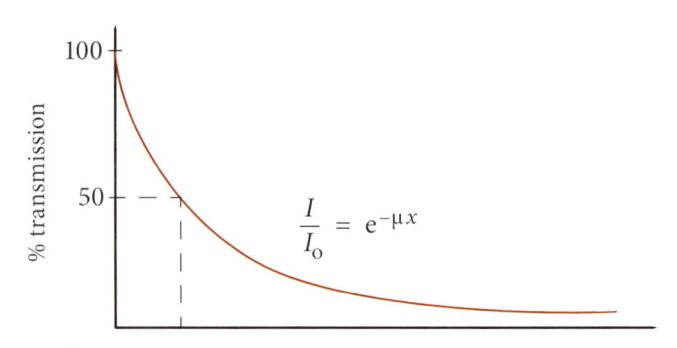

(ii) The quality (penetrating power) of a monoenergetic beam of X-rays can be described in terms of the half-value thickness HVT in a given material. The half-value thickness is the thickness of a material that reduces the intensity of a monoenergetic X-ray beam to half its original value.

IMAGING

X-Ray detection, recording and display techniques

Figure 1540 (a) shows the components of the process of a patient being X-rayed to produce a radiographic image on a photographic plate.

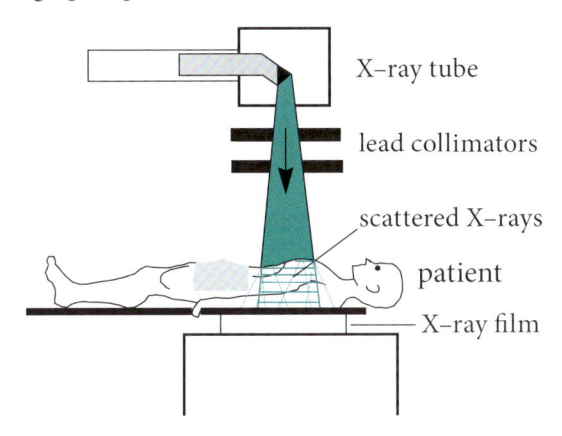

Figure 1540 (a)   Producing a radiographic image.

The **X-ray beam** passes through the glass wall of the X-ray tube, a layer of oil then a 3 mm thick aluminium plate to filter out low energy radiation. It is then collimated by lead plates. The aim is to produce a narrow beam because any random scatter increases the blur of the radiographic image. The amount of exposure time the patient experiences is strictly controlled. The X-rays enter the patient where they are either scattered or absorbed. In order to decrease blurring on the radiograph due to scattering, a lead grid system is inserted before the photogaphic film. Direct X-rays pass between the grid while the scattered X-rays are absorbed by the lead plates, see Figure 1540 (b). The direct X-rays then fall on an intensifying screen cassette containing double-sided film sandwiched between two fluorescent screens, see Figure 1540 (c).

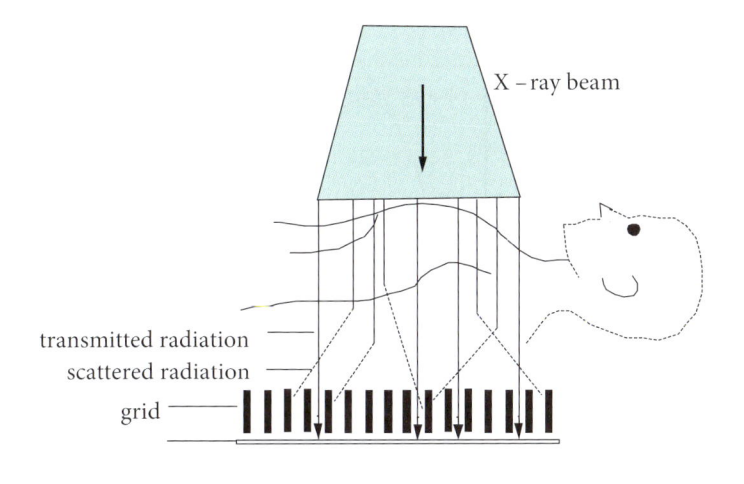

Figure 1540 (b)   Reducing scatter

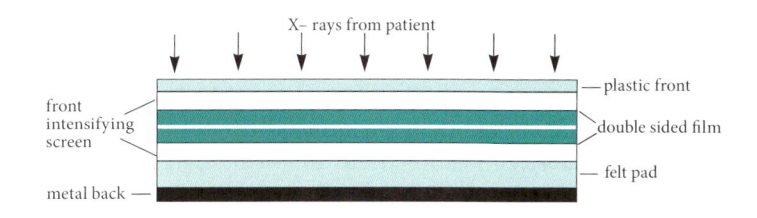

Figure 1540 (c)   Image intensifying screen

The standard X-ray machine used in radiography can produce images of some of the internal organs of the body and the bones. Air pockets, fat and soft tissues can be differentiated from each other because they can attenuate the X-ray beam in different ways. Bones produce a white image because they contain heavier body elements such as calcium and phosphorus in a dense matrix that attenuate the X-ray beam more than the softer tissues. Tissues that contain lighter elements such as hydrogen, carbon, nitrogen and oxygen produce a grey image on the radiograph. It is easy to distinguish a black lung image because it contains lower density air when compared with more dense water that is present in abundance in tissues.

Certain parts of the body are difficult to image against the background of other body parts. In order to improve the contrast of the image, solutions of heavy elements with a large attenuation co-efficient can be introduced into the body. These materials are known as **contrast-enhancing media**. Barium and bismuth can be introduced through the mouth or the rectum for the imaging of the alimentary canal or the appendix. It is common for people with stomach pain or possible gastro-intestinal ulcers to be asked to drink a "barium sulfate meal" before an X-ray is taken. An iodine solution can be introduced intravenously to enhance the image of the cardiovascular system, the kidney and the brain. However, the contrast of the image produced in soft tissue anatomy is not very clear in many situations. Over the past 20 years, this clarity has been greatly improved by using X-rays and electronic detection and display together in computed tomography imaging.

X-Ray quality and imaging techniques

As the relative intensity of the X-rays is increased so too does the electromagnetic spectral range. We say the X-ray quality has increased. The quality of an X-ray beam is a term used to describe its penetrating power.

There are a number of ways that the quality of an X-ray machine can be increased.

455

Chapter 15 (Option C)

1. Increasing the **tube voltage**.
2. Increasing the **tube current**.
3. Using a **target material** with a relatively high atomic number Z.
4. Using **filters**.

Tube voltage

When the accelerating voltage between the cathode and anode of an X-ray tube is increased, the frequency of the X-radiation increases. Therefore, the radiation has more energy and the penetration increases. As the intensity per unit area increases due to the higher potential, so too does the spectral spread as shown in Figure 1541.

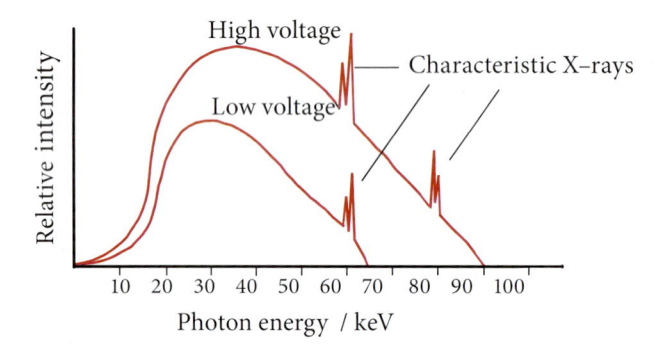

Figure 1541 Effect of tube voltage on X-ray quality.

The following effects can be observed:

(a) E_{max} increases.

(b) λ_{min} decreases.

(c) the peak of the continuous spectrum moves towards higher energies.

(d) the total intensity given by the area under the curve increases, and is $\propto V^2$.

(e) more characteristic line spectrum may appear.

Tube current

Increasing the tube current will increase the rate of thermionic emission from the cathode. Because there are more electrons available to produce X-ray photons, the overall intensity increases. Figure 1542 demonstrates the effects observed when the tube current is increased.

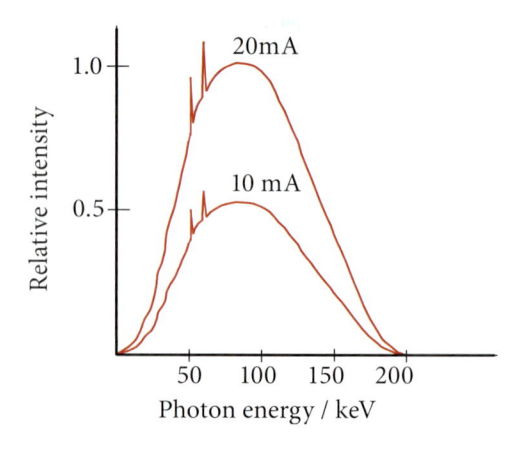

Figure 1542 Effect of tube current on X-ray quality.

The following effects can be observed:

(a) the spectral shape remains the same.

(b) E_{max} remains the same as the voltage is constant.

(c) the total intensity (given by the area under the spectrum) increases as the area under the curve is proportional to I.

Target material

The target material must have a high melting point so that it will not melt with the large heat generated by the accelerated electrons bombarding it. Furthermore, the target material must have a relatively high atomic number so that the mass, size and number of protons in the atoms ensure a greater probability that the bombarding electrons make the necessary collisions to produce X-rays. Common target materials include tungsten (Z = 74) and platinum (Z = 78). Tungsten is more widely used because of its high melting point (3370 °C). See Figure 1543.

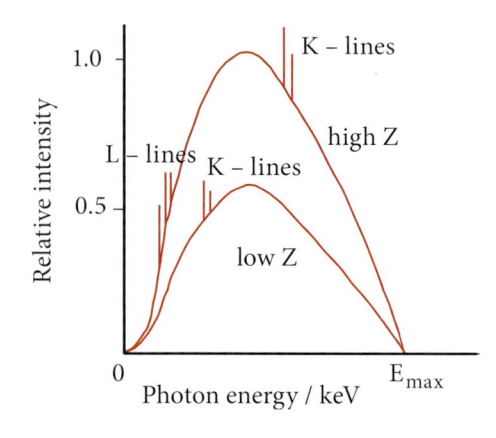

Figure 1543 Effect of target material on X-ray quality

Figure 1542 shows the effects observed when the atomic number is increased.

We then note that:

(a) E_{max} remains constant.

(b) the characteristic line spectra are shifted to higher photon energies.

(c) the X-ray intensity increases as the area under the curve is directly proportional to Z.

Filters

A thin sheet of material is placed in the path of the X-ray beam, and selectively absorbs more lower-energy photons than high-energy photons. The effect of selective filtration is shown in Figure 1544.

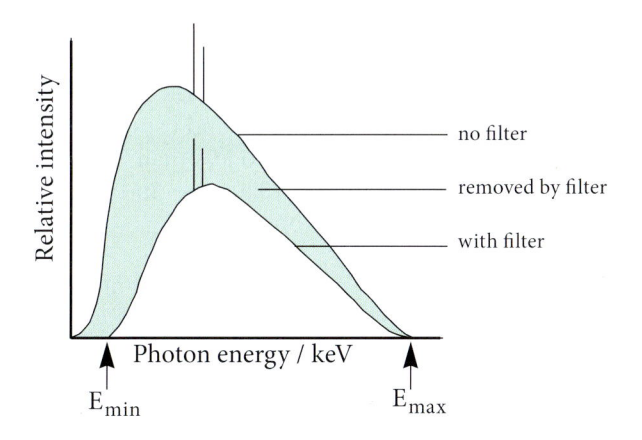

Figure 1544   Effect of filtration on X-ray quality

The following effects can be observed:

(a) E_{max} does not change.

(b) There is a shift in E_{min} towards higher energies.

(c) There is a reduction in X-ray output.

Although the intensity is reduced, the beam is more penetrating because of the removal of lower energy photons. The X-rays are said to be harder.

Example

An X-ray tube has a beam current of 35 mA and it is operated at a voltage of 30 kV.

(a) At what rate does the machine transform energy?

(b) How many electrons reach the target each second?

(c) What is the maximum energy of the X-rays produced? (Assume no thermal energy loss).

(d) What is the minimum wavelength of the X-rays produced?

Solution

(a) $P = VI = (3 \times 10^4 \text{ V}) (3.5 \times 10^{-2} \text{ A}) = 1.05 \text{ kW}$

(b) $q = It = (3.5 \times 10^{-2} \text{ A}) (1 \text{ s}) = 3.5 \times 10^{-2} \text{ C}$

1 C is the charge on 6.25×10^{18} electrons. Thus the number of electrons reaching the target
= $(6.25 \times 10^{18} \text{ e C}^{-1}) \cdot (3.5 \times 10^{-2} \text{ C})$

= 2.2×10^{17} electrons per second

(c) $E = eV = (1.6 \times 10^{-19} \text{ C}) \times (30\,000 \text{ V}) = 4.8 \times 10^{-15} \text{ J}$

(d) $E = \dfrac{hc}{\lambda} \Leftrightarrow \lambda = \dfrac{hc}{E}$

= 6.63×10^{-34} Js $\times 3 \times 10^8$ ms^{-1} / 4.8×10^{-15} J

= 4.143×10^{-11} m

= 0.0414 nm

Computed tomography

Standard X-ray imaging techniques record a longitudinal image on a photographic plate producing 30 shades of grey. Computed tomography (CT) imaging, also called computed axial tomography (CAT) imaging, uses X-rays, scintillation detectors and computer technology to build up an axial scan of a section of an organ or part of the body with 256 grey shades. The tube voltages are about 130 kV, and the exposure times are greater than that of standard radiography. (Tomography means the study of body section radiography).

A patient lies on a table that passes through a circular scanning machine about 60-70 cm in diameter called a gantry. The gantry can be tilted, and the table can be moved in the horizontal and vertical directions. X-rays from the gantry are fired at the organ being scanned and attenuation occurs dependent on the type of tissue

Chapter 15 (Option C)

being investigated. The image produced on the computer monitor is a series of sections or slices of an organ built up to create a three-dimensional image. A schematic diagram of one section is shown in Figure 1545.

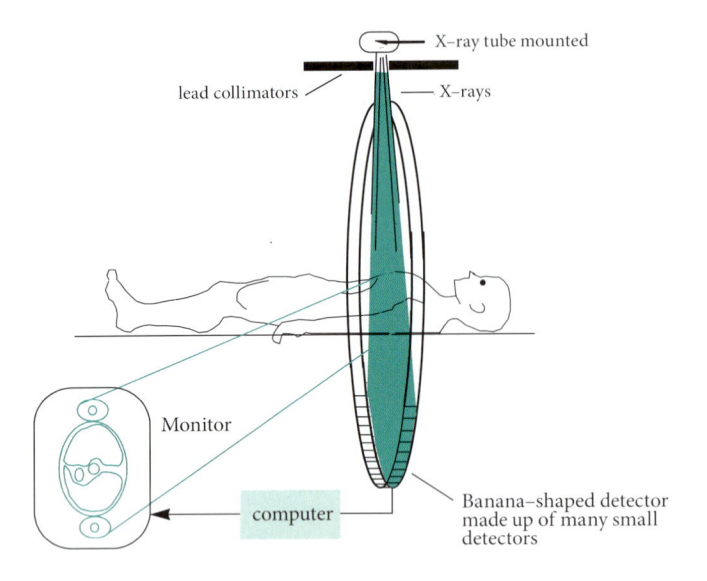

Figure 1545   Section using a CT scanner

A fan beam of around 100 X-ray pulses is produced as the X-ray tube and the photomultiplier detectors around the patient make a 360⁰ rotation. A cross-section or slice of an organ from 1 mm to 10 mm in thickness is obtained with each rotation. The lead collimators control the slice thickness. About 1000 profiles or pictures are obtained in each rotation. A series of slices can be made to produce a 3-dimensional picture of an entire organ. The time required for the complete scan of an entire organ is normally from 3-5 seconds. However, short scanning times of 500 ms can be used when the anatomical region being investigated is affected by the patient's motion and breathing.

The detectors send the information to a series of computers and a host computer oversees the entire operation. The plane of the tomographic image is divided into small pixel areas of about 1 mm^2, each of which can be given a grey shade value from 1 (black) to 256 (white). The thickness of each slice is simultaneously built into a volume pixel called a voxel. The image is produced on a computer monitor, and this image can be manipulated and re-constructed to get rid of interference by subtracting the background. The required well-contrasted image of the organ being investigated is then obtained.

CAT scans provide detailed cross-sectional images for nearly every part of the body including the brain and vessels, the heart and vessels, the spine, abdominal organs such as the liver and kidneys. They are being used in many diagnostic applications including the detection of cancerous tumours and blood clots.

Generation and detection of ultrasound in medical contexts

As already mentioned, ultrasound is sound with frequencies greater than 20 000 Hz.

Just as transverse electromagnetic waves interact with matter as is the case with X-radiation, CAT and MRI, so too ultrasound mechanical waves interact with matter. Three properties they possess are: they can be reflected, refracted and absorbed by a medium.

SONAR (sound navigation and ranging) was developed during World War 1. It is basically the use of sound waves to detect and estimate the range of submerged objects. In the 1930s it had its applications in medical therapy. In the 1940s diagnostic ultrasound developed in parallel with SONAR.

Ultrasound from 20 000Hz to several billion hertz can be produced by ultrasound transducers (a device that converts energy from one form to another) using mechanical, electromagnetic and thermal energy. Normal sound waves are not useful for imaging because their resolution is poor at long wavelengths. Medical ultrasound uses frequencies in the range greater than 1 MHz to less than 20 MHz. In this range with speeds around 1500 m s^{-1} in body tissue the wavelengths are about 1^{-2} mm.

The common transducer used in ultrasound is the piezoelectric crystal transducer. In 1880, Pierre and Jacques Curie observed that when a quartz crystal is subjected to mechanical deformation, a tiny electric potential difference is produced between the faces of the crystal, and conversely the application of an electric potential would deform the crystal and make it vibrate. This is known as the piezoelectric effect. Today, the piezoelectric transducers used in ultrasound have certain ceramic materials in place of crystals. These transducers operate over the entire range of ultrasound frequencies.

When ultrasound meets an interface between two media, the ultrasound wave can undergo reflection, transmission, absorption and scattering. This is similar to when light from air enters glass. The same laws of reflection and refraction occur. For example, with refraction (transmission) the frequency of the ultrasound source will remain constant but a change in the wave's velocity as it crosses the boundary will change the wavelength.

In a typical ultrasound scan, a piezoelectric transducer is placed in close contact with the skin. To minimise the acoustic energy lost due to air being trapped between the transducer and the skin, a gel is applied between the transducer and the skin. The pulse produced by the transducer reflects off various tissue interfaces. The same transducer as a reflected wave or echo again detects the pulse. The electronic representation of the data generated

IMAGING

from the repetition of this process is displayed on an oscilloscope as an ultrasonic image. Thus the distance, size and location of hard and soft tissue structures can be determined.

Acoustic impedance

In ultrasound imaging, it is the reflected portion of the ultrasound beam that is used to produce the image. The greater the difference in the characteristics of the media boundary, the more energy will be reflected to give an echo. The major characteristic is called the **acoustic impedance** of a medium. It is a measure of how easy it is to transmit sound through that medium and it has been found that acoustic impedance depends on both the speed of the sound wave (v) and the density of the medium being considered (r) in the following way:

$Z = rv$

The unit of acoustic impedance is the rayl. This is equivalent to the SI unit kg m^{-2} s^{-1}. Figure 1546 gives the speed, density and acoustic impedance for some biological materials.

Medium	Velocity ms^{-1}	Density kg m^{-3}	Acoustic impedance kg m^{-2}s^{-1} × 10^6
Air (20 °C, 101.3 kPa)	344	1.21	0.0004
water (20 °C)	1482	998	1.48
whole blood (37 °C)	1570	1060	1.66
Brain	1541	1025	1.60
Liver	1549	1065	1.65
Kidney	1561	1038	1.62
Skull bone	4080	1912	7.80
Muscle	1580	1075	1.70
Soft tissue (37 °C)	1540	1060	1.66
Lens of eye	1620	1136	1.84

Figure 1546 The approximate speed of sound in some biological materials

The greater the difference in acoustic impedance between two materials, the greater will be the proportion of the pulse reflected. If I_0 is the initial intensity and I_r is the reflected intensity for normal incidence, it can be shown that:

$$\frac{I_r}{I_0} = \frac{(Z_2 - Z_1)^2}{(Z_2 + Z_1)^2}$$

where Z_1 is the acoustic impedance of material 1 and Z_2 is the acoustic impedance of material 2.

Example

(a) The speed of an ultrasound in blood is 1580 ms^{-1} and the density of the blood is 1060 kgm^{-3}. Calculate the acoustic impedance of blood.

(b) Calculate the thickness of a slice of muscle tissue if its fundamental resonant frequency is 1.5 MHz.

(c) The time delay for a pulse going through fat is 0.133 ms and the speed of ultrasound in the fat is 1450 ms^{-1}. Determine the depth of the fat.

Solution

(a) $Z = rv = 1570$ ms^{-1} × 1060 kgm^{-3} = 1.66 × 10^6 kg m^{-2} s^{-1}.

(b) The wavelength of the resonant frequency
= v / f = 1580 ms^{-1} / 1.5 × 10^6 Hz = 1.05 × 10^{-3} m.

Now the ultrasound has to go down and back. Therefore, the actual wavelength will be half of this = 0.53 mm.

(c) The time taken to reach the boundary of the fat
= ½ × 1.33 × 10^{-4} s = 6.65 × 10^{-5} s

Distance = speed × time = 1450 ms^{-1} × 6.65 × 10^{-5} s = 9.6 cm

A-Scan and B-scan imaging

A-scan

A scan produced by a single transducer when a single bit of information with a one-dimensional base is displayed is called an A-scan (amplitude-modulated mode). The transducer scans along the body and the resulting echoes are plotted as a function of time as shown in Figure 1547. The A-mode measures the time lapsed between when the pulse is sent and the time the echo is received. The first echo is from the skin, the second and third pulses are from either side of the first organ, the fourth and fifth echo are from either side of the second organ. The pulse intensity decreases due to attenuation.

Chapter 15 (Option C)

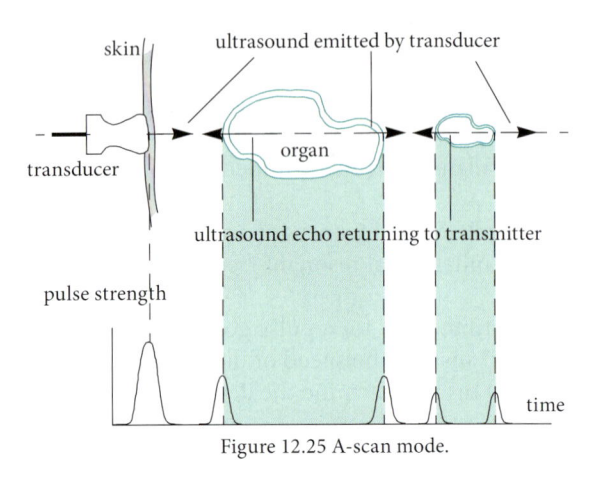

Figure 12.25 A-scan mode.

Figure 1547 A-scan mode

This mode is seldom used, but when it is, it measures the size and distance to internal organs and other organs such as the eye.

B-Scan imaging

In the B-scan mode (brightness-modulated scan), an array of transducers scan a slice in the body. Each echo is represented by a spot of a different shade of grey on an oscilloscope. The brightness of the spot is proportional to the amplitude of the echo as shown in Figure 1548.

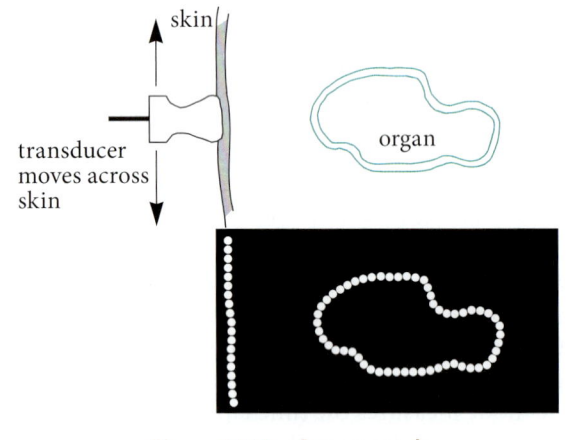

Figure 1548 B-scan mode

The scan head containing many transducers is arrayed so that the individual B-scans can be built up to produce a two-dimensional image. The scan head is rocked back and forth mechanically to increase the probability that the pulse will strike irregular interfaces.

Choice of diagnostic frequency

Resolution (the fine detail that can be detected) is an important factor in ultrasound imaging. A patient may have a small tumour, say in the liver, and the doctor wants to see the photo of the fine detail. The smaller the wavelength, the greater the resolution. That is, higher frequency ultrasound gives more fine detail. If the ultrasound beam is reflected, transmitted, absorbed and scattered, its intensity will decrease. If the frequency of the source is increased too much, the **attenuation** (the reduction in intensity) in fact increases as does the penetration depth. Furthermore, the resolution also decreases if the frequency is increased beyond an optimum point. In a typical pulse-echo diagnostic procedure, the maximum mean ultrasound power delivered is about 10^{-4} W, and the frequency is in the range 1-5 MHz.

Medical imaging techniques (magnetic resonance imaging) involving nuclear magnetic resonance (NMR)

Basic principles of nuclear magnetic resonance (NMR) imaging

The phenomenon known as nuclear magnetic resonance is the basis of the diagnostic tool known as magnetic resonance imaging - MRI. It is a technique used for imaging blood flow and soft tissue in the body. It is the preferred diagnostic imaging technique for studying the brain and the central nervous system.

Rather than using X-rays as the source of radiation, it uses radiation in the radio region of the electromagnetic spectrum and magnetic energy to create cross-sectional slices of the body.

The patient is laid on a table and moved into a chamber containing magnets that can produce a uniform strong magnetic field around 2 T as shown in Figure 1549. Pulses of non-uniform radio-frequency (RF) electromagnetic waves bombard the patient. At particular RF frequencies, the atoms in the tissues absorb and emit energy. This information is sent to a computer that decodes the information and produces a two-dimensional or three-dimensional image on a computer monitor screen.

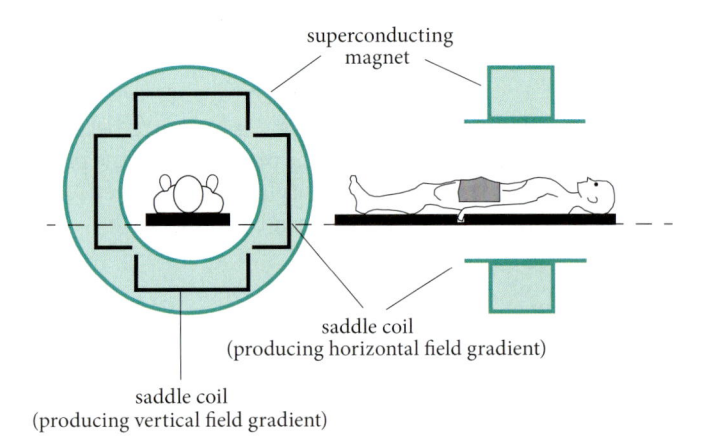

Figure 1549 MRI magnetic fields

IMAGING

Although the principles of nuclear magnetic resonance are beyond the scope of this course, the basic principles of this phenomenon will be outlined in the next few paragraphs.

Recall that when a current is passed through a coiled wire (solenoid), the magnetic field produced is similar to that produced by a simple bar magnet. At the microscopic level, it is known that a charged particle such as a proton or an electron acts like a tiny current loop. As a result, the nuclei of certain atoms and molecules also behave like small magnets due to the rotation or spin of their nuclear protons or neutrons. Spin is in two directions and when nuclei have equal numbers of protons and neutrons, the spin is equal in both directions and there is no net spin. However, if there are different numbers of protons and neutrons, the spins do not cancel and there is a net spin. This happens with hydrogen nuclei.

If hydrogen nuclei are placed in a strong external magnetic field, they will tend to align their rotation axes with the external field direction. However, the laws of quantum mechanics allow certain alignment angles and as a result, the nuclear magnets cannot come into perfect alignment with the external field. Some will align with the magnetic field and others align themselves in the opposite direction to the magnetic field. In fact, they precess like small magnetic tops wobbling at fixed angles around the magnetic field direction.

Now when a weak oscillating magnetic field in the form of pulses of radio waves are superimposed on the strong magnetic field, the oscillating field rotates at right angles to the strong field. If the radio frequency is not a certain frequency, known as the Larmor frequency, the axis of the rotating particle will wobble as described previously. If the applied frequency is equal to the Larmor frequency of precession, the charged particles resonate and absorb energy from the varying radio wave magnetic field. The magnetisation of the material is changed and this is detected by a radio-frequency signal emitted from the sample.

The strength and duration of the radio signals absorbed and emitted are dependent on the properties of the tissue being examined. The proton in the hydrogen atom has a strong resonance signal and its concentration is abundant in body fluids due to the presence of water. Bone shows no MRI signal. MRI is ideal for detecting brain and pituitary tumours, infections in the brain, spine and joints and in diagnosing strokes.

The advantages and disadvantages of diagnostic techniques are summarised with the other imaging techniques in Figure The advantages and disadvantages of diagnostic techniques are summarised with the other imaging techniques in Figure 1550.

	Advantages	Disadvantages
Ultrasound	• Relatively cheap to use. • Abundant ultrasound machines. • No ionising radiation. • Non-invasive. • Good for soft-tissue diagnosis. • Can break down gallstones and kidney stones. • Good for measuring bone density.	• Highly reflective boundaries between bone/tissue and air/tissue prevent effective imaging. • High frequency ultrasound has low penetrative ability. • A limit to the size of objects that can be detected.
X-rays	• Simple to use. • Cheapest alternative. • Abundant X-ray machines. • Good for certain structures.	• Poor at body-function diagnosis. • Not good for differentiating one structure from another. • Resolution not as good as others. • Sometimes enhancing materials need to be ingested. • Radiation dangerous to health.
Computed Tomography	• Good for 3D images showing structure. • Creates cross-sections. • Resolution better than basic X-ray radiography. • Good for tumours and other lesions. • Good for stroke detection.	• Not good for organ function. • Radiation not good for the health. • More expensive than standard X-ray.
Magnetic Resonance Imaging	• No ionising radiation. • Clearest images of the brain. • Best image of the central nervous system.	• Most expensive. • Scan time up to 40 minutes. • Cannot be used with heart pacemakers and metal prostheses.

Figure 1550 Advantages and disadvantages of imaging techniques

CHAPTER 15 (OPTION C)

Exercise 15.5

1. Calculate the minimum wavelength of X-ray photons produced when electrons that have been accelerated through a potential difference of 25 kV strike a heavy metal target.

2. The ground state energy level of a fictitious element is 20 keV and that of the next state is 2.0 keV. Calculate the wavelength of the K_α line associated with this element.

3. Describe the function of the following parts of an X-ray tube:

   (a) The filament
   (b) The potential difference across the electrodes
   (c) The tungsten target

4. Calculate the wavelength and frequency of an X-ray with an energy of 30 000 V.

5. Explain why the target in an X-ray tube is mounted on a disc that rotates at 3600 revolutions per minute.

6. Explain why you would not use hard X-rays for imaging tissues.

7. An X-ray tube has a beam current of 40 mA and it is operated at a voltage of 100 kV.

   (a) Calculate the rate at which the machine transforms energy.
   (b) Determine how many electrons reach the target each second.
   (c) Calculate the maximum energy of the X-rays produced. (Assume no thermal energy loss).
   (d) Calculate the minimum wavelength of the X-rays produced.

8. The half-value thickness of 30 keV X-ray photons in aluminium is 4.8 mm. The initial intensity of the X-ray beam is 2.59×10^2 kW m^{-2}.

   (a) Determine the intensity of the beam after passing through 9.6 mm of aluminium.
   (b) Calculate the linear attenuation coefficient of the aluminium.
   (c) Determine the intensity of the beam after passing through 1.2 mm of aluminium.

9. Describe what is meant by the term X-ray quality, and name two ways in which the quality can be increased.

10. The half-value thickness of 30 keV X-ray photons in aluminium is 2.4 mm. If the initial intensity of the X-ray beam is 4.0×10^2 kW m^{-2}

    (a) Determine the intensity of the beam after passing through 9.6 mm of aluminium
    (b) Calculate the linear attenuation coefficient of the aluminium.
    (c) Determine the intensity of the beam after passing through 1.5 mm of aluminium.

11. CAT and MRI scanners produce tomographic images of parts of the body in diagnostic tests.

    (a) Describe what is meant by the term *tomography*.
    (b) Outline the method by which CAT scans are collected.
    (c) Give two diagnostic applications that CAT scans are used for.
    (d) Discuss the advantages that a CAT scan has when compared to conventional X-ray techniques.

12. MRI is proving to be an extremely useful technique for imaging blood flow and soft tissue in the body. It is the preferred diagnostic imaging technique for studying the brain and the central nervous system.

    (a) Describe the basic principles employed to collect an MRI scan of body tissues
    (b) State the property of the hydrogen atom makes it such a useful atom for MRI diagnosis?
    (c) Give two diagnostic applications that MRI scans are used for.
    (d) Discuss the advantages and disadvantages that a MRI scan has when compared to other diagnostic techniques.

13. Ultrasound is a useful device in medical diagnosis and imaging.

    (a) Describe how ultrasound is different to other types of radiation used in medical diagnosis.
    (b) Explain the SONAR principle used in medical ultrasonic diagnosis.
    (c) Below is data relating to ultrasound transmission in various media.

Medium	Velocity ms⁻¹	Density kg m⁻³	Acoustic Impedance kg m⁻² s⁻¹ × 10⁶
Air (20 °C, 101.3 kPa)	344	1.21	0.0004
Water (20 °C)	1482	998	1.48
Whole blood (37 °C)	1570	1060	1.66
Brain	1541	1025	1.60
Liver	1549	1065	
Kidney	1561	1038	1.62
Skull bone	4080	1912	
Muscle	1580	1075	1.70

(i) Calculate the acoustic impedance for the liver and the skull bone
(ii) Predict whether ultrasound could be used to obtain images of the lung. Explain your prediction.

(d) What is the function of the gel used in ultrasound?
(e) Identify the factors that affect the choice of the diagnostic frequency used in ultrasound.
(f) Distinguish between A-scans and B-scans used in ultrasound diagnosis.
(g) Discuss some of the advantages and disadvantages of ultrasound in medical diagnosis.

14. State and explain which imaging technique is normally used to:

(i) examine the growth of a foetus.
(ii) detect a broken bone.
(iii) detect a tumour in the brain.

15. The following figure shows the variation in intensity I of a parallel beam of X-rays after it has been transmitted through a thickness x of lead.

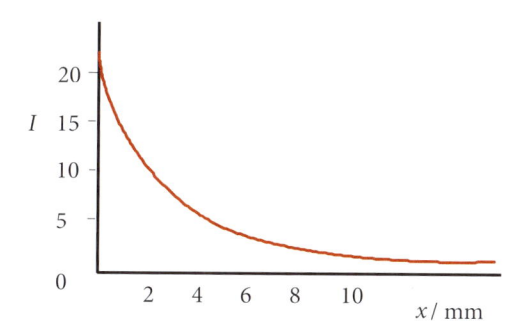

The linear attenuation coefficient was 8 m⁻¹ m for a 1.2 MeV radiation incident on a tissue.

(a) Calculate the thickness of the tissue that is required to reduce the intensity of the radiation by half.

(b) Define half-value thickness.
(c) Estimate the half-value thickness for this beam in lead from the graph.
(d) Determine the thickness of lead that is required to reduce the intensity by 40% of the initial value.
(e) Another sample of lead has a half-value thickness of 4 mm. Determine the thickness of this lead that would reduce the radiation intensity by 80%.

16. (a) State a typical value for the frequency that is used in ultrasound imaging.
(b) The figure below shows an ultrasound transmitter / receiver placed in contact with the skin.

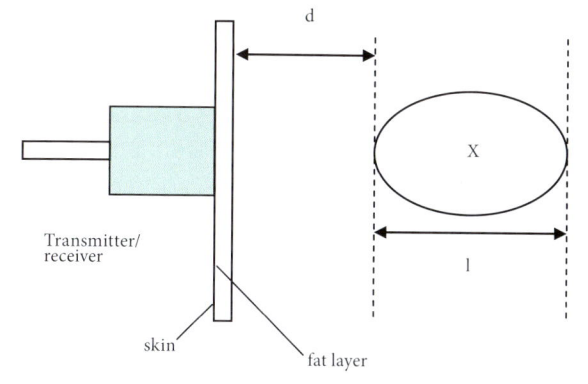

There is a layer of fat and an organ X at distance d from the fat layer. The organs length is l.

On the following graph, the pulse strength of the reflected pulses is plotted against time t where t is the time elapsed between the pulse being transmitted and the time that the pulse is received.

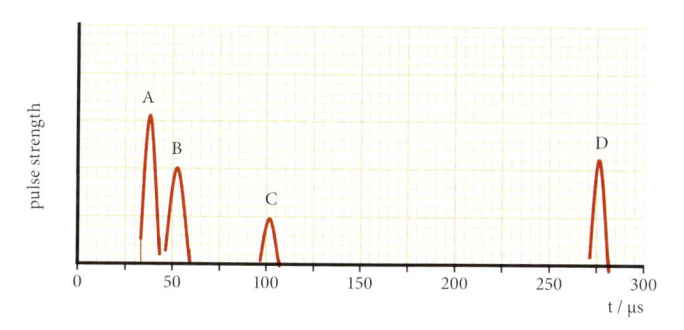

(i) Indicate on the Figure the origin of the reflected pulses shown on the graph.
(ii) The mean speed in tissue and muscle of the ultrasound used in this scan is 2.0×10^3 m s⁻¹. Using data from the graph above, estimate the depth d of the organ beneath the skin and the length l of the organ O.

(c) Is the scan above an A-scan or a B-scan? Explain the difference between these types of scan.

Chapter 15 (Option C)

This page is intentionally left blank

16. Option D: Astrophysics

Contents

D.1 – Stellar quantities
D.2 – Stellar characteristics and stellar evolution
D.3 – Cosmology
D.4 – Stellar processes
D.5 – Further cosmology

Essential Ideas

One of the most difficult problems in astronomy is coming to terms with the vast distances in between stars and galaxies and devising accurate methods for measuring them.

A simple diagram that plots the luminosity versus the surface temperature of stars reveals unusually detailed patterns that help understand the inner workings of stars. Stars follow well-defined patterns from the moment they are created out of collapsing interstellar gas, to their lives on the main sequence and to their eventual death.

The hot big bang model is a theory that describes the origin and expansion of the universe and is supported by extensive experimental evidence.

The laws of nuclear physics applied to nuclear fusion processes inside stars determine the production of all elements up to iron.

The modern field of cosmology uses advanced experimental and observational techniques to collect data with an unprecedented degree of precision and as a result very surprising and detailed conclusions about the structure of the universe have been reached. © IBO 2014

Chapter 16 (Option D)

TOK — THE SCIENCE OF ASTROPHYSICS

'Twinkle, twinkle little star, How I wonder what you are'.

Our planet, Earth, is an insignificant object orbiting an insignificant star, the Sun. The Sun is situated in one arm of an insignificant galaxy, the Milky Way, which contains around 200 billion stars. The Milky Way measures about 105 light years from one end to the other yet this enormous distance is tiny compared to the whole Universe which is about 1010 light years in "diameter". A light year is the distance that light travels in a year. There are about 3.2×10^7 seconds in a year.

The Milky Way is one of about 25 galaxies that make up a so-called "local cluster". Some 50 million light-years from our local cluster is another cluster of galaxies, the Virgo cluster, which contains about a thousand galaxies. There are other clusters that can contain as many as ten thousand galaxies. Amazingly, all these different clusters are grouped into a so-called "super cluster". Between these super clusters are vast voids of empty space. However, interstellar and intergalactic space is not completely empty. It actually contains gas and microscopic dust particles although the density is not very great. The density of interstellar space is estimated to be about 10^{-20} kg m^{-3} and that of inter-galactic space 10^{-25} kg m^{-3}.

Astrophysics is the science that tries to make sense of the Universe by providing a description of the Universe (Astronomy) and by trying to understand its structure and origin (Cosmology). It is a daunting subject for not only does it encompass the whole historical grandeur of physics but it also embraces all of physics as we understand it today, the microscopic and the macroscopic. We cannot understand the structure of stars, their birth and their death unless we understand the very nature of matter itself and the laws that govern its behaviour. It even takes us beyond the realm of physics for, as did our earliest ancestors, we still look up at the stars and ask " who am I and what's it all about?."

This Option can but scrape the surface of this truly vast topic.

D.1 Stellar quantities

NATURE OF SCIENCE:

One of the most difficult problems in astronomy is coming to terms with the vast distances in between stars and galaxies and devising accurate methods for measuring them. (1.1)

IBO 2014

Essential idea: One of the most difficult problems in astronomy is coming to terms with the vast distances in between stars and galaxies and devising accurate methods for measuring them.

Understandings:
- Objects in the universe
- The nature of stars
- Astronomical distances
- Stellar parallax and its limitations
- Luminosity and apparent brightness

Objects in the universe

This section gives a brief summary of what we know to date about the various objects that make up the universe.

Galaxies

As mentioned, **galaxies** are vast collections of stars. There are essentially three types of galaxy and these are discussed in more detail in the AHL secction.

Quasars

Quasars were first discovered in 1960 and their exact nature still remains a mystery. They are extremely bright objects having a luminosity equivalent to that of a 1000 galaxies. They are also very distant (Quasar 3C273 is some 3 billion light years away) and they are also much smaller than any known galaxy.

Nebulae

Nebulae was the name originally given to "misty" type patterns in the night sky. Many such patterns are now recognised as being galaxies. Others are recognised as being the "debris" of a supernova such as the famous Crab Nebula that was first recorded by the ancient Chinese astronomers. Other so-called dark nebulae such as the Horsehead nebula contain a large amount of gas and dust particles and are considered to be the "birth" places of the stars.

ASTROPHYSICS

To conclude this introductory section, Figure 1601 shows the distances as orders of magnitudes from Earth of various astronomical objects.

Object	Distance from Earth/m
Quasar	10^{25}
Nearest galaxy (Andromeda)	10^{22}
Centre of the Milky Way	10^{20}
North Star (Polaris)	10^{19}
Nearest star (Alpha Centauri)	10^{17}
Sun	10^{11}
Moon	10^{8}

Figure 1601   The distance of various astronomical objects

Structure of the solar system

The planets of our solar system orbit the Sun in ellipses with the Sun at one of the foci of the ellipse. Some planets, like our own Earth, have a moon or moons, which orbit the planet. As well as the nine planets many smaller lumps of matter orbit the Sun. Between the orbits of Mars and Jupiter are many millions of lumps of rocks called asteroids. The largest of these Ceres, has a diameter of about 900 km and the smallest are no bigger than about one metre in diameter. This region containing the asteroids is referred to as the asteroid belt.

Another group of objects that orbit the Sun are the comets. Comets are basically lumps of ice and dust only a few kilometres in diameter and their orbits about the Sun are highly elliptical. Because of their small size they are difficult to detect. However, as they pass near the Sun the heat from the Sun starts to vaporise the ice of the comet liberating dust and gases. The liberated gases begin to glow producing a ball of light, the coma, that can be up to 10^6 km in diameter. The coma has a long luminous tail which can be 10^8 km in length. Comets are truly a wonderful and awe-inspiring sight and were often in days gone by considered to be portents of disaster. Perhaps the most famous of the Comets is the Halley comet and was last visible to the naked eye in 1986 and will make another appearance in 2061 which is considerably sooner than a comet called Hyakutake that has an orbital period of about 30,000 years. Figure 1602 summarises some of the details of the planets.

Name of planet	Number of moons	Average distance from the Sun / 10^6 km	Equatorial diameter / 10^3 km	Mass / 10^{23} kg
Mercury	None	57.9	4.9	3.3
Venus	None	108	12.0	49
Earth	1	150	12.8	60
Mars	2	228	6.8	6.4
Jupiter	16	778	143.0	19000
Saturn	18	1430	120.5	5700
Uranus	15	2900	51.2	866
Neptune	8	4500	49.5	103
Pluto	1	5920	2.3	0.13

Figure 1602   Some details of the planets

Stellar cluster

This is a number of stars that is held together in a group by gravitational attraction. The stars in the group were all created at about the same time and there can be many thousands of stars in a group.

Constellation

A **constellation** is a collection of stars that form a recognisable group as viewed from Earth. For example there is a constellation called the Andromeda constellation which contains the galaxy called Andromeda. The ancient Greeks named many of the constellations and perhaps two of the most easily recognisable are the Big Dipper and the Great Bear. Constellations are useful "landmarks" for finding one's way around the night sky.

The relative distances between stars

The stars in a galaxy are not uniformly distributed, however, their separation on average is of the order of 10^{17} m.

In the general introduction we mentioned the vast size of galaxies and the vast number of stars in a single galaxy. We also mentioned that galaxies form clusters. The separation of the galaxies in clusters is of the order 10^{23} m and the separation of clusters is of the order of 10^{24} m. One increase in the power of ten does not sound a lot but at these values it means a huge increase in distance.

The apparent motion of the stars/constellations

As night falls, the stars appear to rise in the East and as the night progresses they move across the night sky until they set in the West. The Earth seems to be at the centre

467

Chapter 16 (Option D)

of a giant celestial sphere and although each individual star appears to keep a fixed position, the whole canopy of the stars appears to rotate in a great circle about an axis through the North and South poles of the Earth.

Because the stars appear to be fixed to a giant celestial sphere they are often referred to as the fixed stars.

The position of the Sun relative to the fixed stars varies slowly over a course of a year. On the first day of spring, the vernal equinox, and the first day of autumn, the autumnal equinox, the Sun rises in the east, traverses the sky, sets in the west and day and night is of equal duration. During summer in the Northern Hemisphere the Sun rises in the Northeast and sets in the Northwest. Daylight hours are longer than the nighttime hours and the further North one goes the longer the daylight hours. In fact, north of the Arctic Circle the Sun never sets. During winter in the Northern Hemisphere the Sun rises in the Southeast, moves close to the southern horizon and sets in the Southwest. Daylight hours are shorter than the nighttime hours and north of the Arctic Circle the Sun never rises above the horizon. The Sun takes about a year to make a journey from West to East around the celestial sphere.

Against the background of the canopy of fixed stars, certain celestial objects do not move in circles. These objects wander back and forth against the backdrop exhibiting what is called retrograde motion. These are the planets and the Greek word for "wanderer" is indeed "planet". The Moon, like the Sun and the stars, traverses the night sky from East to West. The path of the Moon relative to the fixed stars is close to that followed by the Sun. However, it takes only about four weeks to complete a journey round the celestial sphere. The Moon's path around the celestial sphere varies from month to month but remains within about 8° either side of the path followed by the Sun. During the Moon's trip around the celestial sphere it exhibits different phases as seen from Earth, waxing from a new Moon crescent to a full Moon and then waning from the full Moon to a crescent new Moon.

The nature of stars

In an earlier chapter we discussed the topic of nuclear fusion, the process in which two lighter elements such as hydrogen can combine to form a heavier element such as helium and in the process liberate energy. We now know that this conversion of hydrogen to helium is the main source of energy for the stars. Very high temperatures (typically 10^7 K) and pressures are needed for fusion, in order for the nuclei to overcome coulomb repulsion and get close enough to fuse. Stars are formed by interstellar dust coming together through mutual gravitational attraction. The loss in potential energy as this happens can, if the initial mass of the dust collection is sufficient, produce the high temperatures necessary for fusion. The fusion process produces a radiation pressure that can then stop any further gravitational collapse.

In a stable star the thermonuclear processes taking place within the interior of the star do in fact produce a radiation pressure that just balances the gravitational pressure - there is equilibrium between radiation pressure and gravitational pressure. If the initial dust mass is about 80% of the mass of our Sun then the temperature reached by gravitational collapse is not high enough for fusion to take place. In this situation a fully-fledged star is not formed and instead we end up with a hydrogen rich object called a brown dwarf.

Different types of star

Red Giants

These are stars that are considerably larger than our Sun and have a much lower surface temperature than our Sun. The super red giant called Betelgeuse has a diameter equal to that of the distance of Jupiter from the Sun and a surface temperature of about 3000 K.

White Dwarfs

These are stars that are much smaller than the Sun (typically about the volume of the Earth) and have a much higher surface temperature. The white dwarf, Sirius B, has a surface temperature of 20,000 K.

Neutron stars

These are stars which have undergone gravitational collapse to such an extent that their core is effectively made up of just neutrons.

Supernovae

When the core of a star can collapse no further, the outer layers, which are still falling rapidly inwards, will be reflected back causing an enormous shock wave. This shock wave will in turn tear much of the surface of the core away in a colossal explosion. The star has become a supernova. In 1987 the star SK 69202 in the Large Magellanic Cloud went supernova and for a brief instant of time its brilliance was greater than that of the whole Universe by a factor of a 100.

Pulsars

These are rotating neutron stars. As they rotate they emit beams of electromagnetic radiation (usually of radio frequencies) essentially from the poles of the star. Each time a pole lines up with the Earth a pulse of radiation will be detected at the Earth. In 1968 a pulsar was detected in the Crab nebula that has a pulsing frequency of 33 Hz.

Black Holes

It has been suggested that certain stars that undergo gravitational collapse will reach a density and radius such that the gravitational field at the surface of the star will be strong enough to prevent electromagnetic radiation from escaping from the surface. Such stars will not therefore emit any light and are therefore said to be black holes.

Binary stars

Many stars that appear to the naked eye to be a single point of light actually turn out to be two stars rotating about a common centre. Sirius, the brightest star as seen from Earth is in fact a binary star consisting of Sirius A and Sirius B. Sirius A is a main sequence star and Sirius B is a white dwarf.

Cepheid variables

These are stars whose luminosities vary regularly, generally with a period of several days.

Astronomical distances

Introduction

Astronomical distances are vast. For example, it was quoted that the Sun is 25,000 light years (ly) from the centre of our galaxy. This is of the order of 10^{20} m. The distance to our nearest star is of the order of 10^{17} m and the distance to the quasar 3C273 is of the order of 10^{25} m. How are such vast distances measured? Let us first of all look at some of the units that are used in astronomical measurements.

The light year

The **light year** (ly) is defined as the distance that light travels in a year. There are about 3.2×10^7 seconds in a year and the speed of light is very nearly 3.0×10^8 m s^{-1} meaning that one light year is equal to about 9.6×10^{15} m, a more precise value being 9.46×10^{15} m.

The parsec

A useful astronomical distance is that of the average distance between Earth and the Sun, the **astronomical unit** (AU). 1 AU = 1.50×10^{11} m.

We define the **parsec** in terms of the AU. A line of length 1 AU subtends an angle of 1 arcsecond (one second of arc) at a distance of one parsec. In Figure 1603 the object P is 1 pc away from the line AB. Since the angle is so small the lines AP, BP and CP can all be considered to be of the same length.

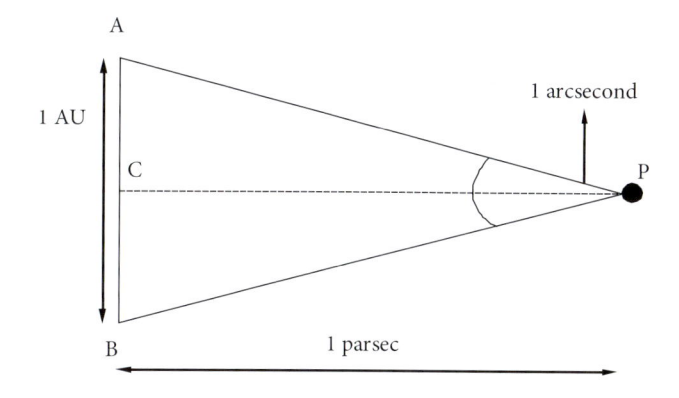

Figure 1603   The parsec

Figure 1604 shows the relationship between the units of distance.

1 AU	= 1.496 10^{11} m
1 ly	= 9.46 10^{15} m
1 ly	= 63 240 AU
1 pc	= 3.086 10^{16} m
1 pc	= 3.26 ly
1 pc	= 206 265 AU

Figure 1604   The relationships between units of distance

Stellar parallax and its limitations

Parallax is the apparent shifting of an object against a distant background when viewed from two different perspectives. Closing one eye and holding an index finger close to your other eye such that it blocks the view of some object at the other end of the room can easily demonstrate parallax. If you move your head, the object will now come into view.

In using the method of parallax to measure stellar distances we must recognise the fact that the Earth is moving through space. When you look out of the window of a moving train or car, objects close to you move past you very rapidly but it takes a long time for the distant landscape to change. In the same way, the stars which are a great distance from the Earth (in astronomical terms) appear to keep their position whereas the nearer stars (and the planets) appear to move against this background of the

469

Chapter 16 (Option D)

so called "fixed stars" as the Earth orbits the Sun. For this reason, astronomers regard the fixed stars as a reference point against which to measure the direction of stars that are closer to the Earth. The position of these stars relative to the fixed stars will depend from which point in the Earth's orbit about the Sun that they are observed. (Incidentally, if we regard the Earth to be fixed in space then the fixed stars appear to rotate about the Earth). Figure 1605 shows how we can use the parallax method to measure the distance from Earth of near stars.

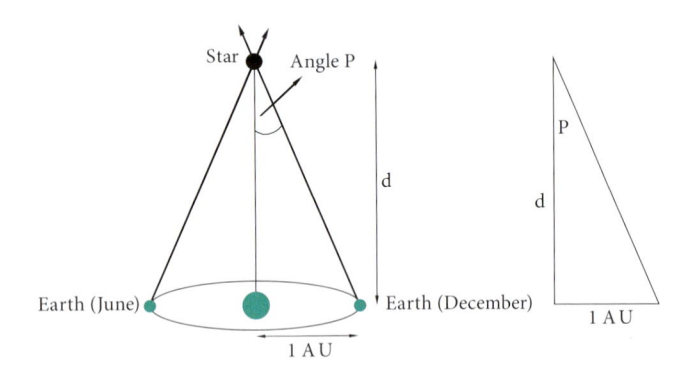

Figure 1605 Using the parallax method

The Earth is shown at two different points in its orbit about the Sun separated by a time period of six months. The angular position (p) of the star is measured against the fixed stars both in December and June.

Clearly we have $\tan p^c = \dfrac{1\,AU}{d}$.

However d is very much larger than 1 AU so the angle p is very small and therefore $\tan p^c \approx p$.

Therefore, we have $p \approx \dfrac{1\,AU}{d} \Leftrightarrow d \approx \dfrac{1\,AU}{p}$.

The parallax method is used to measure stellar distances of up to 100 pc (parallax angle 0.01 arcsec). Beyond this distance the parallax angle becomes too small to be measured with sufficient degree of accuracy. The angle becomes comparable to the error in measurement produced by distortions caused by the Earth's atmosphere. However, orbiting telescopes such as the Hubble avoid this distortion and enable slightly larger stellar distances to be measured by parallax.

Example

The parallax angle for the star Sirius A is 0.37 arcsecond. Calculate the distance of the star from Earth in

(a) metres
(b) parsecs
(c) astronomical units
(d) light years

Solution

(a) 1 arcsecond = $\dfrac{1}{3\,600}$ degrees of arc.

Hence $p = 0.37$ arcs $= \dfrac{0.37}{3\,600} = 1.03 \times 10^{-4} \times \dfrac{2\pi}{360}$ rad.

1 AU = 1.5×10^{11} m, hence

$d = \dfrac{1.5 \times 10^{11}}{1.8 \times 10^{-6}} = 8.36 \times 10^{16}$ m.

(b) If the parallax angle of a star is 1 arcsecond then it is said to be at a distance of 1 pc and hence:

$\dfrac{1}{p(measured\ in\ arc\ seconds)} = \dfrac{1}{0.37} = 2.7$ pc.

(c) 1 pc = 2.06×10^5 AU. Therefore
$d = 2.7 \times 2.06 \times 10^5 = 5.6 \times 10^5$ AU.

(d) 1 pc = 3.26 ly. Therefore $d = 2.7 \times 3.26 = 8.8$ ly.

Luminosity and apparent brightness

The energy radiated by a star is emitted uniformly in all directions as shown in Figure 1606. The total energy emitted by the star per unit time (i.e. the power) is called the Luminosity of the star, L.

For example our sun has a luminosity of 3.90×10^{26} W. (Sometimes written as L☉)

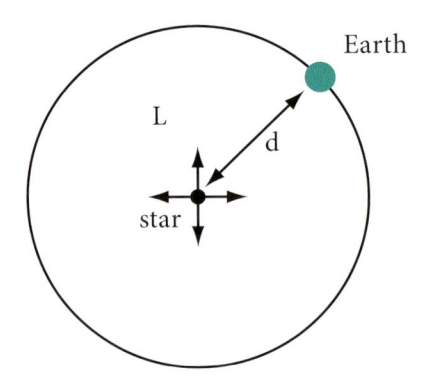

Figure 1606 The luminosity of a star

By the time the energy arrives at the Earth it will be spread out over a sphere of radius d. The energy received per unit time per unit area at the Earth is called the **apparent brightness** of the star b where

$$b = \dfrac{L}{4\pi d^2}$$

The apparent brightness of a star can be measured by attaching a radiation sensitive instrument known as a bolometer to a telescope. If d can be measured then the luminosity of the star can be determined. This is a very important property to know as it gives clues to the internal structure of the star, its age and its future evolution.

If all stars were equally bright, then the further away from the Earth a star is, the less its apparent brightness would be. For example, if the distance of a star A is measured by the parallax method (see below) and found to be at distance $2d$, it would have a quarter of the apparent brightness of star A. A star that is at a distance $4d$ would have one-sixteenth the apparent brightness of star A. The apparent brightness falls off as the inverse square of distance.

However, stars are not all of the same brightness so unless we know a star's luminosity we cannot use a measurement of its apparent brightness to find its distance from the Earth.

Exercise 16.1

1. Using the value for 1 AU given above and taking 1 year = 3.2×10^7 s, verify the conversions between parsecs and light years given in the above table.

2. Our nearest star, Alpha-Centuari, is 4.3 light years from Earth. Calculate the value of the parallax angle that gives this distance?

3. The Sun is 1.5×10^{11} m from the Earth. Estimate how much energy falls on a surface area of 1 m² in a year? State any assumptions that you have made.

D.2 Stellar characteristics and stellar evolution

NATURE OF SCIENCE:

Evidence: The simple light spectra of a gas on Earth can be compared to the light spectra of distant stars. This has allowed us to determine the velocity, composition and structure of stars and confirmed hypotheses about the expansion of the universe. (1.11)

© IBO 2014

Essential idea: A simple diagram that plots the luminosity versus the surface temperature of stars reveals unusually detailed patterns that help understand the inner workings of stars. Stars follow well-defined patterns from the moment they are created out of collapsing interstellar gas, to their lives on the main sequence and to their eventual death.

Understandings:
- Stellar spectra
- Hertzsprung-Russell (HR) diagram
- Mass-luminosity relation for main sequence stars
- Cepheid variables
- Stellar evolution on HR diagrams
- Red giants, white dwarfs, neutron stars and black holes
- Chandrasekhar and Oppenheimer–Volkoff limits

Stellar spectra

Chapter 8 introduced the Stefan-Boltzmann law for a black body. If we regard stars to be black body radiators,

then luminosity L of a star is given by the expression.

$$L = 4\pi R^2 \sigma T^4$$

where R is the radius of the star, T it's surface temperature and σ is the Stefan-Boltzmann constant. If we know the surface temperature and the radius of two stars then we can use the above equation to compare their luminosity. However, in practice we usually use the law to compare stellar radii as explained below.

Wien's (displacement) law

Wien discovered an empirical relation (which he later derived) between the maximum value of the wavelength emitted by a black body and its temperature. The so-called **Wien Displacement Law** is written as

$\lambda_{max} T$ = constant

From measurements of black body spectra the value of the constant is found to be 2.9×10^{-3} m K so that we can write

$\lambda_{max} = \dfrac{2.9 \times 10^{-3}}{T}$

The spectrum of stars is very similar to the spectrum emitted by a black body. We can therefore use the Wien Law to find the temperature of a star from its spectrum. If we know its temperature and its luminosity then its radius can be found from the Stefan law. This is shown in the example below.

Example

The wavelength maximum in the spectrum of Betelgeuse is 9.6×10^{-7} m. The luminosity of Betelgeuse is 104 times the luminosity of the Sun. Estimate the surface temperature of Betelgeuse and also its radius in terms of the radius of the Sun.

Solution

Using the Wien law we have $T = \dfrac{2.9 \times 10^{-3}}{9.6 \times 10^{-7}} \approx 3000$ K.

From the Stefan-Boltzmann law $L = \sigma A T^4 = 4\pi R^2 \sigma T^4$

From which we can write

$\dfrac{L_\odot}{L_{Betel}} = \dfrac{R_\odot^2 T_\odot^4}{R_{Betel}^2 T_{Betel}^4}$ or $\dfrac{R_{Betel}}{R_\odot} = \dfrac{T_\odot^2}{T_{Betel}^2} \times \left(\dfrac{L_{Betel}}{L_\odot}\right)^{1/2}$

Given that the surface temperature of the Sun is 5800 K, substitution gives the radius of Betelgeuse to be 370 times that of the Sun.

Stellar spectra

If a sufficiently high potential is applied between the anode and cathode of a discharge tube that contains a small amount of mercury vapour, the tube will glow. This is the basis of fluorescent lighting tubes. We can arrange for the radiation emitted from the tube to pass through a slit and hence onto a dispersive medium such as a prism or diffraction grating. The radiation can then be focused onto a screen. Images of the slit will be formed on the screen for every wavelength present in the radiation from the tube. Unlike an incandescent source, the mercury source produces a discrete line spectrum and a continuous spectrum in the ultra–violet region. In the visible region, mercury produces three distinct lines- red, green and blue. The wavelength of these lines is unique to mercury. In fact every element can be identified by its characteristic line spectrum. See Figure 1607.

Figure 1607 An example of atomic emission

We can alter the arrangement described above such that we now shine radiation from an incandescent source through the tube containing the mercury vapour when there is no potential difference across the tube. The radiation that has passed through the tube is analysed as above. On the screen we will now observe a continuous spectrum that has three dark lines crossing it. Each line will correspond to the blue, red and green emission lines of mercury. This is what is called an **absorption spectrum**.

Absorption spectra can be used to find out the chemical composition of the atmosphere of a star. The continuous spectrum from a star will be found to contain absorption lines. These lines are formed as the radiation from the surface of the star passes through the cooler, less dense upper atmosphere of the star. The absorption lines will correspond to the emission lines of the elements in the upper atmosphere of the star. Different stars have different spectra. In some stars the lines corresponding to the visible hydrogen spectrum are prominent. Other stars, like our sun, have lines corresponding to the emission lines of elements such as iron, sodium and calcium.

The overall classification system of spectral classes

Stars with similar appearing spectra are grouped together into spectral classes, each class being related to surface temperature. For historical reasons the classes, are labelled OBAFGKM. The stars with the highest temperatures (< 30000 K) are in the O class and stars with the lowest temperatures (3000 K) are in the M class. Our Sun with a surface temperature of 5800 K is in the G class.

The absorption spectrum of a particular star will depend on its surface temperature. If the surface temperature for example is above 10,000 K the photons, leaving the surface will have sufficient energy to ionise any hydrogen atoms in the star's atmosphere. Hence the absorption spectrum will show little evidence of hydrogen being present.

However, if the temperature is about 9000 K the photons will cause excitation in the hydrogen atoms rather than ionisation. Hence, in these stars the spectrum will show strong hydrogen absorption lines. At temperatures above 30,000 K the photons have sufficient energy to produce singly ionised helium and the spectrum of singly ionised helium is different from neutral helium. Hence, if a stellar absorption spectrum has lines corresponding to the emission lines of single ionised helium, we know that the star has a surface temperature in excess of 30,000 K. For every element there is a characteristic temperature range of the source that produces strong absorption lines.

When the effects of temperature are taken into account we find that the composition of all stars is essentially the same, about 74% hydrogen, 25% helium and 1% other elements.

It should also be noted that stellar absorption spectra are another check as to the value for the surface temperature of a star that can be deduced from Wien's law. Figure 1608 below summarises the spectral classes.

Spectral class	Approximate Temperature range /K	Colour	Main absorption lines	Example
O	30000-50000	Blue violet	Ionised helium	Mintaka
B	10000-30000	Blue white	Neutral helium	Rigel
A	7500-10000	White	Hydrogen	Sirius A
F	6000-7500	Yellow white	Ionised metals	Canopus
G	5000-6000	Yellow	Ionised calcium	Sun
K	3500-5000	Orange	Neutral metals	Aldebaran
M	2500-3500	Red orange	Titanium oxide	Betelgeuse

Figure 1608 The spectral classes of stars

The characteristics of spectroscopic and eclipsing binary stars

About half the stars visible from Earth actually consist of two stars orbiting about a common centre. Such systems are called **binary stars**. Stars that can actually be observed as two stars orbiting each other are called visual binaries. In principle it is a relatively easy matter to measure the masses of two such stars. The period of rotation depends on the sum of the masses of the two stars and the separation of the stars. (Actually the orbit is usually elliptical so the separation is in fact the length of the semi-major axis of one star's orbit about the other). By measuring the angular separation of the two stars as seen from Earth and knowing how far they are from Earth, we can calculate their linear separation. The period of revolution can be measured directly (mind you, this might take more than a lifetime since some visual binaries have very long orbital periods) and hence the sum of the masses can be found. The stars actually orbit about their centre of mass and the position of the centre of mass depends on the ratio of the individual star masses. The centre of mass can be found by plotting the orbit of each star separately and so the ratio of the masses can be computed. Knowing the sum of the masses means that the individual masses can be found.

Over many years the masses of many stars in binary systems have been measured. This has enabled a very important relationship to be established between the mass of main sequence stars and their luminosities. Hence we can determine the masses of single main sequence stars by knowing their luminosities.

Eclipsing binaries

Some binary stars cannot be resolved visually as two separate stars. However, the binary nature of the system can be deduced from the fact that the stars periodically eclipse each other. The orientation of the orbit of the stars with respect to the Earth is such that as the stars orbit each other, one will block light from the other. As seen from Earth the brightness of the system will vary periodically. This variation in brightness yields information as to the ratio of the surface temperature of the stars and also the relative size of the stars and the size of their orbit.

Spectroscopic binaries

The binary nature of a system can, in many cases, be deduced from its spectrum. Figure 1609 shows the observed spectrum of a possible binary system taken at different times.

Figure 1609 The spectrum of a binary system

The two spectra A and C are identical. However, corresponding to each line in these spectra there are two lines in spectrum B. One of the lines is of a slightly longer wavelength than the corresponding line in A and C and the other is of a slightly shorter wavelength. They are red–shifted and blue–shifted respectively. We can see how this comes about by looking at Figure 1610.

Chapter 16 (Option D)

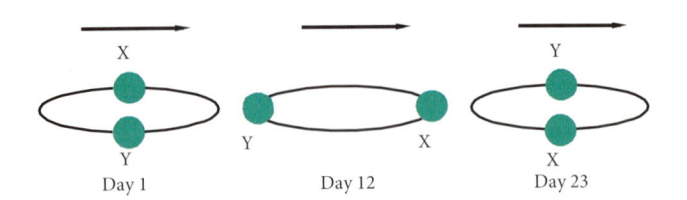

Figure 1610   The red- and blue-shift

The two stars X and Y are orbiting each other as shown. The stars are of the same spectral class and have identical spectra. On day 1 the orbital plane is aligned with respect to the Earth in such a way that star Y completely blocks star X and so only the spectrum of star Y is observed. On day 12 star Y is moving away from the Earth and star X is moving towards the Earth. Because of the Doppler effect (see Chapter 6) the light from Y as observed on Earth will be red-shifted. Similarly because X is moving towards the Earth. light from this star will be observed to be blue-shifted. Hence spectrum B above will be observed. On day 23, star X now completely blocks star Y so only the spectrum from star X is observed.

This is an ideal situation and although it does occur, it is more the exception than the rule. In many systems the stars are not of the same spectral class and so the spectra A and C will not be the same. However, spectrum C will still show red and blue shift. For example one of the stars might be so dim that its spectrum cannot be detected on Earth. However, the single spectrum will shift back and forth as the two stars orbit each other. Such systems are called single–line spectroscopic binaries. A spectroscopic binary might also be an eclipsing binary. These systems, although not common, are very useful since it is possible to calculate the mass and radius of each star from the information that such systems give.

The Hertzsprung-Russell Diagram

In 1911 the Danish astronomer *Ejnar Hertzsprung* noticed that a regular pattern is produced if the absolute magnitude (see next Section) – (or luminosity) – of stars is plotted against their colour (surface temperature). Two years later the American astronomer *Henry Russell* discovered a similar pattern if the luminosity is plotted against spectral class. (*Effectively this is another plot against temperature*). In recognition of the work of these two men, such diagrams are called **Hertzsprung-Russell diagrams**. A typical H-R diagram is shown in Figure 1611. You should note that neither the absolute magnitude scale nor the temperature scale is linear. They are in fact both log scales. For historical reasons the temperature scale is plotted from high to low.

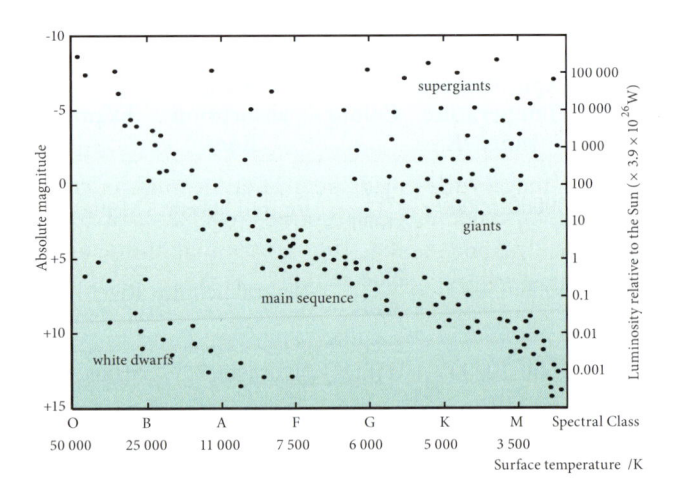

Figure 1611   The Hertzsprung-Russell Diagram

The striking feature about this diagram is that the stars are grouped in several distinct regions with a main diagonal band that contains the majority of stars. For this reason, stars which lie in this band are called main sequence stars. The sequence runs from large luminosity and high surface temperature (top left) to small luminosity and low surface temperature (bottom right hand corner). All stars in the main sequence derive their energy from hydrogen burning (fusion) in the core of the star.

There is another grouping of stars towards the top right-hand corner that have a large luminosity and relatively low surface temperature. To have such a large luminosity at low surface temperatures means that these stars must be huge. For this reason they are called giant stars. Cooler members of this class have a distinctive red appearance and are therefore called red giants. A few stars at low surface temperatures have a very large luminosity (we have already met one such star – Betelgeuse in the constellation of Orion) and these are called supergiants.

There is another grouping of stars towards the bottom left hand corner that have a low luminosity but very high surface temperatures. This means these stars are relatively small (typically the size of the Earth) and because of the low luminosity are called white dwarfs.

Mass-luminosity relation for main sequence stars

Apparent magnitude

An attempt to classify stars by their visual brightness was made 2000 years ago by the Greek astronomer Hipparchus. He assigned a magnitude of 1 to the stars that appeared to be the brightest and to stars that were just visible to the naked eye, he assigned a magnitude of 6. Values between 1 and 6 were assigned to stars with intermediate brightness. This is an awkward scale but is essentially the one still in use today. However, measurement shows that a magnitude 1 star has an apparent brightness 100 times that of a magnitude 6 star. So we now define the **apparent magnitude** scale such that a difference in apparent magnitude of 5 corresponds to a factor of 100 in brightness. This means that 100 stars of magnitude 6 will produce as much power per unit area at the surface of the Earth as a single star of apparent magnitude 1.

To see how the scale works in practice let us work out the ratio of the brightness of a magnitude 1 star to that of a magnitude 2 star.

Let this ratio be r. A change from magnitude 6 to magnitude 1 is a change of 5 magnitudes corresponding to a change in brightness of 100. Hence the factor r^5 that gives a change of 5 magnitudes is equal to 100.

Suppose a star has an apparent magnitude m and an apparent brightness b and it is at a distance d measured in parsecs from the Earth. Its brightness if it were at a distance of 10 pc from the Earth would be, by the inverse square law, $\frac{b}{(d/10)^2}$. Hence if we use the equation relating apparent magnitudes and brightness we have

$$\frac{b_1}{b_2} = \frac{100}{d^2} = (2.521)^{M-m}$$

This equation can only be solved using logarithms. In logarithmic form the equation actually becomes

$$M - m = 2.5\log\left(\frac{100}{d^2}\right) = 5 - 5\log d$$

or $m - M = 5\log d - 5$

The nearest star to the Sun is Alpha Centauri at a distance of 1.3 pc. With an apparent magnitude of 0.1, Alpha Centauri has an absolute magnitude of 4.5. This means that the Sun and Alpha Centauri have very nearly the same luminosity.

Betelgeuse on the other hand, at a distance of 130 pc has an apparent magnitude of +0.50 that gives an absolute magnitude of –5.14 E. This means that if Betelgeuse where at 10 pc from the Earth its brightness would be $(2.512)^{9.96}$ greater than the Sun. (The differences in the absolute magnitudes is 9.96). This means that Betelgeuse has luminosity some 10000 times that of the Sun.

Example

The apparent magnitude of the Sun is -26.7 and that of Betelgeuse 0.50. Calculate how much brighter the Sun is than Betelgeuse.

Solution

We have m_2 for Betelgeuse = 0.50 and m_1 for the Sun = –26.7 to give $m_2 - m_1$ the difference in the apparent magnitudes equal to 27.2.

We therefore have that the ratio, $\frac{b_1}{b_2}$, of the brightness of that of the Sun to that of Betelgeuse to equal

$2.512^{27.2} = 7.6 \times 10^{10}$.

That is the Sun is approximately 80000 million times brighter than Betelgeuse (if both of them are viewed from the Earth).

Example

Calculate the absolute magnitude of the Sun.

Solution

The apparent magnitude of the Sun is –26.7

The distance of the Sun from Earth = 1 AU = 4.9×10^{-6} pc

Therefore $m - M = 5\log(4.9 \times 10^{-6}) - 5$ or

$m - M = -26.5 - 5$.

To give $M = +4.8$.

Chapter 16 (Option D)

The luminosity and stellar distance of a star

We have seen that determining the distance to stars using the method of parallax is only of use up to distances of about 100 pc. Beyond this distance the parallax angle is too small to determine with any degree of accuracy. However, when we discussed apparent brightness and luminosity we found that, from the inverse square law, if the luminosity of a star and its apparent brightness are known, then its distance from Earth can be determined.

The apparent brightness can be measured directly but how do we measure the luminosity? The key to this is the HR diagram and the spectral class of the star. The spectral class of the star is in fact determined from the absorption line spectrum of the star. The surface temperature can also be determined from its spectrum (Wien's law) and this then enables us to plot the star on the HR diagram and hence determine its luminosity.

This method of determining the distance to a star from information gleaned from its spectra is known somewhat misleadingly as spectroscopic parallax, misleading since no parallax is involved in the method.

In theory there is no limit to the stellar distances that can be obtained using the method of spectroscopic parallax. However, in practice beyond 10 Mpc, the error in the determination of the luminosity becomes too large to compute the distance to within a sensible degree of accuracy.

Example

The star Regulus in the constellation of Leo has an apparent brightness of 5.2×10^{-12} that of the Sun and a luminosity 140 times that of the Sun. If the distance from the Earth to the Sun is 4.9×10^{-6} pc, how far from the Earth is Regulus?

Solution

From the inverse square law we know:

$$b = \frac{L}{4\pi d^2}$$

Therefore: $\dfrac{L_{sun}}{L_{reg}} = \dfrac{d^2_{sun}}{d^2_{reg}} \times \dfrac{b_{sun}}{b_{reg}}$

From which: $d^2_{reg} = \left(\dfrac{L_{reg}}{L_{sun}}\right) \times \left(\dfrac{b_{sun}}{b_{reg}}\right) \times d^2_{sun}$

$= 140 \times \left(\dfrac{1}{5.2 \times 10^{-12}}\right) \times (4.9 \times 10^{-6})^2$

To give, $d_{reg} = 25\,pc$.

The mass–luminosity relation

An interesting relationship exists between the luminosity of a main sequence star and its mass. If the define the luminosity L of a star and its mass M in terms of solar units, then it is found that for all Main Sequence stars averaged over the whole sequence that

$$L = M^{3.5}$$

However, it must be borne in mind that this is an average relationship and that the power n to which M is raised is to some extent mass dependant. Generally n is greater than 3 and less than 4.

For example a star that is 5 times more massive than the Sun will be $5^{3.5}$ ($= 280$) times more luminous.

Cepheid variables

The nature of a Cepheid variable

Clearly there are problems with measuring stellar distance beyond 100 pc, the distance limit of the parallax method. (As mentioned above, the Hubble telescope enables this limit to be extended somewhat). However, we have seen that if we can measure the apparent brightness of star and its luminosity then we can determine its distance from the relation $L = 4\pi d^2 b$. In the preceding section we saw that the method of spectroscopic parallax is a powerful tool for determining the luminosity of stars and hence the distance to stars in galaxies beyond the Milky Way. However, there is another method that enables distances greater than 10 Mpc (the limit of spectroscopic parallax) to be determined with great accuracy.

Relationship between period and absolute magnitude for Cepheid variables

In 1784 the amateur astronomer John Goodricke noted that the luminosity of the star Delta Cephei varied regularly. He recorded the apparent magnitude as reaching a maximum of 4.4 and then falling to a minimum of 3.5 in four days, rising to the maximum again in the following 1.5 days. We now know that this periodic change in luminosity is due to the outer layers of the star undergoing periodic contractions and expansions. There are however, two types of Cepheid imaginatively called Type-I and Type-II.

Many other variable stars have since been discovered and they are given the general name Cepheid variables. The position of the Cepheids on the HR diagram is shown in Figure 1612.

ASTROPHYSICS

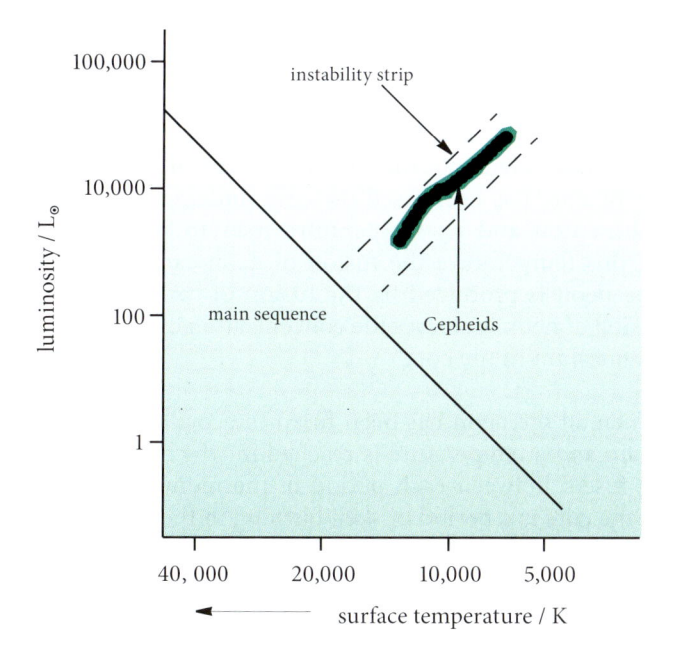

Figure 1612 The position of Type II Cepheid variables on the HR diagram

Cepheid variables are extremely important in determining distances to galaxies. This is because they are extremely luminous (typically 10^4 $L_\odot$) and therefore relatively easily located and also because of the so-called period-luminosity relationship.

The American astronomer *Henrietta Leavitt* showed that there is a actually a linear relationship between the luminosity and period of Cepheid variables.

The graph in Figure 1613 shows the period-luminosity relationship for Type-II Cepheids.

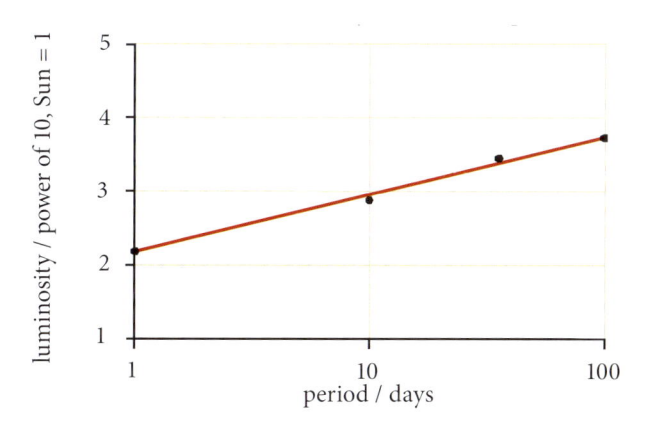

Figure 1613 The period-luminosity relationship for Cepheids

When a Cepheid is located then, by measuring the period, we can determine its luminosity from the period-luminosity relationship. If the apparent brightness of the star is measured then its distance can be computed from the inverse square law.

Cepheid variables act as a sort of "standard candle" in as much as they can be used to check distance measurements made using parallax and/or spectroscopic parallax. However, beyond about 60 Mpc Cepheids are too faint and measurement of their period becomes unreliable. Other techniques have to be used to measure distances greater than 60 Mpc. A summary is given at the end of this chapter outlining these techniques.

Example

The star δ-Cepheid is 300 pc from the Earth (This distance can be determined by parallax from the Hubble telescope). Another variable star is detected in a distant galaxy that has the same period as δ-Cepheid but with an apparent brightness of 10^{-9} of that of δ-Cepheid. How far is the galaxy from Earth?

Solution

The two stars have the same luminosity hence

$$L_{star} = L_\delta = 4\pi d_{star}^2 b_{star} = 4\pi d_\delta^2 b_\delta$$

$$\frac{d_{star}}{d_\delta} = \sqrt{\frac{b_\delta}{b_{star}}} = 3.2 \times 10^4$$

The galaxy is therefore $3.2 \times 10^4 \times 300$ pc ≈ 10 Mpc from Earth.

Summary

The two 'flow' charts in Figures 1614 and 1615 summarise the ideas that we have met so far in this chapter as to how we gain information about the nature of stars and their distance from Earth.

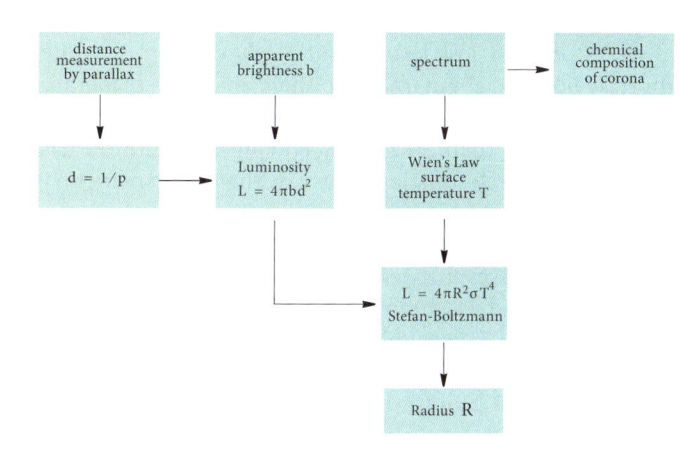

Figure 1614 Distance measured by parallax

477

Chapter 16 (Option D)

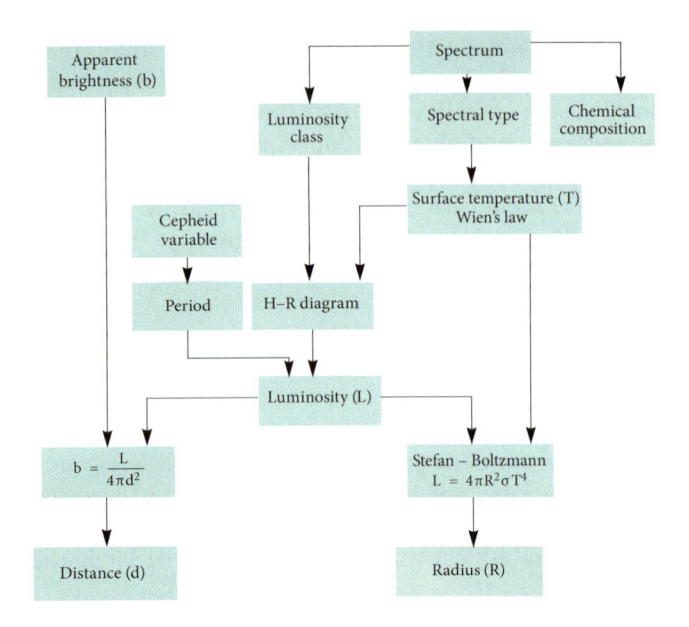

Figure 1615 *Distance determined by spectroscopic parallax - cepheid variables*

Chandrasekhar and Oppenheimer–Volkoff limits

The core of a star like the Sun does not keep contracting under gravity since there is a high density limit set by a quantum mechanical effect known as electron degeneracy. Essentially a point is reached where the electrons cannot be packed any closer. The more massive a white dwarf the greater will be the gravitational force of contraction, hence electron degeneracy sets an upper limit on the mass of a white dwarf. This is known as the **Chandrasekhar limit** after its discoverer and is equal to $1.4\,M_\odot$.

The positions of the red giants and white dwarfs on the H–R diagram are shown in Figure 1611. It is thought that all stars of mass less than about $8\,M_\odot$ end up as white dwarfs ejecting about 60% of their mass as planetary nebulae. However, even stars with masses equal to, or slightly greater than, $8M_\odot$ may end up as white dwarfs should they eject sufficient mass during their planetary nebula phase.

Stars with masses between $4M_\odot$ and $8M_\odot$ are able to fuse carbon and in this process produce neon, sodium, magnesium and oxygen during their final red giant phase.

Whereas the Chandrasekhar limit applies to main sequence stars, the so-called **Oppenheimer–Volkoff limit** applies to neutron stars.

Red giants, white dwarfs, neutron stars and black holes

Most Stars with a mass of $8M_\odot$ or more are able to fuse even more elements than carbon. After all the carbon in the core has been used the core undergoes a further contraction and its temperature rises to some 10^9 K. At this temperature the fusion of neon can take place. The neon is produced by the fusion of carbon and the fusion of neon increases the concentration of oxygen and magnesium in the core.

When all the neon has been fused the core contracts yet again and a temperature is reached in which oxygen can be fused. Between each period of thermonuclear fusion in the core is a period of shell burning in the outer layers and the star enters a new red giant phase. When only shell burning is taking place, the radius and luminosity of the star increases such that the result is a supergiant with a luminosity and radius very much greater than that of a lower mass red giant. Some super giants have a radius several thousand times larger than the Sun and are some of the brightest visible stars. Betelgeuse and Rigel in the Orion constellation are example of supergiants.

Eventually a temperature is reached in the core of a supergiant at which the fusion of silicon can take place. The product of silicon burning is iron.

Figure 1616 shows the structure of the core of a supergiant as it nears the end of its life. The energy of the star comes from six concentric burning shells.

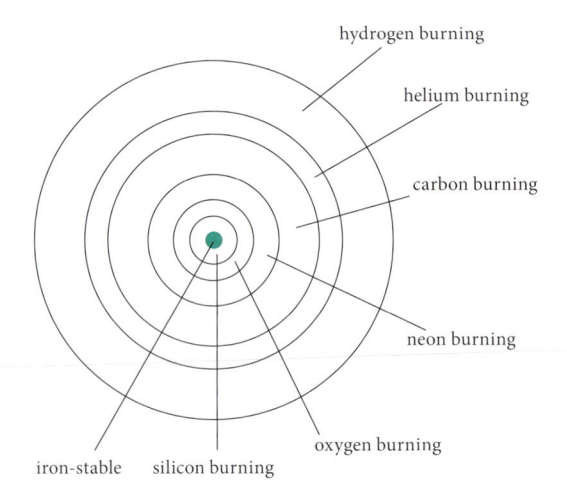

Figure 1616 *The structure of a supergiant*

Because of a very large coulomb repulsion, elements with a proton number of 26 or greater cannot undergo fusion. (Except if there is an enormous energy input). Iron has a proton number of 26 so when the entire core is iron, fusion within the core must cease. The star has reached a critical state.

When the entire inner core is iron it contracts very rapidly and reaches an enormously high temperature ($\approx 6 \times 10^9$ K). The high energy gamma photons emitted at this temperature collide with the iron nuclei breaking the nuclei into alpha-particles. This takes place in a very short time and in the next fraction of a second the core becomes so dense that negative electrons combine with positive protons producing neutrons and a vast flux of neutrinos. This flux of neutrinos carries a large amount of energy from the core causing it to cool and further contract.

The rapid contraction produces an outward moving pressure wave. At this point, because of the contraction of the inner core, material from the other shells is collapsing inward, when this material meets the outward moving pressure wave, it is forced back. The pressure wave continues to accelerate as it moves outward until it is moving faster than the speed of sound and becomes a colossal shock wave that rips the material of the star's outer layers apart. The inner cores are now exposed and a vast amount of radiation floods out into space. The star has become a supernova.

It has been estimated that some 10^{46} J of energy is liberated in such an event and the star loses about 96% of its mass. The energy produced when a star becomes a supernova is sufficiently high to produce all the elements with atomic numbers higher than iron. The material that is flung out in to space will eventually form dark nebulae from which new stars may be formed. And so the process repeats itself. The core material that is left is thought to contract to form a neutron star or a black hole (see next section).

(*On February 23rd, 1987 a supernova was detected in the Large Magellanic Cloud and was so bright that it could be seen in the Southern Hemisphere with the naked eye*).

The sequences of the birth and death of stars is summarised in Figure 1617.

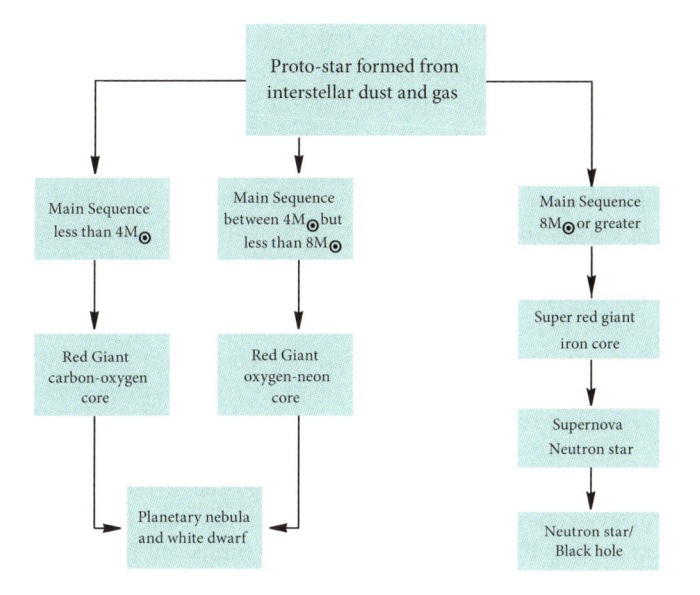

Figure 1617   The sequences of the birth and death of stars

A summary of the stages through which stars like our Sun pass on the way to becoming a white dwarf is shown in Figure 1618.

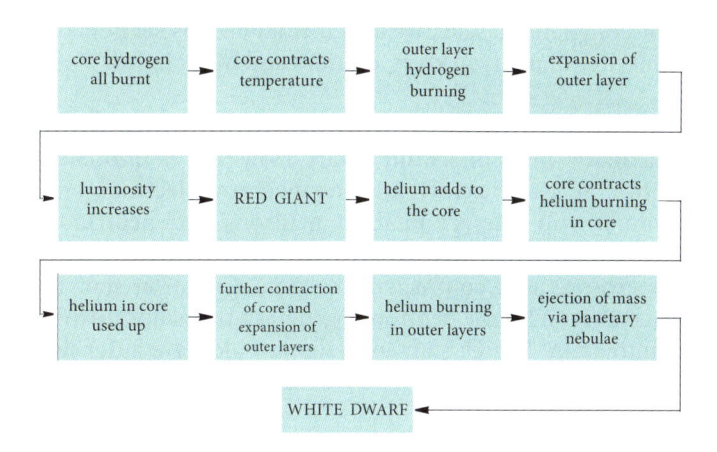

Figure 1618   The life cycle of a star

main sequence stars, the so-called Oppenheimer–Volkoff limit applies to neutron stars.

Figure 1619 and Figure 1620 show the appropriate evolutionary paths on HR diagrams of a low mas star and a high mass star respectively.

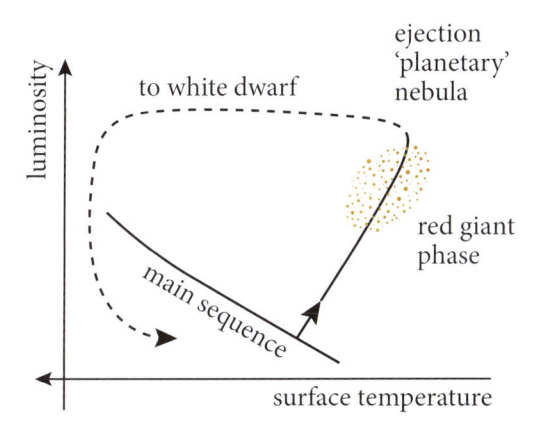

Figure 1619   HR diagram for the evolutionary path of a low mass star.

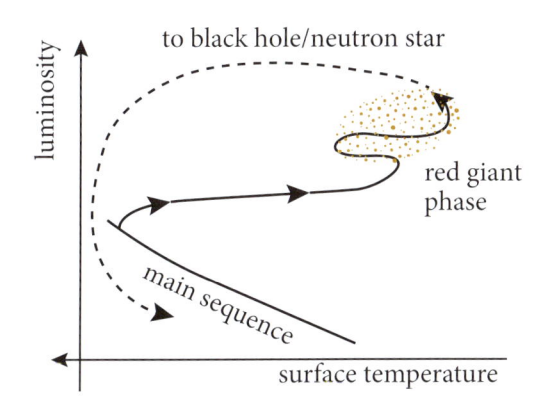

Figure 1620   HR diagram for the evolutionary path of a high mass star.

479

Chapter 16 (Option D)

It is left as an exercise for you to draw appropriate evolutionary paths on an HR diagram of stars of different masses.

For completion's sake, we also include in this section, an overview of **neutron stars, quasars and black holes**.

We have seen that electron degeneracy sets a limit to the maximum mass of a white dwarf. However, as intimated above, at a sufficiently high temperature electrons and protons will interact to form neutrons and neutrinos. It is thought that this is what happens to the core of a supernova, as it contracts to become a neutron star. In such a star it is now neutron degeneracy that stops further contraction. If such stars exist they would have a density of about 4×10^{17} kg m^{-3} and would have a radius of only some 15 km.

The Oppenheimer–Volkoff limit suggests that there is an upper limit to the mass of a neutron star beyond which it will collapse to a black hole. This limit is thought to be about 2-3 solar masses.

Evidence for the existence of neutron stars came in 1967 when Jocelyn Bell as an undergraduate at Cambridge University detected rapidly varying radio pulses from one particular location in the sky. Since then many more such sources have been discovered with periods ranging from about 30 pulses per second to about 1 pulse every 1.5 seconds. This is far faster than the pulses from an eclipsing binary or variable star. Nor could the source be a rotating white dwarf since, at such speeds of rotation, the white¬– dwarf would tear itself apart.

It is now thought that these so-called **pulsars** are in fact rotating neutron stars. Neutron stars are by necessity small and therefore, to conserve angular momentum, they must rotate rapidly. Also, supergiants have a magnetic field and as they shrink to a neutron star the magnetic field strength will become very large. As this field rotates with the star it will generate radio waves. Strong electric fields created by the rotating magnetic field could also create electron-positron pairs and the acceleration of these charges would also be a source of radio waves.

The neutron star model is now firmly accepted by astronomers. The Crab nebula in the constellation Taurus has a pulsar close to its centre with a radio frequency of 33 Hz. It is pretty certain that the nebula is the remains of a supernova.

Quasars

In 1944 Grote Reber, an amateur astronomer, detected strong radio signals from the constellations Sagittarius, Cassiopeia and Cygnus. The first two of these sources were found to lie within the Milky Way. However, in 1951 an odd looking galaxy was found to be the source of the Cygnus signals. The galaxy was subsequently called Cygnus A. One of the extraordinary things about this galaxy is that, unlike all hitherto known galaxies, it exhibits an emission spectrum. Furthermore the emission spectrum showed a very large red-shift indicating that the galaxy was some 220 Mpc from the Earth.

This was the furthest known object in the Universe at this time meaning that

Cygnus A must be one of the most luminous radio sources in the Universe. Because of its star like appearance and strong radio emission, Cygnus A was called a quasar (quasi-stellar radio source). To date about 10,000 quasars have been detected and the most distant is some 3600 Mpc from the Earth. Not only are quasars strong radio emitters but they also have enormous luminosities with the most luminous being some 10000 times more luminous than the Milky Way.

In the 1960s even stranger objects than quasars were detected which emit strong bursts of gamma radiation. It would seem that these objects are even more distant and more luminous than quasars.

Black holes

We have seen that neutron stars have enormous densities and small radii. This means that the gravitational potential at the surface of such a star will also be enormous. The escape velocity from the surface of a planet or star is related to the gravitational potential at its surface. So what would happen if the potential were such that the escape velocity was equal to the speed of light? It would in fact mean that no electromagnetic radiation could leave the surface of the star; it would have become a **black hole**.

The correct way in which to think of the formation of a black hole is in terms of the General Theory of Relativity. This theory predicts that space is warped by the presence of mass. Hence the path of light travelling close to large masses will be curved. Near to a black hole the space is so severely warped that the path of any light leaving the surface will be bent back in on itself. The General Theory also predicts that time slows in a gravitational field. At a point close to a black hole, where the escape velocity just equals the speed of light, time will cease. This point is known as the event horizon of the black hole. When a dying star contracts within its event horizon the entire mass of the star will shrink to a mathematical point at which its density will be infinite. Such a point is known as a singularity. A black hole therefore consists of an event horizon and a singularity

So do black holes exist? We cannot see a black hole, however, its existence might be inferred from the effect that its gravitational field would have on its surroundings. Some stellar objects have been detected as sources of X-ray radiation such as

Cygnus X-1. Spectroscopic observations revealed that close to the location of Cygnus is a supergiant with a mass of about 30M☉. This star itself cannot be the source of X-ray radiation and it has been surmised that the system Cygnus X-1 is in fact a binary, the companion star being a black hole. The intense gravitational field of the black hole draws material from the supergiant and as this material spirals into the black hole it reaches a temperature at which it emits X-rays. Other potential candidates for black holes have been found and some theorists think that there is a black hole at the centre of all galaxies including our own Milky Way.

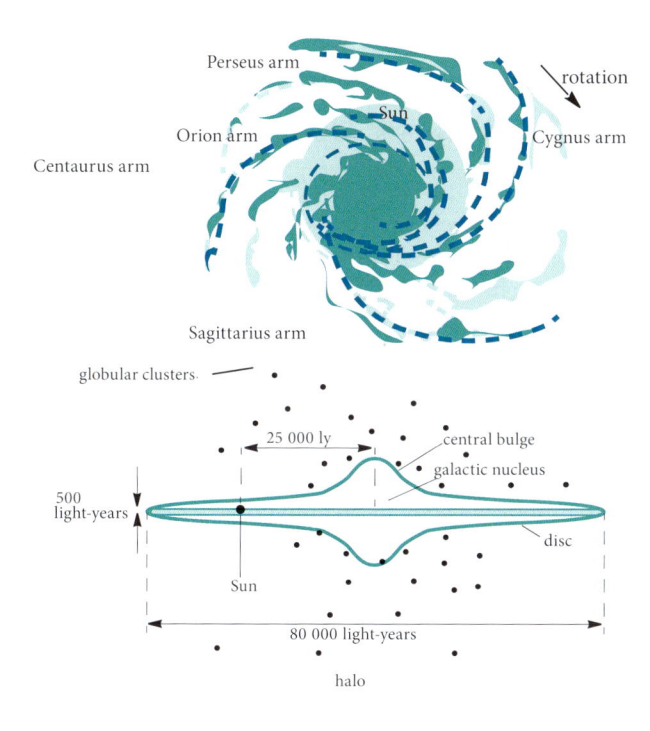

Figure 1621 The Milky Way

Spiral galaxies

There are many spiral galaxies in the universe and the Milky Way is one of them.

All such galaxies are characterised by three main components, a thin disc, a central bulge and a halo. The sketches in Figure 1621 shows a side view and plan view of the Milky Way.

Andromeda is the nearest spiral galaxy to the Milky Way (about 2.2 million light years away) and is just about visible to the naked eye. If it is viewed with binoculars then its spiral shape can be distinguished.

The halo of the Milky Way contains many globular clusters of stars each containing up to a million stars and seemingly very little interstellar dust or very bright stars. This suggests that the halo contains some of the oldest stars in our galaxy. Their age has been estimated at between 10 and 15 billion years.

The bulge at the centre of the Milky Way has the greatest density of stars and many of these would seem to be very hot, young stars. However, direct visible evidence is difficult to collect since the galactic centre contains many dust clouds. Such evidence that we have comes from observations made in the infrared and radio regions of the electromagnetic spectrum.

Elliptical galaxies

As the name implies, these galaxies have an elliptical cross-section and no spiral arms. Some are highly elliptical and some are nearly circular.

Irregular galaxies

These galaxies seem to have no specific structure. Our two nearest galactic neighbours are irregular, the Large and small Magellanic Clouds. The large Magellanic Cloud lies at a distance of about 160,000 light years from the Milky Way.

All galaxies rotate. If they did not, they would collapse under gravitational attraction. Our Sun is about 25,000 light years from the centre of the Milky Way and Doppler shift measurements show it to be moving through space with a speed of 230 km s^{-1}. You can use these values to show that the orbital period of rotation of the Milky Way is about 2.0×10^8 years.

Chapter 16 (Option D)

D.3 Cosmology

> **NATURE OF SCIENCE:**
>
> Occam's Razor: The big bang model was purely speculative until it was confirmed by the discovery of the cosmic microwave background radiation. The model, while correctly describing many aspects of the universe as we observe it today, still cannot explain what happened at time zero. (2.7)
>
> © IBO 2014

Essential idea: The hot big bang model is a theory that describes the origin and expansion of the universe and is supported by extensive experimental evidence.

Understandings:
- The Big Bang model
- Cosmic microwave background (CMB) radiation
- Hubble's law
- The accelerating universe and redshift (z)
- The cosmic scale factor (R)

Newton's model of the universe

Newton's view of the Universe was that it is infinite in extent, contains an infinite number of stars, is static and exists forever. He argued that if this were not the case then the force of gravity between a finite number of stars in a finite universe would draw them together such that eventually the Universe would become a single spherical lump of matter. In the early 1800s a German amateur astronomer, *Henrich Olber*, pointed out that Newton's view leads to a paradox. If the universe is infinite and contains an infinite number of stars then no matter in which direction you looked from the Earth you would see a star. The sky should be ablaze with stars. **Olber's paradox** is then "How come the night sky is black". Or as Kepler put it 30 years before Newton was born, "Why is the sky dark at night?"

Olber's paradox

A simple mathematical argument shows that an infinite universe leads to the view that the sky should be equally bright in all directions to all observers at any place in the Universe.

In Figure 1616 we suppose that the stars are distributed evenly with a density ρ. To an observer the number of stars in a thin shell of thickness d at a distance R away will be $4\pi R^2 \rho d$. (If the shell is very thin then the volume of the shell is $4\pi R^2 d$).

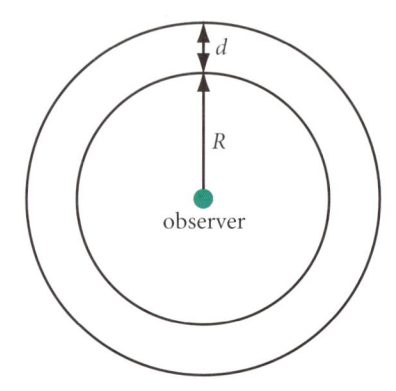

Figure 1616   Olber's paradox

The volume of such shells therefore increases with R^2, which means that the number of stars observed increase as R^2. But the luminosity of any shell falls off as $\frac{1}{R^2}$. Hence the Universe should be uniformly bright in any direction since the luminosity depends only on the density distribution of the stars.

Clearly the night sky is dark and so therefore we must conclude that the universe is not infinite nor are there an infinite number of stars in the universe evenly distributed. For an explanation as to why the night sky is dark we must propose a different model of the universe to that put forward by Newton.

The Big Bang model

In chapter 9 we discussed the apparent shift in frequency of a moving source – the so-called Doppler Effect. Essentially, if a sound source is moving away from an observer then the observer will measure the source as emitting a lower frequency (and longer wavelength) than when the source is stationary. The same effect arises with light sources. Suppose for example an observer measured the visible line spectrum of atomic hydrogen and then compared the wavelengths of the lines with that of a spectrum produced by a hydrogen source that is moving away from him. The lines in the spectrum of the moving source will be measured as having a longer wavelength than those in the spectrum of the stationary source. We say that the spectrum from the moving source is red-shifted. The faster the source is moving away from the observer, the greater will be the cosmological **red-shift**.

In 1920s Edwin Hubble and Milton Humanson photographed the spectra of a large number of galaxies and found that most of the spectra showed a red-shift (nine years earlier, the astronomer Vesto Slipher had detected the red-shift in several nebulae. However, at this time it was not known that the nebulae were galaxies). This means that the galaxies must be moving away from the Earth.

If all the galaxies are rushing away from each other then it is feasible to assume that in the past they were much closer together. It is possible to imagine that sometime long ago all the matter in the Universe was concentrated into a smaller volume. An "explosion" then occurred that threw the matter apart.

The prevailing view of the creation of the Universe is that some 10-20 billion years ago all the matter of the Universe was concentrated into a point of infinite density. Then a cataclysmic explosion initiated the expansion of the Universe. The explosion is called the **Big Bang**. However, we must not think of this like an exploding bomb. When a bomb explodes the shrapnel flies off into space. When the Big Bang happened matter did not fly off into space but space and time itself were created. Before the Big Bang there was no time and there was no space. As the universe expands more and more space is created. In this sense it is wrong to think of the galaxies rushing away from each other. It is the space in which the galaxies are situated that is expanding. There is simple analogy that helps understand this and one that you can easily demonstrate for yourself. Partially inflate a toy balloon and then stick little bits of paper on to it at different places. The surface of the balloon represents space and the bits of paper represent galaxies. As you now further inflate the balloon you will see the "galaxies" move away from each other. The galaxies move even though they are fixed in space because space is expanding.

We must not take the balloon analogy too literally since the balloon is expanding into space whereas the expanding Universe is creating space. There is no outside of the Universe into which it expands. So to ask the question "what is outside the Universe?" is just as meaningless as the question "what was there before the Big Bang?" However, the analogy does help us appreciate that there is no centre to the Universe. Any one of the bits of paper can be chosen as a reference point and all other bits of paper will be appear to be receding form this reference point. This is exactly the same for the galaxies in our Universe.

The expanding Universe also helps us understand the origin of the red-shift observed from distant galaxies. Strictly speaking this is not due to the Doppler effect since the galaxies themselves are not receding from us, it is the space between them that is expanding. When a photon leaves a galaxy and travels through space as the space through which it travels expands, the wavelength associated with the photon will increase in length. The longer it spends on its journey through space the more its wavelength will increase. Hence when the photon reaches our eyes it will be red-shifted and the further from whence it originated, the greater will be the red-shift.

Cosmic microwave background (CMB) radiation

Perhaps the most conclusive piece of evidence for the Big Bang is the existence of an isotropic radiation bath that permeates the entire Universe known as the cosmic microwave background (CMB) radiation. The word isotropic means the same in all directions; the degree of anisotropy of the CMB is about one part in a thousand.

We know that about 24% of the Universe is made up of helium. Calculation shows that the helium produced by nuclear fusion within stars cannot account for this amount. In 1960 two physicists, Dicke and Peebles proposed that sometime during the early history of the Universe it was at a sufficiently high temperature to produce helium by fusion. In this process many high-energy photons would be produced. The photons would have a black body spectrum corresponding to the then temperature of the Universe. As the Universe expanded and cooled the photon spectrum would also change with their maximum wavelength shifting in accordance with Wien's law. It is estimated that, at the present time, the photons should have a maximum wavelength corresponding to a black body spectrum of 3 K.

Shortly after this prediction, Penzias and Wilson were working with a microwave aerial and they found that, no matter in what direction they pointed the aerial, it picked up a steady, continuous background radiation. They soon realised that they had discovered the remnants of the radiation predicted by Dicke and Peebles. This was further evidence that the Universe started life with a bang.

The most modern day measurements confirm all space is filled with radiation corresponding to a black body spectrum of 2.76 K.

The Big Bang model leads to the idea that space is expanding and that within this space, the distribution of the galaxies (and the stars in them) is not uniform. The model is therefore completely at odds with Newton's model of a uniform, static universe, thereby providing a resolution to Olber's paradox.

Hubble's law

We have seen that the Doppler red-shift indicates that the galaxies are moving away from the Earth. By estimating the distance to a number of galaxies Hubble showed that the speed with which the galaxies are moving away from the Earth, the recession speed, is directly proportional to the distance of the galaxy from the Earth. Remember that the galaxies are not actually rushing away from Earth, but that it is actually space that is expanding.

Chapter 16 (Option D)

Hubble published his discovery in 1929 and **Hubble's law** can be written.

$$v = H_0 d$$

where d is the distance to the galaxy and v is its recession speed. H_0 is known as Hubble's constant. An accurate value for Hubble's constant is difficult to measure (as will be shown) but an 'average' value is about 65 km s^{-1} Mpc^{-1}.

For example a galaxy at a distance of 50 Mpc from the Earth will be rushing away from the Earth with a speed of 3250 km s^{-1} (about 0.01 the speed of light). The further away from the Earth, the greater will be the recession speed of a galaxy. The expanding balloon analogy of the expanding universe fits in with the data that enabled Hubble to arrive at his law.

If you concentrate on one galaxy you will see that galaxies that are further away move faster than those nearer to the galaxy that you have chosen. If the surface of the balloon is expanding at a constant rate then the speed with which a galaxy moves relative to another galaxy will be proportional to how far away it is from the other galaxy. The diagram below shows this for a very simple arrangement of five beads on an elastic string.

In Figure 1623 (a) the beads are 10 cm apart. In Figure 1623 (b) the string is stretched at a steady rate until the beads are 15 cm apart. Bead E will now be 60 cm from bead A instead of 40 cm and if the stretching of the string took 1 second, the speed of E relative to A will be 20 cm s^{-1}. On the other hand bead C will now be 30 cm from bead A instead of 20 cm and its speed relative to A will be 10 cm s^{-1}. Doubling the separation between any two beads doubles their relative speed; speed is proportional to separation that is just Hubble's law.

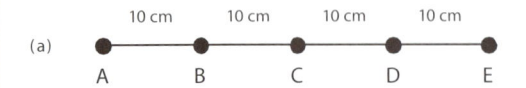

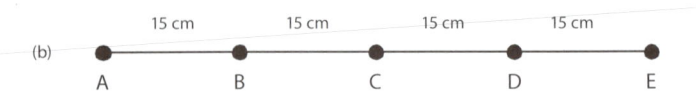

Figure 1623 (a) and (b)   An illustration of Hubble's Law

Hubble was the first astronomer to settle a long-standing debate as to the nature of galaxies. It had been known since 1845 (thanks to the then most powerful telescope in the world, built by *William Parsons*) that some of the nebulae showed a spiral structure. Parsons himself suggested that such nebulae could be "island universes" far beyond the Milky Way. In the 1920s opinions were still divided as to their nature with some astronomers of the opinion that they were relatively small objects scattered about the Milky Way.

In 1923 Hubble took a photograph of the Andromeda nebula that on close examination showed a bright object that he recognised as a Cepheid variable. From measurements of the variable's luminosity he was able to show that the Andromeda nebula is 900 kpc from the Earth and that it has a diameter of some 70 kpc, a diameter much greater than that of the Milky Way. The debate was settled. The universe was far greater in size than had previously been thought and contained many galaxies, some much larger than our own Milky Way.

The limitations of Hubble's law

Determining galactic distances is not easy but it is of key importance. What is needed are "standard candles" which are bright enough to be detected within a particular galaxy. Cepheid variables provide a fairly reliable method for distances up to about 60 Mpc but beyond this they become too faint to be relied upon as a standard candle. Beyond 60 Mpc Red and blue super giants can be used to measure distances of up to about 250 Mpc and the brightest globular clusters can be used as standards up to about 900 Mpc. Beyond 900 Mpc astronomers have to hopefully detect a supernova within the galaxy. To date supernovae have been detected at distances of up to 1000 Mpc.

As astronomers attempt to measure greater and greater galactic distances so the error in their measurements increases. This is because the further away the galaxy, the fewer the independent measurement checks that are available. For instance, if a distance can be measured by parallax it can probably also be checked by using a Cepheid variable, a super red-giant and spectroscopic parallax. At distances beyond 900 Mpc the only method available is to use the luminosity of a supernova. On the other hand if we have an accurate value of Hubble's constant then we can use the Hubble law and red-shift to measure large galactic distances. And there's the rub. Astronomers who use different methods of measuring galactic distances compute different values of the Hubble constant. For example, measurement by supernovae leads to a value between 40 and 65 km s^{-1} Mpc^{-1} whereas measurement by spectroscopic parallax places it between 80 and 100 km s^{-1} Mpc^{-1}. More recent methods using the observation of Cepheid variables from the Hubble telescope yield a value 60 and 90 km s^{-1} Mpc^{-1}.

An accurate value of the Hubble constant would also enable us to calculate the age of the Universe with some degree of confidence.

Consider, for example, two galaxies separated by a distance d. How long did they take to get this distance apart?

If we assume that the universe is expanding at a constant rate, the age or expansion time of the universe is known as the Hubble time T, and it is equal to the reciprocal the Hubble constant as.

Time = distance/speed

From Hubble's Law: $v = H_0 d$

Therefore $T = \dfrac{d}{H_0 d}$

$T \approx \dfrac{1}{H_0}$

This means that the universe could have an age ('could' because we are assuming that the expansion of the Universe is constant) somewhere between

$\dfrac{1}{30}$ km^{-1} s Mpc and $\dfrac{1}{45}$ km^{-1} s Mpc.

Bearing in mind that 1 Mpc = 3×10^{19} km and 1 year = 3×10^7 s this gives an age between 10 and 20 billion years.

Figure 1624 shows the Hubble time if the galaxies have receded to their present distances. Assuming that they have been travelling at constant speed. If the rate of expansion is slowing down, as shown by the deceleration curve, then the actual age of the universe would be less than the Hubble time. Figure 1624 shows the Hubble time if the galaxies have receded to their present distances. Assuming that they have been travelling at constant speed. If the rate of expansion is slowing down, as shown by the deceleration curve, then the actual age of the universe would be less than the Hubble time.

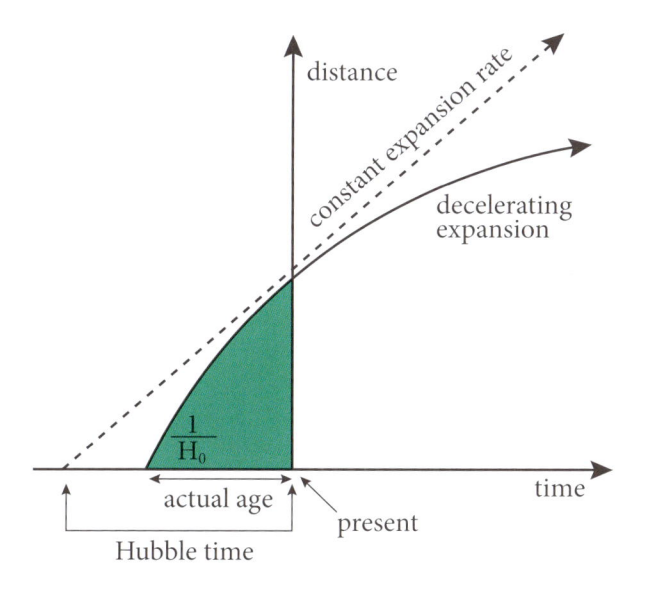

Figure 1624   Hubble time

Example

The wavelength of the blue line in the spectrum of atomic hydrogen as measured in a laboratory on Earth is 486 nm. In the spectrum from a distant galaxy the wavelength of this line is measure as 498 nm. Estimate the recession speed of the galaxy and its distance from Earth.

Solution

From the red-shift we can calculate the recession speed of the galaxy.

$\Delta \lambda = 498 - 486$ nm = 12 nm

Therefore

$\dfrac{12}{486} = \dfrac{v}{3 \times 10^8} = 0.025$

Hence $v = 7400$ km s^{-1}.

Using Hubble's law (see following section) we have:

$d = \dfrac{7400}{65} = 114$ Mpc.

The accelerating universe and redshift (z)

All of the best world telescopes are attempting to trace the expansion history of the universe. Galaxies are not useful for these studies because their properties vary greatly and their size or brightness are not good indicators of distance. Cephieds are good standard "candles" but it is hard to find them beyond 100 million light years, and this is still the local universe. The favoured tool for far away measurements is supernovae that occur in binary star systems. A nova appears to be an event that occurs on the surface of a compact white dwarf in a close binary system. When material from the neighbouring star undergoes thermonuclear reactions on the white dwarf's surface, it triggers violent explosions, and the star becomes between a few thousand and a million times more brighter than before.

In the 1990s, two groups of researchers started to look for supernovae to measure their peak brightness and decay rate as a measure of their absolute brightness. By combining their absolute brightness and apparent brightness, they could find their distance because their light diminishes as the inverse square of their distance.

They were amazed to find that the supernovae at redshifts of 0.5 were fainter than expected for a constant expansion rate. This must mean that they were accelerating as they

485

Chapter 16 (Option D)

moved further and further away. Type Ia supernovae offer a unique opportunity for the consistent measurement of distance out to perhaps 1000 Mpc. The acceleration implies an energy density that acts in opposition to gravity that would cause the expansion to accelerate.

Furthermore, supernovae found at redshifts between 0.5 and 1 with greater distances in space were decelerating.

So what causes the acceleration and the deceleration, and will the universe forever expanding or will it one day collapse? These are questions that are trying to be answered by cosmologists in terms of "dark energy".

Redshift refers to the lengthening of the wavelength of light at observation compared with its length at emission. Several processes can produce redshifts:

- Due to the relative motion in the sources (the simple Doppler shift).

- Due to a relativistic time dilation.

- Due to a photon losing energy as it climbs out of a gravitational field.

The term cosmological redshift z is reserved for the redshifts produced by the overall expansion of space expressed as:

$$z = \frac{\lambda_{obs} - \lambda_{rest}}{\lambda_{rest}}$$

or

$$z = \frac{\Delta \lambda}{\lambda_0} \approx \frac{v}{c}$$

where v is the velocity of the source and c is the speed of light.

It is important to keep the various causes of redshift distinct.

It can be shown that if λ is the wavelength of a spectral line emitted from a stationary source and λ' is the wavelength measured by an observer when the source is moving away from the observer with speed v then

$$\lambda' = \lambda \sqrt{\frac{1 + \frac{v}{c}}{1 - \frac{v}{c}}}$$

where c is the free space velocity of light.

This expression in parenthesis can be written as

$$\frac{\left(1 + \frac{v}{c}\right)\left(1 - \frac{v}{c}\right)}{\left(1 - \frac{v}{c}\right)^2} = \frac{\left(1 - \frac{v^2}{c^2}\right)}{\left(1 - \frac{v}{c}\right)^2}$$

Taking the square root we have,

$$\lambda' = \lambda \sqrt{\frac{\left(1 - \frac{v^2}{c^2}\right)}{\left(1 - \frac{v}{c}\right)^2}} = \frac{\lambda}{\left(1 - \frac{v}{c}\right)} \sqrt{\left(1 - \frac{v^2}{c^2}\right)}$$

If we consider the situation when $v \ll c$ then we can expand this expression by the binomial theorem and ignore second order and higher terms to get

$$\lambda' = \lambda\left(1 + \frac{v}{c}\right) \quad \text{or} \quad \lambda' - \lambda = \frac{\lambda v}{c}$$

such that if the spectral line has been shifted by an amount $\Delta \lambda = \lambda' - \lambda$ then

$$\frac{\Delta \lambda}{\lambda} = \frac{v}{c}$$

Example

A characteristic absorption line found due to ionized helium has a wavelength of 468.6 nm on earth. If the line is shifted to 499.3 nm in a star, determine the recession speed of the star.

Solution

$$\frac{\Delta \lambda}{\lambda_0} = \frac{(499.3 - 468.6)}{468.6} = \frac{v}{c}$$

$$= 6.55 \times 10^{-2}$$

Therefore $v = 6.55 \times 10^{-2} \times 3 \times 10^8 = 1.97 \times 10^7$ m s^{-1}

ASTROPHYSICS

The cosmic scale factor (R)

The cosmological red-shift z that is observed in the spectrum of galaxies can be related to the cosmic scale factor R by which the universe has expanded since the radiation which we are now receiving since it was emitted.

The exact relationship is given by:

$$1 + z = \frac{R}{R_0}$$

$$z = \frac{R}{R_0} - 1$$

where R is the present scale factor, R0 is the scale factor at the time the radiation was emitted.

$\frac{R}{R_0}$ is the factor by which the universe has expanded since

the time at which presently observable radiation was emitted. For example, with a quasar of red-shift 4, we are seeing it, as it was when the universe was one-fifth its present size.

$$\left(\frac{1}{1+z} = \frac{1}{1+4} = \frac{1}{5}\right)$$

Summary

Let us summarise what we know so far about stellar evolution.

We have learnt that galaxies are grouped into clusters that in turn are grouped into super-clusters. We also learned that due to the expansion of the Universe the light that we receive from galaxies is red–shifted. We now see how we can calculate the recession speed of galaxies from their red-shift.

There were a number of important facts that had been established about the Universe. Firstly, it was known that the approximate age of the universe was 20×10^9 years old as calculated from Hubble's Law.

Secondly, Arno Penzias and Robert Wilson discovered cosmic microwave background radiation. The intensity of this radiation at a wavelength of 7.35 cm corresponded to blackbody radiation (Stefan's Law) of 2.76 K, the present temperature of the universe.

With the further knowledge gained from particle physics, there has been some convincing evidence as to how the universe began and how it has evolved in what is termed the standard model. Before the Big Bang, there was no time and space but just a point of extremely dense matter known as a singularity. There was a huge release of energy that was accompanied with temperatures of the order of magnitude greater than 10^{32} K, and the four forces in nature would have been unified into a single force.

Because the quantum effect of gravity has not yet been solved, we begin the journey at 10^{-43} s when the temperature was 10^{32} K. At this temperature, no atoms could have formed. Instead, it is thought that universe consisted of photons and other elementary particles. The universe would have been opaque because as soon as the photons emitted radiation, they would have been scattered and absorbed by electrons and positrons. This energy would have spread out in all space dimensions and the temperature would have dropped.

Let us examine the evolution of the universe from its beginning to now in more detail. It has already been established that we can only surmise what happened after 10^{-43} s. Figure 1625 demonstrates the variation of temperature with time in the evolutionary process.

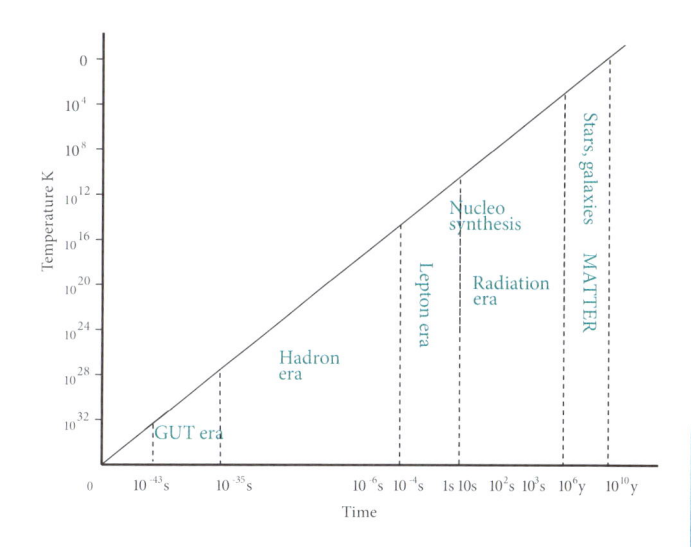

Figure 1625 *Variation of temperature with time of the universe.*

The average kinetic energy of a particle is given by the expression:

$$E = \frac{3}{2} kT$$

where k is **Boltzmann's constant** that has a value of 1.38×10^{-23} JK^{-1}. Because we are dealing with such large energies this equation can be stated as $E = kT$.

Prior to 10^{-43} s, it is speculated that the four forces were unified into only one force and it is believed that the temperature was around 10^{32} K. This means that the energy of a particle would be approximately:

$$E = (1.38 \times 10^{-23} \text{ JK}^{-1} \times 10^{32} \text{ K}) \div (1.6 \times 10^{-19} \text{ JeV}^{-1}) = 10^{19} \text{ GeV}.$$

The particles had so much energy that the symmetry of the four forces was disrupted and the strong, electromagnetic and weak forces created an array of quarks, leptons, photons, W and Z plus gluons. The gravitational force

487

Chapter 16 (Option D)

"condensed out". There was a slight imbalance between the matter and anti-matter already occurring.

At about 10^{27} K and 10^{-35} s, it is thought that the strong force separated out and because the quarks were too close to each other, the strong force could not bind them to form hadrons. There was a sea of quarks, gluons and leptons. However, as the quarks started to separate, quark confinement occurred and hadrons began to form. During the hadron era, the excess of matter over antimatter of the GUT era meant that there was a slight excess of quarks versus antiquarks creating an excess of baryons versus anti-baryons.

At about 10^{15} K and 10^{-6} s, it is believed the weak force separated from the electromagnetic force.

At about 10^{13} K and 10^{-4} s, it is thought that hadrons with an energy of about 1 GeV started to annihilate each other and there existed pair-production/annihilation equilibrium. However, once the energy had dropped below 1 GeV, the pair production of nucleons could not take place. There was a slight excess of energy from the annihilations, enough to form some leftover nucleons, and it is this mass that is in the universe today. Light particles such as the photon and lighter leptons thus dominated the universe in equal numbers and thus began the lepton era.

At about 10^{10} K and 1 s, it is believed that the lighter leptons with energy about 1 MeV were still able to create electrons, positrons and photons in equal numbers and equilibrium between pair-production and annihilation still existed. However, within a few seconds, electrons and positrons started to annihilate each other in larger numbers and their numbers dropped and there was a slight excess of electrons over positrons. At about 10s, there was a large excess of photons and neutrinos and the radiation era began.

At about 10^9 K and 3.2 s, crucial events began to occur as atoms of hydrogen, helium, deuterium and lithium started to form in what is known as **nucleosynthesis**. Let us do a quick calculation of the temperature for average kinetic energy of particles around 500 keV.

$$E = \tfrac{3}{2} kT \text{ and } T = \tfrac{2/3 \, E}{k}$$

$$= \frac{0.67 \times 500 \times 10^3 \text{ eV} \times 1.6 \times 10^{-19} \text{ JeV}^{-1}}{1.38 \times 10^{-23} \text{ JK}^{-1}} = 3.9 \times 10^9 \text{ K}$$

The universe was cooling too quickly for heavier atoms to form and nucleosynthesis stopped.

At about 3000 K and 300 000 years, free electrons with energies of a few electron-volts were bound to nuclei to form atoms. The photons that were previously ionizing atoms were free to spread throughout the universe. The universe became transparent. The photon radiation became red-shifted and the radiation cooled to 2.7 K and they formed the cosmic background radiation that we detect from every part of the universe. So from this point the universe became matter-dominated.

The early universe contained almost equal numbers of particles and antiparticles with a small imbalance favouring matter over anti-matter. However, once photons had reached energies at which they could not undergo pair production from their interaction with particles and anti-particles, matter dominated over anti-matter.

Measuring Astronomical distances – A summary

We conclude this chapter with a summary of the methods available for measuring astronomical distances.

Figure 1626 summarises the methods for measuring astronomical distances.

You should bear in mind that, apart from the parallax method, all the other methods rely on determining the luminosity of some object and then measuring its apparent brightness.

Distance	Method
up to 100 pc	Parallax and Cepheid variables and spectroscopic parallax
up to 10 Mpc	Cepheid variables and spectroscopic parallax
up to 60 Mpc	Cepheid and spectroscopic parallax
up to 250 Mpc	Super red giants and super blue giants and supernovae
up to 900 Mpc	Globular clusters and supernovae
beyond 900 Mpc	Supernovae

Figure 1626 Methods for measuring astronomical distances (table)

ASTROPHYSICS

Exercise 16.2

1. The apparent magnitude of the Andromeda galaxy is 4.8 and that of the Crab Nebula is 8.4. Determine which of these is the brightest and by how much.

2. Estimate the age of the Universe for a value of the Hubble constant that is equal to 80 km s^{-1} Mpc^{-1}.

3. The star Alpha–Centauri B is 1.21 pc from Earth. Calculate

    (a) this distance in AU.
    (b) its parallax angle.

4. Two stars A and B are respectively at distances 50 pc and 500 pc from the Earth. Both have equal brightness. Determine which star is the most luminous and by how much.

5. The diagram below shows the Apparent magnitude scale used by Astronomers.

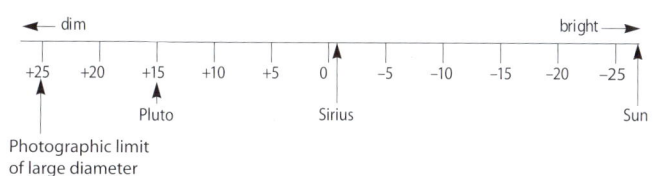

    ← dim   bright →
    +25 +20 +15 +10 +5   0   -5 -10 -15 -20 -25
                        Pluto   Sirius   Sun
    Photographic limit
    of large diameter
    lens telescope

    (a) Explain whether the Sun would be visible if it were at a distance of 10 pc from the Earth.

    (b) Estimate by how much brighter is the Sun than Sirius A.

    (c) The absolute magnitude of Sirius A is less than that of the Sun. Explain whether it is more or less luminous than the Sun. (Sirius A: m = –0.7, Sun: m = –26.7).

6. Explain the difference between an eclipsing binary and a spectroscopic binary.

7. Outline the evidence on which the idea of an expanding Universe is based.

8. A certain line in the spectrum of atomic hydrogen has a wavelength of 121.6 nm as measured in the laboratory. The same line as detected in a distant galaxy has a wavelength of 147.9 nm. Determine the recession speed of the galaxy.

9. State the property of a main sequence star that determines its final outcome. Describe the evolution of a main sequence star to a neutron star.

10. Calculate the wavelength at which the Sun emits most of its energy? (T$_{Sun}$ = 5800)

11. Stars can be assigned to certain spectral classes. The classes are given in the table below.

Spectral class	Colour	Temperature (K)
M	Redorange	2500-3500
K	Orange	3500-5000
G	Yellow	5000-6000
F	Yellowwhite	6000-7500
A	White	7500-10,000
B	Bluewhite	10,000-28,000
O	Bluewhite	28,000-50,000

The table below gives the spectral class and absolute magnitude of some well-known stars.

Name of star	Spectral class	Absolute magnitude (approximate)
Rigel	A	7.0
Vega	A	0
Sun	G	5
Alderbaran	K	0
Pollux	K	+ 2
Sirius B	B	+ 12
Procyon B	F	+ 14
Barnard's star	M	+ 13

(a) Use this table to place the stars on a Hertzprung–Russell diagram in which absolute magnitude is plotted against spectral class (temperature).

(b) For each of the stars identify to which category it could belong.

(c) Identify a star that is hotter and more luminous than the Sun and a star that is cooler and less luminous than the Sun.

489

Chapter 16 (Option D)

D.4 Stellar processes

> **NATURE OF SCIENCE:**
>
> Observation and deduction: Observations of stellar spectra showed the existence of different elements in stars. Deductions from nuclear fusion theory were able to explain this. (1.8)
>
> © IBO 2014

Essential idea: The laws of nuclear physics applied to nuclear fusion processes inside stars determine the production of all elements up to iron.

Understandings:
- The Jeans criterion
- Nuclear fusion
- Nucleosynthesis off the main sequence
- Type Ia and II supernovae

The Jeans criterion

In 1903, Sir James Jean in his article " The stability of a Spherical Nebula" proposed that a gas cloud would not change its shape so long as the potential energy of the internal gravitational force is twice the kinetic energy of motion of the gas particles as proposed by the virial theorem. The virial theorem says that half the change in gravitational energy stays with the star (it heats the star). The other half is radiated away.

If the mass is too high or the temperature is too low, for the equilibrium condition to be satisfied, the cloud will collapse due to gravity. Jean calculated the mass that this happens in the nebula, and this is the physical basis for star formation.

The collapse of an interstellar cloud can only happen if the mass is greater than the Jean mass.

$M > M_J$

Nuclear fusion

The conditions that initiate fusion in a star

We have briefly looked at the way in which stars are formed and also seen that the main source of energy in stars is nuclear fusion. In this section we look at both these processes in more detail and also trace the evolution of stars.

If you view the constellation of Orion with binoculars or a telescope you will see that the middle star of the three stars that makes up Orion's "sword" is very fuzzy. In fact it is not a star at all but a gas cloud – the Orion nebula. This nebula emits its own light and has a temperature of some 10,000 K. Radiation from the two stars in the "sword" excite the hydrogen ions in the nebula and, when the excited atoms de-excite, they emit light. Whereas the Orion nebula consists mainly of hydrogen gas (and some helium) other nebulae can be found that contain dust particles. These particles scatter any starlight and so appear dark against the background of emission nebulae. The most celebrated of the dark nebulae is the Horsehead nebula, so-called because of its distinctive shape. This can be seen against the emission nebula of Orion. (The nebulae associated with Orion's sword are in fact some 450 pc beyond the other two stars in the sword). A typical dark nebula has a temperature of about 100 K and contains between 10^{10} and 10^{19} particles. The particles consist of hydrogen (75%) and helium molecules (24%) and dust (1%). The dust consists of atoms and molecules of many different elements.

It would seem that the dark nebulae are the birthplaces of the stars. Their temperature is low enough and their density high enough for gravity to pull the individual particles together. As the particles move together under their mutual gravitational attraction they lose gravitational potential energy and gain kinetic energy. In other words the temperature of the system increases and as the temperature increases ionisation of the molecules will take place and the system will acquire its own luminosity. At this point the so-called protostar is still very large and might have a surface temperature of some 3000 K and therefore has considerable luminosity. A protostar of mass equal to the Sun can have a surface area some 5000 times greater than the Sun and be 100 times as luminous.

As the gravitational contraction continues the temperature of the core of the protostar continues to rise until it is at a sufficiently high temperature for all the electrons to be stripped from the atoms making up the core. The core has now become plasma, and nuclear fusion now takes place in which hydrogen is converted into helium (sometimes referred to by astronomers as hydrogen burning) and the protostar has become a main sequence star on the Hertzsprung-Russell diagram. The nuclear fusion process will eventually stop any further gravitational contraction and the star will have reached hydrostatic equilibrium in which gravitational pressure is balanced by the pressure created by the nuclear fusion processes.

Whereabouts a protostar "lands" on the main sequence is determined by its initial mass. The greater the initial mass the higher will be the final surface temperature and the greater will be its luminosity. This is illustrated in the Figure 1626.

Astrophysics

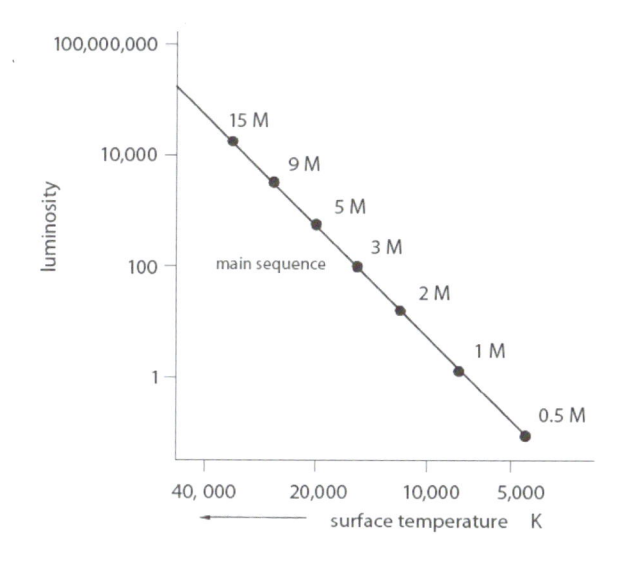

Figure 1626   The relationship of mass and luminosity

The more massive a protostar (more than about 4M☉) the more quickly its core will reach a temperature at which fusion takes place. (Protostars with a mass of about

15 M☉ will reach the main sequence in about 10^4 years whereas protostars with a mass of about 1 M☉ will take about 10^7 years). Its luminosity quickly stabilises but its surface temperature will continue to increase as it further contracts. For protostars with less than this mass, the outer layers are relatively opaque (due to the presence of a large number of negative hydrogen ions) so little energy is lost from the core by radiation. Energy is actually transferred from the core to the surface by convection ensuring that the surface temperature stays reasonably constant. The luminosity therefore will decrease as the protostar contracts.

Gravitational collapse puts a lower and upper limit on the mass of matter that can form a star. As we saw in a previous section, a protostar with a mass less than about 0.08 M☉ will not develop the pressure and temperature necessary to initiate nuclear fusion and will contract to a brown dwarf. If a protostar has a mass greater than about 100 M☉ then the internal pressure created by contraction will overcome the gravitational pressure and vast amounts of matter will be ejected from the outer layer of the protostar thereby disrupting the evolution of the star.

Star's mass on the end product of nuclear fusion

At the end of its lifetime as a main sequence star, all the hydrogen in its core has been used up. How long this takes and a star's ultimate fate depends upon its initial mass. Our Sun, for instance, has been converting hydrogen to helium within its core for some 5×10^9 years and will continue to do so for at least another 5×10^9 years. On the other hand a star with a mass of about 25 M☉ will use up all its hydrogen in about 10^6 years.

Post Main Sequence evolution

Solar mass stars

A star cannot stay in the main sequence forever, as its hydrogen will eventually run out. The time that it spends there is depends on its mass. We have seen that the more mass it has, the more luminous it is and as it uses up its fuel at a faster rate, its lifetime is shorter.

When the core of a star runs out of hydrogen it is no longer in equilibrium and the gravitational force will once again cause the core to contract. The temperature rises in the shell of gas surrounding the core and hydrogen burning commences in the shell. The temperature continues to increase and reactions occur more rapidly and the luminosity increases. The star's envelope expands and the surface area increases as a function of $4\pi r^2$ as shown in Figure 1627. The total energy radiated per square metre increases and as a result the temperature decreases. The star moves up (increased luminosity) and to the right (decreased temperature) on the HR diagram.

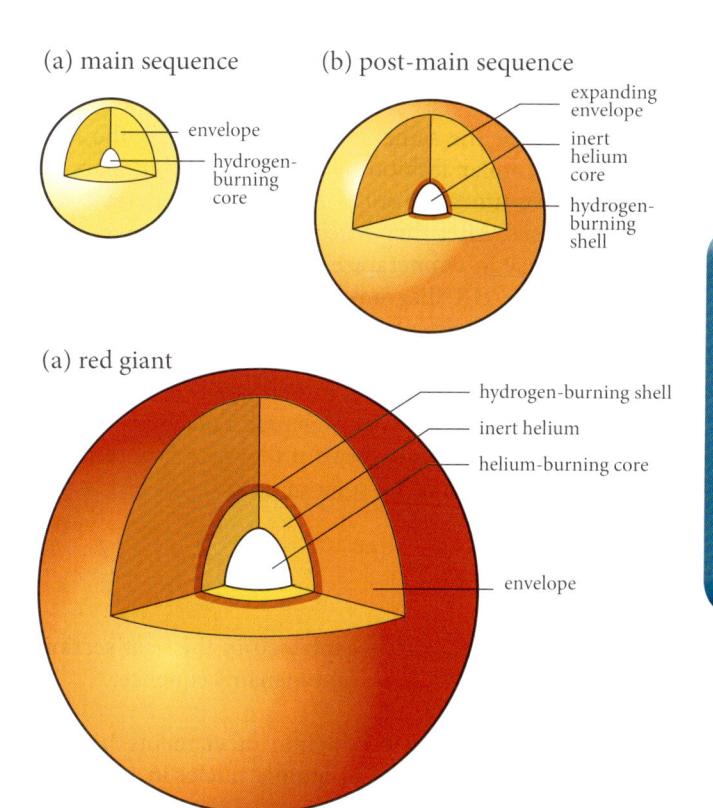

Figure 1627   The evolution of a red giant star

During the time (a few hundred million years) the star is evolving from the main sequence to the red giant phase, helium produced from the hydrogen shell is being added to the helium core. The core contracts and temperature progressively reaches about 10^8 K. Triple alpha reactions commence:

491

He-4 + He-4 → Be-8

Be-8 + He-4 → C-12 + γ

C-12 + He-4 → O-16 + γ

As a consequence, the core releases energy in large amounts and the red giant expands and cools down with the luminosity decreasing. Consequently, it gradually moves lower and to the left of the main sequence of the HR diagram. It still has more luminosity and a higher temperature than it did when it was a main sequence star. It settles into a stable phase and the hydrogen continues to form helium in the core. The core contract as the helium builds up, and the star expands into what is called an AGB (asymptotic giant branch). Again it becomes unstable due to the contraction of the core, and it emits huge amounts of light and gases in what is called a planetary nebula. The hydrogen and helium shells are exposed and they cease to generate energy. The shrinking core cannot contract enough to increase its temperature and the shrunken, cool remnant of the star becomes a white dwarf.

Medium mass stars

Stars smaller or the same approximate mass as the Sun evolve in a similar fashion as just outlined but stars of mass, say five times the solar mass, evolve differently in the later stages. After they reach the red giant stage, they slowly increase in temperature and luminosity and reach a stage on the HR diagram called the instability strip. They undergo a series of expansions and contractions in a cyclic fashion and form a pulsating variable similar to the present Cephied variable. It then swells up again and expels its envelope of hydrogen forming a planetary nebula and evolving into a white dwarf if the mass is within the Chandrasekhar limit of 1.4 solar masses.

High mass stars

The high mass stars will be explained in the next section concerning nucleosynthesis off the main sequence.

Nucleosynthesis is the production of elements heavier than helium by the fusion of atomic nuclei in stars and during supernovae explosions.

From an understanding of stellar structure and stellar evolution, we can understand the origin of elements heavier than helium. When the universe began it only contained the elements hydrogen and helium. All the remaining elements have been produced by nucleosynthesis.

When our galaxy, the Milky Way, formed some time after the beginning of the universe, it contained around 90% of hydrogen atoms and 10% of helium atoms.

The first stars that formed were metal-poor. Later generation stars manufactured heavier atoms and the metal abundance increased.

The most massive stars fused helium into carbon, nitrogen, oxygen, and heavier elements up to iron. When these stars die in supernova explosions they can fuse even heavier elements, and all these are dispersed back into the interstellar medium. Figure 1628 shows the abundance of elements in the universe on an exponential scale 1628(a) and a linear scale 1628(b).

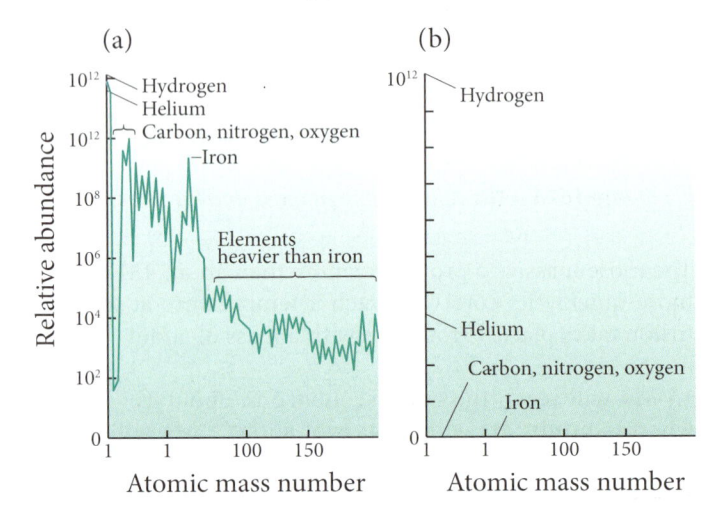

Figure 1628   Abundance of the elements in the universe

We will take the Sun as an example of the changes that take place in nucleosynthesis when a star leaves the main sequence and becomes a red giant. As the Sun ages the continuous energy flow from the core heats the material surrounding the core such that, when all the hydrogen in the core has been used up, hydrogen burning can continue in the surrounding material. There are no fusion processes in the core to counteract gravitational contraction so the core will now start to contract. As the core contracts its temperature will rise and the energy flow from the core will further heat up the outer layers of the Sun. Hydrogen burning now extends further and further into the outer regions and so, as the core contracts, the Sun as a whole actually expands.

This expansion causes the Sun's surface temperature to drop and its luminosity to increase. At a surface temperature of about 3500 K the surface will take on a reddish hue (Wien's law) and the Sun will have a diameter of about 1 AU, sufficient to engulf Mercury and nearly reach Venus. It will also have luminosity about 2000 times that of its present day luminosity.

The helium created by hydrogen burning in the outer layers of the Sun adds to the mass of the core causing the core to further contract. The core temperature will rise and eventually reach a temperature high enough for the fusion of helium to take place. Two of the products of the helium fusion process are carbon-12 and oxygen-16. Most of the carbon in living tissue originated in the core of stars like the Sun in their death throes.

When all the helium in the core has been used up, the core further contracts and its temperature rises such that the energy radiated from the core will now cause helium burning in the outer layers. The Sun has entered a second red giant phase. When it enters this phase its outer layers will reach out and engulf Earth, and it will have luminosity some 10,000 times that of its present luminosity. When it enters this phase it undergoes bursts of luminosity in which a shell of its outer layers is ejected into space. As the Sun ejects its outer layers its very hot core will be exposed.

This core will have a surface temperature of about 100,000 K and the radiation that it emits will ionise the outer gas layers causing them to emit visible radiation producing an, inappropriately named, planetary nebula. The radius of the core will be about that of the Earth and with no fusion reaction taking place within the core it will just simply cool down. The Sun has become a white dwarf star, and as it continues to cool it will eventually fade from sight.

The deaths of high mass stars

In high mass stars with masses between ten and a hundred solar masses, the temperatures of the core are extremely high ranging from around 600 million K and one billion K. There occurs successive burning of a series of elements starting with carbon and oxygen triple alpha reactions. In stars around 4 solar masses at 600 million K, carbon burning nuclear reactions can occur and neon, magnesium, oxygen and more helium are produced for a relatively short period of 600 years.

If the star is initially above nine solar masses, it core contacts greatly and the temperature of the central core increases rapidly to around a billion degrees and neon burning can commence in the supergiant star for one year. With further rises in temperature, the star goes through stages of oxygen burning, and then silicon burning very quickly. The silicon burning eventually gives rise to the forming of iron. Once the iron core is formed, no further energy generation can be produced by thermonuclear reactions, and the core collapses. Figure 1629 shows the series of thermonuclear reactions taking place in concentric shells deep down into the interior.

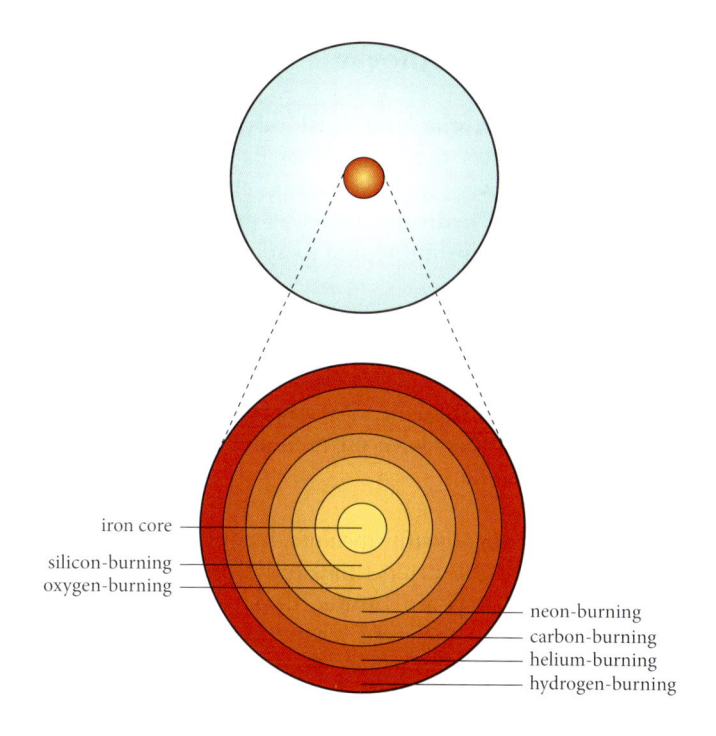

Figure 1629   Thermonuclear reactions in a supergiant.

If the iron core exceeds the Chandrasekhar limit, it cannot form a white dwarf. Instead the iron core collapses on itself due to gravitational self-attraction and the density of the core becomes so great that electrons are forced to combine with protons to form neutrons and huge numbers of neutrinos. These neutrons are packed so closely together that the neutron degeneracy pressure will halt further collapsing. The neutrons are captured – neutron capture. The end result is the formation of a neutron star.

Neutron capture can occur by two main processes called the s and r processes and depends on neutron flux. The slow neutron capture process (s process) is responsible for the production of half the abundances of elements heavier than iron in a galaxy. In the s process, there is low neutron density of the order of 10^8 neutrons per cubic centimetre. If neutrons are added slowly, the unstable nuclei have time to undergo

β-decay. This results in the formation of elements such as strontium, zirconium and barium. If fact, some isotopes can only be produced by slow neutron capture such as strontium-86, molybdenum-96, palladium-104 and tin-116. In the rapid neutron capture process (r process), the neutron flux is high and the nucleus captures many neutrons before it can decay. The r process occurs in supernovas.

In the neutron star, these materials fall inwards to the core and rebounds setting up shock waves that blast the elements formed into space in a cloud of debris and this mixes with surrounding interstellar gas clouds thus enriching them with heavy elements.

CHAPTER 16 (OPTION D)

Type Ia and II supernovae

The type of high-mass star just described where the destruction follows the collapse of its core produces a type II supernova. The expanding envelope contains large amounts of hydrogen that was not consumed and hydrogen lines are prominent in its spectrum.

Supernovas in which the hydrogen lines are weak or absent are known a type 1 supernovas. Type 1 supernovas are further divided into 3 sub-classes- Ia, Ib and Ic. Types Ib and Ic are also thought to result from the collapse of large mass stars that drive off their hydrogen envelopes before they explode.

Type Ia supernovas with absolute magnitudes between -19 and -20 are about ten times more brilliant tan type II supernova. They are produced by a total different mechanism. It is believed that they occur due to thermonuclear carbon burning deep inside a white dwarf that has carbon and oxygen envelopes but no hydrogen envelope. This is likely to happen when the white dwarf has a mass close to the Chandrasekhar limit and is part of a binary system where its companion secretes hydrogen onto it. Helium is produced and the mass slowly builds up. The white dwarf starts to collapse and heats the carbon in its core that leads to a huge surge in energy generation that leads to the star being blown to pieces.

The cloud of debris sweeps out and compresses the interstellar gases in which it is expanding and creates a growing hole that is surrounded by the expanding shell of compressed gas. Fast moving electrons in the cloud emit synchrotron radiation from in all areas of the electromagnetic spectrum. Radio waves are not affected by the clouds of dust that affect the resolution of optical telescopes, and radio observations can be made to reveal the supernova remnants.

Type Ia supernova can be used as standard candles along with Cephied variables. Type 1a supernovae are very bright – often as bright as all the stars in a whole galaxy put together. Because they are so bright, we can see them at very great distances. The disadvantage of supernovae as standard candles is that they don't hang around - you have to spot them when they go off, or shortly afterwards.

D.5 Further cosmology

> **NATURE OF SCIENCE:**
>
> Observation and deduction: Observations of stellar spectra showed the existence of different elements in stars. Deductions from nuclear fusion theory were able to explain this. (1.8)
>
> © IBO 2014

Essential idea: The modern field of cosmology uses advanced experimental and observational techniques to collect data with an unprecedented degree of precision and as a result very surprising and detailed conclusions about the structure of the universe have been reached.

Understandings:
- The cosmological principle
- Rotation curves and the mass of galaxies
- Dark matter
- Fluctuations in the CMB
- The cosmological origin of redshift
- Critical density
- Dark energy

The cosmological principle

The cosmological principle is basically an extension of the Copernican principle that states that the Earth is not a special place but just one of the planets that orbits the Sun. The cosmological principle states that there are no special places in the Universe. When applied to cosmology and the structure of the Universe, the cosmological principle basically asks the question as to whether the Universe is isotropic and homogeneous. These two terms are not equivalent and have a special meaning in cosmology.

Isotropy is the assumption that the Universe looks the same in every direction, that it is isotropic. On a small scale this is not true but if the Universe is isotropic then you will see no difference in the structure of the Universe as you look in different directions. When viewed on the largest scales, the Universe looks the same to all observers and the Universe looks the same in all directions as viewed by a particular observer.

Homogeneity is the assumption that matter is uniformly spread throughout space. Again, this is not true on the small scale because we can see matter concentrated in planets, stars and galaxies. Homogeneity, when viewed on the largest scales, means that the average density of matter is about the same in all places in the Universe and the Universe is fairly smooth on large scales.

For cosmology, we only consider the isotropy and homogeneity of the Universe on scales of millions of light-years in size.

The clearest modern evidence for the cosmological principle is measurements of the cosmic microwave background. Briefly, the CMB is an image of the photons emitted from the early Universe. Isotropy and homogeneity are seen in its random appearance.

Rotation curves and the mass of galaxies

To find the mass of an object we have to view its orbital motion as, for example in a binary star system. Our lifetime is not long enough to view our galaxy rotating significantly, but we can observe radial velocities, proper motions and distances to stars in order to calculate their orbits. From this we can calculate the mass of our galaxy and give us clues about its origin.

Stars in the galactic plane or galactic disc follow nearly circular orbits. For example, the Sun moves about 230 km s-1 in the direction of Cygnus following an orbit 8.5 kpc in radius with a period of orbit of about 240 million years. Halo stars and globular clusters follow highly elongated tipped steeply to the plane of the disc.

In order to describe the rotation of our galaxy we must plot a graph of orbital velocity versus radius from the galactic centre. The graph obtained is called a rotation curve as shown in Figure 1630. This rotation curve demonstrates how the orbital speed of stars and gas clouds varies with distance from the galactic centre.

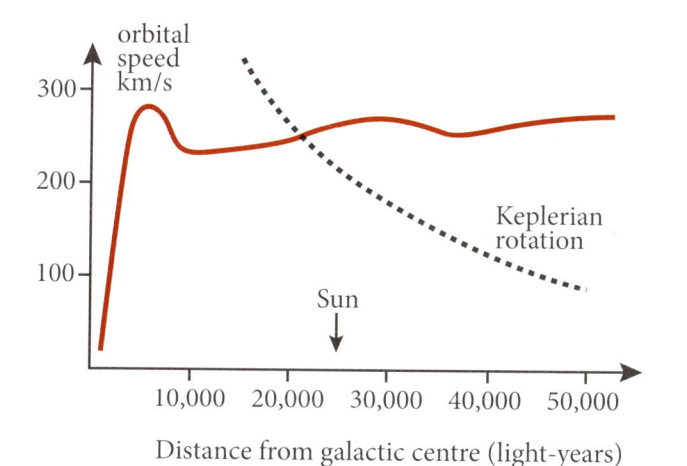

Figure 1630   Rotation curve of our galaxy

We can calculate the mass of the galaxy inside the Sun's galactic centre using Newtonian gravitation.

$$\frac{mv^2}{r} = \frac{GmM}{r^2}$$

$$v^2 = \frac{GM}{r}$$

If the Sun moves in a circular orbit of 2.4 x 1020 m (25 000 light years) at a speed of 2.3 x 105 ms-1 around the galactic centre, the mass inside the galactic centre will therefore be:

$$M = \frac{v^2 r}{G} = \frac{(2.3 \times 10^5)\,2 \times 2.4 \times 10^{20}}{6.67 \times 10^{-11}} = 1.9 \times 10^{41}\ kg$$

The Sun has a mass of 2.0 x 10^{30} kg, so there are just under 10^{11} solar masses inside the radius of the Sun's orbit.

Most of the mass in the galaxy appears to be in and around the nuclear bulge, so stars and gas clouds farther out ought to behave rather like planets revolving around the Sun. Their speeds ought to decrease with increasing distance in a Keplerian fashion as seen in the rotation curve of Figure 1630.

If we want to look at the density of the Universe, we can treat the Universe as a sphere of matter with a certain overall density. Since density equals mass per unit volume, then the mass of the Universe equals the density multiplied by its volume.

$$\frac{v^2}{r} = \frac{4}{3}\frac{\pi r^3 G \rho}{r^2} = \frac{4}{3}\pi r G \rho$$

$$v^2 = \frac{4}{3}\pi G \rho\, r^2$$

$$v = \sqrt{\frac{4}{3}\pi G \rho} \times r$$

Dark matter

Measuring the density of the Universe is no mean feat and is compounded by the fact that the majority of matter in the Universe cannot be seen. We know that galaxies congregate in clusters. However, measurement of the mass of luminous matter in galaxies shows that this mass is not sufficient to keep the galaxies in orbit about the cluster centre. In fact about ten times as much mass is needed. Astronomers therefore have postulated the existence of so-called **dark matter**. They are pretty sure that it has to be there but what it actually consists of is as yet not known. Several theories have been put forward.

One suggestion is that the dark matter consists of neutrinos.

Another is that it is made of MACHOs – massive compact halo objects. These are thought to be dim stars or black holes of between 0.01M☉ and 1 M☉. It is thought that the observed brightening of certain stars over a period of a few days is due to a Macho passing close to the star and thereby bending the light from the star.

Another suggestion is that dark matter is made up of particles called WIMPS – weakly interacting massive particles. The existence of these particles is suggested by certain theories of particle physics but their existence has yet to be confirmed experimentally.

Even if we could measure the density of the universe to a reasonable degree of accuracy we would still not know the fate of the universe since we are still not actually sure of what value the density needs to be to just make the universe flat. Our best theoretical calculations are only accurate to within about 40%.

Fluctuations in the CMB

According to Big Bang theory, temperatures and pressures for the first ~300,000 years of the Universe were such that atoms could not exist. Matter was instead distributed as ionised plasma that was very efficient at scattering radiation. As such, photons were effectively trapped in an impenetrable 'fog' that hid these early times in Universe history.

When the Universe expanded, its temperature and density dropped to a point where the atomic nuclei and electrons were able to combine to form atoms. This is known as the epoch of recombination, and it is at this time that photons were finally able to escape the fog of the early Universe and travel freely. The cosmic microwave background radiation' (CMB) is the record of these photons at the moment of their escape.

The multiple scattering of photons by hot plasma in the early Universe should result in a blackbody spectrum for the photons once they have escaped at the epoch of reionisation.

The photons of the CMB were emitted at the epoch of recombination when the Universe had a temperature of about 3,000 Kelvin. However, they have been cosmological redshifted to longer wavelengths during their ~13 billion year journey through the expanding Universe, and are now detected in the microwave region of the electromagnetic spectrum at an average temperature of 2.725 Kelvin. This agrees well with what Big Bang theory predicts.

In 1989, the Cosmic Background Explorer (COBE) satellite was launched and by 1990 results of the cosmic background radiation studied from space were published.

Figure 1631 shows the final results. The primordial background radiation follows a black body curve with a temperature of 2.735 K.

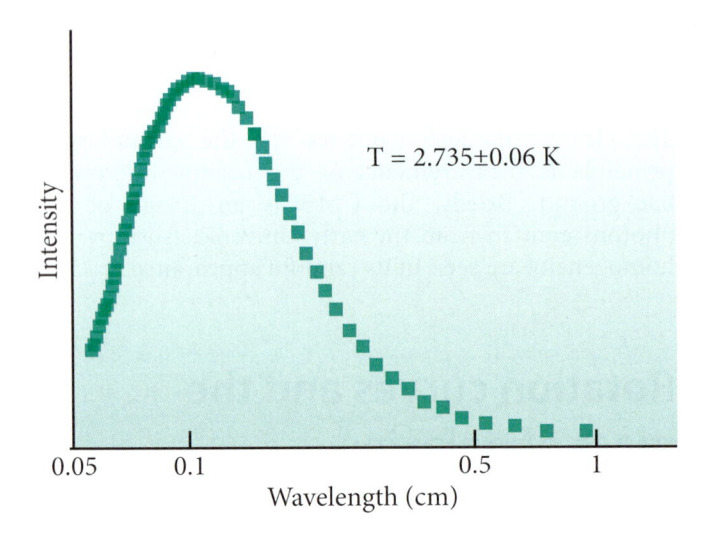

Figure 1631 Background radiation and black body radiation

Studies from the European Space Agency Planck space telescope also confirmed the COBE findings. Although the radiation is almost perfectly isotropic, observation show slight variations in temperature. The anisotropies of the cosmic microwave background were observed by Planck telescope. It shows tiny temperature fluctuations that correspond to regions of slightly different densities. It has slightly shorter wavelengths in the direction of the constellation Leo (hotter). In the opposite direction, the radiation is slightly cooler. Anisotropy is difference in the property of a system with changes in direction. In this case, anisotropy refers to the difference in the temperature of the cosmic microwave background radiation with direction.

The cosmological origin of redshift

Red shifts produced by the expansion of the Universe are referred to as cosmological red shifts. The wavelength of the emitted radiation is lengthened due to the expansion of the Universe. Let us say that one galaxy was formed a long time ago, while another galaxy was formed more recently. Although each galaxy emits the same wavelength of the light, the light from the older galaxy has spent longer travelling through the expanding Universe, and has therefore experienced a greater 'stretching' (redshift). Astronomers are able to determine how far away distant objects are by measuring this wavelength expansion.

ASTROPHYSICS

Critical density

The eventual fate of the Universe is determined by the amount of mass in the Universe.

The Universe could be closed. This means that the density of the Universe is such that gravity will stop the universe expanding and then cause it to contract. Eventually the contraction will result in a 'Big Crunch' after which the whole creation process could start again.

The Universe could be open. This means that the density is such that gravity is too weak to stop the Universe expanding forever.

Another possible development of the Universe is that it could be flat. This means that the density is at a critical value whereby the Universe will only start to contract after an infinite amount of time. The critical density is the average density of matter required for the Universe to just halt its expansion, but only after an infinite time. A Universe with the critical density is said to be flat.

In his theory of general relativity, Einstein demonstrated that the gravitational effect of matter is to curve the surrounding space. In a Universe full of matter, the density of the matter within it controls both its overall geometry and its fate.

The critical density for the Universe is approximately 10^{-26} kg/m³ and is given by:

$$\rho_C = \frac{3H^2}{8\pi G}$$

where H is the Hubble constant and G is Newton's gravitational constant.

Figure 1632 shows the possible development of the Universe depending on the values of the critical density. The straight line shows the development of the Universe if there was no matter in it i.e. the density equals zero. The development of the flat Universe lies somewhere between this line, and the curve for the open Universe. To summarise, if ρ is the density of the Universe and ρ_C is the critical density, then the fate of the Universe will be:

flat if $\rho = \rho_C$

open if $\rho < \rho_C$

closed if $\rho > \rho_C$

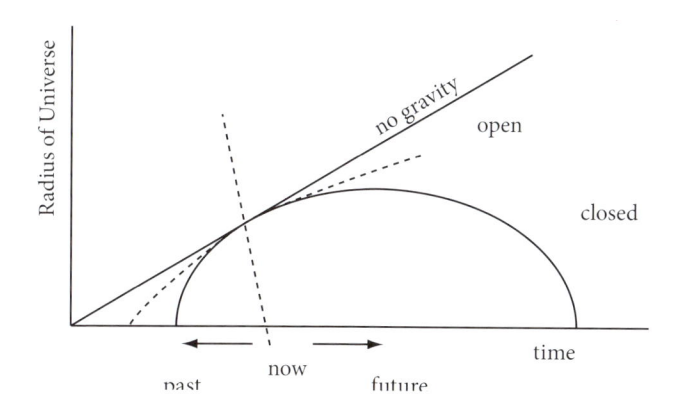

Figure 1632   The fate of the Universe

Some physicists have argued that it must be flat or open but, if this is the case, then there is about 5000 times more dark matter in the Universe than there is luminous matter. Recent measurements suggest that this might indeed be the case and that the Universe is in fact open.

Dark energy

Scientists have found that there is an unknown force that is causing galaxies to move further apart and stretching the fabric of space faster. This force has the potential to pull atoms apart and it could lead to the death of the Universe.

Astronomers from the Berkeley observatory in the USA and the Mount Stromlo observatory in Australia were looking at the data from type Ia supernovas. We know that these supernovas are important tools for measuring large distances in space and are essential for deciphering the expansion of the Universe.

The results they found were quite disturbing. They found that the supernovas were dimmer than expected meaning that they were further away than expected. If that were the case, they would be speeding up, not slowing down. What was causing the acceleration as the gravitational force pulls things together. Perhaps there was some "dark energy" causing this phenomenon.

Further experiments were done with the Hubble telescope on more supernovas, and the same results were found.

So what is "dark energy"? The answer at this time is nobody knows.

497

Chapter 16 (Option D)

This page is intentionally left blank

A

a.f (audio frequency) amplifier
an amplifier that amplifies signals in the approximate range 10 Hz to 20 Hz

aberration
an image defect of which blurring and distortion are the most common image defects. Aberrations can occur with the use of both lenses and mirrors.

absolute magnitude (M)
the apparent magnitude of a star if it were at a distance of 10 pc from Earth.

absolute zero
the point where molecular motion becomes a minimum – the molecules have minimum kinetic energy but molecular motion does not cease.

absorbed dose (D)
the amount of energy E transferred to a particular unit mass m. The SI unit of absorbed dose is J kg^{-1} otherwise known as the Gray (Gy).

absorption spectrum
occurs when white light passes through a substance in the gaseous phase. Dark lines in the white light correspond to the wavelengths characteristic of the emission spectrum of the particular substance.

AC transformer
a device that can be used for increasing or decreasing ac voltages and currents.

acceleration
see average acceleration and instantaneous acceleration

accommodation
the ability of the eye to focus over this range is called accommodation and this is controlled by the ciliary muscles pulling or relaxing in order to change the focal length of the flexible eye lens. The eye has most accommodation for prolonged viewing when viewing at the far point.

accuracy
is an indication of how close a measurement is to the accepted value indicated by the relative or percentage error in the measurement. An accurate experiment has a low systematic error.

acoustic impedance
a measure of how easy it is to transmit sound through a aprticular medium. The unit of acoustic impedance is the rayl.

active solar heating
the use of solar collectors to convert solar energy into heat energy.

adiabatic
expansion or contraction is one in which no thermal energy Q is allowed to flow into or out of the system. For the entire adiabatic process, $Q = 0$.

aerial
a conductor designed to detect a transmitted EM signal

air resistance
a term that refers to the drag force exerted on object as they move through the atmosphere.

albedo (α)
at a surface, is the ratio between the incoming radiation and the amount reflected expressed as a coefficient or as a percentage. (Latin for white)

alpha-particle
a doubly ionised helium atom, that is a helium nucleus.

AM
see amplitude modulation

ammeter
an instrument used to measure the current flowing in an electric circuit and is always connected in series.

A-mode scan
measures the time lapsed between when the pulse is sent and the time the echo is received. The first echo is from the skin, the second and third pulses are from either side of the first organ, the fourth and fifth echo are from either side of the second organ. The pulse intensity decreases due to attenuation.

ampere
defined in terms of the force per unit length between parallel current-carrying conductors.

amplifier
any device that amplifies a signal

amplitude
the maximum displacement of a particle from its equilibrium position when executing SHM For wave motion it is the maximum displacement of the medium through which the wave travels.

amplitude modulation (AM)
the encoding of information on to a carrier wave by producing variations in the amplitude of the carrier wave.

angle of incidence
the angle between the direction of travel of the incident wave and the normal to the boundary

Glossary

angle of refraction
the angle between the direction of travel of the refracted wave and the normal to the boundary

angular frequency (ω)
2π times the linear frequency.

angular magnification
the ratio θ/θ_0 is called the angular magnification M or magnifying power of the lens.

antineutrino
a particle with zero rest mass and zero charge that results from beta-minus decay and decay of a free neutron.

antinode
a point on a stationary wave where the displacement is zero.

antiparticles
all particles have antiparticles which are identical to the particle in mass and half-integral spin but are opposite in charge to their corresponding particle. Although antiparticles have the same mass as their particle pair, they have opposite charge, lepton number, baryon number and strangeness. Some electrically neutral bosons and mesons are their own antiparticle.

aperture
the length of the refracting surface on which the incident rays can be refracted.

apparent brightness
the apparent brightness of a star (b) is the energy received from the star per unit time per unit area of the Earth's surface.

apparent magnitude (m)
a measure of how bright a star appears. The scale is defined such that a difference in apparent magnitude of 5 corresponds to a factor of 100 in brightness. This means that 100 stars of magnitude 6 will produce as much power per unit area at the surface of the Earth as a single star of apparent magnitude 1. The higher the value of m the less bright is the star.

artificial transmutation
a process by which nuclei of an element can be induced to from nuclei of a different element often by the bombardment with neutrons.

APPCDC
Asia-Pacific Partnership for Clean Development and Climate, an organisation that proposed that, rather than imposing compulsory emission cuts, it would work in partnership to complement the Kyoto protocol. The six countries involved were Australia, China, India, Japan, South Korea and the USA.

astronomical unit (AU)
the average distance between Earth and the Sun.
1 AU = 1.50×10^{11} m.

atomic mass unit (u)
this is 1/12th of the mass of an atom of carbon-12.

attenuation
of an X-ray beam is the reduction in its intensity due to its passage through matter.

average acceleration
change in velocity over an interval of time divided by the time interval

average speed
change in distance over an interval of time divided by the time interval

average velocity
change in displacement over an interval of time divided by the time interval

Avogadro's number
one mole of a substance at 0°C and 101.3 kPa pressure (STP) contains 6.02×10^{23} particles.

B

bandwidth
the frequency range covered by the sideband frequencies

baryons
the 'heavyweights' amongst particles that make up matter, including the proton and the neutron. Other baryons include Lamda Λ0, Sigma Σ+, Σ0 and Σ-, Cascade Ξ0 and Ξ- and Omega Ω- particles to name but a few.

becquerel
this is 1 nuclear disintegration per second.

beta particle
a negative or a positive electron associated with radioactive decay.

Big Bang Theory
postulates that the Universe emerged from an enormously dense and hot state about 14 billion years ago. The size of the universe at its beginning was assumed to be extremely small with enormous temperature and pressure. It is assumed that a gigantic "explosion" occurred that created space, time and matter.

binary stars
two stars that orbit a common centre of gravity.

GLOSSARY

biological half-life (T_B)
of a material is the time taken for half the radioactive substance to be removed from the body by biological processes.

black hole
an object whose gravitational field strength at it surface is large enough to prevent light escaping from its surface/ an object whose escape velocity at its surface is equal to or greater than the free space speed of light.

black-body radiation
the radiation emitted by a 'perfect' emitter. The radiation is sometimes called 'temperature radiation' because the relative intensities of the emitted wavelengths are dependant only on the temperature of the black body.

breeder reactor
a nuclear fission reactor that creates or 'breeds' more fissionable material than consumed.

bremsstrahlung
when a fast-moving particle is rapidly decelerated or deflected by another target particle, it radiates most of its energy in the form of photons in what is known as bremsstrahlung or braking radiation in the X-ray region of the electromagnetic spectrum.

Brewster angle (ϕ)
the angle to the normal at which reflected light is completely plane polarized.

Brewster's law
the refractive index n of a substance is related to the Brewster angle *(ϕ) by n = tanϕ*.

Brownian motion
the random, zig-zag motion observed when larger molecules or particles in motion collide with smaller molecules.

B-scan mode
(brightness-modulated scan), an array of transducers scan a slice in the body. Each echo is represented by a spot of a different shade of grey on an oscilloscope.

C

carrier wave
the name given to the wave that is altered by the superposition of the signal wave

cell phones
another name for mobile phones

centre of curvature C
the centre of the sphere of which the lens is made.

centripetal acceleration
the acceleration of a particle traveling in a circle.

centripetal force
the general name given to the force causing a particle to travel in a circle.

cepheid variables
stars whose luminosity varies with a regular frequency.

Chandrasekhar limit
the maximum mass of a star for it to become a white dwarf. ($1.4 M_{sun}$)

change of state (of an ideal gas)
if some macroscopic property of the system has changed eg. phase, temperature, pressure, volume, mass, internal energy.

chemical energy
energy associated with chemical reactions.

chromatic aberration
produces coloured edges around an image. It can be minimised by using an achromatic doublet. It is made from converging crown glass lens and a diverging flint glass lens that are adhered together by canada balsam

coal
an organic material made up primarily of carbon, along with varying amounts of hydrogen, oxygen, nitrogen and sulfur. It is a sedimentary rock.

Coaxial cable
consists of a thin copper wire surrounded by an insulator which in turn is surrounded by a copper grid. This grid is also surrounded by an insulator.

cochlea
the most delicate organ in the hearing process and it contains many intricate structures that will not be fully investigated at this level. It consists of three chambers - two outer chambers, the scala vestibuli (top) and the scala typani (bottom), and an inner chamber called the scala media.

coefficient of volume (or cubical expansion) (β)
the fractional change in volume per degree change in temperature and is given by the relation:

coherent
when the filament of a light globe emits light, the atoms on the filament do not maintain a constant phase relationship because the filament atoms act independantly from each other. The light emitted is incoherent. However, in a laser, each photon of light is in phase with all the other photons. Laser light is coherent.

Glossary

combined cycle gas turbines (CCGT)
a jet engine is used in place of the turbine to turn the generator. Natural gas is used to power the jet engine and the exhaust fumes from the jet engine are used to produce steam which turns the generator.

compression
digital data can be compressed enabling the same bandwidth to be used by several different broadcasting channels

computed tomography (CT) imaging
also called computed axial tomography (CAT) imaging, uses X-rays, scintillation detectors and computer technology to build up an axial scan of a section of an organ or part of the body with 256 grey shades.

conduction
the process by which a temperature difference causes the transfer of thermal energy from the hotter region of the body to the colder region by particle collision without there being any net movement of the substance itself.

conductor
have a low electrical resistance and are therefore able to carry an electric current withour much energy dissipation as heat.

cones
photoreceptors that have slow response rates, and are insensitive at low light levels but are sensitive to particular wavelengths of light, and give us our colour vision. There are around 6.5 million of them. It is thought that the cones can be divided into three colour groups - red cones (64%), green cones (32%), and blue cones (2%).

Conservation of energy
states that energy cannot be created or destroyed but only transformed into different forms. (See conservation of mass-energy and first law of thermodynamics)

Conservation of mass-energy
states that mass and energy are interchangeable and in any interaction mass-energy is conserved.

constellation
a collection of stars that form a recognisable group as viewed from Earth (e.g the Plough)

constructive interference
occurs when two or more waves overlap and their individual displacements add to give a displacement that is greater than any of the individual displacements.

control rods
the rate of nuclear fission in the reactor core can be controlled by inserting or removing the control rods. The control rods are constructed of materials that absorb neutrons.

convection
the process in which a temperature difference causes the mass movement of fluid particles from areas of high thermal energy to areas of low thermal energy (the colder region).

conventional current
flows from the positive to negative terminal.

coolant
a material that circulates through the reactor core and removes thermal energy transferring it to where it can do useful work by converting water into steam.

Coulomb's Law
the force F between two point charges q_1 and q_2 was directly proportional to the product of the two point charges and inversely proportional to the square of the distance between them r^2.

crest
the maximum displacement of a medium through which a wave travels.

critical angle
the angle, measured to the normal, at which a ray incident on a boundary between two media, will undergo total internal reflection in the more dense medium.

critical mass
the smallest possible amount of fissionable material that will sustain a chain reaction.

crude oil
a product of the decomposition of marine plants and animals that were rapidly buried in sedimentary basins where there was a lack of oxygen.

cyclotron
basically like a linac that has been wrapped into a tight spiral.

D

damping
the decrease with time of the amplitude of oscillations.

data transfer rate
the number of bits transmitted per second also called bit rate.

DC amplifier
another name for an operational amplifier

Glossary

de Broglie hypothesis
Any particle with momentum can exhibit wave-like properties and its wavelength is given by the de Broglie formula .

degree of uncertainty
of a measurement is equal to half the limit of reading.

demodulator
removes the carrier wave leaving only the signal waves.

derived quantity
a quantity involving the measurement of two or more fundamental quantities.

destructive interference
occurs when two or more waves overlap and their individual displacements add to give a displacement that is less than any of the individual displacements.

differential amplifier
another term for an operational amplifier

diffraction
the bending and/or spreading of waves when they meet an obstruction or pass through an aperture.

diffusion
a property observed in solids, liquids and gases as something spreads out.

dioptre
the unit for the lens power is the dioptre D with the unit m^{-1}.

dispersion
when a narrow beam of white light undergoes refraction on entering a prism, the light spreads out into a spectrum of colours. The colours range from red at one side of the band, through orange, yellow, green, indigo, to violet at the other side of the band. The separation of the white light into its component colours is due to dispersion.

displacement
distance traveled in a specified direction

Doppler Effect
the phenomenon of the change in frequency that arises from the relative motion between a source and observer.

dosimetry
the study of radiation.

drag force
see air resistance

drift velocity
electrons entering at one end of the metal cause a similar number of electrons to be displaced from the other end, and the metal conducts. Even though they are accelerated along their path, it is estimated that the drift velocity is only a small fraction of a metre each second (about 10^{-4} m s^{-1}).

E

eccentricity
the earth's orbit around the Sun is not circular but rather elliptical and this will affect its orbit every 100 000 and 400 000 years which in turn leads to climate change.

eddy currents
any conductor that moves in a magnetic field has emf induced in it, and as such current, called eddy currents, will also be induced in the conductor. This current has a heating effect in the soft iron core of the transformer which causes a power loss termed an iron loss.

effective half-life (T_E) of the radioactive substance will be less than the physical half-life due to the biological half-life component.

efficiency
of an energy conversion process is the ratio of the useful energy output to the total energy input, usually expressed as a percentage.

Einstein photoelectric equation
relates the maximum kinetic energy of the emitted electrons, f is the frequency of the incident light, f_0 is the threshold frequency and h is the Planck constant

Einstein Principle of Equivalence
states that it is impossible to distinguish between gravitational and inertial effects.

elastic potential energy
the energy associated with a system subject to stress e.g. a stretched spring

electric current
the rate at which charge flows past a given cross-section.

electric field strength (electric field intensity)
at any point in space, E is equal to the force per unit charge exerted on a positive test charge, it is a vector quantity.

electric potential difference
between two points in a conductor is defined as the power dissipated per unit curretn in moving from one point to another.

Glossary

electric potential energy
defined in terms of a point charge moving in an electric field as 'The electric potential at a point in an electric field is defined as being the work done per unit charge in bringing a small positive point charge from infinity to that point'.

electric potential energy
the energy associated with a particle due to its position in an electric field.

electrical energy
this is energy that is usually associated with an electric current and that is sometimes referred to incorrectly as electricity.

electrical resistance
the ratio of the potential difference across the material to the current that flows through it. The units of resistance are volts per ampere (V A^{-1}). However, a separate SI unit called the ohm Ω is defined as the resistance through which a current of 1 A flows when a potential difference of 1 V is applied.

electrical strain gauge
when a metal conducting wire is put under vertical strain, it will become longer and thinner and as a result its resistance will increase. An electrical strain gauge is a device that employs this principle.

electromagnetic waves
waves that consist of oscillating electric and magnetic fields. They are produced by the accelerated motion of electric charge.

electromotive force (emf)
the work per unit charge made available by an electrical source.

electron flow
flows from the negative to the positive terminal.

electron microscope
a microscope that utilizes the wave properties of electrons.

electron-volt (eV)
the energy acquired by an electron as a result of moving through a potential difference of one volt.

electrostatics
the study of stationary electric charges.

elementary particles
particles that have no internal structure, that is, they are not made out of any smaller constituents. The elementary particles are the leptons, quarks and exchange particles.

emission spectra
the spectra produced by excited gaseous atoms or molecules

emissivity
the ratio of the amount of energy radiated from a material at a certain temperature and the energy that would come from a blackbody at the same temperature and as such would be a number between 0 and 1.

energy
the capacity to do work

energy balance climate model
the word "balance" infers that the system is in equilibrium with no energy being accumulated in the earth's surface and atmosphere. This model attempts to account for the difference between the incoming radiation intensity and the outgoing radiation intensity, and the simplest energy balance model chooses temperature as the only variable to be considered.

energy degradation
when energy is transferred from one form to other forms, the energy before the transformation is equal to the energy after (Law of conservation of energy). However, some of the energy after the transformation may be in a less useful form, usually heat. We say that the energy has been degraded.

energy density
the amount of potential energy stored in a fuel per unit mass, or per unit volume depending on the fuel being discussed.

entropy
a thermodynamic function of the state of the system and can be interpreted as the amount of order or disorder of a system.

equipotential lines
lines that join points of equal potential in a gravitational or electric field.

equipotential surface
all points on an equipotential surface at the same potential.

equipotentials
regions in space where the electric potential of a charge distribution has a constant value.

ether
a substance that was thought to permeate the whole of space and that was at absolute rest.

evaporation
a change from the liquid state to the gaseous state that occurs at a temperature below the boiling point.

evaporative cooling
as a substance evaporates, it needs thermal energy input to replace its lost latent heat of vaporisation and this thermal energy can be obtained from the remaining liquid and its surroundings.

exchange particles
elementary particles that transmit the forces of nature.

exponential decay
when a quantity continuously halves in value in equal intervals of time, the quantity is said to decay exponentially.

exposure
is defined for X-radiation and γ-radiation as the total charge (Q) of ions of one sign (either electrons or positrons) produced in air when all the β-particles liberated by photons in a volume of air of mass m are completely stopped in air.

extrapolation
extending the line of best fit outside the plotted points of a graph.

F

far point
the position of the furthest object that can be brought into focus by the unaided eye. The far point of a normal eye is at infinity.

Faraday's Law
can be stated as 'the magnitude of the induced emf in a circuit is directly proportional to the rate of change of magneitc flux of flux linkage.

feedback resistance
the value of the resistance that feeds the output signal of a operational amplifier back to the input.

Feynman diagrams
so named for their inventor, the American physicist *Richard Feynman* (1918–1988). They were developed by Feynmann as a graphical tool to examine the conservation laws that govern particle interactions according to quantum electrodynamic theory.

film badge
a double emulsion photographic film that is placed inside a holder with an area of 3 cm by 5 cm that contains different thicknesses of plastic, an open window and 3 different metal plates. It is pinned to clothing and over a period of time the exposure to radiation results in a darkening of specific areas of the photographic film.

first harmonic (also fundamental)
the first possible mode of vibration of a stationary wave.

first law of thermodynamics
a statement of the Law of Conservation of Energy in which the equivalence of work and thermal energy transfer is taken into account. It can be stated as the heat added to a closed system equals the change in the internal energy of the system plus the work done by the system.

flux linkage (Φ)
If is the flux density through a cross-sectional area of a conductor with coils

focal length (f)
the distance between the principal focus and the centre of the refractingsurface.

forced oscillations
oscillations resulting from the application of an external, usually periodic force.

fossil fuels
naturally occurring fuels that have been formed from the remains of plants and animals over millions of years. The common fossil fuels are peat, coal, crude oil, oil shale, oil tar and natural gas.

fractional uncertainty
see relative uncertainty.

frame of reference
a set of coordinates used to define position

Fraunhofer diffraction
diffraction resulting from the source of light and the screen on which the diffraction pattern is produced being an infinite distance from the diffracting aperture.

frequency
linear frequency (f) is the number of complete oscillations a system makes in unit time.

frequency modulation (FM)
the encoding of information on to a carrier wave by producing variations in the frequency of the carrier wave.

Fresnel diffraction
diffraction resulting from either or both the source of light and the screen on which the diffraction pattern is produced being a finite distance from the diffracting aperture.

frictional force
the force that arises between two bodies in contact.

fundamental
(see first harmonic)

Glossary

fundamental interactions/forces
all forces that appear in nature may be identified as one of four fundamental interactions, either the gravitational, weak, electromagnetic or strong interaction.

fundamental units
kilogram, metre, second, ampere, mole and Kelvin.

G

galaxies
A collection of stars held together by gravity.

gamma ray bursters
astronomical objects that emit intense bursts of gamma radiation thought to be due to the collapse of a rapidly rotating neutron star

gamma ray(s)
high frequency electromagnetic radiation, that is high energy photons.

generator
is essentially a device for producing electrical energy from mechanical energy.

geodesic
the shortest path followed by an object moving in space-time

geostationary satellite
a satellite that orbits Earth in a circular orbit above the equator and has an orbital period of one sidereal day

gluons
the exchange particle that is responsible the quark colour. Just as the positive and negative charges are associated with the electromagnetic force, a three colour charge are associated with quarks and gluons that bind the quarks together.

gravitational lensing
the bending of light by a gravitational field

gravitational mass
the mass that gives rise to the gravitational attraction between bodies as defined by Newton's law of gravity.

gravitational potential
the gravitational potential at a point in a gravitational field is defined as the work done per unit mass in moving a point mass from infinity to the point.

gravitational potential energy
the energy associated with a particle due to its position in a gravitational field.

gravitational red-shift
the observed frequency of light emitted from a source depends upon the position of the source in a gravitational field.

gravitational time dilation
the slowing of time due to a gravitational field

graviton
the exchange particle for the gravitational force. It is an inverse square force with an infinite range that affects all particles and acts on all mass/energy and it has a rest mass of zero.

H

hadrons
are not elementary particles because they are composed of quarks. Mesons consist of a quark and an antiquark. Baryons have three quarks. A proton has 2 up and 1 down quarks - uud, and the neutron has 2 down and I up quarks – ddu. Hadrons interact predominantly via the strong nuclear force, although they can also interact via the other forces.

half-life
see radioactive half-life

half-value thickness
is the thickness of a material that reduces the intensity of a monoenergetic X-ray beam to half its original value.

harmonic series
a series of musical notes arising from a particular fundamental frequency.

harmonics
the different possible modes of vibration of a stationary wave.

heat
the thermal energy that is absorbed, given up or transferred from one object to another.

heat capacity
see thermal capacity

heat engine
any device that converts thermal energy into work.

heat exchanger
a system basically acting as a heat engine driven by chemical reactions (the combustion of fossil fuels) or by nuclear reactions. The working fluid is water heated in a boiler that is converted to steam at high pressure.

GLOSSARY

heat pump
any device that can pump heat from a low-temperature reservoir to a high-temperature reservoir is called a heat pump.

heat
the non-mechanical transfer of energy between a system and its surroundings

Heisenberg Uncertainty Principle
The Uncertainty Principle was proposed by Werner Heisenberg in 1927 as explained in the text

Hertzsprung-Russell diagrams
a plot of the luminosity (or absolute magnitude) against temperature (or spectral class).

Hubble's law
The law states that the relative recession speed between galaxies is proportional to their separation.

I

ideal gas
a theoretical gas that obeys the equation of state of an ideal gas exactly.

ideal gases
obey the equation $pV = nRT$ when there are no forces between molecules at all pressures, volumes and temperatures.

induced current
if the conductor is moved across the magnetic field, then a deflection occurs in the needle of the galvanometer in one direction. After a very short period of time, the needle returns to zero on the scale. The current produced is called an induced current.

inertia
a body's reluctance to change its state of motion.

inertial mass
the mass referred to in Newton's second law

inertial reference frame
a reference frame in which Newton's first law holds true

insolation
incoming solar radiation, it is mainly in the visible region of the electromagnetic spectrum (0.4 μm to 0.7 μm) and short-wave infra-red radiation.

instantaneous acceleration
the rate of change of velocity with time

instantaneous speed
the rate of change of distance with time

instantaneous velocity
the rate of change of displacement with time

insulator
the electrons are held tightly by the atomic nuclei and are not as free to move through a material. They can accumulate on the surface of the insulator but they are not conducting.

intensity
the energy that a wave transports per unit time across unit area of the medium through which it is travelling

interference pattern
the overall pattern produced by interfering waves

Intergovernmental Panel on Climate Change (IPCC)
in the 1980s, the United Nations Environment Programme in conjunction with the World Meteorological Organization set up a panel of government representatives and scientists to determine the factors that may contribute to climate change. The panel was known as the Intergovernmental Panel on Climate Change (IPCC).

internal energy
the sum total of the potential energy and the random kinetic energy of the molecules of the substance making up the system.

internal resistance
the resistance inside a source of electrical energy.

interpolation
drawing the line of best fit between the plotted points of a graph.

inverting amplifier
an operational amplifier in which the non-inverting input is connected to earth.

ionising radiation
when radiation causes ions to form it is called ionising radiation.

ionization current
the current in a gas that results from the ionization of the atoms or molecules of the gas.

ionization
the removal of an electron or electrons from an atom.

isobaric
a graph of pressure as a function of volume change when the pressure is kept constant. Such a process is said to be isobaric. Note that the work done by the gas is equal to the area under the curve.

Glossary

isochoric
a graph of pressure as a function of volume change when the volume is kept constant. Such a process is said to be isochoric. When the volume is kept fixed, the curve of the transformation is said to be an isochore.

isolated system
a system where no energy of any kind enters or leaves the system.

isothermal process
a thermodynamic process in which the pressure and the volume are varied while the temperature is kept constant. In other words, when an ideal gas expands or is compressed at constant temperature, then the gas is said to undergo an isothermal expansion or compression.

isotopes
atoms of the same element with different numbers of neutrons in their nuclei.

K

Kelvin temperature
a fundamental quantity. It is the SI unit of thermodynamic temperature of the triple point of water. One degree Celsius is equal to $1 + 273 = 274$ K.

Kepler's third law
this is the law of periods and states that that the average orbital radius R of a planet about the Sun is related to the period T of rotation of the plane by $R^3 = kT^2$ where k is a constant.

kilogram
the mass of a particular piece of platinum-iridium alloy that is kept in Sèvres, France.

kilowatt-hour (kW h)
the energy consumed when 1 kW of power is used for one hour.

kinetic energy
energy associated with motion

kinetic theory of a gas
when the moving particle theory is applied to gases it is generally called the kinetic theory of gases.

Kirchoff's current law – junction rule
the sum of the currents flowing into a point in a circuit equals the sum of the currents flowing out at that point.

Kirchoff's voltage law – loop rule
in a closed loop the sum of the emfs equals the sum of the potential drops.

Kyoto Protocol
this agreement required industrialized countries to reduce their emissions by 2012 to an average of 5 percent below 1990 levels. A system was developed to allow countries who had met this target to sell or trade their extra quota to countries having difficulty meeting their reduction deadlines.

L

laminations
to reduce the heating effect due to eddy currents, the soft-iron core is made of sheets of iron called laminations that are insulated from each other by an oxide layer on each lamination. This insulation prevents currents from moving from one lamination to the next.

laser
is actually an acronym 'light amplification by stimulated emission of radiation'. A laser is an instrument that has a power source and a light-amplifying substance. There are a variety of solid, liquid and gas lasers available on the market. The common laser used in the laboratory uses a helium- neon gas mixture as the light-amplifying substance.

latent heat of fusion
the quantiy of thermal energy required to change a substance from a solid at its melting point completely to a liquid at its melting point.

latent heat of vaporisation
the quantiy of thermal energy required to change a substance from a liquid at its boiling point completely to a gas at its boiling point.

Law of conservation of electric charge
in a closed system, the amount of charge is constant.

laws of reflection
the angle at which the waves are reflected from a barrier is equal to the angle at which they are incident on the barrier (the angles are measured to the normal to the barrier). All waves, including light, sound, water obey this rule. The normal and the rays associated with the incident and reflected rays all lie in the same plane.

lens
a transparent object with at least one curved surface but more commonly two curved faces. Most lenses are made of glass but perspex (lucite) and quartz lenses are common. They are used to correct defects of vision using spectacles and in optical instruments such as cameras, microscopes and refracting telescopes.

Lenz's Law
  also known as the Second Law of Electromagnetic Induction and it can be stated as 'the direction of the induced emf is such that the current it causes to flow opposes the change producing it'.

leptons
  particles that can travel on their own meaning that they are not trapped inside larger particles. Six distinct types called flavors have been identified along with their antiparticles.

light dependant resistor (LDR)
  is a photo-condutive cell whose resistance changes with the intensity of the incident light.

light year
  the distance that light travels in one year. 1 light year (ly) = 9.46×10^{15} m

limit of reading
  of a measurement is equal to the smallest graduation of the scale of an instrument.

line spectrum
  produced when the spectrum produced by excited gaseous atoms or molecules is passed through a slit and then through a dispersive medium such as a prism of diffraction grating and then brought to a focus on a screen.

linear accelerator (linac)
  is a device that accelerates charged particles in a straight line inside a long evacuated tube.

linear attenuation coefficient
  a beam of homogeneous, monoenergetic X-rays contains photons of only one energy and thus only one wavelength.

linear or lateral magnification m
  (of a lens) is given by the ratio of the height of an image to the height of its object or the ratio of the image distance to the object distance. Linear magnification has no units.

longitudinal waves
  in these types of wave, the source that produces the wave vibrates in the same direction as the direction of travel of the wave i.e. the direction in which the energy carried by the wave is propagated. The particles of the medium through which the wave travels vibrate in the same direction of travel of the wave (direction of energy propagation).

loudspeaker
  a transducer that converts an amplified electrical signal into sound.

luminosity (L)
  the total power radiated by a star.

M

Mach's Principle
  states that inertial and gravitational mass are identical

macroscopic property
  a property that can be observed. Physical properties such as melting point, boiling point, density, thermal conductivity, thermal expansion and electrical conductivity can be observed and measured.

magnetic flux (Φ)
  through a small plane surface is the product of the flux density normal to the surface and the area of the surface. The unit of magnetic flux is the weber Wb.

magnetic force
  a force experienced when a moving charge or a beam of moving charges is placed in a magnetic field.

magnifying power
  see angular magnification.

main sequence stars
  a grouping of stars on a Hertzsprung-Russell diagram that extends diagonally across the graph from high temperature, high luminosity to low temperature low luminosity. Stars on the main sequence derive the energy from hydrogen burning in the core of the star.

Malus' law
  when light of intensity I_0 is incident on an analyzer whose transmission axis makes angle θ to the electric field vector, the intensity I of the transmitted light is given by $I = I_0 cos^2\theta$

mass
  see gravitational mass and inertial mass

mass defect
  The difference in mass between a nucleus and the sum of the mass of its constituent nucleons. The mass of a nucleus is always less than the sum of the mass of its constituent nucleons.

material dispersion
  the spreading out of pulses as they travel along an optic fibre

matter waves
  See de Broglie hypothesis

Glossary

Maxwell's theory
states that electromagnetic radiation consists of oscillating electric and magnetic fields.

mesons
hadrons that can mediate the strong nuclear force. Like the first and second generation leptons, mesons only exist for a short time and they are thus very unstable.

metal structure
positive ions in a 'sea' of delocalised electrons.

method of mixtures
a common indirect method to determine the specific heat capacity of a solid and liquids is called the method of mixtures.

metre
the length of path traveled by light in a vacuum during a time interval of 1/299 792 453 second.

minimum angle of resolution
see Rayleigh criterion

mobile phone
a phone that is not connected by a landline to a telephone exchange

modal dispersion
a situation in which pulses associated with different waves in an optic fibre arrive at the detector at different times

moderator
a material that will slow down the fast neutrons to the speed of the slow thermal neutrons needed for a self-sustained reaction without absorbing the neutrons when they collide with the moderator material.

modes
the name given to the different paths followed by different waves in an optic fibre

modulation
the alteration of a wave form

mole
is the amount of substance that contains as many elementary particles as there are in 0.012 kg of carbon–12. The mole is a fundamental unit.

momentum
the product of mass and velocity

monochromatic
source of radiation is that has a extremely narrow band of frequencies or extremely small narrow wavelength band (or colour in the case of visible light). Most sources of light emit many different wavelengths. Laser light is monochromatic.

monomode fibres
a fibre in which there is only one transmission axis thereby eliminating modal dispersion

Morse code
an electronic communication system that used individual groups electrical pulses to represent letters and that were transmitted along wires

moving particle theory
the basic assumptions of this moving particle theory relevant to thermal energy are:

multiplexing
a means of increasing the bit rate by sending different sets of data apparently simultaneously.

N

natural frequency
the frequency of oscillation of a system that is not subjected to a periodic external force.

natural gas
a product of the decomposition of marine plants and animals that were rapidly buried in sedimentary basins where there was a lack of oxygen.

natural greenhouse effect
a phenomenon in which the natural greenhouse gases absorb the outgoing long wave radiation from the earth and re-radiate some of it back to the earth.

natural radioactivity
a property associated with certain naturally occurring elements in which they emit ionizing radiations.

near point
the position of the closest object that can be brought into focus by the unaided eye. The near point varies from person to person but it has been given an arbitrary value of 25 cm.

nebulae
a cloud of interstellar dust and gas.

nematic liquid crystal
a liquid crystal whose molecules are in the shape of a twisted helix.

neutron number
the number of neutrons in a nucleus

neutrons
an uncharged nucleon

nibble
a 4-bit binary word

node
a point on a stationary wave where the displacement is a maximum.

non-renewable source
one that is considered to be a temporary source that is depleted when it is used.

NTC thermistor
(negative temperature coefficient) the resistance decreases when the temperature rises and theytherefore pass more current.

nuclear binding energy
the energy required to separate the nucleus into it individual nucleons or the energy that would be released in assembling a nucleus from its individual nucleons.

nuclear energy
energy associated with nuclear reactions

nuclear fission
the splitting of a nucleus into two other nuclei.

nuclear fusion
the combining of two nuclei into a single nucleus

nuclear magnetic resonance
the basis of the diagnostic tool known as magnetic resonance imaging (MRI). It is a technique used for imaging blood flow and soft tissue in the body and is the preferred diagnostic imaging technique for studying the brain and the central nervous system. Rather than using X-rays as the source of radiation, it uses radiation in the radio region of the electromagnetic spectrum and magnetic energy to create cross-sectional slices of the body.

nucleon
a proton or a neutron.

nucleon number
the number of nucleons in a nucleus

nucleosynthesis
the different nuclear processes that take place in stars.

nuclide
the general term for a unique nucleus

numerical aperture
is related to the resolution of a lens, and the wavelength of the light (see text for formula)

Nyquist Theorem
states that the sampling signal must be equal to or greater than twice the signal frequency.

O

Ohm's Law
provided the physical conditions such as temperature are kept constant, the resistance is constant over a wide range of applied potential differences, and therefore the potential difference is directly proportional to the current flowing.

Olber's paradox
if Newton's model of a uniform, infinite Universe were correct, then the sky would always be bright. This paradox was first proposed by Henrich Olber in 1823.

operational amplifier
an amplifier with two inputs, very high input impedance and very high gain.

Oppenheimer–Volkoff limit
the maximum mass of a neutron star beyond which it will collapse to a black hole

optic fibres
a fibre in which the carrier wave is light.

optical microscope
a microscope using visible light and lenses to magnify small objects (usually used in biology and medicine)

order of magnitude
the power of ten closest to a number.

oscillating water column (OWC)
wave energy devices that convert wave energy to electrical energy. These can be moored to the ocean floor or built into cliffs or ocean retainer walls.

oscillations
another word for vibrations.

ossicles
a chain of three bones in the ear that transmit vibration form the ear drum to the cochlea. They are called the malleus, incus and stapes, more commonly known as the hammer, anvil and stirrup.

Glossary

P

pair annihilation
when matter (such as an electron) collides with its corresponding antimatter (such as a positron), both particles are annihilated, and 2 gamma rays with the same energy but with a direction at 180⁰ to each other are produced. This is called pair annihilation.

pair production
particle–antiparticle pairs can also be produced when a gamma ray with sufficient energy passes close by a nucleus. The process is the reverse of annihilation and is called pair production.

parallax
the apparent displacement of an object due to the motion of the observer.

parsec
a line of length1 AU subtends an angle of 1 arcsecond (one second of arc /4.8×10^{-6} rad) at a distance of one parsec.

Pauli exclusion principle
states that an orbital can only contain a maximum of two electrons and when the 2 electrons occupy an orbital they have opposite spin.

peak current
an alternating current varies sinusoidally and the maximum current called the peak current.

peat
a brownish material that looks like wood. Although it can be burnt as a fuel, it contains a lot of water, and is very smoky when burnt. Under pressure and over time it will be converted to other forms of coal.

percentage uncertainty
is the relative uncertainty multiplied by 100 to produce a percentage.

period
the time taken for an oscillating system to make one complete oscillation.

periodicity
repetition of motion both in space and in time

phase change
a substance can undergo changes of state or phase changes at different temperatures. Pure substances (elements and compounds) have definite melting and boiling points which are characteristic of the particular pure substance being examined.

phase difference
the time interval or phase angle by which one wave leads or lags another.

photo-electric effect
The emission of electrons from a metal surface that is illuminated with light above a certain frequency

photoelectric work function
The minimum energy ϕ required to remove an electron from the surface of a metal by photo-emission. It is related to the threshold frequency by $\phi = hf_0$.

photon
The existence of the photon was postulated by Einstein in 1905 as being a quantum of electromagnetic energy, regarded as a discrete particle having zero mass, no electric charge, and an indefinitely long lifetime. The energy E of a photon associated with light of frequency f is given by the Planck equation $E = hf$.

photopic vision
cones are responsible for photopic vision or high light-level vision, that is, colour vision under normal light conditions during the day. The pigments of the cones are of three types – long wavelength red, medium wavelength green and short wavelength blue.

photovoltaic devices
use the photoelectric effect. Photons from radiant energy excite electrons in a doped semi-conducting material such as silicon or germanium, and the element becomes conducting allowing electrons to flow in an external circuit to produce electrical energy.

physical half-life (T_R)
of a radioactive nuclide is the time taken for half the nuclei present to disintegrate radioactively.

pixels
the smallest element of an image on a LCD or CCD

Planck constant
Max Planck postulated that energy associated with oscillating atoms is proportional to the frequency of oscillation of the atom. The constant (h) relates the energy (E) of a photon to its associated frequency (f). ($E = hf$) (h = $6.2660693 \times 10^{-34}$ J s)

plasma
a super heated gas.

plasma confinement
plasma has to be confined for 1 second with a density of about 500 trillion atoms per cubic centimetre. Because fusion is not a chain reaction, thes temperature and density conditions have to be maintained for future fusions to occur.

Glossary

polarimeter
essentially a tube that is bounded at both ends with polarizing materials.

polarization
the rotation of the plane of vibration of the electric vector of an electromagnetic wave.

pole (P)
central point of the refracting surface.

pollutants
substances that have undesirable effects on living things and property. Air pollution occurs when these pollutants are introduced into the atmosphere.

population inversion
in the ruby laser, light of energy equivalent to 2.25 eV is absorbed from the flash tube, and this raises the electrons of chromium from the ground state E_1 to an excited state E_3. These electrons quickly undergo spontaneous emission and fall to level E_2 known as the metastable energy state. If the incident radiation from the flash tube is intense enough more electrons are transferred to the E_2 energy level than remain in the ground state – a condition known as population inversion.

positron
a positively charged electron

potential divider
a device that produces the required voltage for a component from a larger voltage.

potential energy
see elastic potential energy, electric potential energy and gravitational potential energy

potential gradient
the rate of change of potential ΔV at a point with respect to distance Δx in the direction in which the change is maximum is called the potential gradient.

power
the rate of working

power of a convex lens (P)
is the reciprocal of the focal length. It is a measure of the strength of a lens as used by optometrists and opthalmologists.

power stations
usually rely on thermal energy, gravitational potential energy or wind power to supply the kinetic energy to rotate a turbine. The turbine contains blades that are made to rotate by the force of water, gas, steam or wind. As the turbine rotates, it turns the shaft of a generator. The electrical energy can be produced by rotating coils in a magnetic field.

precision
is an indication of the agreement among a number of measurements made in the same way indicated by the absolute error. A precise experiment has a low random error.

preferential absorption
the phenomenon in which certain crystals only transmit the vertical or horizontal component of the electric vector of an electromagnetic wave.

pressure
it is defined as the force exerted over an area. The SI unit of pressure is the pascal (Pa).

principal axis
line that passes through the centre of curvature and the centre of the refracting surface.

principal focal plane
the plane that passes through the principal focus and is perpendicular to the principal axis.

principal focus (F)
point through which rays parallel and close to the principal axis pass after refraction if the lens is convex, or appear to come from if the lens is concave.

principle of superposition
the principle of superposition as applied to wave motion states the displacement at a point where two or more wave meet is the vector sum of the individual displacements of each wave at that point.

proper length
the length of an object as measured by an observer at rest with respect to the object

proper time
the time interval between two events as measured by an observer that sees the events take place at the same point in space.

proton number
the number of protons in a nucleus

protostar
a stage in the formation of a star in which the star is self-luminous but in which nuclear fusion as not yet started.

public switched telephone network (PSTN)
land based telephone exchange

pulsar(s)
a pulsating radio source believed to be a rapidly rotating neutron star.

Glossary

pulse oximetry
a non-invasive technique used to monitor the oxygen content of haemoglobin.

pump storage systems
used in off-peak eleticity demand periods. The water is pumped from low resevoirs to higher resevoirs during this period.

Q

quality
of an X-ray beam is a term used to describe its penetrating power.

quality factor
this is approximately equal in value to the number of oscillations that occur before all the energy of an oscillator is dissipated.

quantum
A discrete packet of energy associated with electromagnetic radiation. (see "photon"). Literally from the Latin "how much".

quantum mechanics
The theory proposed in 1926/7 that replaced Newtonian physics.

quantum numbers
the different states in which an electron can exist are determined by four quantum numbers: principal, orbital, magnetic and spin

quark confinement
the property that quarks are always found in groups that are colourless is called quark confinement.

quarks
with a size of less than 10^{-18} m can never be found in isolation as they are trapped inside other composite particles called hadrons of which the proton, the neutron and mesons are examples.

quasars
very distant and very luminous stellar like objects.

R

r.f (radio frequency) amplifier
an amplifier that amplifies signals in the radio frequency range (several kHz to about 100 Mhz)

radiation
the energy produced by a source because of its temperature that travels as electromagnetic waves. It does not need the presence of matter for its transfer.

radiation shielding
ensures the safety of personnel working inside and around the reactor from suffering the ill effects of radiation exposure. There are usually two shields: several metres of high-density concrete to protect the walls of the reactor core from radiation leakage and to help reflect neutrons back into the core and a biological shield to protect personnel made of several centimetres of high density concrete.

radioactive decay
The spontaneous emission by the nuclei of certain atoms, of radiation in the form of alpha particles or beta particles and/or gamma radiation. The decay process cannot be controlled by chemical and physical means.

radioactivity
see natural radioactivity

radius of curvature (R)
the radius of the sphere from which the lens is made.

random uncertainties
are due to variations in the performance of the instrument and the operator. Even when systematic errors have been allowed for, there exists error.

rank advance
as peat became buried beneath more plant matter, the pressure and temperature increased and the water was squeezed out of it. As the material became compacted the peat is converted to lignite, then to sub-bituminous coal and finally bituminous coal. At each stage in the rank advance, the coal has a higher carbon content and a higher energy content per unit mass.

rarefaction
in a sound wave this refers to regions of minimum pressure.

Rayleigh criterion
the images of two sources will be just be resolved by an image forming system if the central maximum of one diffraction pattern image coincides with the first minima of the other diffraction pattern image.

real image
an image that can be seen on a screen that has been put at the point where the rays intersect at a single point.

red giant star
An evolutionary phase of main sequence stars usually with mass less than about $4M_{Sun}$ characterized by low temperature and high luminosity.

red-shift
the Doppler shift of light observed from receding objects.

reflection
occurs when a wave is incident at a boundary between two different media and results in some of the energy of the wave being returned into the medium in which it is travelling before incidence.

refraction
occurs when a wave is incident at a boundary between two different media and results in some of the energy of the incident wave being transmitted across the boundary. If the wavefronts are not parallel to the boundary, the direction of travel of the wave is changed.

refractive index (*n*)
This is defined using the angle of incidence of light in a vacuum and the angle of refraction in the medium whose refractive index is n.

relative uncertainty
equals the absolute uncertainty divided by the measurement. It has no units.

renewable energy source
one that is permanent or one that can be replenished as it is used. Renewable sources being developed for commercial use include solar energy, biomass, wind energy, tidal energy, wave energy, hydro-electric energy and geothermal energy.

reshapers
a device used to re-shape pulses in an optic fibre

resolving power
the minimum angle of resolution

resonance
this occurs when the frequency of forced oscillations is equal to the natural frequency of the system that is being forced.

rest mass-energy
the energy that is equivalent to a body's rest mass

rest mass
the mass of an object as measured by an observer at rest with respect to the object.

rods
photoreceptors that have fast response rates, and are sensitive at low light levels but they are insensitive to colour. There are around 120 million of them.

root-mean-square (r.m.s.) value
the current dissipated in a resistor in an a.c. circuit that varies between I_0 and $-I_0$ would be equal to a current $I_0/\sqrt{2}$ dissipated in a d.c circuit. This d.c current is known as r.m.s. equivalent current to the alternating current.

S

Sankey diagram
in a Sankey diagram, the thickness of each arrow gives an indication of the scale of each energy transformation. The total energy before the energy transfer is equal to the total energy after the transfer otherwise the Law of conservation of energy would be violated.

scalar
a quantity that has only magnitude

scattering
the deflection of EM radiation from its original path due to its collisions with particles in a medium.

Schmitt trigger
a circuit designed to re-shape digital electrical signals

scientific notation
expressing numbers to the power of ten

scotopic vision
rods are responsible for scotopic vision which is the ability to see at low light levels or vision "in the dark" or light levels below 0.034 candela per square metre (-0.034 c dm^{-2}). They do not mediate colour and are sometimes termed "colour blind". Because they do not mediate colour, they are said to have low spatial resolution (acuity).

second
the time for 9 192 631 770 vibrations of the cesium-133 atom.

second law of thermodynamics
implies that thermal energy cannot spontaneously transfer from a region of low temperature to a region of high temperature.

sensors
an input transducers that allows for the transfer of energy from one form to another.

SI unit
an international system of units including the metric system. SI units are those of Le Système International d'Unités adopted in 1960 by the Conférence Générale des Poids et Mesures.

Glossary

sideband frequencies
a modulated wave consists of the carrier wave plus two waves one of frequency ($f_c - f_s$) and the other of frequency ($f_c + f_s$). The frequencies are called the sideband frequencies.

signal wave
the name given to the wave that carries information

significant figures/digits
(sf/sd) are those digits that are known with certainty followed by the first digit which is uncertain.

simple harmonic motion
occurs when the force acting on a system is directed towards the equilibrium position of the system and is proportional to the displacement of the system from equilibrium

Snell's law
is usually applied to light waves and states that when light travels from one medium into another

solar constant
the average radiant power radiated to an area placed perpendicular to the outer surface of the earth's atmosphere while the earth is at its mean distance from the Sun.

SONAR (sound navigation and ranging)
the use of sound waves to detect and estimate the range of submerged objects. In the 1930s it had its applications in medical therapy.

sound intensity
the average power per unit area of a sound wave that is incident perpendicular to the direction of propagation is called the sound intensity. The units of sound intensity are watts per square metre, W m^{-2}. As the sound intensity spreads out from its source, the intensity I is reduced as the inverse square of the distance d from the source.

source independence
the name given to the phenomenon in which audio and visual digital data can be transmitted using the same channel.

space-time
a coordinate system consisting of three dimensions of space and one of time

space–time diagram
the representation of the motion of an object in space-time

specific heat
see specific heat capacity.

specific heat capacity
is the heat capacity per unit mass. It is defined as the quantity of thermal energy required to raise the temperature of one kilogram of a substance by one degree Kelvin.

spectral classes
a classification of stars according to their observed spectrum

speed
see average speed and instantaneous speed

spherical aberration
occurs because the rays that refract at the outer edges of a lens will have a different focal length to those rays that refract near the principal focus. To put it another way, spherical aberration occurs because the rays incident near the edges of a converging lens are refracted more than the paraxial rays

spring constant
the constant k relating the extension x of a spring to the force F causing the extension $F = kx$

standard form
see scientific notation.

standard notation
see scientific notation.

stationary waves
sometimes also referred to as standing waves. Waves in which there is no propagation of energy between points along the wave. The amplitude of a stationary wave varies with position along the wave.

steam engine
an example of external combustion engines. The fuel is burnt outside the engine and the thermal energy is transferred to a piston or a turbine chamber by means of steam.

Stefan's law
the total area under a spectral emission curve for a certain temperature T represents the total energy radiated per metre^2 per unit time E and for that assigned temperature it has been found to be directly proportional to the fourth power T^4.

Stefan-Boltzmann law
A law that relates the luminosity of an object to its absolute temperature and area

Stellar cluster
this is a number of stars that were all created about the same time and that is held together in a group by gravitational attraction.

GLOSSARY

stellar interferometer
a radio telescope that consists of two or more parabolic receiving dishes

step-down transformer
a transformer that if N_s is less than N_p it will be a step-down transformer.

step-index fibre
an optic fibre in which the refractive index of the different materials comprising the fibre change by discrete amounts.

step-up transformer
a transformer that if N_s is greater than N_p then the transformer is a step-up transformer.

strain viewer
a device that use polarized light to view the stress produced in materials subject to strain. It consists of two polaroids with the material under strain placed between them.

string theory
an alternative to quantum theory that proposes that each fundamental particle consists of an oscillating string of a small size compared with the proton. Rather than talking about mathematical particles, string theory talks about oscillating strings that are lines or loops of about 10^{-35} m, and membranes in small dimensions other than the three dimensions that we presently use.

strong nuclear interaction
the short range force of attraction between nucleons.

super red-giant star
an evolutionary phase of main sequence stars usually with mass greater than about $8M_{Sun}$ characterized by low temperature and very high luminosity

surface heat capacity C_s
the energy required to raise the temperature of a unit area of a planet's surface by one degree Kelvin and is measured in J m^{-2} K^{-1}.

synchrotrons
the most powerful members of the accelerator family

system
any object or set of objects that is being investigated. The surroundings will then be everything in the Universe apart from the system.

systematic error
causes a random set of measurements to be spread about a value rather than being spread about the accepted value. It is a system or instrument error.

T

temperature
a scalar quantity that gives an indication of the degree of hotness or coldness of a body. Alternatively, temperature is a macroscopic property that measures the average kinetic energy of particles on a defined scale such as the Celsius or Kelvin scales. At the microscopic level, temperature is regarded as the measure of the average random kinetic energy per molecule associated with its movements.

tension force
this arises when a system is subjected to two equal and opposite forces.

terminal velocity
the velocity reached when the magnitude of the frictional force acting on a body is equal to the magnitude of the driving force.

thermal (heat) capacity
the change in thermal energy for a given change in temperature.

thermal energy (heat)
If a system and its surroundings are at different temperatures and the system undergoes a process, the energy transferred by non-mechanical means is referred to as thermal energy (heat). It is measured in joules.

thermistors
resistors that change resistance with temperature (word derived from thermal resistors).

thermodynamic cycle
a process in which the system is returned to the same state from which it started. That is, the initial and final states are the same in the cyclic process.

thermodynamic engine
device that transforms thermal energy to mechanical energy (work) as in an engine, or mechanical energy to thermal energy such as in refrigeration and air-conditioning systems.

thermodynamics
the name given to the study of processes in which thermal energy is transferred as heat and as work.

Glossary

three phase power
There are 3 conductors on a transmission line to maximize the amount of power that can be generated. Each high voltage circuit has three phases. The generators at the power station supplying the power system have their coils connected through terminals at 120° to each other. When each generator at the power station rotates through a full rotation, the voltages and the currents rise and fall in each terminal in a synchronized manner.

threshold frequency
The frequency below which photoelectric emission will not take place.

threshold intensity of hearing
the minimum detectable intensity for a given frequency is called the threshold intensity of hearing.

time dilation
the slowing of time as observed by an inertial observer who assumes to be at rest with respect to another, moving inertial reference system

total internal reflection
reflection in which all the light incident at a boundary between two media undergoes reflection

transmission rate
another name for bit-rate

transmutation
see artificial transmutation

transverse waves
in these types of wave the source that produces the wave vibrates at right angles to the direction of travel of the wave i.e. the direction in which the energy carried by the wave is propagated. The particles of the medium through which the wave travels vibrate at right angles to the direction of travel of the wave (direction of energy propagation).

travelling wave
a wave that propagates energy

trough
the minimum displacement of a medium through which a wave travels.

tuning circuit
a circuit designed to respond to signals of a certain frequency

U

Uncertainty principle
See 'Heisenberg Uncertainty Principle'

unit of current
is the coulomb per second C s^{-1} and this unit is called the ampere (A).

V

variable
a quantity that varies when another quantity is changed. A variable can be an independent variable, a dependent variable or a controlled variable. An independent variable is altered while the dependent variable is measured. Controlled variables are the other variables that may be present but are kept constant.

vector
a quantity that has both magnitude and direction

vector resolution
giving the x and y components of a vector.

velocity
see average velocity and instantaneous velocity

virtual earth
a point in a circuit that is effectively at earth potential (zero volts)

virtual image
an image that appears to come from a single point when rays are extrapolated to that point.

virtual particle
a particle that cannot be observed during an interaction. A virtual photon is said to be the carrier of the electromagnetic force.

voltmeter
is used to measure the voltage drop across part of an electric circuit and is always connected in parallel.

W

W⁺, W⁻ and Z₀
the exchange particles involved in the weak nuclear interaction.

wave number
the number of waves per centimeter (cm⁻¹)

wave speed
is the speed with which energy is carried in the medium by the wave. A very important fact is that wave speed depends only on the nature and properties of the medium

wavelength
is the distance along the medium between two successive particles that have the same displacement

wave-mechanics
another name for quantum mechanics

weight
another term for the force of gravity acting on an object

weightlessness
if the weight of an object is defined in terms of a 'weighing' process such as the reading on a set of bathroom scales, which in effect measures the contact force between the object and the scales, then objects in free fall are weightless

Wien Displacement Law
a law that relates the maximum wavelength in the blackbody spectrum of an object to the absolute temperature of the object

work
the product of force and displacement in the direction of the force

Index

A

aberration	436
absolute temperature	87, 412
absolute uncertainty	10
absolute zero	89
absorption spectrum	472
acceleration	36
acceleration-time graph	39
achromatic doublet	437
acoustic impedance	459
AC transformer	334
Adams, J. C.	320
adiabatic compression	404
adiabatic expansion	404
A.H.L. Fizeau	112
air resistance	44, 50
Albedo	276
Albert A. Michelson	113
Albert Einstein	114, 222, 225, 350
Alessandro Volta	178
Alpha Centauri	475
alpha particles	214
alternating current	248, 331
Amadeo Avogadro	104
ampere	161
Ampère, André-Marie	161
amplifier	449
amplitude	114, 124
amplitude and intensity	140
Anders Celsius	89
Anderson, Carl	215
André-Marie Ampère	161
angular acceleration	396
angular displacement	195
angular frequency	116
angular magnification	433, 435
angular measure	28
angular momentum	398
angular motion	397
angular velocity	116, 195
anode	179
anthracite	253
antimatter	356
antineutrino	216
antinodes	140, 148
antiparticles	232
antiquarks	231
Antoine Henri Becquerel	216
Antoine Lavoisier	74, 84, 206
aperture	429
apparent brightness	470
apparent magnitude	475
Archimedes' principle	416
arcsecond	444
Aristotle	84, 206
Arthur Eddington	56
A-Scan imaging	459
asteroid	467
astronomical telescope	438
astronomical unit	469
Astronomy	466
Astrophysics	466
asymptotic freedom	241
atmospheric extinction	445
atmospheric turbulence	445
atmospheric turbulence refraction	445
atom	206
atomic emission	472
atomic mass unit	221
attenuation	448, 460
attraction and repulsion	153
auditory canal	129
Australia Telescope Compact Array	444
Avogadro, Amadeo	104
Avogadro constant	103
Avogadro number	104
Avogadro's hypothesis	401

B

Balmer, Johann Jakob	210
baryon number	235
baryons	232
battery	179
Becker, Herbert	212
Becquerel, Antoine Henri	216
Bell, Jocelyn	480
bending of light	384
Benjamin Thompson	84
Bernoulli effect	419
Bernoulli's principle	419
best-fit lines	20
beta particles	214
Betelgeuse	468
Big Bang	228
binary stars	469, 473
biomass	251
black body radiation	274
black hole	386, 480
body at rest	56
Bohr model	348
Bohr, Neils	231
Bohr's quantum condition	354
boiling	95
Boltzmann's constant	487
Boyle, Robert	101
Boyle's Law	101
Bothe, Walther	212
Brewster angle	143
Brewster, David	143
Brewster's law	142, 143
Brownian Motion.	86
B-scan imaging	459
buoyancy force	416

C

calculating displacement	34
calorimeter	92
capacitance	341
capacitor	341
carbon dioxide	257, 279
Carl Anderson	215
Cartesian co-ordinates	375
Cassegrain reflector	441
CAT	458
cathode	179
cathode ray	206
cathode ray oscilloscope	331
CAT scan	458
Cecil Frank Powell	238
Celsius	88
Celsius, Anders	89
centrifugal force	196
centripetal acceleration	196
centripetal force	196
Cepheid variables	469, 476
Ceres	467
CERN	228
Chadwick, James	212
Chandrasekhar limit	478
charge	152
Charles Augustin Coulomb	153
Charles, Jacques	102
Charles' Law	102
chemical energy	246
chemical potential energy	250
Chernobyl	225
Christian Doppler	303
Christian Huygens	112
chromatic aberration	436
circle of least confusion	436
Classical Physics	32, 33, 349
classical theory	351
climate models	280
Clinton Davisson	353
coal	252
coefficient of kinetic friction	61
coefficient of static friction	61
coherent sources	140
comet	467
compound microscope	438
compression	133
computed tomography	457
concave	428
conduction	273
conductor	153

INDEX

conservation of energy	48
conservation of linear momentum	76
conservation of mass-energy	380
constant	418
constant accelerations	42
constant refractive indices	446
constellation	467
constructive interference	139
contact force	52
continuity equation	418
contrast-enhancing media	455
controlled variables	14
convection	273
conventional current	187
converging lenses	428
convex	428
convex lens	429
convex mirror	433
coolant	260
cosmic rays	228
cosmology	466
coulomb	154
Coulomb, Charles Augustin	153
Coulomb force	361
Coulomb repulsion	230
Coulomb scattering	361
Coulomb's law	153
Count Rumford of Bavaria	84
Crab Nebula	480
crest	133
critical angle	447, 137
critical angle reflection	136
critical damping	423
critical mass	224
Crookes, Sir William	206
crude oil	253
CT scanner	458
Curie, Marie	214
Curie, Pierre	214
current	161
curvature	428
cyclic process	405
Cygnus	480, 481

D

Dalton, John	206
damped oscillations	120, 423
damping	120, 422
David Brewster	143
David Gross	241
David Payne	449
David Politzer	241
Davisson, Clinton	353
Davy, Humphry	324
de Brahe, Tycho	319
de Broglie hypothesis	353
de Broglie, Louis	354
de Broglie wavelength	355
decay constant	362
Delta Cephei	476
Democritus	206
density	415
destructive interference	140
diagnostic frequency	460
dielectric	341, 448
dielectric constant	342
difference	314
diffracting aperture	299
diffraction	138
diffraction envelope	294
diffraction grating	294
diode	339
dioptre	429
Dirac, Paul	215
direct current	162
discrete energy	209
dispersion	448
displacement	34, 112, 114
displacement-time graph	286
displacement versus time	34
distance	34
distance-time graph	37
diverging lenses	428
DNA	218
Doppler, Christian	303
Doppler effect	303, 385
Doppler shift	481
dosimetry	217
double convex lens	436
drift velocity	163
driving frequency	424
dry cell	179
dynamic equilibrium	55
dynamic lift	419

E

ear-drum	129
Eddington, Arthur	56
effective resistance	172
efficiency	70, 70
Einstein, Albert	114, 222, 225, 350
Einstein Photoelectric Equation	351
Einstein Principle of Equivalence	384
Ejnar Hertzsprung	474
elastic collision	79
elastic potential energy	68
electrical energy	70, 246
electrical resistance	166
electrical strain gauge	178
electric cells	178
electric current	161
electric fields	152, 194
electric field strength	154, 154
electric potential	312
electric potential difference	163
electric potential energy	311, 312
electrode	179
electrolysis	324
electrolyte	179
electromagnetic force	52
electromagnetic induction	325
electromagnetic spectrum	126
electromagnetic waves	125
electromagnetism	32, 366, 368
electromotive force	325
electron	356
electron antineutrino	231
electron microscope	302
electron-volt	164
electrostatics	152
electroweak	229
elementary particles	231
elliptical galaxies	481
EMF	181, 325
emission rate	277
emissivity	277
EM spectrum	126
energy	65, 123
energy balance climate model	281
energy band theory	153
energy conservation	399
energy density	250
Enrico Fermi	216
entropy	412
equipotential lines	315
equipotentials	310
equipotential surface	310
equivalence principle	382
Ernest Lawrence	228
Ernest Rutherford	208, 357
Ernst Mach	57
error bars	13
Erwin Schrödinger	358
escape speed	310
ether	368
evaporation	95
event horizon	386
exchange particles	238
explosions	79
exponential graphs	18
external forces	52

F

Faraday, Michael	154, 324
Faraday's induction ring	325
Faraday's Law	327
far point	434
fermions	234

521

INDEX

Fermi, Enrico	216
Fermi radius	361
Feynman diagrams	238
Feynman, Richard	238
field lines	310
fields	308
filters	456
First Law of Thermodynamics	402
Fizeau, A.H.L.	112
Fleming's left-hand rule	189
fluid dynamics	415
fluid resistance	50
flux linkage	327
focal length	428
focal plane	429
forced oscillation	121, 422
force on a body	57
forces	51
forces between molecules	58
force-time graphs	75
fossil fuels	251
Foucalt, Jean	113
fractional (relative) uncertainty	10
frame of reference	366
Frank, Wilczek	241
Franklin Roosevelt	225
Fraunhofer diffraction	290
Frederic Joliot	215
free-body diagrams	52
free fall	36, 43
frequency	112, 114, 124
Fresnel diffraction	290
friction of bodies	86
Fritz Strassmann	225
fuel	254
full-wave rectification	340

G

galaxy	466
Galilean relativity	366
Galileo Galilei	56, 84, 112
galvanometer	325
gamma photon	381
gamma radiation	128
gamma ray bursters	375
gamma rays	214
gases	86, 161
Gay-Lussac	102
Geiger-Marsden experiment	360
Geiger-Müller detector	217
Geissler, Heinrich	206
Gell-Mann, Murray	229
General Theory of Relativity	382, 480
generator	331
Georg Simon Ohm	167
geodesic	387
geometry	27
geothermal	251
geothermal power	254
Germer, Lester	353
glass discharge tube	209
gluons	238
God particle	241
Goodricke, John	476
governor	395
graded-index fibres	446
gradient	19
graphical analysis	16
graphs	14
gravitational attraction	387
gravitational field	201
gravitational force	52
gravitational lensing	387
gravitational mass	383
gravitational potential	308
gravitational pull	310
gravitational redshift	382
greek symbols	29
greenhouse effect	278
greenhouse gas	279
G.R. Kirchoff	171
Gross, David	241

H

hadrons	231
Hafele-Keating experiment	378
Hahn, Otto	225
half-life	218, 363
half-value thickness	453
half-wave rectification	340
Halley comet	467
Hans Geiger	217
harmonic	112
harmonic motion	114
harmonic oscillations	114
harmonic oscillator	285
harmonics	149
harmonic series	149
heat energy	246
heat engine	405
heating effect of current	169
heat pump	407
heavily damped	120, 423
Heinrich Geissler	206
Heinrich Hertz	113, 349
Heinrich Lenz	328
Heisenberg Uncertainty Principle	359
Heisenberg, Werner	359
Henrich Olber	482
Henrietta, Leavitt,	477
Henry, Joseph	324
Henry Russell	474
Herbert Becker	212
Hertz, Heinrich	113, 349
Hertzsprung, Ejnar	474
Hertzsprung-Russell diagram	474
Hideki Yukawa	213
Higgs boson	241
Higgs, Peter	241
Hiroshima	225
Hubble's law	484
human ear	130
Humphry Davy	324
Huygens, Christian	112
Hyakutake comet	467
hydroelectric	267
hydroelectric power	266
hydropower	254
hydrostatic equilibrium	417
hyperbola	17
hysteresis	336

I

ideal fluid	417
ideal gas	101, 109, 401
imaging	428
imaging instrumentation	437
imaging techniques	455
impulse	74
incident ray	430
induced current	325
induced EMF	325
inelastic collisions	79
inertial mass	382
inertial reference frame	366
infrared radiation	127, 279
insulator	153
intensity profiles	295
interaction	238
intercepts	19
interference	139, 297
interference pattern	140
two point sources	140
interference patterns	294
internal combustion engine	406
internal energy	89, 402
internal forces	52
internal resistance	182
invariant quantity	380
inverse photoelectric effect	450
ionising radiation	217
Irene Joliot-Curie	214
irregular galaxies	481
Isaac Newton, Sir	84, 112
isobaric	403
isobaric processes	403
isochoric processes	403
isolated system	402
isothermal	403

INDEX

isotopes 212
isovolumetric processes 403

J

Jacques Charles 102
James, Chadwick 212
James Clerk Maxwell 113, 368
James Prescott Joule 65, 85
J. C. Adams 320
Jean Foucalt 113
J. Leverrier 320
Jocelyn Bell 480
Johannes Kepler 319
Johannes Rydberg 210
Johann Jakob Balmer 210
John Dalton 206
John Goodricke 476
Joliot-Curie, Irene 214
Joliot, Frederic 215
Joseph Henry 324
Joseph John (J. J.) Thomson, Sir 152, 207
joule 65
Julius Plücker 206

K

Kelvin 88
Kelvin-Planck statement 411
Kepler, Johannes 319
Kepler's third law 319
kinetic energy 66, 248, 90
kinetic energy changes 119
kinetic theory 86
Kirchhoff's circuit laws 171
Kirchoff, G.R. 171
Kirchoff's current law 171
Kirchoff's voltage law 171

L

laminar 418
laminar flow 420
latent heat 97
latent heat of transformation 98
latitude belt 281
Lavoisier, Antoine 74, 84, 206
law of conservation of electric charge 153
law of conservation of energy 246, 402
law of gravitation 200
law of radioactive decay 362
law of reflection 134
law of universal gravitation 200
Lawrence, Ernest 228

laws of electromagnetic induction 331
laws of mechanics 368
lead-acid battery 180
Leavitt, Henrietta 477
left-hand palm rule 325
length contraction 372
lens 428
Lenz, Heinrich 328
Lenz's Law 328
lepton number 236
leptons 231
Lester Germer 353
Leverrier, J. 320
light dependent resistor 177
light year 469
lignite 252
linear attenuation coefficient 452
linear graph 17
linear magnification 433
linear momentum 74
liquid drop model 224
liquids 86, 161
load factor 255
logarithmic functions 18
logarithmic graphs 18
longitudinal waves 123
Lorentz transformations 369
loudness 130
Louis de Broglie 354
luminosity 476

M

magnetic field 187
magnetic flux 327
magnetic flux density 326
magnetic resonance imaging 460
magnification 433
magnifying power 435
Mach, Ernst 57
Maltese Cross 206
Malus, Etienne 142, 144
Malus' law 144
Manhattan Project, The 225
Marie Curie 214
mass 32, 233, 55
mass defect 222
mass-luminosity 476
material dispersion 448
mathematical sentences 27
matter 356
matter waves 353
Maxwell, James Clerk 113, 368
Maxwell's theory 113, 113, 368
mechanical energy 246
mechanical flexure 445
Melvill, Thomas 210

mesons 232
methane 279
metric multipliers 4
Michael Faraday 154, 206, 324
Michelson, Albert A. 113
Michelson-Morley experiment 374
microwaves 127
middle ear 130
Milky Way 466, 481
Millikan, Robert 208, 351
minimum angle of resolution 301
Minkowski space 377
Minkowski three-space 376
modal dispersion 448
moderator 260
modes 448
modulation 294
molar mass 103
mole 104
molecular behaviour 95
molecular theory 86
moment of inertia 395
momentum 74, 382
monomode fibres 449
motion 33, 37
motor effect 189
Mount Palomar Observatory 441
moving particle theory 86
MRI 458
MRI magnetic fields 460
multipath dispersion 448
muon 231
muon decay experiment 374
muon neutrino 231
Murray Gell-Mann 232
Müller, W. 217

N

Nagasaki 225
natural frequency 121, 130, 121, 422
natural gas 253
near point 434
nebula 466
nebulae 466
Ne'eman, Yuval 232
negative 69
nematic liquid crystal 145
neutrino 216, 362
neutron activation 224
neutron 212
neutron star 468, 480
Newtonian Determinism 348
Newtonian Mechanics 33, 55
newton (N) 57
Newton's First Law of Motion 55
Newton, Sir Isaac 112

523

INDEX

Newton's Law of Cooling	93
Newton's Law of Gravitation	319
Newton's laws of motion	55
Newton's postulates	367
Newton's Second Law of Motion	56, 320, 367, 379, 397
Newton's Third Law of Motion	57
Nicolas Léonard Sadi Carnot	413
Niels Bohr	211
NMR	450
nodes	140, 148
non-mechanical transfer	89
non-renewable energy	252
nuclear binding energy	223
nuclear energy	70, 246, 254
nuclear energy levels	362
nuclear fission	224, 262
nuclear force	52
nuclear fusion	224, 252, 226
nuclear magnetic resonance	460
nuclear power	261
nuclear power station	259
nuclear reactions	221
nucleon	212
nucleon number	212
nucleosynthesis	488
nuclide	212
numerical aperture	302

O

Oersted	324
Oersted, Hans Christian	187
Ohm, Georg Simon	167
Ohm's Law	167
oil shale	253
oil tar	253
Olber, Henrich	482
Olber's paradox	482
Ole Römer	112
Onnes H. Kammerlingh	167
Oppenheimer-Volkoff limit	478
optical compound microscopes	437
optical fibre	137, 447
optical microscope	302
optical path difference	297
optics	368
orbit	467
orbital energy	318
orbital motion	318
orbital speed	318
orders of magnitude	6
oscillation	114, 120, 423
ossicles	130
Otto Hahn	225
outer ear	129
oval window	130

P

pair annihilation	356, 240
pair production	357, 240
parabola	17
parallax method	470
parsec	469
particle acceleration	381
Pascal's principle	416
path difference	140
Paul Dirac	215
Paul Villiard	214
Payne, David	449
peak current	333
peat	252
pendulum	114
percentage uncertainty	10
period	115, 124
periodicity	123
periodic stimulus	424
permeability of free space	162
permittivity constant	154
Peter Higgs	241
phase change	95
phase difference	114, 115, 140
photoelectric effect	348, 452
photon	452
photovoltaic devices	265
Pierre Curie	214
piezoelectric effect	425
piezoelectric transducer	458
pinna	129
Pitot tube	420
pixels	145
planar waves	138
Planck's constant	19, 20, 351
planetary model	230
plasma	262
Plücker, Julius	206
point charges	314
point particle	51
polarimeter	144
polarization	142
polarizers and analysers	143
polaroid	142
Politzer, David	241
pollutants	257
positive	69
positive point charge	312
positron	356
positrons	231
potential	309
potential difference	163
potential divider	175
potential energy	67, 90
potential energy changes	119
potential gradient	313

Pound-Rebka-Snider experiment	382
Powell, Cecil Frank	238
power	65, 69
power dissipation	168
power generation	331
power loss	336
power rating	168
power station	247, 255
power transmission	331
preferential absorption	142
pressure	100, 415
primary cell	179
principal axis	428
principal focal plane	429
principal focus	428
Principia Mathematica	56
principle of conservation of energy	70
principle of superposition	134, 139, 139
probability characteristics	358
projectile	46
projectile motion	45, 50, 367
projectile problems	48
proper length	373
proper time	371
proton number	212
pull	51
pulsar	468, 480
pulse	133
push	51

Q

Q factor	422
quality factor	121
quanta	211
quantum chromodynamics	229
Quantum Mechanics	32, 234, 348
quantum numbers	234
quark confinement	241
quarks	231
quasar	466

R

radian	444
radian measure	115
radiation	273
radiation shielding	260
radioactive decay	229
radioactivity	209
radiography	457
radio interferometry	444
radio telescopes	443
radio waves	127
radius	428

INDEX

radon	218
random errors	9
rarefaction	133
rate of change of momentum	74
ray diagrams	430
Rayleigh Criterion	299, 300
rays	133
real gases	109
real image	432
red giants	468
redshift	382
redshift experiment	388
reference frames	366
reflecting telescopes	441
reflection	133, 134
refracting telescopes	438
refraction	133, 134
refractive index	135
refrigerator	407
regenerators	449
relative density	415
relative motion	41
relativistic mechanics	379
relativistic momentum	379
Relativistic Physics	32
relativity theory	78
renewable energy source	253
reshapers	449
resistance	166
resistivity	166
resistor	166
resolution	460
resonance	122, 122, 425
resonant frequency	425
rest energy	379
rest mass	222, 379
rest mass energy	379
Reynolds number	420
Richard Feynman	238
right-hand palm rule	189, 325
rigid body	52
Robert, Boyle	101
Robert Millikan	208, 351
Robert van der Graaf	228
Römer, Ole	112
Roosevelt, Franklin	225
rotational equilibrium	392
rotational kinetic energy	398
rotational motion	396
Russell, Henry	474
Rutherford, Ernest	208, 357
Rutherford scattering	360
Rydberg, Johannes	210

S

Salter Duck	270
Sankey diagrams	246
satellite energy	320
scalar quantity	312
scalars	22
Schrödinger, Erwin	358
Schwarzschild black holes	386
scientific notation	4
secondary cells	180
second generation muon	231
Second Law of Thermodynamics	246, 411
self-inductance	336
semiconductor	339
sensors	176
SHM	116, 117, 284, 424
significant figures	5
simple harmonic motion	112, 116, 284
single-slit diffraction	289
sinusoidal graph	17
SI units	3
Snell's law	135
Snell, Willebrord	135
solar constant	265, 281
solar energy	254
solar heating	264
solar power	264
solar system	467
solenoid	329
solid friction	60
solids	86, 161
SONAR	458
sound energy	246
sound waves	129
space-time diagram	375
special relativity	353, 369
Special Theory of Relativity	369
specific heat capacity	91
specific latent heat	97
spectra	210
spectral classes	472
speed	35
speed of light	112
sphere	428
spherical aberration	436
spiral galaxies	481
spring constant	54
standing waves	147
static equilibrium	54
stationary waves	147
steam engine	256
Stefan-Boltzmann law	471
Stefan's law	275
stellar interferometer	302
stellar parallax	466, 469
stellar spectra	471
step-index fibres	446
Stokes' law	420
straight-line graph	14
strain viewer	145
strangeness	236
Strassmann, Fritz	225
streamline	418
sublimation	97
sulfur dioxide	257
Sun	468
super cluster	466
supergiant	478
supernova	468
superposition	139
system	402
systematic errors	9

T

target material	456
tau neutrino	231
temperature	87
tension force	53
terminal speed	50
terminal voltage	181
terrestrial telescope	439
Theory of Relativity	233
thermal energy	70, 89
thermal energy transfer	273
thermal fission reactor	260
thermal radiation	273
thermal reactor	225, 260
thermal stability	445
thermistors	177
thermodynamic cycle	405
thermodynamic engine	405
thermodynamics	32, 401
thermoelectric converters	265
thermometer	88
thin film interference	296
thin lens	428
thin lens equation	432
third generation tau	231
Third Law of Thermodynamics	414
Thomas Melvill	210
Thomson, Sir Joseph John (J.J.)	152, 207
Thompson, Benjamin	84
thorium	218
ticker-tape timer	43
tidal power	254, 267
tides	267
timbre	130
time constant	345
time dilation	371, 372
tomography	457
torque	392

525

INDEX

total energy	379
total internal reflection	136, 447, 137
transformer	334
translational equilibrium	54, 392
transmutation	224
transverse waves	123
trap door method	43
travelling wave	148
travelling waves	122
trigonometry	28
trough	133
tube current	456
tube voltage	456
tuning fork	146
tunnelling probability	359
turbulent flow	420
Tycho de Brahe	319

U

ultrasound	458
ultraviolet radiation	128, 279
uncertainties and errors	9
uncertainty principle	348, 359
uniform angular acceleration	396
universal gas constant	401
Universal Gravitational Law	320
universe	466
uranium	218, 248

V

van der Graaf accelerator	228
van der Graaf, Robert	228
vector notation	42
vectors	22
velocity	35
velocity addition	369
velocity-displacement graph	285
velocity-time graph	38
Venturi tube	419
vertex	239
Villiard, Paul	214
Virgo cluster	466
virtual image	432
virtual particle	239
viscosity	420
visible light	128
Volta, Alessandro	178
voltaic pile	179
Volta's crown of cups	179

W

Walther Bothe	212
wave behaviour	132
wave characteristics	132
wavefront	133
wave function	358
waveguide	448
wavelength	124
wave mechanics	358
wave power	270
wave speed	124
weightlessness	321
Werner, Heisenberg	359
white dwarfs	468
Wien's displacement law	471, 472
Wilczek, Frank	241
Willebrord Snell	135
William Crookes, Sir	206
wind generators	269
wind power	251, 254, 268
wind turbines	268
W. Müller	217
work	65, 402
work against a force	65
work function	350
working fluid	405
world line	376

X

X-radiation	128
X-ray	450
X-ray beam	455
X-ray detection	455
X-ray diffraction	87
X-ray radiation	481
X-ray spectrum	450

Y

Young's double-slit experiment	291
Yukawa, Hideki	213
Yuval Ne'eman	232

Z

zero order maximum	291